Metabolic Dynamics in Host-Microbe Interaction

Sameh S. M. Soliman
Mohamed I. Husseiny
Editors

Metabolic Dynamics in Host-Microbe Interaction

Springer

Editors
Sameh S. M. Soliman
Department of Medicinal Chemistry
University of Sharjah
Sharjah, United Arab Emirates

Mohamed I. Husseiny
Beckman Research Institute
City of Hope National Medical Center
Duarte, CA, USA

ISBN 978-981-96-1307-6 ISBN 978-981-96-1305-2 (eBook)
https://doi.org/10.1007/978-981-96-1305-2

This Springer imprint is published by the registered company Springer Nature Singapore Pte Ltd.
The registered company address is: 152 Beach Road, #21-01/04 Gateway East, Singapore 189721, Singapore

Preface

This book is a groundbreaking that delves into the fascinating world of metabolism and its pivotal role in survival, attack, and defense strategies in human-microbe interaction. This book explores the intricate biochemical processes that enable organisms to thrive, adapt, and protect themselves against various challenges they encounter in their environments. The book describes the importance of metabolism in defense and attack by bacteria and fungi in the human environment, and the reverse immunometabolism and gut microbiome responses by the human. Different metabolites from different organisms will be described, in addition to their changes due to interaction with other organisms during infection or defense. The book also describes the methods of detection and identification of metabolites including metabolomics. The use of artificial intelligence in prediction of metabolic pathways following interaction between different organisms and tools used in the identification of therapeutic targets are highlighted. The metabolism in microbiome and their importance are described in relation to infection, metabolic diseases, cancer, and neurological disorders and their effects on therapies.

This book combines the latest scientific research with accessible explanations, providing readers with a deep understanding of how metabolism empowers organisms to thrive and protect themselves in diverse environments. It will be an invaluable resource for researchers, students, physicians, and anyone fascinated by the wonders of biology and the intricate workings of life. The purpose of this book is to provide a comprehensive, interdisciplinary exploration of metabolism in the context of attack and defense strategies. It aims to educate, inspire curiosity, and bridge between scientific knowledge and its practical applications, ultimately contributing to our understanding of the power and complexity of metabolism in living organisms.

This book provides an interdisciplinary approach by integrating concepts from biochemistry, physiology, immunology, ecology, and evolutionary biology. This approach allows readers to gain a comprehensive understanding of metabolism and its role in attack and defense strategies across different organisms. While metabolism is a well-studied topic, this book uniquely focuses on its specific relevance to attack and defense mechanisms in living organisms. It explores how metabolic processes are intricately intertwined with an organism's ability to mount effective defense strategies and execute offensive actions, highlighting the adaptive nature of metabolism in these contexts. The book incorporates the latest research findings and advancements in the field of metabolism and defense mechanisms. It explores emerging research areas, such as the role of the microbiome in metabolism and defense, metabolic interactions between species, use of artificial intelligence, and the application of metabolomics in studying attack and defense. By integrating cutting-edge research, the book offers readers a forward-looking perspective on the topic.

Sharjah, UAE Sameh S. M. Soliman
Duarte, CA, USA Mohamed I. Husseiny

Contents

Editors and Contributors

About the Editors

Sameh S. M. Soliman received his bachelor's and master's degrees in pharmaceutical sciences. He received his PhD from University of Guelph, Canada, with a focus on chemical ecology of natural products and its role in host-microbe interactions. Then he did a postdoctoral fellowship in mycotoxin manipulation and detoxification at Canadian Federal labs followed by another postdoctoral fellowship at the University of California, Los Angeles (UCLA), in chemical and biomolecular engineering of natural products. At UCLA medical School, he received extensive training in infectious diseases, pathogenesis mechanisms, and possible ways in diagnosis and treatment. His research interests include Natural Products as potential medicines and Microbial Products as either toxins or medicines. For >25 years, he got extensive training experience in natural products and including their chemical ecology and their role in nature, in addition to structure optimization modification to understand natural metabolites' ecological roles. He developed a line of research in mycology, fungal infections, and the discovery of effective antimicrobials by optimizing structures and activities against critical proteins in invasive fungal species. He also developed a strategy to combat fungal infection by developing selective anti-pathogenic drugs via deep understanding of the evolved behavior of fungal pathogens. This is by developing neutralizing agents against specific pathogenic treats that are used by the pathogens and do not exist in our microbial flora, thus it weakens the pathogens, while enhancing the human defense system against these pathogens. Currently, as a faculty member at the University of Sharjah, his focus includes teaching and research of natural products and metabolism of different organisms and their ecological roles in nature. Besides, he extensively studies the metabolism of living organisms and their roles in disease development and management.

Mohamed I. Husseiny received his bachelor's and master's degrees in pharmaceutical sciences and medical microbiology. This gave him training in drug development, microbiology, and immunology. Continuing to enlarge his expertise in this area, he completed his PhD at the Institute for Clinical Microbiology, Immunology and Hygiene, Friedrich-Alexander University, Erlangen-Nürnberg, Germany. During his PhD, he established a novel system for gene therapy. These experiences gave him substantial expertise in molecular biology, immunology, gene cloning, protein purification, and various functional analyses. Furthermore, he has extensively trained in murine immunizations and infection. Based on this, he developed an oral vaccine against *Listeriosis* using genetically engineered *Salmonella.* He was employed as a postdoctoral fellow at LA Biomed at Harbor-UCLA Medical Center, Division of Infectious Diseases, Department of Medicine, Torrance, California, USA. He was involved in the project of generating isogenic null mutants of *rFTR1* in *R. oryzae* to study the role of iron-related virulence, and expression and secretion of rFTR1 protein by yeast for vaccination purposes. He employed the genetically engineered *Salmonella* system to develop a vaccine for cancer during his postdoctoral fellowship in the Children's Hospital of Los Angeles. Then, he moved as a postdoctoral fellow into City of Hope, Division of Translational Vaccine Research, Beckman Research Institute. He used a novel strategy to generate a panel of recombinant BK virus (rBKV) expressing various fragments of the HIV-1 gag polypeptide and green fluorescent protein (GFP) under control of the BKV late promoter. He joined Arthur Riggs Diabetes & Metabolism Research Institute, Beckman Research Institute, City of Hope, as postdoctoral fellow in December 2010, moved to staff scientist in 2012, and then promoted to assistant research professor in October 2014. Currently, he is an associate research professor, with over 24 years of experience in vaccination strategy. Throughout his career, his goal remains to employ vaccine strategies to ameliorate and prevent infectious and noninfectious diseases including cancer and autoimmune diseases.

Contributors

Amera A. Abaza Department of Industrial Biotechnology, Genetic Engineering and Biotechnology Research Institute, University of Sadat City, Sadat City, Egypt

Heba Salah Abbas Department of Microbiology and Immunology, Faculty of Pharmaceutical Sciences and Drug Manufacturing, Misr University for Science and Technology (MUST), Giza, Egypt

Department of Microbiology, Egyptian Drug Authority, Previously National Organization for Drug Control and Research, Giza, Egypt

Farah Abdulridha College of Pharmacy, University of Sharjah, Sharjah, UAE

Mays Z. Abu Libdeh Department of Medical Laboratory Sciences, College of Health Sciences, University of Sharjah, Sharjah, UAE

Mohamed Abuelazm Tanta University, School of Medicine, Tanta, Egypt

Ameera Abu-Qiyas Research Institute for Medical and Health Sciences, University of Sharjah, Sharjah, UAE

El-Sayed Akool Department of Pharmacology and Toxicology, Faculty of Pharmacy, Al-Azhar University, Cairo, Egypt

Aseel Aldaihani Department of Medicinal Chemistry, College of Pharmacy, University of Sharjah, Sharjah, UAE

Aya Al Ali College of Pharmacy, Al Ain University, Abu Dhabi, UAE

AAU Health and Biomedical Research Center, Al Ain University, Abu Dhabi, UAE

Rawhia R. Allam Bioinformatics Department, Genetic Engineering and Biotechnology Research Institute, University of Sadat City, Sadat City, Egypt

Fatma Almousa Department of Medicinal Chemistry, College of Pharmacy, University of Sharjah, Sharjah, UAE

Fatimah Almsailem Department of Medicinal Chemistry, College of Pharmacy, University of Sharjah, Sharjah, UAE

Ahmed M. Al-Rawi Research Institute for Medical and Health Sciences, University of Sharjah, Sharjah, UAE

Inas Al-Younis Division of Biological and Environmental Sciences and Engineering (BESE), King Abdullah University of Science and Technology (KAUST), Thuwal, Saudi Arabia

Shahad Arikat Research Institute for Medical and Health Sciences, University of Sharjah, Sharjah, UAE

Amany Balah Department of Pharmacology and Toxicology, Faculty of Pharmacy, Al-Azhar University, Cairo, Egypt

Giulia Bennici Division of Biological and Environmental Sciences and Engineering (BESE), King Abdullah University of Science and Technology (KAUST), Thuwal, Saudi Arabia

Salam Dakalbab Research Institute for Medical and Health Sciences, University of Sharjah, Sharjah, UAE

College of Medicine, University of Sharjah, Sharjah, UAE

Manel Dhahri Biology Department, Faculty of Science, Taibah University, Yanbu Branch, Yanbu, Saudi Arabia

Ashraf El-Baz Department of Industrial Biotechnology, Genetic Engineering and Biotechnology Research Institute, University of Sadat City, Sadat City, Egypt

Mahmoud ElHefnawi Biomedical Informatics and Chemoinformatics Group (BICG), Informatics and Systems Department, National Research Centre, Cairo, Egypt

Tasneem M. Elsayed Division of Cardiology, The Lundquist Institute at Harbor-UCLA Medical Center, Los Angeles, CA, USA

Ashraf S. A. El-Sayed Enzymology and Fungal Biotechnology Lab, Botany and Microbiology Department, Faculty of Science, Zagazig University, Zagazig, Egypt

Mohamed Abd El-Gawad El-Sayed Ahmed Department of Microbiology and Immunology, Faculty of Pharmaceutical Sciences and Drug Manufacturing, Misr University for Science and Technology (MUST), Giza, Egypt

Abdul-Hamid Emwas Core Lab of NMR, King Abdullah University of Science and Technology (KAUST), Thuwal, Saudi Arabia

Muhammed I. Erbay Cerrahpasa Medical School, Istanbul, Turkey

Ahmed K. Ghanem The Loma Linda University Medical Center- Murrieta, Murrieta, CA, USA

Jana H. Ghonaim MIPS, Sharjah, UAE

Mawieh Hamad Department of Medical Laboratory Sciences and the Research Institute for Medical and Health Sciences, University of Sharjah, Sharjah, United Arab Emirates

Rania Hamdy Research Institute for Science and Engineering (RISE), University of Sharjah, Sharjah, UAE

Mohamed E. Hasan Bioinformatics Department, Genetic Engineering and Biotechnology Research Institute, University of Sadat City, Sadat City, Egypt

Zeinab Shawky Hashem Department of Microbiology and Immunology, Faculty of Pharmacy, Minia University, Minia, Egypt

Wael A. H. Hegazy Department of Microbiology and Immunology, Faculty of Pharmacy, Zagazig University, Zagazig, Egypt

Alaa A. Hemeida Bioinformatics Department, Genetic Engineering and Biotechnology Research Institute, University of Sadat City, Sadat City, Egypt

Mohamed I. Husseiny Department of Translational Research & Cellular Therapeutics, Arthur Riggs Diabetes & Metabolism Research Institute, Beckman Research Institute, City of Hope National Medical Center, Duarte, CA, USA

Marija Ivanov Institute for Biological Research "Siniša Stanković", National Institute of the Republic of Serbia, University of Belgrade, Belgrade, Serbia

Mariusz Jaremko Division of Biological and Environmental Sciences and Engineering (BESE), King Abdullah University of Science and Technology (KAUST), Thuwal, Saudi Arabia

Smart-Health Initiative and Red Sea Research Center, Division of Biological and Environmental Sciences and Engineering, King Abdullah University of Science and Technology, Thuwal, Saudi Arabia

Amit Johanis Creighton University, School of Medicine, Phoenix, AZ, USA

Mohammad G. Mohammad Research Institute for Medical and Health Sciences, University of Sharjah, Sharjah, UAE

Department of Medical Laboratory Sciences, College of Health Sciences, University of Sharjah, Sharjah, UAE

Walaa K. Mousa College of Pharmacy, Al Ain University, Abu Dhabi, UAE

AAU Health and Biomedical Research Center, Al Ain University, Abu Dhabi, UAE

College of Pharmacy, Mansoura University, Mansoura, Egypt

Mohamed A. Osman Bioinformatics Department, Genetic Engineering and Biotechnology Research Institute, University of Sadat City, Sadat City, Egypt

Devi Permatasari Division of Biological and Environmental Sciences and Engineering (BESE), King Abdullah University of Science and Technology (KAUST), Thuwal, Saudi Arabia

Frederico G. Pinto Center for Applied and Translational Genomics (CATG), Mohammed Bin Rashid University of Medicine and Health Sciences (MBRU), Dubai Health, Dubai, UAE

Institute of Exact Sciences, Federal University of Viçosa, Rio Paranaíba, Minas Gerais, Brazil

Rand Salamalek College of Pharmacy, University of Sharjah, Sharjah, UAE

Sarra B. Shakartalla Research Institute for Medical and Health Sciences (RIMHS), University of Sharjah, Sharjah, UAE

Faculty of Pharmacy, University of Gezira, Wad Madani, Sudan

Yasmin Shareef College of Pharmacy, University of Sharjah, Sharjah, UAE

Abdel-Aziz S. Shatat Department of Pharmacology and Toxicology, Faculty of Pharmacy, Al-Azhar University, Cairo, Egypt

Yousseria Shetaia Department of Microbiology, Faculty of Science, Ain Shams University, Cairo, Egypt

Mahmoud Singer Department of Radiological Sciences, School of Medicine, University of California Irvine, Irvine, CA, USA

Nelson C. Soares Center for Applied and Translational Genomics (CATG), Mohammed Bin Rashid University of Medicine and Health Sciences (MBRU), Dubai Health, Dubai, UAE

College of Medicine, Mohammed Bin Rashid University of Medicine and Health Sciences (MBRU), Dubai Health, Dubai, United Arab Emirates

Laboratory of Proteomics, Department of Human Genetics, National Institute of Health Doutor Ricardo Jorne (INSA), Lisbon, Portugal

Center for Toxicogenomics and Human Health (toxOMICS), NOVA School-FCM, UNL, Lisbon, Portugal

Marina Soković Institute for Biological Research "Siniša Stanković", National Institute of the Republic of Serbia, University of Belgrade, Belgrade, Serbia

Sameh S. M. Soliman Department of Medicinal Chemistry, College of Pharmacy, University of Sharjah, Sharjah, UAE

Research Institute for Medical and Health Sciences (RIMHS), University of Sharjah, Sharjah, UAE

Dejan Stojković Institute for Biological Research "Siniša Stanković", National Institute of the Republic of Serbia, University of Belgrade, Belgrade, Serbia

Shriraj Susarla Keck School of Medicine at USC, Los Angeles, CA, USA

Mohammed J. Uddin Center for Applied and Translational Genomics (CATG), Mohammed Bin Rashid University of Medicine and Health Sciences (MBRU), Dubai Health, Dubai, UAE

College of Medicine, Mohammed Bin Rashid University of Medicine and Health Sciences (MBRU), Dubai Health, Dubai, UAE

GenomeArc Inc., Toronto, ON, Canada

Aya T. Ziadeh Department of Medical Laboratory Sciences, College of Health Sciences, University of Sharjah, Sharjah, UAE

Metabolism: Fine and Coarse Controls

1

Sameh S. M. Soliman, Rania Hamdy,
and Mohamed I. Husseiny

Abstract

Metabolism encompasses the intricate set of biochemical processes that sustain life by converting nutrients into energy and cellular components. It comprises two primary pathways: catabolism, which breaks down complex molecules to release energy, and anabolism, which builds complex molecules from simpler ones using energy. The regulation of metabolism is achieved through both coarse and fine control mechanisms. Coarse control involves broader adjustments, such as changes in enzyme levels through gene expression and cellular compartmentalization, while fine control entails precise regulation of enzyme activity through allosteric interactions, covalent modifications, and feedback inhibition. These regulatory mechanisms ensure metabolic homeostasis, allowing cells to adapt to varying internal and external conditions. Understanding these controls is crucial for elucidating how metabolic balance is maintained and how metabolic disorders may arise in relation to defense and/or attack.

S. S. M. Soliman (✉)
Department of Medicinal Chemistry, College of Pharmacy, University of Sharjah, Sharjah, UAE

Research Institute for Medical and Health Sciences (RIMHS), University of Sharjah, Sharjah, UAE
e-mail: ssoliman@sharjah.ac.ae

R. Hamdy
Research Institute for Science and Engineering (RISE), University of Sharjah, Sharjah, UAE

M. I. Husseiny
Department of Translational Research & Cellular Therapeutics, Arthur Riggs Diabetes & Metabolism Research Institute, Beckman Research Institute, City of Hope National Medical Center, Duarte, CA, USA

S. S. M. Soliman, M. I. Husseiny (eds.), *Metabolic Dynamics in Host-Microbe Interaction*, https://doi.org/10.1007/978-981-96-1305-2_1

Keywords
Metabolism · Fine control · Coarse control · Host-microbe interaction

1.1 Introduction

Metabolism is a complex network of interdependent metabolic pathways and chemical reactions within the cells that sustain life. These reactions convert nutrients into energy, build cellular structures, and eliminate waste, allowing organisms to grow, reproduce, maintain structural integrity, and adapt to environmental changes. Metabolic pathways are tightly regulated to ensure cells and organisms maintain optimal conditions in response to their surroundings (Brown et al. 2004). The metabolic flux through any pathway must be closely coordinated with the organism's needs of the end product. Metabolic control is the process of adjusting the output of a metabolic pathway in response to an external signal (Zhu and Thompson 2019; Judge and Dodd 2020). Effective control of metabolic pathways is critical for the maintenance of cellular function, growth, and overall homeostasis (Mason and Rathmell 2011).

Enzymes, which are proteins that speed up biochemical reactions, are central to the regulation of metabolism (Chandel 2021). Their functions are controlled through processes like feedback inhibition, allosteric interactions, and covalent changes, allowing cells to adapt their metabolic pathways based on environmental conditions and internal signals (Plaxton 2004; Lorendeau et al. 2015). Hormones, including insulin and glucagon, play a key role in controlling these pathways by adjusting their rates. Organisms must also align their energy needs with available nutrients, a balance achieved through signaling systems that connect external factors like food intake and stress to internal metabolic requirements (Lorendeau et al. 2015). The supply of an active metabolite in an organism is influenced by factors such as growth rate, metabolite localization in specific organs, and degradation of natural resources or end products (Qiu et al. 2023). Understanding these regulatory systems is a key to identifying the causes of metabolic diseases. This chapter explores the regulation and control mechanisms of metabolism within cells.

1.2 Metabolic Regulation Mechanisms

Metabolic regulation mechanisms are the techniques that cells use to control and coordinate the metabolic pathways, ensuring that the production and use of energy and biomolecules are in line with the needs of the cell. This system maintains the cell homeostasis, while responding to changes in its internal and external environment. The regulation of the metabolic process is subjected to important metabolic concepts (Medina-Cleghorn and Nomura 2014). First, metabolic complexity of each organism is the consequence of the cooperation of various enzymes from multiple metabolic pathways. Second, biochemical unity is the idea that although the structure and control of metabolic pathways are identical throughout organisms, there are differences between certain cell types within the same organism that shares the same pathway. Third, the pacemaker enzyme and a pathway's rate of metabolite flux need to be synchronized with the organism's and cells' requirements for the final products. Understanding the actions of the rate-limiting enzyme is a crucial first step toward comprehending the regulation of the process.

1.2.1 Rate-Limiting Enzyme

The rate-limiting step, the slowest stage in a chemical reaction route, is a determinant of the metabolic flux through an entire pathway (Kitamura et al. 2021). In the cell, it is far from equilibrium because of the relatively low activity of the catalyzing enzyme. The rate of this reaction is not limited by substrate availability or the speed of precursor flow but only by the activity of this enzyme, thereby limiting the whole reaction sequence. The rate-limiting step is very exergonic and therefore essentially irreversible under cellular conditions (Zuo et al. 2021). The two basic mechanisms of fine and coarse metabolic regulations allow the cell to modify the rate of metabolic responses (Wegner et al. 2015).

1.3 Fine Metabolic Control

Fine metabolic regulation is the ability of cells to maintain metabolic balance and react rapidly to changes in energy demands, nutrition supply, and environmental variables (Moreno-Sánchez et al. 2008). Instead of depending on energy-intensive processes, this precise regulation modifies preexisting enzymes to help prevent metabolic chaos (Aon et al. 2011). It entails swift, energy-efficient alterations in enzyme activity to regulate the activity of these systems to adapt the cells to any environmental changes and to maintain metabolic efficiency. In order to sustain overall metabolic homeostasis, fine controls frequently interact and cooperate. Fine metabolic control can be understood through the following mechanisms.

1.3.1 Fine Control 1: Substrate/Co-substrate Concentration

Enzyme activity and precursor availability are both important factors in determining metabolic flux. Precursor feeding may enhance the availability of substrates, while biotic or abiotic elicitors may increase the activity of enzymes (Kurata et al. 2007). For instance, determining a pathway's rate-limiting step is essential to manipulating a metabolic pathway.

The rate of pathway flux could be controlled by changes in the substrate concentration for any enzyme that comprises a step in a pathway (Moreno-Sánchez et al. 2008). To obtain a change in the activity of an enzyme, the substrate concentration should be highly increased in hyperbolic kinetics more than sigmoidal kinetics (Robinson 2015). The rate of an enzyme-catalyzed reaction is dependent upon

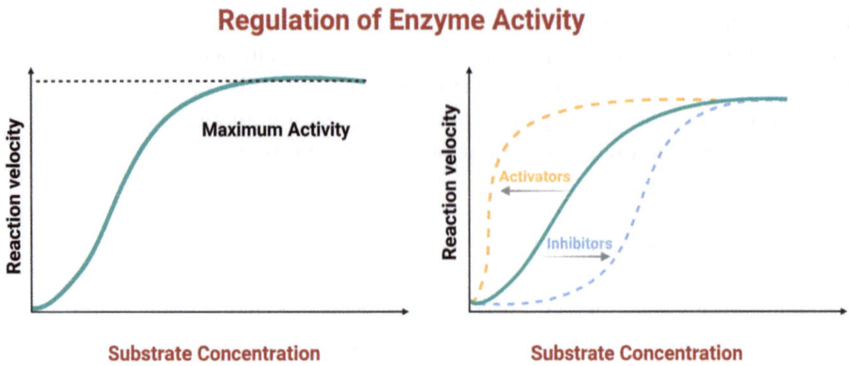

Fig. 1.1 Regulation of enzyme activity. It shows how activators increase enzyme activity (left shift) and inhibitors decrease activity (right shift) relative to the substrate concentration

substrate when it is subsaturating. Substrate concentrations for most enzymes are subsaturating in vivo, and less than or nearly equal to K_m (Michaelis constant) of the enzyme for that substrate. While many multimeric enzymes have numerous substrate binding sites, not all enzymes display simple Michaelis-Menten kinetics. Multimeric enzymes are made up of two or more polypeptides in their initial state (Porter and Miller 2012). The shape of other subunits can be altered, and their ability to bind to substrates enhanced when a substrate attaches to one subunit. This cooperative substrate binding leads to a sigmoidal relationship between enzyme activity and substrate concentration (Freiburger et al. 2011) (Fig. 1.1). These are known as allosteric effects.

1.3.2 Fine Control 2: Allosteric Effector and Enzyme Inhibitions

Allosteric effectors at least regulate all rate-limiting enzymes. Allosteric sites can reversibly bind to certain inhibitor or activator compounds, which are distinct from the active site of multi-subunit regulatory enzymes (Abhishek et al. 2023). The quantity of these nonsubstrate effector molecules can be changed to significantly change an enzyme's activity and, consequently, the flow of metabolites throughout a whole pathway. By attaching to an allosteric site and causing a precise conformational change in the enzyme, allosteric effectors modify the activity of the enzyme; this mechanism is known as heterotropic allosterism (McCullagh et al. 2024). In inactive areas, metabolites bind to enzymes to change their activity. For example, ATP inhibition of phosphofructokinase-1 (PFK-1) reduces glycolysis in case of energy abundance (Mor et al. 2011).

Allosteric effectors themselves may show hyperbolic or sigmoidal (i.e., cooperative) saturation kinetics (Cornish-Bowden 2014). A sigmoidal relationship enables an enzyme's conformation (and hence its activity) to react sensitively to variations in effector concentration. When an allosteric effector binds, a conformational change occurs that either facilitates (activators) or inhibits (inhibitors)

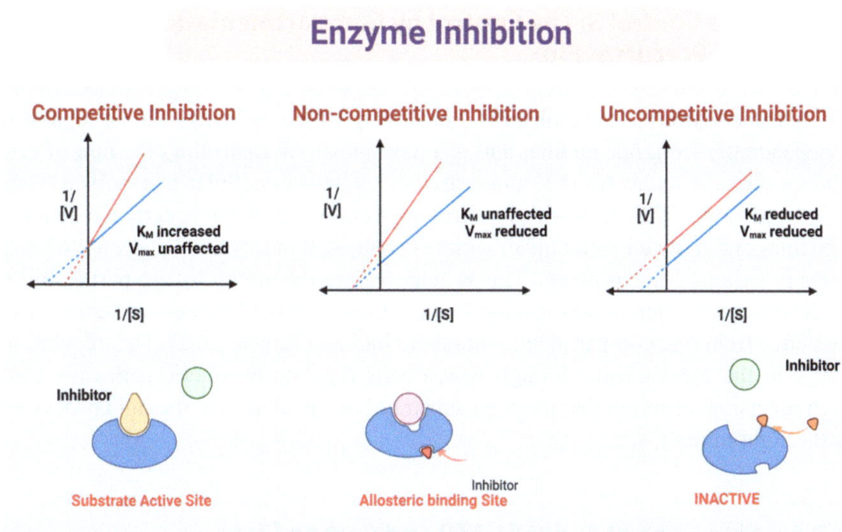

Fig. 1.2 Types of enzyme inhibition. The figure illustrates three types of enzyme inhibition: competitive, noncompetitive, and uncompetitive, and how they affect enzyme activity through different inhibitor interactions. In the Lineweaver-Burk plot, the x-axis represents $1/S$, the reciprocal of the substrate concentration, and the y-axis represents $1/V$, the reciprocal of the reaction velocity, where K_m (Michaelis constant) indicates the enzyme's affinity to the substrate, and V_{max} (maximum velocity) indicates the highest rate at which the enzyme can catalyze the reaction when saturated with substrate

enzyme-substrate interactions. As a result, some or all of the kinetic constants, such as V_{max} (the system's maximum rate), K_m, or S (substrate concentration), may be drastically changed (Lira et al. 2019). In competitive inhibition, K_m increases, while V_{max} remains unchanged; in noncompetitive inhibition, V_{max} decreases without affecting K_m; and in uncompetitive inhibition, both K_m and V_{max} decrease (Fig. 1.2). Furthermore, this regulatory strategy allows metabolites that are remote from a specific reaction to function as feedforward (activators) or feedback (inhibitors) control signals on enzyme activity. All pacemaker enzymes are at least partially controlled by allosteric effectors.

Phosphofructokinase-1 (PFK-1) is an essential gatekeeper enzyme in glycolysis that catalyzes the conversion of fructose 6-phosphate to fructose 1,6-bisphosphate. It is a well-known example of an enzyme that is allosterically controlled (Khan 2009). Low levels of energy in the cell are indicated by both AMP and fructose 2,6-bisphosphate, which activates PFK-1 and promotes glycolysis to create ATP (Bartrons et al. 2018). However, PFK-1 is blocked by ATP and citrate, indicating that there is sufficient energy present and slowing down glycolysis to prevent consuming further glucose (Alfarouk et al. 2014). This regulation ensures that glycolysis only takes place when energy is needed. In cancer cells, PFK-1 plays a critical role in the Warburg effect, where glycolysis is upregulated even in the presence of oxygen, supporting the altered metabolism of cancer (Chen et al. 2016).

1.3.3 Fine Control 3: The Control by Compartmentation and Precursor Flux

Compartmentation of metabolism is a means by which pathways can be regulated independently from one another and is a mechanism of controlling the flux of carbon through the pathways in response to the demands of cellular metabolism (Bar-Peled and Kory 2022). It provides environments more favorable to certain reactions. The integration of cell metabolism requires controlled interaction between the pathways in various compartments. This is achieved via the high selective permeability of the organellar membranes leading to the transport of the limited number of intermediates from one compartment to another (Jain and Zoncu 2022). This regulation refers to the mechanisms through which cells regulate metabolic pathways with high precision, ensuring that the right substrates are available for the right processes at the right time.

1.3.4 Fine Control 4: NADH/ATP and Carbon Flow

The regulation of any pathway depends mainly on providing ATP (the energy currency of cell metabolism), the reducing power (NADH or NADPH), and the building blocks for biosynthesis (Bonora et al. 2012). Their production is controlled by the availability of enzymes (activity and amount) and the substrate in addition to compartmentation and interaction between metabolically distinct organs.

The biosynthetic pathways in nongreen plastids are dependent on a supply of precursors, ATP, and reductants that are generated by the oxidation of carbohydrates (Selinski et al. 2014). The primary function of the plastidic glycolysis and pentose phosphate pathway (PPP) in the nonphotosynthetic tissue is to provide NAD(P)H and ATP for biosynthetic reactions for important processes such as nitrogen assimilation, amino acid synthesis, and fatty acid synthesis. The intermediate of glycolysis may also serve as precursors of acetyl units (Stincone et al. 2015). It is believed that the synthesis of secondary metabolites is associated with the defense reactions. Activation of PPP is a part of plant's defense reactions. Glucose-6-phosphate dehydrogenase (G6PD) is the rate-limiting enzyme of PPP and determines the amount of NADPH by controlling the metabolism of glucose via the PPP (Fig. 1.3).

1.3.5 Fine Control 5: Variation in pH

The majority of enzymes have a distinctive ideal pH where their function is maximal. The activity typically decreases at pH values over or below, albeit to differing extents based on the specific enzyme (Herlet et al. 2017). The pH optimum of an enzyme is not always the same as the pH of its intracellular surroundings, providing the suitable pH switch for other enzymes (Herlet et al. 2017).

Cells contain thousands of enzymes, many of which show different responses to pH; the intracellular pH may represent an important element of fine metabolic

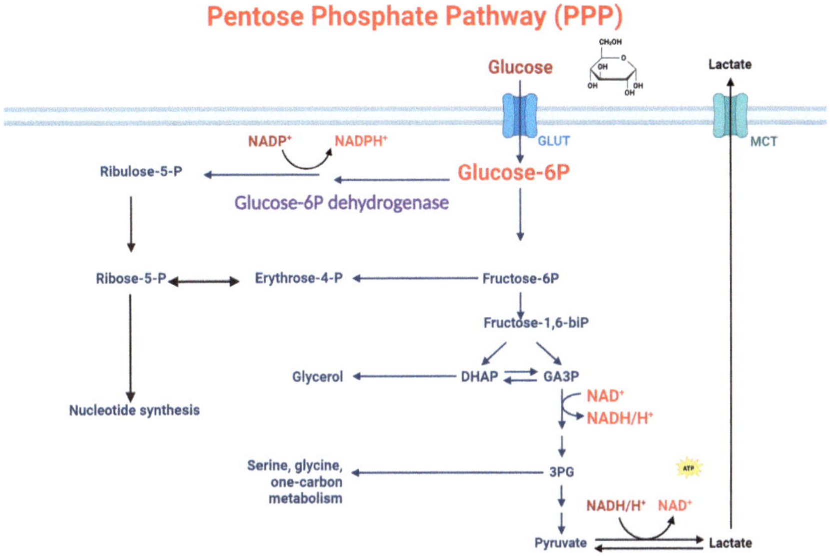

Fig. 1.3 Pentose phosphate pathway. The figure shows that glucose-6-phosphate dehydrogenase (G6PD) is the rate-limiting enzyme in the pentose phosphate pathway (PPP), playing a critical role in regulating the production of NADPH

control (Jin et al. 2022). The light-dependent activation of several enzymes of the reductive pentose phosphate pathway (Calvin-Benson cycle) provides a well-documented example of how changes in pH can contribute to metabolic control within a cell. Photosynthetic electron transport has been linked to H^+ uptake from the stroma into the thylakoid lumen, and thus establishes a proton gradient between the lumen and stroma that drives the photophosphorylation of ADP by inorganic phosphate to form ATP (Foyer et al. 2012).

1.3.6 Fine Control 6: Covalent Modification

The interconversion of an enzyme between a less active (or inactive) and a more active form occurs when two thermodynamically favorable enzyme-catalyzed events result in new, stable covalent bonds on the surface of the target enzyme (Nam et al. 2024). Many different protein post-translational modifications can affect enzymatic function. Changes in the structure of the enzyme caused by covalent modification usually affect the enzyme-substrate interactions, such as V_{max} and K_m. Through covalent modification, it is possible to quickly construct a "new" enzyme form for the cell that has kinetics specific to its transitory metabolic needs (Herman 1980). Enzyme control by reversible covalent modification is the major mechanism whereby extracellular signals can coordinate the control of key enzymes of intracellular pathways. Reversible covalent modification of enzymes typically allows a very marked sensitivity (e.g., "amplification") to signals.

1.3.7 Fine Control 7: The Association of Metabolically Sequential Enzymes

Enzymes that catalyze metabolic processes in a specific order are known as metabolic sequential enzymes. Usually, the next enzyme uses the product of the previous enzyme as a substrate. The rate of flow across soluble metabolic pathways and the degree of enzyme binding are highly correlated (Durek and Walther 2008). Certain sequential metabolic enzyme interactions can affect each other's activity in a reciprocal kinetic way.

1.4 Coarse Control Mechanisms

Coarse metabolic control is an energetically costly, slow process that takes hours to days for eukaryotes and minutes to hours for rapidly developing prokaryotes. It is accomplished by alterations in the overall population of enzyme molecules within the cell (Lane 2011). The relative rates of an enzyme's production and breakdown determine how much of that enzyme is present overall. Therefore, coarse metabolic control can be defined as any change in the rates of proteolysis or gene expression (i.e., transcription, translation, mRNA processing, or degradation). One or all the enzymes in each pathway may be subject to coarse control, which most commonly occurs in response to stimulation, tissue differentiation, or long-term environmental (adaptive) changes. Enzyme proteolysis can be specifically targeted and may be triggered in response to specific stimuli (Bonnans et al. 2014).

The major control of the level of enzyme is determined by the differential expression of their genes with the possibility of translational regulation for the uptake of protein into the organelle (Carthew 2021). It is well known that the control of gene expression and other cellular functions are closely correlated with cell metabolism. In addition to providing the precursor chemicals required for gene expression, metabolic pathways also produce ATP as an energy source (Carthew 2021; Beckwith 2011). Cells must constantly adjust gene expression to their shifting energy and metabolite budgets (Brown 1992).

The transcription is the initial stage in the biological information flow from genome to proteome; transcription regulation is fundamental to the development, complexity, and homeostasis of all species. The crucial role of transcription factors is particularly well illustrated by the link between alterations of their functions and the occurrence of major metabolic diseases (Brynildsen et al. 2005).

Proteolytic regulation of metabolic enzymes is a critical mechanism by which cells control metabolic processes. Enzymatic protein breakdown, or proteolysis, can control the stability, turnover, and activity of enzymes, which can affect a number of different metabolic processes (Motyan et al. 2013). The ubiquitin-proteasome system (UPS) is the common proteolytic system that controls protein degradation in both prokaryotic and eukaryotic cells. It is involved in many physiological functions, including DNA repair, stress responses, cell proliferation (Park et al. 2020), and metabolic pathways (Wang et al. 2023).

1.5 Metabolic Controls in Defense and Attack

Metabolic regulators play a vital role in both defense and attack mechanisms inside organisms (Al-Khayri et al. 2023; Anjali et al. 2023). In defense, cells regulate metabolism to respond to stressors like infections or injury by modifying energy production and synthesis pathways, which aids in generating an effective immune response. For instance, immune cells can switch to glycolysis to rapidly produce energy and maintain their functions during an infection.

Metabolic controls can be manipulated during an attack, like cancer or pathogen invasion, to promote rapid growth or survival. While infections may affect host metabolic pathways to their advantage, tumor cells typically exhibit altered metabolism to support their rapid proliferation. In order to ensure that organisms can respond to threats in an efficient manner while preserving overall metabolic balance, both defense and attack require intricate metabolic balance (Ganeshan and Chawla 2014).

1.6 Microenvironment in Metabolic Control

The cellular microenvironment plays a vital role in shaping the metabolism by influencing factors like oxygen levels, pH, nutrient availability, and signaling interactions with the surrounding cells and the extracellular matrix (Lyssiotis and Kimmelman 2017). These physical and biochemical conditions regulate cellular activity and metabolic pathways, enabling cells to adapt in both healthy and diseased tissues (Simões et al. 2015). The remodeling of tumor microenvironments (TME), influenced by metabolism, leads to pathophysiologic interactions like nutrient sharing, competition, and metabolite role as signaling molecules. These processes are crucial in tumor metabolism and growth, aiding in the design of new drug targets and cancer therapies.

1.6.1 Nutrient Availability

The microenvironment's nutrient availability plays a crucial role in regulating cellular metabolism, influencing whether cells prioritize energy production or biosynthesis, an adaptation commonly seen in rapidly growing tissues like tumors (Arner and Rathmell 2023). Cellular energy status is regulated by two key pathways: AMP-activated protein kinase (AMPK) and the mammalian target of rapamycin (mTOR) (Inoki et al. 2012). AMPK is activated in nutrient-poor microenvironment,

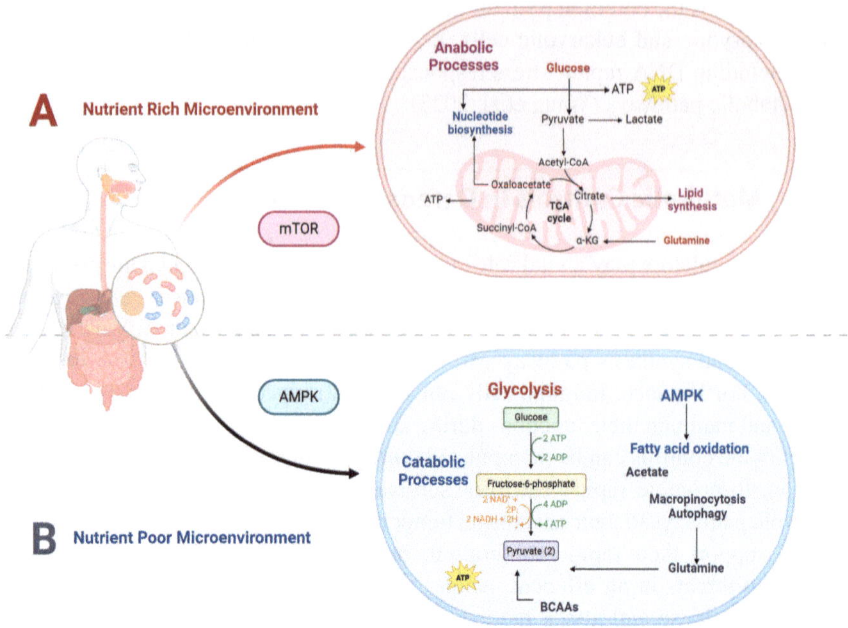

Fig. 1.4 Nutrient availability in metabolic regulation. The figure illustrates metabolic regulation in nutrient-rich (**a**) and nutrient-poor (**b**) microenvironments, highlighting the activation of mTOR in rich conditions to promote anabolic processes like lipid and nucleotide biosynthesis, and AMPK activation in poor conditions to drive catabolic processes such as glycolysis and fatty acid oxidation

promoting catabolic processes like glycolysis and fatty acid oxidation to conserve energy (Shirwany and Zou 2014). On the other hand, mTOR drives anabolic growth and protein synthesis in nutrient-rich microenvironment by integrating signals from glucose and amino acids to promote biomass accumulation (Cuyàs et al. 2014). Cells adapt to nutrient fluctuations by optimizing energy use for survival (Fig. 1.4). In cancer, metabolic reprogramming allows tumor cells to adjust to microenvironmental conditions, balancing energy demands to sustain growth and survival (Faubert et al. 2020).

Microenvironments vary across tissues, with different metabolic profiles tailored to specific functions. For instance, the liver plays a key role in regulating whole-body energy metabolism by controlling glucose synthesis (gluconeogenesis) and storage based on nutrient availability (Alamri 2018).

Microorganisms, like *E. coli*, utilize catabolite repression, where the presence of glucose inhibits the metabolism of alternative carbon sources, ensuring the most efficient use of energy (Kremling et al. 2015). Pathogens manipulate host metabolic pathways to access essential nutrients needed for their survival and replication. Bacteria, viruses, and fungi exploit host resources such as glucose, iron, fatty acids, and amino acids. For example, *E. coli* produces siderophores to scavenge iron (Garénaux et al. 2011), while viruses like HCMV alter glucose and glutamine metabolism for their replication (Yu et al. 2011). In response, the host employs strategies like nutritional immunity to restrict nutrient access, creating a competitive struggle that shapes disease outcomes (Hood and Skaar 2012). Ultimately, these interactions highlight how nutrient availability and metabolic control in both host and pathogens are critical for maintaining homeostasis and influencing health and disease progression.

1.6.2 Oxygen Availability

Oxygen in the microenvironment plays a crucial role in regulating cellular metabolism, particularly through oxidative phosphorylation (De Miguel et al. 2015). Under hypoxic conditions, cells shift to anaerobic glycolysis to maintain energy production (Lee et al. 2021). This shift is driven by hypoxia-inducible factor 1-alpha (HIF-1α), which upregulates key enzymes such as lactate dehydrogenase (LDH) and glucose transporters (GLUT1) to enhance glycolysis and glucose uptake (Lee et al. 2021).

In cancer, HIF-1α stabilizes under hypoxia, promoting the Warburg effect, where cancer cells favor glycolysis even when oxygen is present, supporting rapid growth (Infantino et al. 2021). In the wound microenvironment, hypoxia induces vascular endothelial growth factor (VEGF) production, promoting angiogenesis to restore oxygen and nutrients.

1.6.3 Microenvironment pH Role in the Metabolism Control

pH plays a crucial role in cellular microenvironment by influencing enzyme activity and overall metabolic regulation (Justus et al. 2015). Many metabolic enzymes have an optimal pH range, and deviations can inhibit their function (Tipton et al. 2009). In pathological conditions such as tumors and inflammation, the microenvironment becomes acidic due to lactic acid buildup. This abnormal pH disrupts normal metabolism and weakens immune responses, allowing diseased cells to thrive while impairing regular cellular and immune functions (Asgharzadeh et al. 2017).

Enzymes involved in oxidative phosphorylation, like cytochrome c oxidase, perform better in neutral to slightly alkaline pH, promoting efficient ATP synthesis. In the pentose phosphate pathway (PPP) (Alfarouk et al. 2020), glucose-6-phosphate dehydrogenase (G6PD) is inhibited by low pH, which can impair biosynthesis and antioxidant defenses (Singh et al. 2012). Additionally, low pH impairs immune cell functions, such as those of cytotoxic T cells and natural killer (NK) cells, allowing tumors to evade immune detection and thrive despite immune surveillance (Moinuddin et al. 2024). These examples highlight how pH variations in the microenvironment affect key metabolic pathways, especially in pathological conditions.

1.7 Microenvironment Control in Disease Management

Metabolic control within the microenvironment plays a crucial role in the treatment of various diseases. In cancer therapy, the tumor microenvironment (TME) is characterized by hypoxia and nutrient deprivation, forcing tumors to rely on glycolysis (Warburg's effect) (Chen et al. 2016). Targeting this metabolic pathway by LDH disrupts lactate production, starving cancer cells in low-oxygen conditions (Lee et al. 2021). Additionally, 2-deoxyglucose blocks glycolysis, while G6PD inhibitors target the PPP to reduce NADPH production and increase oxidative stress (Mele et al. 2018). Metformin, a diabetes drug, activates AMPK, suppressing tumor growth by downregulating mTOR, thus decreasing energy demanding biosynthetic pathways (Loubière et al. 2015).

Metabolic control in a microenvironment can enhance wound healing and tissue repair through the shifting of cells toward glycolysis and angiogenesis effect under hypoxic conditions (Ramakrishnan et al. 2014). Metformin enhances insulin sensitivity in diabetes through the influence of inflammation, and glucose uptake, promoting fatty acid oxidation and activation of AMPK (Ford et al. 2015). By restoring metabolic balance, it helps counteract the insulin resistance caused by chronic inflammation in tissues like the liver and adipose tissue (Ford et al. 2015). Accordingly, metabolism is increasingly recognized to play a critical role in immune responses. Infection fuels the pro-inflammatory macrophages (M1) toward glycolysis for rapid energy production, while regulatory T cells (Tregs) prefer oxidative phosphorylation to maintain immunological tolerance (Chauhan and Saha 2018). Targeting such metabolic pathways will allow the establishment of treatments for inflammatory diseases (Chauhan and Saha 2018). Indeed, these metabolic pathways now emerge as a key focus that dramatically expands treatment options for a range of diseases. A deeper comprehension of the microenvironment's influence on metabolic regulation enhances the efficacy of therapeutic interventions in cancer metabolism, inflammation, wound healing, and diabetes management.

1.8 Conclusion

Metabolic control mechanisms, established by both fine and coarse regulatory techniques, are vital for maintaining homeostasis and enabling organisms to adjust to internal and external changes. These regulators function on several levels, ranging from enzyme activity to gene expression, guaranteeing that metabolic flux corresponds to cellular requirements. The microenvironment is crucial in controlling these metabolic processes through factors such as pH, oxygen concentrations, and nutrition availability. Gaining insight into the complex interplay between metabolism and microenvironment dynamics is essential for developing novel therapeutic strategies, especially for diseases like cancer, diabetes, and inflammation where metabolic reprogramming plays a key role in the course of the illness and the effectiveness of treatment.

References

Abhishek S et al (2023) Allosteric crosstalk in modular proteins: function fine-tuning and drug design. Comput Struct Biotechnol J 21:5003–5015

Alamri ZZ (2018) The role of liver in metabolism: an updated review with physiological emphasis. Int J Basic Clin Pharmacol 7(11):2271–2276

Alfarouk KO et al (2014) Glycolysis, tumor metabolism, cancer growth and dissemination. A new pH-based etiopathogenic perspective and therapeutic approach to an old cancer question. Oncoscience 1(12):777

Alfarouk KO et al (2020) The pentose phosphate pathway dynamics in cancer and its dependency on intracellular pH. Metabolites 10(7):285

Al-Khayri JM et al (2023) Plant secondary metabolites: the weapons for biotic stress management. Metabolites 13(6):716

Anjali et al (2023) Role of plant secondary metabolites in defence and transcriptional regulation in response to biotic stress. Plant Stress 8:100154

Aon M, Cortassa S, Lloyd Mbe D (2011) Chaos in biochemistry and physiology. In: Encyclopedia of molecular cell biology and molecular medicine. Wiley, Hoboken

Arner EN, Rathmell JC (2023) Metabolic programming and immune suppression in the tumor microenvironment. Cancer Cell 41(3):421–433

Asgharzadeh MR et al (2017) Molecular machineries of pH dysregulation in tumor microenvironment: potential targets for cancer therapy. Bioimpacts 7(2):115

Bar-Peled L, Kory N (2022) Principles and functions of metabolic compartmentalization. Nat Metab 4(10):1232–1244

Bartrons R et al (2018) Fructose 2,6-bisphosphate in cancer cell metabolism. Front Oncol 8:331

Beckwith J (2011) The operon as paradigm: normal science and the beginning of biological complexity. J Mol Biol 409(1):7–13

Bonnans C, Chou J, Werb Z (2014) Remodelling the extracellular matrix in development and disease. Nat Rev Mol Cell Biol 15(12):786–801

Bonora M et al (2012) ATP synthesis and storage. Purinergic Signal 8(3):343–357

Brown GC (1992) Control of respiration and ATP synthesis in mammalian mitochondria and cells. Biochem J 284(Pt 1):1–13

Brown JH et al (2004) Toward a metabolic theory of ecology. Ecology 85(7):1771–1789

Brynildsen MP, Wong WW, Liao JC (2005) Transcriptional regulation and metabolism. Biochem Soc Trans 33(Pt 6):1423–1426

Carthew RW (2021) Gene regulation and cellular metabolism: an essential partnership. Trends Genet 37(4):389–400

Chandel NS (2021) Basics of metabolic reactions. Cold Spring Harb Perspect Biol 13(8):a040527

Chauhan P, Saha B (2018) Metabolic regulation of infection and inflammation. Cytokine 112:1–11

Chen X-S et al (2016) Anticancer strategies based on the metabolic profile of tumor cells: therapeutic targeting of the Warburg effect. Acta Pharmacol Sin 37(8):1013–1019

Cornish-Bowden A (2014) Understanding allosteric and cooperative interactions in enzymes. FEBS J 281(2):621–632

Cuyàs E et al (2014) Cell cycle regulation by the nutrient-sensing mammalian target of rapamycin (mTOR) pathway. In: Cell cycle control: mechanisms and protocols. Springer, New York, pp 113–144

De Miguel M et al (2015) Cell metabolism under microenvironmental low oxygen tension levels in stemness, proliferation and pluripotency. Curr Mol Med 15(4):343–359

Durek P, Walther D (2008) The integrated analysis of metabolic and protein interaction networks reveals novel molecular organizing principles. BMC Syst Biol 2:100

Faubert B, Solmonson A, DeBerardinis RJ (2020) Metabolic reprogramming and cancer progression. Science 368(6487):eaaw5473

Ford RJ et al (2015) Metformin and salicylate synergistically activate liver AMPK, inhibit lipogenesis and improve insulin sensitivity. Biochem J 468(1):125–132

Foyer CH et al (2012) Photosynthetic control of electron transport and the regulation of gene expression. J Exp Bot 63(4):1637–1661
Freiburger LA et al (2011) Competing allosteric mechanisms modulate substrate binding in a dimeric enzyme. Nat Struct Mol Biol 18(3):288–294
Ganeshan K, Chawla A (2014) Metabolic regulation of immune responses. Annu Rev Immunol 32:609–634
Garénaux A, Caza M, Dozois CM (2011) The Ins and Outs of siderophore mediated iron uptake by extra-intestinal pathogenic Escherichia coli. Vet Microbiol 153(1–2):89–98
Herlet J et al (2017) A new method to evaluate temperature vs. pH activity profiles for biotechnological relevant enzymes. Biotechnol Biofuels 10:234
Herman RH (1980) The principles of metabolic control. In: Herman RH, Cohn RM, McNamara PD (eds) Principles of metabolic control in mammalian systems. Springer US, Boston, pp 1–61
Hood MI, Skaar EP (2012) Nutritional immunity: transition metals at the pathogen–host interface. Nat Rev Microbiol 10(8):525–537
Infantino V et al (2021) Cancer cell metabolism in hypoxia: role of HIF-1 as key regulator and therapeutic target. Int J Mol Sci 22(11):5703
Inoki K, Kim J, Guan K-L (2012) AMPK and mTOR in cellular energy homeostasis and drug targets. Annu Rev Pharmacol Toxicol 52(1):381–400
Jain A, Zoncu R (2022) Organelle transporters and inter-organelle communication as drivers of metabolic regulation and cellular homeostasis. Mol Metab 60:101481
Jin X et al (2022) Effects of pH alterations on stress- and aging-induced protein phase separation. Cell Mol Life Sci 79(7):380
Judge A, Dodd MS (2020) Metabolism. Essays Biochem 64(4):607–647
Justus CR, Sanderlin EJ, Yang LV (2015) Molecular connections between cancer cell metabolism and the tumor microenvironment. Int J Mol Sci 16(5):11055–11086
Khan SA (2009) Beyond the allosteric role of fructose-2,6-bisphosphate: exploring its effects on gene expression and signal transduction. Thesis, University of Minnesota
Kitamura S, Shimizu H, Toya Y (2021) Identification of a rate-limiting step in a metabolic pathway using the kinetic model and in vitro experiment. J Biosci Bioeng 131(3):271–276
Kremling A et al (2015) Understanding carbon catabolite repression in Escherichia coli using quantitative models. Trends Microbiol 23(2):99–109
Kurata H et al (2007) Integration of enzyme activities into metabolic flux distributions by elementary mode analysis. BMC Syst Biol 1:31
Lane N (2011) Energetics and genetics across the prokaryote-eukaryote divide. Biol Direct 6:35
Lee S-H, Golinska M, Griffiths JR (2021) HIF-1-independent mechanisms regulating metabolic adaptation in hypoxic cancer cells. Cells 10(9):2371
Lira AL et al (2019) Allosteric inhibition of α-thrombin enzymatic activity with ultrasmall gold nanoparticles. Nanoscale Adv 1(1):378–388
Lorendeau D et al (2015) Metabolic control of signalling pathways and metabolic auto-regulation. Biol Cell 107(8):251–272
Loubière C et al (2015) Metformin-induced energy deficiency leads to the inhibition of lipogenesis in prostate cancer cells. Oncotarget 6(17):15652
Lyssiotis CA, Kimmelman AC (2017) Metabolic interactions in the tumor microenvironment. Trends Cell Biol 27(11):863–875
Mason EF, Rathmell JC (2011) Cell metabolism: an essential link between cell growth and apoptosis. Biochim Biophys Acta 1813(4):645–654
McCullagh M et al (2024) What is allosteric regulation? Exploring the exceptions that prove the rule! J Biol Chem 300(3):105672
Medina-Cleghorn D, Nomura DK (2014) Exploring metabolic pathways and regulation through functional chemoproteomic and metabolomic platforms. Chem Biol 21(9):1171–1184
Mele L et al (2018) A new inhibitor of glucose-6-phosphate dehydrogenase blocks pentose phosphate pathway and suppresses malignant proliferation and metastasis in vivo. Cell Death Dis 9(5):572

Moinuddin A et al (2024) Metabolic adaptations determine whether natural killer cells fail or thrive within the tumor microenvironment. Immunol Rev 323(1):19–39

Mor I, Cheung E, Vousden K (2011) Control of glycolysis through regulation of PFK1: old friends and recent additions. In: Cold Spring Harbor symposia on quantitative biology. Cold Spring Harbor Laboratory Press, New York

Moreno-Sánchez R et al (2008) Metabolic control analysis: a tool for designing strategies to manipulate metabolic pathways. J Biomed Biotechnol 2008:597913

Motyan JA, Toth F, Tozser J (2013) Research applications of proteolytic enzymes in molecular biology. Biomolecules 3(4):923–942

Nam K et al (2024) Perspectives on computational enzyme modeling: from mechanisms to design and drug development. ACS Omega 9(7):7393–7412

Park J, Cho J, Song EJ (2020) Ubiquitin-proteasome system (UPS) as a target for anticancer treatment. Arch Pharm Res 43(11):1144–1161

Plaxton WC (2004) Principles of metabolic control. In: Functional metabolism: regulation and adaptation. Wiley, Hoboken, pp 1–24

Porter CM, Miller BG (2012) Cooperativity in monomeric enzymes with single ligand-binding sites. Bioorg Chem 43:44–50

Qiu S et al (2023) Small molecule metabolites: discovery of biomarkers and therapeutic targets. Signal Transduct Target Ther 8(1):132

Ramakrishnan S, Anand V, Roy S (2014) Vascular endothelial growth factor signaling in hypoxia and inflammation. J Neuroimmune Pharmacol 9:142–160

Robinson PK (2015) Enzymes: principles and biotechnological applications. Essays Biochem 59:1–41

Selinski J et al (2014) The plastid-localized NAD-dependent malate dehydrogenase is crucial for energy homeostasis in developing Arabidopsis thaliana seeds. Mol Plant 7(1):170–186

Shirwany NA, Zou M-H (2014) AMPK: a cellular metabolic and redox sensor. A minireview. Front Biosci (Landmark Ed) 19:447

Simões RV et al (2015) Metabolic plasticity of metastatic breast cancer cells: adaptation to changes in the microenvironment. Neoplasia 17(8):671–684

Singh S, Anand A, Srivastava PK (2012) Regulation and properties of glucose-6-phosphate dehydrogenase: a review. Int J Plant Physiol Biochem 4(1):1–19

Stincone A et al (2015) The return of metabolism: biochemistry and physiology of the pentose phosphate pathway. Biol Rev Camb Philos Soc 90(3):927–963

Tipton KF, McDonald AG, Dixonw H (2009) Effects of pH on enzymes. In: Contemporary enzyme kinetics and mechanism. Reliable Lab Solutions, p 123

Wang J et al (2023) The ubiquitin-proteasome system in tumor metabolism. Cancers (Basel) 15(8):2385

Wegner A et al (2015) How metabolites modulate metabolic flux. Curr Opin Biotechnol 34:16–22

Yu Y, Clippinger AJ, Alwine JC (2011) Viral effects on metabolism: changes in glucose and glutamine utilization during human cytomegalovirus infection. Trends Microbiol 19(7):360–367

Zhu J, Thompson CB (2019) Metabolic regulation of cell growth and proliferation. Nat Rev Mol Cell Biol 20(7):436–450

Zuo J et al (2021) Glycolysis rate-limiting enzymes: novel potential regulators of rheumatoid arthritis pathogenesis. Front Immunol 12:779787

Bioinformatics and Artificial Intelligence Approaches in Metabolic Pathways

2

Mohamed E. Hasan, Mohamed A. Osman, Alaa A. Hemeida, and Mahmoud ElHefnawi

Abstract

Metabolic pathways, which encompass the series of chemical responses within a cell, are critical for understanding natural processes similar to energy product, growth, and cellular form. This explores the crossroad of metabolic pathway analysis with artificial intelligence (AI), an emerging tool that is transubstantiating the field of biology and biotechnology. AI still marks a shift in how these pathways are studied and modeled. The foundations of AI, particularly its use of machine literacy and data-driven algorithms, offer important tools for prophetic modeling of metabolic pathways. Through AI-grounded models, experimenters can pretend and read metabolic responses more directly, furnishing perceptivity into complaint mechanisms, medicine discovery, and metabolic interventions. This further explores how metabolomics data, which involve the large-scale study of small motes within cells, benefit from AI's advanced analysis ways. AI not only enables the processing of vast datasets but also reveals intricate patterns and correlations that help consolidate our understanding of metabolism in a molecular position. In addition to perfecting exploration styles, AI has opened new possibilities for individualized drug and nutritive interventions.

Keywords

Metabolic pathways · Artificial intelligence · Prophetic modeling · Metabolic responses · Metabolomics data

M. E. Hasan · M. A. Osman · A. A. Hemeida
Bioinformatics Department, Genetic Engineering and Biotechnology Research Institute, University of Sadat City, Sadat City, Egypt

M. ElHefnawi (✉)
Biomedical Informatics and Chemoinformatics Group (BICG), Informatics and Systems Department, National Research Centre, Cairo, Egypt
e-mail: mahef@aucegypt.edu

S. S. M. Soliman, M. I. Husseiny (eds.), *Metabolic Dynamics in Host-Microbe Interaction*, https://doi.org/10.1007/978-981-96-1305-2_2

2.1 Introduction

In biochemistry, a metabolic pathway is defined as a linked group of chemical responses within a cell. The interceders of an enzymatic response are known as metabolites, which are adjusted by a sequence of chemical responses catalyzed by enzymes. In the utmost cases of a metabolic pathway, the product of one enzyme behaves as the substrate for the coming. Still, side products are regarded as waste and removed from the cell. Different metabolic pathways serve the role within a eukaryotic cell and the significance of the pathway in the given cube of the cell. For case, the electron transport chain and oxidative phosphorylation occur in the mitochondrial membrane (Nelson and Cox 2008; Alison et al. 2014; Nicholson 1971; Harvey 2011). In contrast, glycolysis, pentose phosphate pathway, and adipose acid biosynthesis are performed in the cytosol of a cell. Two categories of metabolic pathways are characterized by their capability to either synthesize motes with the application of energy (anabolic pathway) or break down complex motes and emit energy in the process (catabolic pathway). The two pathways round to perform the biosynthesis of an anabolic pathway. In addition to the two separate metabolic pathways is the amphibolic pathway, which can be either catabolic or anabolic grounded on the need for or the vacuity of energy (Voet et al. 2013; Reece and Campbell 2011; Berg et al. 2012a). Pathways are needed for the conservation of homeostasis within an organism, and the flux of metabolites through a pathway is regulated depending on the requirements of the cell and the vacuity of the substrate. The end product of a pathway may be used incontinently, initiate another metabolic pathway, or be stored after use. The metabolism of a cell consists of an elaborate network of connected pathways that enable the conflation and breakdown of motes (anabolism and catabolism).

2.1.1 Catabolic Pathway (Catabolism)

It is a sequence of responses that bring about a net emit of energy in the form of a high-energy phosphate bond formed with the energy carrier's adenosine diphosphate (ADP) and guanosine diphosphate (GDP) to produce adenosine triphosphate (ATP) and guanosine triphosphate (GTP), independently (Harvey 2011).

The net response is, thus, thermodynamically favorable, resulting in lower free energy for the final products (Berg et al. 2002). A catabolic pathway is an exergonic system that produces chemical energy in the form of ATP, GTP, NADH, NADPH, FADH2, etc., from energy-holding sources similar to carbohydrates, fats, and proteins. The end products are frequently carbon dioxide, water, and ammonia. Coupled with an endergonic response of anabolism, the cell can synthesize new macromolecules using the original precursors of the anabolic pathway (Raven et al. 2011). An illustration of a coupled response is the phosphorylation of fructose 6-phosphate to generate the intermediate fructose 1,6-bisphosphate by the enzyme phosphofructokinase together with the hydrolysis of ATP in the pathway of glycolysis. The performing chemical response within the metabolic pathway is largely thermodynamically favorable and, as a result, unrecoverable in the cell.

2.1.2 Anabolic Pathway (Anabolism)

In contrast to catabolic pathways, anabolic pathways bear an energy input to construct macromolecules similar to polypeptides, nucleic acids, proteins, polysaccharides, and lipids. The insulated response of anabolism is inimical in a cell due to a positive Gibbs free energy (ΔG). Therefore, an input of chemical energy through a coupling with an exergonic response is necessary. The coupled response of the catabolic pathway affects the thermodynamics of the response by lowering the overall activation energy of an anabolic pathway and allowing the response to occur. Otherwise, an endogenous response is nonspontaneous synthesis of glucose from non-carbohydrate precursors, primarily in the liver and kidneys. It serves as a critical mechanism to maintain blood glucose levels, especially during fasting or prolonged exercise when glucose supply from dietary sources is insufficient. While gluconeogenesis shares some similarities with the reverse of glycolysis, it involves several unique enzymes, including pyruvate carboxylase, phosphoenolpyruvate carboxykinase, fructose 1,6-bisphosphatase, and glucose 6-phosphatase, which enable the pathway to proceed in the anabolic direction (Berg et al. 2002, 2012b).

2.2 Foundations of Artificial Intelligence

Artificial intelligence (AI) employs machines and software to carry out tasks that are generally associated with human cognitive functions, including natural language understanding, pattern recognition, decision-making, and problem-solving (Tien 2017). When applied to complex datasets, AI has shown exceptional skill in identifying concealed patterns, distinguishing relevant features in noisy environments, and producing precise predictions (Barnes et al. 2020; Li et al. 2020). These strengths render AI particularly advantageous for metabolomics data analysis. The utilization of AI models in metabolomics is highly effective due to the presence of extensive datasets and intricate patterns within the data. The incorporation of AI into metabolomics has significantly enhanced data analysis, biological interpretation, and predictive statistical modeling. Consequently, AI-assisted metabolomics has led to notable advancements in various applications, including disease diagnosis, precision medicine, and drug discovery. Cutting-edge AI algorithms have been created to process unrefined mass spectrometry data, detect chromatographic peaks, identify metabolites, uncover biomarkers, develop predictive models, construct biological pathways, and perform other functions that pose challenges to conventional statistical techniques. Furthermore, AI holds significant potential in integrative multi-omics analysis, thus contributing to advancements in the field of systems biology (Greener et al. 2022; Alajaji et al. 2023). Since the 1990s, AI techniques, mainly including machine learning (ML) and neural networks (NNs), have gained wide applications in data analysis and pattern recognition. For instance, ML was used for sequence analysis and identifying genetic markers in genomics, and clustering algorithms were employed in chemistry for compound classification (Fig. 2.1).

2.2.1 Machine Learning

Machine learning (ML) is employed to develop models that can effectively address large-scale datasets and solve complex problems through a learning process. The foundational principle of ML algorithms lies in their ability to construct mathematical models from a set of sample data (Dhall et al. 2020). Typically, a dataset for developing an ML machine learning model is partitioned into a training subset, which constitutes approximately 70% of the available data and is utilized by the ML algorithm to construct the model and generate predictions, and a testing subset, representing around 30% of the data, which provides an unbiased assessment of the model's performance following training.

Often, an intermediate validation phase is incorporated to assist in identifying the most accurate model and optimizing its hyperparameters. In this scenario, data may be divided according to a 60–20–20 split, with 20% allocated as a validation set. The initial training process necessitates a substantial amount of data, enabling the ML algorithm to have multiple opportunities to learn and refine the model. The learning capability of the algorithm operates through a mathematical function that correlates specific inputs with corresponding outputs. The training dataset informs the algorithms to make predictions without explicit programming, achieved through successive operations that involve adjusting weights and biases to generate predictions within a finite number of iterations (Cohen 2021).

After the algorithm has been trained and developed a generalized model, the next phase involves evaluating the model's performance on an independent dataset containing previously unseen data, yielding sufficiently accurate predictions. Predictions are based on the algorithm's proficiency in associating each input with the statistical representation defined by the user. The success of the algorithm in learning from the input data directly influences the accuracy of its predictions (Antonakoudis et al. 2020).

Constructing a machine learning (ML) model involves a systematic series of steps:

1. *Defining the Training Dataset*: This step entails identifying the appropriate type of data to be utilized as the training dataset, wherein the input data will vary according to the specific problem being addressed.
2. *Gathering the Training Dataset*: A representative dataset that reflects real-world applications must be compiled, containing inputs that effectively respond to the problem under investigation.
3. *Input Feature Representation*: The accuracy of the learned function is significantly influenced by the representation of the input objects. These input objects are converted into feature vectors that encompass various descriptive features. It

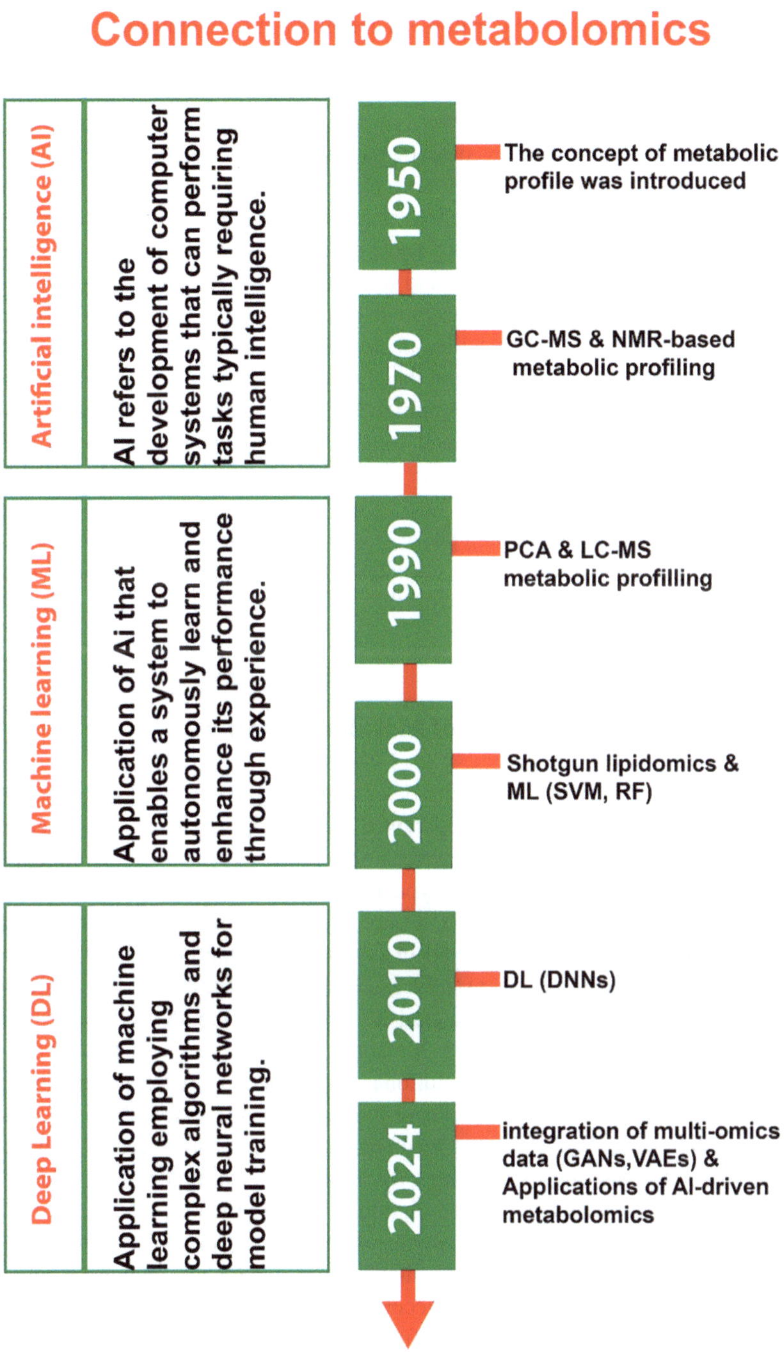

Fig. 2.1 Historical overview of artificial intelligence and its associations with metabolomics

is essential to ensure that the feature set contains a sufficient number of attributes to provide adequate information for accurate output prediction while avoiding excessive dimensionality that could complicate the model.

4. *Determining the Type of Algorithm*: The selection of an appropriate algorithm is crucial, as it will be employed to fit the data during the training and testing phases. This choice depends on multiple factors, including the specific research question being addressed, the characteristics of the dataset, and the category of machine learning being utilized, such as supervised learning, unsupervised learning, reinforcement learning, and semi-supervised learning.
5. *Training the Algorithm*: This phase involves executing the algorithm on the gathered training dataset. The specifics of this process may vary and could require additional user input contingent on the chosen algorithm. Cross-validation techniques may be implemented to optimize hyperparameters—variables that influence how the algorithm is trained (e.g., learning rate, number of branches, clusters, and epochs)—to enhance performance on a subset of the training data.

Validation Following the training phase, a validation step is typically conducted to fine-tune the hyperparameters of the classifiers, thereby improving the model's predictive performance. The general workflow of AI in metabolomics is shown in Fig. 2.2.

2.3 Predictive Modeling of Metabolic Pathways

Mathematical modeling can be used to design specialized convolutional neural networks, trained through a novel unsupervised learning approach to solve linear systems. While computational biology increasingly focuses on large-scale modeling of dynamical systems to predict phenotype from genotype, it remains an emerging field (Tompson et al. 2016). As such, it may not yet consistently deliver the highly precise predictions typically expected in other engineering domains (Lewis et al. 2012). Arguably, the most extensively used and successful modeling fashion in metabolic engineering involves the analysis of internal metabolic fluxes through stoichiometric models of metabolism. Metabolic flux values are constrained by stoichiometry, thermodynamic and evolutionary hypotheticals, or experimental data and are used to suggest inheritable interventions that bring cell metabolism closer to the asked phenotype. While this approach has had significant successes, it has also shown its limitations due to its simplicity. Stoichiometric models are limited for bioengineering purposes because they ignore enzyme kinetics and cannot directly capture dynamic metabolic responses, nor offer a straightforward way to influence ever more abundant proteomics and metabolomics data for increased delicacy. Kinetic models explicitly take into account enzyme kinetics and are suitable to prognosticate the metabolite level as a function of time from the protein level. This type of vaticination is useful to metabolic masterminds to design pathways that have the asked titers, rates, and yields. Metabolomics data analysis using AI Metabolomics

Fig. 2.2 General workflow of AI in metabolomics

is a fairly new discipline within the field of wisdom, arising as a pivotal element in the study of natural systems.

To deal with a large number of variables, point selection, and bracket styles similar to partial least-places discriminant analysis, top element analysis and machine literacy have been largely employed. The use of these styles is especially necessary for individual models. Machine literacy, which is a subdomain of AI, can be divided into three orders: supervised literacy, unsupervised literacy, and semi-supervised literacy. Supervised machine literacy algorithms use intensely statistical approaches to train a model on labeled data and make prognostications about unknown (unlabeled) data. By discrepancy, unsupervised models deal with unlabeled data for training. Machine literacy is the most ideal operation for mass spectrometry data owing to its capability to manage both direct and nonlinear data. Still, using machine literacy for mass spectrometry operations is not a new concept.

2.4 Bioinformatics and Artificial Intelligence Tools Employed in Metabolic Pathways

Several bioinformatics and AI tools are available to study and analyze metabolic pathways. These tools leverage machine learning, systems biology, and computational methods to supply insights into complex biological processes. A few bioinformatics and AI tools commonly used for metabolic pathway analysis are shown in Table 2.1.

2.4.1 Pathway Tools (BioCyc)

The BioCyc collection serves as a distinctive resource for computational systems biology, facilitating global and comparative analyses of genomes and metabolic networks, and acts as a complement to the BioCyc resource of curated PGDBs. The Omics Viewer available on the BioCyc website allows researchers to visualize integrated datasets of gene expression, proteomics, and metabolomics against the metabolic maps of these organisms.

BioCyc.org hosts over 7600 PGDBs for organisms with fully sequenced genomes. Though the majority of these databases are focused on bacterial species, there are PGDBs available for significant experimental organisms such as humans (HumanCyc), yeast (YeastCyc), and Arabidopsis (AraCyc). Each BioCyc database links the genome sequence of a single organism with its predicted metabolic network, encompassing various metabolic pathways, responses, and chemical compounds, all presented through an intuitive graphical interface.

The BioCyc databases also feature newly computed information, including predicted operons and candidate enzymes for catalyzing predicted enzymatic pathways that lack assigned enzymes within the genome annotations. The BioCyc.org website allows users to conduct a variety of queries and data analyses online. Analytical tools provided by BioCyc include the comparison of pathways and genomes among multiple organisms, alignment of orthologous genes across different genomes, and the identification of minimum-cost metabolic network routes connecting two substrates. Additionally, tools are available for displaying omics data—such as metabolomics and gene expression data—on individual pathway maps, on consolidated groups of multiple pathway maps termed pathway collages, and on diagrams representing the complete metabolic network of an organism.

One notable feature of BioCyc is the Smart Tables, which are user-defined collections of metabolites or genes that can be easily transformed into other categories of objects, analyzed, saved in a BioCyc account, and shared with other users. Access to the BioCyc databases is available through the BioCyc.org website, and they can also be downloaded. Furthermore, the Pathway Tools software can be downloaded and installed locally, allowing users to generate BioCyc-like databases for additional genomes, curate existing databases, or perform new analyses that are not accessible online (García-Granados et al. 2019).

Table 2.1 Bioinformatics and artificial intelligence tools used in metabolic pathway analysis

Tool name	Goal	Tool website	Citation
COBRA Toolbox	Modeling and analyzing metabolic networks and flux balance analysis (FBA)	https://opencobra.github.io/cobratoolbox/stable/index.html	Heirendt et al. (2019)
BioModels	Repository for storing mathematical models of biological processes	https://www.ebi.ac.uk/biomodels/	Malik-Sheriff et al. (2020)
MetaboAnalyst	Pathway database for linking genomics, chemical substances, and pathways	https://www.metaboanalyst.ca/	Pang et al. (2021)
Pathway Commons	Integration of biological pathways, including metabolism and signaling	https://www.pathwaycommons.org/	Cerami et al. (2011)
BiGG Models	Database for genome-scale metabolic network models	http://bigg.ucsd.edu/	King et al. (2016)
MINE Database	Database for predicting and exploring metabolites from genome sequences	https://minedatabase.mcs.anl.gov/#/home	Elton et al. (2019)
PathPred	Tool for predicting metabolic pathways and their products using AI	https://www.genome.jp/tools/pathpred/	Takigawa et al. (2010)

2.4.2 MetaCyc

MetaCyc (MetaCyc.org) is a comprehensive metabolic database that is meticulously curated by experts, with its contents derived from experimental studies. Currently, it encompasses over 2400 experimentally determined metabolic pathways linked to more than 2700 organisms, curated from upward of 47,800 scientific publications. The database offers extensive information on metabolic enzymes, reactions, and substrates.

MetaCyc serves as a valuable resource for both teaching biochemistry and facilitating metabolic engineering. The reconfiguration of a metabolic network through genetic engineering can involve several strategies: (i) introducing a new enzyme or pathway into an organism, (ii) substituting an existing enzyme or pathway with an alternative, or (iii) eliminating an enzyme or pathway. Scenarios (i) and (ii) typically require the incorporation of one or more novel enzymes into the host organism, potentially involving enzymes with distinct kinetic properties, different regulatory mechanisms for substrate selection, or new catalytic functions not inherently present in the host organism.

The database provides a searchable encyclopedia of enzymes and pathways, detailing not only the catalytic activities of enzymes but, in many cases, also their regulatory mechanisms regarding substrate selection (to ensure the enzyme's activity under relevant cellular conditions) and cofactor requirements (so that cofactor biosynthesis pathways can be engineered to meet the needs of the introduced enzyme).

MetaCyc aims to present a vast array of pathways from diverse organisms. However, since most pathways have been identified in only a limited number of organisms, and given that MetaCyc does not encompass all relevant literature, there is a recognition that the species information contained within MetaCyc is somewhat limited. Nonetheless, the database is based on experimental data rather than computational predictions. MetaCyc employs the same database schema as EcoCyc, aspiring to provide a similarly rich, literature-based representation for each pathway; however, a minority of pathways currently do not possess the extensive commentary and literature citations that were intended for inclusion. Each pathway in MetaCyc is accompanied by citations referencing the sources from which the information was obtained. Unlike EcoCyc, MetaCyc does not supply genomic data, such as genomic maps or sequences.

2.4.3 COBRA Toolbox

Constraint-grounded reconstruction and analysis (COBRA) provides a molecular mechanistic framework that supports integrative analysis of experimental molecular systems biology data and allows for quantitative predictions of physiochemically and biochemically feasible phenotypic states. The COBRA Toolbox is a comprehensive desktop software suite that includes an array of interoperable COBRA methods. Its broad applications in biology, biomedicine, and biotechnology stem from the flexibility it offers, allowing users to combine functions to implement customized COBRA protocols for any biochemical network.

Version 3.0 of the COBRA Toolbox represents an advancement over the previous versions (1.0 and 2.0) and introduces a variety of new methods for quality-controlled reconstruction, modeling, topological analysis, strain design, experimental design, and network visualization. Furthermore, the toolbox enables the integration of diverse data types, including chemoinformatics, metabolomics, transcriptomics, proteomics, and thermochemical data. A notable enhancement in this version is the inclusion of built-in language integration, which broadens COBRA's applicability by leveraging high-precision, high-performance, and nonlinear numerical optimization solvers for kinetic modeling.

This protocol provides an overview of the new features introduced in COBRA Toolbox v. 3.0 and serves as a guide for generating and analyzing constraint-based models in varied applications. With its unprecedented depth of COBRA methods, the toolbox facilitates in-depth exploration and understanding of complex biochemical networks, making it an invaluable resource for researchers working in systems biology and related fields (Heirendt et al. 2019).

2.4.4 BioModels

Computational modeling has become an essential approach in life sciences research, facilitating the exploration of complex biological systems. To provide a platform for the universal sharing, easy availability, and reproducibility of these models, BioModels (https://www.ebi.ac.uk/biomodels/) was established in 2005 as a repository for high-quality models. The current BioModels platform accommodates the submission of models encoded in various modeling formats, including SBML, CellML, PharmML, the COMBINE library, MATLAB, Mathematica, R, Python, and C.

Submitted models are curated to ensure that they accurately represent natural processes and that their simulation results can be reliably reproduced based on the reference publication. This curation process includes converting models into standard formats and annotating them with controlled vocabularies in compliance with MIRIAM (Minimum Information Required In the Annotation of Models) guidelines.

Over the years, BioModels has experienced steady growth, hosting around 2000 models derived from the published literature, of which approximately 800 are curated models. This makes BioModels the world's largest repository of curated computational models. It is also recognized as the third most utilized data resource among scientists engaged in modeling research, following PubMed and Google Scholar.

BioModels offers significant benefits to modelers by providing access to reliable, semantically enriched curated models in standard formats that are easy to share, reproduce, and utilize, thereby enhancing the overall quality and impact of computational research in the life sciences (Malik-Sheriff et al. 2020).

2.4.5 MetaboAnalyst

The MetaboAnalyst web-based platform has gained widespread usage for the comprehensive analysis and interpretation of metabolomics data. With the introduction of MetaboAnalyst 5.0, the focus has shifted toward bridging the gap from raw metabolomics data to functional insights derived from high-resolution mass spectrometry (HRMS). This version features three key modules designed to facilitate robust analysis:

1. *LC-MS Spectra Processing Module*: This module provides a user-friendly interface capable of performing automated parameter optimization and resumable analysis. It significantly reduces barriers to processing LC-MS1 spectra, making it more accessible for users.
2. *Functional Analysis Module*: Enhancing the previous MS Peaks to Pathways module, this new module allows users to selectively choose specific peak groups of interest and assess their enrichment in potential functions as determined by metabolic pathways and metabolite sets. This enables deeper insights into the biological relevance of the detected metabolites.
3. *Functional Meta-Analysis Module*: This module is designed to integrate multiple global metabolomics datasets obtained under similar conditions or from comparable studies, facilitating a more comprehensive understanding of functional insights across datasets.

In addition to these core modules, MetaboAnalyst 5.0 introduces several new features, including weighted joint-pathway analysis, data-driven network analysis, batch effect correction, the ability to incorporate specialized replicates, and improved composite name matching. The web interface and underlying codebase have also been revamped to enhance performance and user experience.

At the end of an analysis session, users can now easily transition to other compatible modules, streamlining the overall data analysis process and providing a more cohesive user experience. This makes MetaboAnalyst 5.0 a powerful tool for researchers seeking to derive meaningful biological insights from their metabolomics data.

2.4.6 Pathway Commons

Pathway Commons (https://www.pathwaycommons.org) is a comprehensive integrated resource that provides access to publicly available information about biological pathways. This includes details on biochemical responses, the assembly of biomolecular complexes, transport and catalysis events, and interactions involving proteins, DNA, RNA, and small molecules such as metabolites and drug mixtures.

Data in Pathway Commons are sourced from multiple providers and are presented in standard formats, including the Biological Pathway Exchange (BioPAX) language and the Proteomics Standards Initiative Molecular Interactions (PSI-MI) format, ensuring seamless integration of diverse datasets.

Pathway Commons offers the following resources to assist biologists and researchers:

1. *Search Tools*: Users can search this extensive database to find specific biological pathways and related information.
2. *Download Options*: The platform provides a download portal for accessing integrated bulk datasets, including tables of relations and gene sets, which can be invaluable for further research and analysis.
3. *Software Libraries*: Pathway Commons offers applicable software libraries for pathway information processing in several programming languages, including Java, R, Python, and JavaScript, enabling researchers to leverage pathway data in their preferred coding environments.
4. *Web Services*: A web service is available for programmatically querying the entire dataset, allowing for custom data queries and analyses.

Pathway visualization is supported through the Systems Biology Graphical Notation (SBGN), enhancing the interpretability of complex pathways and interactions.

As of now, Pathway Commons includes data from 22 databases, detailing 4794 unique biological biochemical processes (pathways) and approximately 2.3 million relations. To improve the usability of this extensive resource, the platform also develops and maintains interactive web applications and training materials, facilitating pathway exploration and advanced analysis for the research community (Cerami et al. 2011).

2.4.7 KEGG

KEGG (Kyoto Encyclopedia of Genes and Genomes) is an extensively used database and knowledge resource for understanding high-position natural functions and serviceability in terms of molecular commerce networks. These networks include pathways of metabolism, inheritable information processing, environmental information processing, cellular processes, organismal systems, and mortal conditions.

2.4.7.1 Key Components of KEGG Pathways

Metabolic Pathways KEGG's metabolic pathway charts are essential for assaying biochemical processes in organisms. These pathways outline the series of chemical responses being in cells, eased by enzymes. They cover glycolysis, the citric acid cycle, oxidative phosphorylation, and more. Each response step involves metabolites (small motes), enzymes (proteins), and genes garbling these enzymes.

Example: The Glycolysis Pathway in KEGG maps the breakdown of glucose to pyruvate, which is central to cellular respiration and energy product in cells.

Signaling Pathways KEGG includes signaling pathways that represent the communication networks inside the cell and between cells. These pathways help in understanding how cells respond to external signals, similar to hormones, or environmental changes.

Example: The PI3K-Akt signaling pathway is critical in regulating cell survival, growth, and metabolism. KEGG maps this pathway with details on how growth factors spark a waterfall of events leading to cellular responses.

Disease Pathways KEGG includes mortal complaint-related pathways that detail molecular mechanisms underpinning colorful conditions, including cancer, neurodegenerative diseases, contagious conditions, and vulnerable diseases.

Illustration: The Bladder Cancer Pathway in KEGG links specific inheritable mutations and molecular differences to the development and progression of bladder cancer, which is largely applicable for RNA analysis in bladder cancer exploration.

Medicine development and rectifiers: KEGG also supplies medicine and chemical structure databases to help in medicine discovery and development. Experimenters can explore the commerce between different pathways and how medicines affect specific molecular targets.

Example: The Drug-Target Interaction maps show how particular medicines inhibit or enhance the exertion of target proteins, which can be pivotal in designing new treatments.

2.4.7.2 KEGG Pathway Databases

KEGG PATHWAY—visual representations of known pathways that display genes, proteins, metabolites, and responses involved in colorful natural processes.

KEGG MODULE—lower functional units in the pathways, which help in understanding complex biochemical systems.

KEGG DRUG—a database for medicine information, fastening on their relations with pathways and target motes.

2.4.7.3 Applications of KEGG Pathways

Bioinformatics KEGG is extensively used in bioinformatics tools to collude experimental data (like RNA-Seq or proteomics data) onto pathways, helping identify overrepresented pathways in a particular condition (e.g., cancer, infection).

Disease Research KEGG pathways aid in relating implicit biomarkers for conditions and in designing remedial strategies by understanding how certain molecular pathways are altered.

Drug Development Pathway maps help pharmaceutical experimenters in relating molecular targets for medicines and understanding their goods at a systems position.

Functional Genomics KEGG pathways are used to explore the functional counteraccusations of genes, proteins, and enzymes in the environment of systems biology.

2.5 Conclusion

The study of metabolic pathways has long been a foundation of natural and medical exploration, offering perceptivity into the complex biochemical responses that sustain life. Before the arrival of AI, the analysis of these pathways reckoned heavily on homemade trial and traditional computational models. While these styles handed precious information, they were frequently limited in their capability to handle the complexity and scale of metabolic networks, making it grueling to completely understand or prognosticate the issues of metabolic changes, especially in conditions or remedial interventions. With the preface of AI, the field of metabolic pathway exploration has experienced a significant metamorphosis. AI's advanced algorithms and data-processing capabilities have revolutionized the way metabolic data are anatomized, enabling further comprehensive and accurate modeling of these intricate networks. AI allows for the vaticination of metabolic responses under colorful conditions, speeding up discoveries related to complaint mechanisms and medicine development. Likewise, AI-driven analysis of metabolomics data has uncovered retired patterns and biomarkers that were preliminarily delicate to describe. As a result, AI has not only bettered the effectiveness and perfection of metabolic pathway exploration but has also opened new possibilities for individualized drug, synthetic biology, and metabolic engineering, marking a new period in natural exploration and its operations.

References

Alajaji SA et al (2023) Generative adversarial networks in digital histopathology: current applications, limitations, ethical considerations, and future directions. Mod Pathol:100369

Alison S, Papachristodoulou DK, Elliott WH, Elliott DC (2014) Biochemistry and molecular biology, 5th edn, Oxford. ISBN 978-0-19-960949-9

Antonakoudis A, Barbosa R, Kotidis P, Kontoravdi C (2020) The era of big data: genome-scale modelling meets machine learning. Comput Struct Biotechnol J 18:3287–3300. https://doi.org/10.1016/j.csbj.2020.10.011

Barnes EA et al (2020) Indicator patterns of forced change learned by an artificial neural network. J Adv Model Earth Syst 12(9):e2020MS002195

Berg JM, Tymoczko JL, Stryer L (2002) Biochemistry, 5th edn. W. H. Freeman, New York. ISBN 978-0-7167-3051-4

Berg JM, Tymoczko JL, Stryer L, Gatto GJ (2012a) Biochemistry, 7th edn. W.H. Freeman, New York, p 429. ISBN 978-1-4292-2936-4

Berg JM, Tymoczko JL, Stryer L, Gatto GJ (2012b) Biochemistry, 7th edn. W.H. Freeman, New York, pp 480–482. ISBN 978-1-4292-2936-4

Cerami EG, Gross BE, Demir E, Rodchenkov I, Babur O, Anwar N et al (2011) Pathway Commons, a web resource for biological pathway data. Nucleic Acids Res 39(Database issue):685–690

Cohen S (2021) Chapter 1 – The evolution of machine learning: past, present, and future. In: Cohen S (ed) Artificial intelligence and deep learning in pathology. Elsevier, pp 1–12

Dhall D, Kaur R, Juneja M (2020) Machine learning: a review of the algorithms and its applications. In: Proceedings of ICRIC 2019. Springer International Publishing, pp 47–63

Elton DC, Bouwer JC, Travin DY, Aksenov AA, Dorrestein PC, Visel A et al (2019) MINEs: open access databases of computationally predicted enzyme promiscuity products for untargeted metabolomics. J Cheminform 11(1):1–15

García-Granados R, Lerma-Escalera JA, Morones-Ramírez JR (2019) Metabolic engineering and synthetic biology: synergies, future, and challenges. Front Bioeng Biotechnol 7:36

Greener JG et al (2022) A guide to machine learning for biologists. Nat Rev Mol Cell Biol 23(1):40–55

Harvey RA (2011) Biochemistry, 5th edn. Wolters Kluwer, Baltimore. ISBN 978-1-60831-412-6

Heirendt L, Arreckx S, Pfau T, Mendoza SN, Richelle A, Heinken A et al (2019) Creation and analysis of biochemical constraint-based models using the COBRA Toolbox v3.0. Nat Protoc 14(3):639–702

King ZA, Lu J, Dräger A, Miller P, Federowicz S, Lerman JA et al (2016) BiGG Models: a platform for integrating, standardizing and sharing genome-scale models. Nucleic Acids Res 44(D1):515–522

Lewis NE, Nagarajan H, Palsson BO (2012) Constraining the metabolic genotype-phenotype relationship using a phylogeny of in silico methods. Nat Rev Genet 10:291–305

Li RC, Asch SM, Shah NH (2020) Developing a delivery science for artificial intelligence in healthcare. NPJ Digit Med 3(1):107

Malik-Sheriff RS, Glont M, Nguyen TVN, Tiwari K, Roberts MG, Xavier A et al (2020) BioModels—15 years of sharing computational models in life science. Nucleic Acids Res 48(D1):407–415

Nelson DL, Cox MM (2008) Lehninger principles of biochemistry, 5th edn. W.H. Freeman, New York. ISBN 978-0-7167-7108-1

Nicholson DE (1971) An introduction to metabolic pathways by S. Dagley. Sigma Xi, The Scientific Research Society, p 266

Pang Z, Chong J, Zhou G, Morais D, Chang L, Barrette M et al (2021) MetaboAnalyst 5.0: narrowing the gap between raw spectra and functional insights. Nucleic Acids Res 49(W1):388–396

Raven PH, Evert RF, Eichhorn SE (2011) Biology of plants, 8th edn. Freeman, New York, pp 100–106. ISBN 978-1-4292-1961-7

Reece JB, Campbell NA (2011) Campbell biology, 9th edn. Benjamin Cummings/Pearson, Boston, p 143. ISBN 978-0-321-55823-7

Takigawa I, Noguchi T, Kato H, Yamanishi Y, Toh H (2010) PathPred: an enzyme-catalyzed metabolic pathway prediction server. Nucleic Acids Res 38(Web Server issue):138–143

Tien JN (2017) Internet of Things, real-time decision making, and artificial intelligence. AnnData Sci 4:149–178

Tompson J, Schlachter K, Susskind J et al (2016) Accelerating Eulerian fluid simulation with convolutional networks. arXiv preprint arXiv:1607.03597

Voet D, Voet JD, Pratt CW (2013) Fundamentals of biochemistry: life at the molecular level, 4th edn. Wiley, Hoboken. ISBN 978-0-470-54784-7

Methods of Extraction, Detection, and Identification of Different Metabolites—A Metabolomics Approach

3

Giulia Bennici, Devi Permatasari, Inas Al-Younis, Manel Dhahri, Frederico G. Pinto, Mohammed J. Uddin, Abdul-Hamid Emwas, Mariusz Jaremko, and Nelson C. Soares

Abstract

Metabolic dynamics in attack and defense mechanisms are fascinating topics that illustrate the complex ways organisms adapt to their environments. Attack mechanisms involve several functions, including energy production and metabolic adaptation, where some organisms change their appearance or behavior to balance into their environment. Defense mechanisms involve stress response, immune system activation, repair, and recovery. This chapter provides an overview of the methods applied in modern metabolomics, including single-cell metabolomics and targeted/untargeted approaches, emphasizing how metabolite production rapidly shifts in response to environmental and cellular changes. To effectively monitor these spontaneous effects, it is essential to conduct metabolomics studies with a carefully controlled experimental design. This includes meticulous sample collection, efficient metabolite extraction, and optimization of parameters for the most relevant analytical platform. This chapter outlines the most relevant and commonly used methods for metabolite extraction, aiming to maximize the number of metabolites recovered. This is followed by a summary of the advantages and limitations of the most frequently employed analytical tools, with a particular focus on nuclear magnetic resonance (NMR) and various types of mass spectrometry (MS). Finally, we emphasize the importance of integrating different omics sciences to achieve a comprehensive understanding of biological systems.

G. Bennici · D. Permatasari · I. Al-Younis
Division of Biological and Environmental Sciences and Engineering (BESE), King Abdullah University of Science and Technology (KAUST), Thuwal, Saudi Arabia

M. Dhahri
Biology Department, Faculty of Science, Taibah University, Yanbu Branch, Yanbu, Saudi Arabia

S. S. M. Soliman, M. I. Husseiny (eds.), *Metabolic Dynamics in Host-Microbe Interaction*, https://doi.org/10.1007/978-981-96-1305-2_3

Keywords

Metabolomics · Single-cell metabolomics · Metabolite extraction · Metabolite detection · Host-pathogen interaction

3.1 Introduction

The recently emerging field of omics science, known as "metabolomics," is referred to by many as the end point of genomics, transcriptomics, and proteomics and is defined as "the identification and measurement of small molecules resulting from the biochemical activity in the biological system" (Idle and Gonzales 2007). A metabolite has a low molar mass (<1.5 kDa) with key constituent categories comprising lipids, carbohydrates, nucleotides, amino acids and others. There are two main types of metabolomic analysis, called targeted and untargeted approaches.

F. G. Pinto
Center for Applied and Translational Genomics (CATG), Mohammed Bin Rashid University of Medicine and Health Sciences (MBRU), Dubai Health, Dubai, UAE

Institute of Exact Sciences, Federal University of Viçosa, Rio Paranaíba, Minas Gerais, Brazil

M. J. Uddin
Center for Applied and Translational Genomics (CATG), Mohammed Bin Rashid University of Medicine and Health Sciences (MBRU), Dubai Health, Dubai, UAE

College of Medicine, Mohammed Bin Rashid University of Medicine and Health Sciences (MBRU), Dubai Health, Dubai, UAE

GenomeArc Inc, Toronto, ON, Canada

A.-H. Emwas
Core Lab of NMR, King Abdullah University of Science and Technology (KAUST), Thuwal, Saudi Arabia

M. Jaremko
Division of Biological and Environmental Sciences and Engineering (BESE), King Abdullah University of Science and Technology (KAUST), Thuwal, Saudi Arabia

Smart-Health Initiative and Red Sea Research Center, Division of Biological and Environmental Sciences and Engineering, King Abdullah University of Science and Technology, Thuwal, Saudi Arabia

N. C. Soares (✉)
Center for Applied and Translational Genomics (CATG), Mohammed Bin Rashid University of Medicine and Health Sciences (MBRU), Dubai Health, Dubai, UAE

College of Medicine, Mohammed Bin Rashid University of Medicine and Health Sciences (MBRU), Dubai Health, Dubai, UAE

Laboratory of Proteomics, Department of Human Genetics, National Institute of Health Doutor Ricardo Jorne (INSA), Lisbon, Portugal

Center for Toxicogenomics and Human Health (toxOMICS), NOVA School-FCM, UNL, Lisbon, Portugal
e-mail: nelson.soares@dubaihealth.ae

Targeted metabolomics is a hypothesis-driven method in which metabolites are preselected for identification and/or measurement (Dudley et al. 2010). On the other hand, untargeted metabolomics uses full-scan mode to identify and assess metabolites in samples from various populations (e.g., human, animal, plants, bacteria, and others) and/or to monitor variations in how subjects respond to potential internal or external stressors (Oken et al. 2015). Targeted and/or untargeted metabolomics, whether used in combination or independently, are employed for a better understanding of the underlying molecular mechanisms mediating the metabolic responses of cells to changes in their surroundings.

Metabolites are produced at all stages of all biological processes, playing crucial roles in cellular signaling and phenotypic modulation. For example, cells engage in chemical renovation and reparation both in cooperation and competition with other cells in response to the emergence of debris from normal cellular processes, injury to tissue, internal biochemical responses to stressors, and noxious substances of external or internal origin (Wang et al. 2023). Sometimes, these processes fail to clear and repair the tissue, giving rise to pathological conditions (Hajam et al. 2022). In these molecular renovation and repair processes, specific metabolites (Ben Mrid et al. 2021) function as the mediating agents of choice to wipe out, neutralize, attenuate, or simply repel and remove potential antagonists (Mihai et al. 2024; Tomic-Canic et al. 2020). For instance, certain plants elevate the production of reactive oxygen species in response to wounding (Denness et al. 2011; Soares et al. 2009, 2011) or pathogen attack (Hématy et al. 2009). Thus, gaining insight into this apparently well-orchestrated intra- and extra-cellular metabolic exchange is crucial to fully comprehend the mechanisms of the cellular lifecycles and the processes by which cells defend themselves from environmental threats. However, experimentally it is often difficult to distinguish between the metabolites generated in the course of the normal lifecycle of the cell and by the various components of renovation and repair active mechanisms throughout the cell lifecycle. Since most living organisms share a great deal of metabolites, it becomes even more challenging to measure or quantify the inter- or extra-cellular metabolites and identify their respective functions during intracellular and/or interspecies interactions. Consequently, for many years, the data available on the relationship between hosts and pathogens were limited to models of metabolite or gene expression profiles and the qualitative presence or absence of metabolites.

Recent advances in instrumentation and thereto associated methodological developments, continuous refinement of workflows for metabolite extraction and detection, as well as the emergence of comprehensive libraries for the identification of metabolites throughout the last decade (Fig. 3.1) (Träger 1988). These developments have resulted in more reliable and accurate qualitative and quantitative analyses and, consequently, a better understanding of metabolite composition and functions within an organism, for example, the role of metabolites in interactions between host and pathogen. In this context, it is necessary to mention groundbreaking approaches such as ^{13}C-labeling combined with downstream NMR-based metabolomics analysis that generates unique data, specifically regarding pathogen metabolomic flux upon intracellular infection. For example, using this

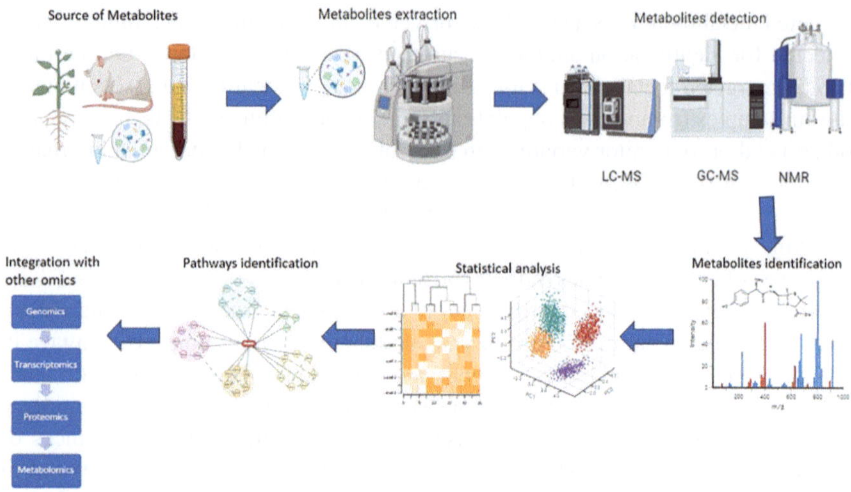

Fig. 3.1 Common design of the metabolomic study and the correlation with multi-omics. First, metabolites were obtained by extraction process according to the types of samples. Then, the extract was analyzed using nuclear magnetic resonance (NMR) and mass spectrometry (MS) to discover metabolites detected and annotated by the library. Statistical analysis was conducted to reveal the important metabolites and then pathway analysis was used to explore the connection deeper with multi-omics integration

methodology, one study has shown that *Shigella*, an intracellular bacterial pathogen, reroutes host cell metabolism to facilitate rapid bacterial proliferation. The study demonstrated that a *Shigella* infection is linked to decreased excretion of lactate and pyruvate and increased excretion of acetate. Further experiments on metabolites mutants confirmed that the *Shigella* metabolism of the host pyruvate is the source of the acetate (Kentner et al. 2014). In another example, ^{13}C-labeling was combined with LC-MS analysis in a study of mice infected with *Listeria monocytogenes* revealing the details of the metabolic dynamics of in vivo activated CD8+T cells and showing the importance of the microenvironment for lymphocyte metabolic responses (Ma et al. 2019).

While metabolic labeling represents a viable option for research based on in vitro infection of cell lines and to some extent in vivo animal studies, for studies involving human patients and/or related samples, then label-free approaches present the more attractive options especially when coupled with either NMR or LC-MS analysis (Letertre et al. 2021). One could argue that quantitative label-based approaches are more accurate than label-free; however, this option provides a pragmatic means toward a more holistically investigation of the metabolomic dynamics of a living system. Label-free approaches have been applied to gain insights into how the immune system reacts to various intruding and obstructive substances, such as infections and cancers (O'Neill et al. 2016), or even to better understand aspects of exaggerated immune cell responses like those widely reported during SARS-CoV-2 infections (Nagana Gowda and Raftery 2023). In fact, there are several examples in the literature demonstrating the utility of label-free-based NMR or LC-MS workflows to understand which metabolites are associated with the pathophysiology of, for example, COVID-19 disease, diagnostics, prognostics (Jalaleddine et al. 2022; Saheb Sharif-Askari et al. 2022; Soares et al. 2023), and therapeutics as well as the impact of various vaccination programs (Abufares et al. 2022). Finally, single-cell metabolomics holds tremendous potential to change the paradigm of metabolic profiling and unveil in detail the role of each metabolite involved in cellular defense and interaction mechanisms (Ali et al. 2022). This chapter summarizes some of the more effective state-of-the-art techniques for metabolite extraction, data processing, and analysis that are used in metabolomics to solve the puzzles of intricate cellular systems biology.

3.2 Extraction Methods

3.2.1 General Methods for Metabolite Extraction

Extraction is the procedure used to isolate the metabolites from the specimen being studied. This process is one of the essential components of the analytical design necessary to ensure the intended outcome. Selecting the appropriate extraction technique depends on the objectives of the overall investigation, as different procedures can lead to different results (Catani et al. 2022). There is a wide variety of extraction techniques available and in use, commonly classified as either conventional or non-conventional extraction techniques. Maceration, Soxhlet, and decoction are examples of commonly used methods in the conventional approach category. Techniques such as pressurized liquid, supercritical fluid, microwave-assisted, ultrasound-assisted, and enzyme-assisted extraction belong to the second category of non-conventional approaches (Pai et al. 2022). All these methods, both conventional and non-conventional, share the following common goals: (1) obtaining targeted bioactive compounds from complex specimens; (2) improving the selectivity of the analytical methods; (3) boosting the concentration of targeted compounds to increase bioassay sensitivity; (4) transforming the bioactive compounds into a form better suited for separation and detection; and (5) providing robust and repeatable methods that are not affected by changes in the sample matrix (Azmir et al. 2013).

The primary goal of any extraction technique is to extract as much of the target compound(s) as possible from a sample matrix while maintaining the integrity of the molecules of interest and minimizing the coextraction of unwanted molecules or contaminants (Cowan 1999; Garcia-Vaquero et al. 2020). The primary conventional method of extraction is solvent-based extraction using either polar or nonpolar solvents, but which solvent to use generally depends on the kind(s) of substance(s) you want to isolate (Azmir et al. 2013). For instance, as one of the polar solvents, water is traditionally utilized as the main extraction solvent, especially when studying components of dried plants that have been used for centuries and have efficacy as medicinal plants (Cowan 1999).

Furthermore, among the conventional methods, there is solid-liquid extraction known as "leaching," which involves the transfer of metabolites from the sample to the organic solvent used. This method will allow the metabolites to dissolve in the solvent and leave the remaining sample. The purpose of the "leaching method" is to segregate and extract molecules of interest from otherwise insoluble high-molecular-weight fractions as well as from other compounds that can obstruct the analysis process at later stages. Maceration is yet another conventional method commonly used for solid-liquid extraction (Luque de Castro and Priego-Capote 2010). By carefully choosing the solvent's polarity and using heat and/or agitation to make the target compounds more soluble, this method allows for the extraction of high-value metabolites (Garcia-Vaquero et al. 2020). Finally, ultrasound-assisted extraction is a novel method combined with solvent extraction procedures. Here, the sample, mixed with the extraction solvent, is submerged in an ultrasonic bath where the mechanical stress caused by the sonication both shortens the extraction time and increases extraction yields (Bucar et al. 2013).

There are numerous challenges associated with the removal of natural products from their natural environment. These challenges are related to the extractive solvent's stability and polarity, toxicity, purity, viscosity, the potential for artifact production during the extraction process, and the bulk material output that should be removed (Abdelmohsen et al. 2022). The primary considerations when selecting solvent are based on factors like cost, availability, toxicity, polarity, boiling point, environmental impact, and inertness. The two key factors that determine how a solute interacts with the solvent are solubility and polarity. The latter is important because, in accordance with the "like-dissolves-like" concept, polar solvents will extract polar natural compounds, which is important for the stability of natural compounds in solvents, while nonpolar solvents allow for the extraction of nonpolar natural compounds. Furthermore, chemical reactions occur when a natural product is dissolved in an organic solvent, sometimes causing unwanted effects that must be dealt with or prevented. Following the removal of the solvent, usually done by evaporation, the natural product may undergo further chemical alterations or stay unaltered. Artifact development stems from the following issues: (1) the creation of novel substances that are not considered to be natural products; (2) degradation of bioactive compounds; (3) the production of active molecules (false positives) from inactive natural products; (4) loss of the overall yield of significant compounds during isolation; (5) toxic components produced from inactive chemicals; and (6) the method's lack of repeatability (Bucar et al. 2013; Maltese et al. 2009).

3.2.2 Methods of Metabolite Extraction for Multi-omics Integrative Analysis

The majority of the existing omics datasets acquired over the past two decades were obtained from a single standalone "omic" approach using specific sample preparation protocols and frequently generated from cells, tissues, or body fluids exposed to completely different conditions than the comparison. For instance, a great deal of the current metabolomic datasets was acquired through NMR spectroscopy-based analysis, which uses sample preparation protocols in many aspects distant from those employed in proteomics workflow, specifically LC-MS/MS-based analysis, the proteomics' gold standard technology (Pusch et al. 2003; Shuken 2023). Therefore, even when subjected to the most sophisticated advanced bioinformatics methodologies, multi-omics downstream statistics and/or bioinformatic integrative analysis must account for the inherent bias and inconsistent interplay among the datasets (Sun and Hu 2016; Krassowski et al. 2020). Thanks to the introduction of LC-MS/MS approaches, proteomics and metabolomics analysis can all be carried out using the same MS equipment (Fig. 3.1) (e.g., (Lin et al. 2024)), resulting in comprehensive multi-omics datasets from a single reference sample. Within this context, it became evident that unified sample preparation protocols that enable the simultaneous extraction of various biomolecule fractions from the same sample would represent an advantageous starting point for further downstream analytical and bioinformatic integrated multi-omics analysis. Recently, Zenati et al. reported a combined multi-omics samples preparation methodology, here the authors compared the coverage and reproducibility of two extraction methods frequently used in plasma: the single-phase methanol precipitation (MeOH ppt) and biphasic chloroform/methanol (MeOH:$CHCl_3$) extraction (Zenati et al. 2023). In the same line, a study by Muelhbauer et al. presented an efficient and unified in-solution approach to prepare lipids, metabolites, and proteins, termed bead-enabled accelerated monophasic multi-omics (Muehlbauer et al. 2023). These extraction protocols have been applied to integrate multi-omics analysis to investigate how different cancer cell lines respond to challenges and/or drug treatment(s) (Haider et al. 2023; Ahmed et al. 2023) and also in the profiling of microbial pathogens (Boucherabine et al. 2024; Alhameed et al. 2023).

3.2.3 Single-Cell Metabolomics

The metabolome of a cell is dynamic, and the production of metabolites quickly changes in response to changes in the environment and cellular activity, effects that may be counteracted by quenching any sample as soon as it has been obtained. Such challenges notwithstanding, advances in single-cell metabolomics now enable us to characterize a cell's metabolome and develop a clear and detailed understanding of its in vivo phenotype and biological and pathological properties (Pandian et al. 2023; Lanekoff et al. 2022).

Using single-cell metabolomics approaches (Fig. 3.2), an overview of the molecular intermediates and products of cellular metabolism within a biological system can be obtained (Ali et al. 2022). Usually, these molecules, which commonly have a molecular weight of less than 1 kDa, are the foundation for biochemical variability inside cells. Numerous factors, including cell cycle, biochemical processes, environmental stress, and pathological states, are associated with the molecular variances across cells of the same kind (Shresta 2020). Well-controlled chemical processes of cell metabolism enable the cell to respond and adjust to a wide range

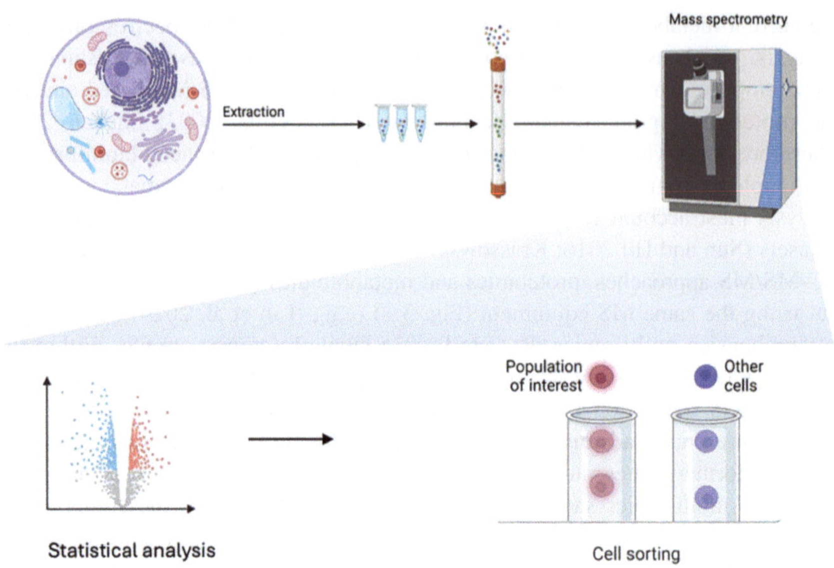

Fig. 3.2 Key steps of single-cell metabolomics, which consist of sample preparation, hence cell isolation; metabolite extraction; metabolite analysis using techniques such as mass spectrometry (MS); data processing and statistical analysis; and finally biological interpretation to find pathways and differentiate between cells

of pathophysiological stimuli. These responses ultimately determine the fate of the cell and the phenotypic result, and while single-cell studies only provide a snapshot of the overall organism's health and functionality, they provide invaluable insights into intracellular functions and intercellular interactions (Ali et al. 2022).

Thanks to the development of new instruments and techniques allowing for detailed mapping and assessment of the functionality of biochemical molecules, single-cell metabolite profiling has become a widely used investigative and analytical approach to uncovering the molecular underpinnings of cellular and metabolic processes. For example, an extensive study of a cell population that is dominated by a single type of cell would usually reveal that all cells share the same properties. On the other hand, a distribution dominated by two cell types may provide different results that represent different cell types. An exact analytical characterization of a certain cell population can only be obtained by single-cell analysis, which will identify distinct metabolic trends among individual cells within the same tissue or culture (Shresta 2020; Prasad et al. 2022). This technique can shed light on various physiological states, responses to environmental changes, pathogen interactions, and the intricate metabolic networks that support cellular functions. Furthermore, advanced techniques such as mass spectrometry (MS) and microfluidics have the capacity to reveal important details about cellular metabolism, triggers of disease and disease progression, and potential targets for treatment. Mass spectrometry has the advantages of high sensitivity, wide dynamic range, and capacity to characterize hundreds to thousands of molecules. MS can also be used to analyze single cells in several ways, from laser capture microdissection to the straightforward extraction of single-cell contents with an electrospray emitter needle. However, single-cell mass spectrometry is still developing, with existing techniques having the capacity mainly to only detect a few abundant molecules present in a single cell (Shresta 2020; Prasad et al. 2022).

Thus, instruments of the current generation have demonstrated remarkable sensitivity in the identification and measurement of hundreds of biomolecules obtained from biological samples. However, single-cell studies create issues related to sampling and preparation. Manual sample manipulation works well for in-depth examination of tiny cell subsets, but it is less useful for characterizing the large-scale cellular heterogeneity found in complex biological structures. Mass spectrometry imaging (MSI) is a potent technique for molecular characterization and chemical mapping. It does not require fluorescent or isotope labeling and can sensitively image thousands of molecules simultaneously. While MSI-based automatic tissue section profiling has the capacity to gather spectra from hundreds of cells, it cannot resolve matrix effects or partition cell contents. However, by limiting matrix impacts from surrounding cells and optimizing analyte extraction, the matrix-assisted laser desorption/ionization (MALDI) matrix technique makes it possible to identify uncommon individuals within a population. Recently, multiple ionization methods, such as MALDI, desorption electrospray ionization (DESI), nanoparticle laser desorption/ionization (nano-LDI), and secondary ion mass spectrometry (SIMS), have been developed for use with MSI techniques (Comi et al. 2017; Zhang et al. 2019).

Overall, recent advances in metabolomics have enabled a more comprehensive analysis of cellular processes and phenotypes. However, absolute quantification and extensive investigation of entire metabolite networks remain constrained. One method with which to overcome these limitations involves stable-isotope labeling, which allows for the detection of mass changes between labeled and unlabeled forms. The incorporation of heavy atoms can reveal metabolic networks and pathways. This approach, when combined with untargeted metabolomics, surpasses traditional metabolomics by depicting pathways and networks of biological processes, accurately identifying metabolites, and quantifying them (Qiu et al. 2023). For instance, this approach has facilitated the discovery of new pathways, such as the ethyl-malonyl-CoA pathway in methylotrophic bacteria and the GABA shunt in *Toxoplasma* parasites. These discoveries exemplify the potential of stable-isotope labeling in elucidating host-pathogen interactions and expanding our understanding of metabolic processes (Chokkathukalam et al. 2014; Olson et al. 2020). The remaining major challenge to overcome to make optimal use of these novel techniques is found at the software level of data interpretation, requiring the development of a holistic approach software capable of integrating all available data and representing them as a functional whole.

3.3 Analytical Detection Methods

Currently, the two major analytical techniques used for metabolomic profiling are nuclear magnetic resonance (NMR) spectroscopy and mass spectrometry (MS) (Emwas et al. 2019; Bingol 2018). While ^{1}H ^{13}C nuclei NMR and 1D and 2D NMR are also used in metabolomics (Irie et al. 2014), the most popular technique is mass spectrometry with more than 80% of published research reporting the use of MS often in combination with a separation technique such as liquid chromatography (LC) or gas chromatography (GC) (Wörheide et al. 2021). Each technique has the ability to separate specific compounds based on their physiochemical properties, with LC-MS being better suited for polar and non-volatile compounds. At the same time, GC-MS is a highly useful tool for the study of nonpolar and volatile organic compounds (Olivier et al. 2019). Despite the high reproducibility and nondestructive nature of NMR, the high sensitivity, selectivity, and resolution, even in the analysis of samples with low concentrations of metabolites, have given rise to an increasing trend toward the use of mass spectrometry-based metabolomics (Wörheide et al. 2021; Yao et al. 2019).

3.3.1 Identification and Quantification Methods

In metabolomics study, few identification and quantification methods can be applied, depending on the experimental design and research questions. As an example, in a targeted metabolomics study, quantification and detection of known metabolites should be identified using reference material with proper analytical tools such as MS. Meanwhile, untargeted metabolomics studies enable the exploration of a wide variety of metabolites present in the sample, allowing the possibility of discovering a novel compound. However, the major limitation of metabolomic studies is that it is often difficult to capture metabolite fluxes and alteration within the sample under specific conditions, which is important to study and observe the interconnected metabolite pathway. This unsteady state analysis cannot be approached by steady analysis. Stable-isotope labeling in metabolomics overcomes the limitations of absolute quantification and flux measurement to deliver precise metabolite identification. Additionally, flux measurement allows comprehensive metabolic regulation, which can uncover the metabolite pathway and novel metabolites. A stable-isotope labeling that can be applied both in NMR and LC-MS can be a suitable tool to tackle the limitations. Stable-isotope labeling can be applied to numerous types of samples like plants, humans, or animals, each with an optimized design depending on the samples' characteristics and the aim of the study. Therefore, the label selection itself must be considered to prevent inaccurate interpretation and allow precise quantification of metabolites (Chokkathukalam et al. 2014).

The complexion in sample preparation and caution label selection made the isotope labeling quite inconvenient. As a result, label-free techniques were developed to simplify previous technique limitations and attain faster quantification results using LC-MS. The identification technique will require a comparison of respected spectra toward the peak intensities between the control and the samples. Like another metabolomic study, the bottleneck issue, which is data analysis, is sometimes tricky and needs assisted methods to approach a reliable result. Therefore, the automated free-label method using certain algorithms can be used as an alternative to gain dependable results. One of the algorithms operated in this technique is usually based on the features correlated with the mass traces, such as metabolite isotope pattern, mass-to-charge spacing, and co-elution format. For instance, this algorithm can be accessed from the openMS FeatureFinderMetabo and found to be effective compared to usual LC-MS feature detection. To improve the accuracy of the detection, this approach and algorithm are still being developed using another potential parameter that can break down any gaps in imprecision within the study and eventually boost the specificity (Kenar et al. 2014).

3.3.1.1 Nuclear Magnetic Resonance (NMR)

Among the most popular metabolomics techniques, NMR and MS have shown a sharp rise in use in recent years. Depending on the particular research issue and metabolites of interest, each technique has advantages in the realm of targeted and untargeted metabolomics studies. NMR is better in targeted metabolomics than MS despite being less sensitive since it can concentrate on specific targeted metabolites and use a few NMR techniques to deliver structural information accurately. Furthermore, this device can accurately and quantitatively quantify metabolites, something that MS is unable to do without a reference substance. In contrast, this tool is less effective in untargeted metabolomics, but it still offers a wide variety of metabolites for thorough profiling and provides specific classes of substances. Additionally, the nondestructive nature of this instrument's conduct toward the sample surpasses that of MS (Emwas et al. 2019; Dayalan et al. 2018).

3.3.1.2 Liquid Chromatography with Tandem Mass Spectrometry (LC-MS/MS)

Mass spectrometry plays a major role in untargeted metabolomics investigations by predicting alterations in the metabolites present in the sample. In some cases, putative molecules from the mass spectra library supplied can be useful when observing how the metabolites change in the treated sample. The objective of untargeted metabolomics is frequently to reveal as many metabolites as possible in order to increase the likelihood of finding new chemicals. Putatively annotated compounds from the MS spectra library are usually purified into single compounds or fractions for further validation using NMR structure elucidation. The hybrid combination of NMR and MS in metabolomics can accelerate the biomarker finding within the sample of interest to gain a profound knowledge of biological living systems and an understanding of the phenotype of organisms. However, the application of LC-MS in targeted metabolomics is also worth considering since it can be used to focus on specific metabolite identification within the sample. To this point, quite complicated sample preparation and destructiveness toward the sample might be the main disadvantages of the instrument (Bingol 2018; Tsoukalas et al. 2024). An in-depth comparison between NMR and MS techniques in metabolomics studies is covered in Table 3.1.

Table 3.1 Comparison of NMR and MS for targeted and untargeted metabolomics study

	Nuclear magnetic resonance		Mass spectrometry	
Indicators	Targeted	Untargeted	Targeted	Untargeted
Sample preparation	Easy and nondestructive	Easy and nondestructive	Sometimes complex and destructive	Sometimes complex and destructive
Sensitivity	Moderate	Moderate	High	High
Reproducibility	High	High	Moderate	Moderate
Quantification	Yes	Yes	Yes, using the reference standard	No
Metabolites coverage	Narrow (focused on specific metabolite)	Broad (comprehensive profiling)	Narrow (focused on specific metabolite)	Broad (comprehensive profiling)
Identified metabolites	Chemical structures	Functional group	Chemical formula with m/z	List of putative compounds

3.3.1.3 Matrix-Assisted Laser Desorption/Ionization Mass Spectrometry (MALDI-MS)

Metabolomics uses MALDI-MS sparingly as an analytical tool. However, as this method involves minimal sample preparation and may offer imaging mass spectrometry, this equipment can be utilized for spatial metabolite identification in tissue. This instrument typically did not standalone in metabolomics since, in comparison to LC-MS, it is less sensitive and repeatable. The LC-MS technique must be added in order to validate the results from the MALDI-MS and attain a greater coverage of metabolites, particularly for the tiny and extremely polar compounds. Mass spectrometry imaging, however, performs better than LC-MS, which has the limitation of requiring metabolite extraction from its location. Without the need to extract metabolites, this device's laser can look at changes in tissue metabolites in response to pathogenic exposure. The incorporation of LC-MS enables the observation of brief changes in metabolites that MALDI-MS could not capture. Therefore, combining the two methods could expedite the investigation of pathogenic processes and metabolic pathways (Irie et al. 2014).

3.3.1.4 Multi-omics

Metabolomics studies have been closely linked with other omics layers studies up to the smallest living system level, such as genomics, transcriptomics, proteomics, and epigenomics, for inclusive observation of mechanisms in the progression of human diseases. Examination of this multi-omics layer affords a holistic understanding of biological living systems and their regulations. Cross-omics interaction enables a profound approach to studying the complexity of each biological layer that can help with biomarker discovery, drug development, and disease mechanisms. Thus, the application of this technique can result in network and pathway analysis, which can lead to personalized medicine. However, the presence of huge data piles must be analyzed prudently to evade any unbiased conclusion. Numerous comprehensive techniques, including machine learning and bioinformatics, are used to raise the accuracy and quality of data processing. It is anticipated that the combination of knowledge-based data integration and data-driven findings from these interdisciplinary approaches will provide recommendations that can be used automatically with necessary modifications rather than solved manually case by case (Wörheide et al. 2021; Olivier et al. 2019).

3.4 Data Processing, Bioinformatics, and Integration

Metabolomics research uses many datasets, which should be handled properly to prevent inaccurate information. Annotation becomes crucial in untargeted metabolomics, which increases the complexity of data collection and processing to observe pharmacological networks. A number of procedures are used, including data normalization and noise reduction, spectral matching with the library, and statistical analysis utilizing multivariant to find the pattern. In order to remove useless data gathering, data normalization should take into account a few crucial parameters: baseline correction, peak shape, peak breadth, and signal-to-noise (S/N) threshold. Metabolite identification, which is achieved by matching the m/z fragment with MS datasets made up of several libraries, is another essential step in this approach. Certain algorithms in each of these libraries use similarity scores to determine how similar the discovered spectra are. To forecast the putatively annotated compounds, several algorithm databases such as MS-Dial, MZmine, RAMSearch, MZ Cloud, MZ Vault, and MetaboAnalyst can be used concurrently. This putatively annotated molecule is categorized as level 2 in metabolite identification by the Metabolomics Standards Initiative (MSI), which states that the detected metabolites should be verified by comparison with a reference standard. However, in NMR, preprocessing methods like phase and baseline correction are usually provided by the machine software, which is often done manually. In addition, metabolite identification uses similar approaches to MS-based, which is through searching in the library, such as Chenomx and AMIX. Compounds identified in NMR should be presented in high abundance and water soluble in the sample. Moreover, for a single compound, data can be characterized and quantitatively measured by using chemical shifts and spin couplings. Statistical analysis can be performed by using supervised and unsupervised methods like PCA and PLS-DA to find the potential biomarkers present in the sample and highlight the answer to the research question. In order to understand the biological mechanisms, the data can also be utilized to observe the route analysis utilizing functional class score (FCS), over-representation analysis (ORA), or other pathway simulation approaches (Yao et al. 2019; Hendriks et al. 2011; Ren et al. 2015; Chen et al. 2022; Want and Masson 2010).

3.5 Future Perspectives and Conclusions

The complex numerous metabolic processes and networks that underlie an organism and, consequently, every single cell and their interactions also reflect in the organism's attack and defense mechanisms, highlighting how diverse living organisms are. All the biochemical pathways for responses to environmental and cellular changes depict the survival abilities of different organisms. Having a full comprehension of these responses by delving deeper into the molecular details of these dynamic systems may allow us to gain a greater understanding of life's adaptability. Consequently, this may widen the current research in biotechnology, drug discovery, and medicine.

These achievements can be reachable with new technologies such as the "omics" sciences, of which metabolomics needs to be highlighted, using advanced analytical methods such as MS and NMR. Further research and development into an integration of omics technologies and computational system biology through deep machine learning models will ultimately lead to a holistic view of cellular metabolism and an understanding of how metabolite dynamics and its fine adjustments will impact cell response to changes in its surroundings. This will potentiate a more comprehensive analysis of cellular processes and phenotypes, providing insights into individual cell processes to closer access to precision molecular intervention across various areas of research, including agriculture, environmental species diversity, drug discovery, and new therapeutics and medicine.

References

Abdelmohsen UR, Sayed AM, Elmaidomy AH (2022) Natural products' extraction and isolation-between conventional and modern techniques. Front Nat Prod 1(May):1–4

Abufares HI, Alsoud LO, Alqudah MAY, Shara M, Soares NC, Alzoubi KH et al (2022) COVID-19 Vaccines, Effectiveness, and Immune Responses. Int J Mol Sci 23(23):15415

Ahmed M, Semreen AM, Giddey AD, Ramadan WS, El-Awady R, Soares NC et al (2023) Proteomic and metabolomic signatures of U87 glioblastoma cells treated with cisplatin and/or paclitaxel. Ann Med [Internet] 55(2) Available from:. https://doi.org/10.1080/07853890.2024.2305308

Alhameed RA, Semreen MH, Hamad M, Giddey AD, Sulaiman A, Al Bataineh MT et al (2023) Multi-omics profiling of Candida albicans grown on solid versus liquid media. Microorganisms 11(12):2831

Ali A, Davidson S, Fraenkel E, Gilmore I, Hankemeier T, Kirwan JA et al (2022) Single cell metabolism: current and future trends. Metabolomics 18(10):77

Azmir J, Zaidul ISM, Rahman MM, Sharif KM, Mohamed A, Sahena F et al (2013) Techniques for extraction of bioactive compounds from plant materials: a review. J Food Eng [Internet] 117(4):426–436. Available from:. https://doi.org/10.1016/j.jfoodeng.2013.01.014

Ben Mrid R, Benmrid B, Hafsa J, Boukcim H, Sobeh M, Yasri A (2021) Secondary metabolites as biostimulant and bioprotectant agents: a review. Sci Total Environ [Internet] 777:146204Available from:. https://doi.org/10.1016/j.scitotenv.2021.146204

Bingol K (2018) Recent advances in targeted and untargeted metabolomics by NMR and MS/NMR methods. High-Throughput 7(2):9

Boucherabine S, Giddey A, Nassar R, Al-Hroub HM, Mohamed L, Harb M et al (2024) Proteomic and metabolomic profiling of methicillin resistant versus methicillin sensitive Staphylococcus aureus using a simultaneous extraction protocol. Front Microbiol 15(June):1–12

Bucar F, Wube A, Schmid M (2013) Natural product isolation-how to get from biological material to pure compounds. Nat Prod Rep 30(4):525–545

Catani M, Felletti S, Franchina FA (2022) Separation techniques [Internet]. In: Metabolomics perspectives: from theory to practical application. INC, pp 63–108. Available from:. https://doi.org/10.1016/B978-0-323-85062-9.00003-9

Chen Y, Li EM, Xu LY (2022) Guide to metabolomics analysis: a bioinformatics workflow. Meta 12(4):357

Chokkathukalam A, Kim DH, Barrett MP, Breitling R, Creek DJ (2014) Stable isotope-labeling studies in metabolomics: new insights into structure and dynamics of metabolic networks. Bioanalysis 6(4):511–524

Comi TJ, Do TD, Rubakhin SS, Sweedler JV (2017) Categorizing cells on the basis of their chemical profiles: Progress in single-cell mass spectrometry. J Am Chem Soc 139(11):3920–3929

Cowan MM (1999) Plant products as antimicrobial agents. Clin Microbiol 12:564–582

Dayalan S, Xia J, Spicer RA, Salek R, Roessner U (2018) Metabolome analysis. In: Encyclopedia of bioinformatics and computational biology: ABC of bioinformatics, pp 396–409

Denness L, McKenna JF, Segonzac C, Wormit A, Madhou P, Bennett M et al (2011) Cell wall damage-induced lignin biosynthesis is regulated by a reactive oxygen species- and jasmonic acid-dependent process in arabidopsis. Plant Physiol 156(3):1364–1374

Dudley E, Yousef M, Wang Y, Griffiths WJ (2010) Targeted metabolomics and mass spectrometry [Internet]. In: Advances in protein chemistry and structural biology, vol 80, 1st edn. Elsevier Inc, p 45–83. Available from:. https://doi.org/10.1016/B978-0-12-381264-3.00002-3

Emwas AH, Roy R, McKay RT, Tenori L, Saccenti E, Nagana Gowda GA et al (2019) Nmr spectroscopy for metabolomics research. Metabolites 9:7

Garcia-Vaquero M, Rajauria G, Tiwari B (2020) Conventional extraction techniques: Solvent extraction [Internet]. In: Sustainable seaweed technologies: cultivation, biorefinery, and applications. Elsevier Inc, pp 171–189. Available from:. https://doi.org/10.1016/B978-0-12-817943-7.00006-8

Haider M, Jagal J, Bajbouj K, Sharaf BM, Sahnoon L, Okendo J et al (2023) Integrated multi-omics analysis reveals unique signatures of paclitaxel-loaded poly(lactide-co-glycolide) nanoparticles treatment of head and neck cancer cells. Proteomics 23(16):1–15

Hajam YA, Rani R, Ganie SY, Sheikh TA, Javaid D, Qadri SS et al (2022) Oxidative stress in human pathology and aging: molecular mechanisms and perspectives. Cells 11(3):552

Hématy K, Cherk C, Somerville S (2009) Host-pathogen warfare at the plant cell wall. Curr Opin Plant Biol 12(4):406–413

Hendriks MMWB, Eeuwijk FA, va Jellema RH, Westerhuis JA, Reijmers TH, Hoefsloot HCJ et al (2011) Data-processing strategies for metabolomics studies. Trends Anal Chem [Internet] 30(10):1685–1698. Available from:. https://doi.org/10.1016/j.trac.2011.04.019

Idle JR, Gonzales FJ (2007) Metabolomics 6(5):348–351

Irie M, Fujimura Y, Yamato M, Miura D, Wariishi H (2014) Integrated MALDI-MS imaging and LC-MS techniques for visualizing spatiotemporal metabolomic dynamics in a rat stroke model. Metabolomics 10(3):473–483

Jalaleddine N, Hachim M, Al-Hroub H, Saheb Sharif-Askari N, Senok A, Elmoselhi A et al (2022) N6-acetyl-L-lysine and p-cresol as key metabolites in the pathogenesis of COVID-19 in obese patients. Front Immunol 13(May):1–11

Kenar E, Franken H, Forcisi S, Wörmann K, Häring HU, Lehmann R et al (2014) Automated label-free quantification of metabolites from liquid chromatography-mass spectrometry data. Mol Cell Proteomics 13(1):348–359

Kentner D, Martano G, Callon M, Chiquet P, Brodmann M, Burton O et al (2014) Shigella reroutes host cell central metabolism to obtain high-flux nutrient supply for vigorous intracellular growth. Proc Natl Acad Sci USA 111(27):9929–9934

Krassowski M, Das V, Sahu SK, Misra BB (2020) State of the field in multi-omics research: from computational needs to data mining and sharing. Front Genet 11(December):1–17

Lanekoff I, Sharma VV, Marques C (2022) Single-cell metabolomics: where are we and where are we going? Curr Opin Biotechnol 75:1–7

Letertre MPM, Giraudeau P, de Tullio P (2021) Nuclear magnetic resonance spectroscopy in clinical metabolomics and personalized medicine: current challenges and perspectives. Front Mol Biosci 8(September):1–25

Lin W, Mousavi F, Blum BC, Heckendorf CF, Lawton M, Lampl N et al (2024) Panama-enabled high-sensitivity dual nanoflow LC-MS metabolomics and proteomics analysis. Cell Rep Methods [Internet] 4(7):100803. Available from:. https://doi.org/10.1016/j.crmeth.2024.100803

Luque de Castro MD, Priego-Capote F (2010) Soxhlet extraction: past and present panacea. J Chromatogr A [Internet] 1217(16):2383–2389. Available from:. https://doi.org/10.1016/j.chroma.2009.11.027

Ma EH, Verway MJ, Johnson RM, Roy DG, Steadman M, Hayes S et al (2019) Metabolic profiling using stable isotope tracing reveals distinct patterns of glucose utilization by physiologically activated CD8+ T cells. Immunity [Internet] 51(5):856–870.e5. Available from:. https://doi.org/10.1016/j.immuni.2019.09.003

Maltese F, van der Kooy F, Verpoorte R (2009) Solvent derived artifacts in natural products chemistry. Nat Prod Commun 4(3):447–454

Mihai MM, Bălăceanu-Gurău B, Ion A, Holban AM, Gurău CD, Popescu MN et al (2024) Host–microbiome crosstalk in chronic wound healing. Int J Mol Sci 25(9):1–22

Muehlbauer LK, Jen A, Zhu Y, He Y, Shishkova E, Overmyer KA et al (2023) Rapid multi-omics sample preparation for mass spectrometry. Anal Chem 95(2):659–667

Nagana Gowda GA, Raftery D (2023) NMR metabolomics methods for investigating disease. Anal Chem 95(1):83–99

O'Neill LAJ, Kishton RJ, Rathmell J (2016) A guide to Immunometabolism for immunologist. Physiol Behav 16(9):553–565

Oken BS, Chamine I, Wakeland W (2015) A systems approach to stress, stressors, and resilience in humans. National Institute Health [Internet], pp 144–154. Available from: https://www.ncbi.nlm.nih.gov/pmc/articles/PMC3624763/pdf/nihms412728.pdf

Olivier M, Asmis R, Hawkins GA, Howard TD, Cox LA (2019) The need for multi-omics biomarker signatures in precision medicine. Int J Mol Sci 20(19):4781

Olson WJ, Di Genova BM, Gallego-Lopez G, Dawson AR, Stevenson D, Amador-Noguez D et al (2020) Dual metabolomic profiling uncovers toxoplasma manipulation of the host metabolome and the discovery of a novel parasite metabolic capability. PLoS Pathog [Internet] 16(4):1–30. Available from:. https://doi.org/10.1371/journal.ppat.1008432

Pai S, Hebbar A, Selvaraj S (2022) A critical look at challenges and future scopes of bioactive compounds and their incorporations in the food, energy, and pharmaceutical sector. Environ Sci Pollut Res [Internet] 29(24):35518–35541. Available from:. https://doi.org/10.1007/s11356-022-19423-4

Pandian K, Matsui M, Hankemeier T, Ali A, Okubo-Kurihara E (2023) Advances in single-cell metabolomics to unravel cellular heterogeneity in plant biology. Plant Physiol [Internet] 193(2):949–965. Available from:. https://doi.org/10.1093/plphys/kiad357

Prasad M, Ghosh M, Kumar R (2022) Single cell metabolomics [Internet]. Metabolomics perspectives: from theory to practical application. INC, pp 457–513. Available from:. https://doi.org/10.1016/B978-0-323-85062-9.00013-1

Pusch W, Flocco MT, Leung SM, Thiele H, Kostrzewa M (2003) Mass spectrometry-based clinical proteomics. Pharmacogenomics 4(4):463–476

Qiu S, Cai Y, Yao H, Lin C, Xie Y, Tang S et al (2023) Small molecule metabolites: discovery of biomarkers and therapeutic targets. Signal Transduct Target Ther 8(1):1–37

Ren S, Hinzman AA, Kang EL, Szczesniak RD, Lu LJ (2015) Computational and statistical analysis of metabolomics data. Metabolomics 11(6):1492–1513

Saheb Sharif-Askari N, Soares NC, Mohamed HA, Saheb Sharif-Askari F, Alsayed HAH, Al-Hroub H et al (2022) Saliva metabolomic profile of COVID-19 patients associates with disease severity. Metabolomics [Internet] 18(11):1–16. Available from:. https://doi.org/10.1007/s11306-022-01936-1

Shresta B (2020) Single cell metabolism. In: Methods in molecular biology, vol 2064, pp 31–59
Shuken SR (2023) An introduction to mass spectrometry-based proteomics. J Proteome Res 22(7):2151–2171
Soares NC, Francisco R, Vielba JM, Ricardo CP, Jackson PA (2009) Associating wound-related changes in the apoplast proteome of medicago with early steps in the ROS signal-transduction pathway. J Proteome Res 8(5):2298–2309
Soares NC, Wojtkowska J, Jackson PA (2011) A proteomic analysis of the wound response in Medicago leaves reveals the early activation of a ROS-sensitive signal pathway. J Proteom [Internet] 74(8):1411–1420. Available from:. https://doi.org/10.1016/j.jprot.2011.03.017
Soares NC, Hussein A, Muhammad JS, Semreen MH, El Ghazali G, Hamad M (2023) Plasma metabolomics profiling identifies new predictive biomarkers for disease severity in COVID-19 patients. PLoS One [Internet] 18(8 August):1–20. Available from:. https://doi.org/10.1371/journal.pone.0289738
Sun YV, Hu Y-J (2016) Integrative analysis of multi-omics data for discovery and functional studies of complex human diseases. Physiol Behav [Internet] 93:147–190
Tomic-Canic M, Burgess JL, O'Neill KE, Strbo N, Pastar I (2020) Skin microbiota and its interplay with wound healing. Am J Clin Dermatol [Internet] 21(s1):36–43. Available from:. https://doi.org/10.1007/s40257-020-00536-w
Träger D (1988) Pyoderma gangraenosum as an orthopedic problem case. Z Orthop Ihre Grenzgeb 126(2):127–129
Tsoukalas D, Sarandi E, Fragoulakis V, Georgaki S, Tsatsakis A (2024) Metabolomics. In: Reference module in biomedical sciences. Elsevier, pp 161–170
Wang J, Sun S, Deng H (2023) Chemical reprogramming for cell fate manipulation: methods, applications, and perspectives. Cell Stem Cell [Internet] 30(9):1130–1147. Available from:. https://doi.org/10.1016/j.stem.2023.08.001
Want E, Masson P (2010) Processing and analysis of GC/LC-MS-based metabolomics data. Springer Protoc, pp 277–298
Wörheide MA, Krumsiek J, Kastenmüller G, Arnold M (2021) Multi-omics integration in biomedical research – a metabolomics-centric review. Anal Chim Acta 1141:144–162
Yao L, Sheflin AM, Broeckling CD, Prenni JE (2019) Data processing for GC-MS- and LC-MS-based untargeted metabolomics. Springer Protoc, pp 287–299
Zenati RA, Giddey AD, Al-Hroub HM, Hagyousif YA, El-Huneidi W, Bustanji Y et al (2023) Evaluation of two simultaneous metabolomic and proteomic extraction protocols assessed by ultra-high-performance liquid chromatography tandem mass spectrometry. Int J Mol Sci 24(2):1–14
Zhang P, Chen YP, Qiu JH, Dai YZ, Feng B (2019) Imaging the microprocesses in biofilm matrices. Trends Biotechnol 37(2):214–226

Bacterial Metabolites in Attack

4

Wael A. H. Hegazy

Abstract

Pathogenic bacteria have evolved a diversity of interplayed approaches and mechanisms to invade and colonize host cells, leading to infections of varying severity. To successfully establish an infection, bacteria must navigate multiple barriers, starting with attachment to host cells, evading the immune response, acquiring essential nutrients, and disseminating within host tissues. These tasks are accomplished through the expression of a huge number of bacterial metabolites regulated by specialized systems, each contributing to different stages of the infection process. This chapter explores bacterial metabolism, focusing on the virulent metabolites that bacteria utilize to establish infections, providing insights into potential targets for developing novel antibacterial therapeutics and strategies.

Keywords

Bacterial metabolites · Bacterial virulence · Bacterial invasion · Bacterial toxins

4.1 Introduction

Pathogenic bacteria have evolved a myriad of interplayed tactics to conquer and colonize host cells, causing a wide range of infection severity. To establish an infection, invading bacteria must overcome multiple barriers starting with attachment to host cells, evasion of the host immunity, acquisition of essential nutrients, and then

W. A. H. Hegazy (✉)
Department of Microbiology and Immunology, Faculty of Pharmacy, Zagazig University, Zagazig, Egypt
e-mail: waelmhegazy@daad-alumni.de

S. S. M. Soliman, M. I. Husseiny (eds.), *Metabolic Dynamics in Host-Microbe Interaction*, https://doi.org/10.1007/978-981-96-1305-2_4

dissemination within host tissues. To accomplish these tasks, bacteria employ an array of virulence factors, which are expressed under the control of specialized systems, each contributing to different stages of the infection process.

Bacterial adhesins such as fimbriae are virulence factors that enable initial attachment to host cells (Patel et al. 2017). Following the attachment, bacteria employ several toxin and effector protein systems to manipulate host-cell functions, resulting in cellular integrity disruption that facilitates bacterial invasion and immune evasion (Do Vale et al. 2016). Inside the cellular compartments, capsules and other surface structures provide additional protection against host immune defenses, ensuring bacterial survival and persistence (Uribe-Querol and Rosales 2017). Bacterial spread, flagella, and motility systems in parallel to a wide diversity of extracellular enzymes such as protease, collagenase, elastase, and others allow bacteria to disseminate through the host tissues (Dowd et al. 2021). Nutrient uptake, such as iron uptake systems, plays a vital role in obtaining essential nutrients in nutrient-limited host environments (Murdoch and Skaar 2022). Furthermore, the bacterial ability to form biofilms enhances the resistance to environmental stresses, antimicrobial agents, and establishing chronic infections.

Regulatory networks and signal transduction pathways allow bacteria to sense and respond to environmental cues, dynamically adjusting the bacterial gene expression to optimize survival and virulence. These interconnected systems collectively facilitate the complex process of bacterial invasion, enabling pathogens to thrive in diverse and often hostile host environments (Chekabab et al. 2014; Shaw et al. 2022). It is imperative to comprehend how bacteria can initiate and establish an infection to develop effective preventive measures. By understanding the intricate strategies of bacterial pathogens that employ diverse metabolic pathways, the challenges of combating bacterial infections can be deeper understood exploring potential targets for novel therapeutic interventions. In this chapter, we delve into bacterial metabolism to explore those virulent metabolites that bacteria utilize to establish their infections.

4.2 Metabolism Meets Virulence

Bacterial infectivity results from an imbalance between bacterial virulence and host resistance. While the mechanisms of infection may differ between bacteria, the ways in which they cause disease can generally be divided into several stages. Bacterial metabolites play crucial roles in pathogenesis and serve as hidden weapons, facilitating infection and enhancing bacterial virulence (Gripenland et al. 2010; Passalacqua et al. 2016). Bacterial metabolites are produced during their metabolic processes and can be discriminated into primary metabolites, which are essential for microbial growth, such as amino acids and nucleotides. In addition, secondary metabolites such as toxins, hydrolytic enzymes, antibiotics, siderophores, and signaling molecules are not directly involved in growth but play crucial roles in interactions with the environment and host cells (Passalacqua et al. 2016; Rohmer et al. 2011; Heroven and Dersch 2014). In a sophisticated manner, the bacterial

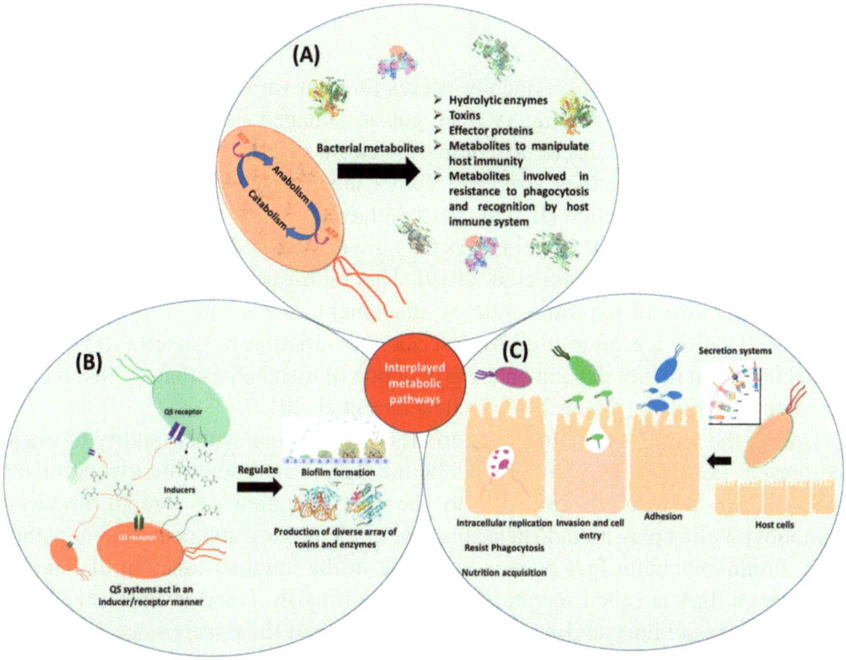

Fig. 4.1 Bacterial metabolites mediate bacterial attacks. In an interplayed manner, metabolites play a crucial role in disrupting host defenses and facilitating bacterial invasion and survival. (**a**) A series of sophisticated metabolic pathways result in the production of huge numbers of metabolites that share in bacterial attack and evasion of host immunity. (**b**) Quorum sensing systems orchestrate the production of various virulence factors and biofilm formation. (**c**) Pathogenic bacteria recruit secretion systems to facilitate their invasion and internalization

metabolites are specialized and play definite roles in the establishment of bacterial infection (Passalacqua et al. 2016; Eisenreich et al. 2010). In the next section, we will shed light on the stages of bacterial infection drawing attention to the most commonly involved metabolites (Fig. 4.1).

4.3 Unveiling the Role of Microbial Metabolites in Host Invasion

Bacterial metabolites are instrumental in the process of host invasion, enabling pathogens to breach host defenses and establish infections. These small molecules facilitate various stages of infection, from initial attachment and entry to evasion of immune responses and establishment of a niche within the host. This section delves into the multifaceted roles of microbial metabolites in host invasion, briefing the mechanisms through which they contribute to pathogenic success.

4.3.1 Entry into the Host

Bacterial entry into the host typically occurs through various routes including the gastrointestinal system, respiratory tract, cut or damaged skin, and mucous membranes. The number of bacteria required to cause disease varies greatly among pathogens depending on their virulence factors and the strength of the host immune response opposing infection (Ji et al. 2015; Velge et al. 2012). In this context, multiplicity of infection (MOI) is defined as the ratio of adsorbed, attached, or infecting bacteria to host cells (Vlazaki et al. 2019). This ratio varies based on the bacterial numbers, duration of exposure, rate of attachment, and nature of opposing host cells. MOI ratios are often used to describe the variation in bacteria-to-host cell interactions that in turn depend on the production of metabolites that are involved in the infection process (Zhang 2013; Bettencourt et al. 2017).

One of the most important characteristics of pathogens is their ability to evade the present host immunity. Immunoglobulins (Ig) or antibodies are glycoproteins produced by B cells in response to the reorganization of foreign invaders. Immunoglobulins play an indispensable role in identifying and neutralizing pathogens. Immunoglobulin IgA plays a vital role in the immune function of mucous membranes. IgA is called secretory Ig as it is primarily found in mucosal tissues such as the respiratory and gastrointestinal tracts and their secretions. In case of bacterial infection through these mucous membranes, IgA neutralizes the bacterial invaders. Interestingly, pathogenic bacterial species such as *Neisseria* spp. and *Staphylococcus aureus* produce IgA protease that facilitates the induction of bacterial infections at their initial stages (Kilian and Russell 2015).

4.3.2 Adherence to Host Cells

Bacteria employ a wide variety of molecules, ranging from simple monomeric metabolites to complex multimeric macromolecules, to adhere to host cells. In the next section, we will address the most involved metabolites and tactics that bacteria employ to adhere to host cells.

4.3.2.1 Adhesins

Pili or fimbriae are hair-like appendages that protrude from the bacterial surface to facilitate adhesion. Besides pili, plenty of non-polymeric bacterial adhesins are produced to recognize various elements on host-cell surfaces. Adhesins could target extracellular matrix components, including elastin, collagens, hyaluronan, proteoglycans, and laminins (Patel et al. 2017). Bacterial pathogens could recognize adhesive glycoproteins like vitronectin, fibrinogen, and particularly fibronectin, which can be found as either secreted or conjugated with the plasma membranes (Stones and Krachler 2015). Furthermore, integrins, selectins, cadherins, and carcinoembryonic antigen-related cell adhesion molecules (CEACAMs) are integral host membrane adhesion receptors to many bacterial pathogens for adhesion, as well as for cell entry (Patel et al. 2017; Kc et al. 2017). As an impressive example, *Streptococcus pyogenes* and *Staphylococcus aureus*, causative agents of a wide range of localized and systemic infections, produce several proteins that bind to fibronectin and collagen and also possess additional enzymatic functions conferring mechanico-functional properties (Foster 2016). For instance, ScpA with serine protease activity and SfbII with lipoproteinase activity are produced by *S. pyogenes* (Kreikemeyer et al. 2004) and FnBP-A of *S. aureus* (Schwarz-Linek et al. 2003). SfbII and FnBP-A bind to ECM-associated fibronectin inducing the aggregation of fibronectin-bound integrin receptors that in turn results in triggering intracellular signaling (Pizarro-Cerdá and Cossart 2006).

4.3.2.2 Autotransporters

Autotransporters are bacterial surface proteins that represent the simplest export mechanisms of proteins. This process, known as type V secretion, involves molecules that have an N-terminal leader peptide for transport through the inner membrane, a C-terminal domain makes a pore in the outer membrane, and a passenger domain is transported across this pore to be displayed on the bacterial surface, where it may be cleaved in certain cases (Henderson et al. 2004; Veiga et al. 2004). It was shown that several autotransporters are associated with bacterial adhesion such as Antigen 43 (Ag43) and TibA in *Escherichia coli* (Ulett et al. 2007; Vo et al. 2017). Another important example of autotransporters is BabA in *Helicobacter pylori*, which distinguishes the Lewis B antigen on red blood cells as well as in gastric mucosa (Doohan et al. 2021). Inflammation of the gastric tissue results in an increase in the expression of other Lewis's antigens, which are recognized by *Helicobacter pylori* sialic acid-binding protein (SabA). This supposes the *Helicobacter pylori* dependence on its own receptors in target tissues to induce serious infections (Mahdavi et al. 2002).

4.3.2.3 Specialized Systems for Bacterial Adhesion

Besides the whole arsenal of adhesins and other metabolites, bacteria have developed specialized systems to ease their adhesion and invasion. However, the systems that regulate the production of virulence metabolites will be described later in this chapter, we preferred to shed light on an example of the translocated intimin receptor (Tir) in *E. coli* and its unique adhesion fashion. Enteropathogenic *E. coli* (EPEC) and enterohemorrhagic *E. coli* (EHEC) have developed their own adhesion system in which they provide both the ligand and the receptor. Following the intimate attachment to intestinal epithelial cells, EPEC and EHEC induce a characteristic attaching and effacing (A/E) lesion. EPEC and EHEC trigger a localized destruction of absorptive microvilli lining the small intestine leading to the accumulation of host cytoskeletal proteins, which in turn results in the formation of pedestal-like structures on which the bacteria perch and flourish (Freeman et al. 2000; Blasche et al. 2014). The required molecules for the induction of A/E lesions are simply metabolites encoded on the locus of enterocyte effacement (Yazdanparast et al. 2012).

4.3.3 Bacterial Invasion

Bacterial pathogens employ a huge diverse metabolite to orchestrate complex strategies for attacking and invading host cells. These metabolites serve as chemical weapons, enabling pathogens to manipulate host biology, breach host defenses, and establish infections. In the following section, we explore how bacteria recruit their metabolite arsenal to successfully invade and cause diseases.

4.3.3.1 Breaching Natural Host Barriers

Following bacterial adhesion, bacterial pathogens will devote a wide array of enzymes and toxins to penetrate the host physical barriers they face. Bacterial pathogens produce enzymes such as mucinases, hyaluronidases, and collagenases to degrade the extracellular matrix and mucosal barriers facilitating initial invasion steps. Then after, enzymes such as proteases, lipases, elastase, and glycosidases are produced to ease the bacterial spread and dissemination into the host tissues (Heinzelmann et al. 2002; Los et al. 2013). *Helicobacter pylori* produces urease to neutralize stomach acid and mucinase to degrade the protective mucus layer of the stomach, facilitating colonization of the gastric epithelium (Mora and Arioli 2014). For deeper infections, other toxins and virulent enzymes are required, for instance, alpha toxin and collagenase that are produced by *Clostridium perfringens* to hydrolyze cell membrane phospholipids, leading to cell lysis and tissue necrosis ending by gas gangrene (Navarro et al. 2018). There is a wide diversity of enzymes and toxins that are involved in breaching the host defenses, which will be more detailed in the coming sections.

4.3.3.2 Invasion of the Host Cells

Bacterial pathogens have developed molecular mechanisms to dynamically facilitate their entry into target cells for replication and dissemination to other host tissues. This invasion can occur through direct interaction with host-cell surface receptors or by injecting bacterial proteins into the host-cell cytoplasm, triggering plasma membrane rearrangements that lead to the engulfment of invader pathogens (Ji et al. 2015; Siegel and Weiser 2015). Numerous pathogens such as *Yersinia enterocolitica*, *Yersinia pseudotuberculosis*, and *E. coli* have utilized their interaction with host-cell receptors to adhere and induce actin cytoskeleton rearrangements leading to cellular invasion. Invasin was the first described bacterial invasion protein (Isberg et al. 1987), it binds to integrins as well as fibronectin (Leo et al. 2015). The engagement of adhesion receptors such as integrin to invasion results in the activation of several small GTPases such as the Rho family of master actin regulators to reshape the bacterial entry site facilitating the bacterial invasion (Wong and Isberg 2005). *Rickettsia* spp. is an intracellular, non-cultivatable bacteria and employs two outer membrane proteins, rOmpA and rOmpB, as putative adhesins favoring the attachment to epithelial cells in vitro (Uchiyama 2012). In a similar way, *Listeria monocytogenes* can traverse the intestinal barrier as well as the blood–brain barrier causing several clinical manifestations (Dussurget et al. 2004). InlA, a prototypic member of the internalin family of leucine-rich repeat molecules, mediates the

Listeria monocytogenes invasion by binding to E-cadherin. The interaction of InlA/E-cadherin is species-specific and recruits two catenins to facilitate the internalization of *Listeria monocytogenes* (Carvalho et al. 2014).

Another invasion strategy employed by enteropathogenic *Salmonella* spp. is recruiting specialized secretion systems to inject effector proteins into the cytoplasm of the host cells. In *Salmonella* enterica spp. the pathogenesis-encoding genes are clustered on certain loci in the chromosome, and these loci are called salmonella pathogenicity islands (SPIs) (Askoura and Hegazy 2020; Hegazy and Hensel 2012). In particular, the SPI-1 encodes type III secretion system-1 (T3SS-1) that possesses its characteristic injectosome and could inject protein effectors that modulate the rearrangement of actin that results in remodeling the host-cell cytoskeleton to facilitate the bacterial invasion (Thabit et al. 2022a; Hegazy et al. 2022). It is worth mentioning that T3SS is expressed in other bacteria such as *Shigella* spp., *Yersinia* spp., and *Pseudomonas* spp. playing significant roles in the bacterial invasion and internalization (Izoré et al. 2011).

4.4 Harnessing Metabolism: Manipulation of Host Immune Responses

Some metabolites modulate the host immune system to create a more favorable environment for the pathogen. For instance, *Helicobacter pylori* secretes urease to neutralize stomach acid conferring a less hostile local environment (Mora and Arioli 2014). In a more complicated manner, bacterial pathogens produce diverse metabolites to employ different mechanisms to evade or suppress host immune responses. The best examples are intracellular replicating bacterial pathogens; for instance, *Mycobacterium tuberculosis* produces mycolic acids to inhibit the maturation of phagosomes in macrophages, allowing the bacteria to survive within these immune cells (Rahlwes et al. 2023). *Streptococcus pneumoniae* produces a polysaccharide capsule that inhibits phagocytosis by macrophages (Bruyn et al. 1992). Other pathogens produce metabolites that disrupt normal host-cell signaling, facilitating infection such as *Bacillus anthracis*, the causative agent of anthrax, interferes with host-cell MAPK signaling pathways inhibiting migration of both antigen-presenting cells (APCs) and T cells that prevent the initiation of adaptive immune responses and recruitment of phagocytes and effector cells to the site of infection (Brossier and Mock 2001). Also, *Listeria monocytogenes* produce pore-forming Listeriolysin O (LLO) induce pores in the phagosome to ease the bacterial escape from phagocytosis (Hamon et al. 2012).

Besides these bacterial metabolites that are involved in evading the host immunity, there are other tactics to manipulate the host immunity. Type III secretion system (T3SS) plays a crucial role in the internalization of bacterial pathogens and resistance to phagocytosis. Following the engulfment of bacterial pathogens into the phagosomes in the involved APCs namely B cells, macrophages, and dendritic cells, pathogenic Gram-negative bacterial cells such as *Salmonella* spp., *Shigella* spp., and *Yersinia* spp. employ their specialized molecular syringe-like apparatus “T3SS injectosome" to translocate effector proteins into the cytoplasm of the host immune cells (Cornelis 2006). These effector proteins are working on disrupting the immune cell signaling to prevent an effective immune response such as Yop in *Yersinia pestis* (Plano and Schesser 2013), promoting the apoptosis of resident immune cells as in enteropathogenic *Shigella flexneri* (Muthuramalingam et al. 2021), or remodeling the host cytoskeleton to ease the thrive and acquisition of nutation such as *Salmonella*-inducing filaments (Sifs) effector proteins in *Salmonella enterica* (Thabit et al. 2022a; Elfaky et al. 2022).

4.4.1 Antimicrobial Peptides (AMPs)

AMPs are small, naturally occurring proteins produced by host cells as part of the innate immune response. AMPs play a crucial role in defending against infections by directly targeting and disrupting the membranes of invading bacteria. AMPs typically have a cationic charge, allowing them to interact with negatively charged bacterial membranes, leading to membrane destabilization and cell death (Bahar and Ren 2013). In addition to their direct antimicrobial effects, AMPs can modulate immune responses, enhance wound healing, and reduce inflammation (Hilchie et al. 2013). In a magnificent adaptation, bacteria could develop several mechanisms to alter the cell envelope composition/charge, degrade AMPs by proteases, and/or extrusion of AMPs by efflux pumps and transport systems (Assoni et al. 2020). These mechanisms are employed by both Gram-negative and Gram-positive bacteria to resist the AMPs (Assoni et al. 2020; Duperthuy 2020).

4.5 Bacterial Colonization and Establishment of a Niche

4.5.1 Nutrient Manipulation and Metabolic Hijacking

Nutrient acquisition from the host environment is a mandatory task for bacterial pathogens to establish their infection. *Salmonella enterica* is equipped with a sophisticated injectosome T3SS-type II that can express *Salmonella*-induced filaments (Sifs). Sifs are tubular structures that emerge from the *Salmonella*-containing vacuole (SCV) and extend throughout the host cell connecting multiple SCVs (Hegazy and Hensel 2012; Thabit et al. 2022a). Sifs maintain the integrity of the SCV to avoid host immunity and facilitate nutrient acquisition to establish a replicative niche inside host cells (Knuff and Finlay 2017). Another tactic for metabolic hijacking, bacteria can manipulate host-cell metabolism to promote their own growth and survival. *Mycobacterium tuberculosis* modifies the host's lipid metabolism to promote lipid accumulation in infected macrophages forming "lipid droplets," which serve as a nutrient reservoir during its intracellular survival (Rahlwes et al. 2023). Another intracellular bacterium, *Chlamydia trachomatis* manipulates host lipid metabolism via redirecting host lipids to its intracellular inclusion, providing essential components for its membrane biogenesis and replication (Rother et al. 2019; Elwell and Engel 2012). *Salmonella* spp. alter host-cell metabolism by inducing glycolysis, which provides the bacteria with nutrients like pyruvate to fuel their intracellular replication (Bumann and Schothorst 2017). *Helicobacter pylori* interferes with cholesterol metabolism in gastric cells, utilizing host cholesterol for its membrane stability (Qaria et al. 2021). *Listeria monocytogenes* increases the synthesis of phospholipids in host cells to build its own membranes, while replicating inside the host (Sauer et al. 2019).

4.5.2 Siderophores

There are a lot of examples that show that successful bacterial pathogens must acquire nutrients from the host environment. However, bacteria could employ specialized metabolites to acquire their nutrition. Iron is essential for many bacterial metabolic processes, but it is often limited in availability due to its insolubility and sequestration by host proteins, particularly during infection. Siderophores are specialized molecules that are produced not only by bacteria but also by fungi, and some plants to scavenge iron from their environment (Hider and Kong 2010). Bacteria release siderophores to bind to iron and transport it back into the cell via specific receptor proteins (Hider and Kong 2010; Ahmed and Holmström 2014). Once inside, the iron is released and utilized for vital cellular functions. Siderophores such as transferrin and lactoferrin play an important role in bacterial virulence, as the ability to acquire iron is essential for the survival and proliferation of pathogens in iron-limited environments such as the host tissues (Ahmed and Holmström 2014). *E. coli* produces enterobactin enabling it to thrive in the urinary tract (Martin et al. 2017). *Mycobacterium tuberculosis* produces mycobactin to maintain its survival and persistence within host macrophages (Ratledge 2004). *Pseudomonas aeruginosa* produces pyoverdine to sequester iron during infection, enhancing its ability to cause disease, especially in immunocompromised patients (Ghssein and Ezzeddine 2022).

4.5.3 Manipulation of Host-Cell Cycle

Bacterial pathogens could successfully manipulate host-cell cycle to enhance their intracellular replication, evade immune responses, and create a favorable niche within host cells. Inhibition of apoptosis, programmed cell death, can prolong the survival of infected cells, providing a stable environment for intracellular bacterial pathogens. For example, *Chlamydia trachomatis* secrets effector proteins to arrest the host-cell cycle at the G2/M phase that results in preventing the host cell from undergoing mitosis and as a consequence the life of the infected host cells is prolonged allowing the intracellular replication of *Chlamydia* for extended periods (Sharma and Rudel 2009; Cheong et al. 2023). Similarly, *Salmonella enterica* can activate the host's checkpoint pathways that lead to cell cycle arrest in the G2 phase (Azimi et al. 2020). *Brucella abortus* produces effector proteins that block the progression of the host-cell cycle at the G1 phase, delaying cell division (Celli 2019). *Helicobacter pylori* secretes the CagA effector protein to disrupt the host-cell cycle by interfering with signaling pathways, leading to abnormal cell proliferation or apoptosis that potentially increases the risk of cancer development (Hatakeyama 2008).

4.5.4 Biofilm Formation

A bacterial biofilm is a structured community of bacteria embedded in a self-produced matrix of extracellular polymeric substances (EPS), which includes proteins, polysaccharides, and nucleic acids. This matrix provides protection and allows the bacteria to adhere to surfaces, resist environmental stresses, evade the host immune response, and increase their tolerance to antibiotics, making biofilms a significant factor in chronic infections and resistance to treatments (Lila et al. 2023; Vestby et al. 2020). Bacterial metabolites play critical roles in biofilm formation, serving as signals and building blocks that regulate the development and stability of biofilms. Principally, metabolites such as polysaccharides, proteins, and lipids donate to the formation of EPS scaffold for biofilm structure (Vestby et al. 2020; Lopez et al. 2010).

The biofilm formation occurs in steps regulated by specialized systems called quorum sensing (QS) systems. QSs orchestrate the production of diverse microbial metabolites that are involved in the different stages of biofilm formation (QS) (Solano et al. 2014; Rajab and Hegazy 2023). Metabolites such as acyl-homoserine lactones (AHLs) and other peptides in Gram-negative and Gram-positive bacteria, respectively, act as signaling molecules that are recognized by QS receptors, enabling bacteria to communicate and coordinate biofilm formation (Li and Tian 2012). QSs regulate the expression of some bacterial virulent hydrolytic exoenzymes, and toxins such as rhamnolipids and pyocyanin in *Pseudomonas aeruginosa* (Bottomley et al. 2007), prodigiosin in *Serratia marcescens* (Van Houdt et al. 2007), violacein in *Chromobacterium violaceum* (Elfaky et al. 2023), and staphyloxanthin in *Staphylococcus aureus* (Almalki et al. 2022a), to modify the biofilm environment, influencing biofilm maturation and resistance. Furthermore, several secondary metabolites such as siderophores and surfactants modulate the availability of nutrients and surface interactions that in turn support biofilm development (Balcazar et al. 2015).

4.6 Microbial Metabolites: The Biochemical Arsenal

The pathogenicity of bacteria is usually a result of a combination of different virulence factors that aim to induce tissue inflammation and increase invasiveness. Bacterial toxins are bioactive molecules released by bacteria that can induce damage to the host. These molecules play crucial roles in bacterial virulence and are usually related to tissue damage and increased invasiveness of the pathogen. In addition, bacterial toxins are commonly responsible for the pathogenicity and clinical manifestations of the illness; hence, the accurate identification of the toxin is sometimes necessary to confirm the diagnosis of some infectious diseases as well as for the selection of patient treatment options. While some pathogenic bacteria secrete one type of toxin, which is responsible for all the symptoms and the disease outcome, others are attacked by a set of multiple toxins that synergistically increase the pathogenic outcomes of the infection.

4.6.1 Bacterial Toxins

Bacterial toxins are generally classified into two main categories: endotoxins and exotoxins (Table 4.1). Endotoxins are Gram-negative bacteria membranal components that trigger inflammatory responses in the host upon bacterial cell lysis, whereas exotoxins are secreted proteins that can affect both nearby and distant sites from the site of bacterial colonization (Lahiri 2000) (Fig. 4.2).

4.6.1.1 Endotoxins

They are Gram-negative bacteria outer membrane integral components, primarily composed of lipopolysaccharides (LPS), to play a crucial role in preserving the structural integrity of the bacterial cell wall and provide protection from hostile environments (Farhana and Khan 2024). The LPS molecule typically comprises three main components: a hydrophobic lipid domain known as "lipid A," a non-repeating "core" oligosaccharide, and a distal polysaccharide with high repetitions known as the "O-antigen" (Raetz and Whitfield 2002).

LPS acquires a vital role in the antigenicity of Gram-negative bacteria mainly caused by the O-antigen moiety, which serves as a major antigenic determinant. The high variability in the O-antigen structure offers a diversity of immune responses elicited by different bacteria (Lerouge and Vanderleyden 2002). Additionally, the lipid A component of LPS is recognized by innate immunity via receptors such as Toll-like receptor 4 (TLR4). This recognition activates strong immune responses leading to the production of pro-inflammatory cytokines that are required for clearance of infection. However, such a response can be potent enough to induce a lethal endotoxic shock characterized by widespread tissue damage and organ failure (Mazgaeen and Gurung 2020). Endotoxins persist even after bacterial death; they are heat stable and not affected by autoclaving. Any pharmaceutical preparation intended for administration by injection must be LPS-free (Shoji and Opal 2021).

Lipoteichoic acid (LTA) is a main component of the Gram-positive bacteria cell wall and plays an important role in their pathogenicity. LTAs are recognized by the host's immunity through Toll-like receptor 2 (TLR2), which triggers an innate immune response (Nilsen et al. 2008). This interaction can lead to the production of pro-inflammatory cytokines and other mediators, contributing to inflammation and sepsis in severe cases. The inflammatory response induced by LTA is similar to the response triggered by LPS in Gram-negative bacteria, while using a different

Table 4.1 Summary of clinically important bacterial toxins

Toxin group	Toxin name	Mechanism	Source	Disease or main symptom
Endotoxins	Lipopolysaccharides (LPS)	Highly antigenic Activates Toll-like receptor 4 (TLR4)	Released from the cell wall of Gram-negative bacteria by lysis	Endotoxic shock
	Lipoteichoic acid	Activation of Toll-like receptor 2 (TLR2)	Released from the outer membrane of Gram-positive bacteria	Inflammation and septic shock
Exotoxins Cytotoxins	Streptococcal pyrogenic exotoxins (SPE)	Superantigens Nonspecific activation of T cells	*Streptococcus pyogenes*	Necrotizing fasciitis and streptococcal toxic shock syndrome (STSS)
	Diphtheria toxin (DT)	A-B type toxin ADP-ribosylation of the elongation factor 2 (EF-2)	*Corynebacterium diphtheria*	Diphtheria
	Anthrax toxins	A-B type toxin Lethal toxin: zinc-dependent metalloprotease/ cleaves mitogen-activated protein kinase kinases (MAPKKs) Odema toxin: Adenylyl cyclase, converting ATP to cyclic AMP (cAMP)	*Bacillus anthracis*	Anthrax
	Shiga toxin (STX1, STX2)	A-B-type toxin Inhibition of protein synthesis by cleaving the 28S rRNA of the ribosome	*Shigella dysenteriae* and *Escherichia coli*	Hemorrhagic colitis Hemolytic uremic syndrome
	Cytotoxic necrotizing factors (CNF1/CNF2)	Rho GTPases activation	*E. coli* *Yersinia pseudotuberculosis*	Tissue necrosis
	Clostridial alpha toxin (α-toxin)	Membrane-damaging toxin Zinc-dependent phospholipase C that hydrolyzes membrane phospholipids	*Clostridium perfringens*	Gas gangrene
	Staphylococcal alpha toxin	Membrane-damaging toxin A pore-forming beta-barrel toxin	*Staph. aureus*	Toxic shock syndrome

	Staphylococcal β-toxin	Membrane-damaging toxin Targets sphingomyelin moiety in cell membrane	*Staph. aureus*	Damage of erythrocytes, fibroblasts, leukocytes, and macrophages
	Streptolysins (SLO, SLS)	Membrane-damaging toxin Targeting cholesterol moiety in membrane forming large pores	*Streptococcus pyogenes*	Necrotizing fasciitis and streptococcal toxic shock syndrome
	Listeriolysins (LLO, LLS)	Membrane-damaging toxin Targets membrane cholesterol of phagocytes and phagosomes	*Listeria monocytogenes*	Listeriosis
	Pneumolysin	Membrane-damaging toxin Targets cholesterol moiety in the cell membrane	*Streptococcus pneumoniae*	Pneumonia Meningitis
	Pertussis toxin (PT)	A-B-type toxin ADP-ribosylation of G-proteins	*Bordetella pertussis*	Whooping cough
	Adenylate cyclase toxin (ACT)	Catalyzes the conversion of ATP to cyclic AMP (cAMP)	*Bordetella pertussis*	Whooping cough
	Exoenzyme Y (ExoY)	Nucleotidyl cyclase	*Pseudomonas aeruginosa*	Critical lung tissue damage
	Panton-Valentine leukocidin (PVL)	Membrane-damaging toxin Form transmembrane pores	*Staphy. aureus*	Recurrent skin and soft tissue infections (SSTIS) and necrotizing pneumonia
	Exfoliatin toxin (ETA)	Serine protease that specifically targets and cleaves desmoglein-1 of the epidermis layer	*Staphy. aureus*	Staphylococcal scalded skin syndrome (SSSS)
Exotoxins Neurotoxins	Botulinum neurotoxin	Inhibiting the release of the neurotransmitter acetylcholine at neuromuscular junctions	*Clostridium botulinum*	Botulism
	Tetanus neurotoxin	Inhibits the release of inhibitory neurotransmitters, such as glycine and gamma-aminobutyric acid (GABA)	*Clostridium tetani*	Lockjaw and tetanus

(continued)

Table 4.1 (continued)

Toxin group	Toxin name	Mechanism	Source	Disease or main symptom
Exotoxins Enterotoxins	Staphylococcal enterotoxin B (SEB)	Superantigen Triggers massive immune response by nonspecific T cell activation	*Staphy. aureus*	Staphylococcal food poisoning
	E. coli heat-labile toxin (LT)	A-B type toxin Activation of adenylate cyclase and increased intracellular cyclic AMP (cAMP)	*E. coli*	Traveler's diarrhea
	B. cereus enterotoxin	Damage to intestinal membrane through pore formation	*Bacillus cereus*	Food poisoning
	C. perfringens enterotoxin	Damage to intestinal membrane through pore formation	*Clostridium perfringens*	Food poisoning
	Clostridial Toxins A and B	Toxin A: Damage to intestinal membrane through pore formation Toxin B: inactivating Rho GTPases	*Clostridium difficile*	*Clostridium difficile* poisoning
	Verotoxin (Shiga-like toxin)	Inhibition of protein synthesis through RNA glycosylation	Enterohemorrhagic *E. coli* (EHEC)	Hemorrhagic colitis Hemolytic uremic syndrome
	Cholera toxin (CT)	A-B type toxin Activates adenylate cyclase, leading to an increase in cAMP levels	*Vibrio cholerae*	Cholera

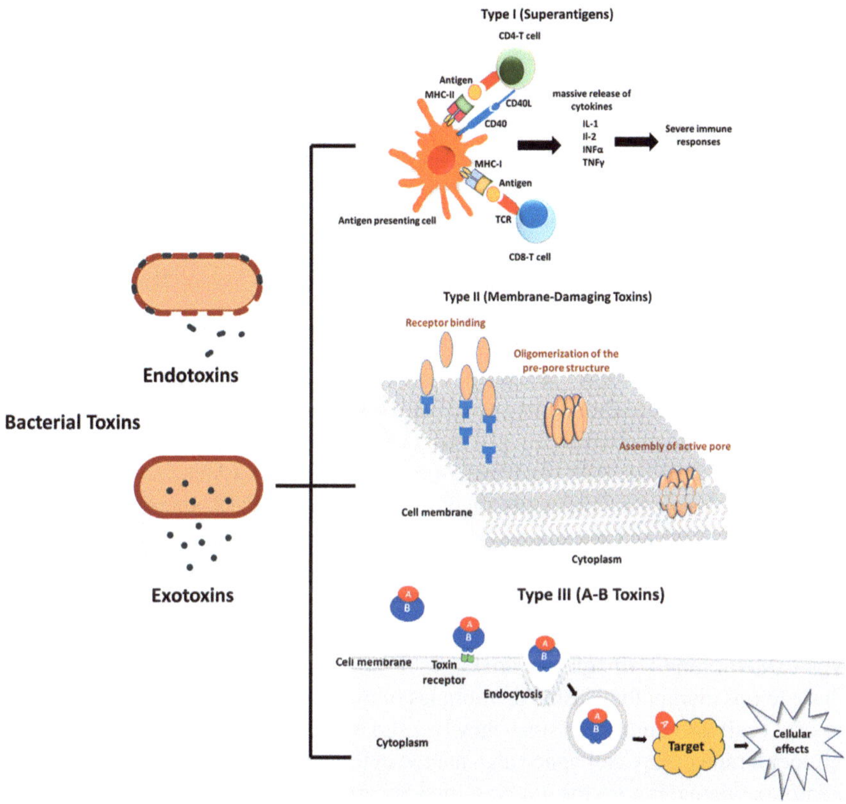

Fig. 4.2 Bacterial toxins are divided mainly into endotoxins and exotoxins. While endotoxins are components of bacterial cells that are released upon bacterial death, exotoxins are produced by living cells. Exotoxins are more specific and more toxic than endotoxins and can be subdivided into three main types (i) superantigens, (ii) pore-forming, and (iii) A-B toxins

pathway. The inflammatory outcomes induced by LTA include arthritis, nephritis, uveitis and encephalomyelitis, multi-organ failure, and septic shock (Ginsburg 2002).

4.6.1.2 Exotoxins

Exotoxins are potent molecules produced by different bacteria, playing a significant role in bacterial pathogenicity. Unlike endotoxins, exotoxins are actively released by bacteria into the surrounding environment (Zhang et al. 2023). Exotoxins are major determinants of bacterial virulence, disrupting normal cellular functions and causing tissue damage, which contributes to further dissemination of the infection. The specificity and high potency of exotoxins make them particularly dangerous, and small amounts are sufficient to cause significant tissue damage (Cavaillon 2018).

Types of Exotoxins Based on Their Mechanism of Action

Type I (Superantigens)

These toxins stimulate the immune system by binding to major histocompatibility complex molecules (MHC) and T cell receptors, leading to massive cytokine release and toxic shock in a manner similar to the septic shock caused by LPS (Unnikrishnan et al. 2002). The most common examples of this type are the staphylococcal superantigens family including the infamous toxic shock syndrome toxin-1 (TSST-1) released by *S. aureus*. TSST-1 can bind directly to MHC class II molecules on antigen-presenting cells (APCs) and T cell receptors, in an antigen-independent manner. This interaction leads to nonspecific stimulation of a large number of T cells, resulting in a massive release of cytokines, which can cause severe immune responses such as life-threatening toxic shock syndrome (Deacy et al. 2021). The syndrome is characterized by a rapid onset of symptoms like high fever, low blood pressure, sunburn-like rash, and multi-organ dysfunction (Atchade et al. 2024). Other examples of superantigens include the staphylococcal enterotoxin B produced by *S. aureus* and associated with food poisoning, and the streptococcal pyrogenic exotoxin produced by *S. pyogenes*, which is known to cause scarlet fever and toxic shock syndrome (Proft and Fraser 2016).

Type II (Membrane-Damaging Toxins)

These toxins disrupt the plasma membranes of the host cells causing cell lysis. They are often called heamolysins since they lyse red blood cells; however, they target the cell membrane of any cell hence also named cytolysins (Braun and Focareta 1991). Membrane-damaging toxins are classified according to their mechanism of action into: (i) Pore-forming toxins, these toxins create pores in the host-cell membranes, causing sudden fluid influx into the cell and cell lysis. Examples include hemolysins produced by *E. coli* and listeriolysin O released by *L. monocytogenes* (Ulhuq and Mariano 2022). (ii) Phospholipases, these enzymes hydrolyze the phospholipids in the cell membrane, hence disrupting the integrity of the phospholipid bilayer and subsequent cell lysis. Notable examples are alpha toxin from *Clostridium perfringens* and beta toxin from *S. aureus* (Monturiol-Gross et al. 2021). (iii) Detergent-like toxins, these toxins are small amphipathic peptides with detergent-like properties which disrupt the hydrophobic regions in the cell membrane such as delta-toxin produced by *S. aureus* (Otto 2014).

Type III (A-B Toxins)

This type of exotoxins is characterized by their two-component structure: the A (active) subunit and the B (binding) subunit. The subunit (B) is responsible for binding to membranal receptors and entry into the host cell, facilitating the entry of the A subunit into the cell. The A subunit is responsible for the pathogenic toxic effects by interfering with essential cellular processes. For example, the A subunit in the diphtheria toxin inhibits protein synthesis by inactivating elongation factor-2, leading to cell death. Similarly, the A subunit in the cholera toxin increases cyclic AMP levels in intestinal cells, causing severe diarrhea and dehydration (Odumosu et al.

2010). The dual-component nature of A-B toxins allows for a high degree of specificity and potency. The B subunit ensures that the toxin targets specific cell types, while the A subunit delivers a precise and often devastating effect. This mechanism not only is efficient but also makes A-B toxins particularly dangerous, as even small amounts can cause significant harm.

Types of Exotoxins According to Their Effects

While some toxins can target all cell types (e.g., cytotoxins), others are specific to certain cells through the recognition of particular receptors on the cell surface or intracellular targets. Upon binding to these receptors, toxins can exert their effects on the cell membrane by disrupting signal transduction pathways, forming pores, or enzymatically modifying membrane components. Alternatively, some toxins enter the cytosol, where they identify and alter specific intracellular targets (Hayes and Van Melderen 2011). Depending on the nature of the target and the type of modification, these intracellular toxins can significantly disrupt cellular functions such as protein synthesis, cell cycle progression, cell homeostasis, vesicular trafficking, and actin cytoskeletal rearrangements. Depending on the target cells and final toxic effects, toxins are classified into three major groups: cytotoxins, neurotoxins, and enterotoxins (Edae and Wabalo 2019).

Cytotoxins

These toxins target essential cell structures such as the cell membrane, the cytoskeleton, intracellular signaling pathways, or protein synthesis pathways, hence inducing significant cell damage and death (Aktories and Barbieri 2005). Most of the bacterial cytotoxins work by modification of GTP-binding proteins, especially the Rho family GTPases. These proteins are essential for maintaining cell shape, motility, and intracellular transport (Mosaddeghzadeh and Ahmadian 2021). Cytotoxins that target GTPases would induce disruption of the actin cytoskeleton and loss of cell integrity and function. Some examples of clinically significant bacterial cytotoxins are listed below.

Streptococcal Pyrogenic Exotoxins (SPE) This group includes multiple exotoxins produced by *S. pyogenes*, which is responsible for a variety of infections from mild pharyngitis to severe invasive diseases like streptococcal toxic shock syndrome (STSS) and necrotizing fasciitis (Ross and Shoff 2024). These exotoxins include several types, with SpeA, SpeB, and SpeC being the most studied. SpeA and SpeC are superantigens, which persuade activation of T cells. The exact mechanism involves binding to MHC-II molecules on APCs and to the variable region of the beta chain of T cell receptors (TCRs) (Proft and Fraser 2016). This binding bypasses the normal antigen processing pathway, leading to the nonspecific activation of a large number of T cells. The result is an immense release of pro-inflammatory cytokines, such as TNF-α, IL-1, and IL-6, which cause severe systemic inflammation, fever, and shock (Brouwer et al. 2023). SpeB, on the other hand, is a cysteine protease that plays a role in pathogenesis by degradation of host proteins, aiding in tissue invasion and immune evasion (Persson et al. 2013). The most common clinical presentation of streptococcal pyrogenic toxicity is the characteristic red rash of scarlet fever, caused by damage to the blood capillaries under the skin.

Diphtheria Toxin It is an A-B type cytotoxin produced by the bacterium *Corynebacterium diphtheria*. The toxin catalyzes the ADP-ribosylation of the elongation factor 2 (EF-2), a critical component of the protein synthesis machinery. This modification inhibits protein synthesis, leading to cell death (Holmes 2000). The toxin is responsible for the clinical presentations of diphtheria infection, including localized inflammation and necrosis of the respiratory epithelium, leading to the formation of a pseudo-membrane that can obstruct the airway. The toxin can be distributed systemically all over the body causing myocarditis, hepatic necrosis, and peripheral neuropathy (Skogmar and Tham 2018). The infection is rare due to mass vaccination programs, which include diphtheria toxoid in the Diphtheria-Tetanus-Pertussis (DPT) childhood vaccination program.

Anthrax Toxins They are A-B-type cytotoxins secreted by the bacterium *Bacillus anthracis*. The anthrax toxins are examples of a tripartite system consisting of three proteins: the protective antigen (PA), the edema factor (EF), and the lethal factor (LF) (Vanden Broeck et al. 2007). These proteins are individually non-toxic, but in combination, they form the two main toxins of *B. anthracis*, the lethal toxin (LT, composed of LF + PA) and the edema toxin (ET, composed of EF + PA) (Liu et al. 2014). The LT starts its activity by binding the PA moiety to host-cell receptors, which facilitates the entry of LF moiety into the cells. Once they enter the cell, LF acts as a zinc-dependent metalloprotease that exactly targets and cleaves mitogen-activated protein kinase kinases (MAPKKs) and disrupts essential cellular signaling pathways (Bardwell et al. 2004). This disruption leads to immune cell dysfunction, massive release of cytokines from macrophages, apoptosis of vascular epithelium, hemorrhage, tissue necrosis, septic shock, and sudden death. On the other hand, the ET initiates its toxic effects by binding the PA moiety to cellular receptors facilitating its entry into host cells. Once inside, the EF moiety acts as an adenylyl cyclase, converting ATP to cyclic AMP (cAMP), resulting in increased intracellular cAMP and disruption of signaling pathways, leading to a range of effects, including impaired immune responses and massive local edema (Tang and Guo 2009). The combined action of both LT and ET results in massive edema, vascular necrosis, and systemic shock. The high lethality of anthrax and the high resistance of its spores to environmental conditions made it one of the most dangerous banned biological weapons (Doganay et al. 2023).

Shiga Toxin (Stx) It is a potent A-B-type cytotoxin produced by certain strains of *Shigella dysenteriae* and *E. coli*. There are two main types of Shiga toxins: Stx1 and Stx2, both of which share a similar structure and mechanism of action. The toxin consists of five B subunits that bind to specific host-cell receptors. Once inside the cell, the A subunit cleaves a specific adenine residue from the 28S rRNA of the ribosome, effectively halting protein synthesis and leading to cell death (Melton-Celsa 2014). The main clinical manifestation of Shiga toxin exposure is hemorrhagic colitis, which is characterized by bloody diarrhea and severe abdominal cramps. If the toxin reaches the systemic circulation, it can cause hemolytic uremic syndrome, which is a life-threatening condition leading to kidney failure, hemolytic anemia, and thrombocytopenia (Liu et al. 2021). The Shiga-producing bacteria are commonly transmitted through contaminated food and water, and so prevention involves proper washing of vegetables and cooking of meat and poultry. There is currently no specific treatment for Shiga toxin related infections except for symptomatic treatment and hydration; antibiotics are generally not recommended as they may increase the release of the toxin (Muhlen and Dersch 2020).

Cytotoxic Necrotizing Factors Cytotoxic necrotizing factors (CNF1/CNF2) are cytotoxins produced by several Gram-negative bacteria including *E. coli* and *Yersinia pseudotuberculosis*. These toxins are known for their ability to activate Rho GTPases, which trigger multiple signaling pathways resulting in prolonged signaling and altered cellular functions. This can lead to a wide range of effects including cytokinesis disruption, modulation of inflammatory and immune responses, increased cell proliferation, apoptosis, and tissue necrosis (Knust and Schmidt 2010).

Clostridial Alpha Toxin Alpha toxin (α-toxin) is a potent type II cytotoxin produced by *C. perfringens*. This toxin is a zinc-dependent phospholipase C enzyme that hydrolyzes phospholipids in cell membranes, leading to cell lysis and tissue destruction. The alpha toxin's enzymatic activity disrupts cell membranes by breaking down phosphatidylcholine, a key component of the cell membrane, resulting in the release of diacylglycerol and other secondary messengers that contribute to inflammation and tissue damage (Monturiol-Gross et al. 2014). The clinical manifestations of the infection include rapid tissue necrosis, severe muscle damage, edema, hemolysis, hypoxia, and systemic toxicity. The α-toxin can trigger platelet aggregation and thrombosis, contributing to vascular damage and further dissemination of the infection (O'Brien and Melville 2004). Gas gangrene is a serious life-threatening condition that requires immediate treatment by a combination of surgical and medical interventions. Surgical debridement and sometimes amputation is crucial to remove all dead and infected tissues. In addition, broad-spectrum antibiotics and hyperbaric oxygen therapy are employed to eradicate the anaerobic pathogen and improve tissue healing (Sison-Martinez et al. 2024).

Staphylococcal Alpha Toxin Also known as α-hemolysin is a type II cytotoxin produced by *S. aureus*. This toxin is a member of the pore-forming β-barrel toxin family, which forms heptameric units on the cellular membrane, creating a barrel pore that disrupts the cell membrane, leading to cell lysis (Moller et al. 2020). At low concentrations, alpha toxin can induce apoptosis in immune cells, such as T cells and monocytes. At higher concentrations, it can cause massive necrosis and other severe cellular reactions by forming large pores that allow uncontrolled calcium influx (Otto 2014). A-hemolysin-related staphylococcal infections can include a range of clinical manifestations ranging from mild skin irritation and fever to keratitis, vascular damage, hypotension, and systemic shock. Treatment by alpha toxin-specific antibodies has shown promise in neutralizing the toxin's effects and providing protection against subsequent infections (Tkaczyk et al. 2012).

Staphylococcal Beta Toxin Staphylococcal β-toxin is a type II cytotoxin produced by *S. aureus*, which targets sphingomyelin moiety in cell membrane into ceramide and phosphorylcholine. This hydrolysis disrupts the integrity of the cell membrane, leading to cell lysis. The most common targets of this toxin are erythrocytes, fibroblasts, leukocytes, and macrophages. Hence, allowing *S. aureus* to evade the host immune response and allow for nutrient acquisition from lysed host cells (Huseby et al. 2007).

Streptolysins Streptolysins are cytotoxic hemolysins secreted by the bacterium *S. pyogenes*. There are two main types of streptolysins: Streptolysin O (SLO) and Streptolysin S (SLS). Both of them belong to type II toxins, which disrupt cell membranes by targeting cholesterol in the host-cell membrane where they oligomerize to form large pores, leading to cell lysis and death. SLO is highly antigenic and oxygen-labile, while SLS is oxygen stable and not antigenic due to its small size (Sierig et al. 2003). SLO contributes to *S. pyogenes* pathogenicity by inducing apoptosis in immune cells, hence contributing to evasion from host immunity. SLS induces significant damage to skin and nasopharyngeal epithelium, increased vascular permeability and inflammation, contributing to severe conditions such as necrotizing fasciitis and streptococcal toxic shock syndrome (Limbago et al. 2000).

Listeriolysins Listeriolysin O (LLO) and listeriolysin S (LLS) are type II cytotoxins released by the bacterium *Listeria monocytogenes*. The latter is the causative agent of listeriosis, a serious foodborne illness caused by ingestion of contaminated food or water. LLO is a cholesterol-dependent cytolysin that is crucial for the bacterium's characteristic ability to escape from phagocytes and survive intracellularly. Once *L. monocytogenes* is phagocytosed, LLO is activated in the acidic environment of the phagosome, where it forms pores in the phagosomal membrane. This allows the bacterium to escape into the cytosol, where it can replicate and spread without being detected by the host's immune system (Nguyen et al. 2019). Interestingly, LLS does not exhibit significant cytotoxic effects on eukaryotic cells, including human cells since it induces only weak hemolysis of red blood cells and does not contribute to the virulence of *L. monocytogenes* in infected tissues. However, studies suggested that LLS's primary role is to enhance the survival and colonization of *L. monocytogenes* in the gastrointestinal environment by eliminating competing cells of microbiota (Quereda et al. 2017). The symptoms of listeriosis can vary widely, ranging from mild flu-like symptoms to severe manifestations like stiff neck, confusion, loss of balance, and convulsions if the infection spreads to the nervous system.

Pneumolysin Pneumolysin is a type II cytotoxin produced by *Streptococcus pneumoniae*, which is a major causative agent of pneumonia, meningitis, and other serious infections. This toxin binds to cholesterol moiety in the cell membrane creating transmembrane pores, leading to cell lysis and death. In addition, pneumolysin can induce immune dysfunction through over-activation of the complement system in a way that can lead to excessive inflammation and tissue damage (Nishimoto et al. 2020). Additionally, pneumolysin can inhibit the function of immune cells such as neutrophils and macrophages, preventing them from effectively responding to the infection. This multifaceted interference with the immune system helps *S. pneumoniae* evade host defenses and establish a persistent infection (Pereira et al. 2022).

Pertussis Toxin (PT) Pertussis toxin (PT) is an A-B-type exotoxin secreted by the bacterium *Bordetella pertussis*, the causative agent of whooping cough. PT is composed of six subunits: five different sequences (S1–S5) that form a holotoxin. The A subunit (S1) possesses enzymatic activity, while the B subunits (S2–S5) facilitate the binding and entry of the toxin into host cells. Once inside the host cells, PT disrupts cellular signaling by ADP-ribosylating the alpha subunits of G-proteins, which are involved in regulating various cellular processes (Locht and Antoine 2021). This disruption leads to an increase in cyclic AMP (cAMP) levels, resulting in altered cellular responses, especially immune responses contributing to the characteristic symptoms of whooping cough, such as severe coughing fits and mucus production. The toxin's ability to modulate the immune system also helps *B. pertussis* to evade host defenses, allowing the bacteria to persist and cause prolonged infection (Decker and Edwards 2021). Pertussis toxin is a key component of the DTP vaccine used to immunize against whooping cough.

Adenylate Cyclase Toxin Adenylate cyclase toxin (ACT) is an exotoxin produced by *B. pertussis*. This toxin plays a crucial role in the pathogenesis of the disease by targeting immune cells, particularly phagocytes like neutrophils and macrophages. ACT binds to these cells through the complement receptor 3 (CD11b/CD18) and, once inside, catalyzes the conversion of ATP to cyclic AMP (cAMP) (Hasan et al. 2015). The resulting increase in cAMP levels disrupts normal cellular functions, impairing the immune response and facilitating bacterial survival and proliferation. High levels of cAMP within phagocytes inhibit their phagocytic ability and impair the production of reactive oxygen species. These effects help *B. pertussis* evade the host immune defenses, which contribute to the persistence of whooping cough infection (Ahmad and Sebo 2020).

Exoenzyme Y (ExoY) It is a cytotoxin produced by *Pseudomonas aeruginosa*. ExoY is a nucleotidyl cyclase, catalyzing the conversion of nucleotides into cyclic nucleotides such as cAMP, cGMP, cUMP, and cCMP. This activity disrupts various cellular processes, leading to altered cell signaling and cytoskeletal rearrangements and ultimately cell death (Beckert et al. 2014). ExoY contributes to the critical lung tissue damage caused by *P. aeruginosa* infection by disrupting the alveolar-capillary barrier, leading to increased vascular permeability, interstitial edema, hemorrhage, infiltration of immune cells, and apoptosis in infected cells (Belyy et al. 2018).

Panton-Valentine Leukocidin (PVL) It is a type II cytotoxin produced by certain strains of *S. aureus*, especially methicillin-resistant *S. aureus*. PVL is composed of two components: LukS-PV and LukF-PV, which work together to form transmembrane pores in host immune cells, particularly neutrophils, monocytes, and macrophages. This pore formation leads to leukocyte destruction and induction of inflammatory responses, making PVL a significant factor in the pathogenesis of severe *S. aureus* infections (Okolie et al. 2013). PVL-positive *S. aureus* strains are known to cause recurrent skin and soft tissue infections (SSTIs), characterized by painful boils and abscesses that are difficult to treat with antibiotics. In addition, PVL-positive *S. aureus* strains are associated with necrotizing pneumonia, characterized by extensive lung tissue damage and high mortality rates (Hoppe et al. 2019).

Exfoliatin Toxin (ETA) It is a cytotoxin produced by *S. aureus*, primarily responsible for causing staphylococcal scalded skin syndrome (SSSS). This syndrome is characterized by widespread blistering and peeling of the skin, resembling burns, and predominantly affects infants and young children. ETA is a serine protease that specifically targets and cleaves desmoglein-1 of the epidermis layer, causing the cells to detach from each other and shedding off the outer skin layer. This disrupts the integrity of the epidermal layer, leading to the characteristic blistering and exfoliation seen in SSSS (Kato et al. 2011).

Neurotoxins

They are potent molecules produced by certain bacteria that specifically target the nervous system, leading to severe neurological effects. These toxins can disrupt normal neuronal function through various mechanisms, such as inhibiting neurotransmitter release, blocking ion channels, or inducing cell death. Examples of bacterial neurotoxins include some of the most potent toxins known to humans, namely botulinum and tetanus toxins. Some clinically significant examples of bacterial neurotoxins are listed below.

Botulinum Neurotoxin Botulinum neurotoxin is an exotoxin secreted by *Clostridium botulinum* and is considered one of the most potent bacterial toxins. It works by inhibiting the release of the neurotransmitter acetylcholine at neuromuscular junctions causing failure of muscle contraction and flaccid muscle paralysis (Choudhury et al. 2021). The exact mechanism involves toxin binding to specific receptors on the motor neurons at the neuromuscular junction followed by entry into the neuron via endocytosis. The light chain of the toxin cleaves SNARE proteins (such as SNAP-25, syntaxin, or VAMP), which are essential for the fusion of synaptic vesicles with the presynaptic membrane. This cleavage prevents the release of acetylcholine into the synaptic cleft and inhibits neurotransmission: Without acetylcholine release, muscle contraction is inhibited, leading to flaccid paralysis (Fabris et al. 2022). The toxin can induce severe respiratory paralysis and death within a few hours if not treated properly. Botulism is a life-threatening condition caused by botulinum toxin. There are three types of botulism: (i) Foodborne Botulism is caused by ingestion of food contaminated with the *C. botulinum* usually due to improper packaging or storage of canned or preserved foods. Symptoms would include severe abdominal cramps, nausea, and vomiting, followed by neurological symptoms such as difficulty swallowing, facial paralysis, and blurred vision. (ii) Wound botulism that is caused by wound infection with *Clostridium botulinum* spores and subsequent germination and release of the toxin into wound tissues. The symptoms are similar to foodborne botulism in addition to fever and wound sepsis. (iii) Infant botulism is caused by ingestion of *Clostridium botulinum* spores, which then grow and produce toxins in the infant's intestines (Rawson et al. 2023). Common sources include honey and contaminated soil. The symptoms are constipation, muscle weakness (floppy baby syndrome), weak cry, irritability, and difficulty feeding and breathing.

The primary treatment for botulism is the administration of botulinum antitoxin, which is an antibody preparation that neutralizes the toxin circulating in the bloodstream. Early administration of the antitoxin is crucial for better treatment outcomes, since it cannot reverse the damage already done by the toxin but can only arrest its progressive toxicity. Despite its potency as a toxin, botulinum neurotoxin has found significant medical and cosmetic applications. In controlled, small doses, it is used to treat various medical conditions such as spastic disorders, chronic migraines, detrusor hyperactivity, and cervical dystonia. It is also widely known for its cosmetic application under the brand name Botox, which helps reduce the appearance of wrinkles by temporarily paralyzing facial muscles.

Tetanus Neurotoxin It is also known as tetanospasmin. It is an extremely potent toxin produced by the bacterium *Clostridium tetani*. The infection results from wound infection with bacterial spores, which are found in soil, dust, or animal feces. The spores germinate under strict anaerobic conditions and secrete the toxin, which is quickly distributed all over the body. The toxin binds irreversibly to peripheral nerve endings and moves along the axons to the central nervous system (CNS) where it inhibits the release of inhibitory neurotransmitters, such as glycine and gamma-aminobutyric acid (GABA((Hassel 2013). This inhibition leads to generalized stiffness, muscle spasm, and lockjaw; the latter is a characteristic symptom of tetanus.

Despite its deadly nature, tetanus can be effectively prevented through vaccination with tetanus toxoid, an inactivated form of the toxin. This vaccine induces an immune response by release of neutralizing antibodies, providing immunity against the toxin. Tetanus is now a rare condition due to the mass immunization program, which includes the tetanus vaccine as part of the DPT childhood vaccination program (Dong et al. 2019).

Enterotoxins

They are a class of exotoxins produced by certain bacteria that specifically target the intestines, leading to gastrointestinal disturbances such as diarrhea, nausea, and vomiting. These toxins are mostly heat-labile, meaning they lose their activity at high temperatures. Enterotoxins can be encoded either chromosomally or on plasmids, which accounts for the high rate of horizontal transfer of such genes among bacterial populations. They often alter the membrane permeability of the intestinal mucosal causing electrolyte imbalance, fluid loss, diarrhea, and dehydration. Clinically significant examples of bacterial enterotoxins are listed below.

Staphylococcal Enterotoxin B (SEB) It is produced by the bacterium *S. aureus*, which is responsible for staphylococcal food poisoning. SEB is also a superantigen, which can trigger a massive immune response by T cell activation. Common symptoms include severe diarrhea, nausea, and intestinal cramping. Unlike most enterotoxins, SEB is heat stable making it particularly dangerous, since they can remain in the food even after boiling (Aguilar et al. 2017).

Escherichia coli **Heat-Labile Toxin (LT)** It is an A-B type enterotoxin produced by enterotoxigenic *E. coli* (ETEC), mostly by *E. coli* O157:H7. This toxin is termed "heat-labile" because it loses its activity when exposed to high temperatures. Structurally, LT is similar to cholera toxin and consists of an AB5 multimeric arrangement, where the B subunits form a pentamer that binds to the host-cell membrane, and the A subunit possesses enzymatic activity. The A subunit catalyzes the ADP-ribosylation of the Gs protein, leading to the activation of adenylate cyclase and an increase in cyclic AMP (cAMP) levels. Elevated cAMP levels result in the secretion of chloride ions and water into the intestinal lumen, causing watery diarrhea (Sheikh et al. 2022).

Bacillus cereus **Food Poisoning** The diarrheal enterotoxins produced by *B. cereus* are heat-labile and cause symptoms such as abdominal cramps and diarrhea. These toxins are typically produced in the small intestine after ingestion of contaminated meat, dairy, or starchy food. In addition to enterotoxin, an emetic toxin named cereulide is also released, heat stable, and causes severe emesis (Dietrich et al. 2021).

Clostridium perfringens **Enterotoxin** It is a type II enterotoxin secreted by *C. perfringens* during the sporulation stage in the intestine. It binds to claudin proteins at the tight junctions between epithelial cells in the gut, forming pores in the cell membranes, disrupting cellular integrity, and causing the characteristic symptoms of abdominal cramps and diarrhea. The spores of *C. perfringens* can survive cooking temperatures and germinate in unrefrigerated food (Freedman et al. 2016).

Clostridial Toxins A and B *Clostridium difficile* produces two main toxins: Toxin A (TcdA) and Toxin B (TcdB). TcdA is an enterotoxin that primarily targets the intestinal lining, causing inflammation, fluid secretion, and diarrhea. TcdB, on the other hand, disrupts the cytoskeletal structure of intestinal epithelial cells by inactivating Rho GTPases, leading to cell rounding and death. The effects of these toxins can extend beyond the intestines to cause systemic effects, including damage to the heart, kidneys, and nervous system. This broader impact is due to the toxins' ability to enter the bloodstream and affect distant organs (Di Bella et al. 2016).

Verotoxin (Shiga-Like Toxin) It is also known as verotoxin. It is a potent toxin secreted by enterohemorrhagic *E. coli* (EHEC), which is responsible for EHEC-induced hemolytic uremic syndrome (HUS). This toxin is structurally and functionally like the Shiga toxin produced by *S. dysenteriae*. The main mechanism of action of verotoxin is protein synthesis inhibition through RNA glycosylation in host cells. The toxin binds to a specific receptor, globotriaosylceramide (Gb3), on the surface of susceptible cells (Melton-Celsa 2014). Once internalized, they cleave a specific adenine residue from the 28S rRNA of the 60S ribosomal subunit, halting the synthesis of proteins, and in turn lead to cell death. This cytotoxic effect accounts for the severe symptoms associated with STEC infections, including bloody diarrhea and HUS (Joseph et al. 2020).

Cholera Toxin (CT) It is a potent A-B-type enterotoxin secreted by *Vibrio cholerae*, which is responsible for the cholera's characteristic severe diarrhea. Structurally, CT is an AB5 multimeric protein complex, consisting of one A subunit and five B subunits (Bagley et al. 2002). The B subunits bind to GM1 gangliosides on the surface of intestinal epithelial cells, facilitating the entry of the A subunit into the cells. Once inside, the A subunit activates adenylate cyclase, leading to an increase in cAMP levels. This elevation in cAMP causes the secretion of chloride ions and water into the intestinal lumen, resulting in massive fluid loss and severe watery diarrhea (Vanden Broeck et al. 2007). Cholera can be life threatening if not treated promptly, primarily due to severe dehydration. Treatment typically involves rehydration therapy; and in severe cases, antibiotics may be administered to reduce the duration of symptoms (Chowdhury et al. 2022).

4.6.2 Bacterial Hydrolytic Enzymes

Bacterial hydrolytic enzymes are crucial for breaking down complex organic molecules into simpler substances, facilitating nutrient uptake and survival in diverse environments. There are many types of bacterial hydrolytic enzymes, each targeting specific substrates. For example, proteases break down proteins, lipases hydrolyze fats, and cellulases degrade cellulose. Such enzymes are commonly released by gut flora in many animals and usually play a significant role in the digestion of food and the absorption of nutrients. This enzymatic activity not only benefits the host by enhancing nutrient availability but also supports the gut flora by providing a steady supply of substrates for growth and reproduction. Bacterial hydrolytic enzymes have wide industrial applications in food production, paper industry, waste management, and biofuel generation.

4.6.2.1 Bacterial Proteases

These enzymes break down proteins into peptides and amino acids, playing a major role in various biological processes and industrial applications. These enzymes are produced by a wide range of bacteria, including many species of *Bacillus*, *Pseudomonas*, *Proteus*, and *Streptomyces*. They are widely used in the detergent industry to remove protein stains, in the food industry for tenderizing meat, in the pharmaceutical industry for developing treatments for various diseases, and in wound debridement to remove dead tissue and promote healing. Some common examples of bacterial proteases include (i) subtilisin, a serine protease produced by *Bacillus subtilis*, widely used in laundry detergents for its ability to break down protein stains. It is also used in the food industry for protein hydrolysis and in the leather industry for dehairing hides (Azrin et al. 2022). (ii) Collagenases, they specifically break down collagen, a major structural protein in connective tissues. Collagenases are produced by various bacteria, for example, *Clostridium histolyticum*, *Bacillus* spp., and *Pseudomonas* spp. Collagenases play a crucial role in bacterial pathogenesis by degrading the extracellular matrix and facilitating the spread of bacteria through host tissues. Collagenases have many applications such as wound debridement and in the food industry for tenderizing meat (Nitulescu et al. 2022). (iii) Elastase is a proteolytic enzyme that specifically degrades elastin, a key component of connective tissue. Degradation of elastin disrupts the structural integrity of connective tissues, facilitating bacterial invasion and colonization. Elastase also plays a role in modulating the host immune response by degrading cytokines and other immune molecules. Elastase is produced by several bacteria, for example, *Pseudomonas aeruginosa*, *Chryseobacterium indologenes*, and *Bacillus* spp. (AlShaikh-Mubarak et al. 2023).

4.6.2.2 Bacterial Lipases

They are enzymes that catalyze the hydrolysis of triglycerides into free fatty acids and glycerol. These enzymes are produced by various bacteria, including *Bacillus* spp., *Staphylococcus* spp., *Pseudomonas* spp., and *Burkholderia* spp. They help bacterial virulence by promoting adipose and fat tissue damage, which contributes to tissue damage and bacterial invasion (Ali et al. 2023). They also have many applications in the detergent industry and dairy products.

4.6.2.3 Bacterial Amylases

They are enzymes that catalyze the hydrolysis of starch into simpler sugars like maltose and glucose. These enzymes are produced by various bacterial species of *Bacillus*, *Pseudomonas*, and *Streptomyces*. Amylases are extensively used in the food industry for brewing, baking, and producing syrups. They also play a role in the textile industry for desizing fabrics and in the paper industry for paper coating (Mehta and Satyanarayana 2016).

4.6.2.4 Bacterial Cellulases

Cellulases are enzymes that catalyze the hydrolysis of cellulose into simpler sugars like glucose. These enzymes are produced by many species of *Clostridium*, *Bacillus*, *Cellulomonas*, and *Thermomonospora*. Cellulases play an important role in nature by maintaining ecosystems through recycling of nutrients by degradation of plant biomass (Behera et al. 2017).

4.7 Metabolic Mastery: How Bacterial Systems Orchestrate the Expression of Virulent Metabolites

The expression of metabolites that contribute to bacterial virulence and pathogenesis is regulated by several regulatory systems. These regulatory systems orchestrate the expression of virulence metabolites in different stages of infection. In the next section, we give examples of these regulatory systems.

4.7.1 Two-Component Systems (TCS)

TCS are common signaling mechanisms in pathogenic bacteria to facilitate the adaption to environmental changes during infection. TCS generally consists of two main components, a membrane-bound sensor kinase that typically consists of a membrane-spanning region and a cytoplasmic kinase domain (Beier and Gross 2006). Sensor kinase detects specific environmental stimuli (e.g., temperature, pH, osmolarity, presence of antimicrobials, or host signals). Upon sensing a signal, the sensor kinase undergoes autophosphorylation at a conserved histidine residue. The second component is the response regulator, which contains a receiver domain that can be phosphorylated and a DNA-binding domain for gene regulation (Capra and Laub 2012). Phosphorylated response regulator mediates changes in gene expression by binding to DNA and activating or repressing target genes related to virulence, and survival in host niche (Mitrophanov and Groisman 2008). There are several examples of TCS; PhoP/Q in *Salmonella* spp. senses magnesium levels and antimicrobial peptides, regulating genes required for virulence and adaptation to host conditions (Gangathraprabhu et al. 2020). EnvZ/OmpR is widely distributed TCS in bacteria such as *E. coli*. It responds to changes in environmental osmolality by regulating the expression of the outer membrane porins OmpF/C (Kenney and Anand 2020). Interestingly, QseC/B in *E. coli* responds to host-derived signals like adrenaline and regulates bacterial motility and virulence genes (Xie et al. 2010). Indeed, these sensor kinases (Qse) are not only found in *E. coli*, but they are also found in other enteric bacteria to eavesdrop on host signals to enhance their virulence and survival (Moreira and Sperandio 2012). Interestingly, these sensor kinases could sense the inducers of another involved virulence controlling systems, quorum sensing, exaggerating the production of virulence factors (Hegazy et al. 2022; Rasko et al. 2008). This is an example shows how the virulence controlling systems are interplayed and working in cooperation. TCS are vital for bacterial survival and pathogenicity, serving as sophisticated signaling networks that enable bacteria to sense their environment and respond effectively.

4.7.2 Ferric Uptake Regulator (Fur)

Fur system is a key transcriptional regulator in bacteria that plays a vital role in iron homeostasis. It controls the expression of the involved genes in iron acquisition, storage, and utilization, ensuring that bacteria can adapt to fluctuating iron availability in their environment (Troxell and Hassan 2013). Fur is homodimeric protein that consists of two identical subunits. The Fur protein has a metal-binding site that can bind iron (Fe^{2+}) undergoing a conformational change that confers its bind to specific DNA sequences called "Fur boxes" located in the promoters of target genes (Pohl et al. 2003; Askoura et al. 2020). When iron is already available, Fur binds to DNA that results in repression of the transcription of the involved genes in iron uptake, such as those encoding iron transport systems and siderophores. This leads to the prevention of the excess intake of iron and protects the bacterial cells from iron toxicity. In iron-limited conditions, the absence of iron causes Fur to be released from the DNA, leading to the de-repression of iron acquisition genes. This enables bacteria to express genes that enhance their ability to scavenge iron from the environment, such as siderophore synthesis and high-affinity transport systems (Cornelis et al. 2009).

4.7.3 Carbon Storage Regulator

Carbon Storage Regulator (Csr) is a vital regulatory network in pathogenic bacteria that manages carbon metabolism and storage in response to nutrient availability (Weilbacher et al. 2003). It primarily involves CsrA, a regulatory protein, and CsrB/C, small non-coding RNAs that modulate CsrA's activity. CsrA binds to specific mRNA targets, and typically represses genes involved in catabolism while promoting genes related to anabolism and storage, while CsrB/C binds to CsrA to prevent it from interacting with target mRNAs. The expression of CsrB/C increases under conditions of high carbon availability, effectively sequestering CsrA and relieving its repression of carbon metabolism-involved genes (Babitzke and Romeo 2007). In nutrient-rich conditions, CsrB/C is expressed, leading to increased binding of CsrA to these RNAs and reduced regulation of target mRNAs. CsrA binds to mRNAs coding for enzymes involved in the breakdown of carbon sources (e.g., glycolysis), inhibiting their expression. Conversely, CsrA can promote the expression of genes involved in storage and biosynthetic pathways, such as those encoding glycogen synthesis. When carbon sources become limited, CsrB/C levels decrease, allowing CsrA to regulate its target mRNAs effectively (Weilbacher et al. 2003; Babitzke and Romeo 2007). In bacterial pathogens, the Csr system has been shown to influence virulence traits. In *Pseudomonas aeruginosa*, CsrA affects biofilm formation and production of virulence factors that impact its ability to establish infections (Rojo 2010). Csr system in *E. coli* controls the important metabolic pathways involved in its survival in different host environments (Weilbacher et al. 2003).

4.7.4 Secretion Systems

Bacterial secretion systems are specialized molecular machines that transport substances across bacterial membranes into the host environment or directly into host cells. They play several important roles in delivering metabolites, toxins, and effector proteins that contribute to virulence (Green and Mecsas 2016). Gram-negative bacteria acquire more complex cell envelope than Gram-positive bacteria, consisting of an inner membrane, a periplasmic space, and an outer membrane. This mandates Gram-negative bacteria to develop more specialized secretion systems to deliver their proteins and other virulence factors into the cytoplasm of host cells (Costa et al. 2015). Type I secretion system (T1SS) involves a tripartite structure that spans the inner membrane, periplasm, and outer membrane secreting substrates in a single step. *E. coli* employs T1SS to secrete hemolysin (HlyA), to lyse red blood cells. Type II secretion system (T2SS) uses a piston-like mechanism to push folded proteins from the periplasm through a pore in the outer membrane. *Vibrio cholerae* uses T2SS to secrete cholera toxin, a key virulence factor in cholera. Similarly, T2SS in *Pseudomonas aeruginosa* is used to deliver elastases and proteases to degrade host tissues, promoting infection. T3SS spans both bacterial membranes and the host cell membrane, forming a needle-like structure that delivers virulence factors directly into host-cell cytoplasm. *Salmonella* spp., *Pseudomonas* spp., *Yersinia* spp., and *Shigella* spp. use T3SS to inject effector proteins into host cells, modulating immune responses and promoting intracellular survival. Type IV secretion system (T4SS) transports proteins or DNA across the bacterial membrane and, in some cases, into host cells. T4SS is also used for horizontal gene transfer via conjugation (Green and Mecsas 2016; Costa et al. 2015; Thanassi et al. 2012). *Helicobacter pylori* utilizes T4SS to deliver CagA protein into gastric epithelial cells, contributing to gastric ulcer formation (Hatakeyama 2008). Autotransporter system or type V secretion system (T5SS) secretes protein contain a domain to forms a pore in the outer membrane, allowing the passenger domain to pass through. *Neisseria gonorrhoeae* exploits T5SS to secrete IgA protease, which helps it evade the host's immune system (Gawarzewski et al. 2013). Type VI Secretion System (T6SS) acts as a contact-dependent weapon delivering effector proteins directly into competing bacteria or eukaryotic cells. *Pseudomonas aeruginosa* and *Vibrio cholerae* recruit the syringe-like T6SS to attack rival bacteria, inject their toxins into host cells, and enhance survival in competitive environments (Silverman et al. 2012). Type IX secretion system (T9SS) secretes large proteins across the outer membrane using a specialized set of machinery. T9SS in *Porphyromonas gingivalis* is involved in secreting proteases that contribute to tissue destruction in periodontal disease (Lasica et al. 2017).

Unlike Gram-negative bacteria, Gram-positive bacteria lack the outer membrane but have a thick peptidoglycan layer. In spite of these differences, Gram-positive bacteria exploit specialized secretion systems to export proteins, toxins, and other virulence factors across the cell membrane into the external environment or host cells (Green and Mecsas 2016). The general secretory pathway or Sec pathway is the most widespread protein secretion system in Gram-positive bacteria. It translocates unfolded proteins across the cytoplasmic membrane. In *Bacillus subtilis*, the Sec pathway is used to secrete degradative enzymes such as proteases and lipases (Yan and Wu 2017). On the other hand, the Twin-Arginine translocation pathway (Tat pathway) is used to transport fully folded proteins across the cytoplasmic membrane in Gram-positive bacteria. This pathway allows the transportation of larger, complex proteins such as *Listeria monocytogenes* virulence factors that are used to evade host immune defenses (Palmer and Berks 2012). Another system is the sortase-mediated secretion system in which sortase anchors surface proteins to the cell wall of Gram-positive bacteria. This system is vital for anchoring adhesins, enzymes, and other proteins that play a role in pathogenesis. *Staphylococcus aureus* uses sortase to attach surface proteins involved in adherence, immune evasion, and invasion (Schneewind and Missiakas 2014). Type VII secretion system (T7SS) is found in certain Gram-positive bacteria such as *Mycobacterium* genus to facilitate the intracellular survival and immune evasion. T7SS in *Mycobacterium tuberculosis* is used to secrete proteins like EsxA, which help in the survival within host macrophages (Rivera-Calzada et al. 2021). Although T7SS is more commonly associated with Gram-positive, a version of T7SS is also used by Gram-negative bacteria, such as some marine *Vibrio* species Type VII secretion system (T7SS) (Pena et al. 2019). Also, Gram-positive bacteria share Gram-negative bacteria in utilizing T4SS to transfer proteins and DNA. *Clostridium difficile* has been shown to use T4SS to secrete toxins that disrupt host-cell functions and cause tissue damage (Rivera-Calzada et al. 2021; Pena et al. 2019).

Secretion systems play vital roles in the expression and translocation of a wide array of proteins and metabolites determining the bacterial ability to establish a stable infection. First, secretion systems deliver toxins that damage host tissues, leading to disease symptoms and facilitating bacterial spread (Green and Mecsas 2016). Secretion systems deliver metabolites that can manipulate host-cell functions, such as altering metabolism, inducing apoptosis, or promoting pathogen replication. Secretion systems such as T3SS, T4SS, T5SS, T6SS, and T7SS could manipulate the host-cell pathways, suppress immune responses, resist phagocytosis, or kill rival bacteria (Pena et al. 2019). Secretion systems help pathogenic bacteria to acquire essential nutrients from the host, such as iron through siderophore secretion (Izoré et al. 2011), and *Salmonella*-inducing filaments (Sifs) which are expressed by T3SS (Izoré et al. 2011). T4SS mediates the horizontal transfer of plasmids, contributing to the spread of virulence traits as well as antibiotic-resistance genes among bacterial populations (Craig et al. 2019). These systems tightly regulate bacterial metabolism, ensuring the timely production of virulence factors necessary for pathogenesis while adapting to host environments. Bacteria rely on these sophisticated secretion systems to deliver their metabolites to interact with their environment, establish infections, and outcompete competitors, making these systems critical to bacterial survival and pathogenicity.

4.7.5 Quorum Sensing (QS) Systems

QS systems are bacterial communication systems that act by employing chemical compounds as a communication signal that in turn ensure an efficient coordination to the production of virulence factors during the different stages of infection (Rutherford and Bassler 2012). The mechanisms of QS and its roles in the regulation of bacterial virulence have been extensively elaborated, revealing that QS systems work in an inducer/receptor manner to regulate the expression of the virulence encoding genes (Cavalu et al. 2022). QS system requires a certain population threshold to be reached before activation, triggering the onset of the system. The inducers are released in minute quantities that remain undetectable when the bacterial population is low. At higher bacterial population, the concentration of secreted inducers becomes sufficient to bind with their specific QS receptors, forming a QS receptor-inducer complex. Once a threshold concentration is reached, the QS receptor-inducer complexes bind to specific promoter regions on the bacterial DNA regulating the transcription of genes responsible for expression of virulence factors (Winzer and Williams 2001).

Basically, QS systems are used in both Gram-positive and Gram-negative bacteria and work in a similar inducer/receptor manner; however, the signaling pathway is quite different (Nazeih et al. 2023). In Gram-negative bacteria, N-acyl-homoserine lactones (AHLs) are the most predominant inducers. AHLs are synthesized via autoinducer synthase genes and released in the surrounding environment to be recognized by their cognate receptors which are mainly Lux-type QS receptors (Papenfort and Bassler 2016; Al-Rabia et al. 2024a). Similarly, QS systems in Gram-positive bacteria utilize autoinducing peptides (AIPs), which are recognized by membrane-bound two-component signal transduction systems (Lila et al. 2023; Abisado et al. 2018). QS systems play a key role in the formation of bacterial biofilms by regulating the production of extracellular polymeric matrix, enhancing the expression of bacterial virulence genes, and increasing the secretion of bioluminescent compounds, exoenzymes, siderophores, and antibiotics (Li and Tian 2012). Furthermore, QS systems orchestrate the expression of virulence, toxin production and motility encoding genes. Here are some examples of virulence factors, which are controlled by QS systems. *Pseudomonas aeruginosa* recruits at least three types of QS systems, two are Lux-type QS systems LasI/R and RhlI/R and an orphan PqsR besides its specialized non-Lux-type Pseudomonas Quinolone Signal (PQS). In addition, *Pseudomonas aeruginosa* could employ other QS systems such as Iqs QS system which regulate the stress response and regulate virulence behavior under nutrient limitation (Lila et al. 2023; Abisado et al. 2018). QS systems regulate the production of virulence factors in *Pseudomonas aeruginosa* such as pyocyanin (a toxin that damages host cells) and hydrolytic enzymes such as elastase, hemolysins, and protease (Bhargava et al. 2014). *Vibrio cholerae* and enterohemorrhagic *E. coli* (EHEC) employ Lux-type QS systems to regulate the production of cholera toxin and Shiga-like toxin, respectively, that results in induction of severe gastrointestinal infections (Papenfort and Bassler 2016). Similarly in Gram-positive bacteria, *Staphylococcus aureus* utilizes its Agr QS system to control the production of toxins like α-hemolysin and proteases, which contribute to tissue damage and immune system evasion (Lila et al. 2023; Khayat et al. 2022a). These examples illustrate how QS signaling systems regulate the expression of a wide diverse of metabolites, which enable the bacteria to adapt to their environment and optimize the expression of virulence factors.

4.8 Metabolite Targeting in Antibiotic Development: Paving the Way for Next-Gen Antimicrobials

4.8.1 Traditional Antimetabolite Antibiotics

Antimetabolite antibiotics represent a known antibiotic class that inhibits bacterial growth by interfering with metabolic processes essential for cell division and survival. The effectiveness of this mechanism depends on targeting specific metabolic functions that bacteria rely on for growth and proliferation (Hutchings et al. 2019). Sulfonamides and diaminopyrimidine derivatives are well-known examples of antimetabolite antibacterials, which inhibit the synthesis of folate in bacteria. Folate is crucial to produce nucleic acids and ultimately for DNA replication. Sulfonamides compete with para-aminobenzoic acid (PABA), a substrate for the enzyme dihydropteroate synthase, blocking the formation of dihydrofolate. Synergistically, diaminopyrimidine derivatives such as trimethoprim inhibit dihydrofolate reductase enzyme that catalyzes the reduction of dihydrofolate to tetrahydrofolate. As a result, bacterial cells are unable to synthesize folate, leading to impaired DNA synthesis and cell division (Capasso and Supuran 2014). Isoniazid is another example of an antimetabolite antibiotic that interferes with the metabolic pathways involved in the synthesis of bacterial essential structures. Isoniazid inhibits the synthesis of mycolic acid, an essential component of the cell wall of *Mycobacterium tuberculosis*. By interfering with this metabolic process, isoniazid effectively weakens the structural integrity of the mycobacterial cell, rendering it more susceptible to immune clearance (Unissa et al. 2016).

4.8.2 Emerging Strategies for Targeting Bacterial Virulence

There are new horizons in the development of innovative approaches to target the metabolites involved in bacterial virulence. Unlike conventional antibiotics, these strategies focus on mitigation of the effects of pathogenic bacteria rather than killing them, which can help prevent the development of resistance (Fernández-Villa et al. 2019; Chen et al. 2011; Batoni et al. 2011; Abdulaal et al. 2024). These strategies are supposed to develop new novel antimicrobial therapies to complement the traditional antibiotics. Bacterial metabolites, such as siderophores, QS signaling molecules, effectors of secretion systems, and exotoxins, play essential roles in bacterial pathogenesis. By inhibiting the production or function of these metabolites, it becomes possible to render pathogens less virulent and more susceptible to the host immune response (Rutherford and Bassler 2012; Garcia-Contreras 2016; Rasmussen and Givskov 2006). These strategies include the development of small molecules that can inhibit the synthesis or action of bacterial metabolites, as well as the use of enzymes or antibodies to neutralize harmful metabolites directly. Such approaches not only reduce the pathogenic potential of bacteria but also minimize the selective pressure that typically drives antibiotic resistance (Hegazy et al. 2021; Koshak et al. 2023).

One promising approach is that monoclonal antibodies are used to neutralize metabolites involved in virulence directly. These antibodies are engineered to specifically target toxins or other harmful proteins, preventing them from interacting with host cells. This approach helps in neutralizing the pathogenic effects as well as enhances the host's immune response against the invading bacteria (Oleksiewicz et al. 2012). For instance, bezlotoxumab is a monoclonal antibody that targets *Clostridium difficile* toxin B (Johnson and Gerding 2019), and other examples are listed in Table 4.2. Another approach is the use of anti-virulence agents to interfere with the key systems, which regulate the expression of metabolites involved in bacterial adhesion, biofilm formation, and production of toxins and hydrolytic enzymes. QS and secretion systems play vital roles in the expression of a huge array of virulence factors and toxins in both Gram-negative and Gram-positive bacteria, and they are potential targets for therapeutic intervention. Interfering with these virulence-controlling systems can disrupt the bacteria's ability to cause disease without exerting selective pressure that leads to resistance (Lila et al. 2023; Alotaibi et al. 2023; Elfaky et al. 2024a). Bearing in mind the central role of QS in regulating diverse metabolic pathways that are involved in the production of virulence factors and biofilm formation, targeting QS is supposed a promising approach (Rajab and Hegazy 2023; Koshak et al. 2023; Al-Rabia et al. 2024b). In this context, numerous studies have been conducted to evaluate the possible application of natural and synthetic compounds as well as repurposing drugs to serve as anti-QS candidates in order to their use as antibacterial complements (Agha et al. 2016; Khayat et al. 2022b, 2023a). Also, targeting of secretion systems is suggested as promising target to mitigate bacterial virulence and develop new antimetabolite antibacterial agents (Thabit et al. 2022a; Hegazy et al. 2022; Cavalu et al. 2022). Some examples representing these new approaches are listed in Table 4.2.

Table 4.2 Examples of new approaches to develop antimetabolites

Agent	Mode of action	Target microbe	References
Monoclonal antibodies			
Bezlotoxumab	Neutralize bacterial toxin	*Clostridium difficile* toxin B	Johnson and Gerding (2019)
Raxibacumab		*Bacillus anthracis*	Mazumdar (2009)
Tocilizumab	Manipulate the involved immune response	Target interleukin-6 (IL-6) that is involved in the immune response against intracellular bacterial infections	Sheppard et al. (2017)
Secukinumab		Target interleukin-17A (IL-17A) that acquires potential mitigation effects on bacterial infections in chronic inflammatory conditions	Sanford and McKeage (2015)

(continued)

Table 4.2 (continued)

Agent	Mode of action	Target microbe	References
Antimicrobial peptides			
AMP 1018	Disrupt biofilm cell membranes	*S. mutants*, *E. coli*	Lemos et al. (2004)
LL-37		*E. coli*	Anunthawan et al. (2015)
DJK-5		*P. aeruginosa*	de la Fuente-Nunez et al. (2015)
DJK-6		*P. aeruginosa*	de la Fuente-Nunez et al. (2015)
Antimicrobial lipids			
Glycerol monolaurate	Disruption of biofilm matrix	*Haemophilus influenzae*, *S. aureus*	Schlievert and Peterson (2012)
Alginate oligosaccharide		*P. aeruginosa*	Powell et al. (2018)
Antagonize QS system			
FDA approved drugs			
Adrenoreceptor blockers	Downregulate expression of Qs encoding genes and antagonize QS receptors	*P. aeruginosa*, *P. mirabilis*, *S. marcescens*, *S. enterica*	Elfaky et al. (2022, 2023); Cavalu et al. (2022); Alotaibi et al. (2023); Almalki et al. (2022b, c); Thabit et al. (2022b)
Gliptin antidiabetics		*P. aeruginosa*, *S. aureus*, *S. marcescens*, *S. mutans*, *S. mutants*	Khayat et al. (2022a, c, 2023b); De et al. (2018); Abbas and Hegazy (2020)
Azole drugs		*P. aeruginosa*, *P. mirabilis*, *S. marcescens*	Agha et al. (2016); Khayyat et al. (2021a, b)
Antipsychotics	Interfere with biofilm formation molecular system	*P. mirabilis*, *E. faecalis*	Holling et al. (2014); Nzakizwanayo et al. (2017)
Phytochemicals			
Stolon (fenugreek)	Antagonize QS system	*P. aeruginosa*	Aldawsari et al. (2021)
Garlic extract			Harjai et al. (2010)
Carvacrol (Oregano)		*P. aeruginosa*	Tapia-Rodriguez et al. (2019)
Xylitol		*S. marcescens*	Khayyat et al. (2021c)
Phytol (piper)		*S. marcescens*	Srinivasan et al. (2017)
Sulforaphane		*P. aeruginosa*	Bendary et al. (2024)
Sodium citrate		*P. aeruginosa*, *S. marcescens P. mirabilis*	Khayat et al. (2022b, 2023a)
Zingerone, Shogaol, Gingerol (ginger)		*P. aeruginosa*, *S. marcescens P. mirabilis*, *A. baumannii*, and *K. pneumoniae*	Koshak et al. (2023); Elfaky et al. (2024a); Kumar et al. (2015)
Vanillin		*C. violaceum*	Choo et al. (2006)

(continued)

Table 4.2 (continued)

Agent	Mode of action	Target microbe	References
Honokiol (Magnolia grandiflora)		*P. aeruginosa*	Elfaky et al. (2024b)
Piceatannol (Euphorbia seeds)		*P. aeruginosa, S. marcescens, P. mirabilis, A. baumannii*, and *K. pneumoniae*	Koshak et al. (2024a)
Arctigenin (Burdock roots)		*P. aeruginosa*	Koshak et al. (2024b)
Emodin (Polygonum cuspidatum)	Downregulation of biofilm formation molecular system	*S. aureus*	Yan et al. (2017)
Curcumin (curcuma)		*Acinetobacter baumannii, P. mirabilis*	Packiavathy et al. (2014); Badr-Eldin et al. (2024)
Tannic acid	Disruption of biofilm matrix	*E. coli*	Samoilova et al. (2019)
Interfere with secretion system and metabolic pathways			
Catecholamines and adrenergic blockers	Interfere with bacterial sensing to the surrounding environmental signaling	*Gut pathogens such as E. coli and S. enterica*	Hegazy et al. (2022); Moreira and Sperandio (2012); Rasko et al. (2008); Karavolos et al. (2013)
Terazosin, Atenolol, and Prazosin	Interfere with T3SS	*S. enterica*	Thabit et al. (2022a); Hegazy et al. (2022); Elfaky et al. (2022)

References

Abbas HA, Hegazy WAH (2020) Repurposing anti-diabetic drug "Sitagliptin" as a novel virulence attenuating agent in Serratia marcescens. PLoS One 15(4):e0231625. https://doi.org/10.1371/journal.pone.0231625. PubMed PMID: 32298346; PubMed Central PMCID: PMC7162429

Abdulaal WH, Alhakamy NA, Asseri AH, Radwan MF, Ibrahim TS, Okbazghi SZ et al (2024) Redirecting pantoprazole as a metallo-beta-lactamase inhibitor in carbapenem-resistant Klebsiella pneumoniae. Front Pharmacol 15:1366459. https://doi.org/10.3389/fphar.2024.1366459. Epub 2024/03/27. PubMed PMID: 38533260; PubMed Central PMCID: PMCPMC10963397

Abisado RG, Benomar S, Klaus JR, Dandekar AA, Chandler JR (2018) Bacterial quorum sensing and microbial community interactions. MBio 9(3):e02331-17. https://doi.org/10.1128/mBio.02331-17. PubMed PMID: 29789364; PubMed Central PMCID: PMCPMC5964356

Agha KA, Abo-Dya NE, Ibrahim TS, Abdel-Aal EH, Hegazy WA (2016) Benzotriazole-mediated synthesis and antibacterial activity of novel N-acylcephalexins. Sci Pharm 84(3):484–496. https://doi.org/10.3390/scipharm84030484. PubMed PMID: 28117314; PubMed Central PMCID: PMCPMC5064239

Aguilar JL, Varshney AK, Pechuan X, Dutta K, Nosanchuk JD, Fries BC (2017) Monoclonal antibodies protect from Staphylococcal Enterotoxin K (SEK) induced toxic shock and sepsis by USA300 Staphylococcus aureus. Virulence 8(6):741–750. https://doi.org/10.1080/21505594.2016.1231295. PubMed PMID: 27715466; PubMed Central PMCID: PMC5626247

Ahmad JN, Sebo P (2020) Adenylate cyclase toxin tinkering with monocyte-macrophage differentiation. Front Immunol 11:2181. https://doi.org/10.3389/fimmu.2020.02181. PubMed PMID: 33013916; PubMed Central PMCID: PMC7516048

Ahmed E, Holmström SJ (2014) Siderophores in environmental research: roles and applications. Microb Biotechnol 7(3):196–208

Aktories K, Barbieri JT (2005) Bacterial cytotoxins: targeting eukaryotic switches. Nat Rev Microbiol 3(5):397–410. https://doi.org/10.1038/nrmicro1150. PubMed PMID: 15821726

Aldawsari MF, Khafagy ES, Saqr AA, Alalaiwe A, Abbas HA, Shaldam MA et al (2021) Tackling virulence of Pseudomonas aeruginosa by the natural furanone sotolon. Antibiotics (Basel) 10(7):871. https://doi.org/10.3390/antibiotics10070871. PubMed PMID: 34356792; PubMed Central PMCID: PMC8300740

Ali S, Khan SA, Hamayun M, Lee IJ (2023) The recent advances in the utility of microbial lipases: a review. Microorganisms 11(2):510. https://doi.org/10.3390/microorganisms11020510. PubMed PMID: 36838475; PubMed Central PMCID: PMC9959473

Almalki AJ, Ibrahim TS, Taher ES, Mohamed MFA, Youns M, Hegazy WAH et al (2022a) Synthesis, antimicrobial, anti-virulence and anticancer evaluation of new 5(4H)-oxazolone-based sulfonamides. Molecules 27(3):671. https://doi.org/10.3390/molecules27030671

Almalki AJ, Ibrahim TS, Elhady SS, Hegazy WAH, Darwish KM (2022b) Computational and biological evaluation of β-adrenoreceptor blockers as promising bacterial anti-virulence agents. Pharmaceuticals 15(2):110. https://doi.org/10.3390/ph15020110

Almalki AJ, Ibrahim TS, Elhady SS, Darwish KM, Hegazy WAH (2022c) Repurposing α-adrenoreceptor blockers as promising anti-virulence agents in Gram-negative bacteria. Antibiotics (Basel) 11(2):178. https://doi.org/10.3390/antibiotics11020178

Alotaibi HF, Alotaibi H, Darwish KM, Khafagy E-S, Abu Lila AS, Ali MA et al (2023) The anti-virulence activities of the antihypertensive drug propranolol in light of its anti-quorum sensing effects against Pseudomonas aeruginosa and Serratia marcescens. Biomedicines 11(12):3161

Al-Rabia MW, Asfour HZ, Alhakamy NA, Abdulaal WH, Ibrahim TS, Abbas HA et al (2024a) Thymoquinone is a natural antibiofilm and pathogenicity attenuating agent in Pseudomonas aeruginosa. Front Cell Infect Microbiol 14:1382289. https://doi.org/10.3389/fcimb.2024.1382289. Epub 2024/04/19. PubMed PMID: 38638827; PubMed Central PMCID: PMCPMC11024287

Al-Rabia MW, Asfour HZ, Alhakamy NA, Bazuhair MA, Ibrahim TS, Abbas HA et al (2024b) Cilostazol is a promising anti-pseudomonal virulence drug by disruption of quorum sensing. AMB Express 14(1):87. https://doi.org/10.1186/s13568-024-01740-1. Epub 2024/08/02. PubMed PMID: 39090255; PubMed Central PMCID: PMCPMC11294311

AlShaikh-Mubarak GA, Kotb E, Alabdalall AH, Aldayel MF (2023) A survey of elastase-producing bacteria and characteristics of the most potent producer, Priestia megaterium gasm32. PLoS One 18(3):e0282963. https://doi.org/10.1371/journal.pone.0282963. PubMed PMID: 36913358; PubMed Central PMCID: PMC10010523

Anunthawan T, de la Fuente-Nunez C, Hancock RE, Klaynongsruang S (2015) Cationic amphipathic peptides KT2 and RT2 are taken up into bacterial cells and kill planktonic and biofilm bacteria. Biochim Biophys Acta 1848(6):1352–1358. https://doi.org/10.1016/j.bbamem.2015.02.021. PubMed PMID: 25767037

Askoura M, Hegazy WAH (2020) Ciprofloxacin interferes with Salmonella typhimurium intracellular survival and host virulence through repression of Salmonella pathogenicity island-2 (SPI-2) genes expression. Pathog Dis 78(1):ftaa011. https://doi.org/10.1093/femspd/ftaa011. PubMed PMID: 32083661

Askoura M, Youns M, Halim Hegazy WA (2020) Investigating the influence of iron on Campylobacter jejuni transcriptome in response to acid stress. Microb Pathog 138:103777. https://doi.org/10.1016/j.micpath.2019.103777. PubMed PMID: 31600543

Assoni L, Milani B, Carvalho MR, Nepomuceno LN, Waz NT, Guerra MES et al (2020) Resistance mechanisms to antimicrobial peptides in Gram-positive bacteria. Front Microbiol 11:593215

Atchade E, De Tymowski C, Grall N, Tanaka S, Montravers P (2024) Toxic shock syndrome: a literature review. Antibiotics (Basel) 13(1):96. https://doi.org/10.3390/antibiotics13010096. PubMed PMID: 38247655; PubMed Central PMCID: PMC10812596

Azimi T, Zamirnasta M, Sani MA, Soltan Dallal MM, Nasser A (2020) Molecular mechanisms of Salmonella effector proteins: a comprehensive review. Infect Drug Resist 13:11–26

Azrin NAM, Ali MSM, Rahman R, Oslan SN, Noor NDM (2022) Versatility of subtilisin: a review on structure, characteristics, and applications. Biotechnol Appl Biochem 69(6):2599–2616. https://doi.org/10.1002/bab.2309. PubMed PMID: 35019178

Babitzke P, Romeo T (2007) CsrB sRNA family: sequestration of RNA-binding regulatory proteins. Curr Opin Microbiol 10(2):156–163

Badr-Eldin SM, Aldawsari HM, Ahmed OA, Kotta S, Abualsunun W, Eshmawi BA et al (2024) Utilization of zein nano-based system for promoting antibiofilm and anti-virulence activities of curcumin against Pseudomonas aeruginosa. Nanotechnol Rev 13(1):20230212

Bagley KC, Abdelwahab SF, Tuskan RG, Fouts TR, Lewis GK (2002) Cholera toxin and heat-labile enterotoxin activate human monocyte-derived dendritic cells and dominantly inhibit cytokine production through a cyclic AMP-dependent pathway. Infect Immun 70(10):5533–5539. https://doi.org/10.1128/IAI.70.10.5533-5539.2002. PubMed PMID: 12228279; PubMed Central PMCID: PMC128358

Bahar AA, Ren D (2013) Antimicrobial peptides. Pharmaceuticals 6(12):1543–1575

Balcazar JL, Subirats J, Borrego CM (2015) The role of biofilms as environmental reservoirs of antibiotic resistance. Front Microbiol 6:1216. https://doi.org/10.3389/fmicb.2015.01216. PubMed PMID: 26583011; PubMed Central PMCID: PMCPMC4628128

Bardwell AJ, Abdollahi M, Bardwell L (2004) Anthrax lethal factor-cleavage products of MAPK (mitogen-activated protein kinase) kinases exhibit reduced binding to their cognate MAPKs. Biochem J 378(Pt 2):569–577. https://doi.org/10.1042/BJ20031382. PubMed PMID: 14616089; PubMed Central PMCID: PMC1223970

Batoni G, Maisetta G, Brancatisano FL, Esin S, Campa M (2011) Use of antimicrobial peptides against microbial biofilms: advantages and limits. Curr Med Chem 18(2):256–279. https://doi.org/10.2174/092986711794088399. PubMed PMID: 21110801

Beckert U, Wolter S, Hartwig C, Bahre H, Kaever V, Ladant D et al (2014) ExoY from Pseudomonas aeruginosa is a nucleotidyl cyclase with preference for cGMP and cUMP formation. Biochem Biophys Res Commun 450(1):870–874. https://doi.org/10.1016/j.bbrc.2014.06.088. PubMed PMID: 24971548; PubMed Central PMCID: PMC4764875

Behera BC, Sethi BK, Mishra RR, Dutta SK, Thatoi HN (2017) Microbial cellulases – diversity & biotechnology with reference to mangrove environment: a review. J Genet Eng Biotechnol 15(1):197–210. https://doi.org/10.1016/j.jgeb.2016.12.001. PubMed PMID: 30647656; PubMed Central PMCID: PMC6296582

Beier D, Gross R (2006) Regulation of bacterial virulence by two-component systems. Curr Opin Microbiol 9(2):143–152

Belyy A, Mechold U, Renault L, Ladant D (2018) ExoY, an actin-activated nucleotidyl cyclase toxin from P. aeruginosa: a minireview. Toxicon 149:65–71. https://doi.org/10.1016/j.toxicon.2017.12.046. PubMed PMID: 29258848

Bendary MM, Ali MAM, Abdel Halim AS, Boufahja F, Chaudhary AA, Elkelish A et al (2024) Investigating sulforaphane's anti-virulence and anti-quorum sensing properties against Pseudomonas aeruginosa. Front Pharmacol 15:1406653. https://doi.org/10.3389/fphar.2024.1406653. Epub 2024/06/05. PubMed PMID: 38835668; PubMed Central PMCID: PMCPMC11148281

Bettencourt P, Carmo N, Pires D, Timóteo P, Anes E (2017) Mycobacterial infection of macrophages: the effect of the multiplicity of infection. In: Méndez-Vilas A (ed) Antimicrobial research: novel bioknowledge and educational programs. Formatex Research Center S.L., pp 651–664

Bhargava N, Sharma P, Capalash N (2014) Pyocyanin stimulates quorum sensing-mediated tolerance to oxidative stress and increases persister cell populations in Acinetobacter baumannii. Infect Immun 82(8):3417–3425

Blasche S, Arens S, Ceol A, Siszler G, Schmidt MA, Häuser R et al (2014) The EHEC-host interactome reveals novel targets for the translocated intimin receptor. Sci Rep 4(1):7531

Bottomley MJ, Muraglia E, Bazzo R, Carfi A (2007) Molecular insights into quorum sensing in the human pathogen Pseudomonas aeruginosa from the structure of the virulence regulator LasR bound to its autoinducer. J Biol Chem 282(18):13592–13600. https://doi.org/10.1074/jbc.M700556200. PubMed PMID: 17363368

Braun V, Focareta T (1991) Pore-forming bacterial protein hemolysins (cytolysins). Crit Rev Microbiol 18(2):115–158. https://doi.org/10.3109/10408419109113511. PubMed PMID: 1930675

Brossier F, Mock M (2001) Toxins of Bacillus anthracis. Toxicon 39(11):1747–1755

Brouwer S, Rivera-Hernandez T, Curren BF, Harbison-Price N, De Oliveira DMP, Jespersen MG et al (2023) Pathogenesis, epidemiology and control of Group A Streptococcus infection. Nat Rev Microbiol 21(7):431–447. https://doi.org/10.1038/s41579-023-00865-7. PubMed PMID: 36894668; PubMed Central PMCID: PMC9998027

Bruyn G, Zegers B, Van Furth R (1992) Mechanisms of host defense against infection with Streptococcus pneumoniae. Clin Infect Dis 14(1):251–262

Bumann D, Schothorst J (2017) Intracellular salmonella metabolism. Cell Microbiol 19(10):e12766

Capasso C, Supuran CT (2014) Sulfa and trimethoprim-like drugs–antimetabolites acting as carbonic anhydrase, dihydropteroate synthase and dihydrofolate reductase inhibitors. J Enzyme Inhib Med Chem 29(3):379–387

Capra EJ, Laub MT (2012) Evolution of two-component signal transduction systems. Ann Rev Microbiol 66(1):325–347

Carvalho F, Sousa S, Cabanes D (2014) How Listeria monocytogenes organizes its surface for virulence. Front Cell Infect Microbiol 4:48

Cavaillon JM (2018) Exotoxins and endotoxins: inducers of inflammatory cytokines. Toxicon 149:45–53. https://doi.org/10.1016/j.toxicon.2017.10.016. PubMed PMID: 29056305

Cavalu S, Elbaramawi SS, Eissa AG, Radwan MF, Ibrahim TS, Khafagy E-S et al (2022) Characterization of the anti-biofilm and anti-quorum sensing activities of the β-adrenoreceptor antagonist atenolol against Gram-negative bacterial pathogens. Int J Mol Sci 23(21):13088. https://doi.org/10.3390/ijms232113088

Celli J (2019) The intracellular life cycle of Brucella spp. Microbiol Spectr 7(2):BAI-0006-2019. https://doi.org/10.1128/microbiolspec.bai-0006-2019

Chekabab SM, Harel J, Dozois CM (2014) Interplay between genetic regulation of phosphate homeostasis and bacterial virulence. Virulence 5(8):786–793

Chen G, Swem LR, Swem DL, Stauff DL, O'Loughlin CT, Jeffrey PD et al (2011) A strategy for antagonizing quorum sensing. Mol Cell 42(2):199–209. https://doi.org/10.1016/j.molcel.2011.04.003. PubMed PMID: 21504831; PubMed Central PMCID: PMCPMC3092643

Cheong HC, Sulaiman S, Looi CY, Chang L-Y, Wong WF (2023) Chlamydia infection remodels host cell mitochondria to alter energy metabolism and subvert apoptosis. Microorganisms 11(6):1382

Choo JH, Rukayadi Y, Hwang JK (2006) Inhibition of bacterial quorum sensing by vanilla extract. Lett Appl Microbiol 42(6):637–641. https://doi.org/10.1111/j.1472-765X.2006.01928.x. PubMed PMID: 16706905

Choudhury S, Baker MR, Chatterjee S, Kumar H (2021) Botulinum toxin: an update on pharmacology and newer products in development. Toxins 13(1):58. https://doi.org/10.3390/toxins13010058. PubMed PMID: 33466571; PubMed Central PMCID: PMC7828686

Chowdhury F, Ross AG, Islam MT, McMillan NAJ, Qadri F (2022) Diagnosis, management, and future control of cholera. Clin Microbiol Rev 35(3):e0021121. https://doi.org/10.1128/cmr.00211-21. PubMed PMID: 35726607; PubMed Central PMCID: PMC9491185

Cornelis GR (2006) The type III secretion injectisome. Nat Rev Microbiol 4(11):811–825. https://doi.org/10.1038/nrmicro1526. PubMed PMID: 17041629

Cornelis P, Matthijs S, Van Oeffelen L (2009) Iron uptake regulation in Pseudomonas aeruginosa. Biometals 22:15–22

Costa TR, Felisberto-Rodrigues C, Meir A, Prevost MS, Redzej A, Trokter M et al (2015) Secretion systems in Gram-negative bacteria: structural and mechanistic insights. Nat Rev Microbiol 13(6):343–359

Craig L, Forest KT, Maier B (2019) Type IV pili: dynamics, biophysics and functional consequences. Nat Rev Microbiol 17(7):429–440

de la Fuente-Nunez C, Reffuveille F, Mansour SC, Reckseidler-Zenteno SL, Hernandez D, Brackman G et al (2015) D-enantiomeric peptides that eradicate wild-type and multidrug-resistant biofilms and protect against lethal Pseudomonas aeruginosa infections. Chem Biol 22(2):196–205. https://doi.org/10.1016/j.chembiol.2015.01.002. PubMed PMID: 25699603; PubMed Central PMCID: PMCPMC4362967

De A, Pompilio A, Francis J, Sutcliffe IC, Black GW, Lupidi G et al (2018) Antidiabetic "gliptins" affect biofilm formation by Streptococcus mutans. Microbiol Res 209:79–85. https://doi.org/10.1016/j.micres.2018.02.005. PubMed PMID: 29580624

Deacy AM, Gan SK, Derrick JP (2021) Superantigen recognition and interactions: functions, mechanisms and applications. Front Immunol 12:731845. https://doi.org/10.3389/fimmu.2021.731845. PubMed PMID: 34616400; PubMed Central PMCID: PMC8488440

Decker MD, Edwards KM (2021) Pertussis (whooping cough). J Infect Dis 224(12 Suppl 2):S310–SS20. https://doi.org/10.1093/infdis/jiaa469. PubMed PMID: 34590129; PubMed Central PMCID: PMC8482022

Di Bella S, Ascenzi P, Siarakas S, Petrosillo N, di Masi A (2016) Clostridium difficile toxins A and B: insights into pathogenic properties and Extraintestinal effects. Toxins 8(5):134. https://doi.org/10.3390/toxins8050134. PubMed PMID: 27153087; PubMed Central PMCID: PMC4885049

Dietrich R, Jessberger N, Ehling-Schulz M, Martlbauer E, Granum PE (2021) The food poisoning toxins of Bacillus cereus. Toxins 13(2):98. https://doi.org/10.3390/toxins13020098. PubMed PMID: 33525722; PubMed Central PMCID: PMC7911051

Do Vale A, Cabanes D, Sousa S (2016) Bacterial toxins as pathogen weapons against phagocytes. Front Microbiol 7:42

Doganay M, Dinc G, Kutmanova A, Baillie L (2023) Human anthrax: update of the diagnosis and treatment. Diagnostics 13(6):1056. https://doi.org/10.3390/diagnostics13061056. PubMed PMID: 36980364; PubMed Central PMCID: PMC10046981

Dong M, Masuyer G, Stenmark P (2019) Botulinum and tetanus neurotoxins. Annu Rev Biochem 88:811–837. https://doi.org/10.1146/annurev-biochem-013118-111654. PubMed PMID: 30388027; PubMed Central PMCID: PMC7539302

Doohan D, Rezkitha YAA, Waskito LA, Yamaoka Y, Miftahussurur M (2021) Helicobacter pylori BabA–SabA key roles in the adherence phase: the synergic mechanism for successful colonization and disease development. Toxins 13(7):485

Dowd GC, Mortuza R, Ireton K (2021) Molecular mechanisms of intercellular dissemination of bacterial pathogens. Trends Microbiol 29(2):127–141

Duperthuy M (2020) Antimicrobial peptides: virulence and resistance modulation in Gram-negative bacteria. Microorganisms 8(2):280

Dussurget O, Pizarro-Cerda J, Cossart P (2004) Molecular determinants of Listeria monocytogenes virulence. Ann Rev Microbiol 58(1):587–610

Edae MCK, Wabalo EK (2019) Bacterial toxins and their modes of action: a review article. J Med Physiol Biophys 55:11–16

Eisenreich W, Dandekar T, Heesemann J, Goebel W (2010) Carbon metabolism of intracellular bacterial pathogens and possible links to virulence. Nat Rev Microbiol 8(6):401–412

Elfaky MA, Thabit AK, Eljaaly K, Zawawi A, Abdelkhalek AS, Almalki AJ et al (2022) Controlling of bacterial virulence: evaluation of anti-virulence activities of prazosin against Salmonella enterica. Antibiotics (Basel) 11(11):1585. https://doi.org/10.3390/antibiotics11111585. PubMed PMID: 36358239

Elfaky MA, Elbaramawi SS, Eissa AG, Ibrahim TS, Khafagy ES, Ali MAM et al (2023) Drug repositioning: doxazosin attenuates the virulence factors and biofilm formation in Gram-negative bacteria. Appl Microbiol Biotechnol 107(11):3763–3778. https://doi.org/10.1007/s00253-023-12522-3. PubMed PMID: 37079062

Elfaky MA, Okairy HM, Abdallah HM, Koshak AE, Mohamed GA, Ibrahim SR et al (2024a) Assessing the antibacterial potential of 6-gingerol: combined experimental and computational approaches. Saudi Pharm J 32:102041

Elfaky MA, Koshak AE, Radwan MF, Abdallah HM, Mohamed GA, Ibrahim SRM et al (2024b) Honokiol from Magnolia tree exhibits antibacterial and anti-virulence potential against Pseudomonas aeruginosa. Arab J Sci Eng:1–12. https://doi.org/10.1007/s13369-024-09303-z

Elwell CA, Engel JN (2012) Lipid acquisition by intracellular Chlamydiae. Cell Microbiol 14(7):1010–1018

Fabris F, Sostaric P, Matak I, Binz T, Toffan A, Simonato M et al (2022) Detection of VAMP proteolysis by tetanus and botulinum neurotoxin type B in vivo with a cleavage-specific antibody. Int J Mol Sci 23(8):4355. https://doi.org/10.3390/ijms23084355. PubMed PMID: 35457172; PubMed Central PMCID: PMC9024618

Farhana A, Khan YS (2024) Biochemistry, lipopolysaccharide. StatPearls, Treasure Island. Disclosure: Yusuf Khan declares no relevant financial relationships with ineligible companies

Fernández-Villa D, Aguilar MR, Rojo L (2019) Folic acid antagonists: antimicrobial and immunomodulating mechanisms and applications. Int J Mol Sci 20(20):4996

Foster T (2016) The remarkably multifunctional fibronectin binding proteins of Staphylococcus aureus. Eur J Clin Microbiol Infect Dis 35:1923–1931

Freedman JC, Shrestha A, McClane BA (2016) Clostridium perfringens enterotoxin: action, genetics, and translational applications. Toxins 8(3):73. https://doi.org/10.3390/toxins8030073. PubMed PMID: 26999202; PubMed Central PMCID: PMC4810218

Freeman NL, Zurawski DV, Chowrashi P, Ayoob JC, Huang L, Mittal B et al (2000) Interaction of the enteropathogenic Escherichia coli protein, translocated intimin receptor (Tir), with focal adhesion proteins. Cell Motil Cytoskeleton 47(4):307–318

Gangathraprabhu B, Kannan S, Santhanam G, Suryadevara N, Maruthamuthu M (2020) A review on the origin of multidrug-resistant Salmonella and perspective of tailored phoP gene towards avirulence. Microb Pathog 147:104352

Garcia-Contreras R (2016) Is quorum sensing interference a viable alternative to treat Pseudomonas aeruginosa infections? Front Microbiol 7:1454. https://doi.org/10.3389/fmicb.2016.01454. PubMed PMID: 27683577; PubMed Central PMCID: PMCPMC5021973

Gawarzewski I, Smits SH, Schmitt L, Jose J (2013) Structural comparison of the transport units of type V secretion systems. Biol Chem 394(11):1385–1398

Ghssein G, Ezzeddine Z (2022) A review of Pseudomonas aeruginosa metallophores: pyoverdine, pyochelin and pseudopaline. Biology 11(12):1711

Ginsburg I (2002) Role of lipoteichoic acid in infection and inflammation. Lancet Infect Dis 2(3):171–179. https://doi.org/10.1016/s1473-3099(02)00226-8. PubMed PMID: 11944187

Green ER, Mecsas J (2016) Bacterial secretion systems: an overview. In: Virulence mechanisms of bacterial pathogens. ASM Press, Washington, DC, pp 213–239

Gripenland J, Netterling S, Loh E, Tiensuu T, Toledo-Arana A, Johansson J (2010) RNAs: regulators of bacterial virulence. Nat Rev Microbiol 8(12):857–866

Hamon MA, Ribet D, Stavru F, Cossart P (2012) Listeriolysin O: the Swiss army knife of Listeria. Trends Microbiol 20(8):360–368

Harjai K, Kumar R, Singh S (2010) Garlic blocks quorum sensing and attenuates the virulence of Pseudomonas aeruginosa. FEMS Immunol Med Microbiol 58(2):161–168. https://doi.org/10.1111/j.1574-695X.2009.00614.x. PubMed PMID: 19878318

Hasan S, Osickova A, Bumba L, Novak P, Sebo P, Osicka R (2015) Interaction of Bordetella adenylate cyclase toxin with complement receptor 3 involves multivalent glycan binding. FEBS Lett 589(3):374–379. https://doi.org/10.1016/j.febslet.2014.12.023. PubMed PMID: 25554420

Hassel B (2013) Tetanus: pathophysiology, treatment, and the possibility of using botulinum toxin against tetanus-induced rigidity and spasms. Toxins 5(1):73–83. https://doi.org/10.3390/toxins5010073. PubMed PMID: 23299659; PubMed Central PMCID: PMC3564069

Hatakeyama M (2008) SagA of CagA in Helicobacter pylori pathogenesis. Curr Opin Microbiol 11(1):30–37

Hayes F, Van Melderen L (2011) Toxins-antitoxins: diversity, evolution and function. Crit Rev Biochem Mol Biol 46(5):386–408

Hegazy WA, Hensel M (2012) Salmonella enterica as a vaccine carrier. Future Microbiol 7(1):111–127. https://doi.org/10.2217/fmb.11.144. PubMed PMID: 22191450

Hegazy WAH, Rajab AAH, Abu Lila AS, Abbas HA (2021) Anti-diabetics and antimicrobials: harmony of mutual interplay. World J Diabetes 12(11):1832–1855. https://doi.org/10.4239/wjd.v12.i11.1832. Epub 2021/12/11. PubMed PMID: 34888011; PubMed Central PMCID: PMCPMC8613656

Hegazy WAH, Salem IM, Alotaibi HF, Khafagy E-S, Ibrahim D (2022) Terazosin interferes with quorum sensing and type three secretion system and diminishes the bacterial espionage to mitigate the Salmonella typhimurium pathogenesis. Antibiotics (Basel) 11(4):465. https://doi.org/10.3390/antibiotics11040465

Heinzelmann M, Scott M, Lam T (2002) Factors predisposing to bacterial invasion and infection. Am J Surg 183(2):179–190

Henderson IR, Navarro-Garcia F, Desvaux M, Fernandez RC, Ala'Aldeen D (2004) Type V protein secretion pathway: the autotransporter story. Microbiol Mol Biol Rev 68(4):692–744

Heroven AK, Dersch P (2014) Coregulation of host-adapted metabolism and virulence by pathogenic yersiniae. Front Cell Infect Microbiol 4:146

Hider RC, Kong X (2010) Chemistry and biology of siderophores. Nat Prod Rep 27(5):637–657

Hilchie AL, Wuerth K, Hancock RE (2013) Immune modulation by multifaceted cationic host defense (antimicrobial) peptides. Nat Chem Biol 9(12):761–768

Holling N, Lednor D, Tsang S, Bissell A, Campbell L, Nzakizwanayo J et al (2014) Elucidating the genetic basis of crystalline biofilm formation in Proteus mirabilis. Infect Immun 82(4):1616–1626. https://doi.org/10.1128/IAI.01652-13. PubMed PMID: 24470471; PubMed Central PMCID: PMCPMC3993403

Holmes RK (2000) Biology and molecular epidemiology of diphtheria toxin and the tox gene. J Infect Dis 181(Suppl 1):S156–S167. https://doi.org/10.1086/315554. PubMed PMID: 10657208

Hoppe PA, Holzhauer S, Lala B, Buhrer C, Gratopp A, Hanitsch LG et al (2019) Severe infections of Panton-Valentine leukocidin positive Staphylococcus aureus in children. Medicine 98(38):e17185. https://doi.org/10.1097/MD.0000000000017185. PubMed PMID: 31567961; PubMed Central PMCID: PMC6756729

Huseby M, Shi K, Brown CK, Digre J, Mengistu F, Seo KS et al (2007) Structure and biological activities of beta toxin from Staphylococcus aureus. J Bacteriol 189(23):8719–8726. https://doi.org/10.1128/JB.00741-07. PubMed PMID: 17873030; PubMed Central PMCID: PMC2168928

Hutchings MI, Truman AW, Wilkinson B (2019) Antibiotics: past, present and future. Curr Opin Microbiol 51:72–80

Isberg RR, Voorhis DL, Falkow S (1987) Identification of invasin: a protein that allows enteric bacteria to penetrate cultured mammalian cells. Cell 50(5):769–778

Izoré T, Job V, Dessen A (2011) Biogenesis, regulation, and targeting of the type III secretion system. Structure 19(5):603–612

Ji S, Choi Y, Choi Y (2015) Bacterial invasion and persistence: critical events in the pathogenesis of periodontitis? J Periodontal Res 50(5):570–585

Johnson S, Gerding DN (2019) Bezlotoxumab. Clin Infect Dis 68(4):699–704

Joseph A, Cointe A, Mariani Kurkdjian P, Rafat C, Hertig A (2020) Shiga toxin-associated hemolytic uremic syndrome: a narrative review. Toxins 12(2):67. https://doi.org/10.3390/toxins12020067. PubMed PMID: 31973203; PubMed Central PMCID: PMC7076748

Karavolos MH, Winzer K, Williams P, Khan CM (2013) Pathogen espionage: multiple bacterial adrenergic sensors eavesdrop on host communication systems. Mol Microbiol 87(3):455–465. https://doi.org/10.1111/mmi.12110. PubMed PMID: 23231070

Kato F, Kadomoto N, Iwamoto Y, Bunai K, Komatsuzawa H, Sugai M (2011) Regulatory mechanism for exfoliative toxin production in Staphylococcus aureus. Infect Immun 79(4):1660–1670. https://doi.org/10.1128/IAI.00872-10. PubMed PMID: 21282415; PubMed Central PMCID: PMC3067547

Kc R, Shukla SD, Walters EH, O'Toole RF (2017) Temporal upregulation of host surface receptors provides a window of opportunity for bacterial adhesion and disease. Microbiology 163(4):421–430

Kenney LJ, Anand GS (2020) EnvZ/OmpR two-component signaling: an archetype system that can function noncanonically. EcoSal Plus 9(1):ESP-0001-2019. https://doi.org/10.1128/ecosalplus.ESP-0001-2019

Khayat MT, Abbas HA, Ibrahim TS, Khayyat AN, Alharbi M, Darwish KM et al (2022a) Anti-quorum sensing activities of gliptins against Pseudomonas aeruginosa and Staphylococcus aureus. Biomedicines 10(5):1169. https://doi.org/10.3390/biomedicines10051169

Khayat MT, Ibrahim TS, Khayyat AN, Alharbi M, Shaldam MA, Mohammad KA et al (2022b) Sodium citrate alleviates virulence in Pseudomonas aeruginosa. Microorganisms 10(5):1046. https://doi.org/10.3390/microorganisms10051046

Khayat MT, Ibrahim TS, Darwish KM, Khayyat AN, Alharbi M, Khafagy ES et al (2022c) Hiring of the anti-quorum sensing activities of hypoglycemic agent linagliptin to alleviate the Pseudomonas aeruginosa pathogenesis. Microorganisms 10(12):2455. https://doi.org/10.3390/microorganisms10122455. PubMed PMID: 36557708; PubMed Central PMCID: PMCPMC9783625

Khayat MT, Elbaramawi SS, Nazeih SI, Safo MK, Khafagy E-S, Ali MA et al (2023a) Diminishing the pathogenesis of the food-borne pathogen Serratia marcescens by low doses of sodium citrate. Biology 12(4):504

Khayat MT, Abbas HA, Ibrahim TS, Elbaramawi SS, Khayyat AN, Alharbi M et al (2023b) Synergistic benefits: exploring the anti-virulence effects of metformin/vildagliptin antidiabetic combination against Pseudomonas aeruginosa via controlling quorum sensing systems. Biomedicines 11(5):1442. https://doi.org/10.3390/biomedicines11051442. PubMed PMID: 37239113; PubMed Central PMCID: PMCPMC10216381

Khayyat AN, Abbas HA, Mohamed MFA, Asfour HZ, Khayat MT, Ibrahim TS et al (2021a) Not only antimicrobial: metronidazole mitigates the virulence of Proteus mirabilis isolated from macerated diabetic foot ulcer. Appl Sci 11(15):6847. https://doi.org/10.3390/app11156847

Khayyat AN, Abbas HA, Khayat MT, Shaldam MA, Askoura M, Asfour HZ et al (2021b) Secnidazole is a promising imidazole mitigator of Serratia marcescens virulence. Microorganisms 9(11):2333. https://doi.org/10.3390/microorganisms9112333

Khayyat AN, Hegazy WAH, Shaldam MA, Mosbah R, Almalki AJ, Ibrahim TS et al (2021c) Xylitol inhibits growth and blocks virulence in Serratia marcescens. Microorganisms 9(5):1083. https://doi.org/10.3390/microorganisms9051083. PubMed PMID: 34070043; PubMed Central PMCID: PMCPMC8158113

Kilian M, Russell MW (2015) Microbial evasion of IgA functions. In: Mucosal immunology. Elsevier, Amsterdam, pp 455–469

Knuff K, Finlay BB (2017) What the SIF is happening—the role of intracellular Salmonella-induced filaments. Front Cell Infect Microbiol 7:335

Knust Z, Schmidt G (2010) Cytotoxic necrotizing factors (CNFs)-a growing toxin family. Toxins 2(1):116–127. https://doi.org/10.3390/toxins2010116. PubMed PMID: 22069550; PubMed Central PMCID: PMC3206620

Koshak AE, Okairy HM, Elfaky MA, Abdallah HM, Mohamed GA, Ibrahim SM et al (2023) Antimicrobial and anti-virulence activities of 4-shogaol from grain of paradise against Gram-negative bacteria: integration of experimental and computational methods. J Ethnopharmacol 323:117611

Koshak AE, Elfaky MA, Albadawi DAI, Abdallah HM, Mohamed GA, Ibrahim SRM et al (2024a) Piceatannol: a renaissance in antibacterial innovation unveiling synergistic potency and virulence disruption against serious pathogens. Int Microbiol:1–13. https://doi.org/10.1007/s10123-024-00532-8. Epub 2024/05/20. PubMed PMID: 38767683

Koshak AE, Elfaky MA, Abdallah HM, Albadawi DAI, Mohamed GA, Ibrahim SRM et al (2024b) Arctigenin from burdock root exhibits potent antibacterial and anti-virulence properties against Pseudomonas aeruginosa. J Microbiol Biotechnol 34(8):1642–1652. https://doi.org/10.4014/jmb.2403.03003. Epub 2024/07/26. PubMed PMID: 39049476; PubMed Central PMCID: PMCPMC11380511

Kreikemeyer B, Klenk M, Podbielski A (2004) The intracellular status of Streptococcus pyogenes: role of extracellular matrix-binding proteins and their regulation. Int J Med Microbiol 294(2–3):177–188

Kumar L, Chhibber S, Kumar R, Kumar M, Harjai K (2015) Zingerone silences quorum sensing and attenuates virulence of Pseudomonas aeruginosa. Fitoterapia 102:84–95. https://doi.org/10.1016/j.fitote.2015.02.002. PubMed PMID: 25704369

Lahiri SS (2000) Bacterial toxins--an overview. J Nat Toxins 9(4):381–408. PubMed PMID: 11126516

Lasica AM, Ksiazek M, Madej M, Potempa J (2017) The type IX secretion system (T9SS): highlights and recent insights into its structure and function. Front Cell Infect Microbiol 7:215

Lemos JA, Brown TA Jr, Burne RA (2004) Effects of RelA on key virulence properties of planktonic and biofilm populations of Streptococcus mutans. Infect Immun 72(3):1431–1440. https://doi.org/10.1128/IAI.72.3.1431-1440.2004. PubMed PMID: 14977948; PubMed Central PMCID: PMCPMC356000

Leo JC, Oberhettinger P, Schütz M, Linke D (2015) The inverse autotransporter family: intimin, invasin and related proteins. Int J Med Microbiol 305(2):276–282

Lerouge I, Vanderleyden J (2002) O-antigen structural variation: mechanisms and possible roles in animal/plant-microbe interactions. FEMS Microbiol Rev 26(1):17–47. https://doi.org/10.1111/j.1574-6976.2002.tb00597.x. PubMed PMID: 12007641

Li YH, Tian X (2012) Quorum sensing and bacterial social interactions in biofilms. Sensors (Basel) 12(3):2519–2538. https://doi.org/10.3390/s120302519. PubMed PMID: 22736963; PubMed Central PMCID: PMCPMC3376616

Lila ASA, Rajab AA, Abdallah MH, Rizvi SMD, Moin A, Khafagy E-S et al (2023) Biofilm lifestyle in recurrent urinary tract infections. Life 13(1):148

Limbago B, Penumalli V, Weinrick B, Scott JR (2000) Role of streptolysin O in a mouse model of invasive group A streptococcal disease. Infect Immun 68(11):6384–6390. https://doi.org/10.1128/IAI.68.11.6384-6390.2000. PubMed PMID: 11035749; PubMed Central PMCID: PMC97723

Liu S, Moayeri M, Leppla SH (2014) Anthrax lethal and edema toxins in anthrax pathogenesis. Trends Microbiol 22(6):317–325. https://doi.org/10.1016/j.tim.2014.02.012. PubMed PMID: 24684968; PubMed Central PMCID: PMC4041834

Liu Y, Tian S, Thaker H, Dong M (2021) Shiga toxins: an update on host factors and biomedical applications. Toxins 13(3):222. https://doi.org/10.3390/toxins13030222. PubMed PMID: 33803852; PubMed Central PMCID: PMC8003205

Locht C, Antoine R (2021) The history of pertussis toxin. Toxins 13(9):623. https://doi.org/10.3390/toxins13090623. PubMed PMID: 34564627; PubMed Central PMCID: PMC8472871

Lopez D, Vlamakis H, Kolter R (2010) Biofilms. Cold Spring Harb Perspect Biol 2(7):a000398. https://doi.org/10.1101/cshperspect.a000398. PubMed PMID: 20519345; PubMed Central PMCID: PMCPMC2890205

Los FC, Randis TM, Aroian RV, Ratner AJ (2013) Role of pore-forming toxins in bacterial infectious diseases. Microbiol Mol Biol Rev 77(2):173–207

Mahdavi J, Sondén B, Hurtig M, Olfat FO, Forsberg L, Roche N et al (2002) Helicobacter pylori SabA adhesin in persistent infection and chronic inflammation. Science 297(5581):573–578

Martin P, Tronnet S, Garcie C, Oswald E (2017) Interplay between siderophores and colibactin genotoxin in Escherichia coli. IUBMB Life 69(6):435–441

Mazgaeen L, Gurung P (2020) Recent advances in lipopolysaccharide recognition systems. Int J Mol Sci 21(2):379. https://doi.org/10.3390/ijms21020379. PubMed PMID: 31936182; PubMed Central PMCID: PMC7013859

Mazumdar S (2009) Raxibacumab. mAbs 1:531–538

Mehta D, Satyanarayana T (2016) Bacterial and archaeal alpha-amylases: diversity and amelioration of the desirable characteristics for industrial applications. Front Microbiol 7:1129. https://doi.org/10.3389/fmicb.2016.01129. PubMed PMID: 27516755; PubMed Central PMCID: PMC4963412

Melton-Celsa AR (2014) Shiga toxin (Stx) classification, structure, and function. Microbiol Spectr 2(4):EHEC-0024-2013. https://doi.org/10.1128/microbiolspec.EHEC-0024-2013. PubMed PMID: 25530917; PubMed Central PMCID: PMC4270005

Mitrophanov AY, Groisman EA (2008) Signal integration in bacterial two-component regulatory systems. Genes Dev 22(19):2601–2611

Moller N, Ziesemer S, Hildebrandt P, Assenheimer N, Volker U, Hildebrandt JP (2020) S. aureus alpha-toxin monomer binding and heptamer formation in host cell membranes – do they determine sensitivity of airway epithelial cells toward the toxin? PLoS One 15(5):e0233854. https://doi.org/10.1371/journal.pone.0233854. PubMed PMID: 32470006; PubMed Central PMCID: PMC7259691

Monturiol-Gross L, Flores-Diaz M, Pineda-Padilla MJ, Castro-Castro AC, Alape-Giron A (2014) Clostridium perfringens phospholipase C induced ROS production and cytotoxicity require PKC, MEK1 and NFkappaB activation. PLoS One 9(1):e86475. https://doi.org/10.1371/journal.pone.0086475. PubMed PMID: 24466113; PubMed Central PMCID: PMC3900566

Monturiol-Gross L, Villalta-Romero F, Flores-Diaz M, Alape-Giron A (2021) Bacterial phospholipases C with dual activity: phosphatidylcholinesterase and sphingomyelinase. FEBS Open Bio 11(12):3262–3275. https://doi.org/10.1002/2211-5463.13320. PubMed PMID: 34709730; PubMed Central PMCID: PMC8634861

Mora D, Arioli S (2014) Microbial urease in health and disease. PLoS Pathog 10(12):e1004472

Moreira CG, Sperandio V (2012) Interplay between the QseC and QseE bacterial adrenergic sensor kinases in Salmonella enterica serovar Typhimurium pathogenesis. Infect Immun 80(12):4344–4353. https://doi.org/10.1128/IAI.00803-12. PubMed PMID: 23027532; PubMed Central PMCID: PMCPMC3497441

Mosaddeghzadeh N, Ahmadian MR (2021) The RHO family GTPases: mechanisms of regulation and signaling. Cells 10(7):1831. https://doi.org/10.3390/cells10071831. PubMed PMID: 34359999; PubMed Central PMCID: PMC8305018

Muhlen S, Dersch P (2020) Treatment strategies for infections with Shiga toxin-producing Escherichia coli. Front Cell Infect Microbiol 10:169. https://doi.org/10.3389/fcimb.2020.00169. PubMed PMID: 32435624; PubMed Central PMCID: PMC7218068

Murdoch CC, Skaar EP (2022) Nutritional immunity: the battle for nutrient metals at the host–pathogen interface. Nat Rev Microbiol 20(11):657–670

Muthuramalingam M, Whittier SK, Picking WL, Picking WD (2021) The Shigella type III secretion system: an overview from top to bottom. Microorganisms 9(2):451

Navarro MA, McClane BA, Uzal FA (2018) Mechanisms of action and cell death associated with Clostridium perfringens toxins. Toxins 10(5):212

Nazeih SI, Ali MA, Halim ASA, Al-Lawati H, Abbas HA, Al-Zharani M et al (2023) Relocating glyceryl trinitrate as an anti-virulence agent against Pseudomonas aeruginosa and Serratia marcescens: insights from molecular and in vivo investigations. Microorganisms 11(10):2420

Nguyen BN, Peterson BN, Portnoy DA (2019) Listeriolysin O: a phagosome-specific cytolysin revisited. Cell Microbiol 21(3):e12988. https://doi.org/10.1111/cmi.12988. PubMed PMID: 30511471; PubMed Central PMCID: PMC6779163

Nilsen NJ, Deininger S, Nonstad U, Skjeldal F, Husebye H, Rodionov D et al (2008) Cellular trafficking of lipoteichoic acid and Toll-like receptor 2 in relation to signaling: role of CD14 and CD36. J Leukoc Biol 84(1):280–291. https://doi.org/10.1189/jlb.0907656. PubMed PMID: 18458151; PubMed Central PMCID: PMC3178504

Nishimoto AT, Rosch JW, Tuomanen EI (2020) Pneumolysin: pathogenesis and therapeutic target. Front Microbiol 11:1543. https://doi.org/10.3389/fmicb.2020.01543. PubMed PMID: 32714314; PubMed Central PMCID: PMC7343714

Nitulescu G, Mihai DP, Zanfirescu A, Stan MS, Gradinaru D, Nitulescu GM (2022) Discovery of new microbial collagenase inhibitors. Life 12(12):2114. https://doi.org/10.3390/life12122114. PubMed PMID: 36556479; PubMed Central PMCID: PMC9781087

Nzakizwanayo J, Scavone P, Jamshidi S, Hawthorne JA, Pelling H, Dedi C et al (2017) Fluoxetine and thioridazine inhibit efflux and attenuate crystalline biofilm formation by Proteus mirabilis. Sci Rep 7(1):12222. https://doi.org/10.1038/s41598-017-12445-w. PubMed PMID: 28939900; PubMed Central PMCID: PMCPMC5610337

O'Brien DK, Melville SB (2004) Effects of Clostridium perfringens alpha-toxin (PLC) and perfringolysin O (PFO) on cytotoxicity to macrophages, on escape from the phagosomes of macrophages, and on persistence of C. perfringens in host tissues. Infect Immun 72(9):5204–5215. https://doi.org/10.1128/IAI.72.9.5204-5215.2004. PubMed PMID: 15322015; PubMed Central PMCID: PMC517428

Odumosu O, Nicholas D, Yano H, Langridge W (2010) AB toxins: a paradigm switch from deadly to desirable. Toxins 2(7):1612–1645. https://doi.org/10.3390/toxins2071612. PubMed PMID: 22069653; PubMed Central PMCID: PMC3153263

Okolie CE, Cockayne A, Penfold C, James R (2013) Engineering of the LukS-PV and LukF-PV subunits of Staphylococcus aureus Panton-Valentine leukocidin for diagnostic and therapeutic applications. BMC Biotechnol 13:103. https://doi.org/10.1186/1472-6750-13-103. PubMed PMID: 24252611; PubMed Central PMCID: PMC3870988

Oleksiewicz MB, Nagy G, Nagy E (2012) Anti-bacterial monoclonal antibodies: back to the future? Arch Biochem Biophys 526(2):124–131

Otto M (2014) Staphylococcus aureus toxins. Curr Opin Microbiol 17:32–37. https://doi.org/10.1016/j.mib.2013.11.004. PubMed PMID: 24581690; PubMed Central PMCID: PMC3942668

Packiavathy IA, Priya S, Pandian SK, Ravi AV (2014) Inhibition of biofilm development of uropathogens by curcumin – an anti-quorum sensing agent from Curcuma longa. Food Chem 148:453–460. https://doi.org/10.1016/j.foodchem.2012.08.002. PubMed PMID: 24262582

Palmer T, Berks BC (2012) The twin-arginine translocation (Tat) protein export pathway. Nat Rev Microbiol 10(7):483–496

Papenfort K, Bassler BL (2016) Quorum sensing signal-response systems in Gram-negative bacteria. Nat Rev Microbiol 14(9):576–588. https://doi.org/10.1038/nrmicro.2016.89. PubMed PMID: 27510864; PubMed Central PMCID: PMCPMC5056591

Passalacqua KD, Charbonneau M-E, O'Riordan MX (2016) Bacterial metabolism shapes the host–pathogen interface. In: Virulence mechanisms of bacterial pathogens. ASM Press, Washington, DC, pp 15–41

Patel S, Mathivanan N, Goyal A (2017) Bacterial adhesins, the pathogenic weapons to trick host defense arsenal. Biomed Pharmacother 93:763–771

Pena RT, Blasco L, Ambroa A, González-Pedrajo B, Fernández-García L, López M et al (2019) Relationship between quorum sensing and secretion systems. Front Microbiol 10:1100

Pereira JM, Xu S, Leong JM, Sousa S (2022) The Yin and Yang of pneumolysin during pneumococcal infection. Front Immunol 13:878244. https://doi.org/10.3389/fimmu.2022.878244. PubMed PMID: 35529870; PubMed Central PMCID: PMC9074694

Persson H, Vindebro R, von Pawel-Rammingen U (2013) The streptococcal cysteine protease SpeB is not a natural immunoglobulin-cleaving enzyme. Infect Immun 81(6):2236–2241. https://doi.org/10.1128/IAI.00168-13. PubMed PMID: 23569114; PubMed Central PMCID: PMC3676003

Pizarro-Cerdá J, Cossart P (2006) Bacterial adhesion and entry into host cells. Cell 124(4):715–727

Plano GV, Schesser K (2013) The Yersinia pestis type III secretion system: expression, assembly and role in the evasion of host defenses. Immunol Res 57:237–245

Pohl E, Haller JC, Mijovilovich A, Meyer-Klaucke W, Garman E, Vasil ML (2003) Architecture of a protein central to iron homeostasis: crystal structure and spectroscopic analysis of the ferric uptake regulator. Mol Microbiol 47(4):903–915

Powell LC, Pritchard MF, Ferguson EL, Powell KA, Patel SU, Rye PD et al (2018) Targeted disruption of the extracellular polymeric network of Pseudomonas aeruginosa biofilms by alginate oligosaccharides. NPJ Biofilms Microbiomes 4:13. https://doi.org/10.1038/s41522-018-0056-3. PubMed PMID: 29977590; PubMed Central PMCID: PMCPMC6026129 funding from AlgiPharma AS. P.D.R. is a director/owner of AlgiPharma AS. The remaining authors declare no competing interests

Proft T, Fraser JD (2016) Streptococcal superantigens: biological properties and potential role in disease. In: Ferretti JJ, Stevens DL, Fischetti VA (eds) Streptococcus pyogenes: basic biology to clinical manifestations. University of Oklahoma Health Sciences Center, Oklahoma City

Qaria MA, Qumar S, Sepe LP, Ahmed N (2021) Cholesterol glucosylation–based survival strategy in Helicobacter pylori. Helicobacter 26(2):e12777

Quereda JJ, Meza-Torres J, Cossart P, Pizarro-Cerda J (2017) Listeriolysin S: a bacteriocin from epidemic Listeria monocytogenes strains that targets the gut microbiota. Gut Microbes 8(4):384–391. https://doi.org/10.1080/19490976.2017.1290759. PubMed PMID: 28156183; PubMed Central PMCID: PMC5570420

Raetz CR, Whitfield C (2002) Lipopolysaccharide endotoxins. Annu Rev Biochem 71:635–700. https://doi.org/10.1146/annurev.biochem.71.110601.135414. PubMed PMID: 12045108; PubMed Central PMCID: PMC2569852

Rahlwes KC, Dias BR, Campos PC, Alvarez-Arguedas S, Shiloh MU (2023) Pathogenicity and virulence of Mycobacterium tuberculosis. Virulence 14(1):2150449

Rajab AA, Hegazy WA (2023) What's old is new again: insights into diabetic foot microbiome. World J Diabetes 14(6):680–704

Rasko DA, Moreira CG, Li de R, Reading NC, Ritchie JM, Waldor MK et al (2008) Targeting QseC signaling and virulence for antibiotic development. Science 321(5892):1078–1080. https://doi.org/10.1126/science.1160354. PubMed PMID: 18719281; PubMed Central PMCID: PMCPMC2605406

Rasmussen TB, Givskov M (2006) Quorum-sensing inhibitors as anti-pathogenic drugs. Int J Med Microbiol 296(2–3):149–161. https://doi.org/10.1016/j.ijmm.2006.02.005. PubMed PMID: 16503194

Ratledge C (2004) Iron, mycobacteria and tuberculosis. Tuberculosis 84(1–2):110–130

Rawson AM, Dempster AW, Humphreys CM, Minton NP (2023) Pathogenicity and virulence of Clostridium botulinum. Virulence 14(1):2205251. https://doi.org/10.1080/21505594.2023.2205251. PubMed PMID: 37157163; PubMed Central PMCID: PMC10171130

Rivera-Calzada A, Famelis N, Llorca O, Geibel S (2021) Type VII secretion systems: structure, functions and transport models. Nat Rev Microbiol 19(9):567–584

Rohmer L, Hocquet D, Miller SI (2011) Are pathogenic bacteria just looking for food? Metabolism and microbial pathogenesis. Trends Microbiol 19(7):341–348

Rojo F (2010) Carbon catabolite repression in Pseudomonas: optimizing metabolic versatility and interactions with the environment. FEMS Microbiol Rev 34(5):658–684

Ross A, Shoff HW (2024) Toxic shock syndrome. StatPearls, Treasure Island. Disclosure: Hugh Shoff declares no relevant financial relationships with ineligible companies

Rother M, Teixeira da Costa AR, Zietlow R, Meyer TF, Rudel T (2019) Modulation of host cell metabolism by Chlamydia trachomatis. Microbiol Spectr 7(3):BAI-0012-2019. https://doi.org/10.1128/microbiolspec.bai-0012-2019

Rutherford ST, Bassler BL (2012) Bacterial quorum sensing: its role in virulence and possibilities for its control. Cold Spring Harb Perspect Med 2(11):a012427. https://doi.org/10.1101/cshperspect.a012427. PubMed PMID: 23125205; PubMed Central PMCID: PMCPMC3543102

Samoilova Z, Tyulenev A, Muzyka N, Smirnova G, Oktyabrsky O (2019) Tannic and gallic acids alter redox-parameters of the medium and modulate biofilm formation. AIMS Microbiol 5(4):379–392. https://doi.org/10.3934/microbiol.2019.4.379. PubMed PMID: 31915750; PubMed Central PMCID: PMCPMC6946639

Sanford M, McKeage K (2015) Secukinumab: first global approval. Drugs 75:329–338
Sauer JD, Herskovits AA, O'Riordan MX (2019) Metabolism of the Gram-positive bacterial pathogen Listeria monocytogenes. In: Gram-positive pathogens. ASM Press, Washington, DC, pp 864–872
Schlievert PM, Peterson ML (2012) Glycerol monolaurate antibacterial activity in broth and biofilm cultures. PLoS One 7(7):e40350. https://doi.org/10.1371/journal.pone.0040350. PubMed PMID: 22808139; PubMed Central PMCID: PMCPMC3394780
Schneewind O, Missiakas D (2014) Sec-secretion and sortase-mediated anchoring of proteins in Gram-positive bacteria. Biochim Biophys Acta 1843(8):1687–1697
Schwarz-Linek U, Werner JM, Pickford AR, Gurusiddappa S, Kim JH, Pilka ES et al (2003) Pathogenic bacteria attach to human fibronectin through a tandem β-zipper. Nature 423(6936):177–181
Sharma M, Rudel T (2009) Apoptosis resistance in Chlamydia-infected cells: a fate worse than death? FEMS Immunol Med Microbiol 55(2):154–161
Shaw C, Hess M, Weimer BC (2022) Two-component systems regulate bacterial virulence in response to the host gastrointestinal environment and metabolic cues. Virulence 13(1):1666–1680
Sheikh A, Tumala B, Vickers TJ, Martin JC, Rosa BA, Sabui S et al (2022) Enterotoxigenic Escherichia coli heat-labile toxin drives enteropathic changes in small intestinal epithelia. Nat Commun 13(1):6886. https://doi.org/10.1038/s41467-022-34687-7. PubMed PMID: 36371425; PubMed Central PMCID: PMC9653437 developed by R.D.H. and licensed by Washington University to PercayAI. The remaining authors declare no competing interests
Sheppard M, Laskou F, Stapleton PP, Hadavi S, Dasgupta B (2017) Tocilizumab (actemra). Hum Vaccin Immunother 13(9):1972–1988
Shoji H, Opal SM (2021) Therapeutic rationale for endotoxin removal with polymyxin B immobilized fiber column (PMX) for septic shock. Int J Mol Sci 22(4):2228. https://doi.org/10.3390/ijms22042228. PubMed PMID: 33672437; PubMed Central PMCID: PMC7926968
Siegel SJ, Weiser JN (2015) Mechanisms of bacterial colonization of the respiratory tract. Ann Rev Microbiol 69(1):425–444
Sierig G, Cywes C, Wessels MR, Ashbaugh CD (2003) Cytotoxic effects of streptolysin o and streptolysin s enhance the virulence of poorly encapsulated group a streptococci. Infect Immun 71(1):446–455. https://doi.org/10.1128/IAI.71.1.446-455.2003. PubMed PMID: 12496195; PubMed Central PMCID: PMC143243
Silverman JM, Brunet YR, Cascales E, Mougous JD (2012) Structure and regulation of the type VI secretion system. Ann Rev Microbiol 66(1):453–472
Sison-Martinez J, Hendriksen S, Cooper JS (2024) Hyperbaric treatment of clostridial myositis and myonecrosis. StatPearls, Treasure Island. Disclosure: Stephen Hendriksen declares no relevant financial relationships with ineligible companies. Disclosure: Jeffrey Cooper declares no relevant financial relationships with ineligible companies
Skogmar S, Tham J (2018) Severe diphtheria with neurologic and myocardial involvement in a Swedish patient: a case report. BMC Infect Dis 18(1):359. https://doi.org/10.1186/s12879-018-3264-9. PubMed PMID: 30064365; PubMed Central PMCID: PMC6069954
Solano C, Echeverz M, Lasa I (2014) Biofilm dispersion and quorum sensing. Curr Opin Microbiol 18:96–104. https://doi.org/10.1016/j.mib.2014.02.008. PubMed PMID: 24657330
Srinivasan R, Mohankumar R, Kannappan A, Karthick Raja V, Archunan G, Karutha Pandian S et al (2017) Exploring the anti-quorum sensing and antibiofilm efficacy of phytol against Serratia marcescens associated acute pyelonephritis infection in Wistar rats. Front Cell Infect Microbiol 7:498. https://doi.org/10.3389/fcimb.2017.00498. PubMed PMID: 29259923; PubMed Central PMCID: PMC5723315
Stones DH, Krachler A-M (2015) Fatal attraction: how bacterial adhesins affect host signaling and what we can learn from them. Int J Mol Sci 16(2):2626–2640
Tang WJ, Guo Q (2009) The adenylyl cyclase activity of anthrax edema factor. Mol Asp Med 30(6):423–430. https://doi.org/10.1016/j.mam.2009.06.001. PubMed PMID: 19560485; PubMed Central PMCID: PMC2783455

Tapia-Rodriguez MR, Bernal-Mercado AT, Gutierrez-Pacheco MM, Vazquez-Armenta FJ, Hernandez-Mendoza A, Gonzalez-Aguilar GA et al (2019) Virulence of Pseudomonas aeruginosa exposed to carvacrol: alterations of the Quorum sensing at enzymatic and gene levels. J Cell Commun Signal 13(4):531–537. https://doi.org/10.1007/s12079-019-00516-8. PubMed PMID: 30903602; PubMed Central PMCID: PMCPMC6946780

Thabit AK, Eljaaly K, Zawawi A, Ibrahim TS, Eissa AG, Elbaramawi SS et al (2022a) Silencing of Salmonella typhimurium pathogenesis: atenolol acquires efficient anti-virulence activities. Microorganisms 10(10):1976. https://doi.org/10.3390/microorganisms10101976. PubMed PMID: 36296252; PubMed Central PMCID: PMCPMC9612049

Thabit AK, Eljaaly K, Zawawi A, Ibrahim TS, Eissa AG, Elbaramawi SS et al (2022b) Muting bacterial communication: evaluation of prazosin anti-quorum sensing activities against Gram-negative bacteria Pseudomonas aeruginosa, Proteus mirabilis, and Serratia marcescens. Biology (Basel) 11(9):1349. https://doi.org/10.3390/biology11091349. PubMed PMID: 36138828; PubMed Central PMCID: PMCPMC9495718

Thanassi DG, Bliska JB, Christie PJ (2012) Surface organelles assembled by secretion systems of Gram-negative bacteria: diversity in structure and function. FEMS Microbiol Rev 36(6):1046–1082

Tkaczyk C, Hua L, Varkey R, Shi Y, Dettinger L, Woods R et al (2012) Identification of anti-alpha toxin monoclonal antibodies that reduce the severity of Staphylococcus aureus dermonecrosis and exhibit a correlation between affinity and potency. Clin Vaccine Immunol 19(3):377–385. https://doi.org/10.1128/CVI.05589-11. PubMed PMID: 22237895; PubMed Central PMCID: PMC3294626

Troxell B, Hassan HM (2013) Transcriptional regulation by Ferric Uptake Regulator (Fur) in pathogenic bacteria. Front Cell Infect Microbiol 3:59

Uchiyama T (2012) Tropism and pathogenicity of rickettsiae. Front Microbiol 3:230

Ulett GC, Valle J, Beloin C, Sherlock O, Ghigo J-M, Schembri MA (2007) Functional analysis of antigen 43 in uropathogenic Escherichia coli reveals a role in long-term persistence in the urinary tract. Infect Immun 75(7):3233–3244

Ulhuq FR, Mariano G (2022) Bacterial pore-forming toxins. Microbiology 168:3. https://doi.org/10.1099/mic.0.001154. PubMed PMID: 35333704; PubMed Central PMCID: PMC9558359

Unissa AN, Subbian S, Hanna LE, Selvakumar N (2016) Overview on mechanisms of isoniazid action and resistance in Mycobacterium tuberculosis. Infect Genet Evol 45:474–492

Unnikrishnan M, Altmann DM, Proft T, Wahid F, Cohen J, Fraser JD et al (2002) The bacterial superantigen streptococcal mitogenic exotoxin Z is the major immunoactive agent of Streptococcus pyogenes. J Immunol 169(5):2561–2569. https://doi.org/10.4049/jimmunol.169.5.2561. PubMed PMID: 12193726

Uribe-Querol E, Rosales C (2017) Control of phagocytosis by microbial pathogens. Front Immunol 8:1368

Van Houdt R, Givskov M, Michiels CW (2007) Quorum sensing in Serratia. FEMS Microbiol Rev 31(4):407–424. https://doi.org/10.1111/j.1574-6976.2007.00071.x. PubMed PMID: 17459113

Vanden Broeck D, Horvath C, De Wolf MJ (2007) Vibrio cholerae: cholera toxin. Int J Biochem Cell Biol 39(10):1771–1775. https://doi.org/10.1016/j.biocel.2007.07.005. PubMed PMID: 17716938

Veiga E, De Lorenzo V, Fernández LA (2004) Structural tolerance of bacterial autotransporters for folded passenger protein domains. Mol Microbiol 52(4):1069–1080

Velge P, Wiedemann A, Rosselin M, Abed N, Boumart Z, Chaussé A-M et al (2012) Multiplicity of Salmonella entry mechanisms, a new paradigm for Salmonella pathogenesis. Microbiologyopen 1(3):243–258

Vestby LK, Gronseth T, Simm R, Nesse LL (2020) Bacterial biofilm and its role in the pathogenesis of disease. Antibiotics (Basel) 9(2):59. https://doi.org/10.3390/antibiotics9020059. PubMed PMID: 32028684; PubMed Central PMCID: PMCPMC7167820

Vlazaki M, Huber J, Restif O (2019) Integrating mathematical models with experimental data to investigate the within-host dynamics of bacterial infections. Pathog Dis 77(8):ftaa001

Vo JL, Martínez Ortiz GC, Subedi P, Keerthikumar S, Mathivanan S, Paxman JJ et al (2017) Autotransporter adhesins in Escherichia coli pathogenesis. Proteomics 17(23–24):1600431
Weilbacher T, Suzuki K, Dubey AK, Wang X, Gudapaty S, Morozov I et al (2003) A novel sRNA component of the carbon storage regulatory system of Escherichia coli. Mol Microbiol 48(3):657–670
Winzer K, Williams P (2001) Quorum sensing and the regulation of virulence gene expression in pathogenic bacteria. Int J Med Microbiol 291(2):131–143. https://doi.org/10.1078/1438-4221-00110. PubMed PMID: 11437336
Wong K-W, Isberg RR (2005) Emerging views on integrin signaling via Rac1 during invasin-promoted bacterial uptake. Curr Opin Microbiol 8(1):4–9
Xie W, Dickson C, Kwiatkowski W, Choe S (2010) Structure of the cytoplasmic segment of histidine kinase receptor QseC, a key player in bacterial virulence. Protein Pept Lett 17(11):1383–1391. https://doi.org/10.2174/092986651109011383. PubMed PMID: 20594156; PubMed Central PMCID: PMCPMC3526665
Yan S, Wu G (2017) Bottleneck in secretion of α-amylase in Bacillus subtilis. Microb Cell Factories 16:1–8
Yan X, Gu S, Shi Y, Cui X, Wen S, Ge J (2017) The effect of emodin on Staphylococcus aureus strains in planktonic form and biofilm formation in vitro. Arch Microbiol 199(9):1267–1275. https://doi.org/10.1007/s00203-017-1396-8. PubMed PMID: 28616631
Yazdanparast A, Mousavi SL, Rasooli I, Amani J, Jalalinadoushan M (2012) Immunogenical study of chimeric recombinant intimin-Tir of Escherichia coli O157: H7 in mice. Arch Clin Infect Dis 7(2):45–51
Zhang J (2013) Transcriptome analysis reveals novel entry mechanisms and a central role of SRC in host defense during high multiplicity mycobacterial infection. PLoS One 8(6):e65128
Zhang H, Zhang Z, Li J, Qin G (2023) New strategies for biocontrol of bacterial toxins and virulence: focusing on quorum-sensing interference and biofilm inhibition. Toxins 15(9):570. https://doi.org/10.3390/toxins15090570. PubMed PMID: 37755996; PubMed Central PMCID: PMC10536320

Bacterial Metabolites in Defense: A Crucial Aspect of Microbial Interaction and Host Protection

5

Zeinab Shawky Hashem

Abstract

Bacterial metabolites are critical in microbial interactions and host defense, going beyond simply competing with other organisms. These bioactive compounds play a key role in inhibiting rivals, influencing the immune responses of hosts, protecting against environmental stresses, and promoting symbiotic relationships. Bacteriocins and antibiotics illustrate how bacteria use metabolites to establish their position and surpass their competitors. Additionally, bacterial metabolites are essential in controlling immune responses, maintaining internal stability, and safeguarding against environmental challenges. In mutualistic interactions, these metabolites contribute to collective defense, benefiting both the bacteria and their hosts. From a bacterial perspective, metabolites such as antibiotics, bacteriocins, and toxins are crucial for survival and colonization. Meanwhile, in humans, metabolites generated by probiotics and gut microbiota are indispensable for gut health and regulating the immune system. Appreciating these diverse functions of bacterial metabolites underscores their importance in defense mechanisms and creates new opportunities for developing therapies and enhancing health.

Keywords

Bacterial metabolites · Microbial defense · Immune modulation · Probiotics · Gut microbiota

Z. S. Hashem (✉)
Department of Microbiology and Immunology, Faculty of Pharmacy, Minia University, Minia, Egypt
e-mail: zeinab.shawky@mu.edu.eg

S. S. M. Soliman, M. I. Husseiny (eds.), *Metabolic Dynamics in Host-Microbe Interaction*, https://doi.org/10.1007/978-981-96-1305-2_5

5.1 Introduction

Bacteria rely on bacterial metabolites, which are important compounds that play a significant role in their survival and success in different environments. These metabolites are produced through complex biochemical processes and are essential for bacteria to protect themselves, interact with other organisms, and adapt to changing conditions. The defensive capabilities provided by these metabolites are critical not only for individual bacteria but also for maintaining balance within microbial communities and their interactions with hosts. The establishment and maintenance of microbial populations depend greatly on microbial interactions, as bacteria actively engage with their environment, other microbes, and host organisms. These interactions involve environmental recognition and the exchange of molecular and genetic information, allowing microorganisms to establish themselves in a community. The host can be affected in a harmful or beneficial manner depending on the nature of the multi-trophic interactions. For example, in humans, the microbial community plays a key role in protecting against diseases caused by microbial pathogens or physiological disturbances (Braga et al. 2016).

Bacteria exert effects by producing bioactive secondary metabolites, which are crucial in microbial interactions and defense mechanisms. The protection of bacteria and host organisms is significantly influenced by these compounds, which also play important roles in ecological interactions (Abdel-Aziz et al. 2017). In this chapter, we will examine the significance of bacterial metabolites in defense, emphasizing their importance in microbial ecosystems and potential medical applications.

5.2 The Role of Bacterial Metabolites in Defense

Bacteria depend on metabolites to defend themselves and thrive in challenging surroundings. These metabolites hinder competitors, regulate host immune reactions, and promote stability. They also offer defense against environmental pressures like temperature variations and oxidative stress. Furthermore, they foster mutually beneficial relationships that bolster the defenses of both bacteria and their hosts. Bacterial metabolites are vital for the survival and ecological prosperity of microorganisms due to their multifaceted roles.

5.2.1 Inhibition of Competitors

To outperform rivals, microorganisms generate and release secondary chemicals that eliminate nearby cells and capture resources. This form of competition driven by natural products probably developed in intricate microbial populations containing viral invaders. Based on this ecological framework, we theorized that microbes excrete compounds that turn natural enemies (such as bacteriophages) into weapons to destroy their competitors (Zang et al. 2024).

Competition plays a significant role in the life of microorganisms. Microbes often coexist in intricate communities consisting of various bacterial, fungal, and protozoan species. With limited resources and space in their habitats, microbes have developed diverse strategies to thrive amidst competition. The release of secondary metabolites stands as one of the primary ways through which bacteria and fungi vie against neighboring microbes. For instance, antibiotics have the capability to hinder the growth or kill competing microorganisms. Anti-adhesion molecules such as biosurfactants can obstruct potential competitors from invading and colonizing. Siderophores, which scavenge scarce environmental iron, indirectly deprive competitors of resources and inhibit their rapid proliferation (Ghoul and Mitri 2016; Westhoff et al. 2021).

One of the most intriguing characteristics of microorganisms is their ability to produce antibiotics, especially from medical and commercial standpoints. Microbes have been a source of a vast number of antibiotics and bioactive compounds, with hundreds of new classes of low-molecular-weight compounds being found annually. Major classes like cephalosporins, penicillin, and tetracyclines are mainly produced by *Streptomyces*, some actinomycetes, and various *Bacillus* species. While antibiotics are primarily used to treat human infectious diseases, they also have applications in veterinary medicine and agriculture. The creation of secondary metabolites involves utilizing common metabolites in more complex and less specific ways compared to primary metabolism. These secondary metabolic pathways are responsible for their production. Polyketide antibiotics, for instance, originate from simple fatty acids. However, their resulting formations are considerably more complex than the fundamental hydrocarbon frameworks of essential fatty acids. Nonetheless, their resulting structures are much more intricate than the basic hydrocarbon frameworks of essential fatty acids (Stephanopoulos et al. 1998).

Bacteriocins are peptides produced by bacteria through ribosomal synthesis, and they possess the capability to hinder or eliminate other bacteria. These antimicrobial peptides are small in size, typically consisting of 30–60 amino acids, and are created by both Gram-positive and Gram-negative bacteria. Bacteriocins display notable variation in their size, structure, mechanisms of action, range of activity, biochemical properties, and the receptors of target cells. It is widely acknowledged that the majority of bacteria, particularly Gram-positive ones, as well as archaea, generate at least one antimicrobial peptide to ensure their survival and gain a competitive edge in their respective surroundings (Ghodhbane et al. 2015).

Bacteriocins can be categorized into three primary groups (Rea et al. 2011). The initial group, referred to as lantibiotics (Class I), comprises peptides that undergo post-translational modifications and contain distinctive amino acids like lanthionine, β-methyl lanthionine, and dehydrated residues. Nisin, a prominent example of a Class Ia lantibiotic, is produced by specific *Lactococcus lactis* strains and is the sole commercially available semi-purified bacteriocin used to control spoilage and pathogenic microorganisms. The second group, Class II bacteriocins, consists of unmodified small peptides (less than 10 kDa) that demonstrate heat stability. This group is further divided into Class IIa, which encompasses pediocin PA1-AcH targeting *Listeria* effectively; Class IIb, which comprises two-peptide unmodified bacteriocins; and Class IIc, which includes circular bacteriocins. Both Class I and Class II bacteriocins are recognized for their broad-spectrum activity against Gram-positive bacteria, making them valuable in food preservation and therapeutic applications.

Bacteriocins display several impressive characteristics that position them as promising substitutes for traditional antibiotics. They possess the unique capability to efficiently eliminate target cells, making them powerful agents against various human and animal pathogens. This includes infections that impact the skin, throat, bladder, bloodstream, and gut. Furthermore, bacteriocins are recognized for their rapid action and quick effectiveness in killing or inhibiting both dormant and active cells. Their Minimum Inhibitory Concentration values are similar to those of conventional antibiotics, underscoring their effectiveness. Another important benefit is their consistent antimicrobial activity across a broad range of environmental conditions, including variations in temperature and pH. Additionally, bacteriocins do not prompt the development of resistance-associated mutations in multiple species simultaneously, unlike broad-spectrum antibiotics. They have varied structures, and they undergo substantial post-translational changes like cyclization, formation of disulfide linkages, and the addition of non-conventional amino acids, which contribute to their stability. These distinctive characteristics enable them to withstand degradation better under extreme temperatures and pH conditions compared to conventional antibiotics (Riley et al. 2013).

5.2.1.1 Specialized Metabolites in Competitive Strategies

Specialized metabolites (SMs) are bioactive compounds synthesized by bacteria, playing roles beyond primary metabolism, particularly in competition between species. Although often considered secondary metabolites, SMs can be crucial for bacterial survival under environmental or competitive stress (Davies 2013). Antibiotics are a well-known example of SMs, mediating chemical interactions between bacteria, but their roles in natural environments, especially at subinhibitory concentrations, remain debated (Romero et al. 2011). SMs have diverse biological functions, many of which are still unknown. Some bacteria utilize SMs, such as antibiotics, in precisely targeted ways that affect competitors beyond mere growth inhibition.

Bacteria have evolved various complex iron acquisition tactics, such as producing high-affinity siderophores and exploiting siderophores made by other microorganisms, in addition to their role in competition. The production of siderophores is closely tied to iron availability and environmental factors, enabling bacteria to adapt to changing nutrient levels and uphold a competitive edge (Cornelis and Andrews 2010). When bacteria modify their environment or use shared resources more efficiently than their competitors, they engage in exploitation competition. SMs play a critical role in this process. For instance, bacteria produce siderophores, which are SMs that help them capture and absorb iron, an essential element for many cellular functions. By producing siderophores, bacteria can increase their own access to iron while decreasing it for others, as observed in various bacterial interactions in different environments. However, siderophores can also be utilized by other species, leading to intricate competitive dynamics (Stubbendieck and Straight 2016). In settings where iron is scarce, bacteria that generate siderophores have a significant edge by seizing this crucial nutrient, often leading to the exclusion of those that do not produce siderophores. However, certain bacteria can take advantage of siderophores created by other organisms, complicating microbial community dynamics. This occurrence, known as "siderophore piracy," adds further levels of competition, as bacterial species not only compete to produce siderophores but also develop methods to utilize those produced by rivals (Kramer et al. 2020).

5.2.1.2 Competitive Interference Via Specialized Metabolites

Competition between bacteria frequently includes the utilization of SMs to hinder the growth or actions of rivals. These metabolites can have impacts that go beyond mere inhibition, affecting bacterial interactions and the dynamics of communities in various and intricate manners. Actually, SMs have a significant impact not just in direct opposition, but also in aiding intricate communication between different species. They have the ability to affect the actions of nearby microorganisms, prompting physiological reactions that alter the composition of the community and competitive relationships (Davies 2013).

Microorganisms produce an exceptional range of SMs, such as polyketides, nonribosomal peptides, ribosomally synthesized and post-translationally modified peptides, terpenes, saccharides, and alkaloids, showcasing unparalleled chemical diversity. Until the mid-twentieth century, these metabolites were often disregarded or considered as waste products. Specialized metabolites, unlike primary metabolites, are usually specific to particular taxa and serve crucial physiological or predatory roles. They also exhibit diverse biological activities, making them valuable for pharmaceutical and agricultural applications, with the potential for developing antimicrobials, anti-cancer agents, and immunosuppressants. Biosynthetic gene clusters produce these metabolites, often remaining silent in standard laboratory conditions and being activated by specific environmental stimuli. The advancement of genomics and computational biology has facilitated the identification of numerous biosynthetic gene clusters through metagenomic analyses. Key classes of specialized metabolites, including polyketides and non-ribosomal peptides, have the capacity to generate a wide array of molecular structures (Loureiro et al. 2018).

Antibiotic as Signaling Molecules Beyond Inhibition

At subinhibitory concentrations, antibiotics, typically known for halting bacterial growth, can also serve as signaling molecules. Instead of just inhibiting competition, antibiotics can impact microbial communities by triggering various responses, such as premature sporulation, antibiotic synthesis, or the activation of secretion systems (Bernier and Surette 2013; Davies et al. 2006). These low antibiotic concentrations can cause significant cellular changes, including the overall transcriptional responses of bacteria exposed to antibiotics. For example, *Streptomyces coelicolor* produces the pigmented antibiotic prodigiosin when exposed to low concentrations of jadomycin B (Wang et al. 2014). Likewise, exposure to subinhibitory kanamycin leads *Pseudomonas aeruginosa* to activate type VI secretion genes (Jones et al. 2013). These observations challenge the traditional view of antibiotics solely as growth inhibitors and suggest that their natural functions in microbial environments may involve complex roles in signaling and regulating community dynamics. Furthermore, these interactions could offer a mechanism for sustaining microbial diversity in environments where growth-inhibitory concentrations are seldom encountered. Such discoveries prompt researchers to reconsider the ecological roles of antibiotics beyond competitive dominance (Bernier and Surette 2013; Davies et al. 2006).

Studies in the recent past have highlighted that antibiotics can trigger hormetic effects, where lower doses stimulate specific bacterial responses, while higher doses hinder growth. Comprehensive genome-wide transcription analyses have uncovered the broader influences of subinhibitory antibiotics, suggesting that they can regulate gene expression beyond direct target interactions (Bernier and Surette 2013). These subinhibitory effects could potentially act as early warning signs of chemical warfare within microbial communities, prompting bacteria to adapt by producing defensive metabolites or altering their behavior. For instance, at low levels of antibiotics, the activation of stress responses or the production of secondary metabolites might improve bacterial survival under competitive conditions. It is essential to understand these concentration-dependent responses to fully comprehend the ecological roles of antibiotics and small molecules in competitive bacterial interactions. Viewing antibiotics as signaling molecules provides new opportunities for investigating their physiological effects in clinical and environmental contexts. Furthermore, this knowledge holds promise for enhancing antibiotic efficacy by utilizing their dual roles in bacterial communities (Romero et al. 2011).

Multi-functionality of Specialized Metabolites

Bacteria produce a vast range of SMs, reflecting their immense chemical diversity, much of which remains poorly understood (Davies and Ryan 2012). While antibiotic activity is often the primary attribute linked to SMs, many such compounds affect bacterial competitors without necessarily inhibiting their growth. For example, *Pseudomonas protegens* synthesizes 2,4-diacetylphloroglucinol, a metabolite recognized for its antifungal properties and commonly employed in biocontrol methods. This SM can also inhibit the cellular differentiation of *Bacillus subtilis* when co-cultured with *P. protegens* (Haas and Défago 2005). On the contrary, certain compounds like the antifungal nystatin (López et al. 2009) and peptide antibiotics can promote biofilm formation in *B. subtilis (*Bleich et al. 2015*)*. Recent research emphasizes the impact of SMs on interspecies competition, demonstrating that their effects extend beyond mere growth inhibition. The findings reveal that specific SMs from *Streptomyces* species can effectively alter the production of secondary metabolites in neighboring bacterial species. This alteration influences competitive interactions through indirect biochemical signaling rather than through direct antimicrobial effects (Čihák et al. 2017).

Additionally, bacillaene, a metabolite produced by *B. subtilis*, was initially recognized for its role as an antibiotic that inhibits protein synthesis. However, bacillaene also impacts the production of prodigiosin in *Streptomyces lividans* and *Streptomyces coelicolor*, even when it does not cause growth inhibition. These findings suggest that SMs can influence bacterial development and interactions through pathways that extend beyond mere antimicrobial effects (Vargas-Bautista et al. 2014). SMs also have functions in disrupting communication systems in bacteria, like quorum sensing. Bacterial behavior synchronization, which includes virulence and biofilm formation, heavily relies on quorum sensing. Some marine bacteria can adjust quorum sensing by producing specialized metabolites that can interfere with these communication pathways. This interference can greatly impact competitive interactions and community dynamics in microbial environments (Teasdale et al. 2009). Despite advances in understanding, much remains unknown about how these modulatory activities affect the competitive fitness of bacteria in natural environments. As a result, researchers have developed model systems using co-cultured bacteria to explore how, under controlled conditions, competitive interactions are influenced by SMS and other factors (Davies and Ryan 2012).

5.2.2 Modulation of Host Immune Responses

The immune system can be influenced by bacterial metabolites, either by boosting or suppressing immune responses to different pathogens and maintaining the body's equilibrium.

5.2.2.1 Immune System Modulation by Short-Chain Fatty Acids (SCFA)

The gut microbiota in humans is a significant source of bacteria that generate SCFAs by breaking down non-digestible polysaccharides, such as starches that are resistant to digestion and dietary fibers. The levels of SCFAs vary at different stages of life due to changes in diet and the microbiota, leading to high acetate production (Tsukuda et al. 2021; Fusco et al. 2023). As individuals age and their diet becomes more diverse, species from *Firmicutes* like *Lactobacillaceae*, *Ruminococcaceae*, and *Lachnospiraceae* become more prevalent, leading to the generation of butyrate and other SCFAs (Biddle et al. 2013).

In early life, acetate is predominantly produced by bacterial strains such as *Bifidobacterium bifidum*, *Bifidobacterium breve*, and *Bifidobacterium infantis*, which utilize human milk oligosaccharides as a substrate. This acetate production plays a key role in facilitating microbial interactions that foster the growth of bacteria responsible for the synthesis of other SCFAs, including propionate and butyrate. After weaning and the transition to a more diverse diet, there is a notable increase in propionate production. This shift is driven by changes in the gut microbiota, particularly the rise of Firmicutes species from the *Clostridia* class, which further contributes to SCFA production. Propionate is produced through various pathways, including the succinate pathway (involving *Bacteroidetes* like *Prevotella* spp.), the acrylate pathway (involving *Coprococcus* spp.), and the propanediol pathway (observed in *Roseburia* and *Blautia* species). Butyrate, the most physiologically significant SCFA, is mainly produced by the Lachnospiraceae and *Ruminococcaceae* families. It is crucial for maintaining gut health by providing energy to colonocytes and maintaining an anaerobic gut environment, which hinders the growth of harmful pathogens such as *Salmonella* (Fusco et al. 2023).

Additional SCFAs, including lactate and formate, are generated. Formate is a byproduct of *Bifidobacteri*a metabolism of human milk oligosaccharides, while lactate is produced in early life by bacteria like *Bifidobacteria* and *Lactobacilli.* Lactate can be further utilized by other gut bacteria, including *Anaerostipes caccae* and *Eubacterium hallii*, to generate butyrate and propionate (Walsh et al. 2022). Ongoing research utilizing methods like comprehensive genetic analysis and metabolic profiling will continue to provide insights into the intricate interactions between SCFA-producing bacteria and human health.

SCFAs are the most common microbial byproducts found in the colon and have various roles in controlling the immune system. To begin with, they serve as the main fuel source for cells in the colon and impact the activation of important genes that uphold the strength of protective barriers and defense mechanisms. Additionally, they manage the activity of certain types of immune cells like macrophages, neutrophils, and dendritic cells. Lastly, SCFAs have a twofold influence on adaptive immune responses, affecting antigen-specific T cells and B cells (Parada Venegas et al. 2019).

SCFAs, which are derived from dietary fiber, play essential roles in regulating immune functions, especially in the intestinal environment. The production of SCFAs stimulates the growth of the cells lining the intestine, improves the intestinal barrier, and influences the production of cytokines. They modulate immune responses through various mechanisms, including the inhibition of histone deacetylases and the activation of G-protein-coupled receptors like GPR43 and GPR109A. SCFAs support anti-inflammatory responses, regulate myeloid cell activity, and facilitate B cell differentiation and antibody production. However, their effects can vary based on context, concentration, and tissue location, potentially both suppressing and promoting inflammation. Although generally protective, SCFAs can also contribute to pathological conditions under specific circumstances, indicating their complex role in immune regulation (Kim 2023).

5.2.2.2 Immune System Regulation by Amino Acid and Indole-Derived Metabolites

Metabolites related to amino acid and indole, especially those originating from tryptophan, have crucial functions in immune regulation mainly by activating the aryl hydrocarbon receptor. Aryl hydrocarbon receptor activation through various natural and synthetic ligands, like indole-3-acetate and kynurenine, affects T cell differentiation, favoring Th17 responses and inhibiting Tregs. However, the impact of aryl hydrocarbon receptor activation is intricate and reliant on the context, as various ligands and transcription factors result in diverse immune responses. Moreover, polyamines, another category of amino acid-related metabolites, demonstrate anti-inflammatory effects by controlling T cell activity and cytokine production. Despite being recognized for their roles in immune modulation, the exact mechanisms through which polyamines and indole-related metabolites derived from the gut microbiota exert their effects are not fully understood (Ilmarinen et al. 2015; Mulero-Navarro and Fernandez-Salguero 2016).

5.2.2.3 Immune System Modulation Through Bile Acid and Cholesterol Metabolites

Bile acids have critical functions in modulating the immune system, especially in the context of inflammatory conditions. The immune responses are regulated by secondary bile acids like lithocholic acid and deoxycholic acid through receptors such as TGR5, FXR, and PXR. These receptors play a role in promoting immune tolerance, as demonstrated by the heightened susceptibility to colitis in mice lacking these receptors. Activation of TGR5, for example, reduces TNF-α production and hinders NF-κB signaling, thus decreasing inflammation. Nonetheless, prolonged exposure to elevated levels of bile acids may trigger inflammation and lead to cancer, indicating that their effects are dependent on the situation. While bile acid metabolites also impact the activation of NLRP3 inflammasome, their roles are intricate and differ depending on the receptors involved, cell types, and tissue locations. In general, bile acids are crucial in maintaining immune equilibrium, yet their functions are diverse and meticulously regulated (Kim 2018).

5.2.3 Protection Against Environmental Stress

Bacteria rely on their metabolites to endure harsh conditions by creating antioxidants, compatible solutes, pH regulators, heat-shock proteins, and endospores. These tactics bolster their ability to withstand challenging environments, allowing them to thrive in extreme conditions such as extreme heat, cold, salinity, and acidity. Recognizing these approaches underscores bacteria's impressive capacity to adapt to environments that would be fatal to numerous other organisms.

5.2.3.1 Antioxidants: Defending Against Oxidative Stress

Reactive oxygen species (ROS) include a variety of chemically reactive molecules, such as superoxide radicals, hydrogen peroxide, singlet oxygen, and hydroxyl radical, and can inflict substantial harm on large molecules such as DNA, RNA, proteins, and lipids, posing a threat to cell survival. Bacteria have developed various defense mechanisms to counteract ROS-induced oxidative stress. One of the most well-studied defenses is the activity of ROS-scavenging enzymes like superoxide dismutases, peroxidases, and catalases, which work to neutralize ROS and safeguard cellular components. Recent studies indicate that specific terminal oxidases within bacterial respiratory chains might have a role in ROS defense, in addition to these detoxifying enzymes. These oxidases, in combination with other specialized enzymes, constitute a complex protective system that reduces oxidative damage and improves bacterial survival during stressful conditions. The intricate network of defenses highlights the significance of ROS regulation in bacterial physiology (Borisov et al. 2021).

5.2.3.2 Compatible Solutes: Coping with Osmotic Stress

Survival in high-salinity environments relies on the adaptation to osmotic pressure and the accumulation of compatible solutes, with glycine betaine and ectoine playing crucial roles in protecting cell structures and maintaining metabolic processes under extreme conditions (Imhoff et al. 2020). Ectoine is found in bacteria and archaea and is essential for shielding cells from external stresses such as high temperatures, osmolarity, dryness, and radiation. It also has applications in cosmetics due to its moisturizing and UV-protective properties, as well as in the pharmaceutical industry for treating various health conditions (Ma et al. 2022).

A comprehensive genomic study of numerous phototrophic bacteria indicated that the capacity to produce glycine betaine is widespread among marine and salt-loving photosynthetic Proteobacteria, as well as their chemotrophic relatives, and certain members of the *Actinobacteria* and *Pirellulaceae* families. This ability is also present in salt-loving Cyanobacteria and *Chloroherpeton thalassium*, which contributes to their survival in high-salinity habitats. On the contrary, most freshwater bacteria do not possess the mechanisms to manufacture or take in these compounds. The production of ectoine has been detected in various photosynthetic Alpha- and *Gammaproteobacteria*, particularly in species such as *Halorhodospira* and several members of *Rhodobacteraceae*. It is also widespread in *Streptomycetes* and Bacilli. Analysis of the evolutionary relationships of glycine-sarcosine methyltransferase and diaminobutyrate-pyruvate aminotransferase sequences indicates alignment with recognized evolutionary groups, with cyanobacterial sequences forming two separate branches. One branch, represented by *Halorhodospira* species, is different from other members of *Ectothiorhodospiraceae*. Furthermore, these bacteria possess a variety of transport systems for osmolytes, which enhances their capability to withstand osmotic stress (Imhoff et al. 2020).

5.2.3.3 Acid-Base Regulation: Adapting to pH Changes

Microorganisms belonging to the domains of Bacteria, Archaea, and Eukarya all require the ability to detect and react to shifts in proton levels, particularly in acidic environments. In order to survive, these life forms have developed different strategies to deal with acid stress, such as regulating pH, preserving the integrity of cell membranes, and fixing damaged large molecules (Lund et al. 2020).

During various bioprocesses, microorganisms often experience acid stress, resulting in the development of a wide range of resistance mechanisms. These mechanisms encompass maintaining pH balance, modifying membrane properties, controlling metabolism, and repairing macromolecules. The protection of probiotics against gastric acids and the improvement of organic acid biosynthesis relies heavily on these acid tolerance mechanisms. The application of systems and synthetic biology shows potential for advancing the industrial implementation of these mechanisms. In industrial environments, microbial acid tolerance is utilized to enhance probiotic survival and organic acid production. Specific resistance elements are employed by different species, requiring tailored strategies for boosting acid tolerance. The use of systems and synthetic biology presents significant opportunities for discovering and enhancing these mechanisms, ultimately allowing the design of microbial cell factories for the production of valuable chemicals (Guan and Liu 2020).

5.2.3.4 Heat-Shock Proteins: Surviving Temperature Extremes

Bacteria rely on the heat-shock response to safeguard themselves from sudden changes in their surroundings, such as fluctuations in temperature. This defense mechanism involves quickly producing heat-shock proteins (HSPs), which are crucial for protein folding, preventing aggregation, and assisting in either the recovery or degradation of misfolded proteins. Bacteria use specialized molecules, referred to as thermosensors, which consist of nucleic acids and proteins, to detect shifts in temperature and control the activation of heat-shock genes. This process is regulated through either positive regulation, involving alternative sigma factors, or negative regulation via transcriptional repressors. Some bacterial species employ both strategies, leading to the creation of complex regulatory networks. In addition to their stress-related functions, certain HSPs also function as virulence factors, enhancing bacterial pathogenicity (Roncarati and Scarlato 2017).

Under normal conditions, most organisms produce a baseline level of HSPs, but their expression significantly increases in response to various stressors. These stressors include physical factors like heat and radiation, as well as chemical and microbial agents such as bacteria, fungi, viruses, and parasites. Initially known for their role in defending against elevated temperatures and stress, HSPs are now known to have a broader role in regulating physiological processes, particularly by interacting with key cellular signaling pathways. They also influence bacterial virulence (Janmohammadi et al. 2015).

HSPs serve various functions in the immune system, both inside and outside cells, such as assisting in antigen presentation and enhancing immune surveillance against tumors. External administration of HSPs has shown potential in triggering immune responses, particularly in cancer therapies and treatments for infectious diseases and autoimmune conditions. Their interaction with viruses is also crucial, as HSPs assist in various stages of viral infection, including cell entry, replication, gene expression, and protein assembly. Viruses, in turn, manipulate host HSPs through transcriptional, translational, and post-translational modifications, as well as by altering their cellular localization. Certain HSPs are especially effective at activating both innate and adaptive immune responses, stimulating dendritic cells and natural killer cells via toll-like receptors, and enhancing MHC-antigen processing and presentation. Due to their pivotal role in immune activation, HSPs are being studied as potential therapeutic agents or targets for treating infectious diseases. Overall, HSPs are essential for maintaining protein stability, cellular homeostasis, and robust immune responses. Their function is influenced by various factors, including diet, microbial stimuli, environmental conditions, and the immune status of the host. Changes in HSP expression levels are often associated with the development of diseases (Bolhassani and Agi 2019).

5.2.4 Symbiosis and Mutual Defense

Bacteria produce vital metabolites that play a key role in mutual defense within symbiotic relationships. For example, in the human gut, beneficial bacteria create metabolites that impede the growth of harmful bacteria, thereby protecting the host from infections. In return, the host provides a stable environment and nourishment for the bacteria. This mutual relationship is crucial for the host's overall defense system, highlighting the importance of bacterial metabolites in maintaining health.

Symbiosis encompasses living together and involves interactions with pathogens, mutualists, and commensals. Bacteria must be highly adapted to their host environments, regardless of the specific nature of their relationship. Therefore, it is not surprising that both pathogens and mutualists share many factors involved in host interaction. Despite the wide variety of hosts, certain elements of the host environment, such as key metabolites and the innate immune system, are universally present. This creates conditions where mechanisms like immune evasion, immune tolerance, and nutrient competition are crucial for successful host colonization across different hosts and bacterial lifestyles. While extensive research has focused on bacterial virulence or mutualistic mechanisms, the role of these shared strategies in promoting host association across diverse bacterial interactions has received less attention (Wiesmann et al. 2023).

Upon finding a host and a suitable environment within it, a bacterium must have the ability to utilize the nutrients available. Eukaryotic hosts that are autotrophic or phototrophic offer microbes the chance to access nutrients that would otherwise be scarce, but obtaining and competing for these nutrients may necessitate specific evolutionary advancements. In hosts, carbon and nitrogen are more inclined to be found as peptides or complex carbohydrates rather than inorganic nitrogen or simple sugars (Rowland et al. 2018; Vives-Peris et al. 2020). This means that the evolution of catabolic enzymes that acquire specialized and complex carbon and nitrogen sources may be crucial in forming symbiotic relationships.

5.3 The Human Microbiome and Defense

The human microbiota comprises about 10–100 trillion symbiotic microbial cells, mainly bacteria in the gut. The human microbiome refers to the genetic material carried by these microbes (Turnbaugh et al. 2007). These microorganisms create an intricate community that influences human physiology, impacting both health and disease. Through coevolution, these microbes have developed a close relationship with the immune system, contributing to metabolic functions, protection against pathogens, and the provision of essential nutrients and energy. Therefore, the human microbiome represents the genetic potential of these microbial communities to impact various bodily functions (Dekaboruah et al. 2020). Figure 5.1 displays examples and functions of metabolites produced by gut microbiota.

The development and functioning of the human body rely significantly on metabolites produced by both the host and the microbiota, many of which are

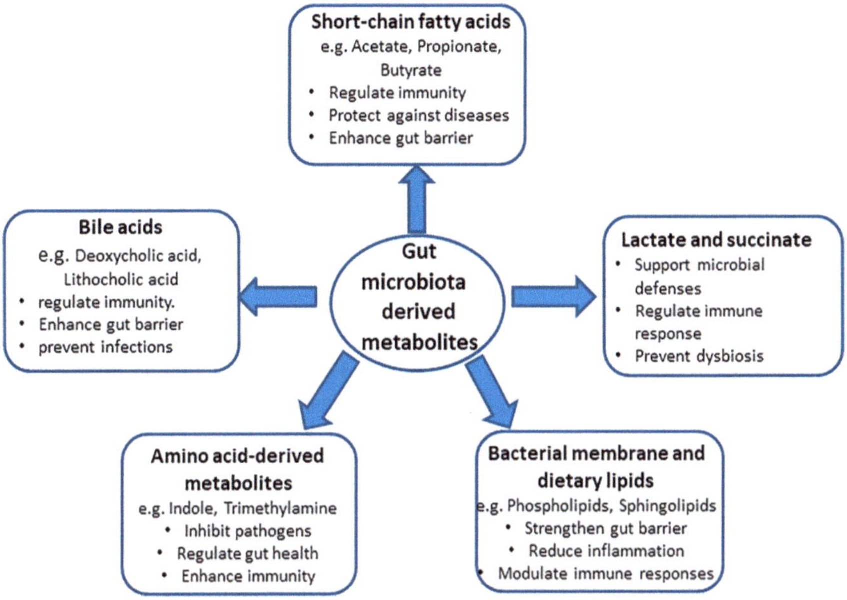

Fig. 5.1 Functions of metabolites derived from gut microbiota

synthesized in the gastrointestinal tract. These gut metabolites play essential roles in energy production, cellular communication, and immune regulation. While the intestines contain host-derived metabolites like free fatty acids, amino acids, and vitamins, the contributions of the gut microbiota, such as SCFAs, are equally critical for maintaining gut balance. Additionally, gut microbiota-produced neurotransmitters and hormones regulate host metabolism, and the microbiota is increasingly recognized as a secondary endocrine system. Disruptions in host-microbiota interactions have been linked to chronic inflammatory conditions, making microbiome-modulated metabolites promising targets for drug development. Understanding these metabolic interactions could pave the way for metabolotherapy, offering new approaches to disease prevention and treatment (Zhang et al. 2023).

The gut microbiota processes dietary components and metabolites produced by the host into substances that can support the microbes or harm host cells. The majority of the compounds present in human feces and urine are either produced or altered by the gut microbiota. These include metabolites derived from the diet such as compound K, metabolites modified by the host as secondary bile acids, and metabolites produced anew by microbes such as SCFAs. Microbial metabolites affect host tissues by entering the small intestine and influencing processes such as intestinal regeneration, immune balance, and barrier integrity (Zheng et al. 2011). Tryptophan metabolism by gut bacteria is significant, with species like *Escherichia coli, Bacteroides* spp., and *Clostridium* spp. converting it into indole and other compounds through the action of the enzyme tryptophanase (Lee and Lee 2010). Health-promoting secondary metabolites often originate from fiber and plant flavonoids, supporting immune responses and reducing inflammation (Saha et al. 2016).

5.3.1 Probiotics and Their Beneficial Metabolites

Probiotics are live microorganisms that, when consumed in sufficient quantities, provide health benefits to the host (Mokoena 2017). Lactic acid bacteria (LAB) are Gram-positive bacteria mainly known for producing lactic acid through carbohydrates metabolism. In addition to lactic acid, LAB release various substances that hinder the growth of other organisms, including bacteriocins, bacteriocin-like inhibitory substances, hydrogen peroxide, diacetyl, and carbon dioxide. The main genera of LAB consist of *Lactobacillus*, *Lactococcus*, *Enterococcus*, *Streptococcus*, and *Pediococcus*, with *Lactobacillus* being the most common probiotic in humans. Apart from LAB, other non-harmful bacterial species, such as the *E. coli* strain Nissle, are crucial for maintaining gut health. This particular strain is commonly utilized as a probiotic to help maintain the balance of intestinal microbiota. Studies have demonstrated its ability to restore the production of human β-defensin 2, which protects the intestinal barrier from the adherence and colonization of pathogenic *E. coli* (Schlee et al. 2007; Liu et al. 2017). Overall, LAB play a vital role in stabilizing and regulating the gut microbiome.

Recent research has identified specific symbiotic bacteria in the intestines, such as *Bacteroides thetaiotaomicron* and *Akkermansia muciniphila*, as potential next-generation probiotics. These bacteria are known to regulate the function of the intestinal epithelium by aiding in the formation of mucous layers, producing antibacterial factors, enhancing the secretion of secretory immunoglobulin A, competitively adhering to intestinal epithelial cells, and promoting the establishment of tight junctions (Liu et al. 2020). Furthermore, metabolites produced by probiotics, such as extracellular proteins, organic acids, indole, bacteriocins, hydrogen peroxide, and nitric oxide, serve a protective role for the intestinal epithelial barrier. They achieve this by stimulating increased mucus secretion from goblet cells, boosting the production of antimicrobial peptides, and enhancing the expression of tight junctions (Kumar et al. 2013). Figure 5.2 illustrates selected health impacts associated with probiotic bioactive metabolites.

5.3.2 Bacteriocins

Bacteria of the microbiota play a crucial role in protecting the human body by producing bacteriocins, which help in outcompeting harmful microorganisms. These antimicrobial peptides are specifically designed to target pathogens in environments where bacteria compete for space and resources. By eliminating harmful bacteria, bacteriocins release nutrients from lysed cells, which in turn benefit beneficial commensal organisms. Some species, such as *Dolosigranulum pigrum*, depend on bacteriocins to obtain essential nutrients from other bacteria. Importantly, bacteriocins

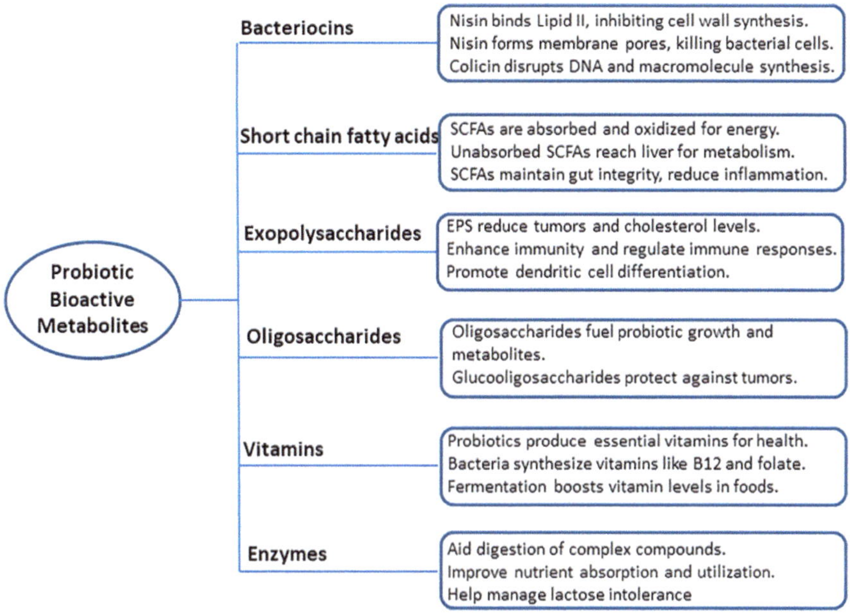

Fig. 5.2 Proposed health benefits of bioactive metabolites from probiotics

selectively attack pathogens while leaving beneficial bacteria unharmed, thus preserving symbiotic relationships and minimizing damage to host cells. Their range of activity varies, with some, like colicins, being specific to certain strains, while others, such as those from *Lactobacillus*, have broader effects across the gut, skin, and vaginal microbiomes. This diversity strengthens the microbiota's defense by targeting multiple pathogens and maintaining a harmonious microbial community (Heilbronner et al. 2021).

Bacteria that produce bacteriocins, especially LAB and intestinal commensals, are commonly employed as probiotics or protective cultures in food production, as well as dietary supplements for humans and animals. The production of bacteriocins has long been acknowledged as a crucial characteristic of these probiotics and protective cultures (Cotter et al. 2013). Although the exact ecological function of bacteriocins is not fully comprehended, they are thought to play a significant part in the efficacy of probiotics within their hosts. Bacteriocins function as adhesion peptides, aiding the producing strain in establishing and sustaining dominance within an existing ecological niche (Anjana and Tiwari 2022).

The advantageous properties of bacteriocins are significantly influenced by their biofunctional characteristics and structural diversity. Ongoing research is dedicated to exploring the bioefficacy of bacteriocins in order to create new probiotics for broader and innovative applications in healthcare. The concerns about using live cells emphasize the importance of considering bacteriocins as a potentially safer alternative in various systems. Many probiotics or protective cultures generate bacteriocins or similar inhibitory substances, which has led researchers to focus on investigating these compounds to address safety concerns regarding the use of live cells. Bacteriocins have been shown to present minimal or negligible risks when applied in animal, human, and food environments, while simultaneously offering a variety of beneficial effects. As a result, they could become the preferred choice for various applications sooner than expected. However, more comprehensive research involving adaptive evolution in diverse systems is essential to fully understand the risks linked to live cell usage and to clarify the practical benefits of bacteriocin-based applications (Wang et al. 2023).

LAB-produced bacteriocins are highly effective in targeting multiple human infectious diseases because of their ability to combat pathogens. For instance, griselimycin, a cyclic bacteriocin, has been proven effective in treating tuberculosis in mice (Kling et al. 2015). Similarly, nisin has been utilized to address infections such as pneumonia, sepsis, and meningitis caused by *Streptococcus pneumonia*. In addition to their role in infectious diseases, bacteriocins like pediocin PA-1 and lactocin AL705 have displayed properties that are both anti-cancer and anti-inflammatory (Huang et al. 2021). Nisin, for instance, hinders cancer cell proliferation by creating ion channels in cell membranes, increasing reactive oxygen species, and disrupting mitochondrial respiration. It has also been observed to have enhanced tumor-clearing effects when used in combination with cancer drugs (Preet et al. 2015). Furthermore, bacteriocins have the ability to regulate the immune response by enhancing the production of anti-inflammatory cytokines and suppressing the levels of pro-inflammatory cytokines through pathways like mitogen-activated protein kinase and Toll-like receptor signaling (Sassone-Corsi et al. 2016).

LAB bacteriocins are not only effective against human pathogens but also against foodborne pathogens. *Pediococcus acidilactici* UL5, for instance, secretes pediocin PA-1, which has demonstrated anti-listerial effects in mice without disturbing the gut's natural microbiota (Dabour et al. 2009). Enterocin CRL35, generated by *Enterococcus mundtii*, has proven effective in pregnant rodents by preventing *Listeria monocytogenes* from spreading to vital organs (Salvucci et al. 2012). Additionally, it has been discovered that nisin enhances immune responses by increasing CD4 and CD8 T cells (Pablo et al. 1999). Finally, LAB strains producing bacteriocins have shown their ability to reduce colonization by vancomycin-resistant enterococci and protect against pneumonia in animal models (De Kwaadsteniet et al. 2009; Millette et al. 2008).

5.4 Conclusion

The essential role of bacterial metabolic products is evident in their significance for microbial survival and human health. Antibiotics, siderophores, and SCFAs are crucial metabolites that serve diverse functions beyond simple defense mechanisms. Bacteria rely on these compounds to compete with other microorganisms, obtain necessary nutrients, and dominate various environments. Production of antibiotics allows bacteria to hinder the growth of rivals, while siderophores enable them to capture vital elements like iron, thereby improving their chances of survival. These interactions are pivotal for microbial ecology and illustrate the sophisticated strategies bacteria employ to maintain their ecological niches.

Bacterial metabolites have far-reaching implications for human health and disease management. SCFAs, for instance, play a significant role in preserving gut health and regulating immune responses. These metabolites help strengthen the intestinal barrier, regulating inflammatory responses, and supporting mental health, underscoring their crucial role in overall well-being. Additionally, bacterial metabolites hold therapeutic potential in managing antibiotic resistance, offering promising avenues for developing new medical interventions. As research progresses, the complex relationships between bacterial metabolites and their effects on human well-being and illness are becoming increasingly evident. Understanding these intricate interactions not only advances our knowledge of microbiology and medicine, but also presents innovative strategies for enhancing health outcomes, addressing disease, and tackling the growing challenge of antibiotic resistance. Further exploration of bacterial metabolic products holds the promise of driving significant advancements in both clinical and environmental health fields.

References

Abdel-Aziz SM, Elsoud MMA, Anise AA (2017) Microbial biosynthesis: a repertory of vital natural products. In: Food biosynthesis. Elsevier, London, pp 25–54

Anjana A, Tiwari SK (2022) Bacteriocin-producing probiotic lactic acid bacteria in controlling dysbiosis of the gut microbiota. Front Cell Infect Microbiol 12:851140

Bernier SP, Surette MG (2013) Concentration-dependent activity of antibiotics in natural environments. Front Microbiol 4:20

Biddle A, Stewart L, Blanchard J, Leschine S (2013) Untangling the genetic basis of fibrolytic specialization by Lachnospiraceae and Ruminococcaceae in diverse gut communities. Diversity 5(3):627–640

Bleich R, Watrous JD, Dorrestein PC, Bowers AA, Shank EA (2015) Thiopeptide antibiotics stimulate biofilm formation in Bacillus subtilis. Proc Natl Acad Sci 112(10):3086–3091

Bolhassani A, Agi E (2019) Heat shock proteins in infection. Clin Chim Acta 498:90–100

Borisov VB, Siletsky SA, Nastasi MR, Forte E (2021) ROS defense systems and terminal oxidases in bacteria. Antioxidants 10(6):839

Braga RM, Dourado MN, Araújo WL (2016) Microbial interactions: ecology in a molecular perspective. Braz J Microbiol 47:86–98

Čihák M, Kameník Z, Šmídová K, Bergman N, Benada O, Kofroňová O et al (2017) Secondary metabolites produced during the germination of Streptomyces coelicolor. Front Microbiol 8:2495

Cornelis P, Andrews SC (2010) Iron uptake and homeostasis in microorganisms. Caister Academic Press, Norfolk

Cotter PD, Ross RP, Hill C (2013) Bacteriocins—a viable alternative to antibiotics? Nat Rev Microbiol 11(2):95–105

Dabour N, Zihler A, Kheadr E, Lacroix C, Fliss I (2009) In vivo study on the effectiveness of pediocin PA-1 and Pediococcus acidilactici UL5 at inhibiting Listeria monocytogenes. Int J Food Microbiol 133(3):225–233

Davies J (2013) Specialized microbial metabolites: functions and origins. J Antibiot 66(7):361–364

Davies J, Ryan KS (2012) Introducing the parvome: bioactive compounds in the microbial world. ACS Chem Biol 7(2):252–259

Davies J, Spiegelman GB, Yim G (2006) The world of subinhibitory antibiotic concentrations. Curr Opin Microbiol 9(5):445–453

De Kwaadsteniet M, Doeschate K, Dicks L (2009) Nisin F in the treatment of respiratory tract infections caused by Staphylococcus aureus. Lett Appl Microbiol 48(1):65–70

Dekaboruah E, Suryavanshi MV, Chettri D, Verma AK (2020) Human microbiome: an academic update on human body site specific surveillance and its possible role. Arch Microbiol 202:2147–2167

Fusco W, Lorenzo MB, Cintoni M, Porcari S, Rinninella E, Kaitsas F et al (2023) Short-chain fatty-acid-producing bacteria: key components of the human gut microbiota. Nutrients 15(9):2211

Ghodhbane H, Elaidi S, Sabatier J-M, Achour S, Benhmida J, Regaya I (2015) Bacteriocins active against multi-resistant gram negative bacteria implicated in nosocomial infections. Infect Disord Drug Targets (Formerly Current Drug Targets-Infectious Disorders) 15(1):2–12

Ghoul M, Mitri S (2016) The ecology and evolution of microbial competition. Trends Microbiol 24(10):833–845

Guan N, Liu L (2020) Microbial response to acid stress: mechanisms and applications. Appl Microbiol Biotechnol 104(1):51–65

Haas D, Défago G (2005) Biological control of soil-borne pathogens by fluorescent pseudomonads. Nat Rev Microbiol 3(4):307–319

Heilbronner S, Krismer B, Brötz-Oesterhelt H, Peschel A (2021) The microbiome-shaping roles of bacteriocins. Nat Rev Microbiol 19(11):726–739

Huang F, Teng K, Liu Y, Cao Y, Wang T, Ma C et al (2021) Bacteriocins: potential for human health. Oxidative Med Cell Longev 2021(1):5518825

Ilmarinen P, Moilanen E, Erjefält JS, Kankaanranta H (2015) The polyamine spermine promotes survival and activation of human eosinophils. J Allergy Clin Immunol 136(2):482–4.e11

Imhoff JF, Rahn T, Künzel S, Keller A, Neulinger SC (2020) Osmotic adaptation and compatible solute biosynthesis of phototrophic bacteria as revealed from genome analyses. Microorganisms 9(1):46

Janmohammadi M, Zolla L, Rinalducci S (2015) Low temperature tolerance in plants: changes at the protein level. Phytochemistry 117:76–89

Jones C, Allsopp L, Horlick J, Kulasekara H, Filloux A (2013) Subinhibitory concentration of kanamycin induces the Pseudomonas aeruginosa type VI secretion system. PLoS One 8(11):e81132

Kim CH (2018) Immune regulation by microbiome metabolites. Immunology 154(2):220–229

Kim CH (2023) Complex regulatory effects of gut microbial short-chain fatty acids on immune tolerance and autoimmunity. Cell Mol Immunol 20(4):341–350

Kling A, Lukat P, Almeida DV, Bauer A, Fontaine E, Sordello S et al (2015) Targeting DnaN for tuberculosis therapy using novel griselimycins. Science 348(6239):1106–1112

Kramer J, Özkaya Ö, Kümmerli R (2020) Bacterial siderophores in community and host interactions. Nat Rev Microbiol 18(3):152–163

Kumar M, Nagpal R, Verma V, Kumar A, Kaur N, Hemalatha R et al (2013) Probiotic metabolites as epigenetic targets in the prevention of colon cancer. Nutr Rev 71(1):23–34

Lee J-H, Lee J (2010) Indole as an intercellular signal in microbial communities. FEMS Microbiol Rev 34(4):426–444

Liu G, Ren W, Fang J, Hu C-AA, Guan G, Al-Dhabi NA et al (2017) L-Glutamine and L-arginine protect against enterotoxigenic Escherichia coli infection via intestinal innate immunity in mice. Amino Acids 49(12):1945–1954

Liu Q, Yu Z, Tian F, Zhao J, Zhang H, Zhai Q et al (2020) Surface components and metabolites of probiotics for regulation of intestinal epithelial barrier. Microb Cell Factories 19:1–11

López D, Fischbach MA, Chu F, Losick R, Kolter R (2009) Structurally diverse natural products that cause potassium leakage trigger multicellularity in Bacillus subtilis. Proc Natl Acad Sci 106(1):280–285

Loureiro C, Medema MH, van der Oost J, Sipkema D (2018) Exploration and exploitation of the environment for novel specialized metabolites. Curr Opin Biotechnol 50:206–213

Lund PA, De Biase D, Liran O, Scheler O, Mira NP, Cetecioglu Z et al (2020) Understanding how microorganisms respond to acid pH is central to their control and successful exploitation. Front Microbiol 11:556140

Ma Z, Wu C, Zhu L, Chang R, Ma W, Deng Y et al (2022) Bioactivity profiling of the extremolyte ectoine as a promising protectant and its heterologous production. 3 Biotech 12(12):331
Millette M, Cornut G, Dupont C, Shareck F, Archambault D, Lacroix M (2008) Capacity of human nisin- and pediocin-producing lactic acid bacteria to reduce intestinal colonization by vancomycin-resistant enterococci. Appl Environ Microbiol 74(7):1997–2003
Mokoena MP (2017) Lactic acid bacteria and their bacteriocins: classification, biosynthesis and applications against uropathogens: a mini-review. Molecules 22(8):1255
Mulero-Navarro S, Fernandez-Salguero PM (2016) New trends in aryl hydrocarbon receptor biology. Front Cell Dev Biol 4:45
Pablo MA, Gaforio JJ, Gallego AM, Ortega E, Gálvez AM, de Cienfuegos A, López G (1999) Evaluation of immunomodulatory effects of nisin-containing diets on mice. FEMS Immunol Med Microbiol 24(1):35–42
Parada Venegas D, De la Fuente MK, Landskron G, González MJ, Quera R, Dijkstra G et al (2019) Short chain fatty acids (SCFAs)-mediated gut epithelial and immune regulation and its relevance for inflammatory bowel diseases. Front Immunol 10:277
Preet S, Bharati S, Panjeta A, Tewari R, Rishi P (2015) Effect of nisin and doxorubicin on DMBA-induced skin carcinogenesis—a possible adjunct therapy. Tumour Biol 36:8301–8308
Rea MC, Ross RP, Cotter PD, Hill C (2011) Classification of bacteriocins from Gram-positive bacteria. In: Prokaryotic antimicrobial peptides: from genes to applications. Springer, New York, pp 29–53
Riley MA, Robinson SM, Roy CM, Dorit RL (2013) Rethinking the composition of a rational antibiotic arsenal for the 21st century. Future Med Chem 5(11):1231–1242
Romero D, Traxler MF, López D, Kolter R (2011) Antibiotics as signal molecules. Chem Rev 111(9):5492–5505
Roncarati D, Scarlato V (2017) Regulation of heat-shock genes in bacteria: from signal sensing to gene expression output. FEMS Microbiol Rev 41(4):549–574
Rowland I, Gibson G, Heinken A, Scott K, Swann J, Thiele I et al (2018) Gut microbiota functions: metabolism of nutrients and other food components. Eur J Nutr 57:1–24
Saha S, Rajpal DK, Brown JR (2016) Human microbial metabolites as a source of new drugs. Drug Discov Today 21(4):692–698
Salvucci E, Saavedra L, Hebert EM, Haro C, Sesma F (2012) Enterocin CRL35 inhibits Listeria monocytogenes in a murine model. Foodborne Pathog Dis 9(1):68–74
Sassone-Corsi M, Nuccio S-P, Liu H, Hernandez D, Vu CT, Takahashi AA et al (2016) Microcins mediate competition among Enterobacteriaceae in the inflamed gut. Nature 540(7632):280–283
Schlee M, Wehkamp J, Altenhoefer A, Oelschlaeger TA, Stange EF, Fellermann K (2007) Induction of human β-defensin 2 by the probiotic Escherichia coli Nissle 1917 is mediated through flagellin. Infect Immun 75(5):2399–2407
Stephanopoulos G, Aristidou A, Nielsen J (1998) Examples of pathway manipulations: metabolic engineering in practice. In: Metabolic engineering. Academic Press, San Diego, pp 203–283
Stubbendieck RM, Straight PD (2016) Multifaceted interfaces of bacterial competition. J Bacteriol 198(16):2145–2155
Teasdale ME, Liu J, Wallace J, Akhlaghi F, Rowley DC (2009) Secondary metabolites produced by the marine bacterium Halobacillus salinus that inhibit quorum sensing-controlled phenotypes in gram-negative bacteria. Appl Environ Microbiol 75(3):567–572
Tsukuda N, Yahagi K, Hara T, Watanabe Y, Matsumoto H, Mori H et al (2021) Key bacterial taxa and metabolic pathways affecting gut short-chain fatty acid profiles in early life. ISME J 15(9):2574–2590
Turnbaugh PJ, Ley RE, Hamady M, Fraser-Liggett CM, Knight R, Gordon JI (2007) The human microbiome project. Nature 449(7164):804–810
Vargas-Bautista C, Rahlwes K, Straight P (2014) Bacterial competition reveals differential regulation of the pks genes by Bacillus subtilis. J Bacteriol 196(4):717–728
Vives-Peris V, De Ollas C, Gómez-Cadenas A, Pérez-Clemente RM (2020) Root exudates: from plant to rhizosphere and beyond. Plant Cell Rep 39(1):3–17

Walsh C, Lane JA, van Sinderen D, Hickey RM (2022) Human milk oligosaccharide-sharing by a consortium of infant derived Bifidobacterium species. Sci Rep 12(1):4143

Wang W, Ji J, Li X, Wang J, Li S, Pan G et al (2014) Angucyclines as signals modulate the behaviors of Streptomyces coelicolor. Proc Natl Acad Sci 111(15):5688–5693

Wang X, Hao G, Zhou M, Chen M, Ling H, Shang Y (2023) Secondary metabolites of Bacillus subtilis L2 show antiviral activity against pseudorabies virus. Front Microbiol 14:1277782

Westhoff S, Kloosterman AM, van Hoesel SF, van Wezel GP, Rozen DE (2021) Competition sensing changes antibiotic production in streptomyces. MBio 12(1):e02729-20. https://doi.org/10.1128/mbio.02729-20

Wiesmann CL, Wang NR, Zhang Y, Liu Z, Haney CH (2023) Origins of symbiosis: shared mechanisms underlying microbial pathogenesis, commensalism and mutualism of plants and animals. FEMS Microbiol Rev 47(6):fuac048

Zang Z, Zhang C, Park KJ, Schwartz DA, Podicheti R, Lennon JT et al (2024) Bacterium secretes chemical inhibitor that sensitizes competitor to bacteriophage infection. bioRxiv. https://doi.org/10.1101/2024.01.31.578241

Zhang Y, Chen R, Zhang D, Qi S, Liu Y (2023) Metabolite interactions between host and microbiota during health and disease: which feeds the other? Biomed Pharmacother 160:114295

Zheng X, Xie G, Zhao A, Zhao L, Yao C, Chiu NH et al (2011) The footprints of gut microbial–mammalian co-metabolism. J Proteome Res 10(12):5512–5522

Fungal Metabolites in Attack

6

Mohamed Abd El-Gawad El-Sayed Ahmed
and Heba Salah Abbas

Abstract

Mycotoxin exposure is a global issue that creates a variety of unpleasant consequences and problems. Mycotoxins are widely detected in Africa, and potential methods for preventing or reducing their contaminating effects are discussed. Mycotoxins are secondary metabolites of fungi that are often poisonous to living beings. Hundreds of mycotoxins have been found thus far, with some regarded as commercially significant, such as aflatoxins, ochratoxins, zearalenone, and fumonisins. Several variables influence the occurrence of mycotoxins in food, including climate circumstances, insect infestation, and inadequate harvesting and storage techniques. Mycotoxin exposure, which happens mostly by consumption, causes a variety of disorders, including mycotoxicoses and mycoses, which can ultimately lead to mortality. Nowadays, nanomaterials with unique capabilities and properties are revolutionary approaches or tools for the quick detection of mycotoxins. Antioxidant additions are the most often utilized agents for mycotoxin removal and effect reduction. Although traditional methodologies are continually advancing, current study patterns are seeking novel answers. Nanotechnology techniques appear to be a viable, successful, and inexpensive method for reducing the health impacts of mycotoxins.

M. A. E.-G. El-Sayed Ahmed (✉)
Department of Microbiology and Immunology, Faculty of Pharmaceutical Sciences and Drug Manufacturing, Misr University for Science and Technology (MUST), Giza, Egypt
e-mail: Mohamed.gawad@must.edu.eg

H. S. Abbas
Department of Microbiology and Immunology, Faculty of Pharmaceutical Sciences and Drug Manufacturing, Misr University for Science and Technology (MUST), Giza, Egypt

Department of Microbiology, Egyptian Drug Authority, Previously National Organization for Drug Control and Research, Giza, Egypt

S. S. M. Soliman, M. I. Husseiny (eds.), *Metabolic Dynamics in Host-Microbe Interaction*, https://doi.org/10.1007/978-981-96-1305-2_6

Keywords
Mycotoxin · Nanotechnology · Human health · Toxicity · Detection · Eradication

6.1 Introduction

Fungal secondary substances are a class of natural compounds that are produced by fungi and have a wide range of biological functions (Boruta 2018; Naranjo-Ortiz and Gabaldón 2020). Based on the biological characteristics of fungi that have been linked to human health, they can be categorized as either safe metabolites or poisonous "mycotoxins" (Eshelli et al. 2018).

Aflatoxin, trichothecene, ochratoxin, fumonisin, zearalenone, and ergot alkaloids are examples of hazardous metabolites from toxic fungi that exhibit liver damage, renal toxicity, and neurological and DNA damages. The virulence of various pathogenic fungi is related to their capacity to synthesize and develop hydrolyzing enzymes. Since siderophore synthesis is a process of iron acquisition, it's being examined. Specifically, these by-products function as molecules that store and chelate iron, allowing dangerous fungi to thrive in human hosts where iron balance is tightly controlled. They also inhibit the production of free radicals, which are a by-product of liberated iron. Further, melanin increases the infectivity of some fungal pathogens. They obstruct phagocytosis's detoxification process, which makes the fungus more resilient to phagocyte assault. Several metabolites have benefits in terms of fungal infection, including pullulan, mannitol, β-(1,3)-glucan, heme-binding proteins, estrogen-binding proteins, farnesol, agglutinin-like sequence proteins, glucuronoxylomannan, and others (Al-Fakih 2014). The discovery of pathogenic pathways and the detection of fungal toxins implicated in pathogenesis may result in the creation of novel, effective antifungal treatments (Al-Fakih 2014). However, since scientists exclusively study mycotoxins with significant economic implications, the amount of mycotoxin exposure in humans is still misunderstood. Consequently, it's critical to step up efforts to identify other harmful fungal toxins that cause diseases for unclear reasons.

The outcomes of the traditional analytical techniques used to identify mycotoxin in food specimens are obtained in just a couple of days or even hours. As a consequence, there is current competition between the food and feed industries, which pushes for lower labor prices, services, and quicker outcomes. Consequently, the importance of quick methods for mycotoxin detection has grown (Singh and Mehta 2020).

6.2 The Influence of Mycotoxin on Human Health

6.2.1 Aflatoxins

Aflatoxin pollution is already a major concern in underdeveloped parts of African and Asian continent; however, it may soon spread to Europe and the North and

South American continent (Sojinrin et al. 2019; Brysse et al. 2013). They are life-threatening carcinogens and mutagenic substances to individuals, as well as animals, generated by fungi (Montesano et al. 1997). Several materials, including maize, peanuts, and wheat, have regularly been discovered to be poisoned with aflatoxins (Eaton and Gallagher 1994; Henry et al. 1999). A significant connection between aflatoxin pollution of agricultural products and basic hepatocellular cancer was established (Henry et al. 1999). Using gene clustering and protein coexpression systems, Cai et al. identified two essential genes, which are histidine-rich glycoprotein and phosphoenolpyruvate carboxykinase 2. Histidine-rich glycoprotein is a protein found in the plasma and produced by the liver that regulates a variety of key biological activities, including vasculogenesis, microbial elimination, attachment of cells, and thrombosis. One of the most directly associated incidences of tumor is the regulatory impact of this protein on vasculogenesis. Also, this protein hinders the production of vascular placental growth factors. Phosphoenolpyruvate carboxykinase 2, a mitochondrial subtype of phosphoenolpyruvate carboxykinase, catalyzes the transformation of oxalic acids to pyruvate in the incidence of guanosine triphosphate as shown in Fig. 6.1. The lifespan curve examinations revealed that a decrease in histidine-rich glycoprotein and phosphoenolpyruvate carboxykinase 2 expressions throughout the progression of liver cancer predicted a bad prognosis. The reduced expression of histidine-rich glycoprotein and phosphoenolpyruvate carboxykinase 2 following aflatoxin exposure to hepatocytes might be linked to the damage of liver cells and oncogenesis (Cai et al. 2020).

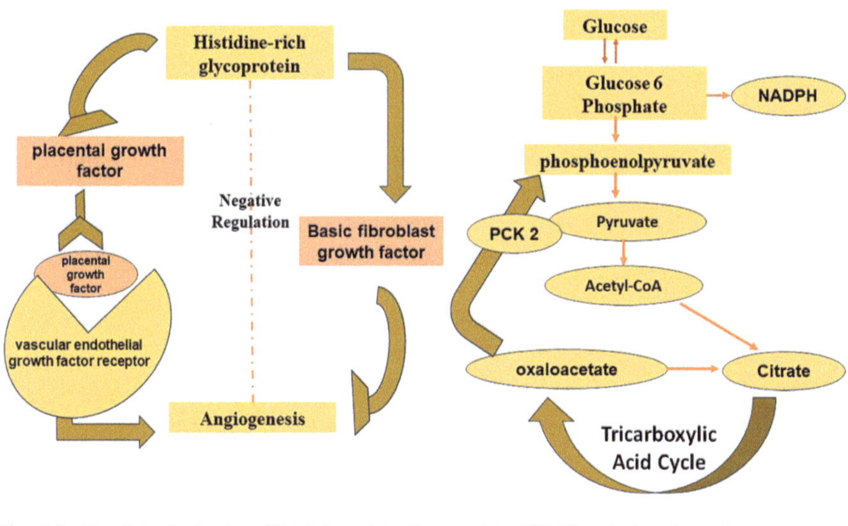

Fig. 6.1 Physiological roles of histidine-rich glycoprotein (HRG) and phosphoenolpyruvate carboxykinase (PCK2) in normal cells. (**a**) HRG prevents vascularization and tumor formation. PIGF, placental growth factor; VEGFR-1, vascular endothelial growth factor receptor 1; bFGF, basic fibroblast growth factor; and (**b**) G6P, glucose 6-phosphate; PEP, phosphoenolpyruvate; OAA, oxaloacetate; and α-KG, α-ketoglutarate. (The figure is created using Microsoft PowerPoint version 2019)

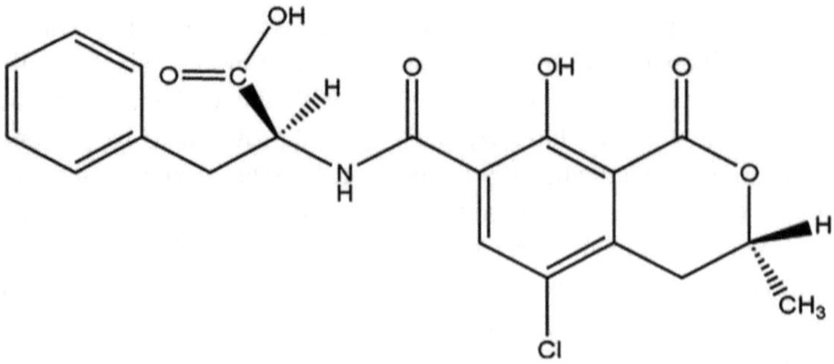

Fig. 6.2 Chemical construction of ochratoxin. (The figure is created using Free Chemical Drawing Software (ACD/Labs), 2024)

6.2.2 Ochratoxin

Ochratoxin is a highly common and hazardous mycotoxin in classes A, B, and C (El Khoury and Atoui 2010). Ochratoxins are pentaketide and generated by several filamentous fungi. Figure 6.2 depicts the chemical construction of ochratoxin (El Khoury and Atoui 2010). According to research, this chemical can have a variety of toxicological consequences, including nephrotoxicity, hepatotoxicity, neurotoxicity, teratogenicity, and immunotoxicity. A role in the causes of Balkan endemic kidney disease, as well as its relationship in urinary tract tumors, has been established (El Khoury and Atoui 2010).

The most serious side effect of ochratoxin is kidney damage. This chemical appears to be a potential kidney-damaging one in all animals other than ruminants (Elling et al. 1985). Research on epidemiology conducted in Denmark revealed that ochratoxin plays an important part in the causes of porcine nephritis (Elling et al. 1985). This mycotoxin has also been linked to human kidney damage, and it is known as Balkan endemic kidney damage, a secondary persistent illness impacting the southeastern region of Europe (Pfohl-Leszkowicz 2009). It is also thought to be the most prevalent manifestation of Tunisian kidney damage, which affects the Tunisian people (Hassen et al. 2004). Further, it has been demonstrated that administering ochratoxin during pregnancy in rats causes several abnormalities in the brain and spinal cord. Soleas et al. (2001) found that ochratoxins may be a source of some tumors and brain injury. Thus, this material appears to be very hazardous to neurological cells and capable of entering the neurons at any moment (Brown et al. 1976).

6.2.3 Fumonisins

Fumonisins are natural mycotoxins generated by the fungus *Fusarium moniliforme*, as well as other *Fusarium* species. In recent years, the fungus *A. niger* has been displayed to create fumonisin compounds, which have been detected in grapes, raisins, wine, and coffee (Scott 2012). Among the various fumonisins discovered too far, fumonisin B1 (FB1) is the main element and a prevalent pollutant in fumonisins. FB1 impacts the nutrient content and sensory qualities of feed, resulting in a decrease in consumption and production efficiency, as well as substantial financial losses. Furthermore, this mycotoxin will contaminate a variety of foods and their products, accounting for 70% of all worldwide food contamination. Those who are exposed to the mycotoxin for an extended period of time will suffer health consequences (Deepthi et al. 2017). Fumonisin has been demonstrated to cause liver dysfunction in various species, including swine, and animals (Gumprecht et al. 1995), as well as particular target organ poisoning, such as lung function in swine (Haschek et al. 1992). Fumonisins have been proven to be potentially hepatocarcinogenic in mice and rat livers (Howard et al. 2001), and the International Agency for Research on Cancer (IARC) has classified fumonisin B1 as Group 2B, which means "probably malignant to humans." Furthermore, FB1 is the primary cause of porcine pulmonary edema (PPE) (Harrison et al. 1990); its cytotoxicity affects the liver and kidney in rats (Gelderblom et al. 1991), and malignancies of the liver and esophagus in humans (Marasas et al. 1988). Fumonisins have a wide range of hazardous effects on organisms, including autophagy, cell death, neurological damage, immune-mediated damage, reproductive harm, tissue and organ toxic effects, and carcinogenesis. They have the potential to cause sickness not just on their own, however in combination with other mycotoxins like aflatoxins. The harmful effects of fumonisins are a complicated process. Prior studies have revealed that fumonisins impact their toxicity via altering sphingolipid breakdown and causing oxidative damage (Kim et al. 2018; Riley and Merrill 2019).

6.2.4 Zearalenone

Zearalenone pollution of cereal grains was recorded around the world, most notably in warmer regions. Zearalenone amounts are usually low in field-contaminated grain but rise during storage when moisture levels exceed 30%–40%. Zearalenone has a significant influence on female reproduction (as indicated by hyperestrogenism); however, it also impacts the reproductive systems of males. Prepubertal pigs are the most vulnerable species. Common clinical indications of hyperestrogenism include vulval swelling, raised size of the uterine and discharges, breast gland overgrowth and secretion, extended estrus, anestrus, a higher probability of pseudopregnancy, impotence, and reduced libido (Gupta et al. 2018).

6.3 Recent Advances in Mycotoxin Detection

6.3.1 Nanomaterials for Mycotoxin Detection

Given the fast-eating habits of food, the recognition of food contaminants (including mycotoxins) must be quick, and capable of meeting the demands of in situ finding and simple popularization. Though advanced massive analytical devices may supply reliable, accurate, and consistent procedures for detecting mycotoxins in complicated dietary matrices, they frequently involve lengthy and complicated handling of sample procedures and demand experienced employees, posing valuable cost, simplicity, and popularity challenges. Aptamer, sensor techniques, or immunoassay that relied on certain detection or adherence of antibodies, aptamers, or synthetic antibodies compensated for the aforementioned inadequacies and became strong instruments for the quick examination and recognition of the presence of mycotoxin (Fan et al. 2023; Pan et al. 2021). Nanomaterials with unique functionalities and qualities are continually being created as carriers of biological molecules, signaling resources, or other uses for the creation of innovative methodologies or technologies for the immediate identification of mycotoxins (Xiong et al. 2022; Nile et al. 2020).

6.3.1.1 Gold Nanoparticles for Aflatoxin Detection

Thin-layer chromatography (TLC) and HPLC are the most used procedures for identifying aflatoxins. Nevertheless, these approaches need considerable preparation of samples, costly devices, and maintenance by trained specialists. Additionally, the enzyme-linked immunosorbent assay (ELISA) has been effectively developed for aflatoxins (Kaittanis et al. 2010), but it requires incubating and cleaning stages, and its use is mostly limited to laboratories (Sojinrin et al. 2019).

Sojinrin et al. developed a sheet of gold nanomaterial immunochromatography to identify aflatoxin B1 in food specimens contaminated with fungi. A small-sized gold nanoparticulate was encapsulated by the combination of bovine serum albumin and the antibody of aflatoxin B1. Thin-layer chromatography (TLC), electrophoresis on gel, and NMR spectroscopic analysis were used to analyze the chemical structures. The gold nanoparticles have the ability to help identify aflatoxin B1 by detecting an obvious purple-blue color, with a limit of detection of 2 ng/mL in an ELISA plate. In just 10 minutes, aflatoxin B11 can be detected using a conjugated gold nanoparticulate immunochromatographic sheet, with a detection limit of 10 ng/g. The smartphone examination of sheets produced findings in 3 seconds, with a limit of detection of 0.3 ng/g for aflatoxin B1 when the amount present was less than 10 ng/g (Sojinrin et al. 2019).

6.3.1.2 Alphatoxin Nanopore for Detection of Mycotoxins Ochratoxin A, Aflatoxin B1, and Fumonisin B1

Previous investigation has demonstrated the capacity of nanopores to identify poisons in aqueous solutions. For instance, mycotoxin differentiation takes place by just one alphatoxin nanopore (Júnior et al. 2019). Alphatoxin, a 33.2-kDa exotoxin generated by *Staphylococcus aureus*, generates a heptameric hole that has been proposed as a prototype protein nanopore for the identification of the chemicals in water-based solutions because of its functional diversity, repeatability, and selectivity. Silva et al. showed that just one alphatoxin nanopore may be used to identify and distinguish between the mycotoxins ochratoxin A, aflatoxin B1, and fumonisin B1 in a fluid with a considerable KCl content and without saline gradient. The addition of different mycotoxins to the high concentrations of KCl solution on alphatoxin nanopore caused changeable and distinctive ionic current blocking events. Each of the three poisons has specific blocking characteristics. The differentiation procedure is based on calculating the remaining ratio of current and analyzing each mycotoxin's duration of stay within the single nanopore. Mycotoxins were identified at the nanomolar scale by just one alphatoxin nanopore, demonstrating that alphatoxin nanopores are a viable molecular instrument for the selective detection of mycotoxins in water-based solutions (da Silva et al. 2023).

6.3.1.3 Copper Oxide Nanoparticle-Based Immunosensor for Zearalenone

An automated sample pretreatment and signal alteration technology were used to create an immunosensor incorporating CuO nanoparticulate-assisted signal magnification.

CuO nanoparticulates, which are very stable and biocompatible, were utilized as carriers to immobilize anti-zearalenone antibodies. The produced CuO nanoparticulates—anti-zearalenone—can keep its capacity to recognize target toxins, while also acting as a signal source and carrier for automatic signal change as presented in Fig. 6.3. Because of the enormous amount of Cu^{2+} ions generated by CuO nanoparticulates in acidic circumstances, toxin recognition is indirectly converted to Cu^{2+} detection. Eventually, a multipurpose microporous plate sensor uses a fluorescent tripeptide molecule to identify Cu in an easy and large-scale fluorescence approach. A linear correlation was found between the intensity of the fluorescence and a logarithmic value of ZEN levels in an amount range of 16.0–1600.0 μg/kg. Furthermore, good precision with a substantial recovery yield of 99.2–104.9% was achieved, which was consistent with the findings gathered through LC-MS/MS of biologically polluted samples (Xuan et al. 2021).

6.4 Nanotechnology and Mycotoxin Eradication

Previous research has shown that real contact with mycotoxins causes increased oxidative damage in organisms (Awad et al. 2014; Hou et al. 2013). Because no effective cure exists, dietary supplements may reduce the toxic effects of mycotoxin, reduce tissue harm caused by oxidative damage, and enable the body to maintain a healthy immune system capable of eradicating the infection. For example, larger dosages of natural antioxidants such as tocopherol, selenium, and zinc are advised in the feeding of animals (Horky et al. 2013, 2016). In recent years, it was discovered that cerium, zinc oxide, silver, gold, selenium, or nanoparticulates formed from carbon had distinct antioxidant capabilities (Nelson et al. 2016; Skalickova et al. 2017).

When evaluating efficiency, it is critical to consider the toxicity of anti-mycotoxin nanomaterials, as well as the harmful effect of mycotoxins. One major issue with evaluating nanomaterial efficacy versus mycotoxin is the employment of simple, single-focus investigations. There was no pertinent proof identified on the

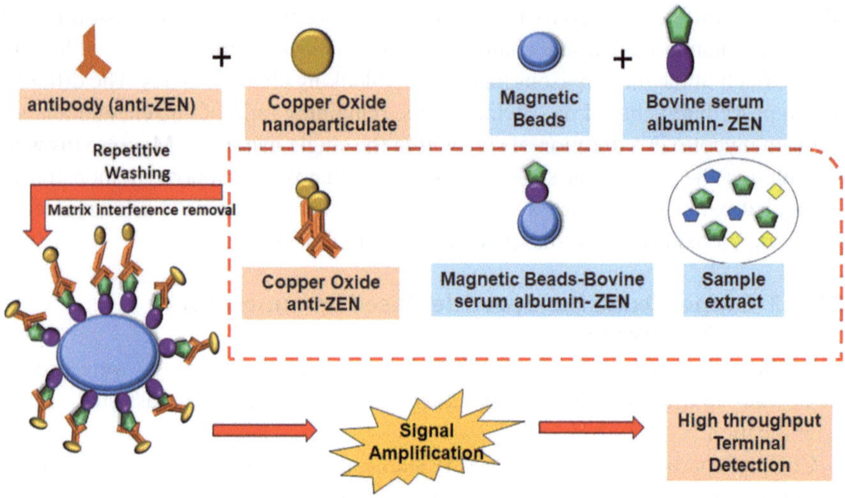

Fig. 6.3 Automated sample preparation and signal converter technology are designed for quick and high-throughput zearalenone antibody identification. (The figure is created using Microsoft PowerPoint version 2019)

relationship between nanoparticulate efficacy, toxicity, and the amount of mycotoxins absorbed from food or feed. Nanoparticle effectiveness against aflatoxin and ochratoxin ranges from 0.065 to 1000 μg/mg, based on the nanomaterial. According to published research, the harmful effects of the most frequent mycotoxins vary from 1 to 30 mg/kg of feed intake. In theory, just 30 mg of nanomaterials per 1 kg of substance feed would be enough to eradicate the hazardous impact (Alshannaq and Yu 2017). In terms of the toxicity of certain nanoparticles, Horky et al. discovered from published research that the safe dosage for mice is on average of 0.3–16,000 mg/kg (Horky et al. 2018). Overall, the findings of nanoadsorbent effectiveness against mycotoxins would have proved significantly more convincing if researchers had taken into account the actual feed dosage of mycotoxins. Graphene has been demonstrated to have a large surface area and a high affinity to mycotoxins. Polymeric nanomaterials have also received interest; they have the potential to replace adsorbents or contain any material, improving the organism's health. In light of these outcomes, this investigation provides fresh insights into possible further studies that might resolve the hurdles connected with using nanoscale to eliminate mycotoxin from crops (Horky et al. 2018).

6.5 Conclusion

Nanoparticulates have the potential use in biosensors for mycotoxin detection rather than eradication. Although the extraordinary qualities of nanoparticles have necessitated the establishment of this field of expertise, the integration of information from toxicological, nanotechnology, and agricultural methods is increasing. These findings provide credence to the concept of a single nanoparticle that might be used for both mycotoxin detection and removal, as well as a nanocage for food delivery. Additional studies should be conducted to examine comprehensive toxicity and efficiency testing, as it is necessary in other nanotechnology areas.

References

Al-Fakih AA (2014) Overview on the fungal metabolites involved in mycopathy. Open J Med Microbiol 04(01):38–63

Alshannaq A, Yu JH (2017) Occurrence, toxicity, and analysis of major mycotoxins in food. Int J Environ Res Public Health 14(6):632

Awad WA, Ghareeb K, Dadak A, Hess M, Böhm J (2014) Single and combined effects of deoxynivalenol mycotoxin and a microbial feed additive on lymphocyte DNA damage and oxidative stress in broiler chickens. PLoS One 9(1):e88028

Boruta T (2018) Uncovering the repertoire of fungal secondary metabolites: from Fleming's laboratory to the International Space Station. Bioengineered 9(1):12–16

Brown MH, Szczech GM, Purmalis BP (1976) Teratogenic and toxic effects of ochratoxin A in rats. Toxicol Appl Pharmacol 37(2):331–338

Brysse K, Oreskes N, O'Reilly J, Oppenheimer M (2013) Climate change prediction: erring on the side of least drama? Glob Environ Chang 23(1):327–337

Cai P, Zheng H, She J, Feng N, Zou H, Gu J et al (2020) Molecular mechanism of aflatoxin-induced hepatocellular carcinoma derived from a bioinformatics analysis. Toxins (Basel) 12(3):203

da Silva AAR, da Silva Júnior JJ, Cavalcanti MI d S, Machado DC, Medeiros PL, Rodrigues CG (2023) Alphatoxin nanopore detection of aflatoxin, ochratoxin and fumonisin in aqueous solution. Toxins (Basel) 15(3):183

Deepthi BV, Somashekaraiah R, Poornachandra Rao K, Deepa N, Dharanesha NK, Girish KS et al (2017) Lactobacillus plantarum MYS6 ameliorates fumonisin B1-induced hepatorenal damage in broilers. Front Microbiol:8

Eaton DL, Gallagher EP (1994) Mechanisms of aflatoxin carcinogenesis. Annu Rev Pharmacol Toxicol 34(1):135–172

El Khoury A, Atoui A (2010) Ochratoxin a: general overview and actual molecular status. Toxins (Basel) 2(4):461–493

Elling F, Nielsen JP, Lillehøj EB, Thomassen MS, Størmer FC (1985) Ochratoxin A-induced porcine nephropathy: enzyme and ultrastructure changes after short-term exposure. Toxicon 23(2):247–254

Eshelli M, Qader MM, Jambi EJ, Hursthouse AS, Rateb ME (2018) Current status and future opportunities of omics tools in mycotoxin research. Toxins (Basel) 10(11):433

Fan Y, Li J, Amin K, Yu H, Yang H, Guo Z et al (2023) Advances in aptamers, and application of mycotoxins detection: a review. Food Res Int 170:113022

Gelderblom WCA, Kriek NPJ, Marasas WFO, Thiel PG (1991) Toxicity and carcinogenicity of the *Fusanum monilzforine* metabolite, fumonisin B_1, in rats. Carcinogenesis 12(7):1247–1251

Gumprecht LA, Marcucci A, Weigel RM, Vesonder RF, Riley RT, Showker JL et al (1995) Effects of intravenous fumonisin B_1 in rabbits: nephrotoxicityxs and sphingolipid alterations. Nat Toxins 3(5):395–403

Gupta RC, Mostrom MS, Evans TJ (2018) Zearalenone. In: Veterinary toxicology. Elsevier, pp 1055–1063

Harrison LR, Colvin BM, Greene JT, Newman LE, Cole JR (1990) Pulmonary edema and hydrothorax in swine produced by fumonisin B1, a toxic metabolite of *Fusarium moniliforme*. J Vet Diagn Invest 2(3):217–221

Haschek WM, Motelin G, Ness DK, Harlin KS, Hall WF, Vesonder RF et al (1992) Characterization of fumonisin toxicity in orally and intravenously dosed swine. Mycopathologia 117(1–2):83–96
Hassen W, Abid-Essafi S, Achour A, Guezzah N, Zakhama A, Ellouz F et al (2004) Karyomegaly of tubular kidney cells in human chronic interstitial nephropathy in Tunisia: respective role of ochratoxin A and possible genetic predisposition. Hum Exp Toxicol 23(7):339–346
Henry SH, Bosch FX, Troxell TC, Bolger PM (1999) Reducing liver cancer – global control of aflatoxin. Science 286(5449):2453–2454
Horky P, Ruttkay-Nedecky B, Kremplova M, Krystofova O, Kensova R, Hynek D et al (2013) Effect of different doses of organically bound selenium on antioxidant status and levels of metal ions in postpartum sows. Int J Electrochem Sci 8(5):6162–6179
Horký P, Skladanka J, Nevrkla P, Slama P (2016) 16. Effect of diet supplemented with antioxidants (selenium, copper, vitamins E and C) on antioxidant status and ejaculate quality of breeding boars. Ann Anim Sci 16(2):521–532
Horky P, Skalickova S, Baholet D, Skladanka J (2018) Nanoparticles as a solution for eliminating the risk of mycotoxins. Nanomaterials (Basel) 8(9)
Hou YJ, Zhao YY, Xiong B, Cui XS, Kim NH, Xu YX et al (2013) Mycotoxin-containing diet causes oxidative stress in the mouse. PLoS One 8(3):e60374
Howard PC, Eppley RM, Stack ME, Warbritton A, Voss KA, Lorentzen RJ et al (2001) Fumonisin b1 carcinogenicity in a two-year feeding study using F344 rats and B6C3F1 mice. Environ Health Perspect 109(suppl 2):277–282
Júnior JJS, Soares TA, Pol-Fachin L, Machado DC, Rusu VH, Aguiar JP et al (2019) Alpha-hemolysin nanopore allows discrimination of the microcystins variants. RSC Adv 9(26):14683–14691
Kaittanis C, Santra S, Perez JM (2010) Emerging nanotechnology-based strategies for the identification of microbial pathogenesis. Adv Drug Deliv Rev 62(4–5):408–423
Kim SH, Singh MP, Sharma C, Kang SC (2018) Fumonisin B1 actuates oxidative stress-associated colonic damage via apoptosis and autophagy activation in murine model. J Biochem Mol Toxicol 32(7)
Marasas WF, Jaskiewicz K, Venter FS, Van Schalkwyk DJ (1988) Fusarium moniliforme contamination of maize in oesophageal cancer areas in Transkei. S Afr Med J 74(3):110–114
Montesano R, Hainaut P, Wild CP (1997) Hepatocellular carcinoma: from gene to public health. JNCI J Natl Cancer Inst 89(24):1844–1851
Naranjo-Ortiz MA, Gabaldón T (2020) Fungal evolution: cellular, genomic and metabolic complexity. Biol Rev 95(5):1198–1232
Nelson B, Johnson M, Walker M, Riley K, Sims C (2016) Antioxidant cerium oxide nanoparticles in biology and medicine. Antioxidants 5(2):15
Nile SH, Baskar V, Selvaraj D, Nile A, Xiao J, Kai G (2020) Nanotechnologies in food science: applications, recent trends, and future perspectives. Nanomicro Lett 12(1):45
Pan R, Li G, Liu S, Zhang X, Liu J, Su Z et al (2021) Emerging nanolabels-based immunoassays: principle and applications in food safety. TrAC Trends Anal Chem 145:116462
Pfohl-Leszkowicz A (2009) Ochratoxin a and aristolochic acid involvement in nephropathies and associated urothelial tract tumours. Arch Ind Hyg Toxicol 60(4):465–483
Riley RT, Merrill AH (2019) Ceramide synthase inhibition by fumonisins: a perfect storm of perturbed sphingolipid metabolism, signaling, and disease. J Lipid Res 60(7):1183–1189
Scott PM (2012) Recent research on fumonisins: a review. Food Addit Contam Part A 29(2):242–248
Singh J, Mehta A (2020) Rapid and sensitive detection of mycotoxins by advanced and emerging analytical methods: a review. Food Sci Nutr 8(5):2183–2204
Skalickova S, Milosavljevic V, Cihalova K, Horky P, Richtera L, Adam V (2017) Selenium nanoparticles as a nutritional supplement. Nutrition 33:83–90
Sojinrin T, Liu K, Wang K, Cui DJ, Byrne H, Curtin JF et al (2019) Developing gold nanoparticles-conjugated aflatoxin B1 antifungal strips. Int J Mol Sci 20(24):6260
Soleas GJ, Yan J, Goldberg DM (2001) Assay of ochratoxin A in wine and beer by high-pressure liquid chromatography photodiode array and gas chromatography mass selective detection. J Agric Food Chem 49(6):2733–2740

Xiong Y, Li W, Wen Q, Xu D, Ren J, Lin Q (2022) Aptamer-engineered nanomaterials to aid in mycotoxin determination. Food Control 135:108661

Xuan Z, Wu Y, Liu H, Li L, Ye J, Wang S (2021) Copper oxide nanoparticle-based immunosensor for zearalenone analysis by combining automated sample pre-processing and high-throughput terminal detection. Sensors 21(19):6538

Fungal Metabolites in Defense

7

Marija Ivanov, Dejan Stojković, and Marina Soković

Abstract

Fungal metabolites play a significant role in defensive systems and contribute to fungal survival, competition, and interaction with their surroundings. This chapter delves into the complex metabolic pathways that fungi rely on to generate a vast range of metabolites, including secondary metabolites such as mycotoxins, antibiotics, and signaling molecules that contribute to intricate defense mechanisms used by fungi. The investigation of fungal metabolites demonstrates their important influence on ecological dynamics, human health, and potential uses in agriculture and industry. The molecular processes behind metabolite synthesis and their interactions with other organisms are critical areas of study for innovative approaches for commercial and medicinal purposes. Understanding fungal metabolites will open new avenues for research and applications, confirming the vital role that fungi play in both human activity and natural ecosystem applications.

Keywords

Fungal secondary metabolism · Defense mechanisms · Metabolite biosynthesis · Molecular processes · Drug resistance · Fungal biology · Fungal infections

M. Ivanov · D. Stojković · M. Soković (✉)
Institute for Biological Research "Siniša Stanković", National Institute of the Republic of Serbia, University of Belgrade, Belgrade, Serbia
e-mail: mris@ibiss.bg.ac.rs

S. S. M. Soliman, M. I. Husseiny (eds.), *Metabolic Dynamics in Host-Microbe Interaction*, https://doi.org/10.1007/978-981-96-1305-2_7

7.1 Introduction

Fungal species are among the most common yet often overlooked human pathogens. As eukaryotic organisms, fungi share many metabolic and physiological pathways with humans, making the development of effective and selective antifungal therapeutics particularly challenging. The complexities of these organisms, and their ability to adapt to various environments, have enabled them to become significant threats to human health (Rodrigues and Nosanchuk 2020).

This chapter aims to shed light on the diverse metabolites and biological features that contribute to fungal defense mechanisms against various stressors. These defensive features have allowed microfungi to emerge as one of the leading causes of infections in the modern era. Fungal metabolites play a crucial role in protecting these organisms from the inhibitory effects of antifungal drugs, the immune responses of their hosts, and competition from other microorganisms in their environment.

Understanding the strategies fungi use to survive in harsh conditions and under various stressors is key to identifying new therapeutic targets. By unraveling the biochemical and molecular weapons fungi deploy, we can pave the way for the development of a new generation of antifungal compounds. These insights could lead to more effective treatments that overcome the current challenges in antifungal therapy, addressing the growing concern of fungal resistance and the limited options available for treating fungal infections.

7.2 Pathogenic Microfungi and Related Infections

Pathogenic fungi are the cause of more than 1.6 million death cases each year (Rokas 2022). Fungi can be the cause of both superficial and invasive infections.

Invasive fungal infections are associated with high rates of morbidity and mortality and often caused by species from the genera *Candida, Cryptococcus,* and *Aspergillus*. Moreover, invasive fungal infections are more frequently diagnosed in immunocompromised patients such as organ transplant and AIDS patients. A concern regarding invasive fungal infections is increasing mainly due to the fact that we are witnessing the increase in the incidence of antifungal-resistant strains, with significant percentage of fungal species being resistant to all four classes of antifungal drugs (Fang et al. 2023).

Superficial fungal infections can affect skin, hair, and nails and are frequently associated with dermatophytes, *Candida* spp., *Malassezia* spp., and dimorphic fungi such as *Sporothrix schenckii*. Skin fungal infections are one of the most common skin diseases that can be suppressed by conventional antifungal treatment, except for onychomycosis that is linked to high treatment failure and recurrence (Shen et al. 2020). Strains belonging to the *Candida* genus are frequently causing oral and vaginal infections. Oral candidiasis or thrush is the infection of oral mucosa linked to fungal overgrowth and invasion of superficial tissues and is the most frequent oral opportunistic infection in individuals with weakened immune systems

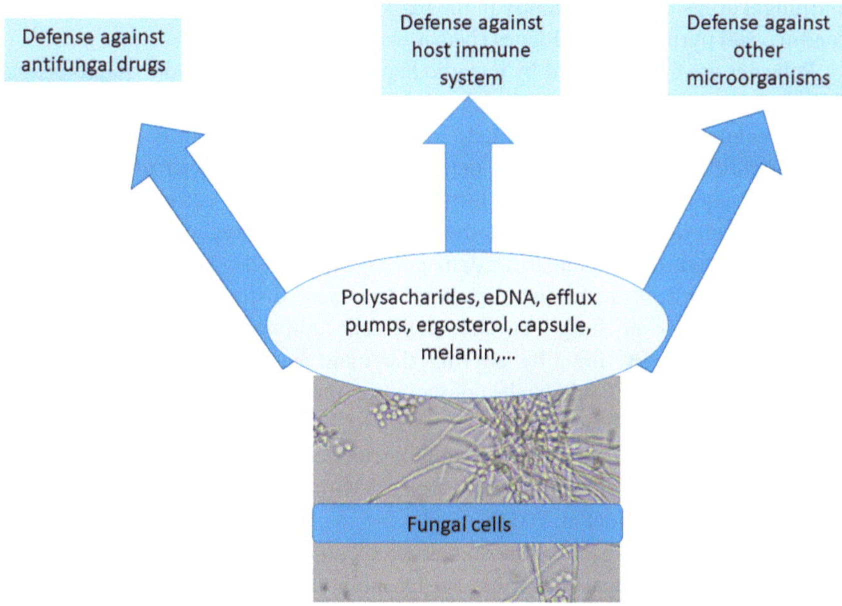

Fig. 7.1 Fungal defense metabolites such as polysaccharides, eDNA, efflux pumps, ergosterol, capsule, and melanin are targeted toward the antifungal drugs, host immune system, and other microorganisms

(Vila et al. 2020). Additionally, vulvovaginal candidiasis in the United States is the second most common vaginal infection, and it is recorded in 70–75% women during their life (Nyirjesy et al. 2022).

7.3 Fungal Defense Techniques

Fungal cells have developed a range of techniques to survive in the competing environments (Fig. 7.1). For example, *Candida albicans* is the species that is successful in causing human infections mostly due to its flexibility—this species has coevolved with its human host and become adapted to the extreme changes in the environment such as the presence of immune cells (Polke et al. 2015). Besides the immune system of the host, fungal cells were forced to gain strategies to fight antifungal treatment and other microorganisms that were competing for the same environment.

7.4 Fungal Defense Techniques Toward Antifungal Drugs

Antifungal drugs are a group of compounds with fungistatic or fungicidal effects toward pathogenic fungal species by specifically targeting some component or process within the fungal cell. According to their unique mode of antifungal action, antifungal drugs that are currently in the market belong to azoles, polyenes, echinocandins, and pyrimidine analogues (Nahar et al. 2024).

Polyenes are characterized by a cyclic heptane or polyketide core macrolactone ring with 20–40 carbon atoms including 3–8 conjugated double bond structure. The most common representatives include nystatin and amphotericin B. Their mechanism of antifungal action starts by incorporation into lipid bilayer of fungal cells, which subsequently leads to pore formation and leakage of cellular material and eventually cell death. Additionally, they induce reactive oxygen species (ROS) to burst in the fungal cell leading to the damage of essential cell components. Moreover, polyenes also adsorb membrane ergosterol and cause disruption in the membrane functioning. However, fungi can alter their cell components in order to become resistant to polyenes, first, by altering the main metabolite target, ergosterol. Ergosterol disturbance could be the consequence of alterations, deletions, and mutations in genes encoding proteins in the ergosterol biosynthesis pathway. Moreover, alterations such as the overexpression of molecular chaperones and modifications in the fungal cell wall composition are associated with polyene resistance in different fungal species (Rabaan et al. 2023).

Echinocandins, cyclic non-ribosomal hexapeptides equipped with a lipophilic side chain, are antifungal drugs that act by inhibiting synthesis of the fungal cell wall. They selectively inhibit activity of enzymes β-1-3-D-glucan synthetase, subsequently increasing susceptibility of fungal cells to osmotic lysis. This drug class includes products such as anidulafungin, caspofungin, and micafungin (Vitiello et al. 2023). Echinocandins are efficient against *Candida* species, with some exceptions due to naturally occurring polymorphism in the target enzyme glucan synthetase. Moreover, as a consequence of abundant echinocandin usage, we are witnessing an increase in the acquired echinocandin resistance, mainly due to alteration in *FKS* genes encoding for glucan synthetase (Arendrup and Perlin 2014).

Pyrimidine analogues, with flucytosine as the main representative, are a group of antifungal drugs that have structure similar to pyrimidine and act by inhibiting DNA and RNA synthesis. Resistance can occur via point mutations in FCY1, gene encoding for cytidine deaminase (Fisher et al. 2022).

Azoles block the ergosterol biosynthesis pathway by inhibiting enzyme lanosterol 14α-demethylase (Erg11) that catalyzes the removal of the 14α-methyl group from lanosterol. Azole resistance can occur due to alterations in enzymes involved in ergosterol biosynthesis, drug efflux, and chromosomal and mitochondrial abnormalities. The drug target enzyme, lanosterol demethylase, is one of the most frequent causes of drug resistance among *Candida* species due to amino acid substitutions in the active site (Cowen et al. 2015).

The forthcoming paragraph will provide further details regarding the metabolites that are most frequently responsible for antifungal resistance and are summarized in Table 7.1.

Table 7.1 Fungal metabolites involved in defense against antifungal drugs

Fungal metabolite	Role	Drug targeted	Mechanism of resistance	References
Ergosterol	Membrane stability	Polyenes, azoles	Alteration in ergosterol biosynthesis, structure, and amount in membrane	Rabaan et al. (2023)
Fungal efflux pumps	Transport	Azoles	Increased drug efflux	Cowen et al. (2015)
Transcription factor Upc2A	Regulates CDR1 transcription	Azoles	Increased efflux pump level	Lee et al. (2023)
Glucan synthetase	Cell wall synthesis	Echinocandins	Alterations in encoding genes leading to drug target modifications	Arendrup and Perlin (2014)
Protein kinase C	Stress response pathway	Echinocandins	Increased production of cell wall components	Healey and Perlin (2018)
Hsp90	Regulates cell wall stress	Echinocandins, azoles	Activation of calcineurin	Singh et al. (2009)
Cytosine permease	Transport	Pyrimidine analogues	No drug transport to active site	Delma et al. (2021)

7.4.1 Ergosterol

All the eukaryotic cell membranes consist of sterols; they have crucial roles in providing structural stability, fluidity, and permeability of the membranes. Moreover, sterols are involved in different processes such as the control of membrane-bound enzyme activities, cell transport, and cell cycle. In the fungal cells, ergosterol is the main sterol compound (Jordá and Puig 2020). Ergosterol and its biosynthesis pathway are at the same time a target for antifungal drugs and the reason for antifungal drug resistance.

Fungal defense against azole drugs is often based on the disturbance of the ergosterol biosynthesis pathway, mainly the changes in the key ergosterol biosynthesis enzyme—lanosterol 14 α-demethylase, cyp51. There are two isoforms of the enzyme, cyp51A and cyp51B, both encoded by the *CYP51* gene. Cyp51B is not essential; mutations in cyp51A are the most common reason for azole resistance, especially among *Aspergillus fumigatus* strains. Amino acid substitutions in cyp51A have been found among both clinical- and laboratory-isolated *A. fumigatus* strains (Chen et al. 2020). In clinical *Candida* strains, point mutations in hot spot regions of the gene *ERG11* have been observed in azole-resistant strains. These mutations reduce the ability of azoles to bind to their target and inhibit its function. Moreover, *Candida* species can overexpress azole target enzyme, for example, by gain-of-function mutations in the *UPC2* regulator of transcription (Bohner et al. 2022). Recently, it has been shown that upregulation or downregulation of *ERG11* gene could happen even upon exposure of fungal cells to some plant molecules such as flavonoids (Ivanov et al. 2021a), suggesting that there is a wide range of factors that might affect azole efficacy.

Understanding the mechanisms of azole resistance is crucial for developing effective antifungal strategies. Current research is focused on identifying new targets within the ergosterol biosynthesis pathway and exploring alternative pathways that could be exploited for therapeutic purposes. Advances in genomics and molecular biology have facilitated the discovery of novel resistance mechanisms and potential drug targets. In addition to drug development, there is an increasing interest in the role of combination therapies, where azoles are used in conjunction with other antifungal agents or adjuvants to overcome resistance. Such combination approaches can enhance the efficacy of existing drugs, reduce the likelihood of resistance development, and provide broader-spectrum antifungal activity.

The ongoing challenge of antifungal resistance highlights the need for continuous surveillance and monitoring of resistance patterns in clinical isolates. This effort will enable the timely identification of emerging resistance trends and the implementation of appropriate treatment strategies. Moreover, public health initiatives aimed at reducing the misuse and overuse of antifungal agents are essential in curbing the spread of resistant fungal strains.

7.4.2 Fungal Efflux Pumps

Among *C. albicans* strains, one of the most common causes of azole resistance is increased drug efflux. Efflux of azole drugs is happening via two main efflux pump families—major facilitator superfamily (MFS) and ATP-binding cassette (ABC) transporters (Holmes et al. 2016). Superfamily of MFS pumps is limited with the substrates, while on the contrary, ABC transporters have broader specificity and higher clinical significance (Cannon et al. 2009).

Proteins belonging to the ABC family possess two classes of domains—nucleotide-binding domains (NBDs, providing energy for substrate transport by hydrolyzing ATP) and transmembrane domains (TMDs, forming substrate channel) (Holmes et al. 2016). Proteins from the ABC superfamily can be grouped into five subfamilies: PDR (pleiotropic drug resistance), MDR (multidrug resistance), MRP (multidrug resistance-associated protein), RLI (RNase L inhibitor)/ALDP (adrenoleukodystrophy protein), and YEF3 (yeast elongation factor EF-3), and a sixth category for unaligned soluble ABC non-transporter proteins unrelated to the described fungal subfamilies (Fig. 7.2). The most common drug transporters Cdr1p and Cdr2p belong to the PDR family. Cdr1p is recognized as the main *Candida* drug transporter, with the overexpression of the *CDR1* gene being directly linked to the higher drug efflux in azole-resistant strains. Protein cdr1p is an integral plasma membrane protein, with typical NBDs and TMDs. Changes in the conformation are caused by ATP binding and its hydrolysis and are passed on from NBDs to TMDs, leading to events such as drug translocation and drug efflux (Prasad and Rawal 2014). Moreover, while the increased expression of the *CDR1* gene is being linked to the higher incidence of azole resistance, its downregulation is seen as a promising

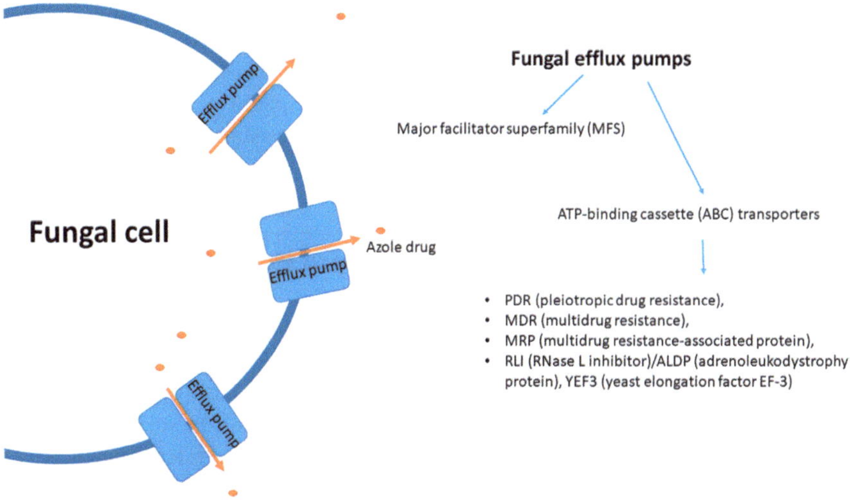

Fig. 7.2 Efflux pumps facilitate faster elimination of azole drugs from the fungal cells, thereby hindering their antifungal effect. There are two major families of fungal efflux pumps

antifungal mechanism (Ivanov et al. 2021b). The zinc cluster transcription factor Upc2A not only regulates *CDR1* expression but is also involved in the regulation of genes that are part of the ergosterol biosynthesis pathway, implicating a wide antifungal resistance role of this regulator (Lee et al. 2023).

Efflux pumps not only contribute to azole resistance but also play a role in the multidrug resistance (MDR) phenotype, where fungal cells become resistant to multiple classes of drugs. This characteristic makes infections caused by resistant *Candida* strains particularly challenging to treat. Understanding the regulation of efflux pump genes and the conditions that trigger their overexpression is crucial for developing strategies to overcome resistance.

Studies have highlighted the role of environmental factors and stress conditions in modulating efflux pump activity. For instance, exposure to sublethal concentrations of azole drugs, oxidative stress, and changes in pH levels can induce the expression of efflux pump genes, thereby contributing to adaptive resistance mechanisms in *C. albicans* (Ivanov et al. 2021b).

Understanding the complex interplay between efflux pump regulation, environmental stimuli, and drug resistance mechanisms is essential for the development of novel therapeutic interventions.

7.4.3 Other Fungal Metabolites Directed Toward Antifungal Drugs

Fungal species can become tolerant to the application of antifungal drugs by employing pathways related to the response to stress. Protein kinase C is part of the stress response pathway and has a role in the increased production of cell wall chitin and mannans. Moreover, it has been shown that cells with lower protein kinase C activity are more sensitive to echinocandin treatment (Healey and Perlin 2018). Moreover, Hsp90 is another part of the stress response pathway but it utilizes calcineurin as the mediator of signaling, and is involved in resistance to both echinocandins and azoles (Singh et al. 2009).

7.5 Fungal Defense Techniques Toward Host Immune Cells

Among the first steps in elimination of fungal pathogens from the host is their recognition by host immune cells. However, despite the diversity of mechanisms immune system utilizes, fungal cells have gained approaches that provide them protection against host recognition. Fungal pathogens have evolved various strategies to evade detection and destruction by the host immune system (Hernández-Chávez et al. 2017) (Fig. 7.3, Table 7.2).

7.5.1 Cell Wall Polysaccharides

The first part of fungal cells that is in contact with environment is the cell wall, and in the majority of fungal pathogens, it is the essential part for provoking host immune response. The composition of the fungal cell wall comprises (i) fibrillar elements that are situated closer to the cell membrane and have a role in mechanical stability, and (ii) hydrophobic and amorphic polymers that are located on the

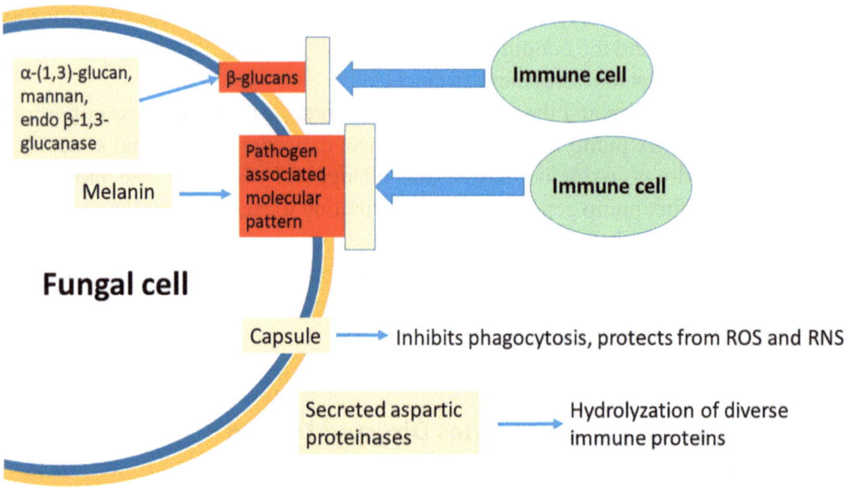

Fig. 7.3 Different parts of fungal cells are important for the fungal protection against human immune cells, including different metabolites involved in blocking β-glucan recognition, melanin, capsule, and hydrolytic enzymes

Table 7.2 Fungal metabolites with a role in defense against host immune system

Fungal metabolite	Role in defense	Immune cell targeted	References
α-(1,3)-Glucan	Blocking surface immune recognition molecules such as β-glucans	Phagocytes, macrophage	Camilli et al. (2018)
Mannan	Blocks β-1,3-glucan recognition	Phagocytes, macrophage	Hernández-Chávez et al. (2017)
Endo-β-1,3-glucanase	Blocks β-1,3-glucan recognition	Phagocytes, macrophage	Hernández-Chávez et al. (2017)
Melanin	Blocking pathogen-associated molecular patterns	Macrophage	Liu et al. (2021)
Polysaccharide capsule	Inhibiting phagocytosis, protecting from ROS and RNS, inducing apoptosis	Macrophage, T cells	Yang et al. (2022)
Rod A	Reduced production of neutrophil extracellular traps	Neutrophils	Bruns et al. (2010)
Secreted aspartic proteinases	Hydrolyze diverse immune proteins	Immune proteins	Naglik et al. (2003)
Urease	Modulates phagolysosomal pH	Macrophage	Fu et al. (2018)

outside and have a role in fungal interaction with the host and its immune system. Fungal cells have used their cell wall to develop traits that would enable them to go unrecognized by host immune cells, mainly by changing the amounts of some components in the cell wall or their location (Hernández-Chávez et al. 2017). Cell wall dominant components are polysaccharides, with the highest percentage of β-glucans. β-Glucans are essential pathogen-associated molecular pattern (PAMP) that initializes response of host immunity upon recognition by membrane-bound immune receptors. However, fungal pathogens have developed some features that could provide them protection against immune response triggered by β-glucans. One of such pathogens is *Histoplasma capsulatum.* This fungal species forms a less immunogenic layer on the cell surface, composed of α-(1,3)-glucan making the fungi unrecognizable to phagocytes. Moreover, another fungus, *Cryptococcus neoformans*, uses a polysaccharide capsule to cover the cell and hide the β-glucans and other immune triggers, which subsequently blocks phagocytosis and macrophage cytokine production. Additionally, in *C. albicans* strains we recognize yeast and the hyphal phase of growth, and the hyphal phase of growth is recognized as a pathogenic morphological form, with β-glucans not exposed on the cell surface during the hyphal growth (Camilli et al. 2018). Dectin-1, a pattern recognition receptor on host immune cells, has shown that it is not efficient against *C. albicans, C. guilliermondii, C. parapsilosis*, and *C. glabrata*, due to β-1,3-glucan shielding by an outer mannan layer. Mannan is protecting these yeast species from being recognized by host immune cells, since it prevents recognition of β-1,3-glucan by host immune receptors (Hernández-Chávez et al. 2017). Moreover, fungal cells have the enzyme, endo-β-1,3-glucanase, with the role in trimming the β-1,3-glucan segments that are on the fungal cell surface and subsequently blocking their recognition by host immune cells. This is a defense mechanism for *Histoplasma* cells (Hernández-Chávez et al. 2017).

7.5.2 Fungal Capsule

Glucuronoxylomannan (GXM) and galactoxylomannan (GalXM) are virulence factors and dominant polysaccharides of *Cryptococcus* capsule. GalXM displays significant immunomodulatory potential (Decote-Ricardo et al. 2019). Capsule polysaccharides have roles in phagocytosis, antigen presentation, and production of antibody and cytokines. Capsule provides escape from the host immunity by different pathways including inhibiting phagocytosis by macrophages. Capsule presence protects the antigens on the cell surface, and it downregulates levels of inflammatory cytokines. Even if the capsule gets taken by macrophages, it can still block their function since it protects fungal cell damage from reactive oxygen and reactive nitrogen species (ROS and RNS, respectively). Additionally, capsule polysaccharide GXM can induce apoptosis of macrophages, blocking subsequent immune protection pathway, while another polysaccharide component GalXM induces T cell apoptosis (Yang et al. 2022).

7.5.3 Melanin

Melanin is a pigment molecule with high molecular weight. It is found within fungal cell walls, providing mechanical stability, protecting against enzyme degradation and UV light. By comparing melanized and non-melanized fungal species, it was concluded that melanized ones are more resistant to phagocytosis (Hernández-Chávez et al. 2017). Melanin's role is to block the host proinflammatory cytokine response since the pigment hides fungal pathogen-associated molecular patterns in species such as *Aspergillus fumigatus*. It was shown that fungal conidia that were non-melanized were able to induce higher IL-6, IL-10, and TNF-α levels compared to the ones with melanin (Liu et al. 2021). Conidia of *A. fumigatus* can inhibit acidification of phagolysosomes of phagocytes that are part of immune defense, with the study suggesting one type of melanin that is found in *A. fumigatus* can inhibit activity of vacuolar ATPase (Thywißen et al. 2011).

Strains of *Cryptococcus* that produce melanin are proven to be more efficient in suppressing host immune defense and inflammatory response and are associated with fatal infections in the central nervous system. These melanin-producing strains could disturb endogenous antimicrobial proteins and are resistant to phagocytic cells. Additionally, *Cryptococcus* strains with melanin are able to inhibit T cell immunity and macrophage production of tumor necrosis factor α and specific lymphocytes (Tajima et al. 2019).

7.5.4 Proteins and Enzymes

Hydrophobin RodA is a surface protein on *Aspergillus fumigatus* conidia that is making the fungi immunologically inert. This protein can reduce formation of neutrophil extracellular traps (NETs) in neutrophils. These nuclear NETs are made of DNA and fungicidal proteins and have the role to prevent *Aspergillus* spreading (Bruns et al. 2010).

Secreted aspartic proteases have different functions during the infective process including hydrolyzation of cells and molecules of the host immune system that helps fungi to survive antimicrobial attack by the host (Naglik et al. 2003). Another fungal enzyme, urease from *Cryptococcus neoformans*, has the role in protecting fungal cells from the elimination by host immune cells. By modulating pH in phagolysosomes, ureases help *C. neoformans* to both survive within and exit from macrophages, which subsequently might lead to the dissemination of the fungus to brain via the macrophage-dependent transport pathway (Fu et al. 2018).

7.6 Biofilm Formation as a Fungal Defense Strategy

Biofilms have a role in protection against both antimicrobial drugs and host immune system (Table 7.3), with increasing research in the area of novel antibiofilm control agents that could be used to treat infections (Ivanov et al. 2022; Carević et al. 2023). Biofilms are formed by cells grouped together and surrounded by polymeric matrix consisting of elements such as polysaccharides, proteins, nucleic acids, and lipids (Ćirić et al. 2019). They can be formed by fungal species, but in nature, we can frequently find polymicrobial biofilms, such as the ones formed between *C. albicans* and *Staphylococcus aureus* (Carević et al. 2022).

One of the resistance mechanisms in biofilms is the formation of polysaccharide net between fungal cells. This sugar complex is in *Candida* species mainly composed of mannan and glucan and is associated with drug resistance (Dominguez et al. 2019). In *Candida auris,* mannan-glucan biofilm matrix sequesters almost 70% of the accessible triazole antifungal drugs (Dominguez et al. 2019).

Table 7.3 Some of the biofilm metabolites are involved in fungal defense

Biofilm Metabolite	Role	References
Polysaccharides	Drug sequestration	Dominguez et al. (2019)
eDNA	Possibly limiting drug penetration, structural maintenance	Pierce et al. (2017)
Tyrosol	Quorum sensing molecule	Kaur and Nobile (2023)
Glucanases	Delivery of glucan to the extracellular matrix	Taff et al. (2012)
Extracellular vesicles	Drug sequestration	Kaur and Nobile (2023)

The extracellular matrix is composed of β-1,3-glucan, which can bind amphotericin B, while on the other hand, it lacks being linked to the higher sensitivity of fungal cells to fluconazole and amphotericin B. Biofilm formation is also accompanied by some changes in the molecular level such as upregulation of *CDR* and *MDR* drug transporter genes. These changes are being linked with the resistance at the onset of biofilm forming procedure. However, in the later steps biofilm resistance is being linked to the changes in the composition of sterols (Cavalheiro and Teixeira 2018). Glucans of the extracellular matrix are essential for the biofilm structure though the role of the glucan modification enzymes is crucial for biofilms. These enzymes have the role in delivering glucans from the fungal cells to the extracellular biofilm matrix. It was shown that mutants without these enzymes are more susceptible to fluconazole treatment during the biofilm growth phase (Taff et al. 2012).

It has been previously observed that cell membranes of cells in biofilm have lower levels of ergosterol than the planktonic cells. Moreover, fungal cells that are in the mature phase of biofilm growth have around one-half of the early-phase cell ergosterol content, implying that mature biofilms use lack of ergosterol to avoid damage caused by the antifungal therapeutics that target ergosterol (Taff et al. 2013).

Another essential component of fungal biofilms is extracellular DNA (eDNA). It is phased-dependently released into biofilms, and its importance was demonstrated upon the addition of DNase, since this enzyme disrupted the integrity of biofilm (Rajendran et al. 2013). eDNA amount reaches up to 5% of matrix weight, and it is secreted from the cells via lysis, quorum sensing, and vesicle excretion. In. *C. albicans*, it contains mainly random non-coding sequence, but is important as an essential structural integrity provider, and it is speculated that it contributes to biofilm antifungal resistance by limiting penetration of the drug (Pierce et al. 2017).

Candida cells have quorum sensing (QS) molecules with different functions. Unlike farnesol, a QS molecule that inhibits hyphal growth and biofilm formation, tyrosol is actively involved in the process of adherence and hyphal growth during biofilm formation. Additionally, incubation of *Candida parapsilosis* cells with tyrosol induced the expression of their genes encoding for drug efflux pumps, *CDR1* and *MDR1* (Kaur and Nobile 2023). Moreover, *Candida* cells in biofilm produce extracellular vesicles; they transport components of the extracellular matrix. These vesicles are essential for biofilm resistance—it was shown that the mutants lacking these transporters have higher susceptibility to the antifungal treatment (Kaur and Nobile 2023).

7.7 Fungal Metabolites Involved in Inhibition of Competing Microorganisms

Among the many secondary metabolites that fungi make are those that suppress competing microorganisms, which is one of the key functions that fungi play in their ecological interactions. These metabolites can limit bacterial and fungal growth, giving the generating fungus a competitive advantage. Examples of these metabolites include antibiotics, mycotoxins, and other bioactive chemicals.

Antibiotics are one of the best-known classes of fungal metabolites having antibacterial properties. As a result of the discovery of penicillin produced by mold *Penicillium notatum*, the antibiotic era began. Antibiotics produced from fungi have been the subject of ongoing research in recent times. For example, newer derivatives of penicillin and other fungal antibiotics have been found and described for their distinct modes of action and their applications in medicine (Sułkowska-Ziaja et al. 2023). Another significant group of fungal metabolites is the cephalosporins, produced by *Cephalosporium* species. New cephalosporin derivatives with enhanced activity against resistant bacterial strains have been developed, showing continued relevance in clinical settings (Ye and Chen 2023).

Mycotoxins are powerfully bioactive secondary metabolites that are also produced by fungi. Because *Aspergillus* species produce aflatoxins, which are poisonous and carcinogenic to both people and animals, they are extensively researched. The formation of aflatoxins and their interactions with host organisms have been better understood (Tiwari et al. 2023). Mycotoxins, despite being harmful, contribute to the ecological fitness of the fungi that produce them by preventing rival microbes from growing in their native environments. Moreover, some fungi produce compounds that specifically target other fungi. For example, gliotoxin, produced by *Aspergillus fumigatus*, exhibits strong antifungal activity, and can induce apoptosis in competing fungal cells. New findings have elucidated the complex mechanisms of gliotoxin action and its dual role in fungal pathogenicity and immune modulation (Bérdy 2012).

In addition to antibiotics and mycotoxins, fungi produce a variety of other bioactive compounds with antimicrobial properties. For instance, the lipopeptides produced by *Trichoderma* species have shown strong antibacterial and antifungal activities, making them promising candidates for biocontrol applications in agriculture (Asad 2022). The recent advances in understanding their biosynthesis and mode of action have opened new avenues for their application in crop protection.

The production of these antimicrobial metabolites is often regulated by complex genetic and environmental factors. Advances in genomics and metabolomics have provided new insights into the diversity and function of fungal secondary metabolites. For example, high-throughput sequencing and metabolite profiling have enabled the discovery of novel antimicrobial compounds and the elucidation of their biosynthetic pathways (Keller et al. 2005). Fungal secondary metabolites play a vital role in inhibiting competing microorganisms, thereby enhancing the ecological fitness of the producing fungi. Furthermore, a wide variety of edible fungi, containing different groups of hydrophilic and lipophilic compounds (Stojković et al. 2013), have shown that these species possess bioactive features against plant fungal pathogens and food contaminants (Vieira et al. 2016; Stojković et al. 2014) and human fungal pathogens (Stojković et al. 2017; Petrović et al. 2013). All these compounds have significant implications for human health, agriculture, and industry, highlighting the importance of continued research in fungal biology and natural product chemistry.

7.8 Concluding Remarks

We have explored the various and intricate functions of fungal metabolites in defensive systems in this chapter. Fungi generate a vast range of metabolites via complex metabolic pathways, which are essential to their ability to survive, compete, and interact with their surroundings. These metabolites include secondary metabolites that contribute to the intricate defense mechanisms used by fungi, such as mycotoxins, antibiotics, and signaling molecules. The investigation of fungal metabolites demonstrates their important influence on ecological dynamics, as well as their potential uses in a number of different contexts. For example, the antifungal and antibacterial substances that fungi create have played a significant role in the development of agricultural products and medications, demonstrating their usefulness in preventing infections and improving crop protection. Furthermore, fungi's significance as a source of bioactive chemicals with possible therapeutic advantages is highlighted by their capacity to create novel metabolites with distinctive architectures.

It is becoming clear that the metabolites of fungus provide interesting opportunities for environmental management and medicinal discovery as the study into their enormous chemical arsenal progresses. It is clear from the complex interplay between fungal defensive mechanisms and their ecological roles that more research is required to completely comprehend the range and potential of fungal metabolites. To develop creative methods for using these substances for commercial and medicinal purposes, future studies should clarify the molecular processes behind metabolite synthesis and their interactions with other organisms. Fungal metabolites are an abundant and varied defense mechanism that affects not only the fungal kingdom but also the environment and human health. Understanding these metabolites will surely open new avenues for research and applications as our understanding and technical prowess grow, confirming the vital role that fungi play in both human activities and natural ecosystems.

Acknowledgments This research was supported by the Ministry of Science, Technological Development and Innovation of the Republic of Serbia, Grant No. 451-03-66/2024-03/200007.

References

Arendrup MC, Perlin DS (2014) Echinocandin resistance: an emerging clinical problem? Curr Opin Infect Dis 27(6):484.

Asad SA (2022) Mechanisms of action and biocontrol potential of Trichoderma against fungal plant diseases – a review. Ecol Complex 49:100978

Bérdy J (2012) Thoughts and facts about antibiotics: where we are now and where we are heading. J Antibiot 65(8):385–395.

Bohner F, Papp C, Gácser A (2022) The effect of antifungal resistance development on the virulence of Candida species. FEMS Yeast Res 22(1):1–10.

Bruns S, Kniemeyer O, Hasenberg M, Aimanianda V, Nietzsche S, Thywien A et al (2010) Production of extracellular traps against Aspergillus fumigatus in vitro and in infected lung tis-

sue is dependent on invading neutrophils and influenced by hydrophobin RodA. PLoS Pathog 6(4):1–18.
Camilli G, Tabouret G, Quintin J (2018) The complexity of fungal β-glucan in health and disease: effects on the mononuclear phagocyte system. Front Immunol 9:343771
Cannon RD, Lamping E, Holmes AR, Niimi K, Baret PV, Keniya MV et al (2009) Efflux-mediated antifungal drug resistance. Clin Microbiol Rev 22(2):291–321.
Carević T, Kostić M, Nikolić B, Stojković D, Soković M, Ivanov M (2022) Hesperetin-between the ability to diminish mono- and polymicrobial biofilms and toxicity. Molecules 27(20).
Carević T, Stojković D, Ivanov M (2023) Plant flavonoids as reservoirs of therapeutics against microbial virulence traits: a comprehensive review update. Curr Pharm Des 29(12):914–927.
Cavalheiro M, Teixeira MC (2018) Candida biofilms: threats, challenges, and promising strategies. Front Med (Lausanne) 5:28.
Chen P, Liu J, Zeng M, Sang H (2020) Exploring the molecular mechanism of azole resistance in Aspergillus fumigatus. J Mycol Med 30(1):100915
Ćirić AD, Petrović JD, Glamočlija JM, Smiljković MS, Nikolić MM, Stojković DS et al (2019) Natural products as biofilm formation antagonists and regulators of quorum sensing functions: a comprehensive review update and future trends. S Afr J Bot 120:65–80
Cowen LE, Sanglard D, Howard SJ, Rogers PD, Perlin DS (2015) Mechanisms of antifungal drug resistance. Cold Spring Harb Perspect Med 5(7):a019752.
Decote-Ricardo D, LaRocque-de-Freitas IF, Rocha JDB, Nascimento DO, Nunes MP, Morrot A et al (2019) Immunomodulatory role of capsular polysaccharides constituents of Cryptococcus neoformans. Front Med (Lausanne) 6:129.
Delma FZ, Al-Hatmi AMS, Brüggemann RJM, Melchers WJG, de Hoog S, Verweij PE et al (2021) Molecular mechanisms of 5-fluorocytosine resistance in yeasts and filamentous fungi. J Fungi 7(11):909.
Dominguez EG, Zarnowski R, Choy HL, Zhao M, Sanchez H, Nett JE et al (2019) Conserved role for biofilm matrix polysaccharides in Candida auris drug resistance. mSphere 2;4(1):e00680-18.
Fang W, Wu J, Cheng M, Zhu X, Du M, Chen C et al (2023) Diagnosis of invasive fungal infections: challenges and recent developments. J Biomed Sci 30(1):1–35.
Fisher MC, Alastruey-Izquierdo A, Berman J, Bicanic T, Bignell EM, Bowyer P et al (2022) Tackling the emerging threat of antifungal resistance to human health. Nat Rev Microbiol 20(9):557–571.
Fu MS, Coelho C, De Leon-Rodriguez CM, Rossi DCP, Camacho E, Jung EH et al (2018) Cryptococcus neoformans urease affects the outcome of intracellular pathogenesis by modulating phagolysosomal pH. PLoS Pathog 14(6):e1007144.
Healey KR, Perlin DS (2018) Fungal resistance to echinocandins and the MDR phenomenon in Candida glabrata. J Fungi (Basel) 4(3):105.
Hernández-Chávez MJ, Pérez-García LA, Niño-Vega GA, Mora-Montes HM (2017) Fungal strategies to evade the host immune recognition. J Fungi 3(4):51.
Holmes AR, Cardno TS, Strouse JJ, Ivnitski-Steele I, Keniya MV, Lackovic K et al (2016) Targeting efflux pumps to overcome antifungal drug resistance. Future Med Chem 8(12):1485–1501.
Ivanov M, Kannan A, Stojković DS, Glamočlija J, Calhelha RC, Ferreira ICFR et al (2021a) Flavones, flavonols, and glycosylated derivatives—impact on Candida albicans growth and virulence, expression of cdr1 and erg11, cytotoxicity. Pharmaceuticals 14(1):1–12.
Ivanov M, Kannan A, Stojković DS, Glamočlija J, Calhelha RC, Ferreira ICFR et al (2021b) Camphor and eucalyptol-anticandidal spectrum, antivirulence effect, efflux pumps interference and cytotoxicity. Int J Mol Sci 22(2):1–14.
Ivanov M, Ćirić A, Stojković D (2022) Emerging antifungal targets and strategies. Int J Mol Sci 23(5):2756.
Jordá T, Puig S (2020) Regulation of ergosterol biosynthesis in Saccharomyces cerevisiae. Genes (Basel) 11(7):1–18.
Kaur J, Nobile CJ (2023) Antifungal drug-resistance mechanisms in Candida biofilms. Curr Opin Microbiol 71:102237.

Keller NP, Turner G, Bennett JW (2005) Fungal secondary metabolism — from biochemistry to genomics. Nat Rev Microbiol 3(12):937–947.

Lee Y, Robbins N, Cowen LE (2023) Molecular mechanisms governing antifungal drug resistance. NPJ Antimicrob Resist 1(1):5.

Liu S, Youngchim S, Zamith-Miranda D, Nosanchuk JD (2021) Fungal melanin and the mammalian immune system. J Fungi (Basel) 7(4):264.

Naglik JR, Challacombe SJ, Hube B (2003) Candida albicans secreted aspartyl proteinases in virulence and pathogenesis. Microbiol Mol Biol Rev 67(3):400.

Nahar D, Mohite P, Lonkar A, Chidrawar VR, Dodiya R, Uddin MJ et al (2024) An insight into new strategies and targets to combat antifungal resistance: a comprehensive review. Eur J Med Chem Rep 10:100120

Nyirjesy P, Brookhart C, Lazenby G, Schwebke J, Sobel JD (2022) Vulvovaginal candidiasis: a review of the evidence for the 2021 centers for disease control and prevention of sexually transmitted infections treatment guidelines. Clin Infect Dis 74(Supplement_2):S162–S168.

Petrović J, Glamočlija J, Stojković DS, Ćirić A, Nikolić M, Bukvički D et al (2013) Laetiporus sulphureus, edible mushroom from Serbia: investigation on volatile compounds, in vitro antimicrobial activity and in situ control of Aspergillus flavus in tomato paste. Food Chem Toxicol 59:297–302.

Pierce CG, Vila T, Romo JA, Montelongo-Jauregui D, Wall G, Ramasubramanian A et al (2017) The Candida albicans biofilm matrix: composition, structure and function. J Fungi 3(1):14.

Polke M, Hube B, Jacobsen ID (2015) Candida Survival Strategies. Adv Appl Microbiol 91:139–235

Prasad R, Rawal MK (2014) Efflux pump proteins in antifungal resistance. Front Pharmacol 29;5:202.

Rabaan AA, Sulaiman T, Al-Ahmed SH, Buhaliqah ZA, Buhaliqah AA, AlYuosof B et al (2023) Potential strategies to control the risk of antifungal resistance in humans: a comprehensive review. Antibiotics (Basel) 18;12(3):608.

Rajendran R, Williams C, Lappin DF, Millington O, Martins M, Ramage G (2013) Extracellular DNA release acts as an antifungal resistance mechanism in mature Aspergillus fumigatus biofilms. Eukaryot Cell 12(3):420.

Rodrigues ML, Nosanchuk JD (2020) Fungal diseases as neglected pathogens: a wake-up call to public health officials. PLoS Negl Trop Dis 14(2):e0007964.

Rokas A (2022) Evolution of the human pathogenic lifestyle in fungi. Nat Microbiol 7(5):607–619.

Shen JJ, Jemec GBE, Arendrup MC, Saunte DML (2020) Photodynamic therapy treatment of superficial fungal infections: a systematic review. Photodiagnosis Photodyn Ther 31:101774.

Singh SD, Robbins N, Zaas AK, Schell WA, Perfect JR, Cowen LE (2009) Hsp90 governs echinocandin resistance in the pathogenic yeast Candida albicans via calcineurin. PLoS Pathog 5(7):e1000532

Stojković D, Reis FS, Barros L, Glamočlija J, Ćirić A, van Griensven LJID et al (2013) Nutrients and non-nutrients composition and bioactivity of wild and cultivated Coprinus comatus (O.F.Müll.) Pers. Food Chem Toxicol 59:289–296.

Stojković DS, Barros L, Calhelha RC, Glamočlija J, Ćirić A, Van Griensven LJLD et al (2014) A detailed comparative study between chemical and bioactive properties of Ganoderma lucidum from different origins. Int J Food Sci Nutr 65(1):42–47.

Stojković DS, Kovačević-Grujičić N, Reis FS, Davidović S, Barros L, Popović J et al (2017) Chemical composition of the mushroom Meripilus giganteus Karst. and bioactive properties of its methanolic extract. LWT Food Sci Technol 79:454–462

Sułkowska-Ziaja K, Trepa M, Olechowska-Jarząb A, Nowak P, Ziaja M, Kała K et al (2023) Natural compounds of fungal origin with antimicrobial activity—potential cosmetics applications. Pharmaceuticals 16(9):1200.

Taff HT, Nett JE, Zarnowski R, Ross KM, Sanchez H, Cain MT et al (2012) A Candida biofilm-induced pathway for matrix glucan delivery: implications for drug resistance. PLoS Pathog 8(8):e1002848.

Taff HT, Mitchell KF, Edward JA, Andes DR (2013) Mechanisms of Candida biofilm drug resistance. Future Microbiol 8(10):1325–1337.
Tajima K, Yamanaka D, Ishibashi K-I, Adachi Y, Ohno N (2019) Solubilized melanin suppresses macrophage function. FEBS Open Bio 9(4):791.
Thywißen A, Heinekamp T, Dahse HM, Schmaler-Ripcke J, Nietzsche S, Zipfel PF et al (2011) Conidial dihydroxynaphthalene melanin of the human pathogenic fungus Aspergillus fumigatus interferes with the host endocytosis pathway. Front Microbiol 3:2:96.
Tiwari S, Singh BK, Dubey NK (2023) Aflatoxins in food systems: recent advances in toxicology, biosynthesis, regulation and mitigation through green nanoformulations. J Sci Food Agric 103(4):1621–1630.
Vieira V, Fernandes Â, Barros L, Glamočlija J, Ćirić A, Stojković D et al (2016) Wild Morchella conica Pers. from different origins: a comparative study of nutritional and bioactive properties. J Sci Food Agric 96(1):90–98.
Vila T, Sultan AS, Montelongo-Jauregui D, Jabra-Rizk MA (2020) Oral candidiasis: a disease of opportunity. J Fungi. 6:15.
Vitiello A, Ferrara F, Boccellino M, Ponzo A, Cimmino C, Comberiati E et al (2023) Antifungal drug resistance: an emergent health threat. Biomedicines 11(4):1063.
Yang C, Huang Y, Zhou Y, Zang X, Deng H, Liu Y et al (2022) Cryptococcus escapes host immunity: what do we know? Front Cell Infect Microbiol 13;12:1041036.
Ye J, Chen X (2023) Current promising strategies against antibiotic-resistant bacterial infections. Antibiotics 12(1):67.

Role of Metabolism in Supporting Immune Responses

8

Shriraj Susarla, Muhammed I. Erbay, Amit Johanis, Mohamed Abuelazm, and Ahmed K. Ghanem

Abstract

Immunometabolism is an emerging field that delves into the complex relationship between the immune system and metabolic processes. This chapter describes the fundamental aspects of immunometabolism, highlighting its significance in explaining the complex mechanisms underlying immune system function and its potential for therapeutic interventions. The long-held belief that immunity and metabolism are distinct systems with separate roles has been challenged in recent years, with growing evidence suggesting a strong link between the two systems. We discuss the bioenergetically demanding nature of immune processes and their distinctive metabolic pathways. Furthermore, we examine signaling and transcription pathways that regulate the activation of both the immune system and the metabolic pathways, underscoring the importance of these common pathways in primary immune regulation. We further explain how dysregulation of normal metabolic pathways contributes to cancer, autoimmunity, and other disease states, and possible treatment approaches based on metabolic intervention in immune cells. Overall, this chapter provides a comprehensive overview of

S. Susarla (✉)
Keck School of Medicine at USC, Los Angeles, CA, USA
e-mail: shriraj.susarla@lundquist.org

M. I. Erbay
Cerrahpasa Medical School, Istanbul, Turkey

A. Johanis
Creighton University, School of Medicine, Phoenix, AZ, USA

M. Abuelazm
Tanta University, School of Medicine, Tanta, Egypt

A. K. Ghanem
The Loma Linda University Medical Center- Murrieta, Murrieta, CA, USA

S. S. M. Soliman, M. I. Husseiny (eds.), *Metabolic Dynamics in Host-Microbe Interaction*, https://doi.org/10.1007/978-981-96-1305-2_8

the emerging field of immunometabolism, highlighting the intricate interplay between immunity and metabolism and its potential to enhance our understanding of the immune system and develop therapeutic applications for clinical intervention in the future.

Keywords

Immunometabolism · Immune · Metabolism · Regulation · Pathways · Dysregulation · Autoimmune

8.1 Introduction

The belief that immunity and metabolism are two distinct systems with distinct roles has persisted for many years. In recent years, it has become evident that there is indeed a strong link between the two systems, prompting the emergence of the study of immunometabolism. Understanding the interactions between metabolic pathways and their critical significance in immune system function is the focus of immunometabolism (Chavakis 2021).

The immune system is a sophisticated and dynamic system that has persisted throughout most of evolutionary history. It allows organisms to respond to a host of local threats, from pathogens and allergens to environmental toxins and cancer cells. It consists of a nonspecific and rapid innate branch and a discerning and measured adaptive branch, both of which cooperate to generate a strong immune response and guard against all manner of insults. Immune processes and their distinct cells are bioenergetically demanding, and thus have distinctive phenotypes of metabolic activity (Yatim and Lakkis 2015).

Metabolic processes maintain homeostasis and the internal environment of all cells through various mechanisms. Numerous pathways maintain a balance between anabolic and catabolic reactions—these may be opposing or synergistic and include all major macromolecules. These reactions are driven by physiological cues and hormonal regulation, and catalyzed by key enzymes and cofactors, including essential vitamins.

Metabolism and immune activity are intrinsically linked, and each of these processes plays a role in supporting immune function. Immune processes require large amounts of energy to support the clonal proliferation of immune cells and the generation of inflammasomes, and thus serve as a form of regulation for metabolic processes. Several specific metabolic pathways are key for normal immune function, such as the respiratory burst in neutrophilic phagocytosis. Furthermore, several signaling and transcription pathways regulate the activation of the immune system, and these may overlap with metabolic pathways and substrates. The study of such interactions is part of the emerging concept called immunometabolism (Chavakis 2021; Hotamisligil 2017).

Various immune cell lines interact differently with metabolic pathways. A constellation of context-dependent forms of immune activation exists, resulting in a plethora of cell fates and metabolic phenotypes. Knowledge of these interactions may provide a unique opportunity for immunomodulation and even therapeutic intervention and is a promising avenue for clinical intervention in the future. In this chapter, we will describe in detail the intersections between immune function and metabolism, highlighting specific interactions across each cell type and response scenario.

8.2 Overview of Immune Cell Types

The immune system is a complex network of substances, specialized cells, and defense mechanisms designed to protect organisms from all manner of external insults. It consists of both the innate and adaptive immune systems, which play differing but essential roles in this response. The innate immune system is responsible for the initial reaction to foreign pathogens and produces a response that is often rapid, nonspecific, and consistent regardless of the properties of the invading substance. On the other hand, the adaptive immune system mounts a slower and much more deliberate defense, consisting of specialized cells that generate circulating antibodies designed to target and isolate the immunogen. Unlike innate immunity, adaptive immunity results in the generation of memory cells, which allows resistance to persist through to future responses to similar pathogens. However, the distinct advantage of innate immunity is its speed, which allows it to provide considerable protection during the lengthy period of clonal proliferation required for adaptive immune function, which may last up to 5 days (Chaplin 2010).

It was initially thought that innate and adaptive immunity worked independently of each other, with adaptive immune cells picking up where the innate immune cells left off after a few days. However, our current understanding of immunity now suggests that these systems are truly synergistic, with overlap between many cellular enzymes and processes at key points in time during an immune response. Ultimately, both innate and adaptive immunity are essential to producing a robust immune response, and in this section, we will discuss each of the specialized cell lines that comprise these systems and the interface between them (Chaplin 2010).

8.2.1 Innate Immune System

Of the two major divisions of the immune system, the innate branch is the more ancient and evolutionarily significant. Adaptive immunity is a late evolutionary adaptation found only in vertebrate organisms, meaning that the innate immune system serves as the primary defense mechanism against environmental threats for these species. Among vertebrates, key components of the innate immune system have been conserved, including external protective structures, antimicrobial peptides, antibody and complement systems, and certain myeloid and lymphoid phagocytic cells. The persistence of these traits highlights their effectiveness and utility in the rapid response to pathogens (Riera Romo et al. 2016).

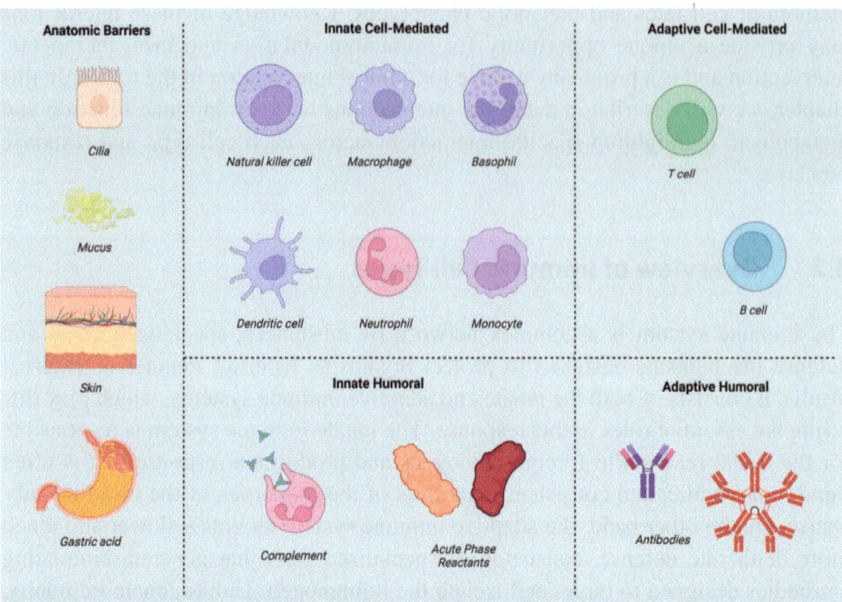

Fig. 8.1 Divisions of the immune system. It is first split into anatomic and physiologic branches; physiologic branches of both innate and adaptive immunity contain cell-mediated and humoral components. (Created in BioRender. Erbay, M. (2024) BioRender.com/t37s752)

The innate immune system is divided into anatomic and physiologic components. The anatomic component includes natural chemical and temperature-based barriers to infection and disease such as skin, mucous membranes, cilia, gastric acid, and enzymes in secretions such as sweat and saliva. The physiologic division contains both cell-mediated and humoral components, which respond directly to pathogens. Hematopoietic cells involved in the cellular response include natural killer (NK) cells, macrophages, and neutrophils; humoral response mechanisms include complement and acute-phase reactants such as C-reactive protein (Fig. 8.1) (Turvey and Broide 2010).

8.2.1.1 Pattern Recognition and Activation

The innate immune system operates based on a series of pattern recognition mechanisms that allow it to detect invading microorganisms. One of the main mechanisms is detecting pathogen-associated molecular patterns, or PAMPs. PAMPs are structures specific to microbes, such as lipopolysaccharide (LPS) in bacterial cell membranes, or flagellin typically found in flagella. These PAMPs bind to specific receptors such as toll-like receptors (TLRs) or nucleotide-binding oligomerization domain-like receptors (NLRs) and result in the activation of the innate immune response. TLRs are specialized to detect extracellular or endocytosed PAMPs, whereas NLRs monitor the intracellular environment (Turvey and Broide 2010; Aristizábal and González 2013).

Receptors in the innate immune system are germline-encoded, which means they are constitutively active and can be produced in large numbers before the onset of an injury or insult, thus generating a much quicker response. The tradeoff for this rapid response is an inability to provide a tailored response to pathogen antigens like the adaptive immune system would; nevertheless, there are nearly a dozen TLR variants and over double this amount of NLRs. Each member of these receptor families binds to a specific type of PAMP, allowing the system to boast a wide range of reactivity to many microbes, including most Gram-positive and Gram-negative bacteria, viruses, and several fungi (Turvey and Broide 2010; Aristizábal and González 2013).

Other pattern recognition mechanisms include recognition of DAMPs or damage-associated molecular patterns. These are typical signs of injury, infection, or inflammation; and DAMP-associated receptors include NLRs and receptors from the RAGE (receptor for advanced glycation end products) family. Typical ligands for these receptors are ATP, K+, and uric acid, which are signs of excessive energy demand, membrane damage, or increased metabolic activity. There is a great deal of overlap between DAMP and PAMP receptors, and they are mechanistically relevant for several chronic diseases and malignancies (Turvey and Broide 2010; Aristizábal and González 2013).

A third recognition strategy is the identification of "missing self" in foreign or infected cells. This strategy involves searching for a lack of molecules expressed by healthy cells; a notable example is the programming of NK cells to respond only to cells lacking major histocompatibility complex 1 (MHC-I). MHC-I is expressed by all normal cells and is a critical marker of self vs. non-self. Disruptions in such processes are the etiology of several autoimmune pathologies such as Crohn's disease and anaphylactic shock, and loss of self-tolerance often occurs through failure, deficiency, or mutation of immune receptor subtypes. However, the risk of self-reactivity is relatively low among innate cells compared to the adaptive immune system. Failures of immune recognition and autoimmunity will be discussed in further detail later in this chapter (Turvey and Broide 2010; Aristizábal and González 2013; Gonzalez et al. 2011).

8.2.1.2 Humoral Innate Immunity

Activation of receptors through any one of these pattern recognition paradigms results in the initiation of the innate immune cascade. One of the first response patterns is the generation of acute-phase reactant proteins, primarily from the liver. These include inflammatory cytokines (interleukin-1, interleukin-6, tumor necrosis factor), prostaglandins, and leukocyte adhesion molecules, which generate a pro-inflammatory environment and assist in opsonization. Below is a brief overview of a few key types of acute-phase reactants and humoral immune effector molecules (Mantovani and Garlanda 2023; Shishido et al. 2012).

Pentraxins Pentraxins are proteins with a conserved pentameric cyclic structure. Notable pentraxins include C-reactive protein and serum amyloid P, which are rapidly produced in amounts up to 1000x upward of normal levels in response to immune activation and released from neutrophils and endothelial cells. They bind ligands on several types of pathogens and opsonize them for clearance via phagocytosis and complement system activation. Disruption in pentraxin activity results in increased susceptibility to bacterial and fungal infection and is a significant risk factor for several inflammatory vascular diseases (Mantovani and Garlanda 2023; Shishido et al. 2012; Manfredi et al. 2008).

Complement The complement system is a series of activating molecules produced by hepatocytes, with intermediates and final products serving a variety of functions. This series may be activated via any one of three pathways: in the classical pathway, C1q binds to antigen-antibody complexes in adaptive immunity; the alternative pathway involves spontaneous production of C3b and binding to a pathogen; and the mannose-binding lectin (MBL) pathway results from MBL binding to mannose residues on a pathogen. All pathways converge at the activation of the C3 convertase, which branches into several downstream products. C3a, C4a, and C5a are potent vasodilators and promoters of inflammation and neutrophil chemotaxis. C3b and C4b act as opsonins for phagocytosis when binding to Factor B on microbial cells. Binding of complement to target cells and generation of intermediate monomers C5-9 result in the formation of a membrane attack complex (MAC), which induces direct cellular damage and subsequent lysis via pore formation at the bacterial cell membrane (Mantovani and Garlanda 2023; Shishido et al. 2012; Sarma and Ward 2011).

Other Acute-Phase Reactants Naturally occurring antibodies (NAb) are produced by B lymphocytes constitutively and are typically IgM variants due to a lack of impetus for class switching under normal circumstances. They mediate debris and damaged cell clearance and are excellent opsonins. Serum amyloid A (SAA) is a major acute-phase reactant that has cytokine-like functions and promotes the induction of anti-inflammatory M2 phenotypes for macrophages, promoting tissue healing and repair. Tissue damage also results in the activation of primary hemostasis and the coagulation cascade, including downstream products of fibrinogen and fibronectin. In addition to serving as markers for inflammation and injury, these can also bind microbes and assist in phagocytosis (Mantovani and Garlanda 2023; Shishido et al. 2012; Grover et al. 2016).

8.2.1.3 Cell-Mediated Innate Immunity

Specific immune cell types also provide a natural nonspecific defense, and each plays a critical role in rapid response to pathogens. These typically include cells from myeloid lineages such as mononuclear mast cells and macrophages, and multilobed nuclear neutrophils, eosinophils, and basophils. Although lymphocytic lineages are predominantly associated with the adaptive immune response, cells such as NK cells also play a role in innate immune defense (Ganeshan and Chawla 2014; Pearce and Pearce 2013).

Granulocytes Key granulocytes involved in innate immunity are mononuclear mast cells and multilobed neutrophils, eosinophils, and basophils. Neutrophils are the primary responders of this system and makeup 50 to 60% of leukocytes at any given time. They are short-lived cells that phagocytose invading pathogens through the release of defensins and microbicidal oxygen products produced in their respiratory burst. They may also trap and destroy microbes using neutrophil extracellular traps (NETs), which are fibers of chromatin and proteases. Neutrophils are recruited to sites of infection via extravasation, which involves chemoattractants and several chemotactic agents (Ganeshan and Chawla 2014; Pearce and Pearce 2013; Mayadas et al. 2014).

Eosinophils are specialized granulocytes that respond to parasites and are professional antigen-presenting cells (APCs) with limited phagocytic ability. Basophils are less commonly found and are primarily responsible for degranulation and release of histamine and prostaglandins to facilitate inflammation and promote clot formation. Like basophils, mast cells release histamine, prostaglandins, and leukotrienes; they play a strong role in anaphylaxis and allergic reactions (Ganeshan and Chawla 2014; Pearce and Pearce 2013; Palmblad 1984).

Macrophages Macrophages are found throughout nearly every tissue type in the body and are essential phagocytes. They have two main activation states; M1 macrophages secrete pro-inflammatory factors such as IL-12 and TNF-α, whereas M2 macrophages secrete actors such as IL-10 and TGF-β to drive tissue repair and immunomodulation, making them especially effective against parasites. The differentiation of macrophages is dependent on their microenvironment, and both subtypes have very different metabolic states and requirements. The M1 (classical) phenotype is selected in response to factors such as PAMPs and IFN-γ, whereas macrophages are recruited to the M2 (alternative) phenotype in response to IL-4 and IL-13 (Ganeshan and Chawla 2014; Pearce and Pearce 2013; Gordon and Martinez-Pomares 2017).

Dendritic Cells Dendritic cells are professional APCs and serve a critical role in bridging innate and adaptive immunity. Their branch-like structure allows them to sample antigens over a large area and to present to a large number of secondary immune cells. Although they may present antigens to innate responders such as macrophages, their primary function lies in the activation of the adaptive immune response through presentation to T lymphocytes; a single dendritic cell may activate up to 3000 T cells (Ganeshan and Chawla 2014; Pearce and Pearce 2013; Bell et al. 1999).

Natural Killer Cells NK cells are lymphocytes that participate in cell-mediated innate immune destruction. They primarily destroy cells via a process known as antibody-dependent cellular cytotoxicity (ADCC), which is discussed in more detail in the following section on adaptive immunity. NK cells have no antigen-specific receptors and respond to all cells lacking MHC-I. Binding of killer cell immunoglobulin receptor (KIR) and other surface inhibitory receptors prevents NK response to normal cells (Ganeshan and Chawla 2014; Pearce and Pearce 2013; Vivier et al. 2008).

8.2.2 Adaptive Immune System

Adaptive immunity refers to the immune system that recognizes and eliminates foreign invaders through antigen-specific antibodies. T cells and B cells are specialized immune cells that identify antigens and launch antigen-specific responses, making them essential components of the adaptive immune response. The adaptive immune system also provides immunologic memory, which ensures the longevity of immune protection by the formation of memory cells containing formerly encountered antigen-specific antibodies. When triggered, these memory cells are able to generate a quicker and more extensive response to the same antigen. T and B cell deficiency causes several diseases that will later be discussed as immunodeficiencies under this section (McComb et al. 2019).

8.2.2.1 T Lymphocytes

T lymphocytes are a pivotal component of adaptive immunity and the generation of cell-mediated responses. All T cells contain specific surface receptors called T cell receptors (TCRs), allowing them to be distinguished from other lymphocytes and to recognize specific antigens presented by MHC molecules of antigen-presenting cells (APCs) (Buck et al. 2015) (Fig. 8.2).

T cells are produced by lymphoid progenitor cells within the bone marrow and are sent to the thymus for maturation. They pass through several checkpoints to ensure an appropriate immune response to foreign pathogens (Thapa and Farber 2019). Migration from the thymic cortex to the medulla occurs during the maturation of T cells via two steps:

Positive Selection Precursor T cells are exposed to MHC class I and class II antigens expressed by the thymic cortex. Over- or under-reaction by precursor T cells leads them to apoptosis, thereby eliminating dysfunctional T cells (Thapa and Farber 2019).

Negative Selection Following positive selection, precursor T cells are carried into the thymic medulla to further eliminate self-reacting T cells via mTEC expression of self-antigens. Self-antigens are expressed by the autoimmune regulator protein (AIRE protein). A deficiency of this regulatory protein can lead to autoimmune polyglandular syndrome (APS), leading to severe autoimmune disease with symptoms such as adrenal insufficiency, hypoparathyroidism, and chronic mucocutaneous candidiasis (Takaba and Takayanagi 2017).

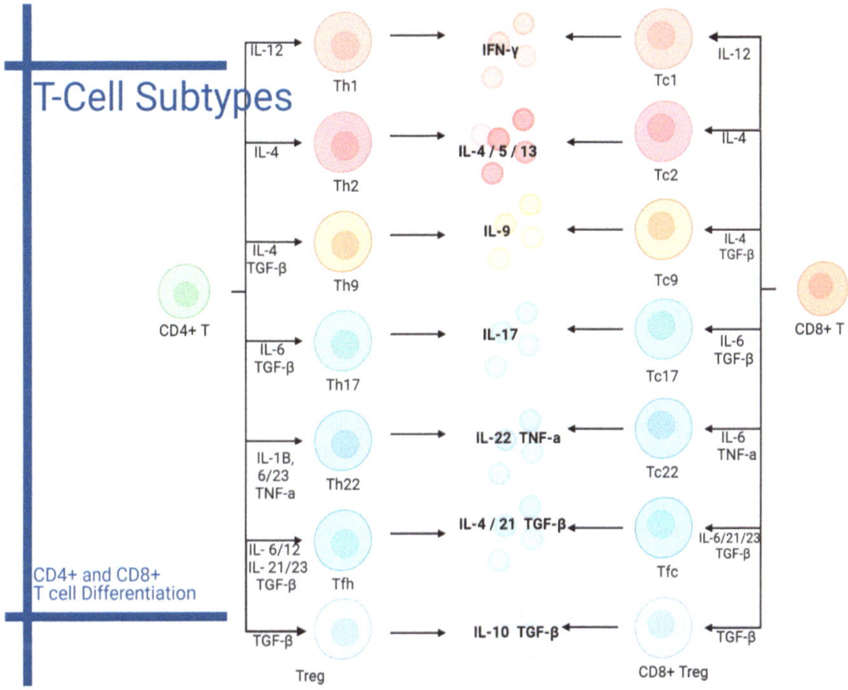

Fig. 8.2 T cell subtypes. Overview of cytokines and the different T cell subtypes that stimulate each other. (Created in BioRender. Erbay, M. (2024) BioRender.com/o43f629)

Processed T cells are regulated to bind to one MHC molecule type by their specific cluster of differentiation (CD). These T cells are also known as single thymocytes, as they are positive for only one CD. $CD8^+/CD4^-$ T cells bind to MHC class I molecules while $CD8^-/CD4^+$ T cells bind to MHC class II molecules (Klein et al. 2014).

Antigen-presenting cells (APCs) activate cellular immune responses by processing antigens and presenting them to T lymphocytes in secondary lymphoid organs such as lymph nodes. Such APCs include dendritic cells, macrophages, and B lymphocytes (Fig. 8.3). This antigen presentation involves exogenous antigens presented on MHC II to CD4+ T cells, and endogenous antigens presented on MHC-I to CD8+ T cells, a process known as cross-presentation (Embgenbroich and Burgdorf 2018; Theisen and Murphy 2017).

Activation requires two signals: the antigen-specific signal through TCR recognition of the antigen-MHC complex, and a co-stimulatory signal mediated by the interaction between B7 proteins (CD80 or CD86) on the antigen-presenting cell and CD28 on the T cells (Chen and Flies 2013). Without this co-stimulatory signal, T cells may become anergic, a mechanism of self-tolerance preventing unnecessary immune responses to self-antigens. Superantigens found in certain bacteria can bypass this requirement, directly activating T cells by linking MHC II on antigen-presenting cells to TCRs on T cells (Chen and Flies 2013).

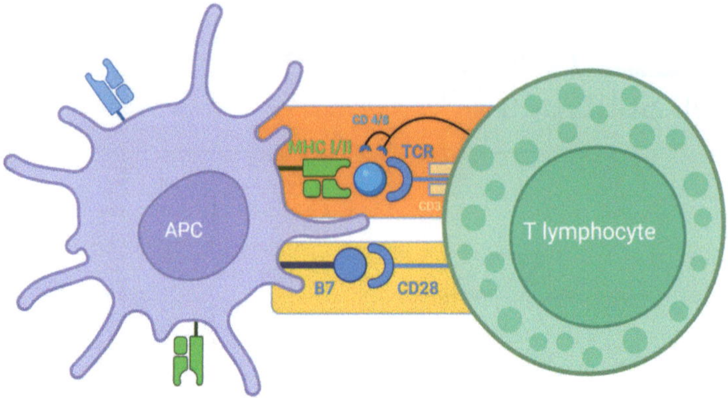

Fig. 8.3 Antigen presentation by dendritic cells to T lymphocytes. The orange box represents the first signal building a connection between an antigen-presenting cell (APC) and a T cell. Antigens attached to MHC class I molecules are attached to T cell receptors (TCR). MHC class I molecules are recognized by $CD8^+$ T cells while MHC class II molecules are matched with $CD4^+$ T cells. The yellow box represents the second signal that mediates interaction between B7 on an APC and CD28 on T cells. (Created in BioRender. Erbay, M. (2024) BioRender.com/n66k818)

Upon activation, T-helper cells produce cytokines that stimulate the production of other immune cells like B cells, macrophages, and cytotoxic T cells; note that cytotoxic T cells directly target and eliminate cells displaying the antigen, and are important in direct defense against infections and malignancies (Hosseini et al. 2021). Each T cell subtype has unique functions and activates a specific type of immune response (Fig. 8.2) (Golubovskaya and Wu 2016; Koh et al. 2023).

Cytokines play an important role both in the conversion of T cell subtypes and in activating immune responses after being released by a T cell subtype. Cytotoxic $CD8^+/CD4^-$ T cells recognize the antigens presented on MHC class II molecules and engage in cell-mediated toxicity, which involves the intracellular destruction of pathogens or neoplasms by inducing apoptosis in the presenting cell. $CD8^-/CD4^+$ T-helper cells can differentiate into five major subtypes: Th1, Th2, Th17, Treg, and Tfh cells. Naive T cells transform into one of these subtypes based on the cytokines released by stimulating cells such as macrophages, thus triggering the appropriate immune response (Fig. 8.3) (Golubovskaya and Wu 2016; Koh et al. 2023).

8.2.2.2 B Lymphocytes

B cells are another pivotal component of adaptive immunity, mediating humoral immune responses by producing antibodies. They are produced and mature in the bone marrow and become active when they encounter antigens. Mature B cells are transmitted to the bloodstream and lymphatic organs such as lymph nodes and the spleen. After maturation, B cells differentiate into plasma cells, producing antibodies or immunoglobulins (Ig) in the bloodstream at a constant rate, and thus mediating long-lived humoral immunity. Naive B cells express IgM and IgD immunoglobulins on their surface (Seifert and Küppers 2016) (Fig. 8.4).

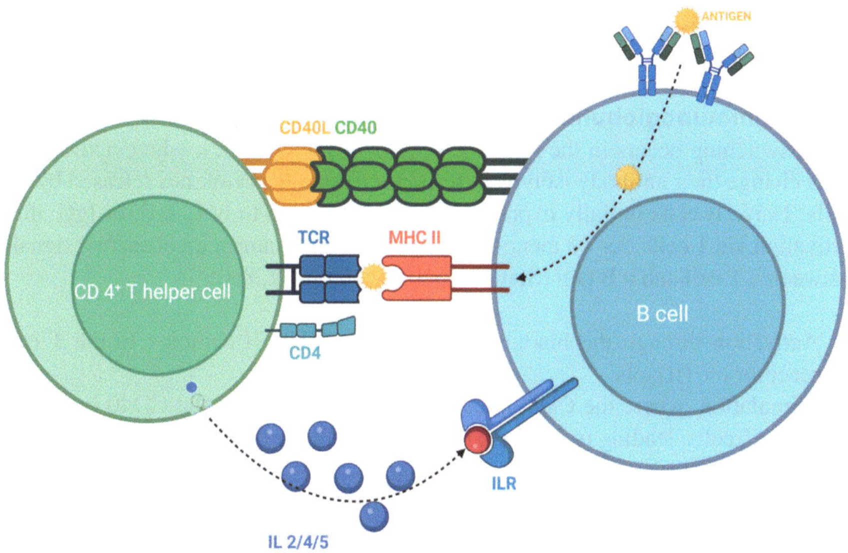

Fig. 8.4 T cell-dependent B cell activation. CD4+ T-helper cells release cytokines to activate B cells following CD40L-CD40 and TCR-Antigen-MHCII complex formation between two cells. (Created in BioRender. Erbay, M. (2024) BioRender.com/n66k818)

8.2.2.3 B Lymphocyte Activation and Affinity Maturation

The activation of B lymphocytes can be categorized as either T cell dependent or T cell independent. Both pathways result in the stimulation and differentiation of B lymphocytes into plasma cells (LeBien and Tedder 2008).

8.2.2.4 T Cell-Dependent B Cell Activation

After encountering protein antigens, a T cell receptor binds to the MHC II antigen complex on the surface of a naive B cell, followed by secondary binding between CD40L on the T cells and CD40 on the B cells. This leads to secretion of cytokines such as interleukin-4 (IL-4) that induce plasma cell differentiation. Within the germinal centers of lymph nodes, plasma cells cycle between the dark zone, where they proliferate and mutate, and the light zone, where they undergo selection. In the light zone, they compete for higher antigen affinity, which produces mature B cells with a better ability to present antigens to T cells, thereby improving survival regulation. This type of activation induces memory B cell formation, providing a long-lasting immunity (Doria-Rose and Joyce 2015).

8.2.2.5 T Cell-Independent B Cell Activation

Nonprotein antigens such as lipopolysaccharides can also initiate T cell independent B cell activation. Cross-linkage of B cell receptors by repetitive nonprotein antigen epitopes such as those found in lipopolysaccharides stimulates B lymphocytes to secrete only IgM. Immunoglobulin class switching requires CD40L-CD40 interaction between B cells and T cells, and cytokines are released as a result of this interaction. This type of activation does not form memory B cells (Lakkis and Weir 2020).

8.2.2.6 Immunoglobulin Class Switching

Class switching occurs in the germinal centers of lymph nodes, where activated B cells change their antibody isotype in response to specific cytokines released by Th cells. Naïve B cells initially express IgM but can switch to IgA, IgE, or IgG after activation via T cells. As we mentioned above, this mechanism involves two signals between a T cell and a B cell (Stavnezer et al. 2008) (Fig. 8.4):

1. Activation through the binding of an antigen-MHC II complex to the T cell receptor on Th cells.
2. Stimulation, where the CD40 receptor on B cells binds to the CD40 ligand on $CD4^+$ T cells, leading to cytokine release and gene rearrangement.

Specific cytokines released from T cells guide class switching from IgM. These cytokines and immunoglobulin class changes are listed below:

IL-4 and IL-13 induce switching to IgE.
IL-5 and TGF-β induce switching to IgA.
IFN-γ, IL-4, and IL-21 induce switching to IgG.

8.2.2.7 Antibody-Dependent Cellular Cytotoxicity

Antibody-dependent cellular cytotoxicity (ADCC) refers to a type of humoral immune response in which effector immune cells such as natural killer (NK) cells and eosinophils bind to and lyse target cells with specific antibodies such as IgG (Fig. 8.5). This mechanism enables innate immune cells to recognize and attack pathogens that do not express pathogen-associated molecular patterns (PAMPs), and it also plays a critical role in the immune response to parasites mediated by eosinophils (see Sect. 8.2.1) (Gómez Román et al. 2014; Wicherska-Pawłowska et al. 2021).

The mechanism begins with antibodies produced by B cells binding to antigens on target cells, which may be tumor cells, viruses, or parasites. Effector cells like NK cells and eosinophils recognize and bind the Fc portion of these antibodies through their Fc receptors, such as CD16 Fc receptors on NK cells (Fig. 8.5). This binding activates signaling pathways within effector cells, leading to the release of cytotoxic substances like perforins, granzymes, or major basic proteins, ultimately killing the target cells. Clinically, ADCC is used in monoclonal antibody therapy, where therapeutic antibodies are used to specifically target and eliminate cancer cells (Hubert and Amigorena 2012).

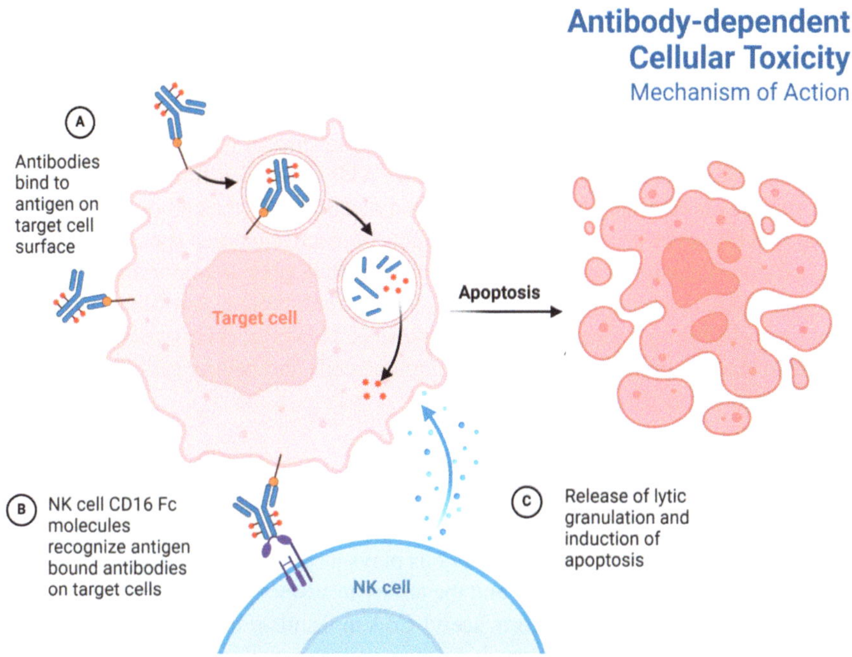

Fig. 8.5 Antibody-dependent cellular toxicity (ADCC). Cancer cells are driven to apoptosis by natural killer (NK) cells after being recognized by antibodies. This mechanism is leveraged in targeted therapy against several cancers (e.g., Trastuzumab). (Created in BioRender. Erbay, M. (2024) BioRender.com/n66k818)

8.3 Metabolic Pathways in Immune Cells

8.3.1 Glycolysis

Glycolysis is the main route of metabolism for most carbohydrates present in the cytosol and is used by all cells (including immune cells) to generate a net of 2 adenosine triphosphate (ATP) molecules promptly. Glycolysis is the breakdown of a glucose molecule into two pyruvate molecules, which can later be converted into four different molecules via four distinct metabolic reactions. These conversion molecules are lactate (used in gluconeogenesis), acetyl-CoA and/or oxaloacetate (used in the citric acid/TCA cycle), and alanine (used in transamination reactions). Within the immune system, the vast majority of cell types primarily rely on glycolysis, as they require short bursts of energy (i.e., M1 macrophages and neutrophils). For this reason, they are often compared with tumor cells, as their metabolism displays Warburg-like properties (Donnelly and Finlay 2015).

8.3.2 Oxidative Phosphorylation

Oxidative phosphorylation is the main source of energy for intracellular reactions, allowing cells to release the energy stored in carbohydrates, fats, and proteins to produce ATP. Within the mitochondria, an electrochemical gradient between inner and in-between membrane spaces generates ATP molecules through an electron transport chain involving oxidation/reduction reactions. By the end of this reaction, up to 36 ATP can be produced. Major amounts of reactive oxygen species (ROS) are also produced during oxidative phosphorylation. ROS play an important role in the immune system, regulating the proliferation, differentiation, and functionality of immune cells through several signaling pathways. However, excessive ROS production also leads to programmed death of immune cells and halting cellular proliferation (Steinert et al. 2021; Liu et al. 2023).

8.3.3 Fatty Acid Oxidation

Fatty acid oxidation (FAO) in immune cells plays a crucial role in tissue repair, pro- and anti-inflammatory responses, and the ability to sustain prolonged immune function. Fatty acids break down to form acetyl-CoA molecules, and ATP is subsequently generated. Fatty acid metabolism facilitates several functions in various immune cells (Kemp et al. 2024):

Innate lymphoid cells (ILCs)—ILCs are tissue-resident and have metabolic profiles reflecting nutrient availability in their local environments. FAO may play a role in modulating ILC function and differentiation (Kemp et al. 2024).

T cells—FAO has been implicated in modulating $CD8^+$ T cell memory formation and regulating T cell quiescence and activation. It can also influence T cell differentiation into specific subsets, such as Th1, Th2, and Th17 cells (Kemp et al. 2024).

B cells—While comparably less studied, FAO may contribute to B cell activation, proliferation, and antibody production. Metabolic reprogramming during B cell activation likely involves both glycolysis and FAO (Kemp et al. 2024).

Dendritic cells (DCs)—FAO can influence DC activation, with saturated fatty acids promoting a pro-inflammatory phenotype and polyunsaturated fatty acids inducing a more tolerogenic state. FAO may also contribute to lipid accumulation in DCs, which has been observed in cancer patients (Kemp et al. 2024).

Macrophages—FAO is important for the alternative activation of macrophages, which promotes tissue repair and resolution of inflammation. It can also influence macrophage polarization and function in response to different environmental cues (Kemp et al. 2024).

8.3.4 Metabolic Pathways and Regulation of Macrophage Polarization

Macrophage differentiation is also regulated by arginine metabolism and glycolysis. Macrophages are highly plastic and can rapidly polarize in response to environmental signals (Kemp et al. 2024). M1 macrophages are induced by IFN-γ and tumor necrosis factor (TNF) and rely predominantly on glycolysis and glucose-driven pathways. In contrast, M2 macrophages are involved in wound healing and tissue repair and predominantly utilize fatty acid oxidation (FAO) and oxidative phosphorylation (O'Neill et al. 2016). In M1 macrophages, glycolysis fuels the pentose phosphate pathway (PPP) to produce nucleotide precursors and NADPH, which are used to generate reactive oxygen species (ROS) for microbial killing. M2 macrophages upregulate the enzyme arginase-1, which breaks down arginine, thereby depriving T cells of this essential nutrient and limiting their proliferation (Sun et al. 2022).

8.3.5 Glutaminolysis

Glutamine is the most abundant amino acid in the human body (Feng et al. 2022). Although it is a nonessential amino acid, it is classified as "conditionally" essential during inflammatory conditions such as infection and injury (Newsholme 2001). This is because glutamine is essential for cell proliferation, as it generates NADPH from $NADP^+$ and boosts the function of triggered immune cells. Rapidly dividing immune cells, including T lymphocytes, consume glutamine at a rate similar to or higher than glucose, even when glucose is abundant. The innate immune system and cells like macrophages and dendritic cells rely on glutaminolysis for their initial defense against pathogens. In the skin, glutamine synthetase (GS) is stored in epidermal keratinocytes and is involved in the synthesis of biologically active compounds necessary for rapid cell division (Xu et al. 2024). The intestinal innate immune system, the largest immune organ, utilizes glutamine to maintain mucosal metabolism, structure, and function during injury or stress. Glutaminolysis has a major role in the adaptive immune system, particularly in the proliferation, differentiation, and effector functions of $CD4^+$ T cells.

8.3.6 Iron and Immune System

Iron plays a vital role in fighting against infections, facilitating functions of innate and adaptive immunity, and is also required for oxygen transport. The human body is able to store non-heme iron in the form of ferritin to be used promptly when in need of oxygen transport or energy metabolism. This mechanism prevents iron toxicity that would otherwise cause excess ROS production (Johnson and Wessling-Resnick 2012). During infection, iron flux is controlled by genes/proteins that prevent bacteria from using iron for their own consumption. Moreover, immune cells regulate their own iron flux while fighting against bacteria (Bao et al. 2024).

Interestingly, iron plays a vital role in neutrophils by inhibiting the formation of neutrophil extracellular traps, thus leading to a more potent antibacterial environment with lower iron levels (Ni et al. 2022). Iron also restores innate lymphoid cells (ILCs) within the gut to fight against bacteria and helps maintain their function (Xiong et al. 2023). Hepcidin has major roles in both iron metabolism and inflammation. Upon activation of the innate immune system, IL-6 induces its expression and release from the liver. Then, hepcidin binds to the ferroprotein efflux protein, leading to its internalization and subsequently decreasing iron levels. This iron retention prevents bacteria from using otherwise available iron that would increase their growth. Hepcidin can also cause changes in the transcription of acute cytokine-induced inflammatory responses (Johnson and Wessling-Resnick 2012).

8.4 Metabolic Reprogramming During Immune Responses

Recent evidence suggests that modifying cellular metabolic processes can impact immune cell mechanisms, including proliferation, differentiation, and effector function. This may occur through modulation of various mechanisms, including regulation of energy balance, macromolecule breakdown, biosynthesis, redox balance, epigenetic modifications, and signal transduction (O'Neill et al. 2016; Murray et al. 2015; Sun et al. 2020). Understanding innate or adaptive immune cells' metabolic reprogramming is crucial for identifying new therapeutic targets in various conditions (Sun et al. 2020; Dai et al. 2017).

Immune responses involve metabolic reprogramming, altering cellular metabolic processes in immune cells to ensure optimal function (Sun et al. 2020). This process is highly dynamic and complex, with 2749 metabolic pathways identified in 3067 diverse organisms (including 477 in humans) according to data from the MetaCyc database (MetaCyc n.d.). These each result in metabolic optimization, where a unique metabolic signature determines the differentiation of specific immune cell subsets as either pro-inflammatory or anti-inflammatory (Sun et al. 2020). Metabolic reprogramming is a highly flexible process that varies by immune cell function.

8.4.1 Innate Cells

8.4.1.1 Monocytes

Monocytes are a crucial component of the innate immune system, monitoring environmental changes and replenishing the pool of tissue macrophages and dendritic cells (DCs) (Yang et al. 2014). The activation of CD14+ Monocytes by lipopolysaccharides (LPS) triggers a metabolic shift, favoring glycolysis over oxidative phosphorylation (OXPHOS). This alteration in energy production allows for a swift host defense mechanism by promoting the production of inflammatory cytokines such as TNFα, IL6, and IL1β, as well as facilitating phagocytosis (Lachmandas et al. 2016).

Under glucose-deprived conditions, monocytes can also boost pro-inflammatory activity through increased fatty acid oxidation (FAO), supporting OXPHOS to compensate for the Warburg effect and fulfilling heightened energy requirements during LPS stimulation (Raulien et al. 2017). In contrast to LPS, synthetic bacterial lipopeptide P3C or *Candida albicans* stimulate both glycolysis and OXPHOS, leading to increased production of pro-inflammatory cytokines and reactive oxygen species (ROS) through toll-like receptor 2 or C-type lectin-signaling pathways (Domínguez-Andrés et al. 2017). Overall, activated monocytes may favor increased glycolysis, which could be counterbalanced by oxidative processes (Sun et al. 2020).

8.4.1.2 Macrophages

Macrophages play a crucial role in maintaining homeostasis, promoting tissue repair, and fighting infection. The innate immune system relies on macrophages, incredibly flexible cells that act as the primary defense against external infections (Wynn et al. 2013). The role of LPS/IFNγ-induced M1 macrophages is significant in host defense and tissue inflammation due to their secretion of inflammatory cytokines, generation of ROS, and amplification of Th1 responses (Koelwyn et al. 2018). The M1 macrophage exhibits increased glycolysis and pentose phosphate pathway (PPP) activity while simultaneously promoting OXPHOS and reduced ATP production in the mitochondria. These effects are attributed to a disrupted tricarboxylic acid (TCA) cycle, which is regulated by Akt (protein kinase B), mTOR (mammalian target of rapamycin), and HIF-1α (hypoxia-inducible factor 1α) (Sun et al. 2020; Koelwyn et al. 2018; Williams and O'Neill 2018).

IL4-triggered M2 macrophages depend heavily on optimal oxidative metabolism, including fatty acid oxidation (FAO) and OXPHOS, to ensure their viability and effectively perform their anti-inflammatory duties during immune resolution (Mehta et al. 2017). Signal transduction and activation of the Stat6 transcription signaling pathway are linked to M2 macrophage polarization and increased mitochondria oxidative activity (Koelwyn et al. 2018). Macrophages may also differentiate into Mox subtypes and exhibit anti-inflammatory and antioxidant properties in response to oxidized phospholipids (Kadl et al. 2010).

8.4.1.3 Dendritic Cells

As the primary antigen-presenting cells, DCs are crucial in bridging innate and adaptive immunity (Worbs et al. 2017). The activation of DCs involves metabolic reprogramming regulated by the PI3K/Akt pathway and counteracted by the AMPK pathway, resembling the Warburg effect in cancer cells (Krawczyk et al. 2010). The switch from lipid oxidation to aerobic glycolysis signifies the metabolic transformation of mitochondrial OXPHOS (Worbs et al. 2017). Evidence suggests early glycolytic bursts in bone marrow-derived DCs can partially bypass glucose requirements during activation (Giovanelli et al. 2019). Furthermore, the formation of DCs from progenitor cells is linked to increased mitochondrial production, facilitated by PGC-1α and supported by PPARγ, mTOR, and MYC (Pearce and Everts 2015). Additionally, recent research has discovered a correlation between fatty acid metabolism and various functions of DCs (De Lima et al. 2018). The inhibition of fatty acid synthesis directly results in decreased production of DCs from their precursors (He et al. 2019). The inhibition of human PBMC development and the induction of DC precursor apoptosis are both consequences of blocking fatty acid synthesis (De Lima et al. 2018).

8.4.1.4 Natural Killer Cells

Natural killer (NK) cells link innate and adaptive immunity. The antiviral and antitumor properties of NK cells are outstanding. They generate IFN cytokines and effectively eliminate target cells through cytotoxic mechanisms (Hu et al. 2022). NK cells rely on glucose as their primary fuel, modulating glycolysis and OXPHOS rates (O'Brien and Finlay 2019). When inactive, NK cells favor OXPHOS as their primary energy source (Keppel et al. 2015). Despite levels of glycolysis and OXPHOS being relatively low in stationary NK cells of mice, they were still adequate to sustain a rapid immune response (O'Brien and Finlay 2019). The application of cytokines to stimulate NK cells results in a significant rise in rates of glycolysis and OXPHOS, which is necessary for NK cells to secrete IFN-γ upon receptor stimulation (Hu et al. 2022).

The activation and metabolism of NK cells are tightly regulated by transcription factors, highlighting their significant role (Hu et al. 2022). Increasing rates of glycolysis and OXPHOS require the activity of transcription factor SREBP (Hu et al. 2022). The regulation of SLC25A1 and ACLY15 increases rates of glycolysis by affecting the expression of the citric acid and malic acid reverse transporter (Assmann et al. 2017). MYC regulates metabolic pathways in NK cells, enhancing glycolysis and mitosis through increased expression of glucose transporters and glycolytic enzymes, thereby promoting efficient OXPHOS (Loftus et al. 2018). mTORC1 signal transduction is also necessary for driving increased glycolysis in cytokine-stimulated NK cells (Hu et al. 2022; Mah et al. 2017).

8.4.2 Adaptive Cells

8.4.2.1 T Lymphocytes

The immune response against foreign antigens heavily relies on T lymphocytes, a highly heterogeneous group of cells encompassing various subsets (Sharabi and Tsokos 2020; Masopust and Schenkel 2013). The primary source of ATP production for peripheral naïve T cells is OXPHOS, which enables them to sustain weak interactions between their T cell receptors and self-peptides displayed on the major histocompatibility complex (MHC) molecules for survival and homeostatic proliferation (Sun et al. 2020; Klein Geltink et al. 2018). The activation of T cells occurs when they encounter antigens on an antigen-presenting cell, which is accompanied by the receipt of co-stimulatory signals and stimulation from cytokines (Dai et al. 2017; Masopust and Schenkel 2013). Despite the presence of enough oxygen to support glucose catabolism through the TCA cycle and OXPHOS, recent studies have suggested that glycolysis remains a crucial metabolic feature of activated effector T cells (Michalek et al. 2011).

The activation of Th1 cells results in lactate dehydrogenase (LDHA) induction to support aerobic glycolysis. This induction of LDHA helps maintain high levels of acetyl-CoA, thereby enhancing histone acetylation and promoting the transcription of IFNγ (Sun et al. 2020; Peng et al. 2016). Also, it was reported that glycolysis plays a crucial role in the effector function of activated T cells, particularly in the generation of the pro-inflammatory cytokine IFNγ (Sun et al. 2020). Although glycolysis did not contribute significantly to these cells' proliferation and survival, OXPHOS was essential for their growth and viability (Chang et al. 2013). On the other hand, Treg cells demonstrated suppression of glycolysis while enhancing levels of OXPHOS. This metabolic adaptation might provide them with a competitive edge in environments characterized by low glucose and high lactate levels, as it enables them to resist the lactate-mediated suppression of T cell function and proliferation (Sun et al. 2020; Newton et al. 2016; Galgani et al. 2016).

8.4.2.2 B Lymphocytes

B lymphocytes play several essential roles in adaptive immunity, such as the production of antibodies targeted toward specific diseases, switching antibody classes, and enhancing affinity (Dai et al. 2017; Victora et al. 2010). B cells diversify their immunoglobulin genes in germinal centers during the immune response, resulting in high-affinity antibodies that play a role in humoral immunity (Victora et al. 2010). Similar to activated T cells, an increase in glycolysis was also observed in B cells when their B cell receptor was engaged or stimulated with cytokines (Sun et al. 2020; Doughty et al. 2006). Glycolysis has been found to have a dual role in these cases, as it not only supports the growth of B cells through the PI3K (phosphatidylinositol 3-kinase) signaling pathway but also acts as a protective mechanism against apoptosis via signal transduction and activation of the Stat6 transcription signaling pathway (Dufort et al. 2007).

A recent study explored the metabolic changes in purified B and T cells when challenged with antigens. The results showed that activated B cells had increased the rate of oxygen consumption and glycolysis, as evidenced by the extracellular acidification rate. These cells also maintained a balanced ratio of oxygen consumption rate to extracellular acidification rate (Caro-Maldonado et al. 2014). B cell proliferation and antibody production rely on increased glycolysis, so inhibition of glycolysis through Glut1-depletion or dichloroacetate significantly reduces B cell proliferation and antibody secretion (Caro-Maldonado et al. 2014).

Short-lived plasma cells can differentiate into long-lived memory B cells, which ensures the production of durable antibodies (Sun et al. 2020). According to metabolic studies, long-lived memory B cells sustain themselves through an elevated capacity for mitochondrial respiration. This increased capacity is believed to result from heightened levels of respiratory substrates, particularly pyruvate. Notably, these cells also demonstrate a higher uptake of glucose in comparison to short-lived plasma cells (Lam Wing et al. 2016). The utilization of cellular glucose for antibody glycosylation is quite extensive; therefore, if the import of pyruvate is disrupted by knocking down mitochondrial pyruvate carrier (Mpc1) or Mpc2, it would lead to a significant decline in the numbers of long-lived memory B cells and antibody secretion (Sun et al. 2020; Lam Wing et al. 2016; Corcoran Lynn and Nutt 2016).

8.5 Metabolism and Immune Signaling

8.5.1 Immune Cells and Metabolic Pathways

Immune system functions are dependent on energy sources from multiple metabolic pathways; and therefore, their regulation is closely linked to various signaling pathways between metabolic and immune cells. Gene expression, transcription factors, and signaling mechanisms are all vital to this interaction and will be discussed in this section.

8.5.1.1 T Cells and Metabolic Pathways

T cell receptor activation often requires two signaling interactions. Prior studies have shown that T cell co-stimulation through the CD28 receptor enhances glucose transporter (GLUT1) expression and subsequent glycolytic pathways and mitochondrial activity (Frauwirth et al. 2002; Klein Geltink et al. 2017; Buck et al. 2016). This transition from oxidative phosphorylation to glycolysis reflects the higher energy requirements of T cells during times of stress, inflammation, or otherwise increased immune activity. Notably, effector T cells rely primarily on aerobic glycolysis (Araki et al. 2009; Pearce et al. 2009). Memory T cells generally utilize lipid oxidation and mitochondrial metabolism; however, they are also able to activate glycolysis by interacting with endoplasmic reticulum mitochondria (Araki et al. 2009; Pearce et al. 2009; Bantug et al. 2018).

On the contrary, regulatory T cells (Tregs) depend mainly on lactate, pyruvate, and lipid oxidation in the mitochondria. FoxP3, a transcription factor for Tregs, is important in suppressing glycolysis to maintain Treg function (Angelin et al. 2017; Beier et al. 2015; Gerriets et al. 2016). Tregs are important for producing immunosuppressive molecules such as interleukin-10 (IL-10) and TGF-β, thus playing a major inhibitory role. Tregs share characteristics with M2 macrophages, which also utilize oxidative metabolism and secrete anti-inflammatory molecules (Loftus and Finlay 2016).

Programmed cell death-1 (PD-1) and cytotoxic T-lymphocyte-associated protein 4 (CTLA4) instead play a role in downregulating the glycolytic pathway, and thus inhibit T cell co-stimulation and activity (Parry et al. 2005). Both PD-1 and CTLA4 are often utilized as targets in cancer therapies. These findings illustrate how an intricate relationship between immune cells, metabolic enzymes, intermediates, and transcription factors helps maintain a regulatory balance in relation to the cell environment and energy requirements (Makowski et al. 2020; Patsoukis et al. 2015).

8.5.1.2 Maintaining Homeostasis

Many tissues such as those found in the gastrointestinal tract and skin are exposed to the extracellular environment and thus are in contact with various pathogens and commensal microorganisms. As a result, maintaining homeostasis while simultaneously defending against pathogens poses a challenge for immune cells. Energy requirements and nutrition are vital for immune cells to maintain their function and are often the premise of regulatory loops (Liang et al. 2020).

8.5.1.3 Nutrient Sensing and Regulation Through mTOR Pathway

In order to maintain optimal immune responses, immune cells must be able to monitor the availability of nutrients, especially during times of high energy demand. One key signaling pathway that allows immune cells to sense nutrients involves the mechanistic target of the rapamycin (mTOR) pathway. Specifically, mTOR complex I (mTORC1) is activated through the phosphatidylinositol 3-kinase (PI3K)/Akt/mTORC1 pathway, which is utilized to monitor the availability of nutrients (Huang et al. 2020). Furthermore, mTORC1 activation is localized at lysosomes through T cell activation and PI3k/Akt signaling and subsequently stimulates Rheb GTPase (Shen and Sabatini 2018; Wyant et al. 2017). This downstream pathway then enables the sestrin/Gator complex to monitor nutrients such as essential amino acids. Additionally, mTORC1 is important for activating AMP-activated protein kinase (AMPK) through phosphorylation of ULK1; this interaction is important to upregulate oxidative metabolic pathways and enhance autophagy when cells are under stress (MacIver et al. 2011; Blagih et al. 2015). Although the activity of mTORC1 is common among all effector T cells, there is still precise regulation of the mTORC1 pathway in different effector T cell subsets, leading to distinct roles (Makowski et al. 2020).

Another regulatory loop involves the epithelial layer of the gastrointestinal tract, which is comprised of immune cells, protective mucus, and epithelial cells that produce antimicrobial proteins (AMPs). Regulation of AMPs is crucial to maintaining homeostasis, and the IL-22-STAT3 signaling and Toll-like receptor (TLR)-MyD88 pathway is important for AMP production (Mukherjee and Hooper 2015). However, studies have determined that levels of AMPs (i.e., Reg3 and Saa1) are substantially decreased with limited nutrient availability (Ji et al. 2019). On the contrary, α-defensin is another AMP whose levels are maintained even during limited nutrient availability, through regulatory mechanisms involving matrix metalloproteinase-7 (MMP7). MMP7 is sustained by Hes1 suppression and de-repression of Atoh1. This regulatory pathway is stimulated by mTOR via the mTOR-Hes1--MP7-α-defensin axis based on the nutritional microenvironment (Liang et al. 2019). Collectively, these findings display the intricate regulation and interplay that enable immune cells to effectively utilize the available energy supply, while also maintaining immune activity when energy stores are limited by activating specific pathways (Liang et al. 2020).

8.5.2 Crosstalk Between Immune Function and Cellular Metabolism

8.5.2.1 Immune System and Metabolic Interactions in Adipose Tissue

Crosstalk between immune cells and metabolic components is very evident in the setting of chronic inflammation and metabolic syndrome; specifically, it occurs in inflammatory processes originating in adipose tissue in the setting of obesity. Metabolic stress experienced by parenchymal cells in various organs leads to tissue remodeling and poor organ function. The innate immune system and cellular interactions among immune cell receptors are relevant in metabolic syndrome. Notably, toll-like receptor 4 (TLR4) and macrophage-inducible C-type lectin (Mincle) have demonstrated crosstalk with metabolic processes in obesity (Tanaka et al. 2014; Saberi et al. 2009; Suganami et al. 2007). Research has shown that pro-inflammatory cytokines, tumor necrosis factor alpha (TNFα), and interleukin-6 (IL-6) are all released through stimulation of TLR4 located on macrophages by saturated fatty acids (SFA) originating from cells in adipose tissue (Hachiya et al. 2022) (Fig. 8.6). Conversely, polyunsaturated fatty acids antagonize the influence of SFAs on

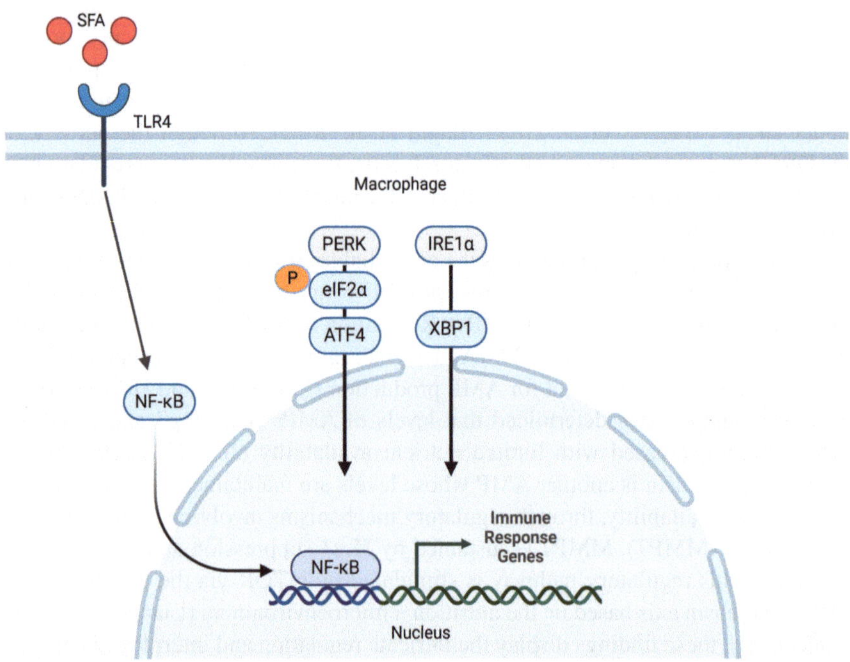

Fig. 8.6 Activation of immune response genes via various pathways mediated through TLR4 in macrophages. Toll-like receptor 4 (TLR4), saturated fatty acids (SFA), nuclear factor kappa-light-chain-enhancer of activated B cells (NF-κB), protein kinase R (PKR)-like endoplasmic reticulum kinase (PERK), eukaryotic initiation factor-2α (eIF2α), activating transcription factor 4 (ATF4), inositol-requiring enzyme 1 alpha (IRE1α), X-box binding protein 1 (XBP1). (Created in BioRender. Erbay, M. (2024) BioRender.com/w01g581)

inflammatory processes and are an example of how the amount and type of lipids in adipocytes regulate the inflammatory response (Hachiya et al. 2022).

8.5.2.2 Saturated Fatty Acids and Integrated Stress Response

Research has shown that SFAs are also able to activate inflammatory cytokine expression in macrophages that lack TLR4, which suggests that metabolic processes play a role in cytokine activation outside of TLR4 interactions (Iwasaki et al. 2014). The integrated stress response (ISR) is often activated by cells and is mediated by eukaryotic initiation factor-2α and activating transcription factor 4 (ATF4) (Costa-Mattioli and Walter 2020). It has been demonstrated that SFAs utilize the ISR pathway to activate pro-inflammatory cytokines; ATF4 stimulates the IL-6 promoter and promotes the localization of NF-κB into the nucleus (Iwasaki et al. 2014). This signaling mechanism explains how the ISR pathway strengthens the inflammatory response that is normally mediated by TLR4. Additionally, this response is often linked to endoplasmic reticulum (ER) stress, as transporters (i.e., CD36) facilitate the metabolism of fatty acids and their accumulation in the endoplasmic reticulum (Leamy et al. 2014; Akoumi et al. 2017; Volmer et al. 2013). As a result, alteration in ER membrane composition results in unfolded protein responses such as the ISR (Hachiya et al. 2022).

8.5.2.3 Inositol-Requiring Enzyme 1α (IRE1α)-X-Box Binding Protein 1 (XBP1) Signaling Pathway

Another example of crosstalk between immune cells and metabolic processes involves the IRE1α-XBP1 signaling pathway, which is important for inflammatory cytokine expression mediated by TLR4 (Martinon et al. 2010). Furthermore, macrophages also depend on extracellular serine to inhibit abnormal TLR4-mediated cytokine production. Overall, these findings showcase the important regulation of immune responses that is maintained by crosstalk between immune signaling and metabolism pathways (Hachiya et al. 2022).

8.5.2.4 Cellular Metabolism and Epigenetic Control of Immune Cells

Gene expression is often regulated by various epigenetic mechanisms that involve covalent alterations of DNA molecules. These methods of gene regulation are important in immune function and are dependent on different metabolic intermediates. One such method involves the TCA cycle intermediate α-ketoglutarate, which serves as a substrate that enables TET enzymes and JmjC domain-containing histone demethylases to induce DNA demethylation and histone demethylation, respectively. NAD+ is a required substrate for histone deacetylation, a process that is carried out by sirtuin (SIRT) histone deacetylases; conversely, SIRT enzymatic activity is inhibited by NADH (Cheng et al. 2014; Lin et al. 2004). Levels of NAD+ and NADH are influenced by glycolytic activity and oxidative phosphorylation; as a result, histone and DNA methylation are balanced by demethylation/methylation rates (Cheng et al. 2014; Lin et al. 2004). Another component of the TCA cycle, succinate instead serves to inhibit both processes. Furthermore, post-translational histone acetylation results in the activation of gene expression and requires acetyl-CoA derived from citrate (Loftus and Finlay 2016). Although these are just a few examples, it is clear that epigenetic regulation of gene expression is vital for optimal immune responses, including both activation and suppression. Moreover, this illustrates the interplay between cellular metabolism and precise immune cell function in accordance with the surrounding environment and conditions (Loftus and Finlay 2016).

8.6 Metabolic Dysregulation in Immune-Mediated Diseases

Disruption of metabolic function can lead to a host of disease states, including autoimmunity, cancer, infection, and chronic inflammation. Metabolic disorders such as obesity cause widespread inflammation and produce abnormal cellular microenvironments. For instance, hyperglycemia associated with type 2 diabetes mellitus disrupts normal T cell function, generating an immunosuppressive environment, which is favorable for tumor growth. Dyslipidemia commonly found in metabolic syndrome patients results in the disruption of cell membranes, impairing normal cell trafficking functions; it also may promote the generation of regulatory T cells. Chronic inflammation and metabolic dysfunction associated with these disease states may even impede gut microbiome composition, leading to gastrointestinal symptoms (Aderinto et al. 2023).

One common phenomenon of metabolic reprogramming is a preferential increase of glucose uptake and transition to aerobic glycolysis in favor of OXPHOS. This shift is also observed in cancer cells and is known as the "Warburg effect"; it is a finding underlying several autoimmune diseases. Another finding that goes hand in hand with the Warburg effect is the generation of a hypoxic microenvironment, which sustains T cells via activation of the signaling molecule hypoxia-inducible factor (HIF). The hypoxia also generates stress which drives the production of mitochondrial ROS, further increasing oxygen consumption. Preferential glycolysis may lead to excess lactate production, and both lactate and oxidative stress contribute to the amplification of damage in several autoimmune disorders. However, it is important to note that many diseases, including those discussed later in this section, may not follow these patterns (Blanco and Kaplan 2023).

8.6.1 Autoimmunity and Metabolism

Autoimmune diseases result from a failure of natural immune function and a pathologic reactionary immune response to self-tissue or antigens. These occur due to a breakdown of innate or adaptive immune self-tolerance mechanisms and lead to a number of acute or chronic disease states. Some typical pathophysiologic mechanisms include failure of thymic negative selection, loss of regulatory T cell function, and immune complex formation. Typically, underlying each of these mechanisms is a metabolic dysfunction in key immune cells due to genetic or environmental factors. Included below is a brief discussion of a few key autoimmune disorders and the metabolic dysregulation responsible for their primary pathophysiology (Theofilopoulos et al. 2017).

8.6.1.1 Systemic Lupus Erythematosus

Systemic lupus erythematosus (SLE) is an autoimmune disorder characterized by a broad pathological phenotype; patients may have a characteristic malar rash, lymphadenopathy, and vasculitis, as well as a number of cardiovascular, renal, and musculoskeletal manifestations. Innate and adaptive immune function are both dysregulated in this disorder. TLRs are activated in response to extracellular genetic material in damaged or apoptotic cells, which serve as self-antigens. This results in further production and binding of autoantibodies, such as anti-Sm, anti-Ro, and anti-La, which form immune complexes that can deposit in several tissue types and activate the complement system and neutrophils to deal further damage. There is also a failure to effectively clear autoreactive B cells and hyperactive T cells, leading to cyclic antigen presentation and activation between dysfunctional B and T cells (Robinson et al. 2022; Justiz Vaillant and Varacallo 2024).

Metabolically, B and T cells show enhanced mitochondrial membrane hyperpolarization, displaying increased excitability. This leads to increased generation of reactive oxygen species intermediates, inflammatory cytokines, and decreased apoptosis of autoreactive T cells. These T cells contain reactively hypertrophic mitochondria with greater cytosolic free calcium and decreased ATP reserves. The cells display signs of increased oxidative stress, and thus a transition to aerobic glycolysis and glutaminolysis, as well as a shift from mitochondrial fusion to fission. Furthermore, abnormalities within mTOR signaling pathways contribute to increased autophagy among B and T cells and disrupt the natural balance between Th17 and regulatory T cells, leading to an autoimmune state. Neutrophils display signs of failed mitophagy in response to the increased amount of oxidized mitochondrial DNA and are unable to keep up with their increased energetic demands. Macrophages are skewed toward an M2 phenotype and preferentially utilize OXPHOS to meet their increased metabolic requirements. The constellation of these metabolic aberrations leads to mitochondrial dysfunction and ultimately produces the hallmark systemic signs of SLE (Sharabi and Tsokos 2020; Robinson et al. 2022).

8.6.1.2 Rheumatoid Arthritis

Rheumatoid arthritis (RA) is a systemic autoimmune disorder that presents with inflammatory arthritis and extraarticular symptoms of lung or ophthalmologic disease. RA arises in response to environmental triggers (i.e., smoking) in patients with genetic predispositions, resulting in the formation of autoantibodies to various native proteins, notably citrullinated and carbamylated peptides. The environmental trigger is believed to cause repeated activation of innate immunity, causing widespread activation and recruitment of both innate and adaptive cells in later stages. The lymphocytic infiltrate and large numbers of cytokines overlaid with chronic inflammation produce progressive destruction of joints and cartilage, presenting earliest in smaller extremity joints and later spreading to larger joints (Weyand and Goronzy 2020; Chauhan et al. 2024).

Cells in RA patients display an anabolic phenotype, with elevated storage of fatty acids as cytoplasmic lipid droplets, decreased oxygen consumption, and decreased levels of free glucose and amino acids. Increased lipogenesis results in the formation of abnormally protruding cell membranes, which leave the cell vulnerable to destruction via pyroptosis. T cells also preferentially shunt toward the pentose phosphate pathway, resulting in decreased ATP, pyruvate, and lactate levels and increased NADPH and glutathione. This shift enables CD4+ T cells to bypass G/M2 checkpoints and proliferate in large numbers; however, cells also age prematurely as the uncharacteristically low levels of reactive oxygen species (ROS) downregulate protective ataxia-telangiectasia mutated (ATM) kinase and leave DNA vulnerable to damage by oxidative stress. Accelerated aging of immune cells in RA is further exacerbated by the loss of MRE11A, a key DNA repair nuclease. This transitions cells into a metabolically downregulated state with reduced oxygen consumption and ATP production. Macrophages in RA are hypermetabolic and display increased glucose and ATP consumption, causing mitochondrial stress. A similar inability to respond to ROS results in ROS-mediated joint destruction, via a phenotype similar to macrophages seen in coronary artery disease. Lastly, RA T cells display abnormal responses to mTOR, resulting in insufficient activation of AMPK. The combination of underactive AMPK and overactive mTOR results in underactive mitochondria, which undergo considerable stress in the RA hyperproliferative T cell environment (Weyand and Goronzy 2020; Chimenti et al. 2015).

8.6.1.3 Multiple Sclerosis

Multiple sclerosis (MS) is a chronic autoimmune condition of the central nervous system, characterized by widespread demyelination and neural damage. Areas of focal inflammation called plaques emerge in waves around microvasculature; these contain a mononuclear infiltrate of microglia and monocytes which bolster T cell invasion. Autoreactive T cells are the primary damage dealers, although recent research also suggests a role for B cells in humoral autoimmunity. Progression of plaques to scar tissue causes the permanent neurological dysfunctions that MS is best known for. Cells in MS show greatly decreased amounts of ROS due to defects in NADPH oxidase activity. They also demonstrate increased leptin activity; leptin, in addition to its relevance in hunger suppression, also acts as a regulatory T cell modulator via the mTOR pathway. Increased leptin results in decreased regulatory T cell proliferation and thus unimpeded immune activity and consequently disease progression. The MS disease state also results in increased concentrations of glutamate and glutamine, which cause autoimmune activation and cytokine release from T cells, excitotoxicity, and neuroinflammation via Th17 T cells (Freitag et al. 2016; Ghasemi et al. 2017).

8.6.1.4 Type 1 Diabetes Mellitus

Type 1 diabetes mellitus (T1DM) arises from the autoimmune destruction of pancreatic beta cells, resulting in a lack of viable insulin. The exact etiology of T1DM is largely undetermined but a human leukocyte antigen (HLA)-related genetic underpinning is strongly suspected. T1DM develops in stages, causing progressively worse hyperglycemia until treatment is warranted to avoid long-term microvascular damage. Pancreatic cell damage is wrought primarily by autoreactive T lymphocytes, although innate cells also infiltrate and play a role. Aberrant B cells secrete autoantibodies to key metabolic substrates such as insulin and accelerate disease progression; regulatory T cells also show relative dysfunction. ROS have been established to have a role in the pathogenesis of T1DM as well; however, unlike in MS and RA, disease progression is due to increased amounts of ROS. The hyperglycemia of T1DM is accompanied by unregulated OXPHOS and FAS; a byproduct of these metabolic pathways is increased production of gamma-aminobutyric acid (GABA), which serves a protective function in preventing NF-κB-mediated apoptosis of pancreatic islet cells (Freitag et al. 2016; Stathopoulou et al. 2019; Lucier and Weinstock 2024).

8.6.2 Therapeutic Approaches Targeting Immune Cell Metabolism

Several of the pathways discussed thus far can serve as targets for therapeutic intervention. Below are examples of some notable treatments which showcase immunometabolic approaches to therapy.

8.6.2.1 Rapamycin

One of the more established medications developed thus far is rapamycin, which acts on the PI3K/Akt/mTOR pathway. Inhibition of mTOR via this pathway results in improved T cell tolerance and downregulation of NK cell cytokine productivity and glycolytic metabolism. Improved T cell tolerance allows rapamycin to be an effective prevention tool for transplant rejection; NK activity is highly relevant in autoimmune diseases, and rapamycin has already been shown to be effective in mitigating disease in several such diseases, including SLE and RA (Teng et al. 2019; Pålsson-McDermott and O'Neill 2020; Wang et al. 2021).

8.6.2.2 Metformin

Metformin is an established treatment for type 2 diabetes mellitus due to its ability to improve glucose metabolism. It also has anti-inflammatory mechanisms—via inhibition of Complex 1 of the electron transport chain, metformin prevents ROS formation and thus limits pro-inflammatory macrophage activity and oxidative stress. The resulting decrease in ATP output also upregulates the AMPK pathway, increasing fatty acid metabolism and leading to a decrease in visceral fat. Furthermore, metformin directly inhibits mTORC1 and improves faulty B cell function. So far, the benefits of metformin have been demonstrated in SLE patients, with further investigations in progress (Stathopoulou et al. 2019; Pålsson-McDermott and O'Neill 2020; Wang et al. 2021; Abdallah et al. 2021).

8.6.2.3 Dimethyl Fumarate

Dimethyl fumarate (DMF) is a current treatment option for MS which promotes a range of anti-inflammatory effects in several cell types via an unclear mechanism. Administration of DMF results in depletion of B cells, an increase in NK cell populations, and a subsequent decrease in cytotoxic T cell phenotypes. From a metabolic standpoint, DMF interacts with several proteins to inhibit glucose-6-phosphate dehydrogenase and limit pentose phosphate pathway shunting, thus decreasing overall cell size and inducing apoptosis in more cells. DMF also increases fatty acid oxidation in mitochondria and decreases overall lipid synthesis, downstream effects of the increased activity of the antioxidant gene Nuclear Factor like 2 (Nrf2) (Stathopoulou et al. 2019; Pålsson-McDermott and O'Neill 2020; Linker and Gold 2013).

8.6.2.4 Methotrexate

Methotrexate is a notable antineoplastic drug, which inhibits folate metabolism. Methotrexate is also effective at reducing inflammation and immune proliferation via an unclear secondary mechanism in chronic diseases such as RA, psoriasis, and Crohn's disease. One suggested mechanism is upregulation of amido-imidazole-carbox-amidoribonucleotide (AICAR), an AMPK activator, which would leading to downstream T cell immunosuppression. This causes an anti-inflammatory phenotypic switch to M2 macrophages, decreases neutrophil activity, and increases regulatory T cell differentiation. Methotrexate also may reduce oxidative stress via free radical scavenging, altering ROS-mediated regulation to favor apoptosis of T lymphocytes (Stathopoulou et al. 2019; Pålsson-McDermott and O'Neill 2020; Maksimovic et al. 2020).

8.6.2.5 2-Deoxy-D-Glucose

2-Deoxy-D-glucose (2-DG) is a glucose analog, which functions as an inhibitor of glycolysis. It competitively binds to hexokinase to prevent glucose binding and accumulates in metabolically active tissues such as macrophages; it has thus been useful for visualization of areas of inflammation for diagnostic purposes. However, it also affects immune regulation, upregulating IL-17 production and formation of regulatory T cells. It is also essential for NK and dendritic cell function and B cell proliferation; and therefore, low levels of 2-DG contribute to weakened immune activity and promotion of tolerance in autoimmune states. In combination with metformin, 2-DG was effective in reducing signs of inflammation in mouse models of SLE. 2-DG injections also slowed disease progression in rat models of Guillain-Barré syndrome. Given its preliminary success, 2-DG may be a strong candidate for treatment of autoimmune and inflammatory conditions (Stathopoulou et al. 2019; Pålsson-McDermott and O'Neill 2020; Wijayasinghe et al. 2022).

8.6.2.6 Promising Therapeutic Targets

While existing therapies have shown promise for their anti-inflammatory properties, novel and promising targets exist in the glycolytic and OXPHOS pathways. GLUT1 is a primary glucose transporter, which is upregulated during proglycolytic states such as inflammation, and is present in macrophages, NK cells, and T effector cells. It also plays a role in wound healing to allow keratinocyte proliferation; a recent study has shown that inhibition of GLUT1 can inhibit inflammation in skin samples from psoriasis patients (Liu et al. 2012). PKM2 is a pyruvate kinase isoform, which exists in multiple subunit states. When in its monomeric or dimeric form, it can exert pro-inflammatory effects, promoting aerobic glycolysis and even cancer cell proliferation. However, studies have shown that tetramerization of PKM2 causes it to revert to its default function as part of the glycolytic pathway, and this form as a treatment has demonstrated improvement of inflammation in coronary artery disease patients (Shirai et al. 2016).

Several enzymes and intermediates in the OXPHOS pathway also demonstrate possible clinical utility. Succinate is an intermediate produced as part of the TCA cycle and exists at the intersection between several metabolic pathways. It is released extracellularly during damaging conditions such as autoimmune disease, ischemia, or extreme temperature. Succinate binding to its receptor promotes a pro-inflammatory phenotype, suggesting a target for inhibition in this receptor. Itaconate is an immunomodulator produced through the TCA cycle that has anti-inflammatory properties, working through a mechanism similar to DMF. Citrate is an essential TCA cycle metabolite, which builds up in active immune cells due to the inhibition of isocitrate dehydrogenase (IDH). IDH is downregulated in cells undergoing active inflammation, preventing normal feedback inhibition of glycolysis, and allowing citrate to reenter metabolism through various pathways. Targeted inhibition of these pathways may provide treatment avenues for several cancers and inflammatory conditions (Pålsson-McDermott and O'Neill 2020; Zotta et al. 2024).

8.7 Conclusion

In this chapter, we have discussed the intersection between immune function and metabolism. In order to carry out their roles, cells of our highly complex and diverse immune system consume large amounts of energy derived through an equally intricate set of metabolic processes. Differential regulation of these processes and intersecting signaling pathways result in multiple immune phenotypes. When this balance is disrupted due to metabolic dysregulation, a number of disease states can emerge, ranging from cancer to autoimmunity. Underlying each of these disease states are points of failure in these key metabolic processes. These represent valuable targets for therapeutic intervention, and several treatments have already been developed which show significant promise in clinical settings. However, ongoing research is essential to fully unlock the potential of immunometabolism—further study of this field may provide opportunities for groundbreaking interventions that could significantly impact patient lives.

References

Abdallah MS, Alarfaj SJ, Saif DS, El-Naggar ME, Elsokary MA, Elsawah HK et al (2021) The AMPK modulator metformin as adjunct to methotrexate in patients with rheumatoid arthritis: a proof-of-concept, randomized, double-blind, placebo-controlled trial. Int Immunopharmacol 95:107575

Aderinto N, Abdulbasit MO, Tangmi ADE, Okesanya JO, Mubarak JM (2023) Unveiling the growing significance of metabolism in modulating immune cell function: exploring mechanisms and implications; a review. Ann Med Surg 85(11):5511

Akoumi A, Haffar T, Mousterji M, Kiss RS, Bousette N (2017) Palmitate mediated diacylglycerol accumulation causes endoplasmic reticulum stress, Plin2 degradation, and cell death in H9C2 cardiomyoblasts. Exp Cell Res 354(2):85–94

Angelin A, Gil-de-Gómez L, Dahiya S, Jiao J, Guo L, Levine MH et al (2017) Foxp3 reprograms T cell metabolism to function in low-glucose, high-lactate environments. Cell Metab 25(6):1282–93.e7

Araki K, Turner AP, Shaffer VO, Gangappa S, Keller SA, Bachmann MF et al (2009) mTOR regulates memory CD8 T-cell differentiation. Nature 460(7251):108–112

Aristizábal B, González Á (2013) Innate immune system. Autoimmunity: from bench to bedside [Internet]. El Rosario University Press

Assmann N, O'Brien KL, Donnelly RP, Dyck L, Zaiatz-Bittencourt V, Loftus RM et al (2017) Srebp-controlled glucose metabolism is essential for NK cell functional responses. Nat Immunol 18(11):1197–1206

Bantug GR, Fischer M, Grählert J, Balmer ML, Unterstab G, Develioglu L et al (2018) Mitochondria-endoplasmic reticulum contact sites function as immunometabolic hubs that orchestrate the rapid recall response of memory CD8(+) T cells. Immunity 48(3):542–55.e6

Bao H, Wang Y, Xiong H, Xia Y, Cui Z, Liu L (2024) Mechanism of iron ion homeostasis in intestinal immunity and gut microbiota remodeling. Int J Mol Sci 25(2):727

Beier UH, Angelin A, Akimova T, Wang L, Liu Y, Xiao H et al (2015) Essential role of mitochondrial energy metabolism in Foxp3^{+} T-regulatory cell function and allograft survival. FASEB J 29(6):2315–2326

Bell D, Young JW, Banchereau J (1999) Dendritic cells. In: Dixon FJ, Alt F, Austen KF, Kishimoto T, Melchers F, Uhr JW (eds) Advances in immunology. 72. Academic Press, pp 255–324

Blagih J, Coulombe F, Vincent EE, Dupuy F, Galicia-Vázquez G, Yurchenko E et al (2015) The energy sensor AMPK regulates T cell metabolic adaptation and effector responses in vivo. Immunity 42(1):41–54

Blanco LP, Kaplan MJ (2023) Metabolic alterations of the immune system in the pathogenesis of autoimmune diseases. PLoS Biol 21(4):e3002084

Buck MD, O'Sullivan D, Pearce EL (2015) T cell metabolism drives immunity. J Exp Med 212(9):1345–1360

Buck MD, O'Sullivan D, Klein Geltink RI, Curtis JD, Chang CH, Sanin DE et al (2016) Mitochondrial dynamics controls T cell fate through metabolic programming. Cell 166(1):63–76

Caro-Maldonado A, Wang R, Nichols AG, Kuraoka M, Milasta S, Sun LD et al (2014) Metabolic reprogramming is required for antibody production that is suppressed in Anergic but exaggerated in chronically BAFF-exposed B cells. J Immunol 192(8):3626–3636

Chang C-H, Curtis Jonathan D, Maggi Leonard B, Faubert B, Villarino Alejandro V, O'Sullivan D et al (2013) Posttranscriptional control of T cell effector function by aerobic glycolysis. Cell 153(6):1239–1251

Chaplin DD (2010) Overview of the immune response. J Allergy Clin Immunol 125(2 Suppl 2):S3–23

Chauhan K, Jandu JS, Brent LH, Al-Dhahir MA (2024) Rheumatoid arthritis. StatPearls. StatPearls Publishing Copyright © 2024, StatPearls Publishing LLC, Treasure Island

Chavakis T (2021) Immunometabolism: where immunology and metabolism meet. J Innate Immun 14(1):1–3

Chen L, Flies DB (2013) Molecular mechanisms of T cell co-stimulation and co-inhibition. Nat Rev Immunol 13(4):227–242

Cheng SC, Quintin J, Cramer RA, Shepardson KM, Saeed S, Kumar V et al (2014) mTOR- and HIF-1α-mediated aerobic glycolysis as metabolic basis for trained immunity. Science 345(6204):1250684

Chimenti MS, Triggianese P, Conigliaro P, Candi E, Melino G, Perricone R (2015) The interplay between inflammation and metabolism in rheumatoid arthritis. Cell Death Dis 6(9):e1887-e

Corcoran Lynn M, Nutt SL (2016) Long-lived plasma cells have a sweet tooth. Immunity 45(1):3–5

Costa-Mattioli M, Walter P (2020) The integrated stress response: from mechanism to disease. Science 368(6489):eaat5314

Dai J, Fang P, Saredy J, Xi H, Ramon C, Yang W et al (2017) Metabolism-associated danger signal-induced immune response and reverse immune checkpoint-activated CD40+ monocyte differentiation. J Hematol Oncol 10(1):141

De Lima TL, Peron G, Oliveira J, Da Rosa LC, Thomé R, Verinaud L (2018) The impact of metabolic reprogramming on dendritic cell function. Int Immunopharmacol 63:84–93

Domínguez-Andrés J, Arts RJW, Ter Horst R, Gresnigt MS, Smeekens SP, Ratter JM et al (2017) Rewiring monocyte glucose metabolism via C-type lectin signaling protects against disseminated candidiasis. PLoS Pathog 13(9):e1006632

Donnelly RP, Finlay DK (2015) Glucose, glycolysis and lymphocyte responses. Mol Immunol 68(2):513–519

Doria-Rose NA, Joyce MG (2015) Strategies to guide the antibody affinity maturation process. Curr Opin Virol 11:137–147

Doughty CA, Bleiman BF, Wagner DJ, Dufort FJ, Mataraza JM, Roberts MF et al (2006) Antigen receptor–mediated changes in glucose metabolism in B lymphocytes: role of phosphatidylinositol 3-kinase signaling in the glycolytic control of growth. Blood 107(11):4458–4465

Dufort FJ, Bleiman BF, Gumina MR, Blair D, Wagner DJ, Roberts MF et al (2007) Cutting edge: IL-4-mediated protection of primary B lymphocytes from apoptosis via Stat6-dependent regulation of glycolytic metabolism. J Immunol 179(8):4953–4957

Embgenbroich M, Burgdorf S (2018) Current concepts of antigen cross-presentation. Front Immunol 9:1643

Feng X, Li X, Liu N, Hou N, Sun X, Liu Y (2022) Glutaminolysis and CD4+ T-cell metabolism in autoimmunity: from pathogenesis to therapy prospects. Front Immunol 13:986847

Frauwirth KA, Riley JL, Harris MH, Parry RV, Rathmell JC, Plas DR et al (2002) The CD28 signaling pathway regulates glucose metabolism. Immunity 16(6):769–777

Freitag J, Berod L, Kamradt T, Sparwasser T (2016) Immunometabolism and autoimmunity. Immunol Cell Biol 94(10):925–934

Galgani M, De Rosa V, La Cava A, Matarese G (2016) Role of metabolism in the immunobiology of regulatory T cells. J Immunol 197(7):2567–2575

Ganeshan K, Chawla A (2014) Metabolic regulation of immune responses. Annu Rev Immunol 32:609–634

Gerriets VA, Kishton RJ, Johnson MO, Cohen S, Siska PJ, Nichols AG et al (2016) Foxp3 and Toll-like receptor signaling balance T(reg) cell anabolic metabolism for suppression. Nat Immunol 17(12):1459–1466

Ghasemi N, Razavi S, Nikzad E (2017) Multiple sclerosis: pathogenesis, symptoms. Diagn Cell-Based Ther Cell J 19(1):1–10

Giovanelli P, Sandoval TA, Cubillos-Ruiz JR (2019) Dendritic cell metabolism and function in tumors. Trends Immunol 40(8):699–718

Golubovskaya V, Wu L (2016) Different subsets of T cells, memory, effector functions, and CAR-T immunotherapy. Cancers (Basel) 8(3):36

Gómez Román VR, Murray JC, Weiner LM (2014) Chapter 1 - Antibody-Dependent Cellular Cytotoxicity (ADCC). In: Ackerman ME, Nimmerjahn F (eds) Antibody Fc. Academic Press, Boston, pp 1–27

Gonzalez S, González-Rodríguez AP, Suárez-Álvarez B, López-Soto A, Huergo-Zapico L, Lopez-Larrea C (2011) Conceptual aspects of self and nonself discrimination. Self Nonself 2(1):19–25

Gordon S, Martinez-Pomares L (2017) Physiological roles of macrophages. Pflugers Arch - Eur J Physiol 469(3):365–374

Grover HS, Saini R, Bhardwaj P, Bhardwaj A (2016) Acute-phase reactants. J Oral Res Rev 8(1):32–35

Hachiya R, Tanaka M, Itoh M, Suganami T (2022) Molecular mechanism of crosstalk between immune and metabolic systems in metabolic syndrome. Inflamm Regener 42(1):13

He Z, Zhu X, Shi Z, Wu T, Wu L (2019) Metabolic regulation of dendritic cell differentiation. Front Immunol 10:410

Hosseini SS, Khalili S, Baradaran B, Bidar N, Shahbazi M-A, Mosafer J et al (2021) Bispecific monoclonal antibodies for targeted immunotherapy of solid tumors: recent advances and clinical trials. Int J Biol Macromol 167:1030–1047

Hotamisligil GS (2017) Foundations of immunometabolism and implications for metabolic health and disease. Immunity 47(3):406–420

Hu C, Xuan Y, Zhang X, Liu Y, Yang S, Yang K (2022) Immune cell metabolism and metabolic reprogramming. Mol Biol Rep 49(10):9783–9795

Huang H, Long L, Zhou P, Chapman NM, Chi H (2020) mTOR signaling at the crossroads of environmental signals and T-cell fate decisions. Immunol Rev 295(1):15–38

Hubert P, Amigorena S (2012) Antibody-dependent cell cytotoxicity in monoclonal antibody-mediated tumor immunotherapy. Onco Targets Ther 1(1):103–105

Iwasaki Y, Suganami T, Hachiya R, Shirakawa I, Kim-Saijo M, Tanaka M et al (2014) Activating transcription factor 4 links metabolic stress to interleukin-6 expression in macrophages. Diabetes 63(1):152–161

Ji L, Zhao X, Zhang B, Kang L, Song W, Zhao B et al (2019) Slc6a8-mediated Creatine uptake and accumulation reprogram macrophage polarization via regulating cytokine responses. Immunity 51(2):272–84.e7

Johnson EE, Wessling-Resnick M (2012) Iron metabolism and the innate immune response to infection. Microbes Infect 14(3):207–216

Justiz Vaillant AAGA, Varacallo M (2024) Systemic lupus erythematosus StatPearls [Internet]. StatPearls Publishing. [updated 2024 Jan-. Available from: https://www.ncbi.nlm.nih.gov/books/NBK535405/

Kadl A, Meher AK, Sharma PR, Lee MY, Doran AC, Johnstone SR et al (2010) Identification of a novel macrophage phenotype that develops in response to Atherogenic phospholipids via Nrf2. Circ Res 107(6):737–746

Kemp F, Braverman EL, Byersdorfer CA (2024) Fatty acid oxidation in immune function. Front Immunol 15:1420336

Keppel MP, Saucier N, Mah AY, Vogel TP, Cooper MA (2015) Activation-specific metabolic requirements for NK cell IFN-γ production. J Immunol 194(4):1954–1962

Klein Geltink RI, O'Sullivan D, Corrado M, Bremser A, Buck MD, Buescher JM et al (2017) Mitochondrial priming by CD28. Cell 171(2):385–97.e11

Klein Geltink RI, Kyle RL, Pearce EL (2018) Unraveling the complex interplay between T cell metabolism and function. Annu Rev Immunol 36(1):461–488

Klein L, Kyewski B, Allen PM, Hogquist KA (2014) Positive and negative selection of the T cell repertoire: what thymocytes see (and don't see). Nat Rev Immunol 14(6):377–391

Koelwyn GJ, Corr EM, Erbay E, Moore KJ (2018) Regulation of macrophage immunometabolism in atherosclerosis. Nat Immunol 19(6):526–537

Koh C-H, Lee S, Kwak M, Kim B-S, Chung Y (2023) CD8 T-cell subsets: heterogeneity, functions, and therapeutic potential. Exp Mol Med 55(11):2287–2299

Krawczyk CM, Holowka T, Sun J, Blagih J, Amiel E, DeBerardinis RJ et al (2010) Toll-like receptor–induced changes in glycolytic metabolism regulate dendritic cell activation. Blood 115(23):4742–4749

Lachmandas E, Boutens L, Ratter JM, Hijmans A, Hooiveld GJ, Joosten LAB et al (2016) Microbial stimulation of different Toll-like receptor signalling pathways induces diverse metabolic programmes in human monocytes. Nat Microbiol 2(3):16246

Lakkis JI, Weir MR (2020) Chapter 31 - Hematologic and infectious complications of chronic kidney disease. In: Kimmel PL, Rosenberg ME (eds) Chronic renal disease, 2nd edn. Academic Press, pp 477–502

Lam Wing Y, Becker Amy M, Kennerly Krista M, Wong R, Curtis Jonathan D, Llufrio Elizabeth M et al (2016) Mitochondrial pyruvate import promotes long-term survival of antibody-secreting plasma cells. Immunity 45(1):60–73

Leamy AK, Egnatchik RA, Shiota M, Ivanova PT, Myers DS, Brown HA et al (2014) Enhanced synthesis of saturated phospholipids is associated with ER stress and lipotoxicity in palmitate treated hepatic cells. J Lipid Res 55(7):1478–1488

LeBien TW, Tedder TF (2008) B lymphocytes: how they develop and function. Blood 112(5):1570–1580

Liang S, Guo XK, Ou J, Huang R, Xue Q, Zhang B et al (2019) Nutrient sensing by the intestinal epithelium orchestrates mucosal antimicrobial defense via translational control of Hes1. Cell Host Microbe 25(5):706–18.e7

Liang S, Ji L, Kang L, Hu X (2020) Chapter Five - Metabolic regulation of innate immunity. In: Dong C, Jiang Z (eds) Advances in immunology. Advances in immunology in China - part B. 145. Academic Press, pp 129–157

Lin SJ, Ford E, Haigis M, Liszt G, Guarente L (2004) Calorie restriction extends yeast life span by lowering the level of NADH. Genes Dev 18(1):12–16

Linker RA, Gold R (2013) Dimethyl fumarate for treatment of multiple sclerosis: mechanism of action, effectiveness, and side effects. Curr Neurol Neurosci Rep 13(11):394

Liu Y, Cao Y, Zhang W, Bergmeier S, Qian Y, Akbar H et al (2012) A small-molecule inhibitor of glucose transporter 1 downregulates glycolysis, induces cell-cycle arrest, and inhibits cancer cell growth in vitro and in vivo. Mol Cancer Ther 11(8):1672–1682

Liu S, Huang B, Cao J, Wang Y, Xiao H, Zhu Y et al (2023) ROS fine-tunes the function and fate of immune cells. Int Immunopharmacol 119:110069

Loftus RM, Finlay DK (2016) Immunometabolism: cellular metabolism turns immune regulator *. J Biol Chem 291(1):1–10

Loftus RM, Assmann N, Kedia-Mehta N, O'Brien KL, Garcia A, Gillespie C et al (2018) Amino acid-dependent cMyc expression is essential for NK cell metabolic and functional responses in mice. Nat Commun 9(1):2341

Lucier J, Weinstock RS (2024) Type 1 diabetes. StatPearls. StatPearls Publishing Copyright © 2024, StatPearls Publishing LLC, Treasure Island

MacIver NJ, Blagih J, Saucillo DC, Tonelli L, Griss T, Rathmell JC et al (2011) The liver kinase B1 is a central regulator of T cell development, activation, and metabolism. J Immunol 187(8):4187–4198

Mah AY, Rashidi A, Keppel MP, Saucier N, Moore EK, Alinger JB et al (2017) Glycolytic requirement for NK cell cytotoxicity and cytomegalovirus control. JCI Insight 2(23):e95128

Makowski L, Chaib M, Rathmell JC (2020) Immunometabolism: from basic mechanisms to translation. Immunol Rev 295(1):5–14

Maksimovic V, Pavlovic-Popovic Z, Vukmirovic S, Cvejic J, Mooranian A, Al-Salami H et al (2020) Molecular mechanism of action and pharmacokinetic properties of methotrexate. Mol Biol Rep 47(6):4699–4708

Manfredi AA, Rovere-Querini P, Bottazzi B, Garlanda C, Mantovani A (2008) Pentraxins, humoral innate immunity and tissue injury. Curr Opin Immunol 20(5):538–544

Mantovani A, Garlanda C (2023) Humoral innate immunity and acute-phase proteins. N Engl J Med 388(5):439–452

Martinon F, Chen X, Lee AH, Glimcher LH (2010) TLR activation of the transcription factor XBP1 regulates innate immune responses in macrophages. Nat Immunol 11(5):411–418

Masopust D, Schenkel JM (2013) The integration of T cell migration, differentiation and function. Nat Rev Immunol 13(5):309–320

Mayadas TN, Cullere X, Lowell CA (2014) The multifaceted functions of neutrophils. Ann Rev Pathol Mech Dis 9:181–218

McComb S, Thiriot A, Akache B, Krishnan L, Stark F (2019) Introduction to the immune system. In: Fulton KM, Twine SM (eds) Immunoproteomics. 2024. Springer New York, New York, pp 1–24
Mehta MM, Weinberg SE, Chandel NS (2017) Mitochondrial control of immunity: beyond ATP. Nat Rev Immunol 17(10):608–620
MetaCyc: Metabolic pathways from all domains of life
Michalek RD, Gerriets VA, Jacobs SR, Macintyre AN, MacIver NJ, Mason EF et al (2011) Cutting edge: distinct glycolytic and lipid oxidative metabolic programs are essential for effector and regulatory CD4+ T cell subsets. J Immunol 186(6):3299–3303
Mukherjee S, Hooper LV (2015) Antimicrobial defense of the intestine. Immunity 42(1):28–39
Murray PJ, Rathmell J, Pearce E (2015) SnapShot: Immunometabolism. Cell Metab 22(1):190-.e1
Newsholme P (2001) Why is L-Glutamine metabolism important to cells of the immune system in health, postinjury, surgery or infection? J Nutr 131(9):2515S–2522S
Newton R, Priyadharshini B, Turka LA (2016) Immunometabolism of regulatory T cells. Nat Immunol 17(6):618–625
Ni S, Yuan Y, Kuang Y, Li X (2022) Iron metabolism and immune regulation. Front Immunol 13:816282
O'Brien KL, Finlay DK (2019) Immunometabolism and natural killer cell responses. Nat Rev Immunol 19(5):282–290
O'Neill LAJ, Kishton RJ, Rathmell J (2016) A guide to immunometabolism for immunologists. Nat Rev Immunol 16(9):553–565
Palmblad J (1984) The role of granulocytes in inflammation. Scand J Rheumatol 13(2):163–172
Pålsson-McDermott EM, O'Neill LAJ (2020) Targeting immunometabolism as an anti-inflammatory strategy. Cell Res 30(4):300–314
Parry RV, Chemnitz JM, Frauwirth KA, Lanfranco AR, Braunstein I, Kobayashi SV et al (2005) CTLA-4 and PD-1 receptors inhibit T-cell activation by distinct mechanisms. Mol Cell Biol 25(21):9543–9553
Patsoukis N, Bardhan K, Chatterjee P, Sari D, Liu B, Bell LN et al (2015) PD-1 alters T-cell metabolic reprogramming by inhibiting glycolysis and promoting lipolysis and fatty acid oxidation. Nat Commun 6(1):6692
Pearce EJ, Everts B (2015) Dendritic cell metabolism. Nat Rev Immunol 15(1):18–29
Pearce EL, Pearce EJ (2013) Metabolic pathways in immune cell activation and quiescence. Immunity 38(4):633–643
Pearce EL, Walsh MC, Cejas PJ, Harms GM, Shen H, Wang L-S et al (2009) Enhancing CD8 T-cell memory by modulating fatty acid metabolism. Nature 460(7251):103–107
Peng M, Yin N, Chhangawala S, Xu K, Leslie CS, Li MO (2016) Aerobic glycolysis promotes T helper 1 cell differentiation through an epigenetic mechanism. Science 354(6311):481–484
Raulien N, Friedrich K, Strobel S, Rubner S, Baumann S, Von Bergen M et al (2017) Fatty acid oxidation compensates for lipopolysaccharide-induced Warburg effect in glucose-deprived monocytes. Front Immunol 8:609
Riera Romo M, Pérez-Martínez D, Castillo FC (2016) Innate immunity in vertebrates: an overview. Immunology 148(2):125–139
Robinson GA, Wilkinson MGL, Wincup C (2022) The role of immunometabolism in the pathogenesis of systemic lupus erythematosus. Front Immunol 12:806560
Saberi M, Woods NB, de Luca C, Schenk S, Lu JC, Bandyopadhyay G et al (2009) Hematopoietic cell-specific deletion of toll-like receptor 4 ameliorates hepatic and adipose tissue insulin resistance in high-fat-fed mice. Cell Metab 10(5):419–429
Sarma JV, Ward PA (2011) The complement system. Cell Tissue Res 343(1):227–235
Seifert M, Küppers R (2016) Human memory B cells. Leukemia 30(12):2283–2292
Sharabi A, Tsokos GC (2020) T cell metabolism: new insights in systemic lupus erythematosus pathogenesis and therapy. Nat Rev Rheumatol 16(2):100–112
Shen K, Sabatini DM (2018) Ragulator and SLC38A9 activate the Rag GTPases through noncanonical GEF mechanisms. Proc Natl Acad Sci USA 115(38):9545–9550

Shirai T, Nazarewicz RR, Wallis BB, Yanes RE, Watanabe R, Hilhorst M et al (2016) The glycolytic enzyme PKM2 bridges metabolic and inflammatory dysfunction in coronary artery disease. J Exp Med 213(3):337–354

Shishido SN, Varahan S, Yuan K, Li X, Fleming SD (2012) Humoral innate immune response and disease. Clin Immunol 144(2):142–158

Stathopoulou C, Nikoleri D, GeorgeBertsias. (2019) Immunometabolism: an overview and therapeutic prospects in autoimmune diseases. Immunotherapy 11(9):813–829

Stavnezer J, Guikema JEJ, Schrader CE (2008) Mechanism and regulation of class switch recombination. Annu Rev Immunol 26(1):261–292

Steinert EM, Vasan K, Chandel NS (2021) Mitochondrial metabolism regulation of T cell–mediated immunity. Annu Rev Immunol 39(1):395–416

Suganami T, Mieda T, Itoh M, Shimoda Y, Kamei Y, Ogawa Y (2007) Attenuation of obesity-induced adipose tissue inflammation in C3H/HeJ mice carrying a Toll-like receptor 4 mutation. Biochem Biophys Res Commun 354(1):45–49

Sun L, Yang X, Yuan Z, Wang H (2020) Metabolic reprogramming in immune response and tissue inflammation. Arterioscler Thromb Vasc Biol 40(9):1990–2001

Sun J-X, Xu X-H, Jin L (2022) Effects of metabolism on macrophage polarization under different disease backgrounds. Front Immunol 13:880286

Takaba H, Takayanagi H (2017) The mechanisms of T cell selection in the thymus. Trends Immunol 38(11):805–816

Tanaka M, Ikeda K, Suganami T, Komiya C, Ochi K, Shirakawa I et al (2014) Macrophage-inducible C-type lectin underlies obesity-induced adipose tissue fibrosis. Nat Commun 5(1):4982

Teng X, Li W, Cornaby C, Morel L (2019) Immune cell metabolism in autoimmunity. Clin Exp Immunol 197(2):181–192

Thapa P, Farber DL (2019) The role of the thymus in the immune response. Thorac Surg Clin 29(2):123–131

Theisen D, Murphy K (2017) The role of cDC1s in vivo: CD8 T cell priming through cross-presentation. F1000Res 6:98

Theofilopoulos AN, Kono DH, Baccala R (2017) The multiple pathways to autoimmunity. Nat Immunol 18(7):716–724

Turvey SE, Broide DH (2010) Innate immunity. J Allergy Clin Immunol 125(2):S24–S32

Victora GD, Schwickert TA, Fooksman DR, Kamphorst AO, Meyer-Hermann M, Dustin ML et al (2010) Germinal center dynamics revealed by multiphoton microscopy with a photoactivatable fluorescent reporter. Cell 143(4):592–605

Vivier E, Tomasello E, Baratin M, Walzer T, Ugolini S (2008) Functions of natural killer cells. Nat Immunol 9(5):503–510

Volmer R, van der Ploeg K, Ron D (2013) Membrane lipid saturation activates endoplasmic reticulum unfolded protein response transducers through their transmembrane domains. Proc Natl Acad Sci USA 110(12):4628–4633

Wang T, Jiao Y, Zhang X (2021) Immunometabolic pathways and its therapeutic implication in autoimmune diseases. Clin Rev Allergy Immunol 60(1):55–67

Weyand CM, Goronzy JJ (2020) Immunometabolism in the development of rheumatoid arthritis. Immunol Rev 294(1):177–187

Wicherska-Pawłowska K, Wróbel T, Rybka J (2021) Toll-Like Receptors (TLRs), NOD-Like Receptors (NLRs), and RIG-I-Like Receptors (RLRs) in innate immunity. TLRs, NLRs, and RLRs ligands as immunotherapeutic agents for hematopoietic diseases. Int J Mol Sci 22(24):13397

Wijayasinghe YS, Bhansali MP, Borkar MR, Chaturbhuj GU, Muntean BS, Viola RE et al (2022) A comprehensive biological and synthetic perspective on 2-Deoxy-d-Glucose (2-DG), a sweet molecule with therapeutic and diagnostic potentials. J Med Chem 65(5):3706–3728

Williams NC, O'Neill LAJ (2018) A role for the Krebs cycle intermediate citrate in metabolic reprogramming in innate immunity and inflammation. Front Immunol 9:141

Worbs T, Hammerschmidt SI, Förster R (2017) Dendritic cell migration in health and disease. Nat Rev Immunol 17(1):30–48

Wyant GA, Abu-Remaileh M, Wolfson RL, Chen WW, Freinkman E, Danai LV et al (2017) mTORC1 activator SLC38A9 is required to efflux essential amino acids from lysosomes and use protein as a nutrient. Cell 171(3):642–54.e12

Wynn TA, Chawla A, Pollard JW (2013) Macrophage biology in development, homeostasis and disease. Nature 496(7446):445–455

Xiong L, Helm EY, Dean JW, Sun N, Jimenez-Rondan FR, Zhou L (2023) Nutrition impact on ILC3 maintenance and function centers on a cell-intrinsic CD71–iron axis. Nat Immunol 24(10):1671–1684

Xu Y, Li M, Lin M, Cui D, Xie J (2024) Glutaminolysis of CD4+ T cells: a potential therapeutic target in viral diseases. J Inflamm Res 17:603–616

Yang J, Zhang L, Yu C, Yang X-F, Wang H (2014) Monocyte and macrophage differentiation: circulation inflammatory monocyte as biomarker for inflammatory diseases. Biomark Res 2(1):1

Yatim KM, Lakkis FG (2015) A brief journey through the immune system. Clin J Am Soc Nephrol 10(7):1274–1281

Zotta A, O'Neill LAJ, Yin M (2024) Unlocking potential: the role of the electron transport chain in immunometabolism. Trends Immunol 45(4):259–273

Metabolic Imbalance in Immune Cells in Relation to Metabolic Disorders, Cancer, and Infections

9

Mahmoud Singer, Rania Hamdy, Jana H. Ghonaim, and Mohamed I. Husseiny

Abstract

The function of the immune system is essential for defending the body against infections and diseases and maintaining overall homeostasis. Metabolism, the set of life-sustaining chemical reactions within cells plays a crucial role in supporting immune responses. In immune cells, metabolic reprogramming is essential for meeting the energetic and biosynthetic demands required for effective immune responses. Various aspects affect the metabolic reprogramming of immune cells and its significance in immune activation and regulation. This chapter delves into the intricate relationship between metabolism and immunity, highlighting how metabolic pathways influence the activation, differentiation, and function of immune cells. By understanding these connections, we can gain insights into potential therapeutic strategies for modulating immune responses in various diseases.

M. Singer (✉)
Department of Radiological Sciences, School of Medicine, University of California Irvine, Irvine, CA, USA
e-mail: singermk@hs.uci.edu

R. Hamdy
Research Institute for Science and Engineering (RISE), University of Sharjah, Sharjah, UAE

J. H. Ghonaim
MIPS, Sharjah, UAE

M. I. Husseiny
Department of Translational Research & Cellular Therapeutics, Arthur Riggs Diabetes & Metabolism Research Institute, Beckman Research Institute, City of Hope National Medical Center, Duarte, CA, USA

S. S. M. Soliman, M. I. Husseiny (eds.), *Metabolic Dynamics in Host-Microbe Interaction*, https://doi.org/10.1007/978-981-96-1305-2_9

Keywords

Immunometabolism · Immune regulation · Metabolic imbalance · Immune dysfunction · Metabolic pathways

9.1 Introduction

Metabolism is the process that sustains life through biochemical reactions. It can be categorized as anabolism—the synthesis of complex molecules from simpler ones, requiring energy input. Catabolism breaks down complex molecules, releasing energy. Immune responses rely heavily on metabolism, influencing how immune cells function and are controlled. Its importance is underscored by key points, including energy supply, cellular function, differentiation, immune cell homeostasis, metabolism, and disease implications (Judge and Dodd 2020). Immune cells' activation status is closely tied to their energy supply. Ample energy is necessary for T-cells, B-cells, and macrophages to complete essential functions, including dividing and developing (Ganeshan and Chawla 2014). These energy needs are fulfilled by ATP generated through the typical metabolic pathways of glycolysis and oxidative phosphorylation. By sensing the level of nutrients, the immune cells can adjust their activation and function accordingly (Al-Khami et al. 2017). Differentiation of immune subsets, such as effector T-cells and regulatory T-cells, is mainly influenced by metabolism and its processes. Effector T-cells quickly generate energy through glycolysis, whereas memory T-cells primarily rely on oxidative phosphorylation and fatty acid oxidation (FAO) (Ma et al. 2024). Similarly, macrophages have the ability to adopt either pro-inflammatory (M1) or anti-inflammatory (M2) phenotypes, depending on metabolic signals. M1 macrophages primarily depend on glycolysis as their main source of energy, whereas M2 macrophages predominantly utilize oxidative metabolism (Viola et al. 2019). Autophagy, a vivacious cellular process responsible for removing damaged components, plays a critical role in maintaining cell health and function and is influenced by metabolic processes (Gomez-Virgilio et al. 2022). Redox balancing affects the production of reactive oxygen species (ROS), key for signaling pathways and microbial killing (Schieber and Chandel 2014). The immune system can be regulated through metabolic signaling molecules like lactate, succinate, and itaconate, which help control inflammation (Lang and Siddique 2024). The complications of diseases can be influenced by disruptions in metabolism. This encompasses conditions such as autoimmune diseases, where the function of immune cells is disrupted, infections that take advantage of the host's metabolism, and cancers that are fueled by alterations in both the tumor microenvironment and metabolism (Aderinto et al. 2023). Exploring the intricate connection between metabolism and immune responses opens the possibility of discovering groundbreaking therapeutic approaches for a broad spectrum of diseases, including infections, autoimmune disorders, and cancer (Huang et al. 2024). The focus of our chapter is to investigate the relationship between metabolic pathways and the immune response. This chapter will explore these perspectives from cellular, biochemical, and immunological angles, aiming to provide a comprehensive explanation.

9.2 Metabolic Pathways in Immune Cells

T-cells and macrophages experience notable metabolic changes to support their growth, development, and optimal performance when they are activated. Resting immune cells primarily generate ATP through oxidative phosphorylation. However, glycolysis still activates even in the presence of oxygen. This phenomenon is commonly known as aerobic glycolysis or the Warburg effect (Balyan et al. 2020).

9.2.1 Glycolysis and Aerobic Glycolysis (the Warburg Effect)

The process of glycolysis involves the breakdown of glucose into pyruvate, resulting in the production of ATP and NADH. To meet the energy and biosynthetic demands of activated immune cells, this pathway is rapidly upregulated, providing quick energy and metabolic intermediates. Glycolytic intermediates play a central role in the synthesis of nucleotides, amino acids, and lipids, which are imperative for the proliferation and function of cells (Soto-Heredero et al. 2020). Glycolysis, an imperative metabolic pathway, takes place in the cytoplasm of cells, breaking down glucose into pyruvate while producing ATP and NADH. Through a sequence of ten enzymatic reactions, the process is divided into two distinct phases: energy investment and energy payoff. Ultimately, every glucose molecule yields a net increase of two ATP molecules. Following glycolysis in normal cells, oxidative phosphorylation occurs under aerobic conditions in order to maximize the production of ATP (Chandel 2021). However, in the case of cancer cells, they frequently demonstrate the Warburg Effect, which refers to their preference for utilizing glycolysis as the primary source of energy, regardless of the presence of oxygen. By transitioning to aerobic glycolysis, cancer cells can efficiently produce ATP and acquire biosynthetic precursors, which are necessary for their rapid proliferation and survival within the hypoxic tumor microenvironment (Vander Heiden et al. 2009).

Energy production and biosynthesis play a role in immune cells for enabling their activation, proliferation, and function. Once activated, immune cells like T-cells and macrophages adjust their metabolism to accommodate heightened energy and biosynthetic requirements. Glycolysis takes on a necessary role as the primary pathway for rapid ATP production, guaranteeing an instantaneous energy supply for cell signaling and effector functions (Domblides et al. 2018). At the same time, glycolytic intermediates play a dual role by providing the necessary building blocks for cell growth and proliferation, such as nucleotides, amino acids, and lipids. In addition, oxidative phosphorylation in mitochondria serves as a continuous source of energy and plays a vital role in the production of ROS, which is necessary for effective pathogen defense. Immune cells possess metabolic flexibility, enabling

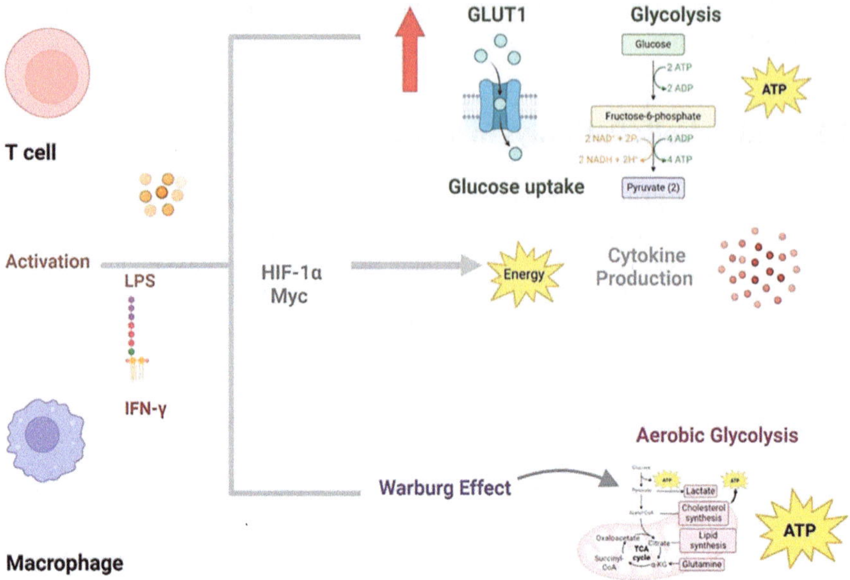

Fig. 9.1 Glucose metabolism and regulation of immune response. The figure illustrates the role of T-cells and macrophages activation in regulating glucose uptake and glycolysis. Upon activation by LPS or IFN-γ, macrophages undergo aerobic glycolysis (Warburg effect), leading to increased ATP production, HIF-1α regulating energy and cytokine production, while T-cells enhance glucose uptake and glycolysis. Where LPS Lipopolysaccharide, IFN-γ Interferon-Gamma, HIF-1α Hypoxia-Inducible Factor 1-alpha, GLUT1 Glucose Transporter 1, ATP Adenosine Triphosphate

them to adapt to different physiological contexts and effectively respond to infections while maintaining immune homeostasis (Lu et al. 2015).

Regulation of metabolism is necessary for immune cells to work properly and react to infections. To support rapid cell growth and proliferation, the phosphatidylinositol-3-kinase (PI3K) -Akt mammalian target of the rapamycin (mTOR) pathway is activated, leading to increased glycolysis and lipid biosynthesis (Mafi et al. 2021). For instance, when T-cells detect antigens, they enhance glucose intake and glycolysis by upregulating glucose transporter 1 (GLUT1) and essential glycolytic enzymes. Likewise, when macrophages are activated by lipopolysaccharide (LPS) or interferon-gamma (IFN-γ), they undergo a metabolic change of the Warburg effect in immune cells (Palmer et al. 2015). This modification entails a move toward aerobic glycolysis, which facilitates the rapid generation of ATP and the production of metabolic intermediates required for biosynthesis (Vander Heiden et al. 2009). Furthermore, the involvement of transcription factors like hypoxia-inducible factor 1 subunit alpha (HIF-1α) and Myc is important in the metabolic reprogramming that occurs during immune activation. This regulation plays a role in ensuring that immune cells are equipped with the energy and biosynthetic resources required to perform essential functions like producing cytokines, multiplying, and eliminating pathogens. As a result, it contributes to the development of a robust and efficient immune response (Infantino et al. 2021) (Fig. 9.1).

9.2.2 Oxidative Phosphorylation (OXPHOS)

Although glycolysis provides immediate energy, oxidative phosphorylation (OXPHOS) remains important for energy production, particularly in memory T-cells and other long-lived immune cells. Maintaining energy balance and supporting cellular longevity and function rely on the process of mitochondrial respiration (van der Windt and Pearce 2012).

Oxidative phosphorylation is key in supporting the long-term activities of immune cells by supplying a constant and efficient energy source, essential for their endurance and effective response over extended periods. In contrast to the quick ATP generation through glycolysis, OXPHOS takes place in the mitochondria and produces a much larger quantity of ATP per glucose molecule by combining the electron transport chain with ATP synthase activity (Kolodziej and O'Halloran 2021). Mitochondrial respiration plays a role in meeting the energy requirements of memory T-cells and other long-lived immune cells, especially during periods of inactivity and when they are reactivated. Memory T-cells, for example, use a combination of FAO and OXPHOS to sustain their survival and preparedness to react to antigens they have encountered before (Angajala et al. 2018). Additionally, OXPHOS plays a spirited role in producing controlled levels of ROS, which are required for signaling pathways associated with immune cell activation and differentiation (Peng et al. 2021a). The ability of immune cells to switch between glycolysis and OXPHOS allows them to adapt to changing energy demands and environmental conditions, ensuring their longevity and functionality. As a result, OXPHOS plays a role in meeting the fundamental energy demands of immune cells and is essential for their ability to swiftly and effectively respond to pathogens upon re-exposure (Delmastro-Greenwood and Piganelli 2013).

OXPHOS is important for the formation of the immune system's memory and its long-lasting protective capabilities. Immune cells of different types employ specialized metabolic strategies that are adapted to their specific functions and states of activation (Angajala et al. 2018). To illustrate, when naïve T-cells are not actively dividing, they primarily depend on OXPHOS and FAO to fulfill their energy requirements. Nevertheless, once activated, they shift toward glycolysis to generate ATP and biosynthetic intermediates rapidly that are decisive for their proliferation and effector functions (Geltink et al. 2018). Signals from the PI3K-Akt-mTOR pathway and transcription factors such as Myc and HIF-1α are responsible for initiating this metabolic reprogramming. Effector T-cells, such as cytotoxic T-lymphocytes (CTLs) and helper T-cells, depend on glycolysis and glutaminolysis to sustain rapid proliferation and cytokine production during their active phases (Dabi et al. 2022). Regulatory T-cells (Tregs) prefer OXPHOS and lipid metabolism, which are essential for their suppressive functions and long-term survival, as they contribute to maintaining immune tolerance (Kurniawan et al. 2020; Singer et al. 2024). Likewise, macrophages demonstrate metabolic adaptability. M1 macrophages, which are identified by their pro-inflammatory characteristics, boost glycolysis and the pentose phosphate pathway to expedite their response to infections and inflammatory signals. In contrast, M2 macrophages, known for their involvement in tissue repair and anti-inflammatory processes, primarily utilize oxidative phosphorylation and FAO (Liu et al. 2022). The metabolism of dendritic cells (DCs) changes according to their activation status. At first, their energy production depends on oxidative phosphorylation, but once activated, they switch to glycolysis. This metabolic adaptation is important in enabling them to present antigens and generate cytokines. The diverse metabolic strategies observed in immune cell types are indicative of the requirement for an adaptable and dynamic immune response. This allows different immune cells to perform their unique function in maintaining immunity (Peng et al. 2021b).

The examination of case studies and clinical implications of OXPHOS highlights the significance of mitochondrial function in both maintaining good health and causing disease, especially in relation to immune responses and metabolic disorders (San-Millan 2023). Studies have demonstrated that abnormalities in OXPHOS can result in various mitochondrial disorders, frequently characterized by symptoms that impact energy-intensive tissues like the brain, muscles, and heart. Mitochondrial myopathies exhibit symptoms like muscle weakness and exercise intolerance resulting from an impaired ability to produce ATP (Chen et al. 2022a). Impaired OXPHOS in the immune system can have a substantial impact on the function of immune cells and result in immunodeficiency. Numerous studies have indicated that patients with mitochondrial diseases commonly display weakened T-cell function, reduced cytokine production, and a compromised ability to mount effective immune responses (Hanaford and Johnson 2022). In addition, the significance of OXPHOS cannot be overstated when it comes to cancer and cancer immunotherapy. The Warburg effect is commonly utilized by tumor cells, as they rely on glycolysis for energy production even in oxygen-rich environments, enabling them to proliferate rapidly (Vander Heiden et al. 2009). Metabolic reprogramming poses challenges for immune cells, specifically T-cells, as they rely on efficient OXPHOS to maintain their activity and lifespan. Inhibitors that specifically target glycolysis or metabolic pathways for cancer cells can potentially alter the tumor microenvironment, creating a more favorable condition for immune cells to function effectively (Chen et al. 2022b). Furthermore, there have been promising results in cancer immunotherapy by improving the efficiency of T-cells through metabolic interventions that enhance mitochondrial function (Aksoylar et al. 2020). For instance, scientific studies have pointed out that enhancing OXPHOS in T-cells can boost their longevity, functionality, and ability to infiltrate tumors (Rangel Rivera et al. 2021). In addition to cancer, modifying the metabolic processes of OXPHOS could be a potential treatment approach for chronic inflammatory diseases and autoimmune disorders, which are characterized by abnormal metabolism contributing to the progression of the conditions (Amo-Aparicio et al. 2024). Through a comprehensive understanding and manipulation of the metabolic pathways that regulate OXPHOS, clinicians can design targeted therapies to rebalance metabolism, boost immune function, and effectively treat various mitochondrial and metabolic diseases (Tian et al. 2020) (Fig. 9.1).

9.2.3 Amino Acid Metabolism

Amino acid metabolism is one of the main pillars of immune cell function and differentiation. Immune cells rely on amino acids for activation and functions (Yang et al. 2023). Glutamine is necessary for immune cell growth and DNA/RNA synthesis during division. Glutamine is used to make other amino acids and helps maintain redox balance. Arginine metabolism is important for T-cell function, as it produces nitric oxide, a signaling molecule that regulates immune responses, T-cell proliferation, and function (Cruzat et al. 2018). Myeloid-derived suppressor cells (MDSC)-mediated arginine depletion in the tumor microenvironment hampers T-cell function and immune responses. The metabolism of branched-chain amino acids (BCAAs) affects immune cell signaling pathways, including mTOR, imperative for T-cell activation. Leucine acts as a signaling molecule, promoting T-cell growth and differentiation (Cortellino and Longo 2023). The kynurenine pathway is essential for immune tolerance and inflammation control. Indoleamine 2,3-dioxygenase (IDO) enzyme is upregulated in inflammatory conditions and tumors, causing tryptophan depletion and immunosuppressive metabolite accumulation. Metabolic reprogramming promotes tumor immune evasion (Krupa and Kowalska 2021). The metabolism of amino acids in the immune system is essential for immune cell function and signaling pathways. Knowledge of metabolic pathways aids in therapeutic interventions for diseases such as cancer and autoimmunity (Yang et al. 2023).

Key amino acids are fundamental for immune cell function, energy production, and immune response regulation (Yang et al. 2023). Glutamine is essential for immune cells, especially dividing cells like lymphocytes and macrophages. It is crucial for cell proliferation and function by supporting nucleotide and protein synthesis. Glutamine helps produce glutathione, shielding immune cells from oxidative stress (Cruzat et al. 2018). Arginine is fundamental for T-cell function as an essential amino acid. It is necessary for the synthesis of nitric oxide (NO) by nitric oxide synthase (NOS) (Marti and Reith 2021). NO functions as a signaling molecule, regulating immune cell communication, blood flow, and pathogen destruction by macrophages. Arginine metabolism affects T-cell differentiation and function (Palmieri et al. 2020). Arginine depletion hampers immune responses and T-cell function (Marti and Reith 2021). Leucine, a BCAA, activates the mTOR pathway decisive for T-cell growth and differentiation. Leucine activates mTOR, enhancing cytokine production and effector T-cell development, boosting immune response. Tryptophan is processed through the kynurenine pathway, influencing immune regulation (Ananieva et al. 2016). IDO enzyme is increased in inflammation and tumors, reducing tryptophan. Depletion and accumulation of kynurenine metabolites can suppress T-cell proliferation and promote the development of regulatory T-cells (Tregs), contributing to immune tolerance and modulation of responses (Zhai et al. 2020). Serine and glycine are needed for cell proliferation and function, synthesizing nucleotides, and biomolecules (Amelio et al. 2014). Cysteine is required for cellular redox balance and immune cell detoxification. Amino acids and their metabolic pathways are required for immune system function, impacting cell health and immune response (Clemente Plaza et al. 2018). Understanding the roles of these amino acids in immune cell function sheds light on potential therapeutic targets for immune enhancement in diseases like cancer, infections, and autoimmune disorders (Cruzat et al. 2014).

Amino acid metabolism and immune responses are reflected in the disease states, from infectious diseases to chronic inflammatory disorders and cancer. Amino acids are essential for immune cell functions (Ling et al. 2023). In infections, immune cells increase amino acid uptake for energy. Pathogens manipulate amino acid availability in their favor. In chronic inflammatory diseases, such as rheumatoid arthritis and Crohn's disease, dysregulated amino acid metabolism causes sustained inflammation and tissue damage (Zheng et al. 2023a). Imbalances in amino acid levels can worsen disease severity and immune activation. In cancer, tumor cells use rewired amino acid metabolism for growth and survival (Wei et al. 2020). Amino acids hijacked by tumors from the microenvironment impair immune cell function, promoting immune evasion and tumor progression. Understanding amino acid metabolism's impact on immune responses is critical for targeted therapies (Balta et al. 2021). Strategies for restoring amino acid balance in immune cells and manipulating availability in the tumor microenvironment can enhance immune responses and mitigate inflammation and autoimmune disorders (Bader et al. 2020).

Amino acid metabolism has significant implications for diverse health conditions, as indicated by meta-analyses and clinical trials. These studies confirm amino acids' impact on metabolic pathways, immune function, and disease progression (Nitz et al. 2019; Zheng et al. 2023b). Meta-analyses on amino acid supplementation in critically ill patients suggest therapeutic potential in intensive care, with improved nitrogen balance, reduced muscle wasting, and potentially enhanced immune function (Arribas-Lopez et al. 2021). Cancer research trials explore new treatment strategies by modifying amino acid availability and targeting metabolic pathways. Restricting dietary methionine has shown promising results in inhibiting tumor growth and enhancing conventional therapies (Regier et al. 2022). Clinical trials for metabolic disorders like phenylketonuria aim to correct defects and enhance patient outcomes. Meta-analyses and clinical trials uncover amino acids' roles in health and disease, informing therapeutic interventions and personalized medicine approaches to improve patient outcomes (Fan et al. 2022) (Fig. 9.2).

9.2.4 Lipid Metabolism

Lipid metabolism is vibrant for the proper functioning of membranes and signaling processes. Cellular compartments rely on lipids for their essential functions, including transport regulation and barrier formation. The bilayer structure found in cell membranes is created through the self-assembly of phospholipids. The inclusion of cholesterol molecules within phospholipids is important for preserving the fluidity and stability of cell membranes, which is essential for maintaining cellular integrity and enabling proper cellular function (Cockcroft 2021). Furthermore, lipid rafts, which are specialized areas in the cell membrane enriched with cholesterol and sphingolipids, act as important platforms for organizing membrane proteins and

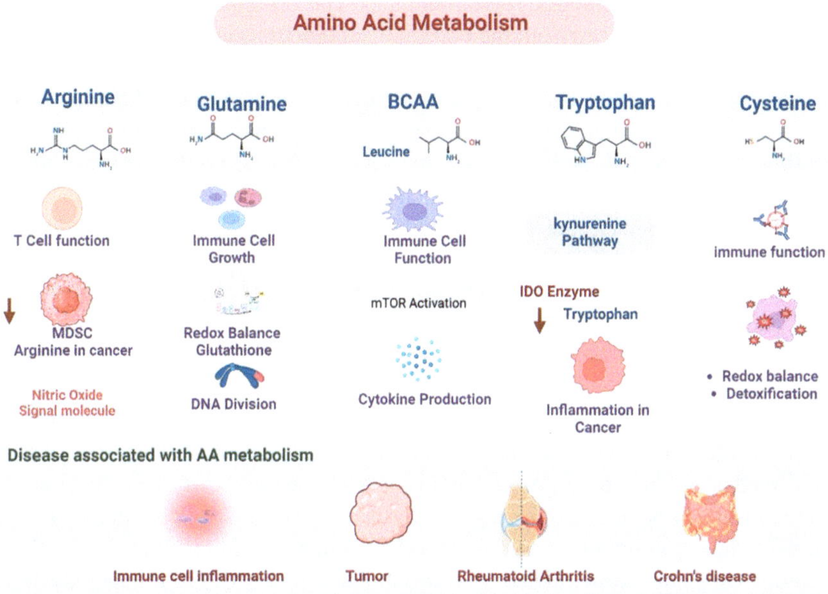

Fig. 9.2 Amino acid metabolism and its effect on the regulatory signals of immune response. This figure highlights the roles of various amino acids in immune function and their links to disease. Arginine, glutamine, branched-chain amino acids (BCAA), tryptophan, and cysteine each play crucial roles in regulating immune cell function, growth, redox balance, cytokine production, and detoxification. Disruptions in these pathways are associated with conditions like immune cell inflammation, tumors, rheumatoid arthritis, and Crohn's disease. Where MDSC Myeloid-Derived Suppressor Cells, mTOR Mechanistic Target of Rapamycin, IDO Indoleamine 2,3-Dioxygenase, AA Amino Acid

facilitating signal transduction. This promotes efficient interaction between signaling molecules and receptors. Cell signaling is influenced by lipids, which generate signaling molecules like prostaglandins, leukotrienes, and sphingosine-1-phosphate (S1P) (Li et al. 2022). These molecules have the responsibility of regulating several physiological processes, including inflammation, immune responses, and cell proliferation. When lipid metabolism is disrupted, it can significantly impact cellular function and overall health. The disruption has the potential to influence the development of multiple diseases, including atherosclerosis, diabetes, and cancer (Garcia et al. 2023). Gaining a comprehensive understanding of how lipids function in membrane formation and signaling pathways can offer valuable knowledge for identifying therapeutic targets. These targets can then be used to regulate lipid metabolism and address diseases that arise from lipid imbalances or disruptions in cellular signaling (Torres et al. 2021).

The critical role of lipid metabolism lies in its ability to regulate immune cell differentiation and polarization, ultimately impacting their functional outcomes and responses to environmental cues (Garcia et al. 2023). During the process of differentiation and activation, immune cells like macrophages, DCs, and T-cells experience distinct metabolic reprogramming. The cells must undergo this reprogramming in order to carry out their specialized functions in both innate and adaptive immunity (Kelly and O'Neill 2015). For example, macrophages have the ability to transform into either classically activated (M1) or activated (M2) phenotypes when exposed to various stimuli, and each phenotype has distinct lipid metabolic profiles (Strizova et al. 2023). When exposed to microbial products or pro-inflammatory cytokines, M1 macrophages enhance their lipid metabolism pathways, particularly FAO, to facilitate their pro-inflammatory functions and antimicrobial responses (Viola et al. 2019). On the other hand, M2 macrophages, which are activated by anti-inflammatory cytokines or signals of tissue repair, increase the absorption and production of lipids, which in turn facilitate tissue restructuring and immune regulation (Chen et al. 2023). Lipid metabolism plays a role in regulating T-cell differentiation and effector functions. T-cell activation leads to metabolic shifts, where the energy production switches from OXPHOS to glycolysis, enabling rapid proliferation and cytokine production (Cao et al. 2023). Moreover, certain lipid mediators, including prostaglandins and leukotrienes that originate from the metabolism of arachidonic acid, regulate the activation and differentiation of T-cells, thereby impacting immune responses and inflammation (Garcia et al. 2023). When immune cells have dysregulated lipid metabolism, it can lead to immune dysfunction and the development of diseases. To give an example, when there is an abnormal buildup of lipids in macrophages, it can cause the formation of foam cells and facilitate the progression of atherosclerosis (Remmerie and Scott 2018). By comprehending the intricate roles of lipid metabolism in immune cell differentiation and polarization, we can gain valuable insights into potential therapeutic strategies that target lipid pathways to manipulate immune responses in diseases like cancer, autoimmune disorders, and chronic inflammatory conditions (Hubler and Kennedy 2016) (Fig. 9.3).

The analysis of longitudinal and epidemiological data on lipid metabolism is important in gaining valuable insights into how lipid profiles, metabolic health, and disease outcomes are interrelated over time in large groups of people (Wazir et al. 2023). These studies observe fluctuations in lipid levels, like cholesterol and triglycerides, within different age groups, genders, and ethnicities, uncovering patterns of lipid metabolism across a wide range of demographic and health characteristics (Frank et al. 2014). Epidemiological research has shown that there are connections between dyslipidemia, which is defined by high levels of low-density lipoprotein (LDL) cholesterol and triglycerides, and a greater risk of cardiovascular diseases, including coronary artery disease and stroke (Alloubani et al. 2021). The use of longitudinal data helps to further understand the relationship between lipid metabolism and aging, as well as how lifestyle choices, obesity, and diabetes can affect disease progression and treatment effectiveness (Zampino et al. 2020). To illustrate, studies conducted over a long period have emphasized the impact of lifestyle changes, like diet and exercise, on lipid metabolism and the potential reduction of cardiovascular risk (Look and Wing 2010). Furthermore, epidemiological information regarding lipid metabolism has greatly influenced public health strategies and clinical guidelines for lipid-lowering therapies, specifically statins. The successful

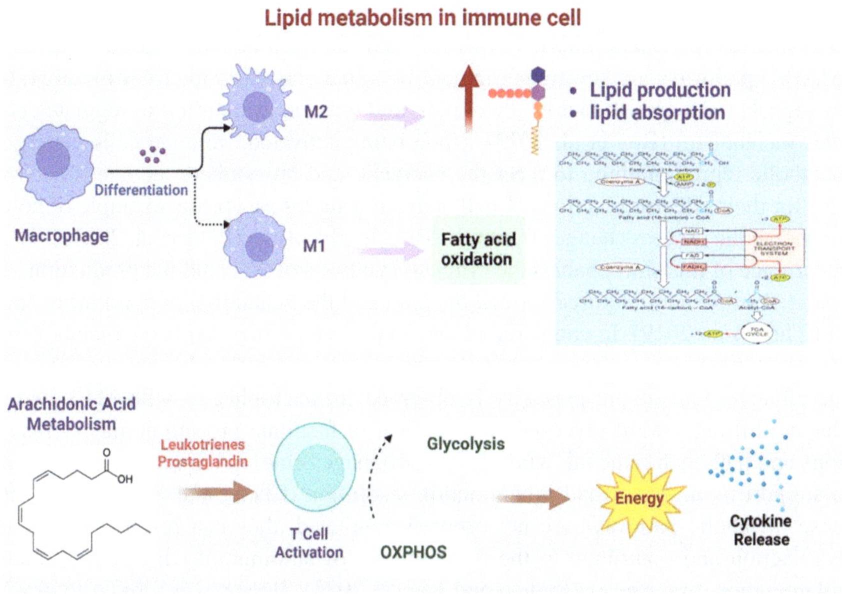

Fig. 9.3 The impact of lipid metabolism on the regulatory signals governing immune response. This figure illustrates the role of lipid metabolism in immune cell function, focusing on macrophage differentiation and T-cell activation. Macrophages can differentiate into M1 or M2 phenotypes, with M2 promoting lipid production and absorption, while M1 is associated with fatty acid oxidation. Arachidonic acid metabolism, where leukotrienes and prostaglandins stimulate T-cell activation, leads to glycolysis and oxidative phosphorylation (OXPHOS), producing energy for cytokine release. Where OXPHOS Oxidative Phosphorylation, ATP Adenosine Triphosphate

reduction of cardiovascular morbidity and mortality rates has been achieved through the targeting of cholesterol synthesis and metabolism with these medications (Morofuji et al. 2022). Lipidomics and high-throughput technologies have allowed us to gain a better understanding by discovering previously unknown lipid biomarkers associated with the risk and progression of diseases (Ahluwalia et al. 2022). In general, having access to longitudinal and epidemiological data on lipid metabolism is energetic to identify risk factors, devise preventive measures, and optimize therapeutic interventions to reduce the impact of dyslipidemia and related metabolic disorders.

9.3 Regulation of Immune Responses by Metabolic Cell Signaling Pathways

Metabolism and cell signaling pathways play a critical role in the intricate and precise regulation of immune responses, which is important for the immune system's proper functioning. In their microenvironment, immune cells, including T-cells, B-cells, macrophages, and DCs, undergo dynamic changes in their metabolic activities as a reaction to various stimuli (Du et al. 2023). The PI3K-Akt-mTOR, AMP-activated protein kinase (AMPK), and mitogen-activated protein kinase (MAPK) pathways are key signaling pathways that play dynamic roles in integrating signals from growth factors, cytokines, and nutrient availability to regulate cellular metabolism (Roy et al. 2023). Upon being activated, immune cells undergo metabolic reprogramming to meet the energetic and biosynthetic demands necessary for their effector functions. T-cell activation serves as a prime example of how the metabolic process changes from OXPHOS to glycolysis (Peng et al. 2021a). The occurrence of this shift enables the efficient synthesis of ATP and the production of essential substances required for cell division and the generation of cytokines (Zhu and Thompson 2019). In comparison, memory T-cells utilize oxidative metabolism to a greater extent in order to sustain their long-term survival and carry out immune surveillance. Phenotypic plasticity is observed in macrophages, with M1 macrophages shifting toward glycolytic metabolism to facilitate pro-inflammatory reactions and pathogen removal, while M2 macrophages employ oxidative metabolism to support tissue repair and inflammation resolution (Cheng and Zheng 2024). If these metabolic pathways are not properly regulated, they can result in immune dysfunction and contribute to the development of autoimmune diseases, chronic inflammation, and cancer (Blanco and Kaplan 2023). Hence, it is vibrant to grasp the intricate relationship between metabolism and cell signaling pathways in immune cells in order to create targeted treatments that can effectively modulate immune responses in various health and disease scenarios (Chi 2022).

The way environmental cues and metabolic reprogramming interact greatly affects the functioning and behavior of the immune system. In response to signals from their surrounding microenvironment, which consist of nutrient availability, oxygen levels, and inflammatory mediators, immune cells regularly fine-tune their metabolic activities (Zhu et al. 2022). The metabolic pathways in immune cells are influenced by environmental cues, which in turn shape their activation, differentiation, and effector functions. Immune cells frequently undergo metabolic changes, such as increased glycolysis, during infection or inflammation. It is crucial because it allows them to produce enough energy for cell proliferation and the generation of cytokines (Mohammadnezhad et al. 2022). The regulation of this metabolic reprogramming is complex and involves important signaling pathways like mTOR, AMPK, and HIF-1α. These pathways identify and combine signals from growth factors, cytokines, and nutrient sensors (Szwed et al. 2021). On the other hand, immune cells may decide switching for OXPHOS to conserve energy and support cellular function in situations of nutrient scarcity or low oxygen levels (hypoxia). If the metabolic adaptations are not controlled, they can disrupt the immune system and potentially contribute to the development of autoimmune diseases, chronic inflammation, and cancer (Piranavan et al. 2020). Gaining a deeper understanding of how environmental cues impact the metabolic reprogramming of immune cells offers valuable insights into the mechanisms of immune regulation. It also presents potential opportunities for therapeutic interventions to modulate immune responses in different disease scenarios, thus paving the way for more efficient treatments that specifically target immune-related disorders (Biswas 2015).

The effects of metabolic reprogramming on immune cell activation and functions are significant, affecting the immune system's capacity to respond effectively to pathogens and maintain immune balance. Immune cells, including T-cells, B-cells, macrophages, and DCs, undergo specific metabolic changes when they encounter antigens or inflammatory signals to fulfill their individual functions (Hu et al. 2022). As an example, the activation of T-cells leads to a transition from OXPHOS to glycolysis, which allows for quick production of ATP and metabolic substances that are necessary for cell growth, division, and the production of cytokines. The regulation of this metabolic switch relies on signaling pathways like mTOR and AMPK. The coordination of metabolic fluxes is achieved by these pathways, which detect nutrient availability and growth factor signals (Shyer et al. 2020). M1 macrophages use glycolysis for inflammation, while M2 macrophages rely on oxidative metabolism for tissue repair. Dysregulation of metabolic pathways can impair immune cell functions, leading to immune dysfunction and susceptibility to diseases (Wang et al. 2022). Comprehending the connection between metabolism and immune cell function helps us understand immune responses and develop effective therapies for immune-related disorders (Aderinto et al. 2023). Experimental models and in vivo studies were important in understanding immune responses. Genetically modified mice are used to investigate the roles of specific genes in immune cell function. Knockout mice lacking mTOR in T-cells have shown the importance of mTOR signaling in T-cell functions. Modifying metabolic pathways pharmacologically reveals immune response effects. Rapamycin, a mTOR inhibitor, can suppress T-cell activation and proliferation, showing the potential for treating immune-mediated diseases (Pollizzi and Powell 2015). In vivo studies validate these findings and show how metabolic alterations affect disease progression and immune cell function. Studies using murine models of sepsis have shown dysregulated metabolism in immune cells, leading to immune suppression and mortality. Experimental models and studies offer insights into how metabolism and cell signaling pathways regulate immune responses, aiding the development of targeted therapies (Nandi et al. 2020).

Metabolism is an active cellular process. Genetically modified mice, cell culture systems, and advanced imaging techniques are used to study these interactions. In vivo studies show tight links between metabolic pathways and immune cell function. T-cells undergo metabolic reprogramming upon activation as a Warburg effect (Heaton et al. 2023). This shift aids the rapid proliferation of T-cells and effector functions. Succinate and itaconate are signaling molecules that affect inflammation. The mTOR and AMPK pathways regulate immune cell activity based on nutrients and energy signals. mTOR supports T-cell growth, while AMPK suppresses immune responses (Weichhart et al. 2015). In vivo models show metabolic dysregulation's role in immune-related diseases. Obesity-induced insulin resistance is linked to chronic low-grade inflammation due to changes in macrophage metabolism (Puschel et al. 2022). Cancer metabolism can impair anti-tumor immunity. Understanding these pathways offers insight into therapeutic strategies for diseases like cancer, autoimmunity, and infections (Ucche and Hayakawa 2024).

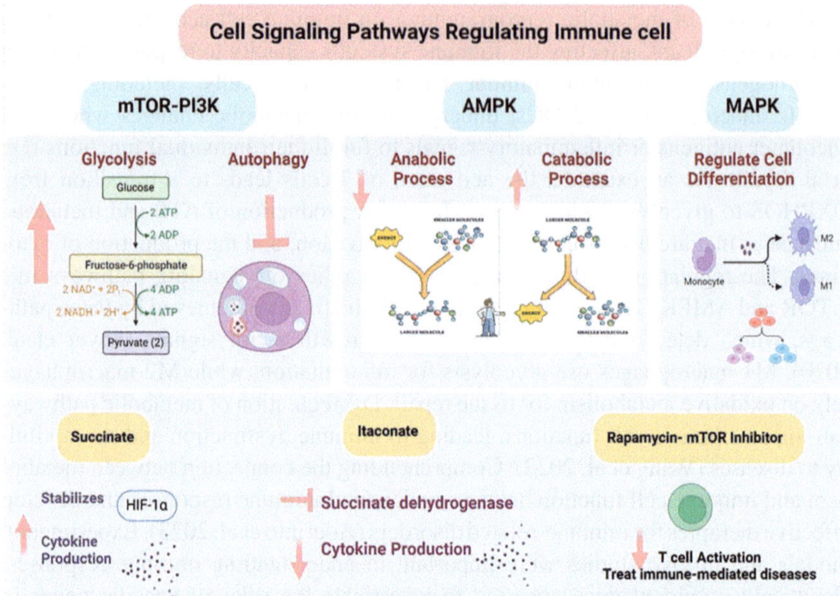

Fig. 9.4 Regulatory signaling for immune response. This figure illustrates the key signaling pathways, including mTOR-PI3K, AMPK, and MAPK pathways. The mTOR-PI3K pathway promotes glycolysis and inhibits autophagy, while AMPK decreases anabolic and increases the catabolic processes, influencing energy production. The MAPK pathway regulates cell differentiation, particularly the transition of monocytes into M1 or M2 macrophages. Additionally, succinate stabilizes HIF-1α to enhance cytokine production, while itaconate inhibits succinate dehydrogenase and reduces cytokine production and rapamycin is an mTOR inhibitor that inhibits T-cell activation, which are key in treating immune-mediated diseases. Where mTOR Mechanistic Target of Rapamycin, PI3K Phosphoinositide 3-Kinase, AMPK AMP-Activated Protein Kinase, MAPK Mitogen-Activated Protein Kinase, HIF-1α Hypoxia-Inducible Factor 1-alpha, ATP Adenosine Triphosphate

Immune responses are regulated by metabolic cell signaling pathways involving immune cells and metabolic processes. T-cells undergo metabolic reprogramming during activation (Shyer et al. 2020). When encountering an antigen, T-cells switch to glycolysis, even in the presence of oxygen, known as the Warburg effect. The mTOR pathway regulates this metabolic shift, supporting T-cell proliferation and functions (Chen et al. 2022b). The AMPK pathway regulates cellular energy balance by activating catabolic pathways and inhibiting anabolic processes (Garcia and Shaw 2017). Certain metabolites act as signaling molecules that influence immune responses. Succinate accumulation in inflammatory macrophages stabilizes HIF-1α, boosting pro-inflammatory cytokine production. Itaconate, from the Krebs cycle, inhibits succinate dehydrogenase and reduces cytokine production (Zhang and Lang 2023) (Fig. 9.4). Lipid metabolism plays a role in immune regulation. S1P regulates lymphocyte trafficking and function by binding to S1P receptors on immune cells. This pathway is essential for lymphocyte egress from lymphoid organs (Bravo et al. 2022). Altered cholesterol metabolism can affect immune cell function.

Cholesterol-rich lipid rafts intervene with immunological synapse formation during T-cell activation (Aguilar-Ballester et al. 2020). Dysregulated cholesterol metabolism impairs immune responses. These examples show how metabolism fuels immune cell functions and shapes immune responses through distinct signaling mechanisms (Cardoso and Perucha 2021).

Metabolite regulation impacts biological processes like development, immune responses, and disease progression. Metabolites can influence gene expression by interacting with gene regulatory components. Acetyl-CoA, a central metabolite, is utilized by histone acetyltransferases (HATs) to activate growth and metabolism-related genes (van der Knaap and Verrijzer 2016). S-adenosyl methionine (SAM) is essential for DNA and histone methylation, which controls gene expression. Fasting or ketogenic diets produce β-hydroxybutyrate, which inhibits histone deacetylases (HDACs) and boosts stress resistance and longevity gene expression (Ungaro et al. 2022). Succinate and fumarate can inhibit prolyl hydroxylases, stabilize HIFs, and activate hypoxia-responsive genes. Metabolites affect gene expression, connecting metabolic states to gene regulatory networks and impacting cellular function and health (Laukka et al. 2016). DNA methylation and histone modification control gene expression through epigenetic changes. Acetyl-CoA, SAM, α-ketoglutarate, and NAD+ are essential for these modifications. Acetyl-CoA is a critical substrate for HATs to acetylate histones, enabling gene transcription (Huo et al. 2021). HDACs remove acetyl groups, leading to gene repression. SAM donates methyl groups for DNA and histone methylation, controlled by DNA methyltransferases (DNMTs) and HMTs, respectively, essential for gene regulation. α-ketoglutarate is very important for ten-eleven translocation (TET) enzymes and Jumonji C (JmjC) domain-containing histone demethylases, involved in DNA and histone demethylation. Sirtuins, a type of histone deacetylase, use NAD+ as a cofactor to connect cellular metabolism with chromatin and gene expression. The availability and fluctuations of these metabolites regulate epigenetic landscapes, impacting cellular processes (Hitchler and Domann 2012).

Metabolites affect immune cell differentiation and memory formation, shaping the immune response. Immune cell differentiation involves reprogramming metabolic pathways for different energy and biosynthetic requirements. Naïve T-cells primarily use OXPHOS but switch to glycolysis when activated for rapid proliferation and effector functions. mTOR and AMPK are key signaling pathways that regulate this metabolic shift by sensing energy and nutrients. Acetyl-CoA and SAM are crucial for T-cell differentiation into effector and regulatory subsets (Shyer et al. 2020). Memory T-cell formation is closely tied to their metabolic state. Memory T-cells use OXPHOS and FAO for long-term survival and rapid response to infections. Metabolites like lactate and itaconate can alter the immune landscape. The interplay between metabolism and immune cell differentiation is decisive for immune homeostasis and memory formation (Corrado and Pearce 2022).

Cohort studies and reviews are essential in research, providing strong evidence on exposure-health relationships. Researchers in prospective cohort studies follow individuals over time, collecting exposure data and tracking specific outcomes. This study design assesses temporal relationships and potential causal links between risk factors and disease. Prospective cohort studies linked smoking to lung cancer, diet to cardiovascular disease, and physical activity to mortality. Systematic reviews synthesize findings from multiple studies on a topic, using rigorous methods to minimize bias and provide comprehensive evidence summaries (Corrado and Pearce 2022). Pooling data from various studies and systematic reviews enhance statistical power and improve estimates of effect sizes. They frequently use meta-analyses to identify overall trends. Cohort studies and systematic reviews contribute to evidence-based practice by providing reliable evidence on health determinants and intervention effectiveness (Koene et al. 2016). Clinical trials are currently focusing on modulating cancer metabolomics to generate an effective therapy (Table 9.1).

Tissue-specific metabolism is the unique metabolic process in different tissues. Tissues have specific metabolic requirements and capabilities. The liver is vital for metabolic homeostasis. During exercise, muscle tissue heavily depends on glycolysis and OXPHOS for ATP generation. Adipose tissue stores and mobilizes lipids, playing a role in energy balance and hormone secretion (Rui 2014). The brain primarily needs glucose but can also use ketone bodies during fasting or low-carb intake. Tissues like the heart primarily use fatty acids as fuel. Various factors, such as hormonal signals, nutrient availability, and gene expression profiles, regulate tissue-specific metabolic pathways. Understanding tissue-specific metabolism is essential for identifying metabolic dysregulation and improving metabolic health (Garcia-Rodriguez and Gimenez-Cassina 2021). Immune cells show remarkable adaptations for diverse functions. The unique conditions and metabolic demands of each tissue drive these adaptations. Alveolar macrophages, Kupffer cells, and microglia are examples of tissue-resident macrophages with specialized roles and phenotypes. Alveolar macrophages clear inhaled pathogens, and Kupffer cells filter blood-borne pathogens (Scur et al. 2022). Gut immune cells create a unique regulatory environment. T-cells adapt metabolism to tissue environments. Cytokines, chemokines, and metabolites influence immune cell function. Tumors induce immune cell adaptations in hypoxic environments, such as the expression of HIFs. Tissue-specific adaptations support immune cells in maintaining homeostasis, defending against infections, and aiding tissue repair across organs (Yan et al. 2022).

The liver is essential for metabolism, including glucose homeostasis, lipid metabolism, and detoxification. In nonalcoholic fatty liver disease (NAFLD), excess triglyceride buildup in liver cells impairs liver function, causing inflammation and potentially advancing to fibrosis and cirrhosis (Rinaldi et al. 2021). Glucose fuels brain neurons, enabling essential functions like neurotransmitter synthesis and synaptic transmission. Disorders like Alzheimer's disease show impaired brain glucose metabolism, causing cognitive decline. Glycolysis and OXPHOS are both utilized by skeletal muscle for ATP generation (Zhang et al. 2021). Muscular dystrophy disrupts muscle metabolism, causing weakness and degeneration. Adipocytes regulate energy storage and release via lipogenesis and lipolysis (Paran et al. 2015). Excessive adipose tissue in obesity affects adipocyte metabolism and adipokine secretion, resulting in insulin resistance and metabolic syndrome. Pancreatic islet cells, including beta cells, secrete insulin in response to high blood glucose levels (Kojta et al. 2020). Dysfunction of pancreatic beta cells in type 2 diabetes leads to

Table 9.1 Current clinical trials on cancer metabolism

NCT Number	Study Title	Phases	Start Date	Locations
NCT06377566	A Study of BV-AVD in People with Bulky Hodgkin Lymphoma	PHASE2	4/17/2024	United States
NCT06472583	Preoperative Immunotherapy Combined with Stereotactic Radiation Therapy Boost in the Treatment of HER2-negative Breast Cancer	PHASE2	4/11/2024	Poland
NCT06327685	Avapritinib With Decitabine in Patients With SM-AHN	PHASE1	3/13/2024	United States
NCT05432856	Impact of Metabolic Health Patterns and Breast Cancer Over Time in Women	NA	3/13/2024	Canada
NCT06261918	Transcriptional and Epimetabolic Profile of Breast Carcinoma with Luminal or HER2+ or Locally Advanced Triple-negative Histotype in Patients With/Without Previous Clinical History of Metabolic Syndrome (PROMETA)	NA	2/1/2024	Italy
NCT06243250	Hyperpolarized 13C-MRI and Metabolomics for Immune-related Adverse Events	NA	2/1/2024	Taiwan
NCT06015659	ZN-c3 + Gemcitabine in Pancreatic Cancer	PHASE2	11/16/2023	United States
NCT05976334	An Absorption, Distribution, Metabolism, Excretion (ADME) Study of [14C] Subasumstat in Adults with Advanced or Metastatic Solid Tumors	PHASE1	11/14/2023	Hungary
NCT06069115	Radiotherapy Exposed Lung Injury: Effect on Functioning—Cohort	NA	7/18/2023	Russian Federation
NCT05912465	Predicting Lung Cancer-Associated Cachexia with PET Imaging	NA	7/1/2023	Italy
NCT05570227	Metabolic Phenotypes in Melanoma	NA	6/27/2023	Australia

(continued)

Table 9.1 (continued)

NCT Number	Study Title	Phases	Start Date	Locations
NCT05903521	BAriatric Surgery Induced Chromosomal Change (BASICC)	NA	6/12/2023	United States
NCT05299125	Amivantamab, Lazertinib, and Pemetrexed for First-line Treatment of Recurrent/Metastatic Non-small Cell Lung Cancers with Epidermal Growth Factor Receptor Mutations	PHASE2	5/24/2023	Brazil
NCT05284487	Clinical Trial to Evaluate Safety and Efficacy of JalosomeÂ® Soothing Gel.	NA	3/27/2023	Italy
NCT05009979	18F-DCFPyL PET/CT in Hepatocellular Carcinoma	PHASE2	1/18/2023	United States
NCT05599048	Hyperpolarized 13C Pyruvate as a Biomarker in Advanced Solid Tumors	PHASE1|PHASE2	12/15/2022	United States
NCT06001619	Prostate Medication, Metabolism, and Gut Microbiota	PHASE4	12/1/2022	Finland
NCT05325281	CPI-613 (Devimistat) in Combination with Chemoradiation in Patients with Pancreatic Adenocarcinoma	PHASE1	10/31/2022	United States
NCT05139784	BeAT1D: Benign Autoimmunity and Type 1 Diabetes	NA	10/24/2022	France
NCT04785989	In Vivo Metabolic Profiling of CLL (Chronic Lymphocytic Leukemia)	NA	6/13/2022	United States
NCT06116851	Prostate Metabolism, Cancer Risk, and Gut Microbiota	NA	6/1/2022	Finland
NCT05296421	Investigating Targetable Metabolic Pathways Sustaining Pancreatic Cancer	NA	3/14/2022	United States
NCT04276272	D4 Choline Breast PET/CT	NA	2/24/2022	United Kingdom
NCT04906746	Ruxolitinib for Cancer Cachexia	EARLY_PHASE1	2/23/2022	United States

(continued)

Table 9.1 (continued)

NCT Number	Study Title	Phases	Start Date	Locations
NCT05207670	A Study Evaluating the Safety, Efficacy, and Pharmacokinetics of Mosunetuzumab Monotherapy in Participants with Select B-Cell Malignancies	PHASE2	2/1/2022	United States, Puerto Rico
NCT06081270	Prospective Study on Resistance-associated Mutations in Metastatic Lung Cancer	NA	12/12/2021	Italy
NCT05081479	A Study of N-Acetylcysteine (N-AC) in People Receiving CAR T-cell Therapy for Lymphoma	PHASE1	12/8/2021	United States
NCT05136846	Papaverine in Combination with Chemoradiation for the Treatment of Stage II-III Non-small Cell Lung Cancer	PHASE1	12/6/2021	United States
NCT04870515	Diet and Physical Activity Intervention for the Prevention of ADT-Induced Metabolic Changes in Patients with Prostate Cancer, TRIPLE-A PILOT Study	NA	11/4/2021	United States
NCT04717050	Reducing Metabolic Dysregulation in Obese Latina Breast Cancer Survivors Using Physical Activity	NA	8/12/2021	United States
NCT05931718	Prospective Evaluation of Diagnosis and Treatment of Patients with Autoimmune Cytopenia Including Autoimmune Hemolytic Anemia, Immune Thrombocytopenia, and Chronic Idiopathic/ Autoimmune Neutropenia	NA	6/1/2021	Italy
NCT04730869	Metabolic Therapy Program with Standard Treatment for Glioblastoma	NA	5/26/2021	New Zealand

(continued)

Table 9.1 (continued)

NCT Number	Study Title	Phases	Start Date	Locations
NCT04667585	Radiotherapy Dose De-escalation in HPV-Associated Cancers of the Oropharynx	NA	4/12/2021	United States
NCT04658303	Evolution of Metabolic and Immune Dysfunction in In-transit Melanoma	NA	2/24/2021	United States
NCT04363983	Interaction Between Host, Microenvironment, and Immunity on Gastrointestinal Neoplasms	NA	1/13/2021	France
NCT04114136	Anti-PD-1 mAb Plus Metabolic Modulator in Solid Tumor Malignancies	PHASE2	9/14/2020	United States
NCT03925675	18F-Fluciclovine PET and Multi-parametric MR Imaging	PHASE2	8/24/2020	United States
NCT04589624	An Investigational Scan (hpMRI) for Monitoring Treatment Response in Patients with Thyroid Cancer and Other Malignancies of the Head and Neck Undergoing Radiation Therapy and/or Systemic Therapy	PHASE1	8/12/2020	United States

www.clinicaltrials.gov
NA not applicable

impaired insulin secretion and dysregulated glucose metabolism. The heart primarily uses FAO for energy (Dludla et al. 2023). Cardiovascular diseases can disrupt cardiac metabolism, resulting in energy depletion and impaired contractile function. The kidneys regulate electrolyte balance, acid-base equilibrium, and glucose reabsorption for metabolic homeostasis (Wang and Chen 2021). Diabetic nephropathy and similar disorders can impair renal metabolism, resulting in electrolyte imbalances and kidney dysfunction. Examples of organ-specific metabolism shed light on the relationship between metabolic pathways and organ function. This knowledge is vigorous for developing targeted therapies to restore metabolic balance and improve outcomes in related diseases (Kumar et al. 2023).

9.4 Metabolism and Immunometabolism in Metabolic Disorders

Immunometabolism is key in metabolic disorders like obesity and diabetes, impacting metabolic balance and immune function (Daryabor et al. 2020). Excessive adipose tissue in obese individuals leads to chronic inflammation with increased cytokine secretion. The inflammatory environment disrupts insulin signaling, causing resistance in peripheral tissues. Obesity disrupts lipid metabolism, leading to the accumulation of lipotoxic intermediates that worsen insulin resistance (Zatterale et al. 2019). Adipose tissue dysfunction disrupts adipokine secretion, causing inflammation and metabolic dysregulation. In type 1 diabetes, pancreatic beta cells are destroyed, causing insulin deficiency (Kawai et al. 2021). Chronic hyperglycemia and insulin resistance in type 2 diabetes are caused by a combination of genetic, environmental, and lifestyle factors. Nutrient excess in diabetes triggers abnormal activation of immune cells, causing chronic inflammation and beta-cell dysfunction. Understanding the links between metabolism and immune responses in metabolic disorders is crucial for targeted therapies (Galicia-Garcia et al. 2020).

Metabolism impacts immune cell function, inflammation, and well-being. Immune cells reprogram their metabolism in response to cues and demands. During infection or inflammation, immune cells rely on glycolysis for energy in effector functions. mTOR and AMPK coordinate nutrient availability and immune signaling to modulate cell activation and differentiation (Ganeshan and Chawla 2014). In metabolic disorders like obesity and diabetes, dysregulated immune cell metabolism leads to chronic inflammation. Chronic inflammation results in insulin resistance and metabolic problems. Caloric restriction and nutrient modulation can enhance metabolic health and alleviate inflammation. Knowing how metabolism, immune cells, and inflammation interact is important for targeting chronic inflammatory diseases (Zatterale et al. 2019). Metabolism's impact on immune cell function and inflammation has promising therapeutic implications for diseases. Targeted interventions can modulate metabolic pathways in immune cells to treat conditions with dysregulated immune responses and chronic inflammation (Aderinto et al. 2023). Restoring metabolic balance through therapies could mitigate inflammation in metabolic disorders. Enhancing mitochondrial function in immune cells may boost immune responses to infections or cancer. Dietary interventions and lifestyle modifications can have therapeutic effects (Clemente-Suarez et al. 2023). Immunometabolism research offers prospects for personalized medicine based on metabolic profiles and immune cell function. These evolving therapeutic strategies promise to improve outcomes in diverse disease contexts with immune dysregulation and chronic inflammation (Ramalho et al. 2020) (Fig. 9.5).

Randomized controlled trials (RCTs) and clinical evaluation reports (CER) are key in clinical research to evaluate interventions across diverse patients. RCTs randomly assign participants to treatment groups, establishing causal relationships between interventions and outcomes. RCTs reduce bias and provide reliable evidence for clinical decisions and guidelines (Zabor et al. 2020). Comparative effectiveness research compares alternative interventions to inform healthcare decisions.

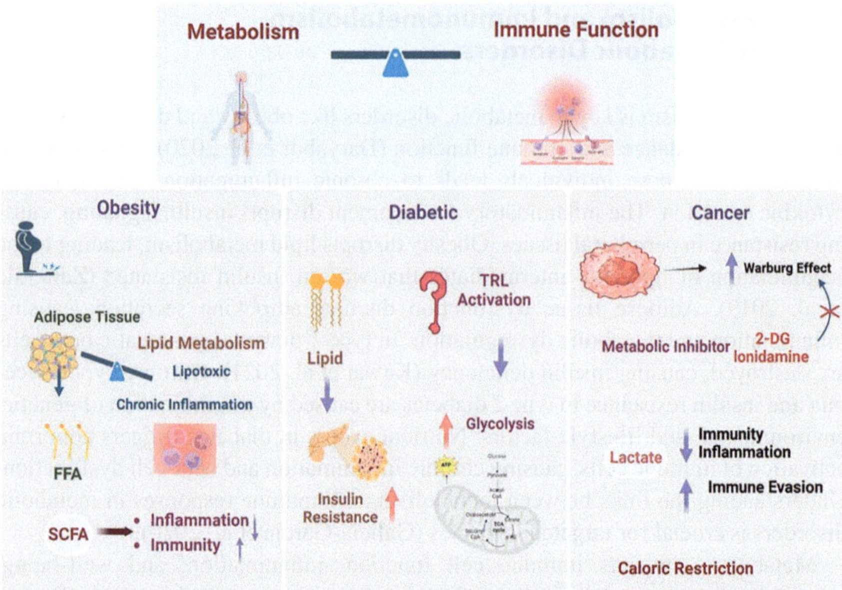

Fig. 9.5 Metabolism imbalance leads to immune dysfunction in chronic diseases. The connection between metabolism and immune function emphasizes how obesity, diabetes, and cancer disrupt this balance. Key metabolic changes like lipotoxicity, glycolysis, and the Warburg effect influence inflammation and immune response, with potential interventions highlighted. Where FFA Free Fatty Acids, SCFA Short-Chain Fatty Acids, TRL Toll-Like Receptor, 2-DG 2-Deoxy-D-glucose

CER utilizes observational data and pragmatic trials to evaluate treatment outcomes. RCTs and CER contribute to evidence-based medicine by guiding informed decisions to improve patient outcomes and optimize healthcare delivery (Dang and Kaur 2016). Metabolic reprogramming is critical for anti-tumor immune responses, showing the connection between cellular metabolism and immune function in tumors. Tumor cells exhibit altered metabolic profiles that promote growth and survival through increased glycolysis (Warburg effect) (Zhang et al. 2023). Immune cells in the tumor undergo metabolic adaptations to maintain their functions. Activated T-cells undergo metabolic changes to support rapid proliferation and cytokine production against tumors. Tregs use oxidative metabolism to maintain their suppressive functions and dampen anti-tumor immunity (Rangel Rivera et al. 2021). Metabolic checkpoints, like the mTOR pathway, impact immune cell metabolism and function in the tumor microenvironment. Understanding metabolic dynamics is energetic for developing immunotherapies that enhance anti-tumor immune responses, benefiting cancer patients (Long et al. 2023). Cancer treatment strategies target cancer cell metabolism and boost immune response. Inhibiting metabolic pathways imperative for cancer cell survival is one approach. Metabolic inhibitors like 2-Deoxy-d-glucose (2-DG) and lonidamine disrupt the Warburg effect in cancer cells, reducing their ability to generate ATP and critical biosynthetic intermediates (Zhang et al. 2022). Targeting mitochondrial metabolism triggers apoptosis in cancer cells that heavily rely on it. Another promising strategy targets metabolic checkpoints that regulate immune cell function in the tumor microenvironment. These drugs can modulate immune cell metabolism and boost T-cell activation (Bader et al. 2020). Dietary interventions, like caloric restriction or ketogenic diets, can enhance conventional cancer therapies by altering tumor cell metabolism and reducing inflammation. Targeting metabolic vulnerabilities in cancer and immune cells shows potential for improving treatment outcomes and patient survival (Talib et al. 2021) (Fig. 9.5).

9.5 Metabolites in Metabolic Imbalance and Immune Dysfunction

Metabolites maintain immune homeostasis. Metabolic disorders like obesity, diabetes, and autoimmune diseases can lead to immune dysfunction and chronic inflammation due to changes in metabolite levels (Andersen et al. 2016). Obesity leads to adipose tissue dysfunction and increased release of free fatty acids, activating toll-like receptors (TLRs) on immune cells and promoting inflammation. Diabetes can lead to oxidative stress, mitochondrial dysfunction, and inflammation due to high levels of metabolites like glucose and lipids (Renovato-Martins et al. 2020). Lactate, a metabolite produced during tumor cell glycolysis, creates an acidic microenvironment that suppresses anti-tumor immunity and aids immune evasion. Short-chain fatty acids (SCFAs) from dietary fiber fermentation can reduce inflammation by influencing immune cell function and promoting T-cell differentiation. Targeted therapeutics to improve immune responses and restore homeostasis in illnesses require an understanding of the role metabolites play in metabolic imbalance and immunological dysfunction (Li et al. 2023).

9.6 Metabolites in Microbiome Interactions

Metabolites mediate host-microbiome interactions. Interactions are crucial for host health and physiological processes. SCFAs are produced by gut bacteria when fermenting dietary fiber. SCFAs such as acetate, propionate, and butyrate energize gut cells and protect barrier integrity. They have immunomodulatory effects, promoting regulatory T-cell differentiation and dampening inflammation (Cong et al. 2022). Gut bacteria modify bile acids to affect lipid metabolism and glucose homeostasis. Microbial enzymes convert primary bile acids into secondary bile acids, which can activate host cell receptors and regulate metabolism and immune responses (Winston and Theriot 2020).

Amino acid metabolites like indole derivatives can affect immune function and gut integrity. An imbalance in gut microbiota composition, known as dysbiosis, is linked to diseases like inflammatory bowel disease (IBD), metabolic disorders, and neurological conditions. Studying metabolites in microbiome interactions unveils their effects on host physiology and potential therapeutic targets for gut health and related disorders (Bernardi et al. 2024).

9.7 Metabolic Reprogramming in Immune Responses

Metabolic reprogramming is the change in cellular metabolism in response to stimuli. Immune cells undergo metabolic reprogramming to meet energetic demands for immune responses. This section discusses the significance of metabolic reprogramming in immune cells for immune activation and regulation (Sun et al. 2020). Upon activation, immune cells undergo a metabolic transformation that fuels their growth, specialization, and ability to carry out their functions. ATP production in resting immune cells is predominantly carried out through OXPHOS. However, even in the presence of oxygen, there is a switch to glycolysis upon activation, giving rise to aerobic glycolysis or the Warburg effect (Sun et al. 2020). A promising approach for treating various diseases involves modulating the metabolism of immune cells. Therapeutic strategies can target and enhance specific metabolic pathways to bolster immune responses against infections and cancer, or they can inhibit these pathways to dampen pathological immune responses in autoimmune and inflammatory diseases (Ganjoo et al. 2023).

Cellular responses to low oxygen levels are carefully controlled by hypoxia-inducible factors (HIFs), a group of transcription factors (Morris et al. 2023). During immune cell activation, HIF-1α is specifically increased and significantly contributes to metabolic reprogramming (Corcoran and O'Neill 2016). By upregulating glycolytic enzymes and glucose transporters, HIF-1α stimulates glycolysis and simultaneously inhibits mitochondrial respiration. By making this shift, we can ensure that the immune cells that are activated have an ample supply of energy and the necessary building blocks for their functions (Infantino et al. 2021). Cell growth, proliferation, and metabolism are regulated by mTOR, also known as the mechanistic target of rapamycin. mTOR signaling is triggered by the presence of nutrients, growth factors, and immune signals (Saxton and Sabatini 2017). An essential regulator in immune cells, mTOR encourages the synthesis of proteins and lipids while actively suppressing the degradation of cellular materials via autophagy (Papadopoli et al. 2019). The differentiation and function of various immune cells, such as T-cells, macrophages, and DCs, heavily rely on mTOR signaling. Metabolic checkpoints act as regulatory mechanisms, ensuring that immune cells are equipped with the necessary metabolic resources for their functions (Chi 2012). Key metabolic checkpoints, such as AMPK, mTOR, and HIFs, play roles in regulating metabolism. These checkpoints combine signals from nutrient availability, energy levels, and environmental conditions to regulate the metabolism and function of immune cells (Long and Zierath 2006).

9.8 Metabolic Dysregulation and Immune Diseases

Autoimmune diseases are marked by the immune system attacking the body's own tissues. Immune cells undergoing metabolic reprogramming can play a part in the onset and advancement of autoimmune diseases through the stimulation and persistence of autoreactive cells (Pisetsky 2023). The immune function is greatly affected by diet and nutrition, as they have an impact on metabolism. The immune system relies on nutrients like vitamins, minerals, and amino acids to have optimal responses (Munteanu and Schwartz 2022). Modifying the diet can have a positive effect on immune cell metabolism and boost immune function in various circumstances. Cancer cells undergo changes in their metabolism to support their fast growth and survival (Luby and Alves-Guerra 2021). Furthermore, this modified metabolism has an impact on the tumor microenvironment, causing the immune system to evade and suppress any immune responses against the tumor. There is promise in targeting metabolic pathways in cancer cells and immune cells to enhance the effectiveness of cancer immunotherapy (Bader et al. 2020). Understanding the connection between metabolic therapies and cancer immunotherapy. The combination of metabolic therapies and cancer immunotherapy has the potential to enhance the effectiveness of treatments. Enhancing the efficacy of immune checkpoint inhibitors and adoptive cell therapies can be achieved by targeting metabolic pathways in both cancer cells and immune cells (Luby and Alves-Guerra 2021).

Long-term infections like tuberculosis and HIV can cause metabolic alterations in immune cells, which can hinder their ability to function properly. Gaining an understanding of these metabolic changes can provide insights into how to improve immune responses and outcomes in chronic infections (Liebenberg et al. 2021). The combination of metabolic therapies and cancer immunotherapy shows promise in enhancing treatment results. By addressing metabolic pathways in cancer cells and immune cells, we can potentially boost the effectiveness of immune checkpoint inhibitors and adoptive cell therapies (Gupta et al. 2022). Anomalies in the immune system's response to self-tissues are what define autoimmune diseases. Alterations in the metabolism of immune cells can contribute to the onset and progression of autoimmune diseases by supporting the activation and survival of cells that target self-tissues (Castro and Gourley 2010). Their effects on metabolism, diet, and nutrition have a significant impact on immune function. Vitamins, minerals, and amino acids are critical for the immune system to function optimally. By making changes to one's diet, it is possible to modify immune cell metabolism and enhance immune function in various contexts (Munteanu and Schwartz 2022).

9.9 Conclusion

Metabolic reprogramming is a fundamental aspect of immune cell activation and function. The switch from oxidative phosphorylation to glycolysis, mediated by factors such as HIF-1α and mTOR, supports the energetic and biosynthetic demands of immune responses. Understanding the mechanisms of metabolic reprogramming in immune cells provides insights into immune regulation and potential therapeutic targets for modulating immune responses in diseases. Understanding the role of metabolism in altering immune responses is fundamental for developing new therapeutic strategies to modulate immune function. An essential component of immune cell activation and function, metabolic reprogramming affects the course of infections, malignancies, and autoimmune illnesses. By targeting metabolic pathways, suppression of the immune responses can be enhanced, offering new avenues for treatment and improving patient outcomes.

References

Aderinto N, Abdulbasit MO, Tangmi ADE, Okesanya JO, Mubarak JM (2023) Unveiling the growing significance of metabolism in modulating immune cell function: exploring mechanisms and implications; a review. Ann Med Surg (Lond) 85(11):5511–5522

Aguilar-Ballester M, Herrero-Cervera A, Vinue A, Martinez-Hervas S, Gonzalez-Navarro H (2020) Impact of cholesterol metabolism in immune cell function and atherosclerosis. Nutrients 12(7):2021

Ahluwalia K, Ebright B, Chow K, Dave P, Mead A, Poblete R et al (2022) Lipidomics in understanding pathophysiology and pharmacologic effects in inflammatory diseases: considerations for drug development. Meta 12(4):333

Aksoylar HI, Tijaro-Ovalle NM, Boussiotis VA, Patsoukis N (2020) T cell metabolism in cancer immunotherapy. Immunometabolism. 2(3):e200020

Al-Khami AA, Rodriguez PC, Ochoa AC (2017) Energy metabolic pathways control the fate and function of myeloid immune cells. J Leukoc Biol 102(2):369–380

Alloubani A, Nimer R, Samara R (2021) Relationship between hyperlipidemia, cardiovascular disease and stroke: a systematic review. Curr Cardiol Rev 17(6):e051121189015

Amelio I, Cutruzzola F, Antonov A, Agostini M, Melino G (2014) Serine and glycine metabolism in cancer. Trends Biochem Sci 39(4):191–198

Amo-Aparicio J, Dinarello CA, Lopez-Vales R (2024) Metabolic reprogramming of the inflammatory response in the nervous system: the crossover between inflammation and metabolism. Neural Regen Res 19(10):2189–2201

Ananieva EA, Powell JD, Hutson SM (2016) Leucine metabolism in T cell activation: mTOR signaling and beyond. Adv Nutr 7(4):798S–805S

Andersen CJ, Murphy KE, Fernandez ML (2016) Impact of obesity and metabolic syndrome on immunity. Adv Nutr 7(1):66–75

Angajala A, Lim S, Phillips JB, Kim JH, Yates C, You Z et al (2018) Diverse roles of Mitochondria in immune responses: novel insights into immuno-metabolism. Front Immunol 9:1605

Arribas-Lopez E, Zand N, Ojo O, Snowden MJ, Kochhar T (2021) The effect of amino acids on wound healing: a systematic review and meta-analysis on arginine and glutamine. Nutrients 13(8):2498

Bader JE, Voss K, Rathmell JC (2020) Targeting metabolism to improve the tumor microenvironment for cancer immunotherapy. Mol Cell 78(6):1019–1033

Balta E, Wabnitz GH, Samstag Y (2021) Hijacked immune cells in the tumor microenvironment: molecular mechanisms of immunosuppression and cues to improve T cell-based immunotherapy of solid tumors. Int J Mol Sci 22(11):5736

Balyan R, Gautam N, Gascoigne NRJ (2020) The ups and downs of metabolism during the lifespan of a T cell. Int J Mol Sci 21(21):7972

Bernardi F, D'Amico F, Bencardino S, Faggiani I, Fanizza J, Zilli A et al (2024) Gut microbiota metabolites: unveiling their role in inflammatory bowel diseases and fibrosis. Pharmaceuticals (Basel) 17(3):347

Biswas SK (2015) Metabolic reprogramming of immune cells in cancer progression. Immunity 43(3):435–449
Blanco LP, Kaplan MJ (2023) Metabolic alterations of the immune system in the pathogenesis of autoimmune diseases. PLoS Biol 21(4):e3002084
Bravo GA, Cedeno RR, Casadevall MP, Ramio-Torrenta L (2022) Sphingosine-1-Phosphate (S1P) and S1P signaling pathway modulators, from current insights to future perspectives. Cells 11(13):2058
Cao J, Liao S, Zeng F, Liao Q, Luo G, Zhou Y (2023) Effects of altered glycolysis levels on CD8(+) T cell activation and function. Cell Death Dis 14(7):407
Cardoso D, Perucha E (2021) Cholesterol metabolism: a new molecular switch to control inflammation. Clin Sci (Lond) 135(11):1389–1408
Castro C, Gourley M (2010) Diagnostic testing and interpretation of tests for autoimmunity. J Allergy Clin Immunol 125(2 Suppl 2):S238–S247
Chandel NS (2021) Glycolysis. Cold Spring Harb Perspect Biol 13(5):a040535
Chen TH, Koh KY, Lin KM, Chou CK (2022a) Mitochondrial dysfunction as an underlying cause of skeletal muscle disorders. Int J Mol Sci 23(21):12926
Chen C, Wang Z, Ding Y, Qin Y (2022b) Manipulating T-cell metabolism to enhance immunotherapy in solid tumor. Front Immunol 13:1090429
Chen S, Saeed A, Liu Q, Jiang Q, Xu H, Xiao GG et al (2023) Macrophages in immunoregulation and therapeutics. Signal Transduct Target Ther 8(1):207
Cheng H, Zheng Y (2024) Advances in macrophage and T cell metabolic reprogramming and immunotherapy in the tumor microenvironment. PeerJ 12:e16825
Chi H (2012) Regulation and function of mTOR signalling in T cell fate decisions. Nat Rev Immunol 12(5):325–338
Chi H (2022) Immunometabolism at the intersection of metabolic signaling, cell fate, and systems immunology. Cell Mol Immunol 19(3):299–302
Clemente Plaza N, Reig Garcia-Galbis M, Martinez-Espinosa RM (2018) Effects of the usage of l-cysteine (l-Cys) on human health. Molecules 23(3):575
Clemente-Suarez VJ, Martin-Rodriguez A, Redondo-Florez L, Ruisoto P, Navarro-Jimenez E, Ramos-Campo DJ et al (2023) Metabolic health, mitochondrial fitness, physical activity, and cancer. Cancers (Basel) 15(3):814
Cockcroft S (2021) Mammalian lipids: structure, synthesis and function. Essays Biochem 65(5):813–845
Cong J, Zhou P, Zhang R (2022) Intestinal microbiota-derived short chain fatty acids in host health and disease. Nutrients 14(9):1977
Corcoran SE, O'Neill LA (2016) HIF1alpha and metabolic reprogramming in inflammation. J Clin Invest 126(10):3699–3707
Corrado M, Pearce EL (2022) Targeting memory T cell metabolism to improve immunity. J Clin Invest 132(1):e148546
Cortellino S, Longo VD (2023) Metabolites and immune response in tumor microenvironments. Cancers (Basel) 15(15):3898
Cruzat VF, Krause M, Newsholme P (2014) Amino acid supplementation and impact on immune function in the context of exercise. J Int Soc Sports Nutr 11(1):61
Cruzat V, Macedo Rogero M, Noel Keane K, Curi R, Newsholme P (2018) Glutamine: metabolism and immune function, supplementation and clinical translation. Nutrients 10(11):1564
Dabi YT, Andualem H, Degechisa ST, Gizaw ST (2022) Targeting metabolic reprogramming of T-cells for enhanced anti-tumor response. Biologics 16:35–45
Dang A, Kaur K (2016) Comparative effectiveness research and its utility in in-clinic practice. Perspect Clin Res 7(1):9–14
Daryabor G, Atashzar MR, Kabelitz D, Meri S, Kalantar K (2020) The effects of type 2 diabetes mellitus on organ metabolism and the immune system. Front Immunol 11:1582
Delmastro-Greenwood MM, Piganelli JD (2013) Changing the energy of an immune response. Am J Clin Exp Immunol 2(1):30–54

Dludla PV, Mabhida SE, Ziqubu K, Nkambule BB, Mazibuko-Mbeje SE, Hanser S et al (2023) Pancreatic beta-cell dysfunction in type 2 diabetes: implications of inflammation and oxidative stress. World J Diabetes 14(3):130–146
Domblides C, Lartigue L, Faustin B (2018) Metabolic stress in the immune function of T cells, macrophages and dendritic cells. Cells 7(7):68
Du H, Bartleson JM, Butenko S, Alonso V, Liu WF, Winer DA et al (2023) Tuning immunity through tissue mechanotransduction. Nat Rev Immunol 23(3):174–188
Fan K, Liu Z, Gao M, Tu K, Xu Q, Zhang Y (2022) Targeting nutrient dependency in cancer treatment. Front Oncol 12:820173
Frank AT, Zhao B, Jose PO, Azar KM, Fortmann SP, Palaniappan LP (2014) Racial/ethnic differences in dyslipidemia patterns. Circulation 129(5):570–579
Galicia-Garcia U, Benito-Vicente A, Jebari S, Larrea-Sebal A, Siddiqi H, Uribe KB et al (2020) Pathophysiology of type 2 diabetes mellitus. Int J Mol Sci 21(17):6275
Ganeshan K, Chawla A (2014) Metabolic regulation of immune responses. Annu Rev Immunol 32:609–634
Ganjoo S, Gupta P, Corbali HI, Nanez S, Riad TS, Duong LK et al (2023) The role of tumor metabolism in modulating T-cell activity and in optimizing immunotherapy. Front Immunol 14:1172931
Garcia D, Shaw RJ (2017) AMPK: mechanisms of cellular energy sensing and restoration of metabolic balance. Mol Cell 66(6):789–800
Garcia C, Andersen CJ, Blesso CN (2023) The role of lipids in the regulation of immune responses. Nutrients 15(18):3899
Garcia-Rodriguez D, Gimenez-Cassina A (2021) Ketone bodies in the brain beyond fuel metabolism: from excitability to gene expression and cell signaling. Front Mol Neurosci 14:732120
Geltink RIK, Kyle RL, Pearce EL (2018) Unraveling the complex interplay between T cell metabolism and function. Annu Rev Immunol 36:461–488
Gomez-Virgilio L, Silva-Lucero MD, Flores-Morelos DS, Gallardo-Nieto J, Lopez-Toledo G, Abarca-Fernandez AM et al (2022) Autophagy: a key regulator of homeostasis and disease: an overview of molecular mechanisms and modulators. Cells 11(15):2262
Gupta SL, Basu S, Soni V, Jaiswal RK (2022) Immunotherapy: an alternative promising therapeutic approach against cancers. Mol Biol Rep 49(10):9903–9913
Hanaford A, Johnson SC (2022) The immune system as a driver of mitochondrial disease pathogenesis: a review of evidence. Orphanet J Rare Dis 17(1):335
Heaton AR, Rehani PR, Hoefges A, Lopez AF, Erbe AK, Sondel PM et al (2023) Single cell metabolic imaging of tumor and immune cells in vivo in melanoma bearing mice. Front Oncol 13:1110503
Hitchler MJ, Domann FE (2012) Redox regulation of the epigenetic landscape in cancer: a role for metabolic reprogramming in remodeling the epigenome. Free Radic Biol Med 53(11):2178–2187
Hu C, Xuan Y, Zhang X, Liu Y, Yang S, Yang K (2022) Immune cell metabolism and metabolic reprogramming. Mol Biol Rep 49(10):9783–9795
Huang Y, Ren H, Tan Y (2024) Editorial: new insights into the metabolic processes of immune-related diseases by multi-omics technologies. Front Endocrinol (Lausanne). 15:1397644
Hubler MJ, Kennedy AJ (2016) Role of lipids in the metabolism and activation of immune cells. J Nutr Biochem 34:1–7
Huo M, Zhang J, Huang W, Wang Y (2021) Interplay among metabolism, epigenetic modifications, and gene expression in cancer. Front Cell Dev Biol 9:793428
Infantino V, Santarsiero A, Convertini P, Todisco S, Iacobazzi V (2021) Cancer cell metabolism in hypoxia: role of HIF-1 as key regulator and therapeutic target. Int J Mol Sci 22(11):5703
Judge A, Dodd MS (2020) Metabolism. Essays Biochem 64(4):607–647
Kawai T, Autieri MV, Scalia R (2021) Adipose tissue inflammation and metabolic dysfunction in obesity. Am J Physiol Cell Physiol 320(3):C375–CC91
Kelly B, O'Neill LA (2015) Metabolic reprogramming in macrophages and dendritic cells in innate immunity. Cell Res 25(7):771–784

Koene RJ, Prizment AE, Blaes A, Konety SH (2016) Shared risk factors in cardiovascular disease and cancer. Circulation 133(11):1104–1114

Kojta I, Chacinska M, Blachnio-Zabielska A (2020) Obesity, bioactive lipids, and adipose tissue inflammation in insulin resistance. Nutrients 12(5):1305

Kolodziej F, O'Halloran KD (2021) Re-evaluating the oxidative phenotype: can endurance exercise save the Western world? Antioxidants (Basel). 10(4):609

Krupa A, Kowalska I (2021) The kynurenine pathway-new linkage between innate and adaptive immunity in autoimmune Endocrinopathies. Int J Mol Sci 22(18):9879

Kumar M, Dev S, Khalid MU, Siddenthi SM, Noman M, John C et al (2023) The bidirectional link between diabetes and kidney disease: mechanisms and management. Cureus 15(9):e45615

Kurniawan H, Soriano-Baguet L, Brenner D (2020) Regulatory T cell metabolism at the intersection between autoimmune diseases and cancer. Eur J Immunol 50(11):1626–1642

Lang R, Siddique M (2024) Control of immune cell signaling by the immuno-metabolite itaconate. Front Immunol 15:1352165

Laukka T, Mariani CJ, Ihantola T, Cao JZ, Hokkanen J, Kaelin WG Jr et al (2016) Fumarate and succinate regulate expression of hypoxia-inducible genes via TET enzymes. J Biol Chem 291(8):4256–4265

Li B, Qin Y, Yu X, Xu X, Yu W (2022) Lipid raft involvement in signal transduction in cancer cell survival, cell death and metastasis. Cell Prolif 55(1):e13167

Li Z, Wang Q, Huang X, Yang M, Zhou S, Li Z et al (2023) Lactate in the tumor microenvironment: a rising star for targeted tumor therapy. Front Nutr 10:1113739

Liebenberg C, Luies L, Williams AA (2021) Metabolomics as a tool to investigate HIV/TB co-infection. Front Mol Biosci 8:692823

Ling ZN, Jiang YF, Ru JN, Lu JH, Ding B, Wu J (2023) Amino acid metabolism in health and disease. Signal Transduct Target Ther 8(1):345

Liu J, Gao M, Yang Z, Zhao Y, Guo K, Sun B et al (2022) Macrophages and metabolic reprogramming in the tumor microenvironment. Front Oncol 12:795159

Long YC, Zierath JR (2006) AMP-activated protein kinase signaling in metabolic regulation. J Clin Invest 116(7):1776–1783

Long Y, Shi H, He Y, Qi X (2023) Analyzing the impact of metabolism on immune cells in tumor microenvironment to promote the development of immunotherapy. Front Immunol 14:1307228

Look ARG, Wing RR (2010) Long-term effects of a lifestyle intervention on weight and cardiovascular risk factors in individuals with type 2 diabetes mellitus: four-year results of the Look AHEAD trial. Arch Intern Med 170(17):1566–1575

Lu J, Tan M, Cai Q (2015) The Warburg effect in tumor progression: mitochondrial oxidative metabolism as an anti-metastasis mechanism. Cancer Lett 356(2 Pt A):156–164

Luby A, Alves-Guerra MC (2021) Targeting metabolism to control immune responses in cancer and improve checkpoint blockade immunotherapy. Cancers (Basel) 13(23):5912

Ma S, Ming Y, Wu J, Cui G (2024) Cellular metabolism regulates the differentiation and function of T-cell subsets. Cell Mol Immunol 21(5):419–435

Mafi S, Mansoori B, Taeb S, Sadeghi H, Abbasi R, Cho WC et al (2021) mTOR-mediated regulation of immune responses in cancer and tumor microenvironment. Front Immunol 12:774103

Marti ILAA, Reith W (2021) Arginine-dependent immune responses. Cell Mol Life Sci 78(13):5303–5324

Mohammadnezhad L, Shekarkar Azgomi M, La Manna MP, Sireci G, Rizzo C, Badami GD et al (2022) Metabolic reprogramming of innate immune cells as a possible source of new therapeutic approaches in autoimmunity. Cells 11(10):1663

Morofuji Y, Nakagawa S, Ujifuku K, Fujimoto T, Otsuka K, Niwa M et al (2022) Beyond lipid-lowering: effects of statins on cardiovascular and cerebrovascular diseases and cancer. Pharmaceuticals (Basel) 15(2):151

Morris DR, Qu Y, Jones-Hall YL, Liu T, Ivanciuc T, Garofalo RP et al (2023) Hypoxia-inducible-factors differentially contribute to clinical disease and the control of viral replication during RSV infection. bioRxiv

Munteanu C, Schwartz B (2022) The relationship between nutrition and the immune system. Front Nutr 9:1082500

Nandi M, Jackson SK, Macrae D, Shankar-Hari M, Tremoleda JL, Lilley E (2020) Rethinking animal models of sepsis - working towards improved clinical translation whilst integrating the 3Rs. Clin Sci (Lond) 134(13):1715–1734

Nitz K, Lacy M, Atzler D (2019) Amino acids and their metabolism in atherosclerosis. Arterioscler Thromb Vasc Biol 39(3):319–330

Palmer CS, Ostrowski M, Balderson B, Christian N, Crowe SM (2015) Glucose metabolism regulates T cell activation, differentiation, and functions. Front Immunol 6:1

Palmieri EM, McGinity C, Wink DA, McVicar DW (2020) Nitric oxide in macrophage immunometabolism: hiding in plain sight. Meta 10(11):429

Papadopoli D, Boulay K, Kazak L, Pollak M, Mallette FA, Topisirovic I et al (2019) mTOR as a central regulator of lifespan and aging. F1000Res 8:8

Paran CW, Zou K, Ferrara PJ, Song H, Turk J, Funai K (2015) Lipogenesis mitigates dysregulated sarcoplasmic reticulum calcium uptake in muscular dystrophy. Biochim Biophys Acta 1851(12):1530–1538

Peng HY, Lucavs J, Ballard D, Das JK, Kumar A, Wang L et al (2021a) Metabolic reprogramming and reactive oxygen species in T cell immunity. Front Immunol 12:652687

Peng X, He Y, Huang J, Tao Y, Liu S (2021b) Metabolism of dendritic cells in tumor microenvironment: for immunotherapy. Front Immunol 12:613492

Piranavan P, Bhamra M, Perl A (2020) Metabolic targets for treatment of autoimmune diseases. Immunometabolism 2(2):e200012

Pisetsky DS (2023) Pathogenesis of autoimmune disease. Nat Rev Nephrol 19(8):509–524

Pollizzi KN, Powell JD (2015) Regulation of T cells by mTOR: the known knowns and the known unknowns. Trends Immunol 36(1):13–20

Puschel GP, Klauder J, Henkel J (2022) Macrophages, low-grade inflammation, insulin resistance and hyperinsulinemia: a mutual ambiguous relationship in the development of metabolic diseases. J Clin Med 11(15):4358

Ramalho R, Rao M, Zhang C, Agrati C, Ippolito G, Wang FS et al (2020) Immunometabolism: new insights and lessons from antigen-directed cellular immune responses. Semin Immunopathol 42(3):279–313

Rangel Rivera GO, Knochelmann HM, Dwyer CJ, Smith AS, Wyatt MM, Rivera-Reyes AM et al (2021) Fundamentals of T cell metabolism and strategies to enhance cancer immunotherapy. Front Immunol 12:645242

Regier DS, Bak A, Bausell H, O'Reilly E, Cowsert LM (2022) Starting the conversation on gene therapy for phenylketonuria: current perspectives of patients, caregivers, and advocates. Mol Genet Metab Rep 31:100855

Remmerie A, Scott CL (2018) Macrophages and lipid metabolism. Cell Immunol 330:27–42

Renovato-Martins M, Moreira-Nunes C, Atella GC, Barja-Fidalgo C, Moraes JA (2020) Obese adipose tissue secretion induces inflammation in Preadipocytes: role of Toll-Like Receptor-4. Nutrients 12(9):2828

Rinaldi L, Pafundi PC, Galiero R, Caturano A, Morone MV, Silvestri C et al (2021) Mechanisms of non-alcoholic fatty liver disease in the metabolic syndrome. A narrative review. Antioxidants (Basel). 10(2):270

Roy T, Boateng ST, Uddin MB, Banang-Mbeumi S, Yadav RK, Bock CR et al (2023) The PI3K-Akt-mTOR and associated signaling pathways as molecular drivers of immune-mediated inflammatory skin diseases: update on therapeutic strategy using natural and synthetic compounds. Cells 12(12):1671

Rui L (2014) Energy metabolism in the liver. Compr Physiol 4(1):177–197

San-Millan I (2023) The key role of mitochondrial function in health and disease. Antioxidants (Basel) 12(4):782

Saxton RA, Sabatini DM (2017) mTOR signaling in growth, metabolism, and disease. Cell 168(6):960–976

Schieber M, Chandel NS (2014) ROS function in redox signaling and oxidative stress. Curr Biol 24(10):R453–R462
Scur M, Mahmoud AB, Dey S, Abdalbarri F, Stylianides I, Medina-Luna D et al (2022) Alveolar macrophage metabolic programming via a C-type lectin receptor protects against lipo-toxicity and cell death. Nat Commun 13(1):7272
Shyer JA, Flavell RA, Bailis W (2020) Metabolic signaling in T cells. Cell Res 30(8):649–659
Singer M, Elsayed AM, Husseiny MI, (2024) Regulatory T-cells: The Face-off of the Immune Balance Frontiers in Bioscience-Landmark 29(11) https://doi.org/10.31083/j.fbl2911377
Soto-Heredero G, de Las G, Heras MM, Gabande-Rodriguez E, Oller J, Mittelbrunn M (2020) Glycolysis - a key player in the inflammatory response. FEBS J 287(16):3350–3369
Strizova Z, Benesova I, Bartolini R, Novysedlak R, Cecrdlova E, Foley LK et al (2023) M1/M2 macrophages and their overlaps - myth or reality? Clin Sci (Lond) 137(15):1067–1093
Sun L, Yang X, Yuan Z, Wang H (2020) Metabolic reprogramming in immune response and tissue inflammation. Arterioscler Thromb Vasc Biol 40(9):1990–2001
Szwed A, Kim E, Jacinto E (2021) Regulation and metabolic functions of mTORC1 and mTORC2. Physiol Rev 101(3):1371–1426
Talib WH, Mahmod AI, Kamal A, Rashid HM, Alashqar AMD, Khater S et al (2021) Ketogenic diet in cancer prevention and therapy: molecular targets and therapeutic opportunities. Curr Issues Mol Biol 43(2):558–589
Tian H, Zhang B, Li L, Wang G, Li H, Zheng J (2020) Manipulation of mitochondrial plasticity changes the metabolic competition between "Foe" and "Friend" during tumor malignant transformation. Front Oncol 10:1692
Torres M, Parets S, Fernandez-Diaz J, Beteta-Gobel R, Rodriguez-Lorca R, Roman R et al (2021) Lipids in pathophysiology and development of the membrane lipid therapy: new bioactive lipids. Membranes (Basel) 11(12):919
Ucche S, Hayakawa Y (2024) Immunological aspects of cancer cell metabolism. Int J Mol Sci 25(10):5288
Ungaro P, Nettore IC, Franchini F, Palatucci G, Muscogiuri G, Colao A et al (2022) Epigenome modulation induced by ketogenic diets. Nutrients 14(15):3245
van der Knaap JA, Verrijzer CP (2016) Undercover: gene control by metabolites and metabolic enzymes. Genes Dev 30(21):2345–2369
van der Windt GJ, Pearce EL (2012) Metabolic switching and fuel choice during T-cell differentiation and memory development. Immunol Rev 249(1):27–42
Vander Heiden MG, Cantley LC, Thompson CB (2009) Understanding the Warburg effect: the metabolic requirements of cell proliferation. Science 324(5930):1029–1033
Viola A, Munari F, Sanchez-Rodriguez R, Scolaro T, Castegna A (2019) The metabolic signature of macrophage responses. Front Immunol 10:1462
Wang LY, Chen C (2021) Energy metabolism homeostasis in cardiovascular diseases. J Geriatr Cardiol 18(12):1044–1057
Wang S, Liu G, Li Y, Pan Y (2022) Metabolic reprogramming induces macrophage polarization in the tumor microenvironment. Front Immunol 13:840029
Wazir M, Olanrewaju OA, Yahya M, Kumari J, Kumar N, Singh J et al (2023) Lipid disorders and cardiovascular risk: a comprehensive analysis of current perspectives. Cureus 15(12):e51395
Wei Z, Liu X, Cheng C, Yu W, Yi P (2020) Metabolism of amino acids in cancer. Front Cell Dev Biol 8:603837
Weichhart T, Hengstschlager M, Linke M (2015) Regulation of innate immune cell function by mTOR. Nat Rev Immunol 15(10):599–614
Winston JA, Theriot CM (2020) Diversification of host bile acids by members of the gut microbiota. Gut Microbes 11(2):158–171
Yan Y, Huang L, Liu Y, Yi M, Chu Q, Jiao D et al (2022) Metabolic profiles of regulatory T cells and their adaptations to the tumor microenvironment: implications for antitumor immunity. J Hematol Oncol 15(1):104

Yang L, Chu Z, Liu M, Zou Q, Li J, Liu Q et al (2023) Amino acid metabolism in immune cells: essential regulators of the effector functions, and promising opportunities to enhance cancer immunotherapy. J Hematol Oncol 16(1):59

Zabor EC, Kaizer AM, Hobbs BP (2020) Randomized controlled trials. Chest 158(1S):S79–S87

Zampino M, AlGhatrif M, Kuo PL, Simonsick EM, Ferrucci L (2020) Longitudinal changes in resting metabolic rates with aging are accelerated by diseases. Nutrients 12(10):3061

Zatterale F, Longo M, Naderi J, Raciti GA, Desiderio A, Miele C et al (2019) Chronic adipose tissue inflammation linking obesity to insulin resistance and type 2 diabetes. Front Physiol 10:1607

Zhai L, Bell A, Ladomersky E, Lauing KL, Bollu L, Sosman JA et al (2020) Immunosuppressive IDO in cancer: mechanisms of action, animal models, and targeting strategies. Front Immunol 11:1185

Zhang W, Lang R (2023) Succinate metabolism: a promising therapeutic target for inflammation, ischemia/reperfusion injury and cancer. Front Cell Dev Biol 11:1266973

Zhang X, Alshakhshir N, Zhao L (2021) Glycolytic metabolism, brain resilience, and Alzheimer's disease. Front Neurosci 15:662242

Zhang Y, Li Q, Huang Z, Li B, Nice EC, Huang C et al (2022) Targeting glucose metabolism enzymes in cancer treatment: current and emerging strategies. Cancers (Basel) 14(19):4568

Zhang X, Song W, Gao Y, Zhang Y, Zhao Y, Hao S et al (2023) The role of tumor metabolic reprogramming in tumor immunity. Int J Mol Sci 24(24):17422

Zheng X, Zhu Y, Zhao Z, Chu Y, Yang W (2023a) The role of amino acid metabolism in inflammatory bowel disease and other inflammatory diseases. Front Immunol 14:1284133

Zheng Y, Yao Y, Ge T, Ge S, Jia R, Song X et al (2023b) Amino acid metabolism reprogramming: shedding new light on T cell anti-tumor immunity. J Exp Clin Cancer Res 42(1):291

Zhu J, Thompson CB (2019) Metabolic regulation of cell growth and proliferation. Nat Rev Mol Cell Biol 20(7):436–450

Zhu Y, Li X, Wang L, Hong X, Yang J (2022) Metabolic reprogramming and crosstalk of cancer-related fibroblasts and immune cells in the tumor microenvironment. Front Endocrinol (Lausanne) 13:988295

Metabolism of Cellular Immunity and Its Role in Supporting Responses to Microbial Infection

10

Ameera Abu-Qiyas, Salam Dakalbab, Shahad Arikat, Ahmed M. Al-Rawi, Mays Z. Abu Libdeh, Aya T. Ziadeh, and Mohammad G. Mohammad

Abstract

Cells are constantly fueled by sugar. While increased sugar consumption in humans might cause them to exhibit a burst of activity, this dependence comes at a significant cost to their long-term health. Similarly, certain cells whose modus operandi involve lots of sugar breakdown have rapid responsiveness but tend to lead short lives. This analogy reflects a growing understanding of cellular metabolism—the complex interplay between energy production and cellular processes. Metabolism not only dictates an organism's overall health but also significantly influences the functions of individual cells, particularly those critical to the immune system, making it a discipline that intersects with several fields of research. Among such interdisciplinary fields is the emerging study of immunometabolism, where the influence of metabolic pathways on immune cell fate and

A. Abu-Qiyas · S. Arikat · A. M. Al-Rawi
Research Institute for Medical and Health Sciences, University of Sharjah, Sharjah, UAE

S. Dakalbab
Research Institute for Medical and Health Sciences, University of Sharjah, Sharjah, UAE

College of Medicine, University of Sharjah, Sharjah, UAE

M. Z. Abu Libdeh · A. T. Ziadeh
Department of Medical Laboratory Sciences, College of Health Sciences, University of Sharjah, Sharjah, UAE

M. G. Mohammad (✉)
Research Institute for Medical and Health Sciences, University of Sharjah, Sharjah, UAE

Department of Medical Laboratory Sciences, College of Health Sciences, University of Sharjah, Sharjah, UAE
e-mail: mmohd@sharjah.ac.ae

S. S. M. Soliman, M. I. Husseiny (eds.), *Metabolic Dynamics in Host-Microbe Interaction*, https://doi.org/10.1007/978-981-96-1305-2_10

function is investigated. In this chapter, we explore various avenues of immunometabolism in immune cells, the crosstalk between immune signaling and metabolic pathways, and the impact of these pathways on shaping immune responses against a myriad of infections.

Keywords

Immunometabolism · Macrophages · Natural killer cells · Dendritic cells · Lymphocyte metabolism · Microbial infection

10.1 Introduction

10.1.1 Metabolism and the Immune System: A Brief Overview

Before exploring the interplay between metabolism and immunity, the key players will be discussed independently first. Metabolism is a set of biochemical reactions that enable cellular function and sustain life. It can be dichotomized into catabolism and anabolism (Fig. 10.1). Catabolism is the breakdown of molecules for energy release, while anabolism is the synthesis of molecules by energy consumption. The dynamic equilibrium between these reactions is necessary to maintain a balance in

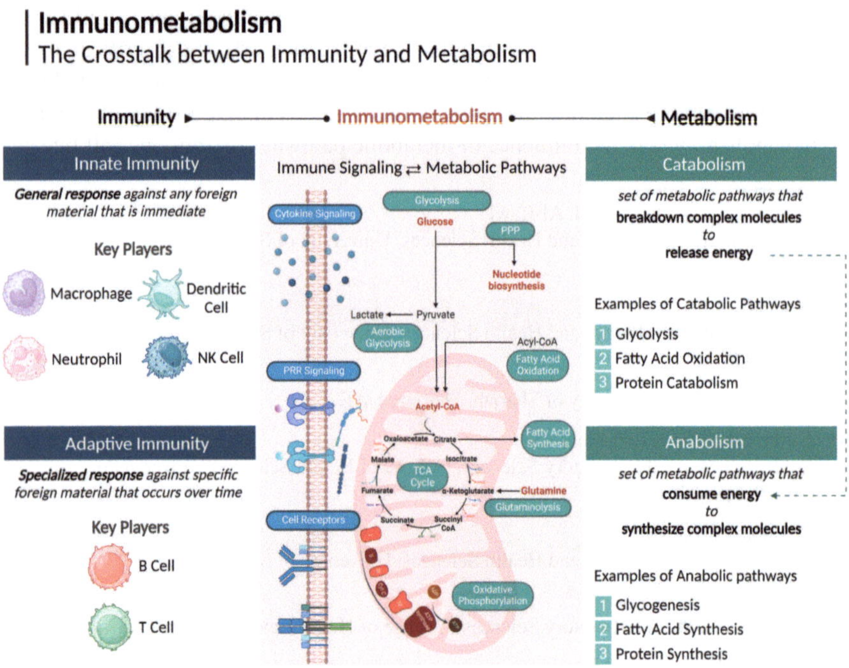

Fig. 10.1 Schematic illustration of the cellular immunity components, metabolic pathways, and intermediary crosstalk

energy utilization and molecule biosynthesis. Thus, dysregulation in these processes not only culminates in the development of metabolic disorders, such as obesity and diabetes, but also impacts immune cells' functions and infection outcomes (Alwarawrah et al. 2018).

Metabolic activities support immunity by fueling its multifaceted defense mechanisms (Olenchock et al. 2017). Immunity is comprised of innate and adaptive components, each safeguarding the body at different stages and in distinct ways. The defense process starts with innate immunity, in which an immediate and general response is launched against any invading foreign substance. Components of innate immunity include the skin, which serves as an anatomical barrier to infection; antimicrobial peptides, which can none-specifically kill a broad range of pathogens; and innate immune cells, which can recognize and engulf foreign bodies in a process known as phagocytosis (Medzhitov and Janeway 2000). Dendritic cells (DCs), macrophages, and neutrophils are the main innate immune cell players that can recognize pathogens by interacting with pathogen-associated molecular patterns (PAMPs) using their pattern recognition receptors (PRRs). Dendritic cells and some subsets of macrophages can then present the antigens of the phagocytosed invaders to T lymphocytes that have the ability to generate specific responses, thereby initiating an adaptive (acquired) immunity. Hence, innate immunity is characterized by non-specific rapid responses that are relatively similar to all pathogens, while adaptive immunity is distinguished by specific responses that take longer to mount as they await other immune cell signals (Hoebe et al. 2004).

10.1.2 Scope of the Chapter and Key Concepts

This chapter is centered on the nexus of immunity and metabolism, where we explore the metabolic landscape governing immune cells involved in innate immune responses such as neutrophils, macrophages, natural killer (NK) cells, and DCs, as well as immune cells mediating adaptive responses, including T and B cells (Fig. 10.1). We also elucidate the bidirectional communication occurring between immune signaling and metabolic checkpoints, and how the metabolic status molds immune responses against a spectrum of microbial infections, including viral, bacterial, fungal, and parasitic infections. Finally, we discuss the burgeoning avenues of research that scientists continue to discover as they unravel these entwined fields and the alluring potential that immunometabolism holds at the forefront in the ongoing hunt for additions to our armamentarium of novel therapeutics. In addressing these points, we hope to provide a holistic understanding of immunometabolism to serve as a valuable resource for students, researchers, and clinicians, as well as inspire nodes for unexplored territory in research and therapeutic interventions.

For a clear understanding of the metabolic intricacies covered in this chapter, it is helpful to review the following key concepts regarding cellular metabolism. The typical metabolic processes that occur within cells include cellular respiration and biosynthesis, which are greatly influenced by oxygen and nutrient availability.

Cellular respiration is a series of catabolic reactions that take place in pursuit of extracting energy from carbon-carbon covalent bonds and transferring it into adenosine triphosphate (ATP). In the presence of oxygen, these catabolic reactions comprise aerobic respiration, which includes glycolysis, the tricarboxylic acid (TCA) cycle, and oxidative phosphorylation (OXPHOS). The metabolic reaction chain is initiated with the production of acetyl-coenzyme A (acetyl-CoA), either by glucose breakdown during *glycolysis* in the cytoplasm or fatty acid breakdown during *β-oxidation* in the mitochondria. Once it is produced, acetyl-CoA enters the *TCA cycle* in the mitochondria and directs the synthesis of electron carriers NADH and FADH2. These electron carriers travel to the inner mitochondrial membrane, where they can donate their electrons to a series of proteins forming an electron transport chain (ETC). *OXPHOS* begins when the electrons are transferred along the ETC, causing a high concentration of protons to be pumped out to the intermembrane space and back in via facilitated diffusion through ATP synthase, a motor enzyme complex that catalyzes ATP synthesis (Pearce Erika and Pearce 2013). In hypoxic conditions, cells resort to anaerobic respiration. During this type of respiration, the pyruvate produced by glycolysis is converted into lactic acid instead of acetyl-CoA, and glycolysis is sustained via regeneration of NAD+ (Krzywinska and Stockmann 2018). Cells may choose to opt for enhanced glycolytic activity even in the presence of oxygen in a process called aerobic glycolysis (Fig. 10.1). This process can have serious downstream effects, including insufficient ATP generation, loss of mitochondrial membrane potential, and toxic lactic acid accumulation (Kelly and O'Neill 2015). Hence, prolonged reliance on glycolysis (as alluded to in the beginning of this chapter) can be detrimental to the health of the cell.

Biosynthesis, on the other hand, is a series of anabolic reactions that utilize ATP to produce proteins, lipids, nucleotides, and sugars required for sustaining cellular functions. Since energy consumption is integral to biosynthetic processes, perturbations in catabolic pathways can adversely impact physiological biosynthesis. In addition to oxygen availability, microenvironments can also impinge on metabolic patterns, as cells will have to adjust if their niche is nutrient-deprived or nutrient-replete (Kaymak et al. 2022). This manifests in the eventual impairment of regular metabolic activities of some immune cells in chronic infections, where they compete with pathogenic invaders for nutrient resources at the infection site (Olive and Sassetti 2016).

10.2 Metabolism and Innate Immunity

Innate immune cells are the body's first line of cellular defense against foreign antigens. Neutrophils, macrophages, NK cells, and DCs are responsible for propagating this response. They express toll-like receptors (TLRs) for pathogen sensing and are responsible for clearing pathogens. Cellular metabolism provides these cells with the energy and resources needed to perform their functions (Cheng et al. 2014). This chapter will explore the metabolic pathways and processes that support the activity of key innate immune cells.

10.2.1 Neutrophils

Neutrophils are the most abundant circulating leukocytes, defending against bacterial and fungal infections. Upon pathogen detection, neutrophils activate their latent cytosolic NADPH oxidase complex and produce microbicidal H_2O_2 (Belambri et al. 2018). Upon pathogen detection and stimulation, they rely on glucose, and in some cases glutamine, for their degranulation of antimicrobial defenses into the extracellular space or directly into the phagosome (Kumar and Dikshit 2019). Their respiratory burst entailing ROS production is also supported by aerobic glycolysis and glutaminolysis mediating NADPH production. Rapid production of NADPH is needed for the NADPH oxidase system and "NETosis," a process by which neutrophils release a net-like mitochondrial DNA in order to trap bacteria (Vorobjeva and Chernyak 2020). These metabolic shifts are induced and maintained by cytokines like IFN-γ and TNF-α, as they prime and enhance the activation of these immune cells, enabling effective defense mechanisms against invading pathogens.

Hypoxia-induced factor 1α (HIF-1α) is an acute-phase inflammatory protein produced in response to several stress conditions, including infection. HIF-1α enhances glycolysis by increasing glucose cellular influx through upregulating glucose transporter 1 (GLUT1) and the glycolytic enzyme phosphoglycerate kinase (PGK). HIF-1α also regulates several antimicrobial factors, including neutrophil elastase, cathepsin G, cathelicidin-related antimicrobial peptide, and inducible nitric oxide (NO) (Table 10.1). After initiating their microbicidal respiratory burst at the site of infection, neutrophils rapidly undergo apoptosis and are cleared from inflammatory sites by macrophages. Their mitochondria depolarize, resulting in the release of cytochrome C at the onset of apoptosis. Therefore, the metabolic status of neutrophils is affected by the surrounding environment that mediates their homing from the bloodstream, activation, and eventual death (Jeon et al. 2020).

10.2.2 Macrophages

Macrophages are specialized immune cells known for their ability to engulf and digest foreign pathogens. They are also important for tissue repair and immune regulation by acting as both phagocytes and antigen-presenting cells (APC). Their activation status and functions can shift dramatically depending on their metabolic state.

Macrophages polarize into different functional states, including pro-inflammatory (M1) or anti-inflammatory (M2) macrophages, each with distinct metabolic profiles (Chinetti-Gbaguidi and Staels 2011; Mohammad et al. 2023). Effector functions of M1 and M2 macrophages are closely regulated by signaling and transcriptional pathways that govern cellular metabolism. Classically activated macrophages (M1) exhibit a pronounced reliance on glycolysis and glutaminolysis to sustain their inflammatory and antimicrobial functions (Ganeshan and Chawla 2014). The pentose phosphate pathway (PPP) and the malate-aspartate shuttle generate NADPH, which is needed for reactive oxygen species (ROS) production. HIF-1α serves as a

Table 10.1 Summary of the metabolic signatures for each immune cell

Immune Cell	Cell Status	Stimuli	Main Energy Source	Metabolic Pathway	Regulation Factors	Outcome	References
Neutrophils	Activated	IFN-γ TNF-α	Glucose and glutamine	↑ Glycolysis ↑ Glutaminolysis	↑ GLUT1, ↑ PGK1, HIF-1α stabilization	NADPH production, NO production, ROS production, Neutrophil extracellular trap (NETosis) formation, Degranulation	Jeon et al. (2020)
Macrophages	M1 (Classical Activation)	LPS IFN-γ TNF-α	Glucose and glutamine	↑ Glycolysis ↑ PPP ↑ Malate-aspartate shuttle ↓ TCA (truncated) ↓ OXPHOS ↑ FAS Arginine catabolism	↑ GLUT1, ↑ PGK1, ↓ AMPK activity, ↑ mTOR activity ↓ PPAR-γ, ↑ PFKFB3, ↑ Citrate, ↑ Itaconate, ↑ Succinate, HIF-1α stabilization iNOS	NADPH production, NO production, ROS production, Pro-inflammatory cytokine production	Zeng et al. (2023), Fan et al. (2018), Phair et al. (2023), Puchalska et al. (2018)
	M2 (Alternative Activation)	IL-4 IL-13 IL-10 TGF-β	Glutamine and fatty acids	↓ Glycolysis ↑ Glutaminolysis ↑ TCA ↑ OXPHOS ↑ FAO Arginine catabolism Ornithine, proline, and polyamine synthesis	↑ AMPK activity ↓ mTOR activity ↑ PPAR-γ ↑ PGC1β ↑ PFKFB1 ↑ Mitochondrial mass Arg1	Anti-inflammatory cytokine production, Tissue repair	Viola et al. (2019), Qing et al. (2019), Wculek et al. (2022)

Natural Killer Cells	Activated	IL-2 IL-12 IL-15 IL-18 IFN-α	Glucose, fatty acids	↑ Glycolysis ↑ PPP	↑ LDHA ↑ PGK1 ↑ mTOR activity ↑ GLUT1 ↑ CD71 ↑ CD98	Cytotoxic degranulation, IFN-γ and TNF-α production	Osuna-Espinoza and Rosas-Taraco (2023)
	Exhausted (persistent infection)	IL-2 IL-12 IL-15 IL-18 IFN-α	Scarce nutrient availability	Impaired glycolytic activity	↑ Tim-3 ↑ PD-1 ↑ NKG2A	Impaired cytotoxicity, Diminished inflammatory function	Osuna-Espinoza and Rosas-Taraco (2023)
Dendritic Cells	mDCs	TLR7/8 ligand TNF-α GM-CSF IL-4	Glucose	↑ Glycolysis ↑ PPP ↓ TCA (truncated) ↓ OXPHOS ↑ FAS	↑ PI3K/Akt ↓ Mitochondrial mass HIF-1α stabilization ↑ mTOR activity	Antigen processing, Pro-inflammatory cytokine production, Co-stimulatory molecule expression	Wu et al. (2023), Basit et al. (2018)
	pDCs	TLR7/8 ligand IFN-α	Glutamine	↑ Glutaminolysis ↑ OXPHOS ↑ FAO	↑ PGC1α ↑ Mitochondrial mass	IFN-α Antiviral response	Basit et al. (2018)

(continued)

Table 10.1 (continued)

Immune Cell	Cell Status	Stimuli	Main Energy Source	Metabolic Pathway	Regulation Factors	Outcome	References
T Cells	Th cells (effector)	TCR ligand IL-1 IL-12	Glucose and glutamine	↑ Glycolysis ↑ FAS Arginine catabolism[a]	↑ GLUT ↑ mTOR activity HIF-1α stabilization ↑ AMPK activity[b]	Clonal expansion, Cytokine secretion, Memory cell transition[c]	Geiger et al. (2016), Wik and Skålhegg (2022)
	Treg cells	IL-10 TGF-β	Fatty acids, lactic acid	↑ OXPHOS ↑ FAO Arginine catabolism	↑ AMPK activity FoxP3 Arg2	Immune suppression, Self-tolerance	Watson et al. (2021), Wik and Skålhegg (2022)
	Tc cells (effector)	TCR ligand	Glucose and glutamine	↑ Glycolysis ↑ Glutaminolysis ↑ Serine metabolism ↑ FAS Arginine catabolism[a]	↑ GLUT ↑ mTOR activity ↑ AMPK activity[b]	Clonal expansion, Cytotoxic degranulation, Memory cell transition[c]	Geiger et al. (2016), Wik and Skålhegg (2022), Reina-Campos et al. (2021)
	Tm cells (quiescent)	IL-7 IL-15	Fatty acids	↑ OXPHOS ↑ FAO	↑ Mitochondrial mass Sustained SRC	Long-term immunological memory formation	Wik and Skålhegg (2022)
	Exhausted (persistent infection)	TCR ligand IL-1 IL-12	Scarce nutrient availability	Impaired metabolic pathways	↑ PD-1 ↑ CTLA-4 Mitochondrial dysfunction	Impaired immune responses	Wik and Skålhegg (2022)

B Cells	GC B cells	BCR ligand	Glucose and fatty acids	↑ Glycolysis ↑ FAO	HIF-1α stabilization ↑ Mitochondrial mass	Proliferation Somatic hypermutation Affinity maturation	Jellusova (2020)
	Plasma cells	BCR ligand IL-21 Il-4	Glucose and glutamine	↑ Glycolysis ↑ OXPHOS	↑ mTOR activity	Nutrient absorption, Autophagy, Antibody secretion	Gaudette et al. (2020), Jellusova (2020)
	B memory cells (quiescent)	IL-4 IL-6 IL-21	Glucose	Glycolysis OXPHOS	↑ Atg5 ↑ Atg7 ↑ Ulk1	Autophagy, Long-term immunological memory formation	Ripperger and Bhattacharya (2021)

[a]Arginine catabolism in Th and Tc cells skews metabolism from glycolysis to OXPHOS, making these cells candidates for Tm formation due to enhanced survival and longevity

[b]AMPK activity is increased in glucose-limited conditions and in the effector-to-memory transition phase due to lack of antigenic stimulation

[c]Memory cell transition is associated with arginine catabolism and AMPK activity

key regulator of glycolytic metabolism in M1 macrophages, controlling the expression of GLUT1 and PGK (Troise et al. 2023). Activation of TLR signaling pathways facilitates mitochondrial recruitment to the phagolysosome, enhancing ROS production and bactericidal activity (Herb and Schramm 2021). Mitochondria-derived signals, such as ROS and succinate generated from the TCA cycle, further amplify the effector phenotype of M1 macrophages. Succinate, primarily derived from glutaminolysis-derived α-ketoglutarate (α-KG) in TCA, stabilizes HIF-1α and enhances cytokine production in response to the Gram-negative bacteria's main cell wall component, lipopolysaccharide (LPS) stimulation (Zeng et al. 2023). Conversely, M2 macrophages rely more on oxidative metabolism, characterized by increased mitochondrial biogenesis and fatty acid oxidation. The metabolic reprogramming in M2 macrophages is initiated by the STAT6 signaling pathway that promotes the expression of genes involved in OXPHOS and mitochondrial function (Viola et al. 2019; Soliman et al. 2022).

LPS inhibits AMP-activated protein kinase (AMPK) activity and upregulates PFKFB3 expression, promoting glycolysis and biosynthetic pathways in highly activated M1 macrophages (Fan et al. 2018; Phair et al. 2023). Additionally, TCA cycle breaks are often observed in these cells, resulting in an accumulation of citrate, succinate, and itaconate. Citrate is redirected to the cytoplasm, where fatty acid synthesis (FAS) can be initiated, another hallmark of M1 macrophages, supporting the production of inflammatory mediators. Altered citrate metabolism can also affect epigenome-related histone acetylation, which results in modulating the expression of glycolytic enzymes, thereby influencing glucose consumption. These metabolic events are essential for the activation and function of macrophages in response to LPS, shaping their immune responses and interactions with other immune cells. Succinate is a TCA cycle intermediate involved in stabilizing HIF-1α, thus promoting glycolysis and inflammation. The increase in succinate levels in LPS-stimulated macrophages is mainly sourced from glutamine metabolism and the glyoxylate shunt. Finally, itaconate accumulation serves M1 macrophages function through its antimicrobial activity, support in succinate accumulation, and ROS generation regulation (Palmieri et al. 2020). On the other hand, the oxidative metabolism in M2 macrophages is induced by IL-4 and IL-13 signaling pathways, inhibiting mTOR and promoting AMPK activity. M2 macrophages express PFKFB1, leading to decreased glycolytic activity. Both PPAR-γ and PGC1β were also shown to direct anti-inflammatory responses and increase mitochondrial biogenesis in M2 macrophages. These macrophage subsets are also distinguished in their differential arginine metabolism methods, as M1 macrophages utilize NOS2, while M2 macrophages employ Arg1. As such, metabolism of arginine in M1 macrophages aids NO generation, while in M2 macrophages, it aids ornithine, proline, and polyamine synthesis in support of tissue repair and wound healing (Líndez and Reith 2021) (Table 10.1).

10.2.3 Natural Killer Cells (NKs)

NK cells are cytotoxic lymphocytes that recognize and eliminate infected or cancerous cells without prior activation. NKs play a prominent role in the resolution of viral infections and are integral to innate immunity.

Resting NK cells primarily utilize oxidative phosphorylation to meet basal energy needs. Activated NK cells shift toward glycolysis, which supports the heightened energy demands for increased cytokine secretion and cytotoxic granule release (Wang et al. 2020). Indeed, the inhibition of LDHA, a central enzyme in aerobic glycolysis, rapidly weakens NK cells' antiviral defenses. Glycolysis provides metabolic intermediates essential for effector functions, enhancing NK cell survival, proliferation, and function. Additionally, the PPP contributes to redox balance and nucleotide synthesis for maintaining NK cell activity and response longevity. These pathways are governed by metabolic regulators, such as mTOR and PGK1, and nutrient-specialized influx transporters, such as CD98, GLUT1, and CD71 (Osuna-Espinoza and Rosas-Taraco 2023). In turn, the nutrient transporters CD98 (amino acid transporter), GLUT1 (glucose transporter), and CD71 (iron transporter) are regulated by stimulation of innate immune cells secreted cytokines, including IFN-α, IL-12, IL-15, and IL-18 and T cell cytokine IL-2 (Table 10.1).

Activated NK cells also alter their lipid composition and utilize fatty acids as their energy source, with the production of lipid mediators that modulate immune responses and cytotoxicity against target cells. Continued activation of NK cells by cytokines due to persistent infection induces an exhausted phenotype characterized by an increased expression of inhibitory receptors Tim-3, PD-1, and NKG2A. This phenotype can lose its cytotoxic and inflammatory functions as metabolic activity becomes impaired during chronic infection (Osuna-Espinoza and Rosas-Taraco 2023).

10.2.4 Dendritic Cells (DCs)

DCs are professional APCs that orchestrate innate and adaptive immune responses. They engulf pathogens, process them, and then present antigens to T helper lymphocytes in a process demanding a lot of energy and, hence, are tightly regulated by metabolic pathways (Wu et al. 2023).

The dynamic nature of metabolic pathways within this cell type is manifested in two broad subsets of DCs: myeloid DCs (mDCs) and plasmacytoid DCs (pDCs). Activated mDCs undergo metabolic reprogramming, characterized by increased glycolysis, to support the energetic demands of antigen processing, cytokine production, and co-stimulatory molecule expression (Wu et al. 2023). The surge in glycolysis also supports FAS through citrate accumulation and metabolism. This shift is regulated by signaling pathways upregulating HIF-1α and PI3K/Akt pathways that are essential for promoting DC maturation and immune activation (Table 10.1). Conversely, stimulated pDCs are primarily fueled by glutamine metabolism and are noted to increase PGC1α expression, which drives OXPHOS and mitochondrial biogenesis, respectively. FAO has also been observed in the pDC subset, which is known for its distinct role in antiviral responses (Basit et al. 2018) (Table 10.1).

10.3 Metabolism and Adaptive Immunity

The adaptive immune system is a complex defense mechanism that identifies and selectively targets specific pathogens, utilizing immune cells such as T and B lymphocytes with specialized recognition receptors. Through this targeted response, the immune system forms effector and memory cells, enabling a quicker and more effective defense upon subsequent encounters with the same pathogen, thereby contributing to long-term immune resilience. In the course of normal development in the bone marrow and later in the thymus (in the case of T cells), lymphocytes undergo a remarkable metabolic remodeling process in response to a wide range of environmental stresses (Patsoukis et al. 2016). The genesis and production of effective immune responses depend on the control of lymphocyte homeostasis. Maintaining homeostasis and cell survival necessitates extrinsic growth factors, and it is increasingly evident that regulating cell metabolism is a crucial component of this control system (Plas et al. 2002). Intriguing insights into the functions of many metabolic pathways and metabolites, with a significant impact on cell growth, function, survival, and even differentiation, have emerged from this field of study. Specifically, when lymphocytes are at rest, they employ an energy-efficient oxidative metabolism; however, upon encountering a pathogen, they undergo a shift to a highly glycolytic metabolism. To sustain cell memory, lymphocytes must revert back to a more quiescent oxidative metabolism following antigen clearance (Hu et al. 2022).

10.3.1 T lymphocytes

T lymphocytes play a central role in the adaptive immune system, orchestrating and executing immune responses. CD8+ T cytotoxic (Tc) cells eliminate pathogen-infected or cancerous cells through the delivery of cytotoxic molecules, while CD4+ T helper (Th) cells influence the immune response by generating cytokines that either enhance or inhibit inflammation (Kishton et al. 2017). Metabolic remodeling is intricately linked to cellular development in T cell biology. The demand for metabolic resources for cell growth, proliferation, and effector chemical production intensifies upon activation of naive T cells (Klein Geltink et al. 2018).

Two primary cooperative processes supplying energy to cells are glycolysis and OXPHOS (Vander Heiden et al. 2009). Glucose transporters (GLUTs) facilitate glucose uptake, driving the TCA cycle and OXPHOS to generate ATP (MacIver et al. 2013). It also supports PPP, yielding pentose sugars crucial for NADPH and nucleic acid synthesis (Sagone Jr et al. 1974; Tuosto et al. 2001). Amino acids, notably serine, play a crucial role in T cell metabolism; serine import facilitates rapid mitochondrial expansion during early T cell activation (Ma et al. 2017), and it is metabolized through the one-carbon metabolism pathway, incorporating into the purine synthesis pathway (Ron-Harel et al. 2016) (Table 10.1). Cysteine availability is vital for protein and glutathione synthesis, impacting T cell activity by regulating osmolality (Sikalidis 2015).

Following T cell activation, a sharp increase in lipid demand occurs (Wang et al. 2011); FAS and absorption are pivotal for the rapid clonal expansion of T cells in the early stages (Muroski et al. 2017). ROS, essential for antigen-specific T cell activation and clonal growth (Pearce Erika and Pearce 2013), primarily results from the ETC (Turrens 2003). The activation and survival of lymphocytes hinge on these metabolic adaptations, as insufficient nutrition can significantly impede T cell proliferation and adaptive immunity (Ozkan et al. 1995).

10.3.1.1 T Helper Lymphocyte Metabolism (Th1, Th2, Th17, Treg)

Within the Th lymphocyte subset, further subcategorization is necessary to distinguish between conventional Th cells and a unique subpopulation identified as T regulatory cells (Tregs). While Th cells are often interchangeably called CD4+ T cells, Tregs display a more defined marker profile through CD3+FoxP3+ expression in addition to the CD4 marker. Inflammatory T cells exhibit remarkably elevated proliferation potential, contrasting with the low proliferation of Tregs (Michalek and Rathmell 2010). The diverse phenotypes of activated T cells necessitate metabolic programs aligned with each cell's functional response.

Glycolysis is integral for complete effector T cell (Te) activation of Th cells, and its absence can impact function (Klein Geltink et al. 2018). Stimuli, including TCR ligands, IL-1, and IL-2, lead to the upregulation of glucose, glutamine, and FAS, initiating CD4+ T cell differentiation into Th1, Th2, and Th17 cells (Michalek et al. 2011; Shi et al. 2011). These factors are regulated through increased GLUT1 expression and mTOR activity. In contrast, high-glucose uptake impairs Tregs' suppressive function and long-term stability. Tregs, instead, predominantly rely on fatty acid oxidation (FAO), driven by AMPK and OXPHOS, for vitality and survival. They exhibit regular AMPK activation, and an in vivo pharmacological AMPK stimulation increases Treg populations (Michalek et al. 2011). Despite poorer activity in high-glucose conditions, Tregs can also maintain a metabolic flexibility that allows them to utilize lactic acid for continued suppressive function and proliferation (Watson et al. 2021). Transcription factors, including HIF-1α and FoxP3, establish the metabolic state and T cell identity of conventional Th and Treg cells, respectively (Gerriets et al. 2016).

The small GTPase RhoA is essential for glycolysis and OXPHOS reprogramming. While Th1 cell differentiation remains unaltered, RhoA-deficient T cells struggle to differentiate into the Th2 cell lineage, suggesting distinct pathways to glycolytic engagement in these T cell subsets (Yang et al. 2016). Th1, Th2, and Th17 subgroups rely on de novo FAS and glycolysis to fulfill lipid needs rather than acquiring external fatty acids (Berod et al. 2014). Conversely, FAO is crucial for the proliferation and suppressive capacity of Tregs (Procaccini et al. 2016). Leptin is essential for Th17 cell differentiation and proliferation, as leptin receptor deletion on T cells reduces Th17 cell function (Reis et al. 2015). Additionally, arginine is a vital requirement for T cells, with Tregs' expression of arginase 2 (Arg2) being linked with mTOR activity attenuation and enhanced suppression. Pathogens can limit T cell activation through the expression of arginase, degrading arginine as an immunosuppressive tactic (Hu et al. 2016). This intricate interplay of metabolic pathways underscores the nuanced regulation essential for T cell functionality and immune responses (Table 10.1).

10.3.1.2 T Cytotoxic Lymphocyte Metabolism

T cytotoxic (Tc) lymphocytes' main role is eliminating abnormal cells and are often interchangeably characterized as CD8+ T cells. Their functions necessitate complex metabolic processes for effective immune responses, including glycolysis and glutaminolysis. Tc cell glycolytic reprogramming and proliferation suggest the activation of mTOR and the involvement of additional GLUT family members in modulating CD8+ T cell glucose uptake (Macintyre et al. 2014). During glucose limitation, AMPK signals drive glutamate transporter gene expression and glutaminolysis to maintain Tc cell metabolism (Blagih et al. 2015). Furthermore, glutamine is essential for Tc cell activation into Teff cells, and these cells express glutamate transporters more frequently, with their loss hindering Tc cell development (Wang et al. 2011; Carr et al. 2010; Sinclair et al. 2013). For Teff cell activation and proliferative ability, mitochondrial ATP synthesis, particularly ATP synthase activity, is crucial within the first 24–48 hours post-T cell activation (Chang et al. 2013). The growth of CD8+ T lymphocytes relies on extracellular serine uptake and its utilization in one-carbon metabolism (Ma et al. 2017; Ron-Harel et al. 2016). Serine is predominantly metabolized for purine biosynthesis in CD8+ T lymphocytes (Kelly and O'Neill 2015). While CD8+ Teff cells utilize exogenous fatty acids for proliferation and accumulation, de novo FAS is essential for optimal expansion and survival (Vander Heiden et al. 2009). Arginine is also implicated in T cell longevity, as inhibiting its uptake significantly reduces T cell survival (Geiger et al. 2016) (Table 10.1).

10.3.1.3 T Memory Lymphocyte Metabolism

Following pathogen elimination, the Te cell population decreases, undergoing apoptosis, leaving behind durable memory T (Tm) cells that exhibit robust responses upon antigen re-exposure (Williams and Bevan 2007). The creation of Tm cells is stimulated and maintained by cytokines, including IL-7 and IL-15. It involves relieving systemic inflammation by blocking the glycolytic flow that caused metabolic stress induced by pathogens in Te cells, thereby enhancing memory cell production (Sukumar et al. 2013; Pearce et al. 2009). Supporting the notion that FAO and spare respiratory capacity (SRC) are interlinked, promoting FAO in T cells improves the formation, long-term survival, and robust response to antigen stimulation of Tm cells (van der Windt et al. 2012). In these cells, OXPHOS is utilized for ATP production to sustain substantial spare respiratory capacity (SRC) and increased mitochondrial mass, crucial for their survival and differentiation (van der Windt et al. 2012; O'Sullivan et al. 2014). SRC plays a critical role in CD8+ Tm cells, not only for growth and persistence but also for rapid recall, relying on mitochondria to supply ATP for quick glycolysis activation upon restimulation. This intricate metabolic orchestration underscores the multifaceted nature of T cell responses and memory formation (Van Der Windt et al. 2013) (Table 10.1).

10.3.1.4 Exhausted T Cell Metabolism

Persistent stimulation of T cells during chronic infections exhausts the metabolic and functional activities of T cells, thereby impairing pathways that are crucial for survival. T cells begin to increase the expression of inhibitory receptors PD-1 and CTLA-4 and adopt apoptotic phenotypes. Chaos ensues as glucose uptake is diminished, mitochondrial function is impaired, and oxidative metabolism is reduced. The depletion of nutrients in the T cell's microenvironment only exacerbates the dysregulation of its metabolic duties (Wik and Skålhegg 2022).

10.3.2 B Cells

Certainly, B cells and T cells share fundamental metabolic characteristics, such as increased glucose influx and the induction of glycolysis upon activation (Doughty et al. 2006; Dufort et al. 2007). As they transition from a naive quiescent state to an anabolic proliferative state, their metabolic profiles undergo corresponding changes. Similar to T cells, activated B cells significantly enhance glucose and glutamine metabolism during clonal proliferation (Garcia-Manteiga et al. 2011; Le et al. 2012).

Following activation, dual programming equips B cell precursors with the necessary metabolic machinery for survival and development. Pro-B and small pre-B cells exhibit lower rates of glucose influx than large B cells. The inner mitochondrial membrane protein EF-Hand Domain 1 (EFHD1)/Swiprosin-2 governs a crucial metabolic switch for pro-B cells to transition into pre-B cells. Pro-B cells express EFHD1, maintaining their capacity for oxidative respiration. The expression of pre-B cell receptor (pre-BCR) in large pre-B cells downregulates EFHD1, prompting these cells to engage more in the glycolytic pathway (Stein et al. 2017). Beyond these metabolic shifts, levels of ROS, influenced by increased mitochondrial ROS generation, also regulate B cell receptor (BCR) activation (Capasso et al. 2010). BCR activation is essential for downstream BCR signaling and cellular proliferation in response to BCR cross-linking (Wheeler and DeFranco 2012) (Table 10.1).

Additionally, increased mitochondrial biogenesis and glucose uptake amplify the metabolic activity of B cells residing in the germinal center (GC) (Waters et al. 2018). Numerous studies demonstrate that GC B + cells exhibit enhanced glycolysis through GLUT1 upregulation and HIF-1α stabilization, promoting anabolism in hypoxic microenvironments (Cho et al. 2016). Enhanced mTOR activity is also a crucial regulator of somatic hypermutation and affinity maturation (Ripperger and Bhattacharya 2021). Nevertheless, recent studies showed that GC B cells predominantly rely on FAO to meet their metabolic needs (Weisel et al. 2020). It is crucial to meticulously assess the metabolic heterogeneity of B cells in GCs, recognizing that glycolysis may only be feasible during the initial stages of clonal proliferation when an energy boost is required (Metur and Klionsky 2021).

Upon exiting the GC, B cells can differentiate terminally into plasma cells that cease to proliferate. The persistence of plasma cells is intricately linked to nutrition absorption, autophagy, and antibody release (Lam et al. 2016; Lam and Bhattacharya 2018). As B cells activated by BCR ligands, IL-21, and IL-4, undergo differentiation into plasma cells, there is a simultaneous increase in both glycolysis and OXPHOS (Price et al. 2018). Glucose, however, is mainly utilized for antibody glycosylation rather than ATP synthesis, and this antibody synthesis process is influenced by mTOR activity. On the other hand, B cells can exit the GC to differentiate into memory B cells, a process that is stimulated and maintained by cytokines such as IL-4, IL-6, and IL-21. The metabolic status of memory B cells is yet to be sufficiently investigated. However, it has been noted that glycolysis, OXPHOS, and autophagy regulated by Atg5, Atg7, and Ulk1 may play a role in the maintenance of metabolic pathways within these cells (Gaudette et al. 2020). The metabolic duality highlights the remarkable adaptability of B cells, allowing them to efficiently direct ATP and metabolic substrates (Metur and Klionsky 2021) (Table 10.1).

10.4 Cytokine Signaling

Cytokines are important not only in orchestrating immune responses but also in modulating the metabolic pathways that mediate them. They play a crucial dual role in the metabolism of cellular immunity, where metabolic reprogramming is required for immune cell activation, differentiation, and function. When immune cells such as macrophages, dendritic cells, and T cells are activated, they undergo significant metabolic reprogramming to meet their energy and biosynthetic needs. Cytokines are key players in this metabolic shift. For example, when macrophages encounter pathogens, they shift from oxidative phosphorylation to glycolysis, a process known as aerobic glycolysis. The production of pro-inflammatory cytokines like TNF-α and IL-1β is increased during this shift, ensuring a rapid supply of ATP and biosynthetic precursors necessary for producing antimicrobial and pro-inflammatory mediators.

Furthermore, T cell activation and differentiation are tightly linked to changes in cellular metabolism. When activated by cytokines like IL-2 and IL-12, naïve T cells switch to glycolysis, allowing for rapid proliferation and effector functions. This metabolic reprogramming is crucial for the differentiation of T cells into various subsets, such as Th1, Th2, Th17, and regulatory T cells. The glycolytic pathway provides energy for antigen engulfing, processing, presentation, and subsequent cytokine production, all of which drive T cell responses (Soto-Heredero et al. 2020). The process can be driven by the production of cytokines such as IL-2 and IL-7, which promote T cell proliferation and differentiation. Similarly, activated DCs also switch to glycolysis, resulting in the modulated production of a cytokine cocktail.

This metabolic shift is influenced by cytokines like GM-CSF and IL-4, which are essential for DC maturation and functionality.

The metabolic state of immune cells is closely linked to their ability to respond to infections. Cytokine-mediated metabolic reprogramming ensures that immune cells are effectively responding to pathogens. Pro-inflammatory cytokines like IFN-γ and TNF-α induce a metabolic shift in macrophages and neutrophils, boosting their antimicrobial functions (Xiao et al. 2021). This shift promotes the production of ROS and NO, which are essential for pathogen killing. These processes are energy-intensive, necessitating a shift to glycolysis to provide the required ATP and metabolic intermediates (Zasłona and O'Neill 2020).

Anti-inflammatory cytokines like IL-10 and TGF-β help resolve inflammation and restore metabolic homeostasis. These cytokines inhibit glycolysis and promote oxidative phosphorylation, which is linked to the anti-inflammatory functions of Tregs and M2 macrophages (Ouyang and O'Garra 2019).

10.5 Metabolic Checkpoints in Immune Signaling

Metabolic checkpoints are points in metabolic pathways by which a cell evaluates metabolic signals in its environment to regulate cellular functions, such as nutrient sensing, cellular energy status, and mitochondrial functions. Immune signaling is the process by which the immune system regulates responses to immune challenges (Marçais et al. 2014). In this section, we aim to discuss the role of metabolic checkpoints in immune signaling.

In macrophages, an enzyme involved in the TCA cycle called fumarate hydratase was discovered to be a metabolic checkpoint, playing a role in the regulation of the macrophage cytokine balance and metabolic state. Consequently, it was shown that inhibition or genetic silencing of fumarate hydratase leads to suppressed mitochondrial respiration, together with increased fumarate levels, indicating the significance of fumarate hydratase in metabolic regulation (Berod et al. 2014).

Metabolic checkpoints in T cells were associated with several functions, such as meeting energy requirements after T cell activation by utilizing glycolysis and glutaminolysis instead of oxidation of free fatty acids and having an interconnecting function between certain metabolites and signaling cascades that affect T lymphocyte responses, therefore influencing immunological function (Roy et al. 2019). These checkpoints may also influence cell cycle progression, differentiation, and apoptosis of T cells, hence altering T cell overall functioning, and their engagement can result in connecting metabolic pathways, influencing metabolite availability, and modulating immune signaling events. Collectively, these functions imply that T cell metabolic checkpoints have crucial roles in the influence of immune responses and cell function (Wang and Green 2012).

A group of cell surface receptors called P2 receptors, (specifically P2X7 receptor), were identified to serve as metabolic checkpoints that have functions in the activation of several immune cells, including T cells, dendritic cells, macrophages, mast cells, and neutrophils. These receptors are activated by extracellular ATP, in a process referred to as purinergic signaling, which results in modulation of T lymphocyte energy metabolism. The deletion or inhibition of the P2X7 receptor has been shown to inhibit T cell activation, impair leukocyte function, and negatively impact the inflammatory response. Additionally, this receptor is shown to be a potential marker of immune-mediated diseases (Vultaggio-Poma and Di Virgilio 2022).

10.6 Immunometabolism and Microbial Infections Interplay

10.6.1 Viral Infections

Given the parasitic nature of viruses, metabolic pathways can greatly differ within the same cell type depending on the infection context. For instance, in the case of HIV, an uninfected activated T cell will increase its glycolytic activity for clonal expansion and effective antiviral responses. In an HIV-infected T cell, however, the virus will hijack the glycolytic flux to cater to viral replication needs (Kavanagh Williamson et al. 2018). As such, the infection status of the cell guides our understanding of the varied metabolic signatures characteristic of responses to viral infections, as well as their exploitation to further viral loads.

Virus-infected cells can recognize their infected state through the detection of viral components by a set of PRRs. Those PRRs include TLR3, TLR7, TLR8, TLR9, RLRs, and NLRs, with the employment of one of these PRRs being dependent on the type of genetic material it can detect, such as dsRNA, ssRNA, or dsDNA. Upon recognition, the PRRs recruit a specific adaptor protein to initiate downstream signaling and mount an antiviral response, mainly through the upregulation of interferons (IFNs) secretion. The secretion of IFNs sounds a siren, provoking a metabolic shift in the stimulated immune cells (Roy et al. 2019; Thompson et al. 2011).

In immune cells like macrophages, dendritic cells, and effector T cells, IFN stimulation is often linked with enhanced glycolysis and interruptions in the TCA cycle. The resulting accumulation of intermediates, including succinate, drives mitochondrial ROS production, thereby promoting effective immune responses via enhanced cytoskeleton remodeling, phagocytosis, and antigen presentation. IFNs can also act on another substrate, L-arginine, by diverting its use for polyamine synthesis, which normally aids viral RNA translation, toward NO production. Beyond energy and amino acid metabolism, IFNs can inhibit cholesterol and FAS as a means of limiting viral replication (Fritsch and Weichhart 2016). In macrophages, the production of an oxysterol called 25-hydroxycholesterol (25-HC) can limit viral spread by reinforcing membrane rigidity and subsequently impeding membrane fusion between the virus and host cell. 25-HC may also be involved in the inhibition of the sterol synthesis pathway, an essential pathway for viral progeny (Fritsch and Weichhart 2016). Moreover, iron availability can also determine the potency of immune responses, as NK cells have been shown to enhance iron uptake during viral infections, and iron deficiency impairs their antiviral activity (Littwitz-Salomon et al. 2021).

In an infected immune cell, the same metabolic pathways can be hijacked to attend to the virus' needs. For instance, HIV-1, DENV, RV, and EBV have been linked with stimulation of glycolytic activity through upregulation of the glucose transporter GLUT1. Augmented HIF-1α stabilization is demonstrated in several viral infections, including HPV, SARS-CoV-2, VSV, KSHV, and NDV, which promotes aerobic glycolysis. Other groups of viruses like DENV, HBV, and IAV (H1N1) have been shown to increase the expression of hexokinase 2 (HK2), an enzyme that catalyzes an irreversible step toward the glycolytic pathway (Li et al. 2023a). Though this pathway was noted for its enhancement in activated immune cells, the machinations that take effect when a virus is in control alter the outcome. Glycolysis is a metabolic pathway that facilitates rapid ATP generation. The ATP amount produced is limited; however, a continuous influx of glucose to fuel glycolysis will provide sufficient energy to swiftly meet the virus' needs.

Though these observations seemingly provide a grim outlook on the chronicles of infected cells, their fate is not sealed. Recent research has demonstrated interesting findings on the potential adaptability of infected cells. Despite enhanced glycolytic activity that is commandeered by an infecting virus, the host cell may still manage to produce interferon-stimulated genes (ISGs) like viperin and OAS-1 to take back control (El Safadi et al. 2023). The interplay between metabolism and the immune system during viral infections is complex and yet to be fully understood. However, research in this area is rapidly advancing, and new insights are emerging into the potential for targeting metabolism to develop new antiviral therapies.

10.6.2 Bacterial Infections

Pathogenic bacteria alter metabolic pathways in the host cells for their survival. On the other hand, the host's immunity activates metabolic pathways in order to suppress the infection. Alterations of metabolic pathways in macrophages by pathogenic intracellular bacteria lead to weakened antibacterial defenses and microbial survival within the macrophage (Sprenger et al. 2018).

Numerous bacteria, like *Yersinia pestis*, express itaconate CoA transferase, itaconyl-CoA hydratase, and (S)-citramalyl-CoA lyase enzymes. These enzymes function in degrading itaconate to inhibit its antibacterial effects (Sasikaran et al. 2014). Other bacteria alter metabolic pathways to induce the availability of glucose and lactate as carbon sources. *Legionella pneumophila* induces aerobic glycolysis in macrophages, which is a metabolic pathway that involves the conversion of glucose into lactate in the presence of oxygen (Galli and Saleh 2021). Bacteria like *Helicobacter pylori* can employ their rocF gene-encoded arginase to compete for arginine metabolism with macrophages, thereby blunting macrophage NOS2 activity and NO generation (Líndez and Reith 2021). A similar and possibly related trend is seen in clinical findings associating severe sepsis patients who did not survive with significantly reduced arginine levels in comparison to sepsis survivors (Su et al. 2015).

On the other hand, host cells can adopt metabolic adaptations that support defensive strategies, thereby limiting bacterial replication and pathogenesis. For instance, IFN-γ can fuel macrophage metabolic adaptation toward enhanced glycolysis by augmenting the expression of the phosphofructokinase muscle isoform (PFK-M). During *Mycobacterium tuberculosis* infections, this metabolic switch supports proinflammatory cytokine production and host defenses (Hackett et al. 2020). Beyond mere support, some metabolic enzymes have been shown to display direct roles in defense. Hexokinase, for example, was shown to act as a PRR in its ability to recognize bacterial peptidoglycans, causing the direct initiation of NLRP3 inflammasome activation (Wolf et al. 2016).

Iron is considered an important micronutrient for bacteria, and its availability plays a key role in bacterial virulence (Carver 2018). In case of an infection, bacteria secrete siderophores, which are high-affinity iron chelators that will scavenge iron and mediate its uptake through surface receptors (Page 2019). Sequestering iron away from bacteria is key for host immunity. The upregulation of hepcidin leads to the degradation of the cellular iron transporter, ferroportin, which reduces iron levels in the plasma. Furthermore, the production of siderocalin as part of innate immunity sequesters iron by binding to siderophores. However, some pathogenic bacterial strains synthesize modified forms of enterobactin, known as "stealth siderophores" that evade siderocalin, hence restoring the pathogen's iron-scavenging ability (Carver 2018). The creative metabolic adaptation strategies manifest in both the host cell and bacterium continues to be studied as they struggle over nutrients and display intriguing metabolic manipulation tactics to triumph.

10.6.3 Fungal Infections

The fungal cell wall forms an interaction point between the host's immunity and the fungus. Metabolic shifts in immune cells occur upon the detection of PAMPs on the fungal surface to induce an antifungal response. The shift includes a reduction in mitochondrial oxidative phosphorylation and induced aerobic glycolysis in immune cells. Along with the activation of mTOR pathway signaling, the shift driving glycolytic activity plays a key role in protection against fungal infections such as *Candida albicans*. This is substantiated by diminished cytokine production following the inhibition of glycolysis and mTOR signaling (Domínguez-Andrés et al. 2017). However, *C. albicans* can manipulate immune cells' metabolism by competing for glucose, leaving immune cells glucose-deprived (Weerasinghe and Traven 2020).

Fungi exhibit the ability of iron modulation in immune cells, specifically macrophages. This is done by different mechanisms, including "iron parasitism," where the fungi use siderophores produced by other microorganisms to chelate iron (Bairwa et al. 2017; Seifert et al. 2008). In addition, fungi, like *C. albicans*, utilize specific receptors to bind to iron-transport and storage proteins, transferrin and ferritin (Ene et al. 2014). Some fungi, like *Rhizopus arrhizus*, secrete peptides that manipulate iron storage in macrophages by upregulating ferroportin and downregulating hepcidin expressions (Soliman et al. 2022). In turn, macrophages can employ iron restriction mechanisms to inhibit the intracellular growth of *Rhizopus* conidia (Andrianaki et al. 2018).

10.6.4 Parasitic Infections

Parasites rely on other organisms to survive and can remodel host metabolism by numerous mechanisms. Indolepyruvate is a metabolite released by some parasites, like *Trypanosoma brucei*, that downregulates HIF-1α levels and glycolysis in order to reduce the production of pro-inflammatory cytokine IL-1β and escape immunity (Chen et al. 2021). The mTOR pathway in dendritic cells is a signaling pathway that controls cell growth and metabolism. Microfilariae of the parasite *Brugia malayi* target the mTOR pathway by downregulating the phosphorylation of mTOR regulatory proteins (Narasimhan et al. 2016).

On the opposite end, host immunity is capable of adapting its own metabolic choices to fight parasitic infections. In *Heligmosomoides polygyrus* infections, M2 macrophages are recruited to infection sites for their Arg1 activity in arginine metabolism. The resulting synthesis of ornithine and polyamines serves to hinder larvae motility and protect host tissue from inflicted damage by the helminth (Líndez and Reith 2021). In addition, neutrophils have been shown to gear their metabolic activity toward glycolysis to support their effector functions against *Leishmania donovani*, with glycolysis inhibition being linked with increased parasitic load and loss of immune control (Ohms et al. 2021).

Parasites utilize iron as a nutrient source. The acquisition of iron is achieved through different effective mechanisms. Some parasites bind directly to ferritin through receptor recognition pathways. Other parasites are capable of phagocytosing erythrocytes and further processing them using hemoglobinase and cysteine proteases (Mach and Sutak 2020). In response to intracellular parasites such as *Plasmodium falciparum*, infected erythrocytes can upregulate ferroportin to deprive parasites of iron supplies through iron export (Zhang et al. 2018). Collectively, these examples highlight the complex nature of host-microbe interactions and their diverse impact on metabolic signatures.

10.7 Emerging Areas of Research and New Therapeutic Approaches

Recent research has revealed that certain metabolites, previously thought to be merely metabolic intermediates, play critical roles as immune response regulators (Li et al. 2023b; Harber et al. 2020; Milanović et al. 2024). Metabolites such as itaconate, succinate, α-KG, and lactate have cytokine-like properties and can modulate immune functions through various mechanisms, offering new therapeutic opportunities.

Itaconate, synthesized from aconitate by the enzyme Immune-responsive gene 1 (IRG1), has emerged as a crucial anti-inflammatory agent. Initially recognized for its antibacterial properties, itaconate significantly suppresses the production of pro-inflammatory cytokines. It achieves this through multiple mechanisms, including the inhibition of succinate dehydrogenase (SDH) to reduce IL-1β production, inducing post-translational modifications that inactivate the redox sensing protein kelch-like ECH-associated protein 1 (KEAP1) and the activation of the anti-inflammatory transcription factor nuclear factor (erythroid-derived 2)-like 2 (Nrf2), and impairing glycolysis by modifying key glycolytic enzymes (Li et al. 2023b; Mills et al. 2016). Derivatives of itaconate, such as dimethyl itaconate (DI) and 4-octyl itaconate (4-OI), have shown promising anti-inflammatory therapeutic potential in models of LPS-induced peritonitis and sepsis (Li et al. 2023b).

Succinate plays a dual role in inflammation, accumulating in lipopolysaccharide (LPS)-activated macrophages to stabilize HIF-1α, thereby promoting the transcription of inflammatory genes like IL-1β. Extracellularly, succinate acts through its receptor, SUCNR1, to activate immune responses. Elevated succinate levels are observed in conditions such as rheumatoid arthritis and type 2 diabetes, making it a potential therapeutic target. Interestingly, succinate can also exhibit anti-inflammatory effects in certain contexts, such as in chronic brain inflammation, underscoring its complex role in immune regulation (Tannahill et al. 2013).

α-KG functions as an anti-inflammatory mediator by influencing T cell differentiation and macrophage activation. It serves as a cofactor for dioxygenases involved in histone demethylation, thus affecting gene expression at the epigenome regulatory level. Also, α-KG activates prolyl hydroxylases that regulate the degradation of HIF-1α, thereby modulating inflammatory responses. The regulatory role of α-KG in inflammation suggests its potential use in treating inflammatory and autoimmune diseases (Liu et al. 2017). Its role as a precursor of glutamine makes it a great candidate for the alleviation of severe burns and infections (Zdzisińska et al. 2017).

Lactate, long considered a toxic byproduct of glycolysis, is now recognized as an important modulator of immune responses. It acts through lactate receptor G protein-coupled receptor (GPR81), a receptor that inhibits inflammatory responses by reducing cyclic AMP levels and influences gene expression related to immune responses through histone lactylation. Modulating lactate levels or its signaling pathways could offer new approaches toward controlling inflammation and immune responses (Hoque et al. 2014; Zhang et al. 2019). Its potential as a target for treating *Mycobacterium tuberculosis* and tumors has been closely studied in recent years (Kiran and Basaraba 2021; Wang et al. 2021).

In summary, immune metabolites extend beyond their conventional metabolic roles to function as key regulators of immune function. Their ability to modulate immune responses through various mechanisms, including influencing transcription factors, enzyme activity, and epigenetic modifications, positions them as promising targets for innovative therapeutic strategies in treating immune and inflammatory diseases (Fig. 10.2).

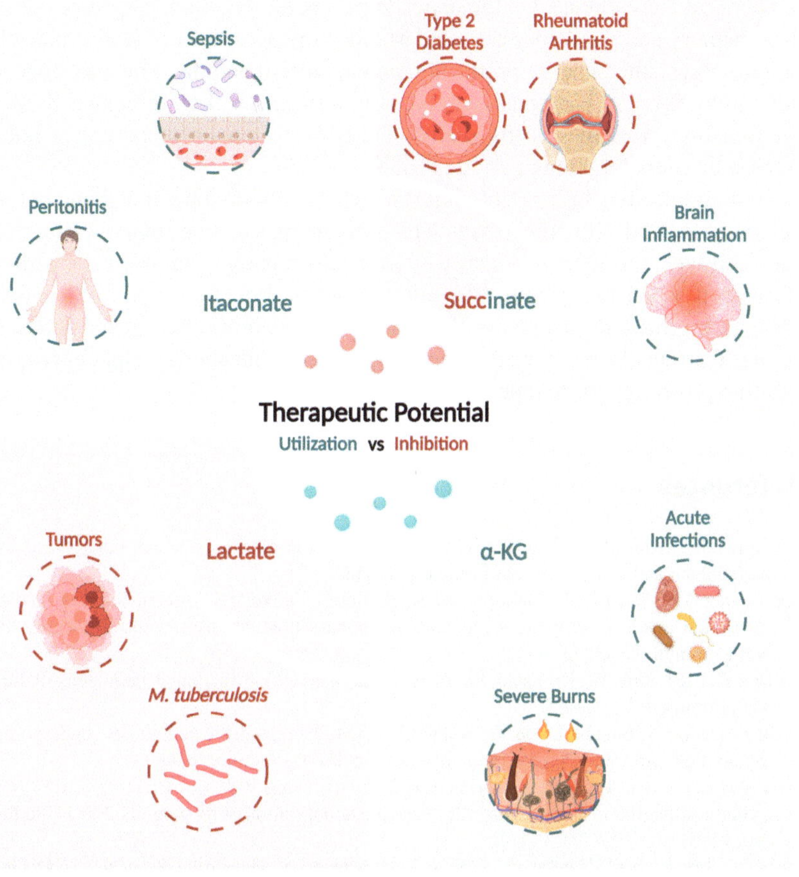

Fig. 10.2 Immunomodulation through utilizing (green) or inhibiting (red) metabolites to treat several diseases highlights the potential therapeutic avenues revealed by immunometabolism research

10.8 Concluding Remarks

In this chapter, we reviewed the metabolic preferences of different types of immune cells while highlighting key features in a variety of infection contexts. These preferences are largely dictated by the energy demands, functional needs, and surroundings of the immune cells. Two main diverging metabolic routes emerge as necessitated by the cells' typical functions. Effector cells, such as T helper and cytotoxic cells, neutrophils, and M1 macrophages favor aerobic glycolysis for rapid energy and biosynthetic precursor production. This preference proves detrimental to their survival, as reliance on glycolysis is inefficient for ATP production and compromises mitochondrial health, in turn revealing why these immune cells are often short-lived. Hence, their functions enable swift defenses against invading pathogens but come at a great price. On the opposite metabolic spectrum, persistent cells such as T memory and regulatory cells and M2 macrophages favor oxidative phosphorylation for sustained energy production and cellular functions. This preference supports mitochondrial biogenesis and the long-term survival of immune cells. Hence, the metabolism of these cells enables the body to remain prepared for potential foreign invaders for a long period of time.

Immunometabolism is a growing research field with many metabolic behaviors yet to be revealed. Although research has demonstrated recurrent metabolic shifts in each cell type, the adaptable nature of metabolism equips immune cells with many stratagems when faced with different opponents. Infection type and context are among the critical determinants of immune cell metabolic choices and defense outcomes. Current efforts in uncovering the nuances of metabolic adaptability promise exciting prospects for therapeutic advancements.

References

Alwarawrah Y, Kiernan K, MacIver NJ (2018) Changes in nutritional status impact immune cell metabolism and function. Front Immunol 9:1055

Andrianaki AM, Kyrmizi I, Thanopoulou K, Baldin C, Drakos E, Soliman SS et al (2018) Iron restriction inside macrophages regulates pulmonary host defense against Rhizopus species. Nat Commun 9(1):3333

Bairwa G, Hee Jung W, Kronstad JW (2017) Iron acquisition in fungal pathogens of humans. Metallomics 9(3):215–227

Basit F, Mathan T, Sancho D, de Vries IJM (2018) Human dendritic cell subsets undergo distinct metabolic reprogramming for immune response. Front Immunol 9:2489

Belambri SA, Rolas L, Raad H, Hurtado-Nedelec M, Dang PMC, El-Benna J (2018) NADPH oxidase activation in neutrophils: role of the phosphorylation of its subunits. Eur J Clin Investig 48:e12951

Berod L, Friedrich C, Nandan A, Freitag J, Hagemann S, Harmrolfs K et al (2014) De novo fatty acid synthesis controls the fate between regulatory T and T helper 17 cells. Nat Med 20(11):1327–1333

Blagih J, Coulombe F, Vincent EE, Dupuy F, Galicia-Vázquez G, Yurchenko E et al (2015) The energy sensor AMPK regulates T cell metabolic adaptation and effector responses in vivo. Immunity 42(1):41–54

Capasso M, Bhamrah MK, Henley T, Boyd RS, Langlais C, Cain K et al (2010) HVCN1 modulates BCR signal strength via regulation of BCR-dependent generation of reactive oxygen species. Nat Immunol 11(3):265–272

Carr EL, Kelman A, Wu GS, Gopaul R, Senkevitch E, Aghvanyan A et al (2010) Glutamine uptake and metabolism are coordinately regulated by ERK/MAPK during T lymphocyte activation. J Immunol 185(2):1037–1044

Carver PL (2018) The battle for iron between humans and microbes. Curr Med Chem 25(1):85–96

Chang C-H, Curtis JD, Maggi LB, Faubert B, Villarino AV, O'Sullivan D et al (2013) Posttranscriptional control of T cell effector function by aerobic glycolysis. Cell 153(6):1239–1251

Chen JY, Zhou JK, Pan W (2021) Immunometabolism: towards a better understanding the mechanism of parasitic infection and immunity. Front Immunol 12:661241

Cheng S-C, Joosten LAB, Netea MG (2014) The interplay between central metabolism and innate immune responses. Cytokine Growth Factor Rev 25(6):707–713

Chinetti-Gbaguidi G, Staels B (2011) Macrophage polarization in metabolic disorders: functions and regulation. Curr Opin Lipidol 22(5):365–372

Cho SH, Raybuck AL, Stengel K, Wei M, Beck TC, Volanakis E et al (2016) Germinal centre hypoxia and regulation of antibody qualities by a hypoxia response system. Nature 537(7619):234–238

Domínguez-Andrés J, Arts RJ, Ter Horst R, Gresnigt MS, Smeekens SP, Ratter JM et al (2017) Rewiring monocyte glucose metabolism via C-type lectin signaling protects against disseminated candidiasis. PLoS Pathog 13(9):e1006632

Doughty CA, Bleiman BF, Wagner DJ, Dufort FJ, Mataraza JM, Roberts MF et al (2006) Antigen receptor–mediated changes in glucose metabolism in B lymphocytes: role of phosphatidylinositol 3-kinase signaling in the glycolytic control of growth. Blood 107(11):4458–4465

Dufort FJ, Bleiman BF, Gumina MR, Blair D, Wagner DJ, Roberts MF et al (2007) Cutting edge: IL-4-mediated protection of primary B lymphocytes from apoptosis via Stat6-dependent regulation of glycolytic metabolism. J Immunol 179(8):4953–4957

El Safadi D, Paulo-Ramos A, Hoareau M, Roche M, Krejbich-Trotot P, Viranaicken W et al (2023) The influence of metabolism on immune response: a journey to understand immunometabolism in the context of viral infection. Viruses 15(12):2399

Ene IV, Brunke S, Brown AJ, Hube B (2014) Metabolism in fungal pathogenesis. Cold Spring Harb Perspect Med 4(12):a019695

Fan K, Lin L, Ai Q, Wan J, Dai J, Liu G et al (2018) Lipopolysaccharide-induced dephosphorylation of AMPK-activated protein kinase potentiates inflammatory injury via repression of ULK1-dependent autophagy. Front Immunol 9:1464

Fritsch SD, Weichhart T (2016) Effects of interferons and viruses on metabolism. Front Immunol 7:630

Galli G, Saleh M (2021) Immunometabolism of macrophages in bacterial infections. Front Cell Infect Microbiol 10:607650

Ganeshan K, Chawla A (2014) Metabolic regulation of immune responses. Annu Rev Immunol 32(1):609–634

Garcia-Manteiga JM, Mari S, Godejohann M, Spraul M, Napoli C, Cenci S et al (2011) Metabolomics of B to plasma cell differentiation. J Proteome Res 10(9):4165–4176

Gaudette BT, Jones DD, Bortnick A, Argon Y, Allman D (2020) mTORC1 coordinates an immediate unfolded protein response-related transcriptome in activated B cells preceding antibody secretion. Nat Commun 11(1):723

Geiger R, Rieckmann JC, Wolf T, Basso C, Feng Y, Fuhrer T et al (2016) L-arginine modulates T cell metabolism and enhances survival and anti-tumor activity. Cell 167(3):829–42.e13

Gerriets VA, Kishton RJ, Johnson MO, Cohen S, Siska PJ, Nichols AG et al (2016) Foxp3 and Toll-like receptor signaling balance Treg cell anabolic metabolism for suppression. Nat Immunol 17(12):1459–1466

Hackett EE, Charles-Messance H, O'Leary SM, Gleeson LE, Muñoz-Wolf N, Case S et al (2020) Mycobacterium tuberculosis limits host glycolysis and IL-1β by restriction of PFK-M via MicroRNA-21. Cell Rep 30(1):124–36.e4

Harber KJ, de Goede KE, Verberk SGS, Meinster E, de Vries HE, van Weeghel M et al (2020) Succinate is an inflammation-induced immunoregulatory metabolite in macrophages. Meta 10(9):372

Herb M, Schramm M (2021) Functions of ROS in macrophages and antimicrobial immunity. Antioxidants 10(2):313

Hoebe K, Janssen E, Beutler B (2004) The interface between innate and adaptive immunity. Nat Immunol 5(10):971–974

Hoque R, Farooq A, Ghani A, Gorelick F, Mehal WZ (2014) Lactate reduces liver and pancreatic injury in Toll-like receptor- and inflammasome-mediated inflammation via GPR81-mediated suppression of innate immunity. Gastroenterology 146(7):1763–1774

Hu H, Juvekar A, Lyssiotis CA, Lien EC, Albeck JG, Oh D et al (2016) Phosphoinositide 3-kinase regulates glycolysis through mobilization of aldolase from the actin cytoskeleton. Cell 164(3):433–446

Hu C, Xuan Y, Zhang X, Liu Y, Yang S, Yang K (2022) Immune cell metabolism and metabolic reprogramming. Mol Biol Rep 49(10):9783–9795
Jellusova J (2020) Metabolic control of B cell immune responses. Curr Opin Immunol 63:21–28
Jeon J-H, Hong C-W, Kim EY, Lee JM (2020) Current understanding on the metabolism of neutrophils. Immune Netw 20(6):e46
Kavanagh Williamson M, Coombes N, Juszczak F, Athanasopoulos M, Khan MB, Eykyn TR et al (2018) Upregulation of glucose uptake and hexokinase activity of primary human CD4+ T cells in response to infection with HIV-1. Viruses 10(3):114
Kaymak I, Luda KM, Duimstra LR, Ma EH, Longo J, Dahabieh MS et al (2022) Carbon source availability drives nutrient utilization in CD8+ T cells. Cell Metab 34(9):1298–1311. e6
Kelly B, O'Neill LAJ (2015) Metabolic reprogramming in macrophages and dendritic cells in innate immunity. Cell Res 25(7):771–784
Kiran D, Basaraba RJ (2021) Lactate metabolism and signaling in tuberculosis and cancer: a comparative review. Front Cell Infect Microbiol 11:624607
Kishton RJ, Sukumar M, Restifo NP (2017) Metabolic regulation of T cell longevity and function in tumor immunotherapy. Cell Metab 26(1):94–109
Klein Geltink RI, Kyle RL, Pearce EL (2018) Unraveling the complex interplay between T cell metabolism and function. Annu Rev Immunol 36:461–488
Krzywinska E, Stockmann C (2018) Hypoxia, metabolism and immune cell function. Biomedicines 6(2):56
Kumar S, Dikshit M (2019) Metabolic insight of neutrophils in health and disease. Front Immunol 10:2099
Lam WY, Bhattacharya D (2018) Metabolic links between plasma cell survival, secretion, and stress. Trends Immunol 39(1):19–27
Lam WY, Becker AM, Kennerly KM, Wong R, Curtis JD, Llufrio EM et al (2016) Mitochondrial pyruvate import promotes long-term survival of antibody-secreting plasma cells. Immunity 45(1):60–73
Le A, Lane AN, Hamaker M, Bose S, Gouw A, Barbi J et al (2012) Glucose-independent glutamine metabolism via TCA cycling for proliferation and survival in B cells. Cell Metab 15(1):110–121
Li J, Wang Y, Deng H, Li S, Qiu H-J (2023a) Cellular metabolism hijacked by viruses for immunoevasion: potential antiviral targets. Front Immunol 14:1228811
Li Z, Zheng W, Kong W, Zeng T (2023b) Itaconate: a potent macrophage Immunomodulator. Inflammation 46(4):1177–1191
Líndez A-AMI, Reith W (2021) Arginine-dependent immune responses. Cell Mol Life Sci 78(13):5303–5324
Littwitz-Salomon E, Moreira D, Frost JN, Choi C, Liou KT, Ahern DK et al (2021) Metabolic requirements of NK cells during the acute response against retroviral infection. Nat Commun 12(1):5376
Liu PS, Wang H, Li X, Chao T, Teav T, Christen S et al (2017) α-ketoglutarate orchestrates macrophage activation through metabolic and epigenetic reprogramming. Nat Immunol 18(9):985–994
Ma EH, Bantug G, Griss T, Condotta S, Johnson RM, Samborska B et al (2017) Serine is an essential metabolite for effector T cell expansion. Cell Metab 25(2):345–357
Mach J, Sutak R (2020) Iron in parasitic protists – from uptake to storage and where we can interfere†. Metallomics 12(9):1335–1347
Macintyre AN, Gerriets VA, Nichols AG, Michalek RD, Rudolph MC, Deoliveira D et al (2014) The glucose transporter Glut1 is selectively essential for CD4 T cell activation and effector function. Cell Metab 20(1):61–72
MacIver NJ, Michalek RD, Rathmell JC (2013) Metabolic regulation of T lymphocytes. Annu Rev Immunol 31:259–283
Marçais A, Cherfils-Vicini J, Viant C, Degouve S, Viel S, Fenis A et al (2014) The metabolic checkpoint kinase mTOR is essential for IL-15 signaling during the development and activation of NK cells. Nat Immunol 15(8):749–757

Medzhitov R, Janeway C (2000) Innate immunity. N Engl J Med 343(5):338–344

Metur SP, Klionsky DJ (2021) Adaptive immunity at the crossroads of autophagy and metabolism. Cell Mol Immunol 18(5):1096–1105

Michalek RD, Rathmell JC (2010) The metabolic life and times of a T-cell. Immunol Rev 236(1):190–202

Michalek RD, Gerriets VA, Jacobs SR, Macintyre AN, MacIver NJ, Mason EF et al (2011) Cutting edge: distinct glycolytic and lipid oxidative metabolic programs are essential for effector and regulatory CD4+ T cell subsets. J Immunol 186(6):3299–3303

Milanović M, Bekić M, Đokić J, Vučević D, Čolić M, Tomić S (2024) Exogenous α-ketoglutarate modulates redox metabolism and functions of human dendritic cells, altering their capacity to polarise T cell response. Int J Biol Sci 20(3):1064–1087

Mills EL, Kelly B, Logan A, Costa ASH, Varma M, Bryant CE et al (2016) Succinate dehydrogenase supports metabolic repurposing of mitochondria to drive inflammatory macrophages. Cell 167(2):457–70.e13

Mohammad MG, Ashmawy NS, Al-Rawi AM, Abu-Qiyas A, Hamoda AM, Hamdy R et al (2023) SARS-CoV-2-free residual proteins mediated phenotypic and metabolic changes in peripheral blood monocytic-derived macrophages in support of viral pathogenesis. PLoS One 18(1):e0280592

Muroski ME, Miska J, Chang AL, Zhang P, Rashidi A, Moore H et al (2017) Fatty acid uptake in T cell subsets using a quantum dot fatty acid conjugate. Sci Rep 7(1):5790

Narasimhan PB, Bennuru S, Meng Z, Cotton RN, Elliott KR, Ganesan S et al (2016) Microfilariae of Brugia malayi inhibit the mTOR pathway and induce autophagy in human dendritic cells. Infect Immun 84(9):2463–2472

O'Sullivan D, van der Windt GJ, Huang SC-C, Curtis JD, Chang C-H, Buck MD et al (2014) Memory CD8+ T cells use cell-intrinsic lipolysis to support the metabolic programming necessary for development. Immunity 41(1):75–88

Ohms M, Ferreira C, Busch H, Wohlers I, Guerra de Souza AC, Silvestre R et al (2021) Enhanced glycolysis is required for antileishmanial functions of neutrophils upon infection with leishmania donovani. Front Immunol 12:632512

Olenchock BA, Rathmell JC, Vander Heiden MG (2017) Biochemical underpinnings of immune cell metabolic phenotypes. Immunity 46(5):703–713

Olive AJ, Sassetti CM (2016) Metabolic crosstalk between host and pathogen: sensing, adapting and competing. Nat Rev Microbiol 14(4):221–234

Osuna-Espinoza KY, Rosas-Taraco AG (2023) Metabolism of NK cells during viral infections. Front Immunol 14:1064101

Ouyang W, O'Garra A (2019) IL-10 family cytokines IL-10 and IL-22: from basic science to clinical translation. Immunity 50(4):871–891

Ozkan H, Olgun N, Sasmaz E (1995) Nutrition, immunity, and infections: T lymphocyte subpopulations in protein-energy malnutrition. Pediatrics 96(1):77

Page MGP (2019) The role of iron and siderophores in infection, and the development of siderophore antibiotics. Clin Infect Dis 69(Suppl 7):S529–Ss37

Palmieri EM, Gonzalez-Cotto M, Baseler WA, Davies LC, Ghesquière B, Maio N et al (2020) Nitric oxide orchestrates metabolic rewiring in M1 macrophages by targeting aconitase 2 and pyruvate dehydrogenase. Nat Commun 11(1):698

Patsoukis N, Bardhan K, Weaver J, Herbel C, Seth P, Li L et al (2016) The role of metabolic reprogramming in T cell fate and function. Curr Trends Immunol 17:1

Pearce Erika L, Pearce EJ (2013) Metabolic pathways in immune cell activation and quiescence. Immunity 38(4):633–643

Pearce EL, Walsh MC, Cejas PJ, Harms GM, Shen H, Wang L-S et al (2009) Enhancing CD8 T-cell memory by modulating fatty acid metabolism. Nature 460(7251):103–107

Phair IR, Nisr RB, Howden AJ, Sovakova M, Alqurashi N, Foretz M et al (2023) AMPK integrates metabolite and kinase-based immunometabolic control in macrophages. Mol Metab 68:101661

Plas DR, Rathmell JC, Thompson CB (2002) Homeostatic control of lymphocyte survival: potential origins and implications. Nat Immunol 3(6):515–521

Price MJ, Patterson DG, Scharer CD, Boss JM (2018) Progressive upregulation of oxidative metabolism facilitates plasmablast differentiation to a T-independent antigen. Cell Rep 23(11):3152–3159

Procaccini C, Carbone F, Di Silvestre D, Brambilla F, De Rosa V, Galgani M et al (2016) The proteomic landscape of human ex vivo regulatory and conventional T cells reveals specific metabolic requirements. Immunity 44(2):406–421

Puchalska P, Huang X, Martin SE, Han X, Patti GJ, Crawford PA (2018) Isotope tracing untargeted metabolomics reveals macrophage polarization-state-specific metabolic coordination across intracellular compartments. iScience 9:298–313

Qing L, Fu J, Wu P, Zhou Z, Yu F, Tang J (2019) Metformin induces the M2 macrophage polarization to accelerate the wound healing via regulating AMPK/mTOR/NLRP3 inflammasome singling pathway. Am J Transl Res 11(2):655

Reina-Campos M, Scharping NE, Goldrath AW (2021) CD8+ T cell metabolism in infection and cancer. Nat Rev Immunol 21(11):718–738

Reis BS, Lee K, Fanok MH, Mascaraque C, Amoury M, Cohn LB et al (2015) Leptin receptor signaling in T cells is required for Th17 differentiation. J Immunol 194(11):5253–5260

Ripperger TJ, Bhattacharya D (2021) Transcriptional and metabolic control of memory B cells and plasma cells. Annu Rev Immunol 39:345–368

Ron-Harel N, Santos D, Ghergurovich JM, Sage PT, Reddy A, Lovitch SB et al (2016) Mitochondrial biogenesis and proteome remodeling promote one-carbon metabolism for T cell activation. Cell Metab 24(1):104–117

Roy S, Rizvi ZA, Awasthi A (2019) Metabolic checkpoints in differentiation of helper T cells in tissue inflammation. Front Immunol 9:3036

Sagone AL Jr, LoBuglio AF, Balcerzak SP (1974) Alterations in hexose monophosphate shunt during lymphoblastic transformation. Cell Immunol 14(3):443–452

Sasikaran J, Ziemski M, Zadora PK, Fleig A, Berg IA (2014) Bacterial itaconate degradation promotes pathogenicity. Nat Chem Biol 10(5):371–377

Seifert M, Nairz M, Schroll A, Schrettl M, Haas H, Weiss G (2008) Effects of the Aspergillus fumigatus siderophore systems on the regulation of macrophage immune effector pathways and iron homeostasis. Immunobiology 213(9):767–778

Shi LZ, Wang R, Huang G, Vogel P, Neale G, Green DR et al (2011) HIF1α–dependent glycolytic pathway orchestrates a metabolic checkpoint for the differentiation of TH17 and Treg cells. J Exp Med 208(7):1367–1376

Sikalidis AK (2015) Amino acids and immune response: a role for cysteine, glutamine, phenylalanine, tryptophan and arginine in T-cell function and cancer? Pathol Oncol Res 21(1):9–17

Sinclair LV, Rolf J, Emslie E, Shi Y-B, Taylor PM, Cantrell DA (2013) Control of amino-acid transport by antigen receptors coordinates the metabolic reprogramming essential for T cell differentiation. Nat Immunol 14(5):500–508

Soliman SSM, El-Labbad EM, Abu-Qiyas A, Fayed B, Hamoda AM, Al-Rawi AM et al (2022) Novel secreted peptides from Rhizopus arrhizus var. delemar with immunomodulatory effects that enhance fungal pathogenesis. Front Microbiol 13:863133

Soto-Heredero G, de Las G, Heras MM, Gabandé-Rodríguez E, Oller J, Mittelbrunn M (2020) Glycolysis – a key player in the inflammatory response. FEBS J 287(16):3350–3369

Sprenger M, Kasper L, Hensel M, Hube B (2018) Metabolic adaptation of intracellular bacteria and fungi to macrophages. Int J Med Microbiol 308(1):215–227

Stein M, Dütting S, Mougiakakos D, Bösl M, Fritsch K, Reimer D et al (2017) A defined metabolic state in pre B cells governs B-cell development and is counterbalanced by Swiprosin-2/EFhd1. Cell Death Differ 24(7):1239–1252

Su L, Li H, Xie A, Liu D, Rao W, Lan L et al (2015) Dynamic changes in amino acid concentration profiles in patients with sepsis. PLoS One 10(4):e0121933

Sukumar M, Liu J, Ji Y, Subramanian M, Crompton JG, Yu Z et al (2013) Inhibiting glycolytic metabolism enhances CD8+ T cell memory and antitumor function. J Clin Invest 123(10):4479–4488

Tannahill GM, Curtis AM, Adamik J, Palsson-McDermott EM, McGettrick AF, Goel G et al (2013) Succinate is an inflammatory signal that induces IL-1β through HIF-1α. Nature 496(7444):238–242
Thompson MR, Kaminski JJ, Kurt-Jones EA, Fitzgerald KA (2011) Pattern recognition receptors and the innate immune response to viral infection. Viruses 3(6):920
Troise D, Infante B, Mercuri S, Netti GS, Ranieri E, Gesualdo L et al (2023) Hypoxic state of cells and immunosenescence: a focus on the role of the HIF signaling pathway. Biomedicines 11(8):2163
Tuosto L, Parolini I, Schröder S, Sargiacomo M, Lanzavecchia A, Viola A (2001) Organization of plasma membrane functional rafts upon T cell activation. Eur J Immunol 31(2):345–349
Turrens JF (2003) Mitochondrial formation of reactive oxygen species. J Physiol 552(2):335–344
van der Windt GJ, Everts B, Chang C-H, Curtis JD, Freitas TC, Amiel E et al (2012) Mitochondrial respiratory capacity is a critical regulator of CD8+ T cell memory development. Immunity 36(1):68–78
Van Der Windt GJ, O'Sullivan D, Everts B, Huang SC-C, Buck MD, Curtis JD et al (2013) CD8 memory T cells have a bioenergetic advantage that underlies their rapid recall ability. Proc Natl Acad Sci 110(35):14336–14341
Vander Heiden MG, Cantley LC, Thompson CB (2009) Understanding the Warburg effect: the metabolic requirements of cell proliferation. Science 324(5930):1029–1033
Viola A, Munari F, Sánchez-Rodríguez R, Scolaro T, Castegna A (2019) The metabolic signature of macrophage responses. Front Immunol 10:1462
Vorobjeva N, Chernyak B (2020) NETosis: molecular mechanisms, role in physiology and pathology. Biochem Mosc 85:1178–1190
Vultaggio-Poma V, Di Virgilio FJB (2022) P2 receptors: novel disease markers and metabolic checkpoints in immune cells. Biomolecules 12(7):983
Wang R, Green DR (2012) Metabolic checkpoints in activated T cells. Nat Immunol 13(10):907–915
Wang R, Dillon CP, Shi LZ, Milasta S, Carter R, Finkelstein D et al (2011) The transcription factor Myc controls metabolic reprogramming upon T lymphocyte activation. Immunity 35(6):871–882
Wang Z, Guan D, Wang S, Chai LYA, Xu S, Lam K-P (2020) Glycolysis and oxidative phosphorylation play critical roles in natural killer cell receptor-mediated natural killer cell functions. Front Immunol 11:202
Wang Z-H, Peng W-B, Zhang P, Yang X-P, Zhou Q (2021) Lactate in the tumour microenvironment: from immune modulation to therapy. EBioMedicine 73:103627
Waters LR, Ahsan FM, Wolf DM, Shirihai O, Teitell MA (2018) Initial B cell activation induces metabolic reprogramming and mitochondrial remodeling. iScience 5:99–109
Watson MJ, Vignali PD, Mullett SJ, Overacre-Delgoffe AE, Peralta RM, Grebinoski S et al (2021) Metabolic support of tumour-infiltrating regulatory T cells by lactic acid. Nature 591(7851):645–651
Wculek SK, Dunphy G, Heras-Murillo I, Mastrangelo A, Sancho D (2022) Metabolism of tissue macrophages in homeostasis and pathology. Cell Mol Immunol 19(3):384–408
Weerasinghe H, Traven A (2020) Immunometabolism in fungal infections: the need to eat to compete. Curr Opin Microbiol 58:32–40
Weisel FJ, Mullett SJ, Elsner RA, Menk AV, Trivedi N, Luo W et al (2020) Germinal center B cells selectively oxidize fatty acids for energy while conducting minimal glycolysis. Nat Immunol 21(3):331–342
Wheeler ML, DeFranco AL (2012) Prolonged production of reactive oxygen species in response to B cell receptor stimulation promotes B cell activation and proliferation. J Immunol 189(9):4405–4416
Wik JA, Skålhegg BS (2022) T cell metabolism in infection. Front Immunol 13:840610
Williams MA, Bevan MJ (2007) Effector and memory CTL differentiation. Annu Rev Immunol 25:171–192
Wolf AJ, Reyes CN, Liang W, Becker C, Shimada K, Wheeler ML et al (2016) Hexokinase is an innate immune receptor for the detection of bacterial peptidoglycan. Cell 166(3):624–636

Wu L, Yan Z, Jiang Y, Chen Y, Du J, Guo L et al (2023) Metabolic regulation of dendritic cell activation and immune function during inflammation. Front Immunol 14:1140749

Xiao N, Nie M, Pang H, Wang B, Hu J, Meng X et al (2021) Integrated cytokine and metabolite analysis reveals immunometabolic reprogramming in COVID-19 patients with therapeutic implications. Nat Commun 12(1):1618

Yang J-Q, Kalim KW, Li Y, Zhang S, Hinge A, Filippi M-D et al (2016) RhoA orchestrates glycolysis for TH2 cell differentiation and allergic airway inflammation. J Allergy Clin Immunol 137(1):231–45.e4

Zasłona Z, O'Neill LAJ (2020) Cytokine-like roles for metabolites in immunity. Mol Cell 78(5):814–823

Zdzisińska B, Żurek A, Kandefer-Szerszeń M (2017) Alpha-ketoglutarate as a molecule with pleiotropic activity: well-known and novel possibilities of therapeutic use. Arch Immunol Ther Exp 65:21–36

Zeng W, Li F, Jin S, Ho P-C, Liu P-S, Xie X (2023) Functional polarization of tumor-associated macrophages dictated by metabolic reprogramming. J Exp Clin Cancer Res 42(1):245

Zhang D-L, Wu J, Shah BN, Greutélaers KC, Ghosh MC, Ollivierre H et al (2018) Erythrocytic ferroportin reduces intracellular iron accumulation, hemolysis, and malaria risk. Science 359(6383):1520–1523

Zhang D, Tang Z, Huang H, Zhou G, Cui C, Weng Y et al (2019) Metabolic regulation of gene expression by histone lactylation. Nature 574(7779):575–580

The Fight for Iron: A Central Theme in Host-Pathogen Interactions

11

Mawieh Hamad

Abstract

Iron is an essential micronutrient for the survival and growth of living matter. In the context of host-pathogen interactions, they both require iron as a cofactor in proteins involved in energy production, DNA replication, and gene transcription, among many metabolic processes. Although iron is essential for life, excess iron is harmful because it generates reactive oxygen species by Fenton chemistry, which leads to oxidative stress, lipid peroxidation, and cell damage and death. Moreover, normal body physiology requires ferrous iron in minimal amounts. These considerations have served as the driving force behind the evolution of intricate mechanisms that tightly maintain iron homeostasis in mammals. Microbial infections trigger an inflammatory response that leads to host iron sequestration and hypoferremia, often called nutritional immunity. The goal of this innate response, along with other measures that help to sequester host iron, is to limit iron availability to the pathogen as a means of inhibiting pathogen survival, growth, and colonization. To counter host-imposed iron sequestration measures, the pathogen initiates a concerted effort to acquire iron rapidly from different sources within the host using several ingenious strategies. This chapter discusses the epic fight between the host as it attempts to sequester iron and the pathogen as it tries to acquire it.

M. Hamad (✉)
Department of Medical Laboratory Sciences and the Research Institute for Medical and Health Sciences, University of Sharjah, Sharjah, United Arab Emirates
e-mail: mabdelhaq@sharjah.ac.ae

S. S. M. Soliman, M. I. Husseiny (eds.), *Metabolic Dynamics in Host-Microbe Interaction*, https://doi.org/10.1007/978-981-96-1305-2_11

Keywords

Iron sequestration · Iron acquisition · Inflammation · Interleukin-6 · Divalent metal transporter lipocalin · Siderophores · Hemophores · Hepcidin · Ferroportin · Virulence · Infectious diseases

11.1 Introduction

Several transition metals are essential for life. About one-third of structurally characterized proteins contain transition metal ions such as zinc, magnesium, and iron (Waldron et al. 2009; Andreini et al. 2008). Iron is ubiquitous in the earth's crust; only aluminum is more abundant than iron. It is one of the most prevalent metal ions in human cells and body fluids and plays crucial roles in human development, physiology, and pathology (Andreini et al. 2008; Beinert et al. 1997). The ability of iron to switch between a reduced electron donor ferrous (Fe^{2+}) state and an oxidized electron acceptor ferric (Fe^{3+}) state makes it suitable as a cofactor in numerous enzymes that catalyze electron transfer-dependent reactions (redox reactions). Iron also serves as a cofactor in a wide range of prosthetic groups and proteins, including heme proteins (hemoglobin, myoglobin, myeloperoxidase, cytochromes, and nitric oxide synthetases), iron-sulfur clusters (mitochondrial aconitase, coenzyme Q10, respiratory complexes I–III), and nonheme proteins (lactoferrin, transferrin, and ferritin, hypoxia-inducible factor prolyl hydroxylases), which directly affect oxygen transport, ATP production, DNA replication and repair, and gene expression, among other metabolic activities.

Under normal conditions, the amount of iron needed to support the totality of body metabolism and physiology is strictly limited. Therefore, any significant decrease or increase in body iron concentration (iron deficiency or iron overload) is harmful to human health. Iron deficiency has long been associated with anemia, compromised immunity, and increased susceptibility to infection, among other adverse consequences. On the other hand, iron overload, or the accumulation of iron as free iron Fe^{2+} or Fe^{3+} ions or as iron present in low-molecular-weight molecules, has long been known to promote the generation and propagation of reactive oxygen species (ROS) such as superoxide anion and hydrogen peroxide (H_2O_2). ROS production leads to oxidative stress, which results in lipid peroxidation, DNA damage, and apoptosis. Iron-induced oxidative stress has been linked to compromised immunity, liver cirrhosis, cancer, cardiovascular complications, diabetes, and hypothyroidism, among other pathologies. At neutral pH (pH 7), soluble ferrous iron is oxidized to insoluble ferric iron, which polymerizes into highly insoluble ferric (oxy)hydroxides. Moreover, although iron is abundant in the environment, its poor bioavailability necessitates strict conservation and recycling. Together, these considerations clearly explain the evolution of intricate mechanisms that tightly maintain mammalian iron homeostasis at the levels of absorption, utilization, storage, and release.

Dietary iron is absorbed, mainly by duodenal enterocytes, through several receptors, including the divalent metal iron transporter DMT1. In the circulation, ferric iron binds to the iron carrier protein transferrin (TF), and the complex binds to the transferrin receptor 1 (TfR1). The whole complex is then endocytosed by target cells; although in theory, all body cells require iron, hepatocytes and macrophages are the prime body cells that handle body iron recycling and storage. The TF-TfR1 pathway is the main route for importing iron into other body cells with high demand for iron, such as bone marrow cells, muscle cells, brain cells, macrophages (MΦs), and various other immune cells. MΦs, especially those of the spleen, internalize significant amounts of iron by phagocytosing senescent red blood cells (RBCs). The release of iron from iron-absorbing enterocytes, iron-storing hepatocytes, and MΦs occurs through the iron-exporter ferroportin (FPN). FPN cell surface expression is negatively regulated by the hormone hepcidin. The binding of hepcidin to FPN induces its ubiquitination, internalization, and degradation and blocks iron release. Hepcidin synthesis is regulated by several factors, including iron levels, anemia, hypoxia, erythropoiesis, inflammation, and sex hormones. Increased iron stores or decreased demand for iron increase hepcidin synthesis to reduce cellular iron release from iron-absorbing enterocytes and iron-storing MΦs and hepatocytes. Sequestered iron is stored in hepatocytes and MΦs as ferritin and hemosiderin. Conversely, increased demand for iron, as in the case of iron deficiency or hypoxia, decreases hepcidin synthesis to enhance iron release via FPN.

Iron is critical for the development and function of host immunity. It is also essential for pathogen growth and virulence. Therefore, iron deficiency may put human hosts at increased risk of microbial infections because it compromises the competence of the immune response. Iron deficiency has long been associated with microbial infections of the respiratory tract, gastrointestinal tract, and the skin. Iron overload, as in the case of hereditary hemochromatosis (HH), promotes pathogen growth and virulence but suppresses host immunity (Marx 2002). Bacterial infections with high demand for iron, such as those caused by *Yersinia enterocolitica* and *Vibrio vulnificus,* occur at a higher frequency in HH patients relative to the general population. As microbial pathogens require iron to establish an infection, one of the goals of innate immunity is to limit host iron availability. Attempts at iron sequestration by host immunity are usually met by aggressive attempts by the pathogen to acquire iron. Soon after the pathogen gains access to the host, the fight for iron quickly escalates into an all-out war. Immune cells sequester iron by upregulating the expression of toll-like receptors (TLRs) such as TLR4 and proinflammatory cytokines such as interleukin 1 beta (IL-1β), IL-6, and tumor necrosis factor-alpha (TNF-α) to enhance hepcidin synthesis and block cellular iron release (Shafarin et al. 2016). This limits iron availability and deprives the pathogen of the iron required for growth and virulence (Ganz 2006). Microbial pathogens counter host iron sequestration through various strategies, including the secretion of potent iron chelators that extract iron from a wide range of iron-containing molecules in the host.

11.2 Iron Homeostasis in the Human Host

Iron homeostasis in humans and other mammals is maintained through intricate mechanisms that regulate iron absorption, utilization, storage, and release (Anderson and Wang 2012). Systemic iron homeostasis is primarily maintained through the interaction between the liver-derived iron regulatory peptide hormone hepcidin and FPN (SLC40a1), the only known cellular iron exporter that exports iron from enterocytes, MΦs, and hepatocytes (Fig. 11.1). Hepcidin blocks iron export by inducing FPN ubiquitination, internalization, and degradation. Among the key triggers that modulate iron metabolism in humans are anemia (iron-deficiency anemia and anemia of chronic disease), erythropoiesis, hypoxia, inflammation, and iron overload. Although these triggers initiate corrective responses at different junctions, they ultimately converge at the hepcidin-FPN axis. While regulating systemic iron homeostasis is seemingly straightforward, regulating cellular iron metabolism is complicated and nuanced.

11.2.1 Systemic Iron Homeostasis

Systemic iron homeostasis in humans and other mammals is maintained at four key stages: intestinal absorption, extracellular fluid trafficking, tissue distribution and utilization, and storage. Interestingly, iron loss through cell shedding, desquamation, and chronic inflammation is not subject to any known regulatory mechanisms.

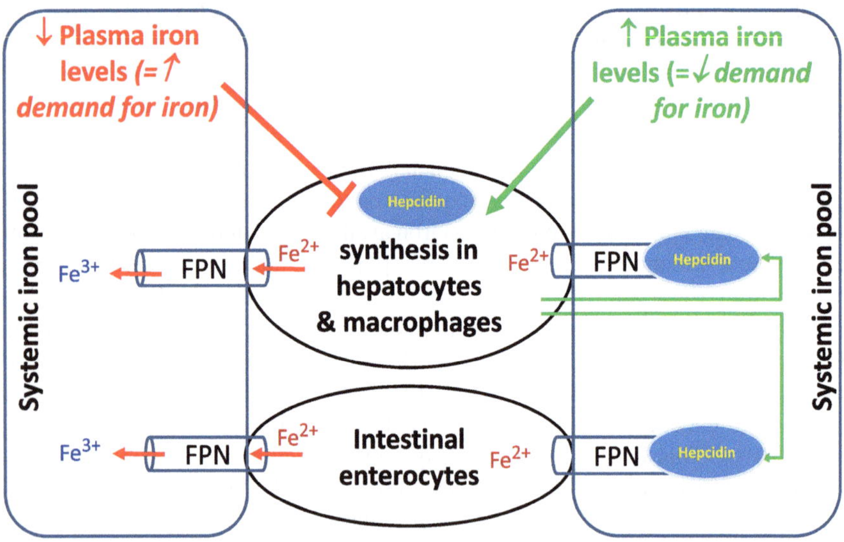

Fig. 11.1 The general layout of systemic iron homeostasis in mammals. Under conditions of low iron availability, hepcidin synthesis and release by hepatocytes and macrophages are blocked to maintain FPN surface expression and promote iron release. In contrast, when iron levels are high, hepcidin synthesis and release are enhanced to minimize FPN surface expression and block iron release

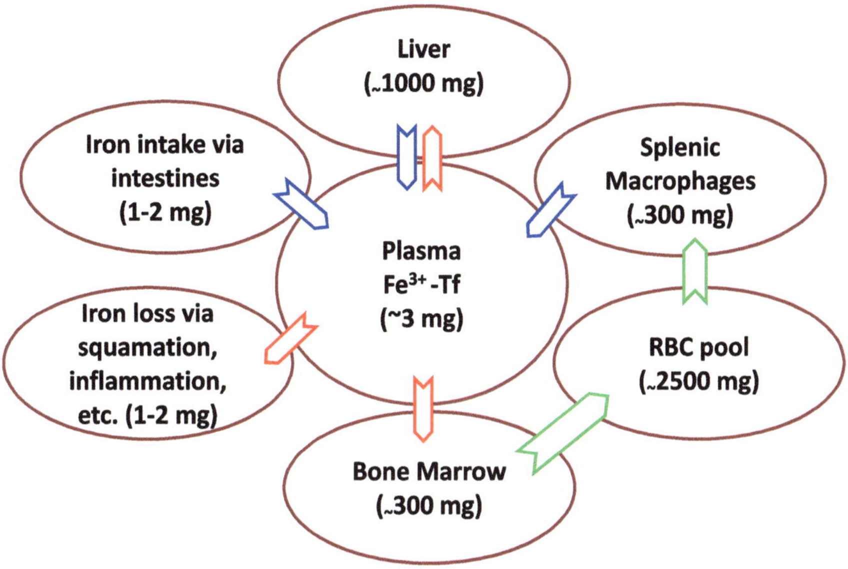

Fig. 11.2 Body iron distribution and recycling. Dietary iron is released into the circulation as ferric iron, which immediately binds to transferrin (TF) to form bi-ferric-TF complexes for trafficking to various utilization and storage sites. Iron in hemoglobin (Hb) is recycled back into the circulation through splenic macrophages, which engulf senescent and damaged RBCs, and through hepatocytes, which take up iron as Hb-haptoglobin complexes, among other forms

Healthy adults contain about 4 grams (g) of iron in their bodies (Fig. 11.2), which is roughly 40 milligrams (mg) of iron per kilogram (kg) body weight in women and 50 mg/kg body weight in men. Homeostatic serum iron concentration in humans ranges between 40 and 160 µg/dl in women and between 60 and 160 µg/dl in men. Serum iron levels outside this range often indicate iron deficiency at the lower end and iron overload at the upper end. Duodenal enterocytes, the point of entry of dietary iron, absorb about 1–2 mg of iron daily. Once absorbed, the major compartments that participate in the maintenance of iron homeostasis are the erythron (RBCs and RBC precursors in bone marrow) and iron stores (hepatocytes and splenic MΦs). The erythron is the largest iron compartment in the body; it contains about 3 g (55–65%) of total body iron in adults. Erythron iron is mostly attached to heme in hemoglobin (Hb). Hepatocytes store up to 1 g (20%) of total body iron in cytoplasmic and mitochondrial forms of ferritin. Hepatocytes also store iron in hemosiderin as a secondary cytoplasmic iron store, especially when ferritin iron-storing capacity is exhausted. It must be noted that the concentration of stored iron could significantly vary depending on how balanced the rates of iron intake, utilization, and loss are. On average, the human body loses 1–2 mg daily, which is roughly equivalent to the daily dietary iron intake. Splenic MΦs, which take up senescent and damaged RBCs, contain small amounts of iron (<50 mg), which could increase by up to six-fold (Morgan and Walters 1963; Korolnek and Hamza 2015; Fillet et al. 1989) due to increased RBC phagocytosis (erythrophagocytosis), decreased iron release, or both. Extraction of iron from senescent RBCs is the major source of systemic iron. Once phagocytosed, RBCs are degraded within the phagolysosome for heme iron to be extracted and released into the cytosol; heme itself may also exit the phagolysosome through the transporter homolog of the Caenorhabditis elegans heme-responsive gene 1 (HRG1) (White et al. 2013). Negligible amounts of iron are present in myoglobin and other iron-containing proteins and enzymes in muscles and other body tissues (Morgan and Walters 1963; Korolnek and Hamza 2015; Fillet et al. 1989).

Iron in the small intestine exists as heme from hemoglobin and myoglobin in meat and as nonheme from meat and plant foods; nonheme ferric iron is highly insoluble. Iron absorption by duodenal enterocytes occurs through DMT1 and the heme carrier protein (HCP1) (Rouault 2005), which transports heme iron. DMT1, a 12 transmembrane domain integral protein on the apical membrane of absorptive enterocytes, transports different divalent cations, including ferrous iron (Gunshin et al. 1997; Fleming et al. 1997). Mutations in DMT1 of rodents (Fleming et al. 1998) and humans (Mims et al. 2005) were shown to precipitate microcytic anemia. Dietary nonheme iron is first reduced to ferrous iron by ferric reductase duodenal cytochrome B (DCYTB) or cytochrome b reductase 1 (CYBRD1) on intestinal apical brush borders (McKie et al. 2001). Once internalized, the activity of heme oxygenase 1 (HO1) helps to release ferrous iron (Poss and Tonegawa 1997), which then crosses the basolateral membrane through FPN into the circulation. Before entering the circulation, ferrous iron is oxidized to ferric iron by the ferroxidases ceruloplasmin and hephaestin. Ferric iron in circulation instantly binds to the plasma iron carrier protein transferrin (TF) to form bi-ferric-TF complexes (holo-transferrin). Under homeostatic conditions, less than 45% of TF is saturated, and the rest remains as iron-free transferrin (apo-transferrin). The TF pool in the circulation carries a total of 2–3 mg of iron per cycle, which can be repeated several times a day. The percentage of apo-transferrin significantly increases when serum iron levels exceed the TF carrying capacity, as in the case of iron overload resulting from hereditary hemochromatosis (HH) (Chua et al. 2004).

Holo-transferrin delivers iron to body cells by binding to transferrin receptors (TfR1 and TfR2), where the complex is then endocytosed, and ferric iron is extracted within acidic endosomes (Fig. 11.3) (West et al. 2000). TfR1 is the main receptor for holo-transferrin; the affinity of TfR1 to holo-transferrin is tens of folds higher than that of TfR2 (West et al. 2000). Deletion of TFRC, the gene that encodes TfR1, induces severe anemia and death in mice (Levy et al. 1999). Previous studies have suggested that glyceraldehyde-3-phosphate dehydrogenase (GAPDH) can also function as a holo-transferrin receptor (Kumar et al. 2012). Once ferric iron is reduced by ferrireductases including the six-transmembrane epithelial antigen of the prostate proteins STEAP2 and STEAP3 (Ohgami et al. 2005), DCYTB (in enterocytes), and glucose/ribitol dehydrogenase family protein (SDR2) (Hentze et al. 2010), it is released into the cytoplasm through DMT1 and the zinc transporter Zrt- and Irt-like protein 14 (ZIP14; solute carrier protein family 39; SLC39A14) (Lane et al. 2015). Depletion of ZIP14 reduces free iron uptake by hepatocytes and pancreatic cells and diminishes parenchymal cell iron loading in HH in mice

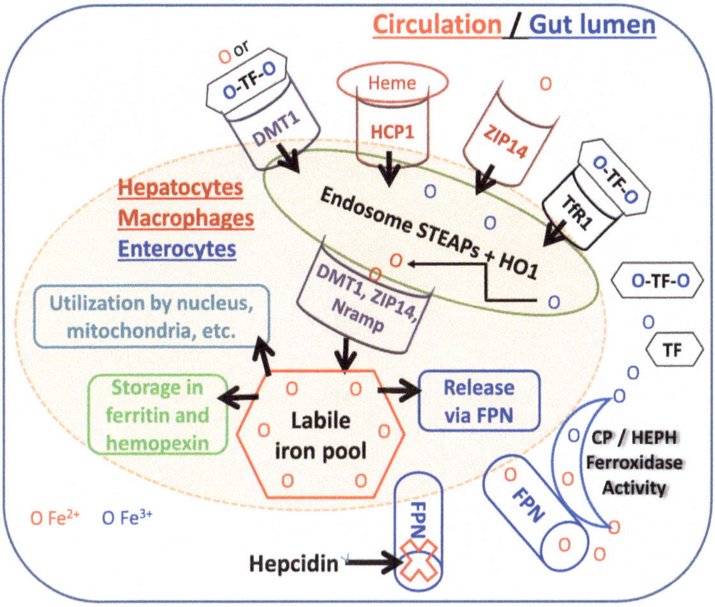

Fig. 11.3 The role of enterocytes, hepatocytes, and macrophages in iron homeostasis. Enterocytes absorb dietary heme and nonheme iron through several receptors, including DMT1, ZAI14, and HCP1, endocytose the receptor-iron complexes, and export iron into the circulation via FPN. Hepatocytes and macrophages import transferrin-bound iron through TfR1, endocytose the receptor-ferric-transferrin complexes, and export ferrous into the cytoplasm for utilization, storage, or release. Within the endosome, iron is extracted via metalloreductases (STEAP2, STEAP3, Dcytb, etc.) with the aid of the heme degrader heme oxygenase 1 (OH1) along with other proteins and released into the cytoplasm via DMT1 and Nramp. Iron release into circulation via FPN is promoted in the absence of hepcidin and blocked in the presence of hepcidin

(Jenkitkasemwong et al. 2015). Extracted iron can also be released into the cytoplasm by the natural resistance-associated macrophage protein 1 (Nramp1) (Soe-Lin et al. 2009) and the type IV mucolipidosis-associated protein (TRPML1) (Lane et al. 2015; Pasquadibisceglie et al. 2023).

11.2.2 Cellular Iron Homeostasis

The expression of iron regulatory genes TFRC (TfR1 gene), solute carrier family 40 member 1 (SLC40A1), which encodes FPN, ferritin heavy and light chain genes, and DMT1 is primarily regulated at the post-transcriptional level (Fig. 11.4). This largely depends on the interaction between the iron regulatory proteins 1 and 2 (IRP1 and IRP2) and the iron-responsive element (IRE) sequences in the 5′ or 3′ untranslated regions (UTRs) of target mRNAs. The main trigger of the IRP/IRE system is cellular iron content. Under conditions of iron deficiency, the IRE stem-loop structure at the 3′UTR of TFRC and DMT1 mRNAs binds to the IRP1/2 complex. This stabilizes these mRNAs and enhances their translation, with the resulting

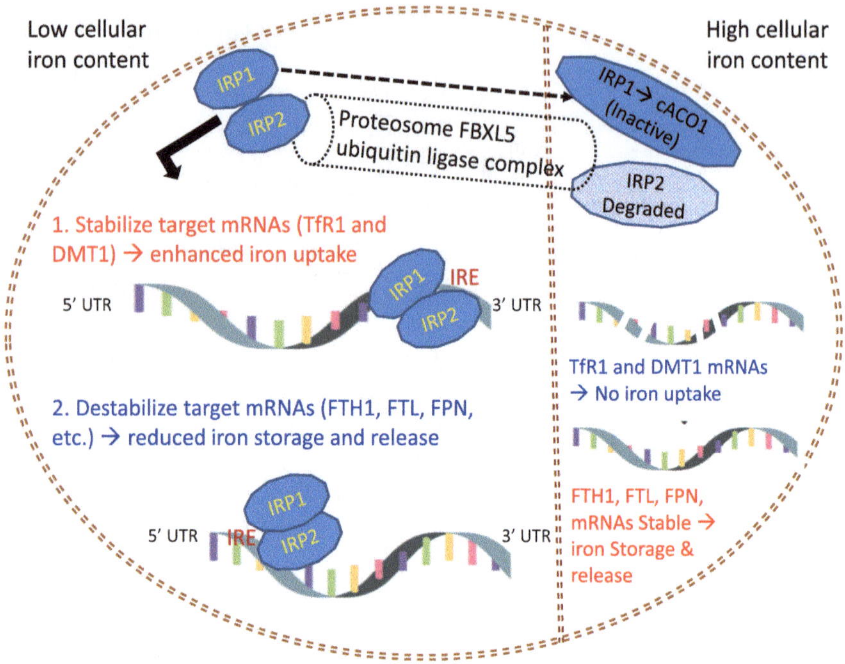

Fig. 11.4 Expression of key iron metabolism-related genes under low and high iron content. The expression of genes that mediate iron uptake (mainly TfR1 and DMT1) and genes that mediate iron storage (FTH1 and FTL; components of ferritin) and release (FPN). Decreased cellular iron content facilitates the complexing of IRP1 and IRP2, which then binds the IREs at the 3′UTRs of TfR1 and DMT1 mRNAs. This stabilizes these mRNAs and permits their translation. At the same time, the IRP1/2 complex binds IREs at the 5′UTRs of FTH1, FTL, aconitase, and FPN mRNAs. This destabilizes these mRNAs and inhibits their translation. Increased cellular iron content, on the other hand, is associated with the conversion (inactivation) of IRP1 into cytoplasmic aconitase 1 (ACO1) and the proteasomal degradation of IRP2. This results in reduced synthesis of TfR1 and DMT1 proteins, leading to reduced iron uptake by cells. This also increases the synthesis of FTH1 and FTL to enhance iron storage and that of FPN to promote cellular iron release

proteins contributing to enhanced iron uptake. In contrast, binding of the IRP1/2 complex to IREs at the 5′UTR of FTH1, FTL, FPN, and the cytoplasmic aconitase (cACO1; also known as the IRE binding protein) destabilizes these mRNAs, inhibits their translation, and minimizes iron storage and export (Zhang et al. 2014). IRP1 and c-ACO1 interconvert depending on cellular iron content; high iron content leads to cACO1 assembly, while low iron content destabilizes the ISCs in cACO1 IRP1, allowing it to convert to IRP1 (Baumgart and Bott 2011).

11.2.3 Regulation of FPN Expression

FPN is a 12 transmembrane domain protein that serves as the main iron exporter in mammalian cells (McKie et al. 2001; Donovan et al. 2000). FPN releases ferrous iron from enterocytes, hepatocytes, and MΦs with the help of the Cu-containing ferroxidases ceruloplasmin (Gromadzka et al. 2020) and hephaestin (Vulpe et al. 1999) that oxidize ferrous iron to stable, nonreactive, and insoluble ferric iron. The expression of FPN is regulated at the transcriptional, post-transcriptional, and post-translational levels. Post-transcriptionally, IRP/IRE binding at 5′UTR of FPN's mRNA represses its translation under conditions of iron deficiency (Abboud and Haile 2000). Transcriptionally, the expression of FPN is upregulated by heme (Marro et al. 2010) and downregulated by inflammation (Liu et al. 2005). Post-translationally, the stability of membrane-bound FPN is regulated by hepcidin, which induces its degradation; binding of hepcidin to FPN induces its ubiquitination, internalization, and degradation (Pigeon et al. 2001; Nicolas et al. 2001, 2002; Ross et al. 2012; Qiao et al. 2012; Nemeth et al. 2004). The binding of hepcidin to FPN induces conformational changes that expose lysine (K) residues to ubiquitination by the E3 ubiquitin ligase RNF217, which promotes FPN proteolysis (Jiang et al. 2021). Iron-dependent epigenetic regulator Ten-eleven translocation methylcytosine dioxygenase 1 (TET1) regulates the ring finger protein 17 (RNF217) cytoplasmic accumulation by demethylating its 5-methylcytosines (5-mC). Reduced cytosolic iron content also induces FPN internalization via a hepcidin-independent mechanism that involves the E3 ubiquitin ligase neural precursor cell expressed developmentally downregulated 4-like (NEDD4-2) and the NEDD4-2 binding protein NEDD family interacting protein 1 (Nfdip-1) (Nemeth and Ganz 2021).

11.2.4 Regulation of Hepcidin Synthesis

The 25 amino-acid peptide hormone hepcidin is primarily produced by hepatocytes (Krause et al. 2000; Park et al. 2001). The spleen, adipose tissues, heart, brain, pancreas, and kidneys also produce considerable amounts of hepcidin in response to changes in systemic iron status (Daher et al. 2019; Sow et al. 2007; Schwarz et al. 2012; Schwartz et al. 2021; Qian et al. 2014; Urrutia et al. 2013). Hepcidin was originally described as an acute-phase antimicrobial peptide whose expression is triggered by infection (Park et al. 2001; Nemeth et al. 2003; Krause et al. 2003; Valore and Ganz 2008). The 84 amino-acid long pre-pro-peptide undergoes post-translational maturation that involves the cleaving of a 24 amino-acid N-terminal signal peptide, followed by the removal of the pro-region by furin-like convertases from the 60-residue long polypeptide (Park et al. 2001; Valore and Ganz 2008). The C-terminus of the mature 25-residue polypeptide consists of four highly conserved disulfide bonds that form a β-hairpin, and the unstructured N-terminus interacts with FPN (Nemeth et al. 2006). Excess hepcidin in the circulation is cleared through urine. Major triggers that modulate hepcidin expression up/down include anemia and erythropoiesis, hypoxia, inflammation, and estrogen (Fig. 11.5).

Liver hepcidin expression is mainly regulated by the bone morphogenetic protein 6 (BMP6)/suppressor of mothers against the decapentaplegic (SMAD) pathway (Andriopoulos Jr et al. 2009; Meynard et al. 2009; Canali et al. 2017a); BMPs 2, 4, and 9 may also contribute to hepcidin expression (Truksa et al. 2006; Canali et al. 2017b). Upregulated expression of BMP6, which occurs in response to elevated

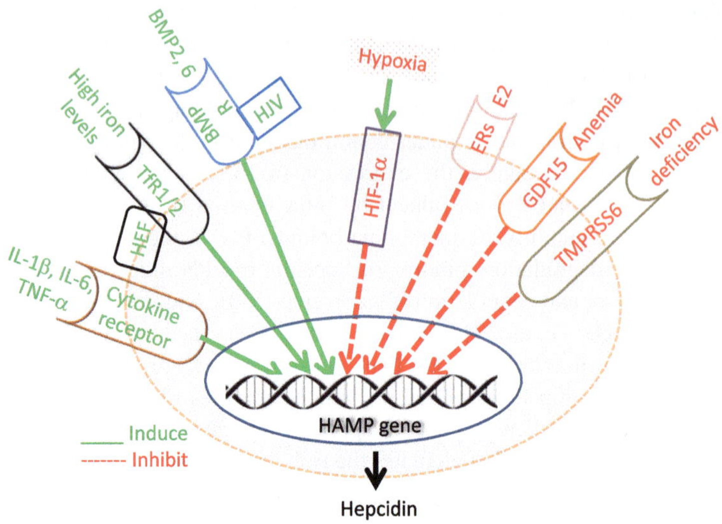

Fig. 11.5 Regulation of hepcidin synthesis. Anemia, hypoxia, iron deficiency, and/or elevated levels of estrogen (E2) block hepcidin synthesis through specific receptors and mediators. On the other hand, inflammation, iron overload, and increased synthesis of BMB2 or BMP6 enhance hepcidin synthesis through their respective receptors and mediators

body iron stores, allows it to engage its cell surface (hepatocyte) BMP receptor (BMPR) along with the coreceptor hemojuvelin (HJV) (Babitt et al. 2006). Type 1 (activin-like kinases ALK2 and ALK3) and type 2 (BMPR2 and ActR2A) kinases as members of the transforming growth factor-beta (TGF-β) receptor superfamily also act as BMPRs in iron homeostasis (Steinbicker et al. 2011; Mayeur et al. 2014). BMP6/receptor-coreceptor engagement triggers the phosphorylation of cytoplasmic SMAD1/5/8 (Miyazono et al. 2010), which recruits SMAD4 before the complex translocates to the nucleus (Wang et al. 2005), where the SMAD complex acts as a transcription factor through its ability to bind to BMP-responsive elements (BMP-RE) in the promoter region of HAMP (the hepcidin gene) (Truksa et al. 2009). The phosphorylation of SMAD1/5/8 is under the negative control of SMAD6 and SMAD7 (An et al. 2018). Moreover, both BMP6 and iron levels can upregulate the expression of the transmembrane serine protease, matriptase-2 (TMPRSS6) (Meynard et al. 2011), which acts as a negative regulator of hepcidin via its ability to cleave HJV and thus disrupt BMPR signaling (Silvestri et al. 2008a). This is further supported by the observation that soluble HJV competes with and disrupts membrane-bound HJV signaling (Lin et al. 2005).

Since erythropoiesis represses hepcidin expression, it has long been postulated that there must exist an erythropoiesis-related mechanism that regulates iron homeostasis. In this regard, the erythroblast-derived cytokines growth differentiation factor 15 (GDF-15) and twisted gastrulation factor 1 (TWSG1) were suggested as candidate inhibitors of hepcidin synthesis in human hepatocytes (Tanno et al. 2007, 2009). However, several observations including the fact that phlebotomy-stimulated suppression of hepcidin is not blocked by GDF-15 depletion in mice (Casanovas et al. 2013) and that TWSG1 expression does not change in different mouse models of anemia (Mirciov et al. 2017) have challenged the ascribed role of GDF15 and TWSG1 in hepcidin synthesis. The possibility that erythroferrone (ERFE) may regulate hepcidin gene expression during erythropoiesis has also been challenged by the finding that stimulated erythropoiesis in ERFE-knockout mice does not substantially suppress HAMP (hepcidin gene) expression (Kautz et al. 2015).

Hypoxia-inducible factor (HIF) family of proteins, mainly HIF1α and HIF2α, regulate hepcidin expression along with several genes involved in hypoxia and anemia (Kaelin and Ratcliffe 2008). Hypoxia, through HIF1α and HIF2α, upregulates the expression of erythropoietin (EPO) (Kapitsinou et al. 2010; Morita et al. 2003), which suppresses hepcidin synthesis (Liu et al. 2012). HIF1α can directly repress hepcidin synthesis by binding to hypoxia response elements (HREs) in the promoter region of the HAMP gene in mice (Peyssonnaux et al. 2007). It can also indirectly repress HAMP gene expression through its involvement with BMP6/SMAD signaling. Hypoxia-induced upregulation of the TMPRSS6 via HIF-HRE binding in the promoter region of TMPRSS6 (Maurer et al. 2012) was reported to cleave HJV and repress HAMP gene expression (Silvestri et al. 2008b).

Inflammation regulates hepcidin synthesis, mainly through the IL-6/signal transducer and activator of the transcription (STAT) signaling pathway (Verga Falzacappa et al. 2007). Engagement of IL-6 with its receptor (IL-6R, CD126) activates the Janus kinase 1 (JAK) protein leading to STAT3 phosphorylation (Ganz and Nemeth 2015) and translocation to the nucleus, where activated STAT3 binds the STAT binding motif in the HAMP gene promoter to induce basal transcription (Verga Falzacappa et al. 2007). Intact SMAD1/5/8 complex, which is related to the BAMP6/SMAD pathway, increases IL-6/STAT3-induced HAMP expression (Canali et al. 2016), possibly through activin B acting as a surrogate ligand for the SMAD1/5/8 complex during infection (Canali et al. 2016; Kanamori et al. 2016). IL-1β, TNF-α, and TLR4, among other inflammatory mediators, can contribute to HAMP gene expression regulation, mostly through the nuclear factor-kappa beta (NF-κβ) pathway (Drakesmith and Prentice 2012).

Several lines of evidence have long suggested that sex hormones, mainly estrogen (17-β estradiol; E2), are involved in regulating iron metabolism. Several studies have established that increased serum levels of E2 correlate negatively with serum hepcidin concertation (Bajbouj et al. 2018; Ulas and Cay 2011) and positively with serum iron levels (Horiguchi et al. 2005; Pentecost and Teng 1987; Stuckey et al. 2006; Harrison-Findik 2010; Qian et al. 2015). Contraceptive use significantly increases serum iron concentration (Akinloye et al. 2011; Huang et al. 2013; Lehtihet et al. 2016; Milman et al. 1992). Ovariectomized rats tend to have significantly reduced serum iron levels relative to controls (Raso et al. 2009). Administration of E2 to such rats was reported to reverse iron deficiency (Horiguchi et al. 2005). Elevated serum E2 levels were also shown to modulate lactoferrin (LTF) expression in mouse uterine tissue (Pentecost and Teng 1987; Stuckey et al. 2006). Moreover, the elevation of endogenous E2 levels during in vitro fertilization in humans was associated with a significant reduction in plasma hepcidin levels (4.85–1.43 ng/mL)

(Ulas and Cay 2011). Knockout of BMP6 in mice was shown to lead to gender-based differences in serum hepcidin levels and body iron concentration (Harrison-Findik 2010). E2-induced uterotrophic responses in rodents were also shown to lead to coordinated expression of several iron regulatory genes including LTF, ceruloplasmin (CP), gelatinase-associated lipocalin (NGAL; LCN2), and FPN (Stuckey et al. 2006). Exposure to the organic pollutants polychlorinated biphenyls (PCBs), which have a high affinity for estrogen receptors (ERs), was reported to inhibit HAMP gene expression in the human liver cancer cell line Hep-G2 (Qian et al. 2015). Treatment of cultured human cancer cell lines MCF-7 (breast), SKOV-3 (ovarian), HuH7, and Hep-G2 (liver) with increasing concentrations of E2 reduced hepcidin synthesis and maintained FPN surface expression (Shafarin et al. 2016; Bajbouj et al. 2017; Yang et al. 2012; Hou et al. 2012; Liu et al. 2009). In contrast, treatment with the E2 antagonist ICI 182,780 reversed the E2-mediated downregulation of HAMP gene expression (Liu et al. 2009). E2 treatment in human ovarian cancer cell lines ES-2 and SKOV-3 was shown to upregulate HIF1α expression and reduce hepcidin production (Kazi et al. 2009).

11.3 Iron and Host Immunity

Iron is essential for the development and function of the immune system. Like other body cells, immune cells use iron for ATP production, DNA replication, and gene expression, among other cellular activities. Therefore, iron deficiency compromises the ability of cells like MΦs, neutrophils, and lymphocytes to proliferate and differentiate. In contrast, iron availability promotes cytokine and chemokine production and antibody synthesis, among other immune responses. Iron is also a significant player in the inflammatory response that ensues during microbial infections. Under homeostatic conditions, while pathogens compete for iron with the host, host immunity remains competent enough to mount protective anti-pathogen responses. However, under conditions of iron deficiency or iron overload, host susceptibility to microbial infections significantly increases. Immune cells sequester iron via proinflammatory mediators that enhance hepcidin synthesis to block cellular iron export. Iron sequestration by host immunity, which is often referred to as nutritional immunity, deprives pathogens of the iron needed for growth, virulence, and dissemination (Ganz 2006). Previous studies have suggested that nutritional immunity, by itself, can reduce iron availability by as much as 30%, leading to what is often called hypoferremia of infection (Fleming et al. 1997).

11.3.1 Role of Iron in Innate Immunity

Systemic iron levels in the host significantly decrease following a microbial infection as part of the anti-infection innate immune response (Johnson and Wessling-Resnick 2012). Proinflammatory cytokines and other innate immunity mediators induce hepcidin synthesis to limit iron release by duodenal enterocytes, MΦs, and hepatocytes. Concurrently, considerable amounts of circulating iron are taken up by MΦs, and hepatocytes for storage (Martínez et al. 2017; Weinberg 1975). MΦs, neutrophils, and natural killer (NK) cells contribute to iron sequestration by releasing specific iron scavengers, as will be discussed later.

Depending on the microenvironment in which they exist, unpolarized MΦs differentiate into proinflammatory M1 cells or anti-inflammatory M2 cells. Classically, lipopolysaccharides (LPS) trigger M1 cell polarization and IL-4 triggers M2 cell polarization. M1 cells secrete several proinflammatory cytokines, including tumor necrosis factor (TNF-α), IL-1β, IL-6, and IL-18, among others. M2 cells produce several anti-inflammatory cytokines, including IL-10, transforming growth factor-beta (TGF-β), and platelet-derived growth factor (PDGF), among others (Amit et al. 2016). M2 cells are further divided into four subsets (M2a-M2d) based on the unique biomarker profile they express and the cytokines they produce. Previous work has shown that iron plays a role in MΦ polarization (Yang et al. 2021; Camiolo et al. 2020). Mounting evidence suggests that M1 cells display a protective iron sequestration phenotype and M2 cells display a nonprotective iron release phenotype. This is based on several findings; for example, while M1 cells upregulate ferritin synthesis for iron storage (Soares and Hamza 2016; Cairo et al. 2011), M2 cells downregulate its expression (McKie et al. 2001). Moreover, the expression of TfR1 (Ganz and Nemeth 2011) and ferritin (Torti and Torti 2002) on/in MΦs and other innate cells was reported to significantly increase during inflammation, further enhancing iron uptake and storage (Ganz and Nemeth 2011). Expression of the heme binding receptor CD163 and the heme-degrading protein heme oxygenase 1 (HO-1) are highly upregulated in M1 but not M2 cells (DeRosa and Leftin 2021).

Previous work has established that the expression of several iron metabolism-related genes including TFRC (TfR1), FTH1, and SLC40A1 (FPN) undergoes significant shifts in different innate cells during microbial infections. For example, under conditions of increased iron availability, neutrophils upregulate the expression of the bacteriostatic factor lipocalin 2 (LCN2) to block bacterial siderophores from extracting host iron (Flo et al. 2004). Neutrophils also upregulate the production of LTF and store it intracellularly in neutrophils' secondary granules to release it at infectious foci to compete with microbe-derived iron-chelating molecules (Rosa et al. 2017). Previous work has demonstrated that LTF promotes the growth of beneficial bacteria with low iron requirements including *Lactobacillus* and *Bifidobacterium spp* (Yang et al. 2022). Reduced iron availability following deferasirox (iron chelator) treatment (Puri et al. 2019) or upon the production of the heme scavenger haptoglobin (Pirschel et al. 2022) reduces neutrophil recruitment during inflammation. Whether iron chelation promotes (Völlger et al. 2016) or inhibits (Van Avondt et al. 2016) the formation of neutrophil extracellular traps (NETs) is still debatable. Iron chelation was reported to impair the production of ROS and hence reduce NET formation in response to bacterial infection (Pritchard et al. 2015), it was also reported to decrease NET release in septic mice (Ren et al. 2017).

The natural killer (NK) cell population is divided into two subsets based on phenotypic characteristics and functional properties (Rebuffet et al. 2024). The first subset represents conventional or killer NKs, which are $CD16^{+/bright}CD56^{dim}$. Their main function is to directly recognize ligands on the surface of tumor cells and virally infected cells and then degranulate to release perforins and granzymes to degrade target cells. Cytotoxic NKs also secrete the soluble tumor necrosis factor-related apoptosis-inducing ligand (TRAIL), the anti-B cell maturation antigen (BCMA), and/or the FasL (CD95L), each to engage its corresponding receptor on target cells and initiate a specific response either in the NK cell and/or the target cell. The second NK cell subset is comprised of $CD16^{dim}CD56^{bright}$ cells, which secrete several inflammatory cytokines such as interferon-gamma (IFN-γ), TNF-α, and granulocyte monocyte colony-stimulating factor (GM-CSF). Increased iron availability promotes NK activation, while iron deficiency hinders it (Sottile et al. 2019; Littwitz-Salomon et al. 2021; Osuna-Espinoza and Rosas-Taraco 2023). Adequate iron supply in NK cells supports energy production, the formation and packaging of granules, and the generation of ROSs as components of the NK cell cytotoxic machinery. It is also involved in regulating IFN-γ gene expression via both transcriptional modulatory effects and changes to NK cell's epigenetic landscape (Soares and Weiss 2015). Notably, the link between virally infected cell susceptibility to NK cytotoxic activity and the expression of TfR was originally made in 1986 (Borysiewicz et al. 1986).

11.3.2 Role of Iron in Lymphocyte Differentiation and Function

Lymphocyte activation and proliferation require iron-containing proteins that regulate gene transcription, DNA replication, and cell activation. Therefore, iron deficiency impairs T-cell differentiation and activation and compromises adaptive immunity (Ni et al. 2022). Numerous illustrative examples of how iron is important for proper lymphocyte functioning have been documented. For instance, increased iron content was reported to inhibit T helper (Th) cell polarization to IFN-γ-producing Th1 cells (Pfeifhofer-Obermair et al. 2021; Knobloch et al. 2020). As lymphocytes import iron through the holo-transferrin-TfR1 pathway, increased expression of TfR1 on lymphocytes, which is promoted by IL-2 signaling, is critical for cell proliferation and activation (Frost et al. 2021; Macedo et al. 2004). Increased cellular iron content in regulatory T (Treg) cells, which express TfR1 at levels higher than that in Th cells, was reported to induce ferroptosis in Tregs (Feng et al. 2021). Hepatic iron overload resulting from MHC-class I deficiency in mice was associated with weak anti-viral responses (Cardoso et al. 2002), likely due to defective antigen processing and presentation (Doherty and O'Farrelly 2000). The positive impact of including iron oxide nanoparticles as adjuvants in viral vaccines, vis-à-vis, enhanced Th1, Th17, and T cytotoxic (Tc) cell responses, further illustrates how important iron is for the induction of protective adaptive immunity (Neto et al. 2018).

Regarding the role of iron in B cell function, previous work has shown that iron deficiency is often associated with reduced T-dependent synthesis of anti-vaccine antibodies (Jiang et al. 2019) and that administration of anti-TfR1 antibodies in multiple myeloma leads to reduced cell growth and survival (Leoh et al. 2018).

Moreover, mutations in TFRC were reported to precipitate defective B cell responses (Jabara et al. 2016). Increased synthesis of the iron scavenger LTF during infection and vaccination was shown to enhance the synthesis of antimicrobial IgA and IgG2b antibodies (Jang et al. 2015).

11.3.3 Immune Cells and Mediators Influence Iron Metabolism

It is worth noting that, just as iron availability is critical for immune cell differentiation and function, various immune cells and mediators can directly or indirectly modulate iron metabolism. The list of immune cells and mediators that differentially contribute to this effort is extensive, but citing key examples should suffice. MΦs play a central role in iron homeostasis through their ability to acquire, recycle, store, and release iron (Hentze et al. 2010). Inflammation-induced HO-1 activity in MΦs facilitates the extraction and release of heme iron (Poss and Tonegawa 1997). Neutrophils also release iron-binding proteins, such as lactoferrin, to restrict microbial growth by sequestering iron and depriving pathogens of this essential micronutrient (Soares and Weiss 2015). Inflammatory cytokines (IL-6, TNF-α, and IL-1β), which upregulate hepcidin synthesis to reduce the release of cellular iron and hence initiate an intricate cascade of events that limit the ability of pathogens to access iron, are key in this regard (Nemeth and Ganz 2006). IFN-γ also modulates iron metabolism by upregulating the expression of ferritin and downregulating the expression of FPN, leading to increased iron sequestration (Weiss and Schett 2013).

11.4 Iron Metabolism in Microbial Pathogens

For microbial pathogens, the primary goal of acquiring iron is to facilitate the production of proteins with ISC attachments (Kim et al. 2017; Miller and Auerbuch 2015; Rodionov et al. 2005). Microbial ISCs facilitate electron transfer reactions, enzymatic catalysis, and the regulation of gene expression. Moreover, the regulation of nitric oxide (NO) metabolism and detoxification is aided by specific ISC-containing proteins that operate under the control of oxidative stress-responsive transcription factors, including the superoxide response regulator protein (SoxR) (Kim et al. 2017). Pathogens also store excess iron in specialized molecules such as bacterioferritin for use under conditions of limited iron availability (Sousa Gerós et al. 2020). For example, *Cryptococcus neoformans* utilizes the ferritin-like cleavage factor two protein 1 (Cft1) to use it for iron storage (Jung et al. 2010). Microbial pathogens also use iron to facilitate the production of redox-responsive proteins that, in many instances, serve as transcription factors. Several bacterial species use ferric iron levels as a cue to continue or halt transcriptional repression of iron acquisition and virulence genes. During the early phase of microbial infections, the pathogen often experiences iron starvation due to limited iron availability in the host microenvironment. As such, the ability to acquire iron from the host is an important virulence factor that gives the pathogen a competitive advantage in establishing an infection.

11.4.1 Host-Induced Iron Starvation

Pathogen access to host iron is hindered by several mechanisms, some of which are part of the normal host physiology while others are triggered by inflammation. (i) The immediate switch of soluble ferrous to insoluble ferric iron and the formation of highly insoluble ferric (oxy)hydroxide polymers restrict iron availability to invading pathogens. This may explain the evolution of free ferric iron uptake mechanisms in many microbial pathogens. (ii) Loading free ferric iron onto TF, which is abundant in host plasma and interacts with free ferric iron with high affinity to form holo-transferrin. This further exacerbates iron scarcity in the host's extracellular fluids. (iii) Inflammation upregulates the expression of hepcidin, which degrades FPN and blocks cellular iron release. (iv) Inflammation also induces neutrophils to produce LTF as a high-affinity iron scavenger at mucosal surfaces; this parallels the action of TF in the cytoplasm. (v) Restricted trafficking of intracellular iron hinders the capacity of intracellular pathogens to access iron. The upregulated expression of Nramp1 on phagosome membranes, which facilitates the pumping of iron out of the phagosome where the pathogen is likely to reside, is a good example of this process. Interestingly, neutrophils also produce the antimicrobial peptide neutrophil LCN2 to scavenge catechol-type siderophores (enterobactin) and deprive *Enterobacteriaceae* of siderophore-extracted iron (Xiao et al. 2017), where Nramp1 facilitates the release of LCN2 in neutrophils (Johnson and Wessling-Resnick 2012). (vi) Classical nutritional immunity is ineffective in the case of intracellular pathogens such as *Salmonella* spp. that reside in the cytoplasm of MΦs. This is because FPN degradation sequesters cellular iron and increases the labile iron pool allowing the pathogen to thrive. In this case, increased production of IFN-γ helps to resolve this issue by upregulating the expression of LCN2 and FPN in MΦs to decrease cellular iron content (Nairz et al. 2008). In short, nutritional immunity and the resulting hypoferremia along with the rest of the iron sequestration mechanisms highlighted above collectively reduce iron concentration in the host to $<10^{-8}$ M, which is at least 100-fold below the optimal 10^{-6} M iron concentration microbial pathogens require for growth and virulence.

11.4.2 Regulation of Iron Metabolism in Microbial Pathogens

As microbial pathogens encounter a brief period of iron starvation upon infecting a mammalian host, they initiate a concerted effort to acquire iron rapidly from different sources within the host. Many bacterial species (*S. aureus*, *Bartonella* spp., *Bacteroides* spp., *Porphyromonas gingivalis*, *Haemophilus influenzae*, etc.) acquire iron by targeting hemoproteins, mostly because of ubiquity and ease of access (Skaar et al. 2004a; Gruss et al. 2012). Since most heme exists in hemoglobin, under conditions of low iron availability, extracellular and facultative intracellular bacterial pathogens secrete hemolysin to lyse RBCs and release heme. Therefore, the expression of hemolysin and similar enzymes during microbial infections is considered a pathogenicity determinant (Los et al. 2013). Free hemoglobin or heme that dissociates from it is captured by secreted and/or cell surface-anchored heme receptors (Rouault 2005). Microbial pathogens can also target heme in serum albumin (*Candida* spp.), heme-hemopexin complexes, or hemoglobin-haptoglobin complexes (*H. influenzae*) (Zambolin et al. 2016).

These and other iron acquisition strategies require the synthesis of specific soluble and cell surface receptors. Therefore, the pathogen starts modulating the expression of a wide range of genes involved in iron acquisition. The starting point of this process is the ability of the microbial pathogen to sense iron scarcity and to interpret it as a cue that it is inside a host and that it needs to assemble its iron acquisition machinery. Depending on cellular iron content, two distinct families of highly conserved iron-responsive regulators (IRRs) control the transcription of several iron acquisition and virulence genes (Fig. 11.6). Regulation of the expression of such genes in Gram-negative bacteria and some Gram-positive bacteria (those with low G/C DNA content including *Staphylococcus* spp., *Bacillus* spp., and *Listeria monocytogenes*, etc.) is mediated by the ferric iron uptake regulator (Fur) family of IRRs (Escolar et al. 1999). In Gram-positive bacteria with high G/C DNA content (e.g., *Corynebacterium diphtheriae*), the regulation of iron metabolism-related and associated virulence genes is mediated by the diphtheria toxin repressor (DtxR) IRR (Hantke 2001). *Mycobacteriaceae* employ a DtxR-like IRR that is specifically referred to as the mycobacterial iron-dependent regulator (IdeR) (Pandey and

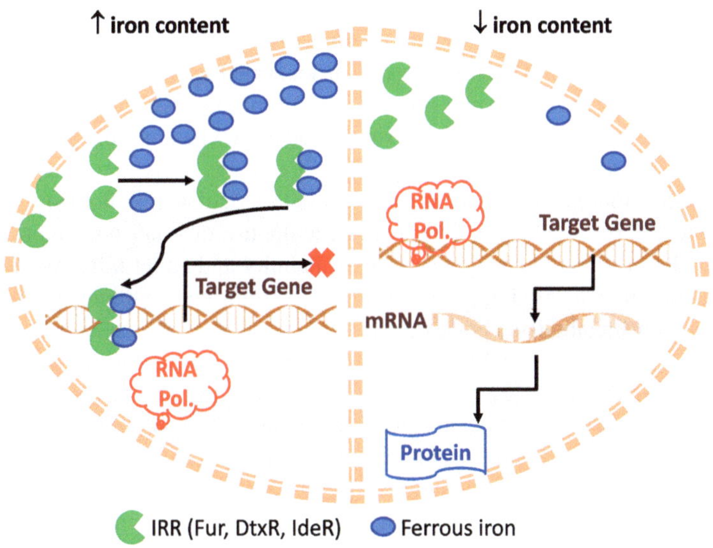

Fig. 11.6 How iron-responsive regulators (IRRs) repress/de-repress target genes. When cellular iron content is high, iron binds to apo-IRR, and monomer holo-IRRs combine to form dimers that then bind to consensus sequences in the promoter region of a target gene. This inhibits RNA polymerases from properly binding to the promoter region. This results in the repression of target gene transcription. In contrast, low cellular iron content keeps IRRs in the apo-IRR state, which allows the promoter region of a target gene to be accessed by RNA polymerases and the subsequent gene transcription

Rodriguez 2014). IRRs (Fur and DtxR or IdeR) contain a metal corepressor binding site in the C-terminus domain and a DNA-binding site in the N-terminus domain (Escolar et al. 1999; White et al. 1998). Ferrous iron binding to the metal corepressor binding site leads to conformational changes that allow the DNA-binding site to engage a consensus sequence within the promoter region of a target gene (De Lorenzo et al. 1987; Boyd et al. 1990; Baichoo and Helmann 2002). Decreased cellular iron content in the pathogen means that no ferrous-IRR (ferrous-Fur or ferrous-DtxR) complexes form and that the expression of iron metabolism- and virulence-related genes is not repressed. In contrast, increased cellular iron content promotes the formation of ferrous-IRR complexes to engage IRR binding sequences in the promoter region of target genes, thereby inhibiting RNA polymerase binding and gene transcription.

Microbial pathogens can also sense heme, which gives them situational awareness. Like iron sensing, heme sensing serves as a cue to the pathogen that it is inside a host and that it needs to assemble its iron acquisition machinery. The sensing of heme in pathogens like *Serratia marcescens*, *S. aureus*, *C. diphtheriae*, and *Bacillus anthracis* among others is mediated by heme-responsive activators (HRAs) (Vanderpool and Armstrong 2003; Bibb et al. 2007; Stauff and Skaar 2009). A well-studied example of bacterial HRAs is the transcriptional regulator hemagglutinin/protease regulator (HapR) in *Vibrio cholerae*.; HapR expression helps transition the pathogen to a virulent state. Heme binds to HapR, leading to conformational changes in the protein that alter its DNA-binding behavior, thereby regulating the expression of various virulence factors (Caigoy et al. 2022). The heme-responsive transcriptional activator (HemAT) has been identified as an HRA in *Shewanella oneidensis*; it regulates the expression of several genes involved in heme biosynthesis and respiration, enabling the bacterium to adapt to changing oxygen and heme levels (Hrusa et al. 2015).

Post-transcriptional mechanisms are also used to regulate iron homeostasis in, at least some, microbes. One such mechanism is the use of small non-coding RNAs (sRNAs) that respond to changes in iron availability and target mRNAs of several iron-related genes to modulate their stability or translation efficiency and hence modulate iron metabolism. For example, *Salmonella enterica* utilizes the sRNA RyhB to post-transcriptionally regulate the expression of some iron acquisition genes (Faner and Feig 2013).

Both iron starvation and heme sensing modulate the expression of several iron acquisition-related genes (siderophores and hemophores among others), which are typically repressed when the pathogen is outside the host or when iron levels are high. In other words, the expression of iron acquisition genes in microbial pathogens is regulated at the pre-transcriptional (repression/de-repression) level. This gene transcription repression/de-repression strategy applies to the expression of the bacterial IsdG and IsdI genes, whose products play an important role in heme degradation and iron extraction in pathogenic bacteria like *S. aureus* (Skaar et al. 2004b). IsdG is a heme oxygenase that catalyzes the oxidative cleavage of heme, which results in the release of iron, carbon monoxide, and biliverdin. IsdI, on the other hand, is a heme-degrading enzyme that breaks down heme into biliverdin and iron without carbon monoxide production (Hammer and Skaar 2011; Torres et al. 2006). Both IsdG and IsdI are part of the iron-regulated surface determinant (Isd) system, which is involved in heme uptake and utilization. Regulation of the expression of IsdG and IsdI is controlled by IRRs like the Fur system. Their expression is also regulated by HRAs such as HapR in *V. cholerae*, which directly interacts with heme and modulates the expression of IsdG and IsdI genes in response to heme availability (Caigoy et al. 2022). In some bacteria, a two-component system comprised of a sensor kinase and a response regulator controls the expression of IsdG and IsdI in response to different environmental cues including iron and heme levels. In *S. aureus,* for example, a two-component system called the staphylococcal respiratory regulator (SrrAB) has been implicated in the regulation of IsdG and IsdI expression in response to oxygen and heme availability (Skaar et al. 2004b; Hammer and Skaar 2011). Besides iron acquisition, several microbial genes that contribute to pathogenicity and virulence are also regulated by IRR-mediated gene repression systems. These include toxin and adhesin genes, genes that promote biofilm formation, and autoinducer genes that respond to perturbations in cell population density (quorum sensing) (Fillat 2014; Carpenter et al. 2009). The positive impact of Fur and DtxR on pathogen growth, survival, and virulence has been documented in several human pathogens, including *S. aureus*, *H. influenzae*, and *M. tuberculosis* (Schmitt and Holmes 1991; Troxell and Hassan 2013).

Pathogenic fungi use multiple mechanisms to acquire iron from the host. For example, *Candida albicans* use the transcription factor suppressor of essential function (Sef1; also known as Cph1) to modulate iron metabolism by regulating the expression of iron acquisition and storage genes (Noble 2013). Orthologs of Sef1, which was first described in *Saccharomyces cerevisiae*, were reported in several fungal species including *C. albicans*. In *C. albicans*, Sef1 regulates the expression of multiple genes involved in iron acquisition including those encoding the high-affinity iron permeases (FTR1 and FTR2), the ferroxidase FET34, and the multi-copper oxidase CIG1, which collectively enhance iron uptake from the surrounding environment. Previous work has demonstrated that Sef1 recognizes and binds to IREs in the promoter region of target genes, which facilitates their coordinated expression (Umeyama et al. 2005). Deletion of Sef1 was shown to impair *C. albicans* growth and virulence, especially under conditions of low iron availability (Umeyama et al. 2005). The expression of Sef1 was reported to be induced by low iron content in the surrounding environment through the activity of the transcription factor Hap43, which binds to the promoter region of Sef1 and activates its transcription (Fourie et al. 2018).

11.4.3 Iron Acquisition Strategies in Microbial Pathogens

Microbial pathogens employ direct and indirect iron acquisition strategies in response to host iron sequestration (Fig. 11.7). The direct strategy involves the acquisition of ferric iron from the surrounding environment using specific receptors and transporters. For example, the ferric-siderophore uptake system uses secreted

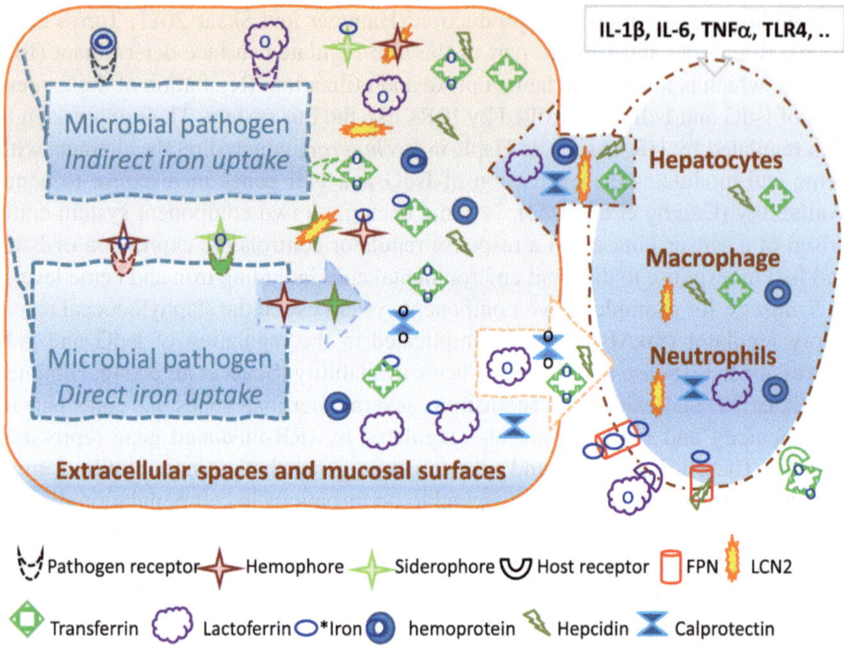

Fig. 11.7 Host and pathogen fight for iron. Host immune cells and hepatocytes produce and release several iron-binding proteins such as transferrin (liver cells, macrophages), lactoferrin, LCN2, and calprotectin (neutrophils and macrophages) into the circulation. These proteins scavenge significant amounts of iron available in the circulation and return via their respective receptors to macrophages and hepatocytes for iron to be internalized and sequestered. Under the influence of proinflammatory cytokines from several immune cells (macrophages, dendritic, etc.), hepatocytes and macrophages release hepcidin, which recirculates to degrade FPN and inhibit iron release. To counter host iron sequestration, microbial pathogens directly acquire iron by expressing receptors for host-released iron-scavenging proteins (transferrin, lactoferrin, and hemoproteins). They also indirectly acquire iron by releasing several iron chelators (siderophores and hemophores), which bind ferric iron, and heme among other iron-containing moieties to be re-captured via siderophore or hemophore receptors on pathogen cell surface. *The blue circle representing iron does not distinguish between free iron and iron bound to heme or other moieties or between ferric and ferrous iron; the figure is only meant to serve as a simplified approximation of host iron sequestration measures and pathogen iron acquisition strategies

siderophores to bind ferric iron for the complex to then be captured by specific cell surface receptors such as the enterobactin receptor FepA and the ferrichrome receptor FhuA on *E. coli*. The heme uptake system involves the capture of heme from host hemoproteins using heme-specific receptors; the *H. influenzae* HxuA hemophore-HxuB receptor is a good example of this strategy. The indirect strategy involves the acquisition of iron from host iron-binding proteins and molecules. *Neisseria gonorrhoeae* uses the TbpA receptor to recognize siderophores that scavenge iron from TF and LTF. Microbes can also acquire iron directly from host heme-containing proteins, such as hemoglobin. Such pathogens express receptors that recognize and bind heme or heme-containing proteins, facilitating their internalization and subsequent release of iron; the Isd system in *S. aureus* is a common example of this strategy.

Siderophores are small, low-molecular-weight (0.5–1.500 KDa), iron-chelating molecules that enable the pathogen to scavenge ferric iron from different hemoproteins and insoluble iron. Currently, more than 250 different siderophores have been described in bacteria and fungi. Siderophores use deprotonated, negatively charged oxygen molecules to capture positively charged ferric iron with high affinity (Hider and Kong 2010). Except for vibriobactin (Vbb) (Li et al. 2012; Allred et al. 2013), siderophores use six oxygen atoms to construct hexadentate octahedral complexes for ferric iron entrapment. The complex then engages specific siderophore receptors for subsequent internalization and esterase hydrolysis (Hider and Kong 2010). Siderophores bind and solubilize iron from various sources and exhibit a diverse range of chemical structures (Miethke and Marahiel 2007). For example, enterobactin is a catecholate siderophore produced by many *Enterobacteriaceae*, including *Escherichia coli* and *Klebsiella pneumoniae* (Hantke 2001). It consists of three 2,3-dihydroxybenzoic acid (DHB) moieties linked by amide bonds. The molecule stably complexes with ferric iron to then be transported into the bacterial cell. Pyoverdine, a fluorescent siderophore produced by *P. aeruginosa*, consists of a peptide backbone with variable side chains. It binds ferric iron, and the complex is then internalized into the bacterial cell via specific receptors. Desferrioxamine B, a hydroxamate siderophore produced by a range of bacteria (*Streptomyces* spp. and *Mycobacterium* spp.) and fungi, contains a hexadentate iron-binding motif. It stably complexes with ferric iron to be internalized via surface receptors (Haas 2014). *V. cholerae* produces vibriobactin as its primary siderophore to facilitate iron acquisition (Sandy and Butler 2009) and *N. gonorrhoeae* expresses TbpA and TbpB receptors to capture iron directly from human transferrin (Schryvers and Stojiljkovic 1999).

Microbial pathogens also secrete hemophores, which are high-affinity heme scavengers that bind heme and facilitate its extraction and transport into the pathogen (Wandersman and Delepelaire 2004; Contreras et al. 2014). For example, *S. marcescens* secretes the marcescens hemophore (HasAsm), which stably complexes with heme to be recognized by its respective surface receptor (HasR) for internalization. Heme uptake from hemopexin (HxuA) hemophore, produced by *H. influenzae*, binds to and extracts heme from hemopexin and haptoglobin and internalizes it via its surface HxuB receptor. PhuS hemophore from *P. aeruginosa* binds heme in hemoglobin and internalizes it via the PhuR receptor (Torres et al. 2006; Wandersman and Delepelaire 2004).

Besides their iron scavenging and chelation potential, siderophores can modulate the activity of iron-sensitive host transcription factors. For example, HIF-1α in the cytosol is inactive in the presence of sufficient O_2 or iron supply. Under normoxic conditions, host-derived prolyl hydroxylases (PHD) hydroxylate specific proline residues in the O_2-dependent degradation domain (ODD) of HIF-1α, thus creating binding sites for the von Hippel-Lindau protein and thereby marking HIF-1α for subsequent proteasomal degradation (Hartmann et al. 2008). This action further deprives the pathogen of iron as the activation of the inhibition of HIF-1α allows for hepcidin to be produced and iron release to be blocked. Inhibition of HIF-1α by PHD is countered by several siderophores, which stabilize HIF-1α and restore its activity in an iron-dependent manner. The benefits the pathogen gains from stabilizing HIF-1α are counterbalanced by the fact that HIF-1α regulates nitric oxide synthase (iNOS), which mediates NO production during the phagocytic oxidative burst. NO also inhibits PHD itself, thus degrading HIF-1α permits PHD to function as a siderophore (Peyssonnaux et al. 2005).

Another example of siderophores interfering with iron-dependent host responses relates to the interaction between enterobactin and myeloperoxidase (MPO), a ferric-dependent heme peroxidase, which uses H_2O_2 to generate HOCl ROS during the oxidative burst in phagocytes. Apo-enterobactin, but not holo-enterobactin, was reported to indirectly disrupt the activity of MPO (Singh et al. 2015), possibly by serving as an enzyme substrate.

With the impending success of the pathogen in establishing an infection following the deployment of its iron acquisition tools such as siderophores and hemophores, innate immunity deploys additional measures to limit or disrupt iron acquisition by the pathogen while enhancing the take-up and killing activities of innate immune cells. Chief among these measures is the production of several antimicrobial peptides and acute-phase reactant proteins that interact with and trap different microbial products. For example, LCN1 and LCN2 from neutrophils, MΦs, and other body cells bind to and sequester apo- or holo-catecholate, and to a lesser extent carboxylate, siderophores. The expression of LCN2 is upregulated in response to inflammation-related toll-like receptor 4 (TLR4)-MyD88-dependent NF-κβ pathway (Flo et al. 2004; Peek et al. 2012). Deficient synthesis of LCN2 has previously been associated with an increased risk of urinary tract infections (UTIs) in humans (Forster et al. 2017) and mice (Liu et al. 2013). Epithelial cells also secrete LCN1 (tear LCN), which has a high binding affinity for lipophilic compounds such as bacterial and fungal catecholate and hydroxamate siderophores (Fluckinger et al. 2004). In relative terms, while the range of siderophores targeted by LCN2 is narrower than that targeted by LCN1, LCN2 has a higher binding affinity for siderophores than LCN1, making it the prime siderophore scavenger in many mammalian systems (Fluckinger et al. 2004). The sensing of microbial siderophores including enterobacter and DFO has been shown to trigger the production of the proinflammatory cytokine IL-8 (Nelson et al. 2007; Choi et al. 2004), which contributes to the recruitment of additional immune cells and mediators to the site of infection.

Limited iron availability in the surrounding environment often triggers a switch from planktonic growth to biofilm formation. Biofilms are structured microbial communities embedded within a self-produced extracellular matrix. Iron sequestered and concentrated within biofilms promotes microbial persistence and resistance to host immune responses and antimicrobial drugs. Several studies have demonstrated that microbial pathogens employ competitive antagonists to resist LCN2-dependent neutralization of siderophores. For example, *K. pneumoniae* releases the enterobactin cleavage product 2,3-dihydroxybenzoylserine (DHBS) as decoys (stealth siderophores) to engage and neutralize LCN2 (Holden et al. 2018).

Microbial pathogens like *S. enteritidis* increase the production of cyclic diguanylate monophosphate (c-di-GMP) to promote biofilm formation and induction of drug resistance (Holden et al. 2018). The logic of this process, vis-à-vis the pathogen, is questionable since c-di-GMP interacts with key danger (DNA) sensors in innate immune cells including the stimulator of interferon genes (STING) and DExD/H-Box Helicase 41 (DDX41), where their activation enhances the production of the antimicrobial type 1 interferons. This potential disadvantage is countervailed by the fact that c-di-GMP can engage LCN2 and possibly block its binding to catecholate siderophores (Li et al. 2015).

Some bacterial species circumvent inflammation-induced iron starvation by growing inside iron-rich compartments like endosomes and phagosomes of iron-scavenging cells; *M. tuberculosis* growing in MΦs is an ideal example of this case. Therefore, upregulated expression of proteins that channel ferrous iron from endosomal compartments to the cytoplasm including Nramp1 (SCLA11) and Nramp2/DMT-1 on late phagosomes promote resistance to intracellular pathogens (Vidal et al. 1995; Jabado et al. 2003). Moreover, pathogens like *Borrelia burgdorferi* rely on manganese (Mn) with multiple oxidation states rather than iron as a cofactor in many of its functional proteins (Kurthkoti et al. 2017).

11.5 Therapeutic Implications

Growing appreciation of the role of iron in infection and immunity has provided valuable insights into new and innovative approaches to manage and treat microbial infections. The list of experimental modalities and potential therapeutic strategies is diverse and still expanding. Given that pathogens need iron for numerous metabolic reactions, enhancing the ability of the host to sequester iron has served as a broad strategy in developing various modalities to help manage microbial infections. For example, the use of modulators that enhance the inflammatory response against pathogens was reported to improve the iron sequestration potential of the host (Nairz et al. 2010). Modulation of iron metabolism by iron regulators like hepcidin has proven useful in limiting microbial growth in animal models (Pigeon et al. 2001). Iron-selective antibacterial agents such as salicylhydrazones and pyrazolones, which selectively extract iron from microbial iron acquisition molecules, were shown to disrupt iron homeostasis and limit bacterial growth (Saha et al. 2013). Iron chelation via deferoxamine and deferiprone has long been shown to exert antimicrobial activity against numerous pathogenic bacterial and fungal species (Saha et al. 2013). The fact that iron availability is a requirement for several antibiotics like beta-lactams, fluoroquinolones, and aminoglycosides to be effective antibiotics helps explain why using iron chelation as an adjunctive therapy significantly enhances the antimicrobial activity of antibiotics both in the case of antibiotic-susceptible and antibiotic-resistant bacteria (Marchetti et al. 2020; Kohanski et al. 2007).

The development of bacterial resistance to antibiotics is a never-ending public health concern. In this respect, a class of experimental agents called sideromycins was introduced over the last two decades to help mitigate this problem. A sideromycin is a conjugate of a siderophore plus a toxin or an antibiotic compound (Chu et al. 2010). The siderophore in this case acts as a vehicle, and the toxin or the antibiotic is the payload to be delivered to target cells. Sideromycins are designed to selectively target and inhibit the growth or kill microbial pathogens expressing the receptor(s) for the siderophore used in the conjugate. This strategy promotes significant improvement over conventional antibiotic therapy as it is reasonably targeted and specific. Previous studies have suggested that sideromycins circumvent antibiotic resistance mechanisms (Kramer et al. 2020) and may help manage and treat antibiotic-resistant infections (Negash et al. 2019). A handful of sideromycins have been through clinical testing including cefiderocol (Zhanel et al. 2019; Mey et al. 2005). The drug is currently used to treat complicated urinary tract and kidney infections; it has demonstrated significant activity against a broad spectrum of Gram-negative multidrug-resistant pathogens including P. aeruginosa (Portsmouth et al. 2018). Its precise mechanism of action involves the binding of ferric iron through the siderophore component and then the delivery of the antibiotic and the siderophore into the bacterial cell cytoplasm via iron transporters (Portsmouth et al. 2018; Ito et al. 2016). Cefiderocol inside the target bacterial cells binds to penicillin-binding protein and blocks cell wall synthesis (Ito et al. 2016).

The iron-chelating properties of siderophores have also inspired the development of several siderophore mimetics that limit iron availability to pathogens. One such memetic is the siderophore analog 2,3-dihydroxybenzoyl isonicotinoyl isoleucine (DIBI), which is a synthetic compound designed as a memetic of the naturally occurring siderophore enterobactin (del Parquet et al. 2018). DIBI has demonstrated significant antimicrobial activity against various bacterial pathogens (del Parquet et al. 2018; Fischbach et al. 2006). The synthetic iron chelator 2,2-dipyridylamine (DPA) has also shown significant antimicrobial activity when used as an approximate functional memetic of lipocalins (Ferreira et al. 2016). Synthetic hemophore-mimetic peptides have also been developed and tested against different microbial pathogens. For example, the P4 peptide, as an effective scavenger of heme, limits bacterial growth (Wencewicz et al. 2009).

The specificity of siderophores has also provided for the design of potential drugs that target specific siderophores/siderophore receptors as a means of selectively depleting iron. For example, the metallic element gallium, a competitive antagonist to siderophore iron, does not directly kill bacterial cells but exploits their iron metabolism to inhibit their growth (Li et al. 2022; Pereira et al. 2024). Since the ionic radius of gallium is similar to that of ferric iron, it can be picked up by bacterial cells through iron transport systems (Bonchi et al. 2014). Internalized gallium has been shown to disrupt iron-dependent reactions and metabolic processes and thereby inhibit cell growth and induce cell death (Qiao et al. 2022). Several studies have also shown that the activity of gallium is wide-ranging; both Gram-positive and Gram-negative species including antibiotic-resistant microbial strains were shown to be susceptible to gallium (Li et al. 2022; Baldoni et al. 2010). The drug precipitates significant antibacterial effects in *P. aeruginosa*, *Acinetobacter baumannii*, and *S. aureus* (Goss et al. 2018; Antunes et al. 2011). Gallium was also shown to be of potential value in managing some difficult-to-treat biofilm-forming bacteria (Valappil et al. 2008).

11.6 Concluding Remarks

The fight for iron between mammalian hosts and microbial pathogens is complex and dynamic. One important aspect of the dynamism of this process is that it has and will continue to drive reciprocal evolutionary processes to improve the host's ability to sequester iron and the pathogen's ability to acquire iron. It is plausible therefore that the fight for iron between host and pathogen is likely to become more intense and acrimonious. In this context, it is worth noting that differences in the time scale of evolution in unicellular bacteria versus complex multicellular mammals are massive. Therefore, any evolutionary changes in the iron acquisition machinery in bacteria are likely to outpace any reciprocal evolutionary changes in the iron sequestration machinery in mammalian hosts. Over time, therefore, the pathogen may gain the upper hand in the fight for iron, at least transiently. The consequences of these evolutionary processes have provided high levels of diversity in the host's iron sequestration machinery and the pathogen's iron acquisition machinery. This diversity has provided a treasure trove of theoretical possibilities for the development of antibacterial agents that focus on enhancing the host's iron sequestration capacity and/or blocking the pathogen's iron acquisition capacity. While some of these potential therapies have shown some promise as antibacterial agents, more work is still needed to fully address their efficacy and safety either as adjunctive or solo therapies.

References

Abboud S, Haile DJ (2000) A novel mammalian iron-regulated protein involved in intracellular iron metabolism. J Biol Chem [Internet] 275(26):19906–19912. [cited 2024 Jul 19]. Available from: https://www.jbc.org/article/S0021-9258(19)80067-4/fulltext

Akinloye O, Adebayo TO, Oguntibeju OO, Oparinde DP, Ogunyemi EO (2011) Effects of contraceptives on serum trace elements, calcium and phosphorus levels. West Indian Med J [Internet]:308–315. [cited 2024 Jul 19]. Available from: https://pesquisa.bvsalud.org/gim/resource/ru/lil-672773

Allred BE, Correnti C, Clifton MC, Strong RK, Raymond KN (2013) Siderocalin outwits the coordination chemistry of vibriobactin, a siderophore of Vibrio cholerae. ACS Chem Biol [Internet] 8(9):1882–1887. [cited 2024 Jul 20]. Available from: https://pubs.acs.org/doi/10.1021/cb4002552

Amit I, Winter DR, Jung S (2016) The role of the local environment and epigenetics in shaping macrophage identity and their effect on tissue homeostasis. Nat Immunol [Internet] 17(1):18–25. [cited 2024 Jul 20]. Available from: https://www.nature.com/articles/ni.3325

An P, Wang H, Wu Q, Wang J, Xia Z, He X, Wang X, Chen Y, Min J, Wang F (2018) Smad7 deficiency decreases iron and haemoglobin through hepcidin up-regulation by multilayer compensatory mechanisms. J Cell Mol Med [Internet] 22(6):3035–3044. Jun [cited 2024 Jul 19]. Available from: https://onlinelibrary.wiley.com/doi/10.1111/jcmm.13546

Anderson GJ, Wang F (2012) Essential but toxic: controlling the flux of iron in the body. Clin Exp Pharmacol Physiol [Internet] 39(8):719–724. [cited 2024 Jul 18]. Available from: https://onlinelibrary.wiley.com/doi/10.1111/j.1440-1681.2011.05661.x

Andreini C, Bertini I, Cavallaro G, Holliday GL, Thornton JM (2008) Metal ions in biological catalysis: from enzyme databases to general principles. J Biol Inorg Chem [Internet] 13(8):1205–1218. [cited 2024 Jul 18]. Available from: http://link.springer.com/10.1007/s00775-008-0404-5

Andriopoulos B Jr, Corradini E, Xia Y, Faasse SA, Chen S, Grgurevic L, Knutson MD, Pietrangelo A, Vukicevic S, Lin HY, Babitt JL (2009) BMP6 is a key endogenous regulator of hepcidin expression and iron metabolism. Nat Genet [Internet] 41(4):482–487. [cited 2024 Jul 19]. Available from: https://www.nature.com/articles/ng.335

Antunes LCM, Han J, Ferreira RBR, Lolić P, Borchers CH, Finlay BB (2011) Effect of antibiotic treatment on the intestinal metabolome. Antimicrob Agents Chemother [Internet] 55(4):1494–503. [cited 2024 Jul 20]. Available from: https://journals.asm.org/doi/10.1128/AAC.01664-10

Babitt JL, Huang FW, Wrighting DM, Xia Y, Sidis Y, Samad TA, Campagna JA, Chung RT, Schneyer AL, Woolf CJ, Andrews NC, Lin HY (2006) Bone morphogenetic protein signaling by hemojuvelin regulates hepcidin expression. Nat Genet [Internet] 38(5):531–539. [cited 2024 Jul 19]. Available from: https://www.nature.com/articles/ng1777

Baichoo N, Helmann JD (2002) Recognition of DNA by fur: a reinterpretation of the fur box consensus sequence. J Bacteriol [Internet] 184(21):5826–5832. [cited 2024 Jul 20]. Available from: https://journals.asm.org/doi/10.1128/JB.184.21.5826-5832.2002

Bajbouj K, Shafarin J, Abdalla MY, Ahmad IM, Hamad M (2017) Estrogen-induced disruption of intracellular iron metabolism leads to oxidative stress, membrane damage, and cell cycle arrest in MCF-7 cells. Tumour Biol 39(10):1010428317726184

Bajbouj K, Shafarin J, Allam H, Madkour M, Awadallah S, El-Serafy A, Sandeep D, Hamad M (2018) Elevated levels of estrogen suppress hepcidin synthesis and enhance serum iron availability in premenopausal women. Exp Clin Endocrinol Diabetes [Internet] 126(07):453–459. [cited 2024 Jul 19]. Available from: http://www.thieme-connect.de/DOI/DOI?10.1055/s-0043-124077

Baldoni D, Steinhuber A, Zimmerli W, Trampuz A (2010) In vitro activity of gallium maltolate against Staphylococci in logarithmic, stationary, and biofilm growth phases: comparison of conventional and calorimetric susceptibility testing methods. Antimicrob Agents Chemother [Internet] 54(1):157–163. [cited 2024 Jul 20]. Available from: https://journals.asm.org/doi/10.1128/AAC.00700-09

Baumgart M, Bott M (2011) Biochemical characterisation of aconitase from Corynebacterium glutamicum. J Biotechnol [Internet] 154(2–3):163–170. [cited 2024 Jul 19]. Available from: https://www.sciencedirect.com/science/article/pii/S0168165610003019

Beinert H, Holm RH, Münck E (1997) Iron-sulfur clusters: nature's modular, multipurpose structures. Science [Internet] 277(5326):653–659. [cited 2024 Jul 18]. Available from: https://www.science.org/doi/10.1126/science.277.5326.653

Bibb LA, Kunkle CA, Schmitt MP (2007) The ChrA-ChrS and HrrA-HrrS signal transduction systems are required for activation of the *hmuO* promoter and repression of the *hemA* promoter in *Corynebacterium diphtheriae*. Infect Immun [Internet] 75(5):2421–2431. [cited 2024 Jul 20]. Available from: https://journals.asm.org/doi/10.1128/IAI.01821-06

Bonchi C, Imperi F, Minandri F, Visca P, Frangipani E (2014) Repurposing of gallium-based drugs for antibacterial therapy. BioFactors [Internet] 40(3):303–312. [cited 2024 Jul 20]. Available from: https://iubmb.onlinelibrary.wiley.com/doi/10.1002/biof.1159

Borysiewicz LK, Graham S, Sissons JGP (1986) Human natural killer cell lysis of virus-infected cells. Relationship to expression of the transferrin receptor. Eur J Immunol [Internet] 16(4):405–411. [cited 2024 Jul 20]. Available from: https://onlinelibrary.wiley.com/doi/10.1002/eji.1830160416

Boyd J, Oza MN, Murphy JR (1990) Molecular cloning and DNA sequence analysis of a diphtheria tox iron-dependent regulatory element (dtxR) from Corynebacterium diphtheriae. Proc Natl Acad Sci USA [Internet] 87(15):5968–5972. [cited 2024 Jul 20]. Available from: https://pnas.org/doi/full/10.1073/pnas.87.15.5968

Caigoy JC, Shimamoto T, Mukhopadhyay AK, Shinoda S, Shimamoto T (2022) Sequence polymorphisms in *Vibrio cholerae* HapR affect biofilm formation under aerobic and anaerobic conditions. Alexandre G, editor. Appl Environ Microbiol [Internet] 88(17):e01044-22. [cited 2024 Jul 20]. Available from: https://journals.asm.org/doi/10.1128/aem.01044-22

Cairo G, Recalcati S, Mantovani A, Locati M (2011) Iron trafficking and metabolism in macrophages: contribution to the polarized phenotype. Trends Immunol [Internet] 32(6):241–247. [cited 2024 Jul 20]. Available from: https://www.cell.com/trends/immunology/fulltext/S1471-4906(11)00050-0

Camiolo G, Barbato A, Giallongo C, Vicario N, Romano A, Parrinello NL, Parenti R, Sandoval JC, García-Moreno D, Lazzarino G, Avola R, Palumbo GA, Mulero V, Li Volti G, Tibullo D, Di Raimondo F (2020) Iron regulates myeloma cell/macrophage interaction and drives resistance to bortezomib. Redox Biol [Internet] 36:101611. [cited 2024 Jul 20]. Available from: https://www.sciencedirect.com/science/article/pii/S2213231720308168

Canali S, Core AB, Zumbrennen-Bullough KB, Merkulova M, Wang CY, Schneyer AL, Pietrangelo A, Babitt JL (2016) Activin B induces noncanonical SMAD1/5/8 signaling via BMP type I receptors in hepatocytes: evidence for a role in hepcidin induction by inflammation in male mice. Endocrinology [Internet] 157(3):1146–1162. [cited 2024 Jul 19]. Available from: https://academic.oup.com/endo/article-abstract/157/3/1146/2422727

Canali S, Zumbrennen-Bullough KB, Core AB, Wang CY, Nairz M, Bouley R, Swirski FK, Babitt JL (2017a) Endothelial cells produce bone morphogenetic protein 6 required for iron homeostasis in mice. Blood J Am Soc Hematol [Internet] 129(4):405–414. [cited 2024 Jul 19]. Available from: https://ashpublications.org/blood/article-abstract/129/4/405/36190

Canali S, Wang C, Zumbrennen-Bullough KB, Bayer A, Babitt JL (2017b) Bone morphogenetic protein 2 controls iron homeostasis in mice independent of Bmp6. Am J Hematol [Internet] 92(11):1204–1213. [cited 2024 Jul 19]. Available from: https://onlinelibrary.wiley.com/doi/10.1002/ajh.24888

Cardoso EM, Macedo MG, Rohrlich P, Ribeiro E, Silva MT, Lemonnier FA, de Sousa M (2002) Increased hepatic iron in mice lacking classical MHC class I molecules. Blood J Am Soc Hematol [Internet] 100(12):4239–4241. [cited 2024 Jul 20]. Available from: https://ashpublications.org/blood/article-abstract/100/12/4239/88309

Carpenter BM, Whitmire JM, Merrell DS (2009) This is not your mother's repressor: the complex role of fur in pathogenesis. Infect Immun [Internet] 77(7):2590–2601. [cited 2024 Jul 20]. Available from: https://journals.asm.org/doi/10.1128/IAI.00116-09

Casanovas G, Spasić MV, Casu C, Rivella S, Strelau J, Unsicker K, Muckenthaler MU (2013) The murine growth differentiation factor 15 is not essential for systemic iron homeostasis in phlebotomized mice. Haematologica [Internet] 98(3):444. [cited 2024 Jul 19]. Available from: https://www.ncbi.nlm.nih.gov/pmc/articles/PMC3659953/

Choi EY, Kim EC, Oh HM, Kim S, Lee HJ, Cho EY, Yoon KH, Kim EA, Han WC, Choi SC, Hwang JY, Park C, Oh BS, Kim Y, Kimm KC, Park KI, Chung HT, Jun CD (2004) Iron chelator triggers inflammatory signals in human intestinal epithelial cells: involvement of p38 and extracellular signal-regulated kinase signaling pathways. J Immunol [Internet] 172(11):7069–7077. [cited 2024 Jul 20]. Available from: https://journals.aai.org/jimmunol/article/172/11/7069/80950

Chu BC, Garcia-Herrero A, Johanson TH, Krewulak KD, Lau CK, Peacock RS, Slavinskaya Z, Vogel HJ (2010) Siderophore uptake in bacteria and the battle for iron with the host; a bird's eye view. Biometals [Internet] 23(4):601–611. [cited 2024 Jul 20]. Available from: http://link.springer.com/10.1007/s10534-010-9361-x

Chua AC, Olynyk JK, Leedman PJ, Trinder D (2004) Nontransferrin-bound iron uptake by hepatocytes is increased in the Hfe knockout mouse model of hereditary hemochromatosis. Blood [Internet] 104(5):1519–1525. [cited 2024 Jul 18]. Available from: https://ashpublications.org/blood/article-abstract/104/5/1519/18924

Contreras H, Chim N, Credali A, Goulding CW (2014) Heme uptake in bacterial pathogens. Curr Opin Chem Biol [Internet] 19:34–41. [cited 2024 Jul 20]. Available from: https://www.sciencedirect.com/science/article/pii/S1367593113002391?casa_token=RAc8xkyJyk8AAAAA:yK38IebgJXZI-K0aIDhG9rOSbkXy-6RIDnS5xjNcYX1pRG8mwOIacm31cbN3-dnXCacGwrtXH7Y

Daher R, Lefebvre T, Puy H, Karim Z (2019) Extrahepatic hepcidin production: the intriguing outcomes of recent years. World J Clin Cases [Internet] 7(15):1926. [cited 2024 Jul 19]. Available from: https://www.ncbi.nlm.nih.gov/pmc/articles/PMC6695539/

De Lorenzo V, Wee S, Herrero M, Neilands J (1987) Operator sequences of the aerobactin operon of plasmid ColV-K30 binding the ferric uptake regulation (fur) repressor. J Bacteriol [Internet] 169(6):2624–2630. [cited 2024 Jul 20]. Available from: https://journals.asm.org/doi/abs/10.1128/jb.169.6.2624-2630.1987

del Parquet MC, Savage KA, Allan DS, Davidson RJ, Holbein BE (2018) Novel iron-chelator DIBI inhibits Staphylococcus aureus growth, suppresses experimental MRSA infection in mice and enhances the activities of diverse antibiotics in vitro. Front Microbiol [Internet] 9:1811. [cited 2024 Jul 20]. Available from: https://www.frontiersin.org/articles/10.3389/fmicb.2018.01811/full

DeRosa A, Leftin A (2021) The iron curtain: macrophages at the interface of systemic and microenvironmental iron metabolism and immune response in cancer. Front Immunol [Internet] 12:614294. [cited 2024 Jul 20]. Available from: https://www.frontiersin.org/articles/10.3389/fimmu.2021.614294/full

Doherty DG, O'Farrelly C (2000) Innate and adaptive lymphoid cells in the human liver. Immunol Rev [Internet] 174:5–20. [cited 2024 Jul 20]. Available from: https://europepmc.org/article/med/10807503

Donovan A, Brownlie A, Zhou YI, Shepard J, Pratt SJ, Moynihan J, Paw BH, Drejer A, Barut B, Zapata A, Law TC, Brugnara C, Lux SE, Pinkus GS, Pinkus JL, Kingsley PD, Palis J, Fleming MD, Andrews NC, Zon LI (2000) Positional cloning of zebrafish ferroportin1 identifies a conserved vertebrate iron exporter. Nature [Internet] 403(6771):776–781. [cited 2024 Jul 19]. Available from: https://www.nature.com/articles/35001596

Drakesmith H, Prentice AM (2012) Hepcidin and the iron-infection axis. Science [Internet] 338(6108):768–772. [cited 2024 Jul 19]. Available from: https://www.science.org/doi/10.1126/science.1224577

Escolar L, Pérez-Martín J, De Lorenzo V (1999) Opening the iron box: transcriptional metalloregulation by the fur protein. J Bacteriol [Internet] 181(20):6223–6229. [cited 2024 Jul 20]. Available from: https://journals.asm.org/doi/10.1128/JB.181.20.6223-6229.1999

Faner MA, Feig AL (2013) Identifying and characterizing Hfq–RNA interactions. Methods [Internet] 63(2):144–159. [cited 2024 Jul 20]. Available from: https://www.sciencedirect.com/science/article/pii/S1046202313001400?casa_token=rvQxBoHpWC8AAAAA:tsL2sSW3h5XijTM9cw1V7Dmg7rn9Toj8ZqdAN1vg_prqWmo8emhKaEIBSRFw9sE2fKMrQT8sj24

Feng P, Yang Q, Luo L, Sun Y, Lv W, Wan S, Guan Z, Xiao Z, Liu F, Li Z, Dong Z, Yang M (2021) The kinase PDK1 regulates regulatory T cell survival via controlling redox homeostasis. Theranostics [Internet] 11(19):9503. [cited 2024 Jul 20]. Available from: https://www.ncbi.nlm.nih.gov/pmc/articles/PMC8490516/

Ferreira D, Seca AM, Diana CGA, Silva AM (2016) Targeting human pathogenic bacteria by siderophores: a proteomics review. J Proteomics [Internet] 145:153–166. [cited 2024 Jul 20]. Available from: https://www.sciencedirect.com/science/article/pii/S1874391916301282?casa_token=SxN9WG1B7-AAAAAA:j5kxo4MZ7aDhQ85wOXvoeXn7YNPic-zhtf-hltcGSJ8HdvjYlx3LGsasuR9k1ym6VW0gV0OgLw0

Fillat MF (2014) The FUR (ferric uptake regulator) superfamily: diversity and versatility of key transcriptional regulators. Arch Biochem Biophys [Internet] 546:41–52. [cited 2024 Jul 20]. Available from: https://www.sciencedirect.com/science/article/pii/S0003986114000563?casa_token=QAhtgldII10AAAAA:XSZgc1hTJ_41xwvMPmtTBgBmVHeSRrRulSIxwCAFNPhgD25gq6yXCwUQXKv1JDdMe4iwSnr1gPw

Fillet G, Beguin Y, Baldelli L (1989) Model of reticuloendothelial iron metabolism in humans: abnormal behavior in idiopathic hemochromatosis and in inflammation. [cited 2024 Jul 18]. Available from: https://ashpublications.org/blood/article-abstract/74/2/844/167195

Fischbach MA, Lin H, Zhou L, Yu Y, Abergel RJ, Liu DR, Raymond KN, Wanner BL, Strong RK, Walsh CT, Aderem A, Smith KD (2006) The pathogen-associated *iroA* gene cluster mediates bacterial evasion of lipocalin 2. Proc Natl Acad Sci USA [Internet] 103(44):16502–16507. [cited 2024 Jul 20]. Available from: https://pnas.org/doi/full/10.1073/pnas.0604636103

Fleming MD, Trenor CC III, Su MA, Foernzler D, Beier DR, Dietrich WF, Andrews NC (1997) Microcytic anaemia mice have a mutation in Nramp2, a candidate iron transporter gene. Nat Genet [Internet] 16(4):383–386. [cited 2024 Jul 18]. Available from: https://www.nature.com/articles/ng0897-383

Fleming MD, Romano MA, Su MA, Garrick LM, Garrick MD, Andrews NC (1998) *Nramp 2* is mutated in the anemic Belgrade (*b*) rat: evidence of a role for Nramp2 in endosomal iron transport. Proc Natl Acad Sci USA [Internet] 95(3):1148–1153. [cited 2024 Jul 18]. Available from: https://pnas.org/doi/full/10.1073/pnas.95.3.1148

Flo TH, Smith KD, Sato S, Rodriguez DJ, Holmes MA, Strong RK, Akira S, Aderem A (2004) Lipocalin 2 mediates an innate immune response to bacterial infection by sequestrating iron. Nature [Internet] 432(7019):917–921. [cited 2024 Jul 20]. Available from: https://www.nature.com/articles/nature03104

Fluckinger M, Haas H, Merschak P, Glasgow BJ, Redl B (2004) Human tear lipocalin exhibits antimicrobial activity by scavenging microbial siderophores. Antimicrob Agents Chemother [Internet] 48(9):3367–3372. [cited 2024 Jul 20]. Available from: https://journals.asm.org/doi/10.1128/AAC.48.9.3367-3372.2004

Forster CS, Johnson K, Patel V, Wax R, Rodig N, Barasch J, Bachur R, Lee RS (2017) Urinary NGAL deficiency in recurrent urinary tract infections. Pediatr Nephrol [Internet] 32(6):1077–1080. [cited 2024 Jul 20]. Available from: http://link.springer.com/10.1007/s00467-017-3607-6

Fourie R, Kuloyo OO, Mochochoko BM, Albertyn J, Pohl CH (2018) Iron at the centre of Candida albicans interactions. Front Cell Infect Microbiol [Internet] 8:185. [cited 2024 Jul 20]. Available from: https://www.frontiersin.org/articles/10.3389/fcimb.2018.00185/full

Frost JN, Tan TK, Abbas M, Wideman SK, Bonadonna M, Stoffel NU, Wray K, Kronsteiner B, Smits G, Campagna DR, Duarte TL, Lopes JM, Shah A, Armitage AE, Arezes J, Lim PJ, Preston AE, Ahern D, Teh M, Naylor C, Salio M, Gileadi U, Andrews SC, Dunachie SJ, Zimmermann MB, van der Klis FRM, Cerundolo V, Bannard O, Draper SJ, Townsend ARM, Galy B, Fleming MD, Lewis MC, Drakesmith H (2021) Hepcidin-mediated hypoferremia disrupts immune responses to vaccination and infection. Med [Internet] 2(2):164–179. [cited 2024 Jul 20]. Available from: https://www.cell.com/med/fulltext/S2666-6340(20)30021-0

Ganz T (2006) Hepcidin—a peptide hormone at the interface of innate immunity and iron metabolism. In: Antimicrobial peptides and human disease [Internet], pp 183–198. [cited 2024 Jul 18]. Available from: https://link.springer.com/chapter/10.1007/3-540-29916-5_7

Ganz T, Nemeth E (2011) Hepcidin and disorders of iron metabolism. Annu Rev Med [Internet] 62(1):347–360. [cited 2024 Jul 20]. Available from: https://www.annualreviews.org/doi/10.1146/annurev-med-050109-142444

Ganz T, Nemeth E (2015) Iron homeostasis in host defence and inflammation. Nat Rev Immunol 15(8):500–510

Goss CH, Kaneko Y, Khuu L, Anderson GD, Ravishankar S, Aitken ML, Lechtzin N, Zhou G, Czyz DM, McLean K, Olakanmi O, Shuman HA, Teresi M, Wilhelm E, Caldwell E, Salipante SJ, Hornick DB, Siehnel RJ, Becker L, Britigan BE, Singh PK (2018) Gallium disrupts bacterial iron metabolism and has therapeutic effects in mice and humans with lung infections. Sci Transl Med [Internet] 10(460):eaat7520. [cited 2024 Jul 20]. Available from: https://www.science.org/doi/10.1126/scitranslmed.aat7520

Gromadzka G, Tarnacka B, Flaga A, Adamczyk A (2020) Copper dyshomeostasis in neurodegenerative diseases—therapeutic implications. Int J Mol Sci [Internet] 21(23):9259. [cited 2024 Jul 19]. Available from: https://www.mdpi.com/1422-0067/21/23/9259

Gruss A, Borezée-Durant E, Lechardeur D (2012) Environmental heme utilization by heme-auxotrophic bacteria. Adv Microb Physiol [Internet] 61:69–124. [cited 2024 Jul 20]. Available from: https://www.sciencedirect.com/science/article/pii/B9780123944238000032

Gunshin H, Mackenzie B, Berger UV, Gunshin Y, Romero MF, Boron WF, Nussberger S, Gollan JL, Hediger MA (1997) Cloning and characterization of a mammalian proton-coupled metal-ion transporter. Nature [Internet] 388(6641):482–488. [cited 2024 Jul 18]. Available from: https://www.nature.com/articles/41343

Haas H (2014) Fungal siderophore metabolism with a focus on aspergillus fumigatus. Nat Prod Rep [Internet] 31(10):1266–1276. [cited 2024 Jul 20]. Available from: https://pubs.rsc.org/en/content/articlehtml/2014/np/c4np00071d

Hammer ND, Skaar EP (2011) Molecular mechanisms of *Staphylococcus aureus* iron acquisition. Annu Rev Microbiol [Internet] 65(1):129–147. [cited 2024 Jul 20]. Available from: https://www.annualreviews.org/doi/10.1146/annurev-micro-090110-102851

Hantke K (2001) Iron and metal regulation in bacteria. Curr Opin Microbiol [Internet] 4(2):172–177. [cited 2024 Jul 20]. Available from: https://www.sciencedirect.com/science/article/pii/S1369527400001843?casa_token=PsG_v_fVLbQAAAAA:JTblLng_H2YfUqqzOjPlLMJjNqeKjpHaksva9aWIcvYjnRnnDk2VbFR7Ofy_aGd4Yh7PspOw6v8

Harrison-Findik DD (2010) Gender-related variations in iron metabolism and liver diseases. World J Hepatol [Internet] 2(8):302. [cited 2024 Jul 19]. Available from: https://www.ncbi.nlm.nih.gov/pmc/articles/PMC2999297/

Hartmann H, Eltzschig HK, Wurz H, Hantke K, Rakin A, Yazdi AS, Matteoli G, Bohn E, Autenrieth IB, Karhausen J, Neumann D, Colgan SP, Kempf VA (2008) Hypoxia-independent activation of HIF-1 by enterobacteriaceae and their siderophores. Gastroenterology [Internet] 134(3):756–767. [cited 2024 Jul 20]. Available from: https://www.sciencedirect.com/science/article/pii/S0016508507021695?casa_token=V72ugZejECUAAAAA:qHZOR10o4XVJPbmgRwCRnTrZZnzXZCXJQxvGVMzR2557Iv0zVWAOaCEAt_sEExhPGHDZEigcR_M

Hentze MW, Muckenthaler MU, Galy B, Camaschella C (2010) Two to tango: regulation of mammalian iron metabolism. Cell 142(1):24–38

Hider RC, Kong X (2010) Chemistry and biology of siderophores. Nat Prod Rep [Internet] 27(5):637–657. [cited 2024 Jul 20]. Available from: https://pubs.rsc.org/en/content/articlehtml/2010/np/b906679a

Holden VI, Wright MS, Houle S, Collingwood A, Dozois CM, Adams MD, Bachman MA (2018) Iron acquisition and siderophore release by carbapenem-resistant sequence type 258 Klebsiella pneumoniae. Castanheira M, editor. mSphere [Internet] 3(2):e00125-18. [cited 2024 Jul 20]. Available from: https://journals.asm.org/doi/10.1128/mSphere.00125-18

Horiguchi H, Oguma E, Kayama F (2005) The effects of iron deficiency on estradiol-induced suppression of erythropoietin induction in rats: implications of pregnancy-related anemia. Blood [Internet] 106(1):67–74. [cited 2024 Jul 19]. Available from: https://ashpublications.org/blood/article-abstract/106/1/67/103193

Hou Y, Zhang S, Wang L, Li J, Qu G, He J, Rong H, Ji H, Liu S (2012) Estrogen regulates iron homeostasis through governing hepatic hepcidin expression via an estrogen response element. Gene [Internet] 511(2):398–403. [cited 2024 Jul 20]. Available from: https://www.sciencedirect.com/science/article/pii/S0378111912011699?casa_token=mbqngUGVTFAAAAAA:ounrngMp9KrIKa4T68pPDWOoKR7M%2D%2DrSI0pw-nkMEMsWLV21Cl9Ugt7IO706S2yzhnjGuD93pSY

Hrusa G, Farmer W, Weiss BL, Applebaum T, Roma JS, Szeto L, Aksoy S, Runyen-Janecky LJ (2015) TonB-dependent heme iron acquisition in the tsetse fly symbiont sodalis glossinidius. Goodrich-Blair H, editor. Appl Environ Microbiol [Internet] 81(8):2900–2909. [cited 2024 Jul 20]. Available from: https://journals.asm.org/doi/10.1128/AEM.04166-14

Huang X, Xu Y, Partridge NC (2013) Dancing with sex hormones, could iron contribute to the gender difference in osteoporosis? Bone [Internet], Elsevier 55:458–460. [cited 2024 Jul 19]. Available from: https://www.sciencedirect.com/science/article/pii/S8756328213001117

Ito A, Nishikawa T, Matsumoto S, Yoshizawa H, Sato T, Nakamura R, Tsuji M, Yamano Y (2016) Siderophore cephalosporin cefiderocol utilizes ferric iron transporter systems for antibacterial activity against Pseudomonas aeruginosa. Antimicrob Agents Chemother [Internet] 60(12):7396–7401. [cited 2024 Jul 20]. Available from: https://journals.asm.org/doi/10.1128/AAC.01405-16

Jabado N, Cuellar-Mata P, Grinstein S, Gros P (2003) Iron chelators modulate the fusogenic properties of Salmonella-containing phagosomes. Proc Natl Acad Sci [Internet] 100(10):6127–6132. [cited 2024 Jul 20]. Available from: https://www.pnas.org/doi/abs/10.1073/pnas.0937287100

Jabara HH, Boyden SE, Chou J, Ramesh N, Massaad MJ, Benson H, Bainter W, Fraulino D, Rahimov F, Sieff C, Liu ZJ, Alshemmari SH, Al-Ramadi BK, Al-Dhekri H, Arnaout R, Abu-Shukair M, Vatsayan A, Silver E, Ahuja S, Davies EG, Sola-Visner M, Ohsumi TK, Andrews NC, Notarangelo LD, Fleming MD, Al-Herz W, Kunkel LM, Geha RS (2016) A missense mutation in TFRC, encoding transferrin receptor 1, causes combined immunodeficiency. Nat Genet [Internet] 48(1):74–78. [cited 2024 Jul 20]. Available from: https://www.nature.com/articles/ng.3465

Jang YS, Seo GY, Lee JM, Seo HY, Han HJ, Kim SJ, Jin BR, Kim HJ, Park SR, Rhee KJ, Kim WS, Kim PH (2015) Lactoferrin causes IgA and IgG2b isotype switching through beta-glycan binding and activation of canonical TGF-β signaling. Mucosal Immunol [Internet] 8(4):906–917. [cited 2024 Jul 20]. Available from: https://www.sciencedirect.com/science/article/pii/S1933021922009746

Jenkitkasemwong S, Wang CY, Coffey R, Zhang W, Chan A, Biel T, Kim JS, Hojyo S, Fukada T, Knutson MD (2015) SLC39A14 is required for the development of hepatocellular iron overload in murine models of hereditary hemochromatosis. Cell Metab [Internet] 22(1):138–150. [cited 2024 Jul 19]. Available from: https://www.cell.com/cell-metabolism/fulltext/S1550-4131(15)00214-4

Jiang Y, Li C, Wu Q, An P, Huang L, Wang J, Chen C, Chen X, Zhang F, Ma L, Liu S, He H, Xie S, Sun Y, Liu H, Zhan Y, Tao Y, Liu Z, Sun X, Hu Y, Wang Q, Ye D, Zhang J, Zou S, Wang Y, Wei G, Liu Y, Shi Y, Eugene Chin Y, Hao Y, Wang F, Zhang X (2019) Iron-dependent histone 3 lysine 9 demethylation controls B cell proliferation and humoral immune responses. Nat Commun [Internet] 10(1):2935. [cited 2024 Jul 20]. Available from: https://www.nature.com/articles/s41467-019-11002-5

Jiang L, Wang J, Wang K, Wang H, Wu Q, Yang C, Yu Y, Ni P, Zhong Y, Song Z, Xie E, Hu R, Min J, Wang F (2021) RNF217 regulates iron homeostasis through its E3 ubiquitin ligase activity by modulating ferroportin degradation. Blood J Am Soc Hematol [Internet] 138(8):689–705. [cited 2024 Jul 19]. Available from: https://ashpublications.org/blood/article-abstract/138/8/689/475832

Johnson EE, Wessling-Resnick M (2012) Iron metabolism and the innate immune response to infection. Microbes Infect [Internet] 14(3):207–216. [cited 2024 Jul 20]. Available from: https://www.sciencedirect.com/science/article/pii/S1286457911002541

Jung WH, Saikia S, Hu G, Wang J, Fung CKY, D'Souza C, White R, Kronstad JW (2010) HapX positively and negatively regulates the transcriptional response to iron deprivation in Cryptococcus neoformans. PLoS Pathog\ [Internet] 6(11):e1001209. [cited 2024 Jul 20]. Available from: https://journals.plos.org/plospathogens/article?id=10.1371/journal.ppat.1001209

Kaelin WG, Ratcliffe PJ (2008) Oxygen sensing by metazoans: the central role of the HIF hydroxylase pathway. Mol Cell [Internet] 30(4):393–402. [cited 2024 Jul 19]. Available from: https://www.cell.com/AJHG/fulltext/S1097-2765(08)00292-X

Kanamori Y, Sugiyama M, Hashimoto O, Murakami M, Matsui T, Funaba M (2016) Regulation of hepcidin expression by inflammation-induced activin B. Sci Rep 6(1):1–12

Kapitsinou PP, Liu Q, Unger TL, Rha J, Davidoff O, Keith B, Epstein JA, Moores SL, Erickson-Miller CL, Haase VH (2010) Hepatic HIF-2 regulates erythropoietic responses to hypoxia in renal anemia. Blood J Am Soc Hematol [Internet] 116(16):3039–3048. [cited 2024 Jul 19]. Available from: https://ashpublications.org/blood/article-abstract/116/16/3039/27825

Kautz L, Jung G, Du X, Gabayan V, Chapman J, Nasoff M, Ganz T (2015) Erythroferrone contributes to hepcidin suppression and iron overload in a mouse model of β-thalassemia. Blood J Am Soc Hematol [Internet] 126(17):2031–2037. [cited 2024 Jul 19]. Available from: https://ashpublications.org/blood/article-abstract/126/17/2031/34444

Kazi AA, Molitoris KH, Koos RD (2009) Estrogen rapidly activates the PI3K/AKT pathway and hypoxia-inducible factor 1 and induces vascular endothelial growth factor A expression in luminal epithelial cells of the rat uterus. Biol Reprod [Internet] 81(2):378–387. [cited 2024 Jul 20]. Available from: https://academic.oup.com/biolreprod/article-abstract/81/2/378/2557897

Kim J, Park C, Imlay JA, Park W (2017) Lineage-specific SoxR-mediated regulation of an endoribonuclease protects non-enteric bacteria from redox-active compounds. J Biol Chem

[Internet] 292(1):121–133. [cited 2024 Jul 20]. Available from: https://www.jbc.org/article/S0021-9258(20)31735-X/fulltext

Knobloch J, Casjens S, Lehnert M, Yanik SD, Körber S, Lotz A, Rupp J, Raulf M, Zschiesche W, Weiss T, Kronsbein J, Koch A, Brüning T, Pesch B (2020) Exposure to welding fumes suppresses the activity of T-helper cells. Environ Res [Internet] 189:109913. [cited 2024 Jul 20]. Available from: https://www.sciencedirect.com/science/article/pii/S0013935120308082?casa_token=jc1XTkHJ1c4AAAAA:iz7hA3vT7bno-xWehJXsNJbuBR4392nwicWMlSjWdFRIu90wsA7uS5cuA6ucS0hseeV9cslSSn8

Kohanski MA, Dwyer DJ, Hayete B, Lawrence CA, Collins JJ (2007) A common mechanism of cellular death induced by bactericidal antibiotics. Cell [Internet] 130(5):797–810. [cited 2024 Jul 20]. Available from: https://www.cell.com/fulltext/S0092-8674(07)00899-9

Korolnek T, Hamza I (2015) Macrophages and iron trafficking at the birth and death of red cells. Blood J Am Soc Hematol [Internet] 125(19):2893–2897. [cited 2024 Jul 18]. Available from: https://ashpublications.org/blood/article-abstract/125/19/2893/34265

Kramer J, Özkaya Ö, Kümmerli R (2020) Bacterial siderophores in community and host interactions. Nat Rev Microbiol [Internet] 18(3):152–163. [cited 2024 Jul 20]. Available from: https://www.nature.com/articles/s41579-019-0284-4

Krause A, Neitz S, Mägert HJ, Schulz A, Forssmann WG, Schulz-Knappe P, Adermann K (2000) LEAP-1, a novel highly disulfide-bonded human peptide, exhibits antimicrobial activity. FEBS Lett [Internet] 480(2–3):147–150. [cited 2024 Jul 19]. Available from: https://www.sciencedirect.com/science/article/pii/S0014579300019207

Krause A, Sillard R, Kleemeier B, Klüver E, Maronde E, Conejo-García JR, Forssmann WG, Schulz-Knappe P, Nehls MC, Wattler F, Wattler S, Adermann K (2003) Isolation and biochemical characterization of LEAP-2, a novel blood peptide expressed in the liver. Prot Sci [Internet] 12(1):143–152. [cited 2024 Jul 19]. Available from: https://onlinelibrary.wiley.com/doi/10.1110/ps.0213603

Kumar S, Sheokand N, Mhadeshwar MA, Raje CI, Raje M (2012) Characterization of glyceraldehyde-3-phosphate dehydrogenase as a novel transferrin receptor. Int J Biochem Cell Biol [Internet] 44(1):189–199. [cited 2024 Jul 18]. Available from: https://www.sciencedirect.com/science/article/pii/S1357272511002895?casa_token=YhYnnsFzCcoAAAAA:y3bEJKeBGfsFmJjZfQjXecUNOzwZ4U_hSrJAAWJTyvrQru8ijmeDX_kDXEoZjj1TSaT8bhrE3f0

Kurthkoti K, Amin H, Marakalala MJ, Ghanny S, Subbian S, Sakatos A, Livny J, Fortune SM, Berney M, Rodriguez GM (2017) The capacity of *Mycobacterium tuberculosis* to survive iron starvation might enable it to persist in iron-deprived microenvironments of human granulomas. Stallings CL, editor. mBio [Internet] 8(4):e01092-17. [cited 2024 Jul 20]. Available from: https://journals.asm.org/doi/10.1128/mBio.01092-17

Lane DJR, Merlot AM, Huang MH, Bae DH, Jansson PJ, Sahni S, Kalinowski DS, Richardson DR (2015) Cellular iron uptake, trafficking and metabolism: key molecules and mechanisms and their roles in disease. Biochim Biophys Acta (BBA) Mol Cell Res [Internet] 1853(5):1130–1144. [cited 2024 Jul 19]. Available from: https://www.sciencedirect.com/science/article/pii/S0167488915000361

Lehtihet M, Bonde Y, Beckman L, Berinder K, Hoybye C, Rudling M, Sloan JH, Konrad RJ, Angelin B (2016) Circulating hepcidin-25 is reduced by endogenous estrogen in humans. PloS One [Internet] 11(2):e0148802. [cited 2024 Jul 19]. Available from: https://journals.plos.org/plosone/article?id=10.1371/journal.pone.0148802

Leoh LS, Kim YK, Candelaria PV, Martínez-Maza O, Daniels-Wells TR, Penichet ML (2018) Efficacy and mechanism of antitumor activity of an antibody targeting transferrin receptor 1 in mouse models of human multiple myeloma. J Immunol [Internet] 200(10):3485–3494. [cited 2024 Jul 20]. Available from: https://journals.aai.org/jimmunol/article/200/10/3485/106464

Levy JE, Jin OU, Fujiwara Y, Kuo F, Andrews N (1999) Transferrin receptor is necessary for development of erythrocytes and the nervous system. Nat Genet [Internet] 21(4):396–399. [cited 2024 Jul 18]. Available from: https://www.nature.com/articles/ng0499_396

Li N, Zhang C, Li B, Liu X, Huang Y, Xu S, Gu L (2012) Unique iron coordination in iron-chelating molecule vibriobactin helps Vibrio cholerae evade mammalian siderocalin-mediated immune response. J Biol Chem [Internet] 287(12):8912–8919. [cited 2024 Jul 20]. Available from: https://www.jbc.org/article/S0021-9258(20)60853-5/fulltext

Li W, Cui T, Hu L, Wang Z, Li Z, He ZG (2015) Cyclic diguanylate monophosphate directly binds to human siderocalin and inhibits its antibacterial activity. Nat Commun [Internet] 6(1):8330. [cited 2024 Jul 20]. Available from: https://www.nature.com/articles/ncomms9330

Li F, Liu F, Huang K, Yang S (2022) Advancement of gallium and gallium-based compounds as antimicrobial agents. Front Bioeng Biotechnol [Internet] 10:827960. [cited 2024 Jul 20]. Available from: https://www.frontiersin.org/articles/10.3389/fbioe.2022.827960/full

Lin L, Goldberg YP, Ganz T (2005) Competitive regulation of hepcidin mRNA by soluble and cell-associated hemojuvelin. Blood [Internet] 106(8):2884–2889. [cited 2024 Jul 19]. Available from: https://ashpublications.org/blood/article-abstract/106/8/2884/21824

Littwitz-Salomon E, Moreira D, Frost JN, Choi C, Liou KT, Ahern DK, O'Shaughnessy S, Wagner B, Biron CA, Drakesmith H, Dittmer U, Finlay DK (2021) Metabolic requirements of NK cells during the acute response against retroviral infection. Nat Commun [Internet] 12(1):5376. [cited 2024 Jul 20]. Available from: https://www.nature.com/articles/s41467-021-25715-z

Liu XB, Nguyen NBH, Marquess KD, Yang F, Haile DJ (2005) Regulation of hepcidin and ferroportin expression by lipopolysaccharide in splenic macrophages. Blood Cells Mol Dis [Internet] 35(1):47–56. [cited 2024 Jul 19]. Available from: https://www.sciencedirect.com/science/article/pii/S1079979605000586

Liu S, Le May C, Wong WP, Ward RD, Clegg DJ, Marcelli M, Korach KS, Mauvais-Jarvis F (2009) Importance of extranuclear estrogen receptor-α and membrane G protein–coupled estrogen receptor in pancreatic islet survival. Diabetes [Internet] 58(10):2292–2302. [cited 2024 Jul 20]. Available from: https://diabetesjournals.org/diabetes/article-abstract/58/10/2292/13740

Liu Q, Davidoff O, Niss K, Haase VH (2012) Hypoxia-inducible factor regulates hepcidin via erythropoietin-induced erythropoiesis. J Clin Investig [Internet] 122(12):4635–4644. [cited 2024 Jul 19]. Available from: https://www.jci.org/articles/view/63924

Liu Z, Petersen R, Devireddy L (2013) Impaired neutrophil function in 24p3 null mice contributes to enhanced susceptibility to bacterial infections. J Immunol [Internet] 190(9):4692–4706. [cited 2024 Jul 20]. Available from: https://journals.aai.org/jimmunol/article/190/9/4692/87020

Los FC, Randis TM, Aroian RV, Ratner AJ (2013) Role of pore-forming toxins in bacterial infectious diseases. Microbiol Mol Biol Rev [Internet] 77(2):173–207. [cited 2024 Jul 20]. Available from: https://journals.asm.org/doi/abs/10.1128/mmbr.00052-12

Macedo MF, De Sousa M, Ned RM, Mascarenhas C, Andrews NC, Correia-Neves M (2004) Transferrin is required for early T-cell differentiation. Immunology [Internet] 112(4):543–549. [cited 2024 Jul 20]. Available from: https://onlinelibrary.wiley.com/doi/10.1111/j.1365-2567.2004.01915.x

Marchetti M, De Bei O, Bettati S, Campanini B, Kovachka S, Gianquinto E, Spyrakis F, Ronda L (2020) Iron metabolism at the interface between host and pathogen: from nutritional immunity to antibacterial development. Int J Mol Sci [Internet] 21(6):2145. [cited 2024 Jul 20]. Available from: https://www.mdpi.com/1422-0067/21/6/2145

Marro S, Chiabrando D, Messana E, Stolte J, Turco E, Tolosano E, Muckenthaler MU (2010) Heme controls ferroportin1 (FPN1) transcription involving Bach1, Nrf2 and a MARE/ARE sequence motif at position- 7007 of the FPN1 promoter. Haematologica [Internet] 95(8):1261. [cited 2024 Jul 19]. Available from: https://www.ncbi.nlm.nih.gov/pmc/articles/PMC2913073/

Martínez D, Oyarzún R, Pontigo JP, Romero A, Yáñez AJ, Vargas-Chacoff L (2017) Nutritional immunity triggers the modulation of iron metabolism genes in the sub-antarctic notothenioid Eleginops maclovinus in response to Piscirickettsia salmonis. Front Immunol [Internet] 8:1153. [cited 2024 Jul 20]. Available from: https://www.frontiersin.org/articles/10.3389/fimmu.2017.01153/full

Marx JJM (2002) Iron and infection: competition between host and microbes for a precious element. Best Pract Res Clin Haematol [Internet] 15(2):411–426. [cited 2024 Jul 18]. Available from: https://www.sciencedirect.com/science/article/pii/S1521692602900010

Maurer E, Gütschow M, Stirnberg M (2012) Matriptase-2 (TMPRSS6) is directly up-regulated by hypoxia inducible factor-1: identification of a hypoxia-responsive element in the *TMPRSS6* promoter region. Biol Chem [Internet] 393(6):535–540. [cited 2024 Jul 19]. Available from: https://www.degruyter.com/document/doi/10.1515/hsz-2011-0221/html

Mayeur C, Leyton PA, Kolodziej SA, Yu B, Bloch KD (2014) BMP type II receptors have redundant roles in the regulation of hepatic hepcidin gene expression and iron metabolism. Blood J Am Soc Hematol [Internet] 124(13):2116–2123. [cited 2024 Jul 19]. Available from: https://ashpublications.org/blood/article-abstract/124/13/2116/33093

McKie AT, Barrow D, Latunde-Dada GO, Rolfs A, Sager G, Mudaly E, Mudaly M, Richardson C, Barlow D, Bomford A, Peters TJ, Raja KB, Shirali S, Hediger MA, Farzaneh F, Simpson RJ (2001) An iron-regulated ferric reductase associated with the absorption of dietary iron. Science [Internet] 291(5509):1755–1759. [cited 2024 Jul 18]. Available from: https://www.science.org/doi/10.1126/science.1057206

Mey AR, Wyckoff EE, Kanukurthy V, Fisher CR, Payne SM (2005) Iron and fur regulation in *Vibrio cholerae* and the role of fur in virulence. Infect Immun [Internet] 73(12):8167–8178. [cited 2024 Jul 20]. Available from: https://journals.asm.org/doi/10.1128/IAI.73.12.8167-8178.2005

Meynard D, Kautz L, Darnaud V, Canonne-Hergaux F, Coppin H, Roth MP (2009) Lack of the bone morphogenetic protein BMP6 induces massive iron overload. Nat Gen [Internet] 41(4):478–481. [cited 2024 Jul 19]. Available from: https://www.nature.com/articles/ng.320

Meynard D, Vaja V, Sun CC, Corradini E, Chen S, López-Otín C, Grgurevic L, Hong CC, Stirnberg M, Gütschow M, Vukicevic S, Babitt JL, Lin HY (2011) Regulation of TMPRSS6 by BMP6 and iron in human cells and mice. Blood J Am Soc Hematol [Internet] 118(3):747–756. [cited 2024 Jul 19]. Available from: https://ashpublications.org/blood/article-abstract/118/3/747/28940

Miethke M, Marahiel MA (2007) Siderophore-based iron acquisition and pathogen control. Microbiol Mol Biol Rev [Internet] 71(3):413–451. [cited 2024 Jul 20]. Available from: https://journals.asm.org/doi/10.1128/MMBR.00012-07

Miller HK, Auerbuch V (2015) Bacterial iron–sulfur cluster sensors in mammalian pathogens. Metallomics [Internet] 7(6):943–956. [cited 2024 Jul 20]. Available from: https://academic.oup.com/metallomics/article-abstract/7/6/943/6015117

Milman N, Kirchhoff M, JØrgensen T (1992) Iron status markers, serum ferritin and hemoglobin in 1359 Danish women in relation to menstruation, hormonal contraception, parity, and postmenopausal hormone treatment. Ann Hematol [Internet] 65(2):96–102. [cited 2024 Jul 19]. Available from: http://link.springer.com/10.1007/BF01698138

Mims MP, Guan Y, Pospisilova D, Priwitzerova M, Indrak K, Ponka P, Divoky V, Prchal JT (2005) Identification of a human mutation of DMT1 in a patient with microcytic anemia and iron overload. Blood [Internet] 105(3):1337–1342. [cited 2024 Jul 18]. Available from: https://ashpublications.org/blood/article-abstract/105/3/1337/19640

Mirciov CS, Wilkins SJ, Dunn LA, Anderson GJ, Frazer DM (2017) Characterization of putative erythroid regulators of hepcidin in mouse models of anemia. PLoS One [Internet] 12(1):e0171054. [cited 2024 Jul 19]. Available from: https://journals.plos.org/plosone/article?id=10.1371/journal.pone.0171054

Miyazono K, Kamiya Y, Morikawa M (2010) Bone morphogenetic protein receptors and signal transduction. J Biochem [Internet] 147(1):35–51. [cited 2024 Jul 19]. Available from: https://academic.oup.com/jb/article-abstract/147/1/35/752057

Morgan EH, Walters MNI (1963) Iron storage in human disease: fractionation of hepatic and splenic iron into ferritin and haemosiderin with histochemical correlations. J Clin Pathol [Internet] 16(2):101–107. [cited 2024 Jul 18]. Available from: https://jcp.bmj.com/content/16/2/101.abstract

Morita M, Ohneda O, Yamashita T, Takahashi S, Suzuki N, Nakajima O, Kawauchi S, Ema M, Shibahara S, Udono T, Tomita K, Tamai M, Sogawa K, Yamamoto M, Fujii-Kuriyama Y (2003) HLF/HIF-2α is a key factor in retinopathy of prematurity in association with erythropoietin. EMBO J [Internet] 22(5):1134–1146. [cited 2024 Jul 19]. Available from: http://emboj.embopress.org/cgi/doi/10.1093/emboj/cdg117

Nairz M, Fritsche G, Brunner P, Talasz H, Hantke K, Weiss G (2008) Interferon-γ limits the availability of iron for intramacrophage *Salmonella typhimurium*. Eur J Immunol [Internet]

38(7):1923–1936. [cited 2024 Jul 20]. Available from: https://onlinelibrary.wiley.com/doi/10.1002/eji.200738056

Nairz M, Schroll A, Sonnweber T, Weiss G (2010) The struggle for iron – a metal at the host-pathogen interface: iron at the host-pathogen interface. Cell Microbiol [Internet] 12(12):1691–1702. [cited 2024 Jul 20]. Available from: https://onlinelibrary.wiley.com/doi/10.1111/j.1462-5822.2010.01529.x

Negash KH, Norris JK, Hodgkinson JT (2019) Siderophore–antibiotic conjugate design: new drugs for bad bugs? Molecules [Internet] 24(18):3314. [cited 2024 Jul 20]. Available from: https://www.mdpi.com/1420-3049/24/18/3314

Nelson AL, Ratner AJ, Barasch J, Weiser JN (2007) Interleukin-8 secretion in response to aferric enterobactin is potentiated by siderocalin. Infect Immun [Internet] 75(6):3160–3168. [cited 2024 Jul 20]. Available from: https://journals.asm.org/doi/10.1128/IAI.01719-06

Nemeth E, Ganz T (2006) Regulation of iron metabolism by hepcidin. Annu Rev Nutr [Internet] 26(1):323–342. [cited 2024 Jul 20]. Available from: https://www.annualreviews.org/doi/10.1146/annurev.nutr.26.061505.111303

Nemeth E, Ganz T (2021) Hepcidin-ferroportin interaction controls systemic iron homeostasis. Int J Mol Sci 22(12):6493

Nemeth E, Valore EV, Territo M, Schiller G, Lichtenstein A, Ganz T (2003) Hepcidin, a putative mediator of anemia of inflammation, is a type II acute-phase protein. Blood J Am Soc Hematol [Internet] 101(7):2461–2463. [cited 2024 Jul 19]. Available from: https://ashpublications.org/blood/article-abstract/101/7/2461/106644

Nemeth E, Tuttle MS, Powelson J, Vaughn MB, Donovan A, Ward DM, Ganz T, Kaplan J (2004) Hepcidin regulates cellular iron efflux by binding to ferroportin and inducing its internalization. Science [Internet] 306(5704):2090–2093. [cited 2024 Jul 19]. Available from: https://www.science.org/doi/10.1126/science.1104742

Nemeth E, Preza GC, Jung CL, Kaplan J, Waring AJ, Ganz T (2006) The N-terminus of hepcidin is essential for its interaction with ferroportin: structure-function study. Blood [Internet] 107(1):328–333. [cited 2024 Jul 19]. Available from: https://ashpublications.org/blood/article-abstract/107/1/328/21711

Neto LMM, Zufelato N, De Sousa-Júnior AA, Trentini MM, Da Costa AC, Bakuzis AF, Kipnis A, Junqueira-Kipnis AP (2018) Specific T cell induction using iron oxide based nanoparticles as subunit vaccine adjuvant. Hum Vaccin Immunother [Internet]:1–16. [cited 2024 Jul 20]. Available from: https://www.tandfonline.com/doi/full/10.1080/21645515.2018.1489192

Ni S, Yuan Y, Kuang Y, Li X (2022) Iron metabolism and immune regulation. Front Immunol [Internet] 13:816282. [cited 2024 Jul 20]. Available from: https://www.frontiersin.org/articles/10.3389/fimmu.2022.816282/full

Nicolas G, Bennoun M, Devaux I, Beaumont C, Grandchamp B, Kahn A, Vaulont S (2001) Lack of hepcidin gene expression and severe tissue iron overload in upstream stimulatory factor 2 (*USF2*) knockout mice. Proc Natl Acad Sci USA [Internet] 98(15):8780–8785. [cited 2024 Jul 19]. Available from: https://pnas.org/doi/full/10.1073/pnas.151179498

Nicolas G, Bennoun M, Porteu A, Mativet S, Beaumont C, Grandchamp B, Sirito M, Sawadogo M, Kahn A, Vaulont S (2002) Severe iron deficiency anemia in transgenic mice expressing liver hepcidin. Proc Natl Acad Sci USA [Internet] 99(7):4596–4601. [cited 2024 Jul 19]. Available from: https://pnas.org/doi/full/10.1073/pnas.072632499

Noble SM (2013) Candida albicans specializations for iron homeostasis: from commensalism to virulence. Curr Opin Microbiol [Internet] 16(6):708–715. [cited 2024 Jul 20]. Available from: https://www.sciencedirect.com/science/article/pii/S1369527413001574?casa_token=9GspchD72TUAAAAA:CzOUP68gJIJqxOMG4Jt-Mf1KREQli03L-0QlBb9Twbd3048dtCHWOETXLnn7zL5H4jxXwbKuH4c

Ohgami RS, Campagna DR, Greer EL, Antiochos B, McDonald A, Chen J, Sharp JJ, Fujiwara Y, Barker JE, Fleming MD (2005) Identification of a ferrireductase required for efficient transferrin-dependent iron uptake in erythroid cells. Nat Genet [Internet] 37(11):1264–1269. [cited 2024 Jul 18]. Available from: https://www.nature.com/articles/ng1658

Osuna-Espinoza KY, Rosas-Taraco AG (2023) Metabolism of NK cells during viral infections. Front Immunol [Internet] 14:1064101. [cited 2024 Jul 20]. Available from: https://www.frontiersin.org/articles/10.3389/fimmu.2023.1064101/full

Pandey R, Rodriguez GM (2014) IdeR is required for iron homeostasis and virulence in M *ycobacterium tuberculosis*. Mol Microbiol [Internet] 91(1):98–109. [cited 2024 Jul 20]. Available from: https://onlinelibrary.wiley.com/doi/10.1111/mmi.12441

Park CH, Valore EV, Waring AJ, Ganz T (2001) Hepcidin, a urinary antimicrobial peptide synthesized in the liver. J Biol Chem [Internet] 276(11):7806–7810. [cited 2024 Jul 19]. Available from: https://www.jbc.org/article/S0021-9258(19)32119-2/abstract

Pasquadibisceglie A, Bonaccorsi di Patti MC, Musci G, Polticelli F (2023) Membrane transporters involved in iron trafficking: physiological and pathological aspects. Biomolecules [Internet] 13(8):1172. [cited 2024 Jul 19]. Available from: https://www.mdpi.com/2218-273X/13/8/1172

Peek ME, Bhatnagar A, McCarty NA, Zughaier SM (2012) Pyoverdine, the major siderophore in *Pseudomonas aeruginosa*, evades NGAL recognition. Interdiscip Perspect Infect Dis [Internet] 2012:1–10. [cited 2024 Jul 20]. Available from: http://www.hindawi.com/journals/ipid/2012/843509/

Pentecost BT, Teng CT (1987) Lactotransferrin is the major estrogen inducible protein of mouse uterine secretions. J Biol Chem [Internet] 262(21):10134–10139. [cited 2024 Jul 19]. Available from: https://www.sciencedirect.com/science/article/pii/S0021925818610889

Pereira D, Ferreira S, Ramírez-Rodríguez GB, Alves N, Sousa Â, Valente JF (2024) Silver and antimicrobial polymer nanocomplexes to enhance biocidal effects. Int J Mol Sci [Internet] 25(2):1256. [cited 2024 Jul 20]. Available from: https://www.mdpi.com/1422-0067/25/2/1256

Peyssonnaux C, Datta V, Cramer T, Doedens A, Theodorakis EA, Gallo RL, Hurtado-Ziola N, Nizet V, Johnson RS (2005) HIF-1α expression regulates the bactericidal capacity of phagocytes. J Clin Investig [Internet] 115(7):1806–1815. [cited 2024 Jul 20]. Available from: https://www.jci.org/articles/view/23865

Peyssonnaux C, Zinkernagel AS, Schuepbach RA, Rankin E, Vaulont S, Haase VH, Nizet V, Johnson RS (2007) Regulation of iron homeostasis by the hypoxia-inducible transcription factors (HIFs). J Clin Investig [Internet] 117(7):1926–1932. [cited 2024 Jul 19]. Available from: https://www.jci.org/articles/view/31370

Pfeifhofer-Obermair C, Tymoszuk P, Nairz M, Schroll A, Klais G, Demetz E, Engl S, Brigo N, Weiss G (2021) Regulation of Th1 T cell differentiation by iron via upregulation of T cell immunoglobulin and mucin containing protein-3 (TIM-3). Front Immunol [Internet] 12:637809. [cited 2024 Jul 20]. Available from: https://www.frontiersin.org/articles/10.3389/fimmu.2021.637809/full

Pigeon C, Ilyin G, Courselaud B, Leroyer P, Turlin B, Brissot P, Loréal O (2001) A new mouse liver-specific gene, encoding a protein homologous to human antimicrobial peptide hepcidin, is overexpressed during iron overload. J Biol Chem [Internet] 276(11):7811–7819. [cited 2024 Jul 19]. Available from: https://www.jbc.org/article/S0021-9258(19)32120-9/abstract

Pirschel W, Mestekemper AN, Wissuwa B, Krieg N, Kröller S, Daniel C, Gunzer F, Tolosano E, Bauer M, Amann K, Heinemann SH, Coldewey SM (2022) Divergent roles of haptoglobin and hemopexin deficiency for disease progression of Shiga-toxin–induced hemolytic-uremic syndrome in mice. Kidney Int [Internet] 101(6):1171–1185. [cited 2024 Jul 20]. Available from: https://www.sciencedirect.com/science/article/pii/S0085253822000163

Portsmouth S, van Veenhuyzen D, Echols R, Machida M, Ferreira JCA, Ariyasu M, Tenke P, Nagata TD (2018) Cefiderocol versus imipenem-cilastatin for the treatment of complicated urinary tract infections caused by Gram-negative uropathogens: a phase 2, randomised, double-blind, non-inferiority trial. Lancet Infect Dis [Internet] 18(12):1319–1328. [cited 2024 Jul 20]. Available from: https://www.thelancet.com/journals/laninf/article/PIIS1473-3099(18)30554-1/abstract

Poss KD, Tonegawa S (1997) Heme oxygenase 1 is required for mammalian iron reutilization. Proc Natl Acad Sci USA [Internet] 94(20):10919–10924. [cited 2024 Jul 18]. Available from: https://pnas.org/doi/full/10.1073/pnas.94.20.10919

Pritchard LK, Spencer DI, Royle L, Bonomelli C, Seabright GE, Behrens AJ, Kulp DW, Menis S, Krumm SA, Dunlop DC, Crispin DJ, Bowden TA, Scanlan CN, Ward AB, Schief WR, Doores

KJ, Crispin M (2015) Glycan clustering stabilizes the mannose patch of HIV-1 and preserves vulnerability to broadly neutralizing antibodies. Nat Commun [Internet] 6(1):7479. [cited 2024 Jul 20]. Available from: https://www.nature.com/articles/ncomms8479

Puri S, Kumar R, Rojas IG, Salvatori O, Edgerton M (2019) Iron chelator deferasirox reduces *Candida albicans* invasion of oral epithelial cells and infection levels in murine oropharyngeal candidiasis. Antimicrob Agents Chemother [Internet] 63(4):e02152-18. [cited 2024 Jul 20]. Available from: https://journals.asm.org/doi/10.1128/AAC.02152-18

Qian ZM, He X, Liang T, Wu KC, Yan YC, Lu LN, Yang G, Luo QQ, Yung WH, Ke Y (2014) Lipopolysaccharides upregulate hepcidin in neuron via microglia and the IL-6/STAT3 signaling pathway. Mol Neurobiol [Internet] 50(3):811–820. [cited 2024 Jul 19]. Available from: http://link.springer.com/10.1007/s12035-014-8671-3

Qian Y, Zhang S, Guo W, Ma J, Chen Y, Wang L, Zhao M, Liu S (2015) Polychlorinated biphenyls (PCBs) inhibit hepcidin expression through an estrogen-like effect associated with disordered systemic iron homeostasis. Chem Res Toxicol [Internet] 28(4):629–640. [cited 2024 Jul 19]. Available from: https://pubs.acs.org/doi/10.1021/tx500428r

Qiao B, Sugianto P, Fung E, del-Castillo-Rueda A, Moran-Jimenez MJ, Ganz T, Nemeth E (2012) Hepcidin-induced endocytosis of ferroportin is dependent on ferroportin ubiquitination. Cell Metab [Internet] 15(6):918–924. [cited 2024 Jul 19]. Available from: https://www.cell.com/cell-metabolism/fulltext/S1550-4131(12)00139-8

Qiao Y, Li Y, Ye Y, Yu Y, Wang W, Yao K, Zhou M (2022) Gallium-based nanoplatform for combating multidrug-resistant *Pseudomonas aeruginosa* and postoperative inflammation in endophthalmitis secondary to cataract surgery. ACS Appl Mater Interfaces [Internet] 14(46):51763–51775. [cited 2024 Jul 20]. Available from: https://pubs.acs.org/doi/10.1021/acsami.2c15834

Raso GM, Irace C, Esposito E, Maffettone C, Iacono A, Di Pascale A, Santamaria R, Colonna A, Meli R (2009) Ovariectomy and estrogen treatment modulate iron metabolism in rat adipose tissue. Biochem Pharmacol [Internet] 78(8):1001–1007. [cited 2024 Jul 20]. Available from: https://www.sciencedirect.com/science/article/pii/S0006295209004444?casa_token=arAV2iTSa2AAAAAA:u4A_1Rd79Xy8ZMizH3FzroxY3fBGhusydSRKQ1EEp44K_cXD6LhWyC5qTiVBiUb8VHf9B0AD0Co

Rebuffet L, Melsen JE, Escalière B, Basurto-Lozada D, Bhandoola A, Björkström NK, Bryceson YT, Castriconi R, Cichocki F, Colonna M, Davis DM, Diefenbach A, Ding Y, Haniffa M, Horowitz A, Lanier LL, Malmberg KJ, Miller JS, Moretta L, Narni-Mancinelli E, O'Neill LAJ, Romagnani C, Ryan DG, Sivori S, Sun D, Vagne C, Vivier E (2024) High-dimensional single-cell analysis of human natural killer cell heterogeneity. Nat Immunol [Internet]:1–15. [cited 2024 Jul 20]. Available from: https://www.nature.com/articles/s41590-024-01883-0

Ren X, Zou L, Zhang X, Branco V, Wang J, Carvalho C, Holmgren A, Lu J (2017) Redox signaling mediated by thioredoxin and glutathione systems in the central nervous system. Antioxid Redox Signal [Internet] 27(13):989–1010. [cited 2024 Jul 20]. Available from: http://www.liebertpub.com/doi/10.1089/ars.2016.6925

Rodionov DA, Dubchak IL, Arkin AP, Alm EJ, Gelfand MS (2005) Dissimilatory metabolism of nitrogen oxides in bacteria: comparative reconstruction of transcriptional networks. PLoS Comput Biol [Internet] 1(5):e55. [cited 2024 Jul 20]. Available from: https://journals.plos.org/ploscompbiol/article?id=10.1371/journal.pcbi.0010055

Rosa L, Cutone A, Lepanto MS, Paesano R, Valenti P (2017) Lactoferrin: a natural glycoprotein involved in iron and inflammatory homeostasis. Int J Mol Sci [Internet] 18(9):1985. [cited 2024 Jul 20]. Available from: https://www.mdpi.com/1422-0067/18/9/1985

Ross SL, Tran L, Winters A, Lee KJ, Plewa C, Foltz I, King C, Miranda LP, Allen J, Beckman H, Cooke KS, Moody G, Sasu BJ, Nemeth E, Ganz T, Molineux G, Arvedson TL (2012) Molecular mechanism of hepcidin-mediated ferroportin internalization requires ferroportin lysines, not tyrosines or JAK-STAT. Cell Metab [Internet] 15(6):905–917. [cited 2024 Jul 19]. Available from: https://www.cell.com/cell-metabolism/fulltext/S1550-4131(12)00138-6

Rouault TA (2005) The intestinal heme transporter revealed. Cell [Internet] 122(5):649–651. [cited 2024 Jul 18]. Available from: https://www.cell.com/AJHG/fulltext/S0092-8674(05)00864-0

Saha R, Saha N, Donofrio RS, Bestervelt LL (2013) Microbial siderophores: a mini review. J Basic Microbiol [Internet] 53(4):303–317. [cited 2024 Jul 20]. Available from: https://onlinelibrary.wiley.com/doi/10.1002/jobm.201100552

Sandy M, Butler A (2009) Microbial iron acquisition: marine and terrestrial siderophores. Chem Rev [Internet] 109(10):4580–4595. [cited 2024 Jul 20]. Available from: https://pubs.acs.org/doi/10.1021/cr9002787

Schmitt MP, Holmes RK (1991) Iron-dependent regulation of diphtheria toxin and siderophore expression by the cloned Corynebacterium diphtheriae repressor gene dtxR in C. diphtheriae C7 strains. Infect Immun [Internet] 59(6):1899–1904. [cited 2024 Jul 20]. Available from: https://journals.asm.org/doi/10.1128/iai.59.6.1899-1904.1991

Schryvers AB, Stojiljkovic I (1999) Iron acquisition systems in the pathogenic *Neisseria*. Mol Microbiol [Internet] 32(6):1117–1123. [cited 2024 Jul 20]. Available from: https://onlinelibrary.wiley.com/doi/10.1046/j.1365-2958.1999.01411.x

Schwartz AJ, Goyert JW, Solanki S, Kerk SA, Chen B, Castillo C, Hsu PP, Do BT, Singhal R, Dame MK, Lee HJ, Spence JR, Lakhal-Littleton S, Vander Heiden MG, Lyssiotis CA, Xue X, Shah YM (2021) Hepcidin sequesters iron to sustain nucleotide metabolism and mitochondrial function in colorectal cancer epithelial cells. Nat Metab 3(7):969–982

Schwarz P, Kübler JA, Strnad P, Müller K, Barth TF, Gerloff A, Feick P, Peyssonnaux C, Vaulont S, Adler G, Kulaksiz H (2012) Hepcidin is localised in gastric parietal cells, regulates acid secretion and is induced by Helicobacter pylori infection. Gut [Internet] 61(2):193–201. [cited 2024 Jul 19]. Available from: https://gut.bmj.com/content/61/2/193.short

Shafarin J, Bajbouj K, El-Serafy A, Sandeep D, Hamad M (2016) Estrogen-dependent downregulation of hepcidin synthesis induces intracellular iron efflux in cancer cells in vitro. Biol Med [Internet] 8(7):1. [cited 2024 Jul 18]. Available from: https://www.researchgate.net/profile/Ahmed-El-Serafi-2/publication/311657636

Silvestri L, Pagani A, Nai A, De Domenico I, Kaplan J, Camaschella C (2008a) The serine protease matriptase-2 (TMPRSS6) inhibits hepcidin activation by cleaving membrane hemojuvelin. Cell Metab [Internet] 8(6):502–511. [cited 2024 Jul 19]. Available from: https://www.cell.com/AJHG/fulltext/S1550-4131(08)00319-7

Silvestri L, Pagani A, Camaschella C (2008b) Furin-mediated release of soluble hemojuvelin: a new link between hypoxia and iron homeostasis. Blood J Am Soc Hematol [Internet] 111(2):924–931. [cited 2024 Jul 19]. Available from: https://ashpublications.org/blood/article-abstract/111/2/924/103758

Singh V, Yeoh BS, Xiao X, Kumar M, Bachman M, Borregaard N, Joe B, Vijay-Kumar M (2015) Interplay between enterobactin, myeloperoxidase and lipocalin 2 regulates E. coli survival in the inflamed gut. Nat Commun [Internet] 6(1):7113. [cited 2024 Jul 20]. Available from: https://www.nature.com/articles/ncomms8113

Skaar EP, Humayun M, Bae T, DeBord KL, Schneewind O (2004a) Iron-source preference of *Staphylococcus aureus* infections. Science [Internet] 305(5690):1626–1628. [cited 2024 Jul 20]. Available from: https://www.science.org/doi/10.1126/science.1099930

Skaar EP, Gaspar AH, Schneewind O (2004b) IsdG and IsdI, heme-degrading enzymes in the cytoplasm of Staphylococcus aureus. J Biol Chem [Internet] 279(1):436–443. [cited 2024 Jul 20]. Available from: https://www.jbc.org/article/S0021-9258(18)52833-7/fulltext

Soares MP, Hamza I (2016) Macrophages and iron metabolism. Immunity [Internet] 44(3):492–504. [cited 2024 Jul 20]. Available from: https://www.cell.com/immunity/fulltext/S1074-7613(16)30054-1

Soares MP, Weiss G (2015) The Iron age of host–microbe interactions. EMBO Rep [Internet] 16(11):1482–1500. [cited 2024 Jul 20]. Available from: https://www.embopress.org/doi/10.15252/embr.201540558

Soe-Lin S, Apte SS, Andriopoulos B, Andrews MC, Schranzhofer M, Kahawita T, Garcia-Santos D, Ponka P (2009) Nramp1 promotes efficient macrophage recycling of iron following erythrophagocytosis in vivo. Proc Natl Acad Sci USA [Internet] 106(14):5960–5965. [cited 2024 Jul 19]. Available from: https://pnas.org/doi/full/10.1073/pnas.0900808106

Sottile R, Federico G, Garofalo C, Tallerico R, Faniello MC, Quaresima B, Cristiani CM, Di Sanzo M, Cuda G, Ventura V, Wagner AK, Contrò G, Perrotti N, Gulletta E, Ferrone S, Kärre K, Costanzo FS, Carlomagno F, Carbone E (2019) Iron and ferritin modulate MHC class I expression and NK cell recognition. Front Immunol [Internet] 10:224. [cited 2024 Jul 20]. Available from: https://www.frontiersin.org/articles/10.3389/fimmu.2019.00224/full

Sousa Gerós A, Simmons A, Drakesmith H, Aulicino A, Frost JN (2020) The battle for iron in enteric infections. Immunology [Internet] 161(3):186–199. [cited 2024 Jul 20]. Available from: https://onlinelibrary.wiley.com/doi/10.1111/imm.13236

Sow FB, Florence WC, Satoskar AR, Schlesinger LS, Zwilling BS, Lafuse WP (2007) Expression and localization of hepcidin in macrophages: a role in host defense against tuberculosis. J Leukoc Biol [Internet] 82(4):934–945. [cited 2024 Jul 19]. Available from: https://academic.oup.com/jleukbio/article-abstract/82/4/934/6975321

Stauff DL, Skaar EP (2009) The heme sensor system of Staphylococcus aureus. Bact Sens Signal [Internet] 16:120–135. [cited 2024 Jul 20]. Available from: https://karger.com/Article/Abstract/219376

Steinbicker AU, Bartnikas TB, Lohmeyer LK, Leyton P, Mayeur C, Kao SM, Pappas AE, Peterson RT, Bloch DB, Yu PB, Fleming MD, Bloch KD (2011) Perturbation of hepcidin expression by BMP type I receptor deletion induces iron overload in mice. Blood J Am Soc Hematol [Internet] 118(15):4224–4230. [cited 2024 Jul 19]. Available from: https://ashpublications.org/blood/article-abstract/118/15/4224/29052

Stuckey R, Aldridge T, Lim FL, Moore DJ, Tinwell H, Doherty N, Davies R, Smith AG, Kimber I, Ashby J, Orphanides G, Moggs JG (2006) Induction of iron homeostasis genes during estrogen-induced uterine growth and differentiation. Mol Cell Endocrinol [Internet] 253(1–2):22–29. [cited 2024 Jul 19]. Available from: https://www.sciencedirect.com/science/article/pii/S0303720706001122

Tanno T, Bhanu NV, Oneal PA, Goh SH, Staker P, Lee YT, Moroney JW, Reed CH, Luban NL, Wang RH, Eling TE, Childs R, Ganz T, Leitman SF, Fucharoen S, Miller JL (2007) High levels of GDF15 in thalassemia suppress expression of the iron regulatory protein hepcidin. Nat Med [Internet] 13(9):1096–1101. [cited 2024 Jul 19]. Available from: https://www.nature.com/articles/nm1629

Tanno T, Porayette P, Sripichai O, Noh SJ, Byrnes C, Bhupatiraju A, Lee YT, Goodnough JB, Harandi O, Ganz T, Paulson RF, Miller JL (2009) Identification of TWSG1 as a second novel erythroid regulator of hepcidin expression in murine and human cells. Blood J Am Soc Hematol [Internet] 114(1):181–186. [cited 2024 Jul 19]. Available from: https://ashpublications.org/blood/article-abstract/114/1/181/26093

Torres VJ, Pishchany G, Humayun M, Schneewind O, Skaar EP (2006) *Staphylococcus aureus* IsdB is a hemoglobin receptor required for heme iron utilization. J Bacteriol [Internet] 188(24):8421–8429. [cited 2024 Jul 20]. Available from: https://journals.asm.org/doi/10.1128/JB.01335-06

Torti FM, Torti SV (2002) Regulation of ferritin genes and protein. Blood J Am Soc Hematol [Internet] 99(10):3505–3516. [cited 2024 Jul 20]. Available from: https://ashpublications.org/blood/article-abstract/99/10/3505/106937

Troxell B, Hassan HM (2013) Transcriptional regulation by Ferric Uptake Regulator (Fur) in pathogenic bacteria. Front Cell Infect Microbiol [Internet] 3:59. [cited 2024 Jul 20]. Available from: https://www.frontiersin.org/articles/10.3389/fcimb.2013.00059/full

Truksa J, Peng H, Lee P, Beutler E (2006) Bone morphogenetic proteins 2, 4, and 9 stimulate murine hepcidin 1 expression independently of Hfe, transferrin receptor 2 (Tfr2), and IL-6. Proc Natl Acad Sci USA [Internet] 103(27):10289–10293. [cited 2024 Jul 19]. Available from: https://pnas.org/doi/full/10.1073/pnas.0603124103

Truksa J, Lee P, Beutler E (2009) Two BMP responsive elements, STAT, and bZIP/HNF4/COUP motifs of the hepcidin promoter are critical for BMP, SMAD1, and HJV responsiveness. Blood J Am Soc Hematol [Internet] 113(3):688–695. [cited 2024 Jul 19]. Available from: https://ashpublications.org/blood/article-abstract/113/3/688/25164

Ulas M, Cay M (2011) Effects of 17β-estradiol and vitamin E treatments on blood trace element and antioxidant enzyme levels in ovariectomized rats. Biol Trace Elem Res [Internet] 139(3):347–355. [cited 2024 Jul 19]. Available from: http://link.springer.com/10.1007/s12011-010-8669-2

Umeyama T, Kaneko A, Nagai Y, Hanaoka N, Tanabe K, Takano Y, Niimi M, Uehara Y (2005) Candida albicans protein kinase CaHsl1p regulates cell elongation and virulence. Mol Microbiol [Internet] 55(2):381–395. [cited 2024 Jul 20]. Available from: https://onlinelibrary.wiley.com/doi/10.1111/j.1365-2958.2004.04405.x

Urrutia P, Aguirre P, Esparza A, Tapia V, Mena NP, Arredondo M, González-Billault C, Núñez MT (2013) Inflammation alters the expression of DMT 1, FPN 1 and hepcidin, and it causes iron accumulation in central nervous system cells. J Neurochem [Internet] 126(4):541–549. [cited 2024 Jul 19]. Available from: https://onlinelibrary.wiley.com/doi/10.1111/jnc.12244

Valappil SP, Ready D, Neel EAA, Pickup DM, Chrzanowski W, O'Dell LA, Chrzanowski W, Pratten J, Newport RJ, Smith ME, Wilson M, Knowles JC (2008) Antimicrobial gallium-doped phosphate-based glasses. Adv Funct Mater [Internet] 18(5):732–741. [cited 2024 Jul 20]. Available from: https://onlinelibrary.wiley.com/doi/10.1002/adfm.200700931

Valore EV, Ganz T (2008) Posttranslational processing of hepcidin in human hepatocytes is mediated by the prohormone convertase furin. Blood Cells Mol Dis [Internet] 40(1):132–138. [cited 2024 Jul 19]. Available from: https://www.sciencedirect.com/science/article/pii/S1079979607001738

Van Avondt K, Schimmel M, Bulder I, van Mierlo G, Nur E, van Bruggen R, Biemond BJ, Luken BM, Zeerleder S (2016) Circulating iron in patients with sickle cell disease mediates the release of neutrophil extracellular traps. Blood [Internet] 128(22):161. [cited 2024 Jul 20]. Available from: https://www.sciencedirect.com/science/article/pii/S0006497119301624

Vanderpool CK, Armstrong SK (2003) Heme-responsive transcriptional activation of *Bordetella bhu* genes. J Bacteriol [Internet] 185(3):909–917. [cited 2024 Jul 20]. Available from: https://journals.asm.org/doi/10.1128/JB.185.3.909-917.2003

Verga Falzacappa MV, Vujic Spasic M, Kessler R, Stolte J, Hentze MW, Muckenthaler MU (2007) STAT3 mediates hepatic hepcidin expression and its inflammatory stimulation. Blood [Internet] 109(1):353–358. [cited 2024 Jul 19]. Available from: https://ashpublications.org/blood/article-abstract/109/1/353/117146

Vidal S, Gros P, Skamene E (1995) Natural resistance to infection with intracellular parasites: molecular genetics identifies *Nramp1* as the *Bcg/Ity/Lsh* locus. J Leukoc Biol [Internet] 58(4):382–390. [cited 2024 Jul 20]. Available from: https://academic.oup.com/jleukbio/article/58/4/382/6979028

Völlger L, Akong-Moore K, Cox L, Goldmann O, Wang Y, Schäfer ST, Naim HY, Nizet V, von Köckritz-Blickwede M (2016) Iron-chelating agent desferrioxamine stimulates formation of neutrophil extracellular traps (NETs) in human blood-derived neutrophils. Biosci Rep [Internet] 36(3):e00333. [cited 2024 Jul 20]. Available from: https://portlandpress.com/bioscirep/article-abstract/36/3/e00333/91699

Vulpe CD, Kuo YM, Murphy TL, Cowley L, Askwith C, Libina N, Gitschier J, Anderson GJ (1999) Hephaestin, a ceruloplasmin homologue implicated in intestinal iron transport, is defective in the sla mouse. Nat Genet [Internet] 21(2):195–199. [cited 2024 Jul 19]. Available from: https://www.nature.com/articles/ng0299_195

Waldron KJ, Rutherford JC, Ford D, Robinson NJ (2009) Metalloproteins and metal sensing. Nature [Internet] 460(7257):823–830. [cited 2024 Jul 18]. Available from: https://www.nature.com/articles/nature08300

Wandersman C, Delepelaire P (2004) Bacterial iron sources: from siderophores to hemophores. Annu Rev Microbiol [Internet] 58(1):611–647. [cited 2024 Jul 20]. Available from: https://www.annualreviews.org/doi/10.1146/annurev.micro.58.030603.123811

Wang RH, Li C, Xu X, Zheng Y, Xiao C, Zerfas P, Cooperman S, Eckhaus M, Rouault T, Mishra L, Deng CX (2005) A role of SMAD4 in iron metabolism through the positive regulation of hepcidin expression. Cell Metab [Internet] 2(6):399–409. [cited 2024 Jul 19]. Available from: https://www.cell.com/AJHG/fulltext/S1550-4131(05)00337-2

Weinberg ED (1975) Iron and susceptibility to infectious diseases. [cited 2024 Jul 20]. Available from: https://www.cabidigitallibrary.org/doi/full/10.5555/19750419639

Weiss G, Schett G (2013) Anaemia in inflammatory rheumatic diseases. Nat Rev Rheumatol [Internet] 9(4):205–215. [cited 2024 Jul 20]. Available from: https://www.nature.com/articles/nrrheum.2012.183

Wencewicz TA, Möllmann U, Long TE, Miller MJ (2009) Is drug release necessary for antimicrobial activity of siderophore-drug conjugates? Syntheses and biological studies of the naturally occurring salmycin "Trojan Horse" antibiotics and synthetic desferridanoxamine-antibiotic conjugates. Biometals [Internet] 22(4):633–648. [cited 2024 Jul 20]. Available from: http://link.springer.com/10.1007/s10534-009-9218-3

West AP, Bennett MJ, Sellers VM, Andrews NC, Enns CA, Bjorkman PJ (2000) Comparison of the interactions of transferrin receptor and transferrin receptor 2 with transferrin and the hereditary hemochromatosis protein HFE. J Biol Chem [Internet] 275(49):38135–38138. [cited 2024 Jul 18]. Available from: https://www.jbc.org/article/S0021-9258(19)55936-1/abstract

White A, Ding X, Vanderspek JC, Murphy JR, Ringe D (1998) Structure of the metal-ion-activated diphtheria toxin repressor/tox operator complex. Nature [Internet] 394(6692):502–506. [cited 2024 Jul 20]. Available from: https://www.nature.com/articles/28893

White C, Yuan X, Schmidt PJ, Bresciani E, Samuel TK, Campagna D, Hall C, Bishop K, Calicchio ML, Lapierre A, Ward DM, Liu P, Fleming MD, Hamza I (2013) HRG1 is essential for heme transport from the phagolysosome of macrophages during erythrophagocytosis. Cell Metab [Internet] 17(2):261–270. [cited 2024 Jul 18]. Available from: https://www.cell.com/cell-metabolism/fulltext/S1550-4131(13)00013-2?code=cell-site&switch=standard&mobileUi=0

Xiao X, Yeoh BS, Vijay-Kumar M (2017) Lipocalin 2: an emerging player in iron homeostasis and inflammation. Annu Rev Nutr [Internet] 37(1):103–130. [cited 2024 Jul 20]. Available from: https://www.annualreviews.org/doi/10.1146/annurev-nutr-071816-064559

Yang Q, Jian J, Katz S, Abramson SB, Huang X (2012) 17β-Estradiol inhibits iron hormone hepcidin through an estrogen responsive element half-site. Endocrinology 153(7):3170–3178

Yang Y, Tian Q, Wu S, Li Y, Yang K, Yan Y, Shang L, Li A, Zhang L (2021) Blue light-triggered Fe2+-release from monodispersed ferrihydrite nanoparticles for cancer iron therapy. Biomaterials [Internet] 271:120739. [cited 2024 Jul 20]. Available from: https://www.sciencedirect.com/science/article/pii/S0142961221000910?casa_token=POwEpOMeq2MAAAAA:B7_36A55-rSpuPuL_ys1zuR3Cgsa3uwNV8CORZ1xLSli-DA0X8sPVFPIk4XC6J6nNCjv6s8KaEWA

Yang N, Venezuela J, Almathami S, Dargusch M (2022) Zinc-nutrient element based alloys for absorbable wound closure devices fabrication: current status, challenges, and future prospects. Biomaterials[Internet]280:121301.[cited2024Jul20].Availablefrom:https://www.sciencedirect.com/science/article/pii/S0142961221006578?casa_token=m939WGlgvTgAAAAA:oeJGjhGLKWXhPBPeSZnuw8s0qLHoT2Fb-Jx7p0eCsJGpMuqzyeG6_ACLqViN-9Q63KI26jVNkuQ

Zambolin S, Clantin B, Chami M, Hoos S, Haouz A, Villeret V, Delepelaire P (2016) Structural basis for haem piracy from host haemopexin by Haemophilus influenzae. Nat Commun [Internet] 7(1):11590. [cited 2024 Jul 20]. Available from: https://www.nature.com/articles/ncomms11590

Zhanel GG, Golden AR, Zelenitsky S, Wiebe K, Lawrence CK, Adam HJ, Idowu T, Domalaon R, Schweizer F, Zhanel MA, Lagacé-Wiens PRS, Walkty AJ, Noreddin A, Lynch Iii JP, Karlowsky JA (2019) Cefiderocol: a siderophore cephalosporin with activity against carbapenem-resistant and multidrug-resistant Gram-negative bacilli. Drugs [Internet] 79(3):271–289. [cited 2024 Jul 20]. Available from: http://link.springer.com/10.1007/s40265-019-1055-2

Zhang DL, Ghosh MC, Rouault TA (2014) The physiological functions of iron regulatory proteins in iron homeostasis-an update. Front Pharmacol [Internet] 5:124. [cited 2024 Jul 19]. Available from: https://www.frontiersin.org/articles/10.3389/fphar.2014.00124/full

Weinberg ED (1975) Iron and susceptibility to infectious disease. [illegible] Available from: [illegible]

Weiss G, Schett G (2013) Anaemia in inflammatory rheumatic diseases. Nat Rev Rheumatol 9(4):205–215 [cited 2024 Jul 26]. Available from: [illegible]

[illegible] DA, Möllmann U, Long TE, Miller MJ (2009) Is drug release necessary for antimicrobial activity of siderophore-drug conjugates? Syntheses and biological studies of the naturally [illegible]

The Mechanisms and Therapeutic Implications of Metabolic Communication in the Tumor-Immune Microenvironment

12

Mahmoud Singer, Rania Hamdy, Tasneem M. Elsayed, and Mohamed I. Husseiny

Abstract

The tumor-immune microenvironment (TME) is a dynamic and complex ecosystem where metabolic communication between tumor cells and immune cells plays a crucial role in shaping tumor progression and therapeutic responses. In the middle of tumor invasion mechanisms and immune defense interactions, metabolites play a mediator role in the fate of TME stability. This chapter provides a comprehensive analysis of the metabolic interactions within the TME, focusing on how tumor cells alter the metabolic landscape to influence immune cell function and vice versa. We will discuss the key metabolic pathways involved, the effects of metabolic changes on immune cell activity, and the implications for developing novel therapeutic strategies.

M. Singer (✉)
Department of Radiological Sciences, School of Medicine, University of California Irvine, Irvine, CA, USA
e-mail: singermk@hs.uci.edu

R. Hamdy
Research Institute for Science and Engineering (RISE), University of Sharjah, Sharjah, UAE

T. M. Elsayed
Division of Cardiology, The Lundquist Institute at Harbor-UCLA Medical Center, Los Angeles, CA, USA

M. I. Husseiny
Department of Translational Research & Cellular Therapeutics, Arthur Riggs Diabetes & Metabolism Research Institute, Beckman Research Institute, City of Hope National Medical Center, Duarte, CA, USA

S. S. M. Soliman, M. I. Husseiny (eds.), *Metabolic Dynamics in Host-Microbe Interaction*, https://doi.org/10.1007/978-981-96-1305-2_12

Keywords

Immunometabolism · Tumor microenvironment · Communication · Cancer metabolism · Metabolic pathways

12.1 Introduction

Cancer is a disease that involves uncontrolled cell proliferation, the ability to spread to different organs, and the ability to escape from the immune responses (Brown et al. 2023). Comprising tumor cells, immune cells, stromal cells, and extracellular matrix components, the tumor microenvironment (TME) has a significant influence on tumor behavior and patient outcomes. The growth and survival of cancer cells, as well as the function and activity of immune cells, are influenced by pivotal metabolic communication within this microenvironment. Delving into these metabolic interactions not only enhancing the understanding of tumor biology but also unveiling exciting opportunities for therapeutic intervention (de Visser and Joyce 2023; Ribeiro Franco et al. 2020).

12.2 Metabolic Pathways in Tumor Cells

12.2.1 The Warburg Effect (Carbohydrate Metabolism)

The Warburg metabolism refers to the phenomenon in which cancer cells primarily rely on glycolysis for energy production, even when there is enough oxygen available. The purpose of this metabolic shift is to support the quick multiplication of cells and to generate the necessary intermediates for biosynthetic processes. Metabolic reprogramming is the change from oxidative phosphorylation to aerobic glycolysis, indicating a metabolic shift (Liberti and Locasale 2016). While oxygen is available, cancer cells often opt for glycolysis, a metabolic pathway that converts glucose into lactate instead of undergoing complete oxidation in the mitochondria. Cancer cells employ aerobic glycolysis to quickly generate ATP and produce the metabolic intermediates needed for their ongoing cell growth and proliferation. Upregulation of key enzymes, including hexokinase, phosphofructokinase-1 (PFK-1), and pyruvate kinase M2 (PKM2), drives the increase in glycolysis (Vander Heiden et al. 2009). As a means of supporting elevated glycolytic activity, cancer cells commonly bolster their glucose intake by overexpressing glucose transporters, with GLUT1 being particularly prominent. This increased uptake guarantees a steady flow of glucose for glycolysis, a crucial process for meeting the intense metabolic needs of rapidly proliferating cancer cells (Cunha et al. 2023). The Warburg effect grants cancer cells the ability to adjust their metabolism in response to different microenvironmental conditions, enhancing their adaptability. Although glycolysis produces fewer ATP molecules per glucose compared to oxidative phosphorylation, it provides crucial intermediates that contribute to the synthesis of nucleotides and lipids in anabolic pathways (Bose et al. 2021). This metabolic adaptation supports tumor growth and can confer resistance to therapies. The shift to glycolysis also influences the tumor microenvironment. Increased lactate production from aerobic glycolysis contributes to acidification of the extracellular environment, which can promote tumor invasion and suppress anti-tumor immune responses, mainly T cells and macrophages. Additionally, metabolic byproducts from glycolysis can affect the behavior of the surrounding cells and tissues (Wang et al. 2021a).

The dependence of cancer cells on Warburg metabolism offers multiple possible approaches for therapy. The exploration of disrupting the glycolytic flux in cancer cells involves the investigation of inhibitors that target important glycolytic enzymes, including hexokinase and PFK-1 inhibitors. As an illustration, preclinical studies have shown that 3-bromopyruvate, which inhibits hexokinase, holds the potential for reducing tumor growth (Casas-Benito et al. 2023; Zhang et al. 2022a). Disrupting cancer cell metabolism can be achieved through the therapeutic strategy of modulating glucose transport, where targeting GLUT1 offers an alternative approach. Blocking GLUT1 inhibitors can decrease glucose absorption, hindering the glycolytic pathway, and potentially decelerating tumor growth (Marchesi et al. 2020; Pliszka and Szablewski 2021). The combination of glycolytic inhibitors with other treatments, like chemotherapy or targeted therapies, has the potential to improve the effectiveness of therapy. By focusing on the metabolic vulnerabilities of cancer cells, this approach aims to minimize the development of resistance (Zhang et al. 2022a; El-Tanani et al. 2024). By exploiting metabolic vulnerabilities and targeting interconnected metabolic pathways with glycolysis, such as those associated with lactate production or mitochondrial function, new therapeutic opportunities may arise (Muluh et al. 2023; Wolpaw and Dang 2018). Cancer biology is marked by the fundamental feature of Warburg metabolism, which involves heightened aerobic glycolysis (Fig. 12.1). Rapid tumor growth and the ability to adapt to the tumor microenvironment are supported by this metabolic shift. Promising cancer therapy strategies involve focusing on different components of the glycolytic pathway (Barba et al. 2024).

Targeting glycolytic enzymes is a promising strategy in cancer therapy, focusing on the metabolic vulnerabilities of tumor cells (Abdel-Wahab et al. 2019). Hexokinase 2 (HK2), inhibitors such as 2-deoxyglucose (2-DG), and lonidamine impede glucose phosphorylation, a key step in glycolysis (Wang et al. 2023a; Garcia et al. 2019). Inhibiting pyruvate kinase M2 (PKM2) with agents like shikonin reduces glycolytic flux, impairing both energy production and biosynthesis required for tumor growth (Wang et al. 2017). Additionally, lactate dehydrogenase A (LDHA) inhibitors, including FX11 and GSK2837808A, prevent the conversion of pyruvate to lactate, further disrupting the metabolic adaptability of cancer cells and mitigating lactate-driven immune suppression (Liu et al. 2021a; Certo et al. 2021). These inhibitors aim to exploit the dependency of cancer cells on glycolysis and represent promising therapeutic strategies.

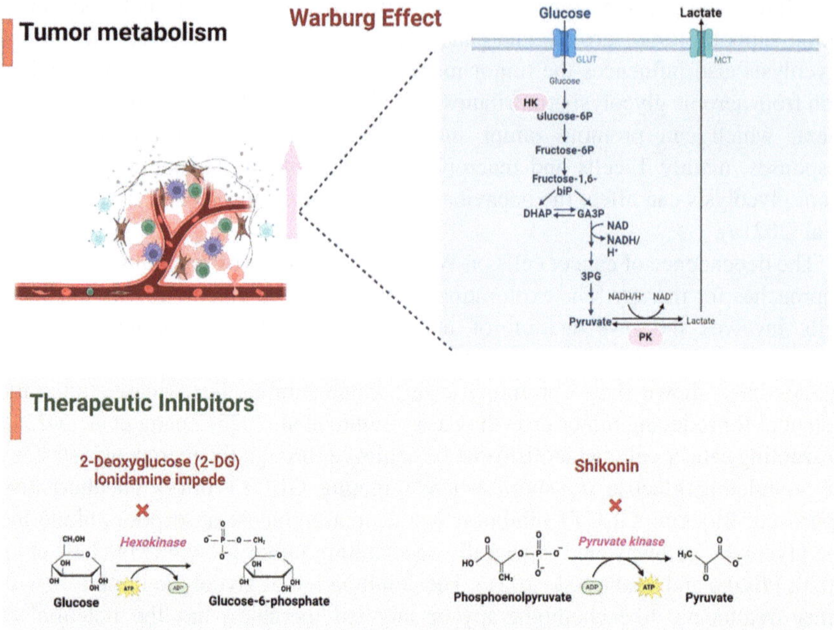

Fig. 12.1 Tumor metabolism and therapeutic inhibitors targeting glycolysis. This figure illustrates the Warburg effect in cancer cells, where glycolysis enzymes such as hexokinase (HK) and pyruvate kinase (PK) are upregulated. Inhibitors like 2-deoxyglucose (2-DG) and lonidamine target HK, while Shikonin inhibits PK, both disrupting glycolysis and reducing cancer cell energy production in cancer cells

12.2.2 Cancer Lipid Metabolism

Cancer lipid metabolism is a dynamic field that offers valuable insights into the ways malignant cells adapt lipid processing to sustain their growth and survival. Unlike typical cells, cancer cells frequently undergo significant metabolic reprogramming, which involves modifications in lipid synthesis, oxidation or degradation, and transport. These changes not only facilitate the growth of tumors but also provide new possibilities for targeted therapy (Fu et al. 2020; Zhang et al. 2022b). Elevated de novo lipogenesis, a metabolic process converting non-lipid precursors into fatty acids and other lipids, is commonly observed in cancer cells. Enzymes such as fatty acid synthase (FASN) and acetyl-CoA carboxylase (ACC) are often upregulated, facilitating the synthesis of lipids necessary for membrane biogenesis and energy storage (Butler et al. 2020; Koundouros and Poulogiannis 2020). The increased lipid synthesis is a defining characteristic of numerous cancers and is associated with a negative prognosis. The decreased activity of fatty acid oxidation (FAO) in various cancers leads to an accumulation of fatty acids within the TME (Gupta et al. 2024). By reducing reliance on FAO for energy, this shift enables cancer cells to allocate fatty acids toward anabolic processes like membrane synthesis and signaling. In tumors, the activity of enzymes essential for FAO, such as carnitine palmitoyl transferase 1 (CPT1), is frequently suppressed (De Oliveira and Liesa 2020). To meet their increased metabolic demands, cancer cells frequently alter lipid transport mechanisms. Enhancing lipid uptake and redistribution can occur through changes in the expression of lipid transport proteins, including fatty acid-binding proteins (FABPs) and scavenger receptor class B type 1 (SR-B1) (Zhang et al. 2022b). Furthermore, the increased storage of lipids in lipid droplets serves as a vital reserve, sustaining cellular functions during phases of accelerated development. Within the tumor microenvironment, lipids play a pivotal role in modulating cell signaling, immune responses, and the progression of tumors. Eicosanoids and sphingolipids, which are bioactive lipids, have the ability to regulate inflammatory responses, promote angiogenesis, and enhance cell survival, thereby facilitating tumor growth and metastasis (Delmas et al. 2023).

The metabolism of lipids is altered in cancer cells to facilitate the synthesis of membranes and signaling pathways that are critical for their proliferation and survival. The upregulation of de novo lipogenesis and fatty acid oxidation is common in cancer cells. Disease progression and metastasis in prostate cancer are linked to the upregulation of FASN expression (Zekovic et al. 2023). The presence of distinctive lipid metabolic alterations in cancer opens multiple avenues for therapeutic intervention. Potential cancer therapies are being explored by targeting lipid synthesis using inhibitors for key enzymes, including FASN inhibitors. By targeting the lipid supply necessary for tumor growth and survival, these inhibitors work to impede the progression of cancer (Fu et al. 2020; Butler et al. 2020). In addition, lipid transport proteins and pathways can be targeted to disrupt the uptake and utilization of lipids in cancer cells, presenting another potential approach. By targeting the metabolic flexibility of tumors, this approach has the potential to significantly enhance the efficacy of treatments. By combining lipid metabolism inhibitors with conventional treatments like chemotherapy or targeted therapies, there is a potential for synergistic effects that can help overcome resistance (Gupta et al. 2024; Beloribi-Djefaflia et al. 2016; Wang et al. 2023b). There is potential for improved therapeutic outcomes and a targeted approach to addressing the metabolic vulnerabilities of cancer cells (Fig. 12.2). Lipid availability in the TME plays a significant role in influencing immune cell function, including their activation, differentiation, polarization, and response to tumor cells (Corn et al. 2020).

12.2.3 Amino Acid Metabolism

Cancer biology heavily relies on amino acid metabolism, as it affects tumor growth, survival, and the development of drug resistance. To fulfill their heightened demands for protein synthesis, energy production, and redox balance, cancer cells demonstrate modified amino acid metabolism (Chen et al. 2024). The altered metabolism of serine and glycine, crucial for one-carbon metabolism, is commonly observed in cancer cells. Serine acts as a precursor for glycine synthesis, and both amino acids are essential to produce purines, pyrimidines, and the antioxidant glutathione (Pan

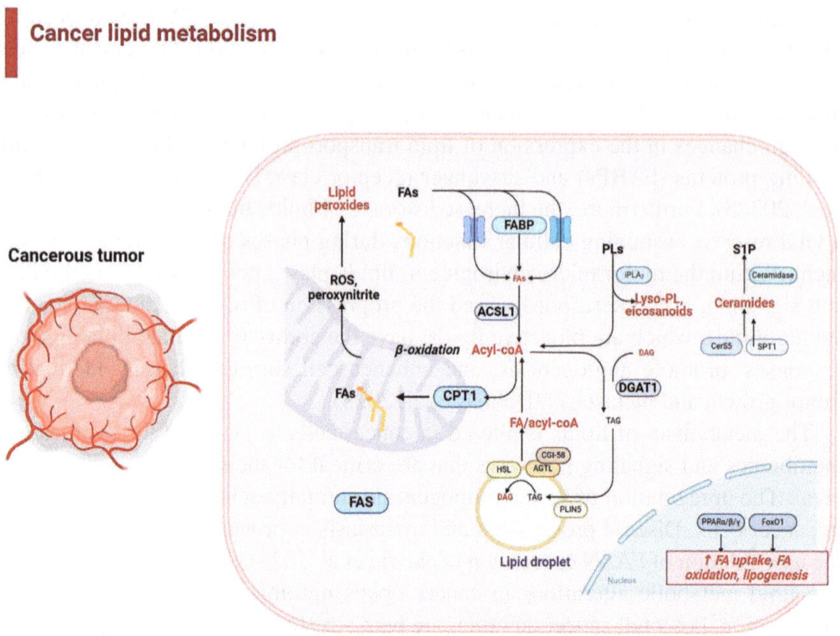

Fig. 12.2 Cancer lipid metabolism pathways and their role in tumor growth. This figure illustrates the key lipid metabolic pathways involved in cancer cell survival and growth. Enzymes like Fatty Acid Synthase (FAS), Acyl-CoA Synthetase Long Chain 1 (ACSL1), Carnitine Palmitoyl Transferase 1 (CPT1), and lipid-binding proteins such as FABP and Ceramides are involved in processing fatty acids (FAs) and other lipid molecules. Lipid droplets act as energy reserves for fast-growing cancer cell and tumor progression

et al. 2021). The serine-glycine pathway is frequently upregulated in cancer, providing essential components for cell proliferation and maintaining redox balance (Amelio et al. 2014). Phosphoglycerate dehydrogenase (PHGDH) is a key enzyme in serine synthesis. Inhibitors of PHGDH aim to block serine production, disrupting nucleotide synthesis and tumor growth (Mullarky et al. 2019). Serine hydroxymethyltransferase (SHMT) is essential for serine-glycine interconversion and one-carbon metabolism. SHMT inhibitors, such as SHIN1, are being investigated for their potential to impair cancer cell proliferation (Sun et al. 2023). One notable feature of cancer cells is their modified amino acid metabolism, which involves elevated consumption of tryptophan and other vital amino acids. The metabolism of tryptophan to kynurenine by indoleamine 2,3-dioxygenase (IDO) leads to immune tolerance and suppression. IDO inhibitors, such as Epacadostat, are being explored to reverse this immune suppression and enhance the efficacy of immunotherapies by restoring T cell function, thus boosting the body's anti-tumor immune response (Tang et al. 2021). 6-Diazo-5-oxo-L-norleucine (DON) is another glutamine antagonist that targets multiple enzymes in glutamine metabolism, reducing its availability for cancer cell proliferation (Pham 2020). Certain cancers, such as hepatocellular carcinoma and melanoma, cannot synthesize arginine due to the loss of arginine-succinate

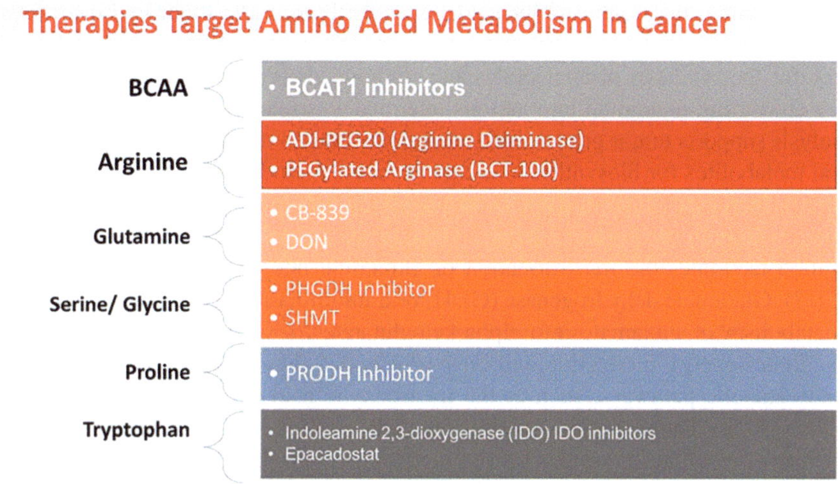

Fig. 12.3 Therapies targeting amino acid metabolism in cancer. The therapies are organized by the metabolic enzymes or pathways they inhibit, with specific drugs listed under each category. Where, BCAT1 Branched-Chain Aminotransferase 1, ADI-PEG20 Arginine Deiminase Polyethylene Glycol-conjugated, PEGylated Arginase (BCT-100) Arginase enzyme attached to polyethylene glycol, CB-839 A selective inhibitor of glutaminase, DON 6-Diazo-5-oxo-L-norleucine, a glutamine antagonist, PHGDH Phosphoglycerate Dehydrogenase, SHMT Serine Hydroxymethyltransferase, PRODH Proline Dehydrogenase, IDO Indoleamine 2,3-Dioxygenase

synthetase 1 (ASS1) and depend on external sources of arginine (Sun and Zhao 2022). ADI-PEG20 (Arginine Deiminase) enzyme depletes extracellular arginine, starving cancer cells that are unable to synthesize it (Kim et al. 2009). PEGylated Arginase (BCT-100) breaks down circulating arginine, making it unavailable for cancer cells reliant on it for growth (Burke 2022).

Essential for protein synthesis and energy production, branched-chain amino acids (BCAAs) including leucine, isoleucine, and valine are of utmost importance (Basson et al. 2023). BCAA metabolism is often disrupted in various cancers, resulting in heightened uptake and altered catabolism (Fig. 12.3). The heightened presence of BCAAs can enhance tumor growth by bolstering mTOR signaling, a key regulator of cellular metabolism and growth. Conversely, certain tumors showcase diminished levels of BCAAs, indicating modifications in their metabolic requirements. Branched-chain amino acid aminotransferase 1 (BCAT1) inhibitors are implicated in tumor progression making them a key target for cancer therapy (Nong et al. 2022).

The methionine cycle, crucial for the conversion of methionine into homocysteine and cysteine, is frequently disturbed in cancer cells (Jung et al. 2021). Involved in epigenetic modifications and cell signaling, the synthesis of S-adenosylmethionine (SAM) heavily depends on this essential cycle. To sustain irregular methylation patterns and perpetuate genetic instability, cancer cells may modify methionine metabolism (Pascale et al. 2022).

Glutamine, an essential amino acid, is frequently over-consumed by cancer cells. Glutamine aids in anabolic processes by supplying the necessary carbon and nitrogen for the synthesis of nucleotides and amino acids (Wise and Thompson 2010). By converting glutamine into various intermediates like α-ketoglutarate, glutaminolysis supports tumor growth and survival. It plays a vital role in providing essential metabolites for biosynthesis, energy production, and redox balance (Yoo et al. 2020). Increased glutamine uptake is a common trait in cancer cells, as it supplies them with both carbon and nitrogen. Once inside the cell, the enzyme glutaminase (GLS) facilitates the transformation of glutamine into glutamate (Cluntun et al. 2017). Glutamate dehydrogenase (GDH) or transaminases play a role in the further metabolism of glutamate into alpha-ketoglutarate (α-KG). The process not only facilitates energy production and biosynthesis but also supports various anabolic and catabolic pathways, such as the tricarboxylic acid (TCA) cycle (Arnold and Finley 2023). The conversion of glutamine to α-KG not only produces ATP but also generates essential intermediates required for cell proliferation. The synthesis of nucleotides, amino acids, and lipids, which are crucial for tumor growth, is supported by the TCA cycle intermediates derived from glutamine. The reliance on glutamine highlights its significance as a vital energy source for cancer cells that rapidly multiply (Altman et al. 2016).

Within cancer cells, glutaminolysis plays a pivotal role in preserving the balance of redox reactions. The metabolic pathway involving the transformation of glutamine to glutamate and α-KG results in the production of reduced nicotinamide adenine dinucleotide (NADH) and various other reducing agents (Jin et al. 2016). These components play a crucial role in managing oxidative stress and aiding in the elimination of reactive oxygen species (ROS). Redox homeostasis is boosted by glutaminolysis, allowing cancer cells to thrive in the challenging tumor microenvironment. In the tumor microenvironment, cancer cells frequently adjust their glutaminolytic activity in response to the specific conditions of the tumor microenvironment. For instance, when faced with hypoxic conditions, the body may rely more heavily on glutamine for its metabolic needs. Furthermore, the accessibility of glutamine can have a profound impact on the immune cell makeup inside tumors, altering immune responses and potentially playing a role in immune evasion (Aboelella et al. 2021; Wang et al. 2024).

Several potential therapeutic targets arise from the metabolic reliance on glutaminolysis in cancer cells. The goal of inhibitors like CB-839 (Telaglenastat) and other inhibitors in clinical development is to interrupt the process of converting glutamine to glutamate by targeting glutaminases (Kao et al. 2022). The presence of these inhibitors can restrict the availability of crucial metabolites, leading to a decline in tumor growth. According to preliminary studies, the targeting of GLS has shown promise in the treatment of multiple cancer types, including breast, lung, and pancreatic cancers (Jin et al. 2023). Ongoing exploration involves therapeutic approaches that aim to interfere with downstream mechanisms involving glutamate or α-KG, including GDH inhibition or manipulation of TCA cycle function. The implementation of these approaches may disrupt the metabolic flexibility of cancer cells, thereby diminishing their ability to proliferate. By combining glutaminolysis inhibitors with conventional treatments like chemotherapy or targeted therapies, there is potential to increase the effectiveness of therapy. These combinations offer a promising approach to improving patient outcomes by targeting cancer metabolism and overcoming resistance mechanisms (Lukey et al. 2013).

The changes in how cancer cells process amino acids present multiple potential treatment options, as they contribute to the activation of cancer-causing signaling pathways and can impact the immune system's response by promoting tumor growth, energy generation, and maintaining redox equilibrium (Wei et al. 2020). When tumor cells take up more glutamine, it can lead to reduced glutamine levels in the TME, which can impact the metabolism of nearby immune cells. Modifying the metabolism of glutamine can have an impact on the functioning of immune cells, leading to immune suppression or tolerance (Wang et al. 2024; Li et al. 2021). The targeting of glutaminolysis presents exciting prospects for cancer treatment. To improve patient outcomes and devise effective treatments, it is vital to delve deeper into the mechanisms that drive glutaminolysis and its role in the advancement of cancer (Shen et al. 2021). Inhibiting specific enzymes involved in the catabolism of serine-glycine pathways, methionine cycle, or BCAA metabolism through pharmacological interventions can modify the epigenetic landscape of tumors, resulting in impaired tumor growth, reduced amino acid availability, and disrupted cancer cell growth and proliferation (Reina-Campos et al. 2020).

12.3 Metabolic Interactions Between Tumor Cells and Immune Cells

12.3.1 Macrophages

A key factor in cancer progression and treatment efficacy is the dynamic relationship between tumor cells and macrophages within the TME. Macrophages, with their ability to switch between pro-inflammatory (M1) and anti-inflammatory (M2) phenotypes, have a profound impact on tumor metabolism and vice versa. Exploring the intricacies of these metabolic interactions unveils crucial insights into tumor biology and potential therapeutic approaches (Di Somma et al. 2021). Tumor cells tend to release substances like cytokines, growth factors, and extracellular matrix components, which prompt macrophages to shift toward the M2 phenotype. The primary role of M2 macrophages is to promote tissue repair and dampen immune activity. Their role in supporting tumor growth involves stimulating the formation of new blood vessels, suppressing the body's immune system's ability to fight tumors, and facilitating the spread of cancer cells (Kerneur et al. 2022). The polarization process is influenced by factors like interleukin-10 (IL-10) and transforming growth factor-beta (TGF-β), which can bring about changes in macrophage metabolism and function. Within the TME, macrophages display metabolic modifications that facilitate their crucial function in driving tumor progression (Wang et al. 2021b). The energy production mechanisms differ between M1 and M2 macrophages, with M2 macrophages favoring oxidative phosphorylation (OXPHOS) and FAO over

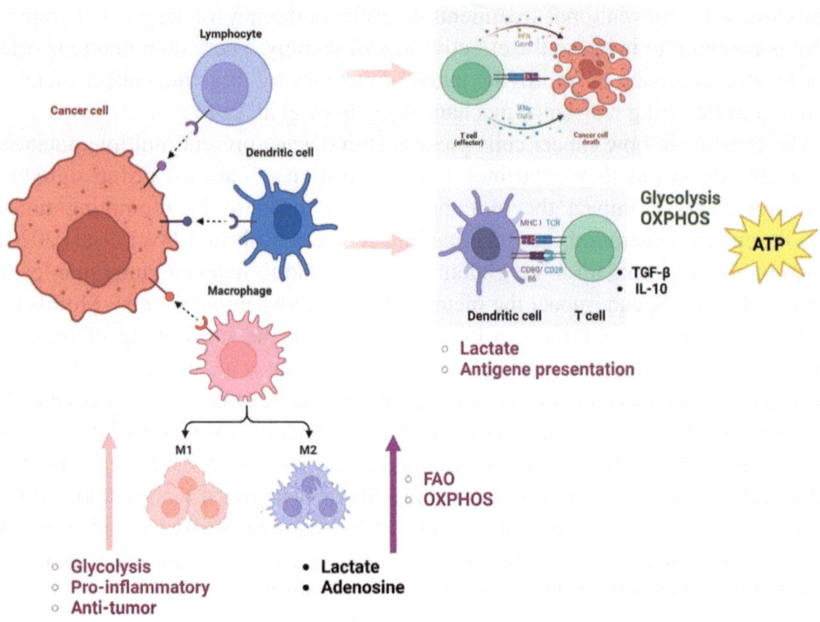

Fig. 12.4 Interplay between cancer cells and immune cells in the tumor microenvironment. This figure illustrates the interactions between cancer cells, immune cells (lymphocytes, dendritic cells, and macrophages), and metabolic processes within the tumor microenvironment. Cancer cells communicate with dendritic cells and macrophages to modulate the immune response, with key shifts between pro-inflammatory and anti-tumor states (M1 macrophages) and immunosuppressive states (M2 macrophages). The metabolic processes of glycolysis and oxidative phosphorylation (OXPHOS) in immune cells and cancer cells are highlighted. Lactate produced by cancer cells can impair antigen presentation and support immune evasion. Adenosine and lactate also play critical roles in shaping the tumor microenvironment, influencing immune cell polarization and metabolic activity. IFN-γ Interferon Gamma, TNF-α Tumor Necrosis Factor Alpha, TGF-β Transforming Growth Factor-Beta, IL-10 Interleukin 10, FAO Fatty Acid Oxidation, OXPHOS Oxidative Phosphorylation

glycolytic metabolism. This change in metabolism enables M2 macrophages to preserve their immune regulatory functions and secret factors that facilitate the progression of tumors. In contrast, M1 macrophages exhibit heightened glycolytic activity and release pro-inflammatory cytokines, bolstering their role in promoting anti-tumor immunity (Viola et al. 2019) (Fig. 12.4).

Metabolites, such as lactate and adenosine, released by tumor cells, can modulate macrophage function by altering their metabolism. Increased glycolysis in tumor cells leads to the release of high levels of lactate, which can acidify the TME and promote macrophage polarization toward the M2 phenotype (Tao et al. 2023). Both tumor cells and macrophages can produce adenosine, a compound that can impede the functioning of M1 macrophages and promote M2 polarization. The metabolic changes that occur create an ideal environment for tumor growth, enabling the tumor to evade the immune system and promote the formation of new blood vessels (Zhang and Sioud 2023). The metabolic profile of the TME can be influenced by the competition for nutrients between tumor cells and macrophages. The scarcity of glucose and glutamine caused by tumor cells' excessive consumption hinders macrophages' access to these vital resources (Elia and Haigis 2021). The metabolic remodeling caused by this competition can significantly influence the polarization and function of macrophages (Fig. 12.4). Moreover, tumor cells can release substances that modify the way macrophages absorb nutrients and metabolize, causing additional effects on the tumor microenvironment (Chen et al. 2023a).

Shifting M2 macrophages toward an M1-like phenotype through modulation of macrophage polarization has the potential to strengthen anti-cancer immunity. In order to address this issue, researchers have explored strategies such as inhibiting M2-associated signaling pathways or introducing agents that induce M1 polarization (Kerneur et al. 2022). Disrupting the supportive role of macrophages in cancer progression can be achieved by targeting key metabolic pathways involved in tumor-macrophage interactions. Inhibitors that specifically target lactate production or adenosine signaling are potential options (Li et al. 2023; You et al. 2023). For example, by blocking lactate transport or adenosine receptors, their influence on macrophage polarization could be reduced. The addition of metabolic inhibitors to immunotherapy or conventional treatments could potentially boost therapeutic efficacy (Noe et al. 2021). These combination strategies aim to disrupt the metabolic interactions between tumor cells and macrophages, leading to improved immune response and reduced tumor growth (Mantovani et al. 2022). Modifying macrophage function and inhibiting tumor progression could be achieved by investigating inhibitors that specifically target enzymes involved in tumor-induced macrophage metabolic changes, including fatty acid oxidation or glycolysis (Zhang et al. 2024).

12.3.2 T Lymphocytes

In the regulation of immune responses and tumor progression, the metabolic interactions between tumor cells and T cells within the TME are of utmost importance. The function and efficacy of T cells, specifically cytotoxic T lymphocytes (CTLs) and regulatory T cells (Tregs), are greatly impacted by their metabolic state, which affects their ability to fight against tumors. In a different vein, tumor cells actively reshape their metabolic terrain to evade immune monitoring and foster their own development (Ganjoo et al. 2023; Jung and Le 2021). In the presence of tumor cells, the environment becomes hostile with low oxygen levels, nutrient scarcity, and the buildup of metabolic byproducts like lactate (Lim et al. 2020). These conditions can compromise the ability of T cells to function properly and survive. In the TME, the presence of elevated lactate levels and low pH can impede the ability of T cells to activate and multiply. In addition, tumor cells can modify the accessibility of important nutrients like glucose and amino acids, leading to alterations in T cell metabolism and function. To function optimally, T cells rely heavily on their metabolic state (Heintzman et al. 2022). Rapid energy production and proliferation upon activation are achieved by effector T cells, such as CTLs, through their heavy reliance on glycolysis. On the other hand, Tregs, which play a role in promoting immune tolerance and suppressing anti-tumor responses, primarily employ FAO and OXPHOS as their main energy sources (Rangel Rivera et al. 2021). The suppressive function of Tregs is associated with a metabolic shift from glycolysis to FAO, which can be exploited by tumor cells to hinder anti-tumor immune responses (Wang et al. 2023c).

Tumor cells often employ the tactic of inhibiting immune responses by expressing immune checkpoint molecules like PD-L1, which interact with T cell receptors like PD-1. The impact of these checkpoints on T cell metabolism can be observed through their modification of the signaling pathways associated with glycolysis and oxidative phosphorylation (He and Xu 2020). PD-1 signaling, for instance, has been demonstrated to alter the metabolism of T cells, diminishing their effectiveness and making them more prone to exhaustion. Nutrient scarcity triggers fierce competition between tumor cells and T cells in the TME, as both vie for essential resources such as glucose, amino acids, and lipids (Patsoukis et al. 2015). Immune cells are often deprived of necessary resources due to the dominance of tumor cells, leading to metabolic starvation. Moreover, tumors can release metabolites, including adenosine, that actively hinder the function of T cells and foster immune tolerance (Ganjoo et al. 2023). The accumulation of these metabolites can skew T cell metabolism toward an exhausted or anergic state, further aiding tumor-immune evasion (Xuekai et al. 2024).

Restoring T cell function could be achieved by targeting the metabolic pathways that tumors have disrupted. For example, by boosting T cell glycolysis or preventing lactate buildup, it is possible to enhance T cell function in the TME (Cao et al. 2023). Enhancing the effectiveness of immune responses against tumors can be achieved by utilizing drugs that selectively target metabolic pathways linked to T cell exhaustion or by restoring the conditions of the TME (Bader et al. 2020). The use of checkpoint inhibitors to obstruct PD-1/PD-L1 interactions or other inhibitory pathways can revive T cell activity and counter the metabolic suppression enforced by tumors. The synergy between these inhibitors and T cell metabolism-modulating agents could significantly boost the immune system's ability to combat tumors. To improve T cell function in the nutrient-depleted TME, it may be beneficial to provide essential nutrients or precursors that are lacking, such as glucose or specific amino acids (Wojtukiewicz et al. 2021). The potential of this approach lies in its ability to synergize with other immunotherapies, resulting in a stronger response from T cells against tumors. The suppression of Treg function, including metabolic pathways like FAO, could be targeted to enhance anti-tumor immunity by reducing their suppressive effects (Bader et al. 2020). A particular focus lies on strategies that aim to preserve the function of effector T cells while specifically targeting Treg metabolism (Fig. 12.2). The interplay of metabolic processes between tumor cells and T cells is crucial in determining the strength of immune responses and has a direct influence on the advancement of tumors (Zeng and Chi 2013). The impaired function of T cells can be attributed to the metabolic changes induced by tumors, and conversely, the metabolic activity of T cells directly impacts their ability to combat tumors. Exploring these interactions can provide valuable insights that can guide the development of therapeutic strategies to enhance immune responses and improve the effectiveness of cancer treatments (Ganjoo et al. 2023).

12.3.3 Dendritic Cells (DCs)

As antigen-presenting cells, dendritic cells (DCs) hold a vital position in the immune system, responsible for initiating and regulating immune responses. Metabolic crosstalk between DCs and tumor cells plays a pivotal role in shaping the anti-tumor immune response within the TME (Del Prete et al. 2023). The metabolic conditions created by tumor cells pose a significant challenge for DCs, with nutrient deprivation, hypoxia, and the presence of metabolic byproducts like lactate. Under these conditions, the metabolism and function of DCs can undergo modifications, potentially influencing their overall capabilities; for example, the presence of high lactate levels in the TME results in acidification of the environment, creating barriers for DC maturation and antigen presentation (Xia et al. 2021). Not only do tumors impair DC function, but they also deplete essential nutrients like glucose and amino acids, which in turn reduces their ability to activate T cells. DCs predominantly utilize OXPHOS and glycolysis as their energy sources, and the equilibrium between these pathways influences how efficiently they process and present antigens (Leone and Powell 2020). Altered metabolic conditions in the TME can drive DC metabolism toward pathways that are less conducive to supporting their antigen-presenting functions. Impaired T cell activation can occur when there is metabolic stress or a buildup of inhibitory metabolites, which can interfere with antigen processing and presentation (Moller et al. 2022). Metabolites including adenosine and prostaglandins are commonly released by tumor cells, causing suppression of DC function and the promotion of immune tolerance. TME experiences an accumulation of adenosine, primarily caused by elevated extracellular ATP breakdown, which leads to the inhibition of DC maturation and antigen presentation. The reduction in DC activity can result in diminished T cell activation, creating an immune-suppressive environment that promotes the growth of tumors (Cortellino and Longo 2023). Tumor cells secrete cytokines, growth factors, and extracellular matrix components that can reprogram DC metabolism and function. For example, TGF-βand IL-10 can drive DCs toward a regulatory or tolerogenic phenotype, impairing their ability to initiate effective anti-tumor immune responses. This reprogramming often involves changes in DC metabolic pathways, such as shifts from glycolysis to FAO, which can further inhibit their function (Zhao et al. 2021) (Fig. 12.2).

A promising therapeutic approach involves targeting the metabolic pathways that tumors alter to restore DC function. By implementing interventions that target lactate accumulation and nutrient replenishment, the maturation and antigen presentation of DCs could be enhanced. Implementing this could potentially strengthen DCs' ability to trigger T cells and initiate a more efficient immune response to combat tumors. The overall immune response can be strengthened by integrating immune checkpoint inhibitors with techniques that regulate the metabolism of DCs. Improving DC function and promoting T cell activation may be achievable by blocking inhibitory checkpoint pathways and addressing metabolic challenges (Hu et al. 2024; Yu et al. 2021). The restoration of DC function and the reinforcement of anti-tumor immunity can be achieved by blocking the production or action of immunosuppressive metabolites, such as adenosine. The immune-suppressive effects of tumors on DCs might be reduced by agents that block adenosine receptors or inhibit their synthesis. Enhancing the ability of DC-based vaccines to present tumor antigens and activating T cells could be achieved by engineering them to overcome metabolic constraints. To address this, researchers can explore methods such as preconditioning DCs or genetically enhancing their antigen-presenting capabilities. The intricate nature of these interactions offers valuable insights into the mechanisms behind tumor-immune evasion, shedding light on potential strategies to bolster the body's ability to fight against cancer (Xing et al. 2023; Zahavi and Hodge 2023) (Fig. 12.2).

12.3.4 Regulatory T Cells (Tregs)

Maintaining immune tolerance and preventing overactive immune responses, regulatory T cells (Tregs) are essential components of the immune system. Tregs tend to accumulate in the TME, where they exert their suppressive effects, hindering the body's ability to mount an anti-tumor immune response. The metabolic interactions between tumor cells and Tregs play a crucial role in shaping their behavior and influencing the progression of the tumor (Oparaugo et al. 2023). The presence of tumor cells can result in an environment that is metabolically demanding, marked by a lack of nutrients, low oxygen levels, and the buildup of immunosuppressive metabolites. Tregs can adjust their metabolism to flourish in this challenging environment. Tregs, unlike effector T cells, predominantly utilize OXPHOS and FAO to produce energy (Luby and Alves-Guerra 2021). Their preference for this metabolic pathway contributes to their long-term survival and strengthens their ability to suppress functions within the TME. Metabolic byproducts like lactate, adenosine, and prostaglandins are released by tumor cells, and these substances play a role in shaping Treg metabolism and function. By creating an environment with elevated lactate levels and low pH, the TME can effectively enhance the activity of Tregs, enabling them to thrive and carry out their suppressive functions (Wang et al. 2021a). Adenosine, a byproduct of breaking down extracellular ATP, not only promotes Treg expansion and function but also suppresses effector T cell activity. By producing these metabolites, the body creates an environment that hinders the immune system, promoting tumor growth and progression. The competition for nutrients between tumor cells and Tregs can shape Treg function and behavior (Ohta and Sitkovsky 2014). The high glucose and amino acid consumption by tumor cells restricts the availability of these resources for Tregs. Tregs can adapt their metabolic pathways, including boosting FAO and utilizing different nutrients, as a strategy to sustain their activity and overcome competition.

Tregs can effectively survive and carry out their functions in the nutrient-limited tumor microenvironment thanks to this metabolic adaptation (El Sayed et al. 2021).

Factors like TGF-β and IL-10 are secreted by tumor cells, contributing to the development and functionality of Treg cells. The adoption of a more suppressive phenotype by reprogrammed Tregs, due to these factors, further strengthens their ability to inhibit anti-tumor immune responses. TGF-β plays a significant role in regulating Treg metabolism by activating metabolic pathways that facilitate their ability to suppress immune responses (Hinshaw and Shevde 2019).

Disrupting the function of Treg cells by targeting their metabolism might have the potential to strengthen the body's ability to fight tumors. Inhibiting FAO or enhancing glycolysis in Tregs. New therapeutic strategies could emerge by targeting specific metabolic enzymes or pathways in Tregs, which are agents that play a crucial role in metabolism (Kurniawan et al. 2020). The reduction of Treg function and restoration of effective anti-tumor immunity could be achieved by inhibiting the production or action of immunosuppressive metabolites, such as adenosine or lactate. To address the immune-suppressive effects of tumors and boost T cell responses, potential strategies include the administration of adenosine receptor antagonists or lactate dehydrogenase inhibitors (Leone and Powell 2021). The addition of metabolic inhibitors to immune-modulating therapies, such as immune checkpoint inhibitors, could result in synergistic effects. A promising approach involves targeting Treg metabolism and immune checkpoints, which could potentially reverse Treg-mediated immune suppression and strengthen the body's anti-tumor immune response. Disrupting Treg induction and function is possible by employing strategies that target tumor-derived factors like TGF-β or IL-10 (Gonzalez-Navajas et al. 2021). In order to combat Treg-mediated immune suppression within the tumor microenvironment, one possible strategy involves the use of antibodies or small molecules to inhibit these factors. Metabolic alterations caused by tumors can amplify Treg function, thereby facilitating immune evasion. Gaining an understanding of these interactions offers valuable insights into potential therapeutic approaches to overcome Treg-induced suppression and enhance the body's ability to fight tumors (Sarkar et al. 2021).

12.4 Therapeutic Implications

12.4.1 Targeting Metabolic Pathways in Tumor Cells

The rapid growth, proliferation, and survival of cancer cells are made possible by their distinct metabolic adaptations. By adjusting glucose metabolism, enhancing lipid synthesis, and modifying amino acid metabolism, these metabolic reprogramming strategies present promising therapeutic targets. Developing strategies to disrupt cancer cell metabolism while sparing normal cells involves targeting the unique metabolic pathways in tumor cells (Navarro et al. 2022). Here are the simple targets:

1. Glycolysis and the Warburg effect are widely recognized as the predominant metabolic changes in cancer. Regardless of the availability of oxygen, tumor cells exhibit the Warburg effect, utilizing glycolysis as their primary energy

source. Cancer cells undergo this shift, enabling them to quickly produce ATP and biosynthetic intermediates, facilitating their rapid proliferation. Enzymes like hexokinase, PFK-1, and PKM2 play a crucial role in glycolysis and are frequently overexpressed in tumors. The glycolytic flux can be disrupted and tumor growth reduced by using inhibitors that target glycolytic enzymes such as 3-bromopyruvate (hexokinase inhibitor) and PFK-1 inhibitors. By inhibiting glycolysis, drugs like 2-deoxy-D-glucose (2-DG) can change the acidic TME, resulting in enhanced immune cell functioning (Liberti and Locasale 2016).
2. Glutaminolysis, the metabolic pathway responsible for converting glutamine to α-ketoglutarate, plays a crucial role in tumor cells' metabolism. In addition to its role in the TCA cycle, this pathway is also involved in nucleotide synthesis and maintaining redox balance. Crucial for this process are enzymes such as GLS and GDH, which play a vital role. Preclinical studies have revealed the potential of GLS inhibitors like CB-839 in impairing tumor cell growth by targeting glutaminolysis. Clinical trials are currently underway for these inhibitors, with scientists and medical professionals closely monitoring their progress. Introducing glutaminase inhibitors, such as CB-839, has revolutionized therapy by targeting glutaminolysis, ultimately influencing tumor growth and immune modulation (Jin et al. 2016).
3. Altered lipid metabolism is a common characteristic of cancer cells, involving increased synthesis and oxidation of fatty acids to support membrane biogenesis and energy production. Tumors frequently show heightened expression of enzymes like FASN, which are responsible for lipid synthesis. The potential of FASN inhibitors, like Orlistat, to disrupt lipid metabolism and suppress the growth of tumor cells is currently under investigation. The potential impact of drugs like statins and fatty acid synthase inhibitors on the TME is currently being studied because of their ability to target lipid synthesis and oxidation (Butler et al. 2020).
4. Supporting their growth and survival, tumor cells undergo changes in amino acid metabolism. Dysregulation of BCAAs and tryptophan metabolism is a common occurrence in cancer. Inhibiting tumor growth and bolstering immune responses is currently being studied through the targeting of amino acid transporters and catabolic pathways (Chen et al. 2024).

Targeting specific metabolic pathways in tumor cells requires precision to ensure that normal cells are not affected by any off-target effects. When metabolic inhibitors are combined with other treatment approaches, such as chemotherapy or immunotherapy, there is a possibility of improving treatment efficacy and overcoming resistance mechanisms. One possible approach to enhance the overall anti-tumor response is to combine glycolysis inhibitors with immune checkpoint inhibitors. By utilizing personalized medicine approaches, scientists can analyze the metabolic profiles of individual tumors to pinpoint their specific vulnerabilities, thus enabling the selection of targeted therapies (Stine et al. 2022).

Various mechanisms, such as the activation of compensatory metabolic pathways or the emergence of genetic mutations, can enable tumor cells to resist metabolic inhibitors. Overcoming these challenges and improving therapeutic outcomes may be possible by combining metabolic inhibitors with agents that specifically target resistance mechanisms. To assess the safety and efficacy of metabolic inhibitors in cancer treatment, ongoing clinical trials are of utmost importance. Advancement in cancer therapy relies heavily on continued research into the mechanisms of metabolic reprogramming in tumors, as well as the discovery of groundbreaking inhibitors. By targeting glycolysis, glutaminolysis, lipid metabolism, and amino acid metabolism, various strategies can provide therapeutic advantages. By integrating these metabolic approaches alongside other therapeutic modalities and prioritizing customized strategies, we may able to overcome cancer resistance to therapies, manipulate with tumor microenvironment, and achieve more therapeutic success (Talib et al. 2021).

12.4.2 Reprogramming Immune Cells

The metabolic processes of immune cells are adapted in the presence of different stimuli, including tumor microenvironments and chronic inflammation. The metabolic state of immune cells plays a critical role in determining their effectiveness against tumor cells in the context of cancer. Reprogramming immune cells by targeting these metabolic pathways shows potential as a therapeutic strategy to enhance anti-tumor immunity and enhance clinical outcomes (Xia et al. 2021).

In the pursuit of cancer therapy, there are several approaches that can be utilized to enhance immune cell reprogramming. The importance of glycolysis in T cell activation and function cannot be overstated, as it involves the breakdown of glucose to generate energy and metabolic intermediates. Glycolysis plays a crucial role in meeting the heightened energy and biosynthetic needs of activated T cells, specifically CTLs. The Warburg effect, observed in cancer cells, is also mirrored in T cells during activation (Cao et al. 2023). The improvement of anti-tumor activity of T cells can be achieved through the enhancement of glycolysis, whereas the inhibition of glycolytic pathways may decrease their effectiveness. To illustrate, the manipulation of T cell responses within the tumor microenvironment can be accomplished by employing inhibitors that specifically target glycolytic enzymes or metabolic regulators. The survival and suppressive functions of Tregs are maintained by their metabolic preference (Bader et al. 2020). Modifying the metabolic pathways of Tregs has the potential to decrease their immunosuppressive effects. As an example, the impairment of Treg function and enhancement of anti-tumor immune responses could be achieved by targeting FAO or OXPHOS pathways using specific inhibitors. Macrophages can flexibly adapt their metabolism, transitioning between glycolysis and OXPHOS in response to changes in their polarization state (97). The energy source for M1 macrophages, which exhibit pro-inflammatory and tumoricidal functions, is mainly glycolysis. Conversely, M2 macrophages, which are accountable for immune suppression and tissue repair, depend on OXPHOS and FAO. Through the manipulation of metabolic processes, it becomes possible to prompt a transformation of macrophages from an M2 to an M1 phenotype, resulting in an enhancement of their anti-tumor activities (Wculek et al. 2022; Kempkes et al. 2019; Liu et al. 2021b). The use of metabolic modulators or inhibitors that specifically affect M2 macrophages can lead to the development of a macrophage phenotype that is more inflammatory and effective against tumors. The altered metabolism caused by tumors can hinder the functioning of DCs. The capacity of DCs to present tumor antigens and activate T cells could potentially be improved by enhancing glycolysis or restoring normal metabolic pathways (Mehta et al. 2021). The efficacy of cancer vaccines and immunotherapies could be improved by metabolic interventions that target and normalize DC function. Immune cells can be made more powerful in their ability to eradicate tumors by strategically altering their metabolic pathways. One way to enhance anti-tumor immune responses is by administering drugs that stimulate glycolysis in T cells, hinder FAO in Tregs, or induce macrophages to adopt an M1 phenotype (Bader et al. 2020).

Personalizing therapeutic strategies based on the metabolic profile of immune cells within individual tumors can optimize the effectiveness of these approaches. The incorporation of metabolic reprogramming alongside immune checkpoint inhibitors or CAR-T cell therapy could lead to synergistic effects. By combining glycolysis-enhancing agents with PD-1/PD-L1 inhibitors, it is possible to augment T cell activity and overcome immune suppression. The emergence of resistance to metabolic inhibitors can be attributed to either the development of compensatory metabolic pathways or adaptive changes in immune cells (Golonko et al. 2024). The effectiveness of reprogramming strategies and patient outcomes can be enhanced by combining metabolic interventions with agents that specifically target these resistance mechanisms. Enhancing anti-tumor immunity shows promise by targeting metabolic pathways and reprogramming immune cells. It is possible to enhance the functionality and effectiveness of immune cells against tumors by manipulating glycolysis, OXPHOS, and FAO in various cell types (Bader et al. 2020). Improving clinical outcomes in cancer treatment will require the integration of these metabolic strategies with other therapeutic modalities, as well as a personalized approach. The advancement of innovative and successful therapies heavily relies on further exploration of how metabolism and immune function interact (Debela et al. 2021).

12.4.3 Combination Therapies

To enhance the effectiveness and address resistance, combination therapy in cancer treatment often combines different therapeutic modalities. To enhance treatment effectiveness, researchers are exploring the simultaneous targeting of metabolic pathways alongside conventional therapies like chemotherapy, immunotherapy, or targeted therapies (Bayat Mokhtari et al. 2017). Combination therapies aim to improve therapeutic outcomes and minimize resistance by disrupting the metabolic processes that tumors rely on, like a well-coordinated attack on their lifeline. This concise review delves into the reasoning behind focusing on metabolic pathways in combination therapy while showcasing existing strategies and potential advantages (Bader et al. 2020).

One common occurrence is that tumors develop resistance to single-agent therapies by adapting metabolically, among other mechanisms. In the case of cancer cells, if a particular pathway is blocked, they can adapt by transitioning to alternative metabolic pathways. By combining metabolic inhibitors with conventional treatments, it is possible to suppress these adaptive responses and improve the effectiveness of the overall treatment (Navarro et al. 2022). Disrupting the energy and biosynthetic demands of tumor cells can render them more sensitive to chemotherapy and radiotherapy, and this can be achieved by inhibiting glycolysis or glutaminolysis. Implementing this approach can boost the effectiveness of these treatments and potentially lessen the required doses, thereby mitigating associated toxicity (Zhao et al. 2013). Along with altering tumor cell behavior, metabolic reprogramming also has consequences for the tumor microenvironment. Immune cell activity can be modified, and the development of new blood vessels can occur as a result of this. To create a more favorable TME for these treatments, metabolic inhibitors can be combined with immunotherapies or anti-angiogenic agents. By modulating metabolism, it is possible to enhance the infiltration and activity of immune cells in the TME, leading to improved efficacy of immune checkpoint inhibitors or CAR-T cell therapies (Chen et al. 2020).

Many tumors exhibit an increase in glycolysis, which is a fundamental metabolic pathway. When glycolysis inhibitors, such as 3-bromopyruvate or 2-deoxyglucose, are added to chemotherapy, it amplifies the cytotoxic effects of the chemotherapeutic agents. These inhibitors target the metabolic support system of cancer cells, amplifying the effectiveness of conventional chemotherapy in treating tumors (Ganapathy-Kanniappan and Geschwind 2013). The conversion of glutamine to α-ketoglutarate, known as glutaminolysis, facilitates tumor growth and allows tumors to evade the immune system. Combining immune checkpoint inhibitors, such as anti-PD-1/PD-L1 antibodies, with glutaminase inhibitors like CB-839 can heighten the immune system's ability to target and fight tumors (Leone et al. 2019). By disrupting glutaminolysis, T cell function can be enhanced, leading to increased effectiveness of immune checkpoint blockade. Increased fatty acid synthesis is a common characteristic of tumor cells, enabling the creation of membranes and the production of energy (Allison et al. 2017). The combination of FASN inhibitors, such as Orlistat, and tyrosine kinase inhibitors (TKIs) can interfere with lipid metabolism, resulting in heightened tumor sensitivity to targeted therapies (Wang et al. 2023b). The goal of this combination approach is to hinder tumor growth and combat TKI resistance. The metabolic state of tumor cells can impact the efficacy of radiotherapy. Radiotherapy outcomes can be improved by incorporating inhibitors of metabolic pathways, such as those that target OXPHOS or the pentose phosphate pathway. By modifying the metabolic profile, these inhibitors can increase the sensitivity of tumor cells to radiation, ultimately boosting the effectiveness of therapy (Chen et al. 2023b).

12.5 Challenges and Future Directions

Metabolic communication within the TME is an area of research that is rapidly progressing. To develop targeted therapies, it is imperative for future research to concentrate on uncovering specific metabolic targets through a comprehensive exploration of the metabolic pathways that are exclusive to cancer cells and immune cells. Simultaneously, there is a need for the development of novel therapeutic strategies to target metabolic changes, reprogram immune cells, and enhance treatment outcomes. The incorporation of metabolic profiling and immune cell analysis in personalized medicine has led to successful outcomes, with treatment strategies customized to each patient showing great potential (Bader et al. 2020; Navarro et al. 2022; Pal et al. 2022; Qiu et al. 2023).

Targeting metabolic pathways offers a promising approach to enhance cancer treatment. By disrupting key metabolic processes and integrating these strategies with chemotherapy, immunotherapy, or targeted therapies, it is possible to improve therapeutic efficacy and overcome resistance (Wu et al. 2022). To advance cancer therapy and enhance patient outcomes, it is vital to continue studying the interplay between metabolic inhibitors and other treatments, as well as adopting personalized approaches (Krzyszczyk et al. 2018).

A key obstacle in employing metabolic inhibitors is to achieving selectivity in reducing off-target effects and toxicity to healthy tissues. Critical steps in achieving therapeutic efficacy and minimizing adverse effects can include developing selective targeted inhibitors and carefully optimizing dosing schedules (Zhang et al. 2022a). Enhancing treatment efficacy can be achieved by personalizing combination therapy according to the metabolic profile of individual tumors (Labrie et al. 2022). The use of metabolic profiling and biomarkers enables the identification of appropriate and optimized therapeutic strategies tailored to individual patients (Massa and Seliger 2023). The exploration of novel metabolic inhibitors and their synergistic effects with other treatments will propel the advancement of successful therapeutic strategies (Bayat Mokhtari et al. 2017).

12.6 Conclusion

Metabolic communication within the TME is a critical determinant of tumor progression and immune responses. Understanding how tumor cells alter the metabolic landscape that influences immune cell function can provide valuable insights into tumor biology and therapeutic development. By targeting these metabolic interactions, we can develop more effective therapies and improve patient outcomes. Herein, we provide a detailed analysis of the metabolic interactions within TME, highlighting the complex interplay between tumor cells and immune cells. By understanding these interactions, novel therapeutic strategies to enhance immune responses and improve cancer treatment can be developed.

References

Abdel-Wahab AF, Mahmoud W, Al-Harizy RM (2019) Targeting glucose metabolism to suppress cancer progression: prospective of anti-glycolytic cancer therapy. Pharmacol Res 150:104511
Aboelella NS et al (2021) Oxidative stress in the tumor microenvironment and its relevance to cancer immunotherapy. Cancers (Basel) 13(5):986
Allison KE, Coomber BL, Bridle BW (2017) Metabolic reprogramming in the tumour microenvironment: a hallmark shared by cancer cells and T lymphocytes. Immunology 152(2):175–184
Altman BJ, Stine ZE, Dang CV (2016) From Krebs to clinic: glutamine metabolism to cancer therapy. Nat Rev Cancer 16(10):619–634
Amelio I et al (2014) Serine and glycine metabolism in cancer. Trends Biochem Sci 39(4):191–198
Arnold PK, Finley LWS (2023) Regulation and function of the mammalian tricarboxylic acid cycle. J Biol Chem 299(2):102838
Bader JE, Voss K, Rathmell JC (2020) Targeting metabolism to improve the tumor microenvironment for cancer immunotherapy. Mol Cell 78(6):1019–1033
Barba I, Carrillo-Bosch L, Seoane J (2024) Targeting the Warburg effect in cancer: where do we stand? Int J Mol Sci 25(6):3142
Basson C et al (2023) The tryptophan-kynurenine pathway in immunomodulation and cancer metastasis. Cancer Med 12(18):18691–18701
Bayat Mokhtari R et al (2017) Combination therapy in combating cancer. Oncotarget 8(23):38022–38043
Beloribi-Djefaflia S, Vasseur S, Guillaumond F (2016) Lipid metabolic reprogramming in cancer cells. Oncogenesis 5(1):e189
Bose S, Zhang C, Le A (2021) Glucose metabolism in cancer: the Warburg effect and beyond. Adv Exp Med Biol 1311:3–15
Brown JS et al (2023) Updating the definition of cancer. Mol Cancer Res 21(11):1142–1147
Burke L (2022) Rewiring of proline metabolism in colorectal carcinogenesis: causes and consequences. University of Leicester
Butler LM et al (2020) Lipids and cancer: emerging roles in pathogenesis, diagnosis and therapeutic intervention. Adv Drug Deliv Rev 159:245–293
Cao J et al (2023) Effects of altered glycolysis levels on CD8(+) T cell activation and function. Cell Death Dis 14(7):407
Casas-Benito A, Martinez-Herrero S, Martinez A (2023) Succinate-directed approaches for Warburg effect-targeted cancer management, an alternative to current treatments? Cancers (Basel) 15(10):2862
Certo M et al (2021) Lactate modulation of immune responses in inflammatory versus tumour microenvironments. Nat Rev Immunol 21(3):151–161
Chen B et al (2020) Metabolic modulation via mTOR pathway and anti-angiogenesis remodels tumor microenvironment using PD-L1-targeting codelivery. Biomaterials 255:120187
Chen S et al (2023a) Macrophages in immunoregulation and therapeutics. Signal Transduct Target Ther 8(1):207
Chen M et al (2023b) Metabolic interventions in tumor immunity: focus on dual pathway inhibitors. Cancers (Basel) 15(7):2043
Chen J et al (2024) Amino acid metabolism in tumor biology and therapy. Cell Death Dis 15(1):42
Cluntun AA et al (2017) Glutamine metabolism in cancer: understanding the heterogeneity. Trends Cancer 3(3):169–180
Corn KC, Windham MA, Rafat M (2020) Lipids in the tumor microenvironment: from cancer progression to treatment. Prog Lipid Res 80:101055
Cortellino S, Longo VD (2023) Metabolites and immune response in tumor microenvironments. Cancers (Basel) 15(15):3898
Cunha A et al (2023) Targeting glucose metabolism in cancer cells as an approach to overcoming drug resistance. Pharmaceutics 15(11):2610

De Oliveira MP, Liesa M (2020) The role of mitochondrial fat oxidation in cancer cell proliferation and survival. Cells 9(12):2600
de Visser KE, Joyce JA (2023) The evolving tumor microenvironment: from cancer initiation to metastatic outgrowth. Cancer Cell 41(3):374–403
Debela DT et al (2021) New approaches and procedures for cancer treatment: current perspectives. SAGE Open Med 9:20503121211034366
Del Prete A et al (2023) Dendritic cell subsets in cancer immunity and tumor antigen sensing. Cell Mol Immunol 20(5):432–447
Delmas D et al (2023) Emergence of lipid droplets in the mechanisms of carcinogenesis and therapeutic responses. Cancers (Basel) 15(16):4100
Di Somma S et al (2021) Cross talk of macrophages with tumor microenvironment cells and modulation of macrophages in cancer by virotherapy. Biomedicines 9(10):1309
El Sayed R et al (2021) Metabolic factors affecting tumor immunogenicity: what is happening at the cellular level? Int J Mol Sci 22(4):2142
Elia I, Haigis MC (2021) Metabolites and the tumour microenvironment: from cellular mechanisms to systemic metabolism. Nat Metab 3(1):21–32
El-Tanani M et al (2024) Metabolic vulnerabilities in cancer: a new therapeutic strategy. Crit Rev Oncol Hematol 201:104438
Fu Y et al (2020) Lipid metabolism in cancer progression and therapeutic strategies. Med Comm 2(1):27–59
Ganapathy-Kanniappan S, Geschwind JF (2013) Tumor glycolysis as a target for cancer therapy: progress and prospects. Mol Cancer 12:152
Ganjoo S et al (2023) The role of tumor metabolism in modulating T-cell activity and in optimizing immunotherapy. Front Immunol 14:1172931
Garcia SN, Guedes RC, Marques MM (2019) Unlocking the potential of HK2 in cancer metabolism and therapeutics. Curr Med Chem 26(41):7285–7322
Golonko A et al (2024) Dietary factors and their influence on immunotherapy strategies in oncology: a comprehensive review. Cell Death Dis 15(4):254
Gonzalez-Navajas JM et al (2021) The impact of Tregs on the anticancer immunity and the efficacy of immune checkpoint inhibitor therapies. Front Immunol 12:625783
Gupta A, Das D, Taneja R (2024) Targeting dysregulated lipid metabolism in cancer with pharmacological inhibitors. Cancers (Basel) 16(7):1313
He X, Xu C (2020) Immune checkpoint signaling and cancer immunotherapy. Cell Res 30(8):660–669
Heintzman DR, Fisher EL, Rathmell JC (2022) Microenvironmental influences on T cell immunity in cancer and inflammation. Cell Mol Immunol 19(3):316–326
Hinshaw DC, Shevde LA (2019) The tumor microenvironment innately modulates cancer progression. Cancer Res 79(18):4557–4566
Hu N et al (2024) The role of metabolic reprogramming in the tumor immune microenvironment: mechanisms and opportunities for immunotherapy in hepatocellular carcinoma. Int J Mol Sci 25(11):5584
Jin L, Alesi GN, Kang S (2016) Glutaminolysis as a target for cancer therapy. Oncogene 35(28):3619–3625
Jin J et al (2023) Targeting glutamine metabolism as a therapeutic strategy for cancer. Exp Mol Med 55(4):706–715
Jung JG, Le A (2021) Metabolism of immune cells in the tumor microenvironment. Adv Exp Med Biol 1311:173–185
Jung MK et al (2021) Role of branched-chain amino acid metabolism in tumor development and progression. J Cancer Prev 26(4):237–243
Kao TW et al (2022) Therapeutic targeting of Glutaminolysis as a novel strategy to combat cancer stem cells. Int J Mol Sci 23(23):15296
Kempkes RWM et al (2019) Metabolic pathways involved in regulatory T cell functionality. Front Immunol 10:2839

Kerneur C, Cano CE, Olive D (2022) Major pathways involved in macrophage polarization in cancer. Front Immunol 13:1026954
Kim RH et al (2009) Arginine deiminase as a novel therapy for prostate cancer induces autophagy and caspase-independent apoptosis. Cancer Res 69(2):700–708
Koundouros N, Poulogiannis G (2020) Reprogramming of fatty acid metabolism in cancer. Br J Cancer 122(1):4–22
Krzyszczyk P et al (2018) The growing role of precision and personalized medicine for cancer treatment. Technology (Singap World Sci) 6(3–4):79–100
Kurniawan H, Soriano-Baguet L, Brenner D (2020) Regulatory T cell metabolism at the intersection between autoimmune diseases and cancer. Eur J Immunol 50(11):1626–1642
Labrie M et al (2022) Therapy resistance: opportunities created by adaptive responses to targeted therapies in cancer. Nat Rev Cancer 22(6):323–339
Leone RD, Powell JD (2020) Metabolism of immune cells in cancer. Nat Rev Cancer 20(9):516–531
Leone RD, Powell JD (2021) Fueling the revolution: targeting metabolism to enhance immunotherapy. Cancer Immunol Res 9(3):255–260
Leone RD et al (2019) Glutamine blockade induces divergent metabolic programs to overcome tumor immune evasion. Science 366(6468):1013–1021
Li T, Copeland C, Le A (2021) Glutamine metabolism in cancer. Adv Exp Med Biol 1311:17–38
Li M et al (2023) Metabolism, metabolites, and macrophages in cancer. J Hematol Oncol 16(1):80
Liberti MV, Locasale JW (2016) The Warburg effect: how does it benefit cancer cells? Trends Biochem Sci 41(3):211–218
Lim AR, Rathmell WK, Rathmell JC (2020) The tumor microenvironment as a metabolic barrier to effector T cells and immunotherapy. elife 9:9
Liu KX et al (2021a) Harnessing lactate metabolism for radiosensitization. Front Oncol 11:672339
Liu Y et al (2021b) Metabolic reprogramming in macrophage responses. Biomark Res 9(1):1
Luby A, Alves-Guerra MC (2021) Targeting metabolism to control immune responses in cancer and improve checkpoint blockade immunotherapy. Cancers (Basel) 13(23):5912
Lukey MJ, Wilson KF, Cerione RA (2013) Therapeutic strategies impacting cancer cell glutamine metabolism. Future Med Chem 5(14):1685–1700
Mantovani A et al (2022) Macrophages as tools and targets in cancer therapy. Nat Rev Drug Discov 21(11):799–820
Marchesi F et al (2020) Manipulation of glucose availability to boost cancer immunotherapies. Cancers (Basel) 12(10):2940
Massa C, Seliger B (2023) Combination of multiple omics techniques for a personalized therapy or treatment selection. Front Immunol 14:1258013
Mehta AK et al (2021) Macrophage biology and mechanisms of immune suppression in breast cancer. Front Immunol 12:643771
Moller SH, Wang L, Ho PC (2022) Metabolic programming in dendritic cells tailors immune responses and homeostasis. Cell Mol Immunol 19(3):370–383
Mullarky E et al (2019) Inhibition of 3-phosphoglycerate dehydrogenase (PHGDH) by indole amides abrogates de novo serine synthesis in cancer cells. Bioorg Med Chem Lett 29(17):2503–2510
Muluh TA, Shu XS, Ying Y (2023) Targeting cancer metabolic vulnerabilities for advanced therapeutic efficacy. Biomed Pharmacother 162:114658
Navarro C et al (2022) Metabolic reprogramming in cancer cells: emerging molecular mechanisms and novel therapeutic approaches. Pharmaceutics 14(6):1303
Noe JT et al (2021) Lactate supports a metabolic-epigenetic link in macrophage polarization. Sci Adv 7(46):eabi8602
Nong X et al (2022) The mechanism of branched-chain amino acid transferases in different diseases: research progress and future prospects. Front Oncol 12:988290
Ohta A, Sitkovsky M (2014) Extracellular adenosine-mediated modulation of regulatory T cells. Front Immunol 5:304
Oparaugo NC et al (2023) Human regulatory T cells: understanding the role of Tregs in select autoimmune skin diseases and post-transplant nonmelanoma skin cancers. Int J Mol Sci 24(2):1527

Pal S et al (2022) Targeting cancer-specific metabolic pathways for developing novel cancer therapeutics. Front Immunol 13:955476
Pan S et al (2021) Serine, glycine and one-carbon metabolism in cancer (review). Int J Oncol 58(2):158–170
Pascale RM et al (2022) S-Adenosylmethionine: from the discovery of its inhibition of tumorigenesis to its use as a therapeutic agent. Cells 11(3):409
Patsoukis N et al (2015) PD-1 alters T-cell metabolic reprogramming by inhibiting glycolysis and promoting lipolysis and fatty acid oxidation. Nat Commun 6:6692
Pham LGC (2020) Glutamine antagonist 6-diazo-5-oxo-L-norleucine, a Hyaluronan synthesis inhibitor, slows triple negative breast cancer growth. Biological Sciences 67
Pliszka M, Szablewski L (2021) Glucose transporters as a target for anticancer therapy. Cancers (Basel) 13(16):4184
Qiu S et al (2023) Small molecule metabolites: discovery of biomarkers and therapeutic targets. Signal Transduct Target Ther 8(1):132
Rangel Rivera GO et al (2021) Fundamentals of T cell metabolism and strategies to enhance cancer immunotherapy. Front Immunol 12:645242
Reina-Campos M, Diaz-Meco MT, Moscat J (2020) The complexity of the serine glycine one-carbon pathway in cancer. J Cell Biol 219(1):e201907022
Ribeiro Franco PI et al (2020) Tumor microenvironment components: Allies of cancer progression. Pathol Res Pract 216(1):152729
Sarkar T, Dhar S, Sa G (2021) Tumor-infiltrating T-regulatory cells adapt to altered metabolism to promote tumor-immune escape. Curr Res Immunol 2:132–141
Shen YA et al (2021) Inhibition of glutaminolysis in combination with other therapies to improve cancer treatment. Curr Opin Chem Biol 62:64–81
Stine ZE et al (2022) Targeting cancer metabolism in the era of precision oncology. Nat Rev Drug Discov 21(2):141–162
Sun N, Zhao X (2022) Argininosuccinate synthase 1, arginine deprivation therapy and cancer management. Front Pharmacol 13:935553
Sun W, Zhao E, Cui H (2023) Target enzymes in serine-glycine-one-carbon metabolic pathway for cancer therapy. Int J Cancer 152(12):2446–2463
Talib WH et al (2021) Targeting drug chemo-resistance in cancer using natural products. Biomedicines 9(10):1353
Tang K et al (2021) Indoleamine 2, 3-dioxygenase 1 (IDO1) inhibitors in clinical trials for cancer immunotherapy. J Hematol Oncol 14:1–21
Tao H et al (2023) Unveiling the veil of lactate in tumor-associated macrophages: a successful strategy for immunometabolic therapy. Front Immunol 14:1208870
Vander Heiden MG, Cantley LC, Thompson CB (2009) Understanding the Warburg effect: the metabolic requirements of cell proliferation. Science 324(5930):1029–1033
Viola A et al (2019) The metabolic signature of macrophage responses. Front Immunol 10:1462
Wang X, Zhang F, Wu X-R (2017) Inhibition of pyruvate kinase M2 markedly reduces chemoresistance of advanced bladder cancer to cisplatin. Sci Rep 7(1):45983
Wang ZH et al (2021a) Lactate in the tumour microenvironment: from immune modulation to therapy. EBioMedicine 73:103627
Wang H et al (2021b) The impact of the tumor microenvironment on macrophage polarization in cancer metastatic progression. Int J Mol Sci 22(12):6560
Wang S et al (2023a) Advances in the study of Hexokinase 2 (HK2) inhibitors. Anti-Cancer Agents Med Chem (Formerly Current Medicinal Chemistry-Anti-Cancer Agents) 23(7):736–746
Wang Z et al (2023b) Lipid metabolism as a target for cancer drug resistance: progress and prospects. Front Pharmacol 14:1274335
Wang Y et al (2023c) Targeting the metabolism of tumor-infiltrating regulatory T cells. Trends Immunol 44(8):598–612
Wang B et al (2024) A glutamine tug-of-war between cancer and immune cells: recent advances in unraveling the ongoing battle. J Exp Clin Cancer Res 43(1):74

Wculek SK et al (2022) Metabolism of tissue macrophages in homeostasis and pathology. Cell Mol Immunol 19(3):384–408
Wei Z et al (2020) Metabolism of amino acids in cancer. Front Cell Dev Biol 8:603837
Wise DR, Thompson CB (2010) Glutamine addiction: a new therapeutic target in cancer. Trends Biochem Sci 35(8):427–433
Wojtukiewicz MZ et al (2021) Inhibitors of immune checkpoints-PD-1, PD-L1, CTLA-4-new opportunities for cancer patients and a new challenge for internists and general practitioners. Cancer Metastasis Rev 40(3):949–982
Wolpaw AJ, Dang CV (2018) Exploiting metabolic vulnerabilities of cancer with precision and accuracy. Trends Cell Biol 28(3):201–212
Wu HL et al (2022) Targeting nucleotide metabolism: a promising approach to enhance cancer immunotherapy. J Hematol Oncol 15(1):45
Xia L et al (2021) The cancer metabolic reprogramming and immune response. Mol Cancer 20(1):28
Xing J, Zhang J, Wang J (2023) The immune regulatory role of adenosine in the tumor microenvironment. Int J Mol Sci 24(19):14928
Xuekai L et al (2024) Advances in reprogramming of energy metabolism in tumor T cells. Front Immunol 15:1347181
Yoo HC et al (2020) Glutamine reliance in cell metabolism. Exp Mol Med 52(9):1496–1516
You M et al (2023) Signaling pathways in cancer metabolism: mechanisms and therapeutic targets. Signal Transduct Target Ther 8(1):196
Yu T et al (2021) Metabolic interventions: a new insight into the cancer immunotherapy. Arch Biochem Biophys 697:108659
Zahavi D, Hodge JW (2023) Targeting immunosuppressive adenosine signaling: a review of potential immunotherapy combination strategies. Int J Mol Sci 24(10):8871
Zekovic M et al (2023) Alteration of lipid metabolism in prostate cancer: multifaceted oncologic implications. Int J Mol Sci 24(2):1391
Zeng H, Chi H (2013) The interplay between regulatory T cells and metabolism in immune regulation. Onco Targets Ther 2(11):e26586
Zhang Q, Sioud M (2023) Tumor-associated macrophage subsets: shaping polarization and targeting. Int J Mol Sci 24(8):7493
Zhang Y et al (2022a) Targeting glucose metabolism enzymes in cancer treatment: current and emerging strategies. Cancers (Basel) 14(19):4568
Zhang C et al (2022b) New dawn for cancer cell death: emerging role of lipid metabolism. Mol Metab 63:101529
Zhang S et al (2024) Fatty acid metabolism of immune cells: a new target of tumour immunotherapy. Cell Death Discov 10(1):39
Zhao Y, Butler EB, Tan M (2013) Targeting cellular metabolism to improve cancer therapeutics. Cell Death Dis 4(3):e532
Zhao H et al (2021) Inflammation and tumor progression: signaling pathways and targeted intervention. Signal Transduct Target Ther 6(1):263

The Role of Microbiome in Metabolic Pattern Changes

13

El-Sayed Akool, Abdel-Aziz S. Shatat, and Amany Balah

Abstract

The human microbiome, comprising 100 trillion bacteria, is vital for health and regulates metabolism, immunity, and neurological signaling; dysbiosis can lead to various diseases. The gut microbiota ferments insoluble dietary fiber and polysaccharides to produce short-chain fatty acids (SCFAs), which are essential for intestinal homeostasis and energy metabolism as well as controlling immune cell development and anti-inflammatory properties. Gut microbes also produce secondary bile acids, which act as ligands for bile acid receptors. Furthermore, the gut microbiota also produces lipids, which are associated with metabolic disorders. In addition, it is essential for cardiovascular health and preventing metabolic syndrome, cardiovascular diseases, and cancer. Moreover, the gut microbiota plays a role in red meat metabolism. The gut microbiome also influences metabolic diseases, including bile acid bioactivity and circadian misalignment. Finally, the microbiome is crucial for metabolism regulation and therapeutic interventions, with current treatments mainly involving fecal microbiota transplantation while advanced techniques like probiotics and postbiotics are being explored.

Keywords

Microbiome · Metabolic syndrome · Systemic Metabolism · Gut microbiota · Nutrient Metabolism

E.-S. Akool (✉) · A.-A. S. Shatat · A. Balah
Department of Pharmacology and Toxicology, Faculty of Pharmacy, Al-Azhar University, Cairo, Egypt
e-mail: elsayedakool@azhar.edu.eg

S. S. M. Soliman, M. I. Husseiny (eds.), *Metabolic Dynamics in Host-Microbe Interaction*, https://doi.org/10.1007/978-981-96-1305-2_13

13.1 Introduction

In recent years, our understanding of the human microbiome, the genome of all our microbes, has expanded dramatically. Moreover, the technological developments in genome sequencing and metagenomic analysis (the genetic study of genomes extracted directly from environmental samples) have made it possible for researchers to examine these microorganisms, their roles, and the interactions between the microbiome and its hosts in both health and disease. Estimated 100 trillion bacteria make up the human microbiome, the majority of which are found in the intestines (Amon and Sanderson 2017).

Complex microbial communities found in our bodies are essential to good health. The majority of human microbes reside in the intestines and are affected by medication, diet, lifestyle, pregnancy, and genetics. In addition to producing a variety of substances that impact the host, the gut microbiota is crucial for controlling metabolism, strengthening immunity, getting rid of toxins, controlling gut endocrine function, and controlling neurological signaling. Numerous acute and chronic diseases, such as immunological, inflammatory, neurodevelopmental, and degenerative disorders, as well as metabolic disorders like obesity, type 2 diabetes (T2D), non-alcoholic liver disease, and cardio-metabolic diseases, are linked to dysbiosis of the microbiome. Ten percent of the metabolites in mammalian blood are microbial metabolites, which reach distant organ systems like the liver and brain and have an impact on them (Mavros and Ronda 2022). In this review, the role of the microbiome in metabolic pattern changes has been summarized.

13.2 Composition of the Microbiome

The microbiome is dynamic and changes with early development, environmental factors like diet and antibiotic use, and especially in response to disease, unlike the host genome, which is relatively constant (Lozupone et al. 2012). Most importantly, the type of feed (either breast milk or formula feeds), the delivery method (either cesarean section or vaginal birth), the mother's nutritional status (either overweight or undernourished), as well as the use of antibiotics all affect the newborn's intestinal microbiome (Meropol and Edwards 2015). It is believed that the intricacy and adaptability of the newborn microbiota during this formative stage of development is crucial for preserving immune system homeostasis and affects health in the long run (Charbonneau et al. 2016). At least 1000 distinct species of bacteria, representing the two main phyla, *Bacteroidetes* and *Firmicutes*, can be found in a healthy human digestive tract. To comprehend the possible impacts on human health and disease susceptibilities of modifications in the composition of the microbiome, it is crucial to clarify the composition of the microbiome in healthy individuals. The gut is also home to other microorganisms, such as fungi, viruses, and archaea. But compared to bacteria, there hasn't been as much research done on them.

13.3 The Microbiome and Systemic Metabolism

Systemic metabolism is the intricate web of interconnected, distinct metabolic mechanisms that occur within organ systems, tissues, and cells and coordinate the movement of metabolites throughout the entire organism. The key to comprehending host-microbiome metabolic interactions in the future is figuring out which of these processes the microbiome influences in detail. Though research in this area is still in its infancy, a great deal has been accomplished in the past 15 years since the first reports of the microbiome's role in metabolic regulation.

13.3.1 Energy Intake

Through changes in the digestion of complex macronutrients, the microbiome can directly regulate energy intake; alternatively, it can influence hunger and satiety indirectly. The microbiome regulates the amount of energy that is excreted after consumption, according to early research in the area that connected energy homeostasis and the microbiome. According to these studies on the connection between obesity and the microbiome, the microbiomes of lean and obese hosts differ in terms of composition and metabolic activity, which may have an effect on the host's capacity to obtain energy from food (Lund et al. 2020). Although the exact mechanisms through which different microbiomes have varying capacities overall to determine the food's calorie content are still unclear, a recent study suggested that the metabolite dimethylglycine plays a crucial role in weight gain (Fluhr et al. 2021).

By interacting with the neuroendocrine axis, the microbiome can affect the signaling of hunger and satiety in addition to directly influencing the extraction of calories from food. For instance, endogenous release of hunger-regulating peptides, including peptide YY (PYY) and glucagon-like peptide 1 (GLP-1), can occur in response to circulating short-chain fatty acids (SCFAs) derived from the microbiome, which are byproducts of commensal bacteria's fermentation of carbohydrates. The SCFA receptors GPR (G-protein coupled receptor)-41 and GPR-43 are expressed by the gastrointestinal tract's enteroendocrine cells (Tazoe et al. 2009) and, when activated, secrete PYY and GLP-1, which induce satiety (Psichas et al. 2015). The negative feedback mechanism triggered by food consumption is exemplified by SCFAs' ability to suppress appetite. This underscores the pivotal function of the microbiota as a regulator of physiological adaptation to environmental factors, thereby facilitating energy balance. Moreover, hunger and satiety may be regulated by the microbiome through mechanisms other than SCFAs. For example, ClpB, a melanocyte-stimulating hormone (α-MSH) mimic produced by Escherichia coli, can stimulate PYY release and induce activity of proopiomelanocortin (POMC) neurons (Breton et al. 2016).

An essential aspect of metabolic homeostasis is the connection between energy intake and the microbiome. As previously mentioned, the microbiota has been associated with the control of food intake and energy absorption. Nevertheless, energy balance involves both energy expenditure and energy intake.

13.3.2 Energy Expenditure

Energy can be used either anabolitically through growth or catabolitically through combustion. This emphasis is on the microbiome's function in controlling catabolic processes that affect energy consumption. It will pay special attention to the non-shivering thermogenesis process, which has attracted a lot of attention because it has been linked to better cardio-metabolic health in people (Becher et al. 2021). To keep the body warm when exposed to cold, brown and beige adipocytes support a specialized type of heat production known as non-shivering thermogenesis (Chouchani et al. 2019). Traditionally, in white adipose depots, prolonged cold exposure induces the production of beige adipocytes, also called "brite" adipocytes, which combine characteristics of brown and white adipocytes, though subcutaneous depots are more effective in this process than visceral depots. Both Beige and brown adipocytes need the protein uncoupling protein 1 (UCP-1) for non-shivering thermogenesis. It accomplishes this by separating respiration from ATP synthesis and dissipating the proton gradient across the inner mitochondrial membrane (Cannon and Nedergaard 2004). Various groups have investigated the possibility of the microbiome influencing non-shivering thermogenesis, owing to the growing interest in the gut microbiome's role in energy expenditure and systemic metabolism. According to early research on this topic, mice given antibiotics or kept germ-free showed enhanced insulin sensitivity as well as increased Ucp1 expression in both subcutaneous and visceral white adipose tissue (Suarez-Zamorano et al. 2015). However, a more recent study discovered the opposite, showing that the depletion of gut microbiota hampered thermoregulation and lowered energy metabolism. The study also found that the depleted microbiota's recolonization restores this reduced effect of thermogenesis in part by using SCFA butyrate (Li et al. 2019). However, a further study showed that the microbiome contributes to the metabolites of amino acids to maximize hepatic TCA cycle fluxes to promote gluconeogenesis (Krisko et al. 2020). It has been reported that depletion of the microbiome had no effect on energy expenditure or the recruitment of thermogenic tissues. Consequently, there is disagreement in the field about how the microbiota controls energy consumption via non-shivering thermogenesis. Crucially, comprehending the subtleties of this connection between non-shivering thermogenesis and the microbiota may offer individuals who are either hypometabolic or hypercaloric different avenues for regulating their metabolism (Thaiss et al. 2017).

13.4 Metabolism and Absorption of Nutrients

13.4.1 Indigestible Polysaccharide Degradation

The gut microbiota creates metabolites called short-chain fatty acids (SCFAs) by fermenting indigestible polysaccharides and insoluble dietary fiber. Colonocytes primarily obtain their nutrients from butyrate. Of all the SCFAs, acetate has the highest intestinal concentration. Through the portal vein, it enters the liver where it is oxidized and mostly utilized by hepatocytes. A component of hepatic gluconeogenesis is propionate. These SCFAs are bioactive molecules involved in energy metabolism, control the development of immune cells, and have anti-inflammatory properties. They also preserve the barrier function of the intestines. The synthesis of SCFA involves multiple pathways (Koh et al. 2016). Through the Wood–Ljungdahl pathway, which yields acetyl-CoA from CO2, acetogenic bacteria and glycolysis both produce acetate from pyruvate. Butyryl-CoA is created from acetoacetyl-CoA, and then butyrate is produced. The phylum Firmicutes contains the *Ruminococcaceae*, *Lachnospiraceae*, *Erysipelotrichaceae*, and *Clostridiaceae*, which are important butyrate-producing bacterial taxa. Butyrate is also produced by Clostridium species (like *C. butyricum*) and *Butyrivibrio* species (like *B. fibrisolvens*). Acetate is produced by acetobacter and *Gluconobacte*r species (Koh et al. 2016; Knip and Siljander 2016). Through the interconversion of SCFAs, butyrate is produced from 24% of all acetate. Propionate synthesis involves three pathways: the lactate-mediated acrylate pathway, the succinate-mediated succinate pathway, and the propanediol-mediated propanediol process, which makes use of deoxy sugars such as rhamnose and fucose (Louis and Flint 2017). Bacteria that break down mucin and produce propionate include *Akkermansia muciniphila* and other species. SCFAs are crucial for maintaining the environment of the intestines as well as for the metabolism of energy and glucose, inflammation, and immunological response. For instance, in order to activate peroxisome proliferator-activated receptor gamma (PPARγ) and maintain an anaerobic environment in the intestinal tract, butyrate is necessary for intestinal epithelial cells (IECs) to promote mitochondrial β-oxidation, maintain hypoxia for epithelial cells, and prevent oxygen transfer into the intestinal lumen (Rivera-Chávez et al. 2016; Byndloss et al. 2017). Most interestingly, the intestinal production of acetate and propionate is also crucial for intestinal homeostasis. They cause colonic Treg to guard against colitis induced experimentally. According to a different study, acetate inhibits intestinal inflammation by signaling through the neutrophil-expressed G-protein coupled receptor (GPR). Additionally, it improves the function of the intestinal barrier and supports the body's defense against infections (Fujisaka et al. 2023).

13.4.2 Lipid and Bile Acids Metabolism

The main bacterial metabolites that impact host physiology are secondary bile acids. Also, hepatocytes generate chenodeoxycholic acid and cholic using cholesterol. During digestion, these lipid-soluble bile acids are connected to taurine or glycine to form these water-soluble bile salts, which are then stored in the gallbladder before being expelled into the gut. Primary bile salts are converted into secondary bile salts by gut microbes. These bile acids are actively reabsorbed into the hepatic portal circulation along the proximal and distal ileum. Additionally, bacteria deconjugate certain primary and secondary conjugated bile salts to produce lipid-soluble bile acids, which are subsequently passively absorbed into the circulation of the hepatic portal. About 95% of the bile acids that are supplied to the duodenum are recycled through the enterohepatic circulation. The dietary lipids are emulsified and made easier to absorb by bile acids. They also serve as ligands for bile acid receptors, such as transmembrane G-protein coupled receptor 5 (TGR5) and farnesoid X receptor (FXR), which control the expression of bile acid transporter genes as well as energy and cholesterol metabolism (Wahlström et al. 2016). Deoxycholic acid inhibits bacterial overgrowth and has been linked to hepatocellular carcinoma (Yoshimoto et al. 2013). It was recently discovered that the gut microbiome of centenarians was unique and enriched in *Odoribacter laneus*, *Odoribacteraceae*, and *Parabacteroides merdae*. *Isoallolithocholic acid*, which possesses antimicrobial properties against Gram-positive bacteria, is produced by these microbiota (Sato et al. 2021). Therefore, bile acid produced by gut bacteria may promote human longevity.

The gut microbiota also produces lipids. Primarily found in eukaryotes are the genes encoding mediators involved in sphingolipid biosynthesis. However, genes that control sphingolipid biosynthesis are also present in some Bacteroides bacterial species. Moreover, α-galactosylceramide, produced by *Bacteroides fragilis*, inhibits the activation of intestinal natural killer T (NKT) cells (An et al. 2014). Bacteroides sphingolipid concentrations in the feces of individuals suffering from inflammatory bowel disease (IBD) are comparatively low and inversely connected with inflammation. Intestinal inflammation and changed host ceramide pools occur when genetically modified *Bacteroides thetaiotaomicron*, which lacks sphingolipid synthesis, colonizes germ-free (GF) mice (Brown et al. 2019). Bacteroides may therefore produce sphingolipids that preserve intestinal homeostasis.

Unsaturated fatty acids in the diet are also saturated by gut bacteria. According to Kishino and colleagues (Kishino et al. 2013), Lactobacillus plantarum uses catalysts of hydration and dehydration to enzymatically metabolize unsaturated fatty acids, resulting in the production of conjugated, hydroxy, and oxo-fatty acids as well as partially saturated trans-fatty acids. Administering linoleic acid triggers the arachidonic acid cascade, causing adipose tissue to become chronically inflamed. *Lactobacillus salivarius* and *Lactobacillus gasseri* hydroxylate linoleic acid to produce 10-hydroxy-cis-12-octadecenoic acid, which enhances GLP-1 secretion, improves glucose metabolism, and reduces obesity (Miyamoto et al. 2019). One of the main factors influencing the microbial composition is dietary nutrients. By using plasma lipidomic analysis, it was possible to determine how dynamically the microbiota affects systemic lipid profiles in GF or colonized mice fed varying nutritional compositions. The degree of this influence is associated with metabolic disorders (Watanabe et al. 2021). Therefore, lipid absorption, bile acid metabolism, inflammatory signals through energy metabolism, and lipid synthesis are all regulated by the gut microbiota.

The gut microbiome plays a role in synthesizing lipids that are crucial for maintaining membrane integrity. Bacterial lipids are essential for supporting various host interactions. The lipid composition of Escherichia coli is substantial, constituting 10% of the cell's total dry weight. Key lipid classes in bacterial membranes include phospholipids (PE, PS, PC, PI, PG), glycerolipids (DAG, TAG), cardiolipins (CL), and saccharolipids (LPS). Various bacterial phyla exhibit distinct lipid classes, like sphingolipids in commensal Bacteroidetes strains (Brown et al. 2023; Johnson et al. 2020).

Recent research has shown that gut microbiome strains, including *Bacteroides*, can sulfonate cholesterol, impacting serum cholesterol levels in mice. The presence of specific gene clusters in these microbes can influence the absorption of dietary cholesterol and impact blood lipid profiles. Variations in cholesterol-metabolizing microbial enzymes among human populations can explain differences in cholesterol and its derivatives in blood lipid panels (Brown et al. 2023; Gérard 2020).

Omega-3 and omega-6 fatty acids are essential for human health, with linoleic and alpha-linolenic acids needing to be obtained from the diet as the body cannot synthesize them. These polyunsaturated fatty acids play crucial roles in membrane structure and inflammation regulation, where omega-3 fatty acids exhibit anti-inflammatory properties while omega-6 fatty acids are pro-inflammatory. Maintaining a balanced ratio of omega-6 to omega-3 fatty acids is vital for cardiovascular health and protection against metabolic syndrome, cardiovascular diseases, and cancer. However, the Western diet often contains an imbalance with excessive omega-6 fatty acids compared to traditional diets, potentially leading to health issues. Changes in dietary patterns over the last century have resulted in increased levels of linoleic acid in human adipose tissue and breast milk, impacting childhood development negatively. Gut microbiome enzymes from species like *Bifidobacterium* and *Lactobacillus* can metabolize lipids, influencing inflammatory signals within the host. These enzymes can also alter polyunsaturated fatty acids, affecting their oxidative potential and contributing to metabolic syndrome and obesity susceptibility. Moreover, the gut microbiome plays a significant role in modulating the uptake and metabolism of polyunsaturated fatty acids, ultimately influencing susceptibility to inflammatory diseases, metabolic syndromes, and cardiovascular issues. Understanding how gut microbial strains and enzymes interact with dietary saturated and monounsaturated fats is crucial for comprehending personalized responses to diet and inflammatory triggers, highlighting the importance of balanced fatty acid intake for overall health and well-being (Brown et al. 2023; Mariamenatu and Abdu 2021).

Westernized societies are facing rising rates of metabolic syndrome and autoimmunity due to chronic inflammation linked to changes in lipid consumption. The gut microbiome responds to dietary shifts, influencing lipid metabolism and inflammation through potent lipid stimulators. In the small intestine, microbial-derived lipids interact with the immune system, balancing immune responses and lipid metabolism. Disrupted lipid absorption can lead to microbiome-mediated inflammation, affecting T cell balance and gut health. Sterol derivatives from the microbiome impact immune pathways, emphasizing the role of early-life exposure to lipids in disease susceptibility. Genetic interactions, dietary profiles, and microbiome products influence chronic inflammation, with microbiome-generated lipids playing a significant but understudied role in systemic inflammation and lipid signaling (Brown et al. 2023; Christ et al. 2019).

13.4.3 Vitamins

Vitamins A, D, E, and K are fat-soluble, while B-complex and C are water-soluble. Water-soluble vitamins function as coenzymes, while fat-soluble vitamins are delivered to different tissues, where they serve as bioactive effectors and signaling molecules. Vitamin supplementation modifies gut microbiota, boosts biodiversity, and raises SCFA levels (Pham et al. 2021). Nevertheless, vitamin synthesis is done by gut bacteria (Steinert et al. 2020). All but vitamin D cannot be synthesized by humans. Thus, nutrition and gut flora determine the availability of vitamins in humans. The B, K, and C complex vitamins are produced by the gut microbiota. By analyzing the microbial genome, the capacity of bacteria to produce thiamine (B1), riboflavin (B2), niacin (B3), pantothenate (B5), pyridoxine (B6), biotin (B7), folate (B9), and cobalamin (B12) was assessed. Forty to sixty-five percent of the 256 human commensal bacteria have a B-complex synthesis pathway (Magnúsdóttir et al. 2015). Gut microbiota and vitamins interact, with the former influencing microbial community structures (Pham et al. 2021). Administering vitamin A to neonates within the first two days of their lives has a sex-dependent effect on the composition of gut bacteria. It raises the abundance of Bifidobacterium in boys, but it raises the abundance of *Actinobacteria* and *A. muciniphila* in girls (Huda et al. 2019). The composition of gut bacteria is influenced by plasma concentrations and dietary vitamin A intake (Wu et al. 2011; Li et al. 2017), which might help provide defense against different infections (Long et al. 2007).

The structure of the microbial community in the gut is also altered by vitamin D, and supplementation is linked to notable alterations in the abundances of *Firmicutes*, *Actinobacteria*, and *Bacteroidetes*. Vitamin D supplementation and 25(OH)-D levels exhibit a negative correlation with the families *Veillonellaceae* and *Oscillospiraceae* (*Firmicutes*) (Bellerba et al. 2021). By reducing immunological activation, fortifying the intestinal barrier, and increasing the production of antibacterial peptides, vitamin D may provide protection against inflammatory bowel disease (IBD) (Pham et al. 2021). Vitamin D receptors (VDRs) are regulated by specific species of bacteria. Wu et al. claim that the probiotics *Lactobacillus rhamnosus* and *Lactobacillus plantarum* increase VDR in intestinal epithelial cells (IECs) and prevent colitis in a way that is dependent on VDR signaling (Wu et al. 2015). Variations in VDR have an impact on both parabacteroides abundance and β-diversity, according to a genome-wide association study (Wang et al. 2016). According to this data, the gut microbiota helps to preserve intestinal homeostasis and regulate vitamin D signaling.

13.4.4 Amino Acid Metabolism

Red meat's choline and L-carnitine are converted to trimethylamine (TMA) by the gut microbiota, which also breaks down proteins and produces amino acids (AAs). After entering the bloodstream, the latter is converted to trimethylamine-N-oxide (TMAO) by the liver. It was discovered that TMAO encouraged atherosclerosis in an animal model. In a dose-dependent manner, plasma TMAO may raise the risk and mortality of cardiovascular disease (CVD) (Schiattarella et al. 2017). Wang et al. demonstrated that microbial TMA production is inhibited by 3,3-dimethyl-1-butanol (DMB), a structural analog of choline. Oral DMB administration suppresses arteriosclerosis development in mice lacking apolipoprotein E (KO) by lowering TMAO levels and blocking the formation of macrophage foam cells induced by the choline diet. Hence, according to Wang et al., microbial intervention may aid in the prevention of arteriosclerosis (Wang et al. 2015).

Other AAs are also metabolized by gut microbiota. Plasma concentrations of the branched-chain amino acids (BCAAs) valine, leucine, and isoleucine are associated with insulin resistance in diabetic patients (Jang et al. 2016). By combining analyses of metagenomics and serum metabolomics, Pedersen and colleagues demonstrated that microbiota, which is rich in pathways involved in BCAA biosynthesis, is linked to elevated BCAA levels in individuals with insulin resistance (Pedersen et al. 2016). The two main organisms that synthesize BCAAs are *Prevotella copri* and *Bacteroides vulgatus*. When *P. copri* is given orally to mice on a high-fat diet, the serum levels of BCAAs rise, exacerbating their glucose intolerance. According to Pedersen et al., the gut microbiota's production of BCAAs may therefore be a factor in insulin resistance (Pedersen et al. 2016).

Tryptophan, phenylalanine, and tyrosine, three aromatic amino acids, have been linked to a higher incidence of diabetes (Chen et al. 2016). Moreover, bacteria contain genes that encode mediators of aromatic AA metabolism. The use of antibiotics has a significant impact on these metabolic pathways. Thus, the host AA profile is influenced by the gut microbiota's metabolism of aromatic amino acids (Fujisaka et al. 2018).

13.4.5 Glucose Metabolism and Gut Microbiota

Studies on the connections between diabetes, obesity, and gut microbiota have advanced. Depending on their race and nationality, type 2 diabetic patients (T2D) have distinct gut microbiomes. Qin and colleagues conducted a metagenomic analysis on a Chinese cohort with T2D. Their findings indicated an upregulation of genes regulating resistance to oxidative stress, a rise in the production of hydrogen sulfide and methane metabolism, and a decrease in the number of bacteria that produce butyrate (Qin et al. 2012). Karlsson et al. used a shotgun metagenomic analysis on a cohort of 145 European females who had T2D, impaired glucose tolerance, and a normal status to characterize functional and compositional changes in the gut microbiota. They developed a model of random forests that individuals with compromised glucose tolerance could use based on the metagenomic data. However, this model was unable to fit the Qin et al. cohort consistently (Karlsson et al. 2013). For this reason, metagenomic profiles may be a helpful indicator of impaired glucose metabolism in both healthy people and diabetics. But there are also racial and regional differences in gut microbiota that need to be considered. However, there is evidence linking impaired glucose metabolism to microbial dysfunction.

Furthermore, gut dysbiosis exists at the impaired glucose tolerance stage even before T2D develops. The gut microbiota and its functional genes, along with a decrease in the number of butyrate-producing bacteria, are features of T2D and prediabetes, according to Wu et al.'s metagenomic analysis of about 1500 Swedish participants (Wu et al. 2020). The authors demonstrated that bacterial compositional/functional changes are linked to insulin resistance at the impaired glucose tolerance stage by using the random forest algorithm to create a machine learning model to identify individuals with prediabetes or diabetes (Wu et al. 2020). Analysis of the gut microbiota may be able to diagnose glucose intolerance at an earlier stage and enhance early intervention. Building a machine learning model that forecasts when healthy and prediabetic patients will develop T2D has proven challenging as a result of these studies (Aasmets et al. 2021). However, it may serve as a prognostic marker for individual metabolic disease risks.

Type 1 diabetes (T1D) is also associated with gut microbiota. Proteomically analyzing stool samples from T1D patients, Gavin et al. found proteins linked to altered gut microbiota, decreased intestinal barrier functions, and intestinal inflammation even prior to the onset of the disease (Gavin et al. 2018). Before the onset of the disease, there were brief rises in *Streptococcus infantarius* and *Ruminococcus gnavus* abundance as well as a decrease in microbial diversity in children's gut microbiota who were genetically predisposed to type 1 diabetes (Kostic et al. 2015). *P. copri* and *Eubacterium siraeum* were more common in T1D patients, but *Firmicutes* and *Faecalibacterium prausnitzii* were more common in controls, according to a metagenomic analysis of 74 T1D patients and 296 healthy controls. This study linked host glucose control to modified bacterial species and metabolic pathways (Shilo et al. 2022). Thus, the development of T1D may be linked to modifications in the gut microbiota and the surrounding environment. The gut bacteria that is commensal as an immunological mechanism, *Parabacteroides distasonis,* has an epitope that resembles the insulin beta chain target protein of the autoimmune response in T1D. These bacteria may produce autoimmune antibodies that contribute to the pathophysiology of T1D (Girdhar et al. 2022). On the other hand, other reports contend that T1D development is inhibited by the microbiota. In a GF environment, but not when intestinal bacteria are present, MyD88-KO NOD mice develop T1D (Wen et al. 2008). According to research by Shimokawa and colleagues, *Ruminococcus* spp. may delay the onset of T1D by stimulating CD8+ regulatory T cells (Shimokawa et al. 2020). This finding suggests the onset of T1D is closely linked to the immune system and microbiota. This data unequivocally suggests the role of microbiota in glucose metabolism.

13.5 The Microbiome and Metabolic Syndrome

Herman Haller initially used the term "metabolic syndrome" in the 1970s while researching the different risks connected to atherosclerosis. But as early as the 1920s, dysmetabolism was linked to cardiovascular risk factors; soon after, the relationship between these risk factors and android adiposity, the phenotype that accompanies dysmetabolism and is defined by the distribution of fat in the trunk and upper body, was reported. Subsequently, Gerald Reaven dubbed MetS "syndrome X" in 1988, emphasizing insulin resistance as one of the clusters of risk factors (Sarafidis and Nilsson 2006).

It has been difficult to come up with a common definition for MetS, even after these clinical findings have been known for more than 80 years. Since then, several organizations have released their own definitions for MetS. Among them are the American Association for Clinical Endocrinology, the World Health Organization, and the International Diabetes Federation (Saklayen 2018). While central obesity, dyslipidemia, insulin resistance, and hypertension are all included in these definitions, there are differences in the definitions and cutoffs of each of these risk factors. As more research is done and data is gathered, more risk factors might surface to further muddle, or perhaps clarify, MetS. For instance, the definition of MetS has completely disregarded the gut; however, data from the past ten years, as the field of the microbiome has gained prominence, have demonstrated that the gut may be the source of mechanisms that result in MetS risk factors.

In 2007, several studies in humans and rodents provided evidence for a gut-centric theory of metastatic syndrome (MetS). These studies demonstrated that a prolonged high-fat diet (HFD) causes defects in the intestinal barrier, which allows intestinal luminal contents (food antigen, bacterial byproducts, bacteria themselves) and bacterial LPS to pass into systemic circulation. Because of the resulting low-grade inflammation's ability to inhibit normal glycemic function, it was named metabolic endotoxemia and continued to persist (Cani et al. 2007). This was the first report of Toll-like receptor (TLR) activation by bacterial LPS causing an innate immune response that reduced insulin sensitivity. The investigation of variables influencing intestinal permeability has thus emerged as one of the most exciting fields in microbiome research. Additionally, it has been found that the gut microbiota and its metabolites affect and influence research areas like circadian misalignment and the function of bile acids as bioactive molecules that have previously been connected to metabolic diseases. More defining MetS phenotypes that take into account defects at the gut level may eventually be produced by the findings of these exciting areas of research.

13.5.1 Obesity Is Associated with Abnormal Gut Microbiota

Since the middle of the twentieth century, the prevalence of obesity and its metabolic co-morbidities has sharply increased in developed nations. The obesity epidemic is thought to be primarily caused by sedentary lifestyles, increased food consumption, and widespread polygenetic susceptibility (McAllister et al. 2009). It may also be made worse by the widespread use of antibiotics. Furthermore, there is growing evidence that the pathophysiology of obesity involves some environmental effects that are mediated by the gut microbiota. It has been reported in 2006 that lean mice could gain weight if they had a transferrable microbiota linked to obesity (Turnbaugh et al. 2006). Subsequent epidemiological research has revealed variations in the gut microbiota between obese and lean individuals. At the species level, twin studies have shown that the abundance of SCFA producers, like *Eubacterium ventriosum* and *Roseburia intestinalis*, is linked to obesity, whereas the methanogenic archaeon *Methanobrevibacter smithii* and the butyrate-producing *Oscillospira* spp. may be linked to leanness (Gophna et al. 2017; Tims et al. 2013). Another metagenome-wide association study of lean people and obese people revealed that the glutamate-fermenting commensal *Bacteroides thetaiotaomicron* was significantly less abundant in obese people and had an inverse relationship with serum glutamate concentration (Liu et al. 2017). Additionally, feeding mice with *B. thetaiotaomicron* prevented them from becoming obese, suggesting that probiotics or other microbial compounds may be used in the future to target the gut microbiota in obesity intervention efforts. Studies on the gene families and gut microbial pathways indicate that obesity is linked to a decrease in superoxide reductase and a reduced ability for unidirectional conjugation, which is the transfer of genetic material between bacteria and may cause intestinal oxidative stress (Thingholm et al. 2019) (Fig. 13.1). In addition to observational studies, a fecal microbiota transplantation study demonstrates that transferring the phenotype of the human donor to the recipient animal in a diet-dependent manner occurs when stools from twins discordant for obesity are transplanted into germ-free mice (Ridaura et al. 2013). The ongoing, intense search for gut microbial messengers that connect the gut microbiota to the body's energy metabolism is justified by this groundbreaking study. However, it's crucial to keep in mind that different medications may have an impact on the gut microbiota when studying the possible role that altered gut microbiota may play a role in the development of chronic disorders, including obesity. For

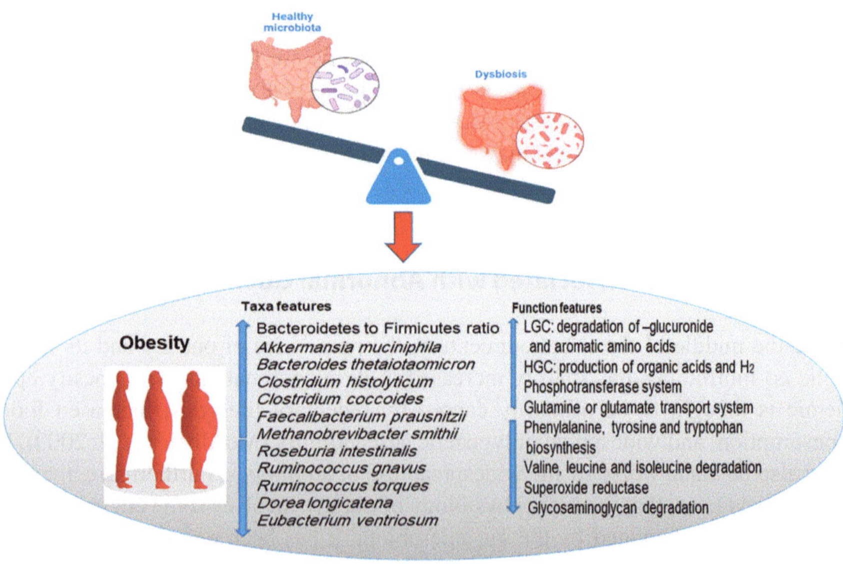

Fig. 13.1 Obesity is associated with abnormal gut microbiota

example, a gut microbiota that promotes health has been associated with the use of statins, which are cholesterol-lowering medications that are frequently prescribed to prevent arteriosclerosis (Vieira-Silva et al. 2020). It goes without saying that observations of any effects of microbiota-modifying drugs should ideally be followed up in interventional studies in order to address any causality of the relationships.

13.5.2 Modified Gut Microbiota Is Associated with Type 2 Diabetes and Prediabetes Levels

Indirect evidence that the intestinal microbiota and the distal gut's hormonal secretion capabilities may be involved in glucose regulation comes from extensive epidemiological studies showing that patients who undergo total colectomy have a higher risk of T2D than those who do not (Jensen et al. 2018). Furthermore, mechanistic research in rodents has demonstrated that hyperglycemia may enhance intestinal barrier permeability by reprogramming intestinal epithelial cells through transcription, altering tight junction integrity, and GLUT2-dependently, resulting in a leaky mucosa (Thaiss et al. 2018). Therefore, it is very important to determine whether an aberrant gut microbiota causes or maintains high blood glucose in T2D and its precursor states. T2D is the most common endocrine disorder, accounting for approximately 90% of all diabetes cases. It is also increasing in incidence and prevalence, similar to obesity, affecting 5–15% of adults (Deshpande et al. 2008). Obesity and T2D are linked to a complex network of hundreds of genes. The majority of people are obese or overweight, and these conditions are still in their early stages. The pathophysiology is characterized by increased insulin biosynthesis and secretion as a compensatory mechanism, primarily in the skeletal muscles, liver, and adipose tissues. Insulin biosynthesis decreases when insulin resistance persists, and hyperglycemia deteriorates as a result (Bornfeldt and Tabas 2011). Other common characteristics of the T2D phenotype include hypertension, a heightened propensity for platelet aggregation, and dyslipidemia, which is defined by high levels of circulating triglycerides and low-density lipoprotein cholesterol in addition to hyperglycemia. Premature arteriosclerosis is a risk factor for all these abnormalities (Laakso and Kuusisto 2014). As a result, people with T2D are treated with a variety of medications that target both elevated blood sugar levels and cardiovascular co-morbidity risk factors, in addition to changing their health-related behaviors. Nonetheless, these people's gut microbiota is impacted by their multidrug regimens in a variety of ways (Vich Vila et al. 2020). Thus, the early-stage drug-naive T2D, known as prediabetes, has been the focus of recent epidemiological studies seeking to clarify connections between the gut microbiota and the disease. Moreover, elevated plasma glucose levels within the range of those without diabetes are characteristic of prediabetes, but they are high enough to put them at risk of developing ischemic cardiovascular disease and overt T2D. The gut microbiota of these drug-naive people with prediabetes shows a decrease in the abundance of bacteria that produce butyrate, an increase in the abundance of bacteria that have the potential to cause inflammation, and a decrease in the abundance of *Akkermansia muciniphila* (Zhong et al. 2019) (Fig. 13.2).

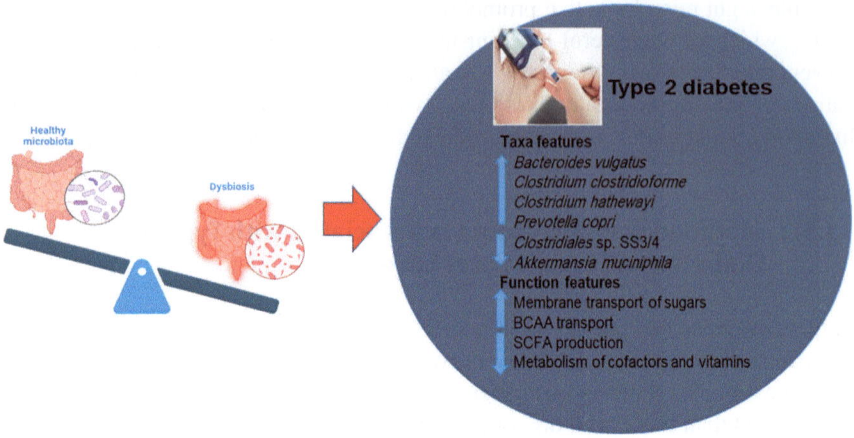

Fig. 13.2 Modified gut microbiota is associated with type 2 diabetes

13.5.3 Cardio-Metabolic Disorders (CMD) Are Associated with Gut Microbial Dysbiosis

Arteriosclerosis is a major cause of morbidity and mortality in the Western world. Arteriosclerosis patients often have a history of long-term, metabolic dysfunctions that are clinically silent, such as low-grade inflammation, insulin resistance, and elevated blood levels of lipids, glucose, and insulin (Michos et al. 2019). As a result, it can be difficult to discern between the possible impact of an aberrant gut microbiota on early-stage metabolic disturbances and an imbalanced gut microbiota that is speeding up the pathophysiology of ischemic heart disease at a stage distal to the dysmetabolic stage. Furthermore, like people with T2D, people with CMD also take a lot of medications, which makes it difficult to distinguish real arteriosclerosis signals in the gut microbiota from signatures brought on by complicated medications, pre-morbidities, and co-morbidities. These issues haven't been methodically handled up to this point. Without accounting for pre- or co-morbidities, a metagenome-wide association study of fecal samples from individuals with CMD, without taking pre- or co-morbidities into consideration, showed that, in comparison to healthy control subjects, the gut microbiome of these individuals is more abundant in oral cavity species and has enriched abundances of *Enterobacteriaceae*, such *as Escherichia coli*, *Enterobacter aerogenes,* and *Klebsiella* spp. (Jie et al. 2017) (Fig. 13.3). On the other hand, anti-inflammatory *Faecalibacterium prausnitzii* and *Bacteroides* species were less common in CMD patients. A decline in certain fermentation capacities was observed at the functional level, but there was also enrichment in the phosphotransferase system, amino acid transporters, and enzymes responsible for the synthesis of TMA and LPS (Jie et al. 2017). It was recently demonstrated that ischemic heart failure is linked to a dysbiotic gut microbiota with lower abundances of *Oscillibacter* spp., *Faecalibacterium*, and *Alistipes* and higher abundances of *Ruminococcus*, *Acinetobacter*, and *Veillonella* spp. (Cui et al. 2018).

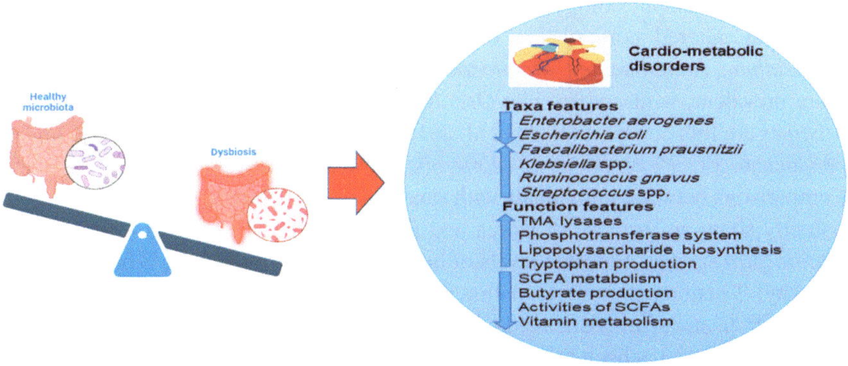

Fig. 13.3 Cardio-metabolic disorders (CMD) are associated with gut microbial dysbiosis

According to Cui and colleagues, these patients' microbiomes were functionally enriched in genes related to the synthesis of TMAO and LPS (Cui et al. 2018). Taken together, these studies on cardiovascular and metabolic diseases (CMD) indicate a shift towards a microbiome that is less focused on fermentation and more inclined towards promoting inflammation. The discovery of the TMAO pathway, which is linked to ischemic vascular conditions and involves the gut microbiota, is especially fascinating (Wang et al. 2011). TMA, a byproduct generated by gut bacteria from compounds like phosphatidylcholine, lecithin, and l-carnitine present in various animal sources including red meat, poultry, seafood, and eggs, enters the bloodstream where it is converted into TMAO in the liver. Research on rodents demonstrates that a diet rich in TMAO or its precursors hastens the progression of arteriosclerosis, promotes blood clot formation, and heightens the risk of thrombosis (Zhu et al. 2016). Moreover, the production of bacterial TMA in the intestines can be suppressed using 3,3-dimethyl-1-butanol, a TMA-lyase inhibitor, which has been shown to reduce arteriosclerosis and thrombotic events. Epidemiological evidence has further validated these findings, indicating a direct link between elevated plasma levels of TMAO and a higher incidence of heart attacks, strokes, and premature death among individuals with stable coronary artery disease. This collective body of research underscores the role of TMAO as a significant biomarker for arteriosclerosis (Haghikia et al. 2018).

13.5.4 Intestinal Microbiota Dysbiosis in Disorders of the Metabolic Liver

Non-alcoholic fatty liver disease (NAFLD) impacts a significant portion, estimated at 20–40%, of the global adult population and is commonly viewed as a hepatic manifestation of metabolic syndrome. NAFLD encompasses a spectrum of conditions, from the milder form known as simple steatosis to the more severe and inflammatory non-alcoholic steatohepatitis (NASH) (Brunt et al. 2011). Recent research is increasingly highlighting the potential role of disrupted gut microbiota as a key factor in the pathophysiology of NAFLD and NASH. Individuals with NAFLD often exhibit an altered composition of gut bacteria, with certain species such as *Clostridium, Anaerobacter, Streptococcus, Escherichia*, and *Lactobacillus* being more prevalent, while others like *Oscillibacter, Flavonifaractor, Odoribacter, and Alistipes* are less common (Jiang et al. 2015). In the case of NASH, there is a higher prevalence of *Proteobacteria, Enterobacteriaceae*, and *Escherichia* species. Comparisons between individuals with steatosis or NASH and healthy controls consistently show an increase in Dorea and *Ruminococcus* species and a decrease in *Oscillospira* species. Elevated levels of hepatotoxic compounds like 2-butanone and 4-methyl-2-pentanone in fecal matter, which are harmful to individuals with metabolic liver diseases, are associated with these microbiota changes (Del Chierico et al. 2017). Moreover, it's been proposed that individuals with NAFLD harbor gut microbiota that produce more ethanol compared to healthy individuals, likely due to an abundance of ethanol-producing bacteria like *E. coli* within their microbiomes (Nair et al. 2001) (Fig. 13.4). This excess ethanol production can lead to gut barrier dysfunction, portal endotoxemia, and tissue damage through the activation of NF-κB signaling pathways (Xu et al. 2011). The compromised detoxification pathways in the livers of NAFLD patients result in an accumulation of reactive oxygen species, causing oxidative damage to hepatocytes and fostering hepatic inflammation and steatohepatitis (de Medeiros and de Lima 2015). Individuals with liver cirrhosis also exhibit significant alterations in their gut microbiota, with an enrichment of *Proteobacteria* and *Fusobacteria*. Functional changes in the gut microbiome of those with liver cirrhosis and failure include alterations in processes related to nitrogen metabolism, GABA biosynthesis, denitrification capacity, and haem biosynthesis, all of which can impact liver health (Schwenger et al. 2019). Research further indicates that an abundance of alcohol-producing bacteria like *Klebsiella pneumoniae* in the gut can exacerbate the progression of NAFLD. Additionally, microbial metabolites such as phenylacetate have been identified as contributors to lipid buildup in the liver and the development of NASH, particularly in obese women without diabetes, as revealed through integrative multi-omics approaches (Yuan et al. 2019).

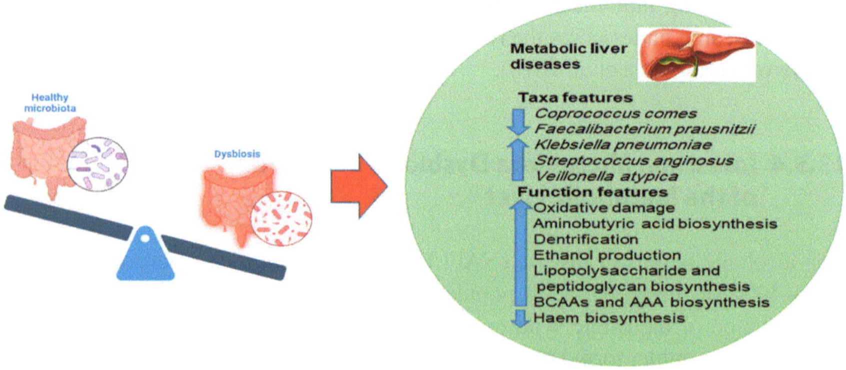

Fig. 13.4 Intestinal microbiota dysbiosis in metabolic liver diseases

13.5.5 Malnutrition-Related Disruption of Gut Microbiota

About 160 million people worldwide suffer from malnutrition, which is the main cause of death for children under the age of five. Breastfeeding, having access to food and water are important protective factors against malnutrition, and maturing a healthy gut microbiota, which is characterized by a brief bloom of bifidobacteria (Million et al. 2017). The early loss of two crucial components of mother milk, *Bifidobacterium longum* and *Bifidobacterium pseudolongum*, is one of the first indicators of intestinal dysbiosis in severe acute malnutrition (SAM) (Smith et al. 2013) (Fig. 13.5). It seems that *B. lungum* and *B. pseudolongum* play a significant role in forming the early gut microbiota in infants, in part because they secrete broad-spectrum lantibiotics (Lee et al. 2011). After that, an abnormal inversion of the ratio of facultative to anaerobes is the main symptom of SAM dysbiosis. Under the right dietary conditions, this inversion could be reversed and eventually result in microbial communities that are richer in anaerobes. But if malnutrition persists, the loss of a mature, healthy anaerobic gut microbiota can eventually result in deficiencies in energy harvesting, immunological responses, and vitamin synthesis (Kelly et al. 2004). It can also be linked to persistent malabsorption, diarrhea, and systemic bacterial invasion (Stecher and Hardt 2008). The intestinal microbiota's developmental biology is not considered when developing the current dietary therapy regimens for SAM. One study used pig and mouse models to create diets that would help the microbiota mature after weaning. This might help children with SAM's microbiota mature and put their growth and metabolic profiles back on a more wholesome path (Gehrig et al. 2019). It is critically necessary to conduct long-term randomized clinical trials to assess the impact of these dietary plans on the weight gain and general physiology of impacted children.

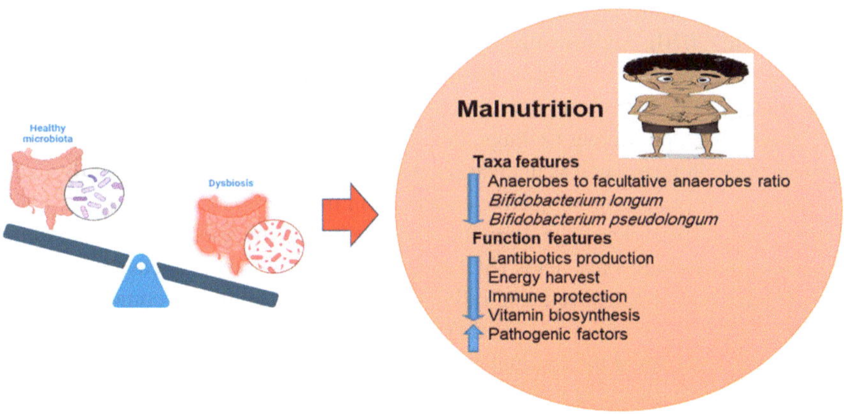

Fig. 13.5 Malnutrition-related disruption of gut microbiota

13.6 Diagnostics, Therapeutics, and Clinical Applications

Identifying the pathways that govern systemic metabolism originating from the microbiome is crucial as it opens up the potential for targeting these pathways for therapeutic purposes. Microbiome-based treatments are attractive due to their non-invasive nature, low risk of toxicity, and ease of administration. Looking forward, there are several promising avenues where knowledge of the microbiota could be harnessed to enhance human metabolic well-being. Currently, the sole clinically approved microbiome-based treatment for *Clostridium difficile* infection is fecal microbiota transplantation (FMT) (Napolitano and Covasa 2020). Various studies have shown that utilizing FMT or antibiotic therapy for community replacement as therapeutic approaches for obesity and metabolic disorders is ineffective, underscoring the necessity for more refined methods to modulate systemic metabolism via the gut. Ongoing research is exploring numerous advanced techniques, many of which rely on live bacteria (probiotics) or their metabolites (postbiotics). Investigation into the potential therapeutic advantages of administering *A. muciniphila* has been prompted by the discovery of a negative association between obesity and *Akkermansia*. This represents a prime example of an intervention based on a specific microbial species (Si et al. 2022). Preliminary results from a recent double-blind randomized controlled trial involving 32 overweight and insulin-resistant human subjects suggest that the administration of pasteurized *A. muciniphila* may enhance insulin levels and plasma cholesterol in overweight individuals (Depommier et al. 2019). In mice fed a high-fat diet, *A. muciniphila* can produce a protein that boosts thermogenesis and GLP-1 secretion, leading to improvements in weight and metabolic parameters (Yoon et al. 2021). While these interventions did not significantly reduce body weight, they did enhance insulin sensitivity and various secondary indicators. Efforts have also been made to transplant a "lean" microbiome into obese individuals, mirroring findings from preclinical mouse studies (Allegretti et al. 2020). Research has indicated that underfeeding as opposed to overfeeding and vancomycin treatment compared to a placebo resulted in greater calorie loss in stool, aligning with preclinical observations on microbiota energy extraction (Basolo et al. 2020). Additionally, bariatric surgery, the most successful weight loss procedure, has been linked to certain side effects mediated by microbial production of secondary bile acid (Chaudhari et al. 2021).

Despite these promising findings, prebiotic and probiotic interventions encounter several challenges, such as unclear mechanisms, inconsistent clinical results, and the reliance on proxies or subjective endpoints in initial clinical studies (Suez et al. 2019). Various theories have been posited regarding the mechanism, including increased short-chain fatty acids (SCFAs), hormone signaling, mucus layer thickness, barrier function, and reduced inflammation. However, the outcomes of different studies have not always aligned. While utilizing a single probiotic strain for treatment is considered the optimal approach for establishing causation, conflicting results have been observed in other studies using the same probiotic blends (Veiga et al. 2020). For example, some research has demonstrated improved body mass index (BMI) in obese children with non-alcoholic fatty liver disease (NAFLD)

(Rajkumar et al. 2014) and metabolic advantages in overweight adults (Alisi et al. 2014). Conversely, in obese adolescents, the probiotic mixture VSL#3, comprising *Lactobacillus, Bifidobacterium* species, and *Streptococcus thermophilus*, unexpectedly led to weight gain and increased adiposity (Jones et al. 2018). It is advisable to avoid subjective secondary metrics like waist circumference. Moreover, despite promising preclinical investigations and early clinical trials, most extensive human trials exploring the impact of probiotics on weight loss have shown minimal to no benefits (Wiciński et al. 2020). Several meta-analyses have indicated no significant weight-related advantages in both adults and children, although comparing probiotic studies directly is challenging due to variations in probiotic strains, dosages, and treatment durations (Perna et al. 2021). Although some meta-analyses have reported statistically significant benefits for overweight adults, none of the observed effect sizes have exceeded 1 kg of weight loss (Koutnikova et al. 2019). Despite the comparatively strong preclinical evidence, a number of factors may have contributed to the challenge of translating probiotic therapies to humans. Preclinical model mice are usually young, housed in carefully regulated conditions, have caged microbiomes that are in balance (Hildebrand et al. 2013), and eat homogenous diets. The human microbiome, on the other hand, differs greatly depending on factors like diet, sex, age, and location (Zhang et al. 2021). Effective probiotic species colonization depends on variations in the human microbiome because certain individuals have a high engraftment resistance (Zmora et al. 2018). Recent research indicates that due to the varied reactions to probiotics and other therapies targeting the microbiome, the microbiome has the potential to serve as a valuable diagnostic tool for precision medicine approaches to combat obesity and metabolic disorders. The predictive ability of the microbiome's composition for post-meal glycemic responses to diets enables the development of personalized nutrition strategies to optimize host metabolism (Zeevi et al. 2015). Comparing these tailored interventions to a Mediterranean diet regimen, recent clinical trials involving patients with T2D have demonstrated reduced HbA1c levels and improved glycemic management in both prediabetic individuals and those newly diagnosed with the condition (Rein et al. 2022). A comprehensive understanding of how microbial species, genetic pathways, and dietary elements interact is crucial for leveraging microbiome data effectively. Two studies investigating the human microbiome's capacity to metabolize various fiber types unveiled upregulated genetic pathways and species that could outcompete others for available nutrients (Delannoy-Bruno et al. 2021). Notably, alterations in the plasma proteomes of human subjects were observed in response to controlled diets rich in these fibers. In a related trial, a high-fiber diet was linked to an increase in genes responsible for glycan degradation by microbes, while a diet rich in fermented foods correlated with enhanced microbial diversity and reduced markers of inflammation (Wastyk et al. 2021). Drawing on data from the PREDICT 1 trial, which encompassed over a thousand UK patients, including twin pairs, along with a validation group in the USA, two recent studies have shed light on the intricate relationships between the microbiome, diet, and metabolic outcomes (Asnicar et al. 2021). They discovered that the makeup of the microbiome was a strong predictor of fasting cardiovascular metabolic markers, postprandial lipid, and, to a lesser extent, glycemic response. A variety of plant-based, raw diets were linked to more robust microbiome species, some of which could only be detected by metagenomic assemblies, underscoring the importance of additional characterization. The collective findings from these studies suggest a potential approach to addressing the variability and occasional unpredictability of individual responses to interventions targeting the microbiome: utilizing both taxonomic and metagenomic data to forecast individual reactions. Beyond the direct impacts of microbiome composition on host well-being, numerous instances highlight the diverse effects of microbial metabolites on human health. Manipulating these metabolites represents an additional therapeutic avenue for leveraging microbiome discoveries. An illustrative example of a metabolite-based treatment is trimethylamine N-oxide (TMAO). TMAO, derived from the breakdown of choline and other TMA-containing compounds by microbes, is linked to conditions such as heart disease, atherosclerosis, T2D, thrombosis, Alzheimer's disease, and stroke (Haghikia et al. 2018; Missailidis et al. 2016; Vogt et al. 2018; Zhuang et al. 2019). Targeting bacterial TMA production in preclinical studies has shown therapeutic benefits in mouse models of atherosclerosis and platelet aggregation (Roberts et al. 2018). Current research is exploring TMAO-targeted treatments for chronic kidney disease and cardiovascular ailments, with TMAO also serving as a biomarker in various clinical trials (Kimber et al. 2020; Yang et al. 2019; Awoyemi et al. 2021). While numerous other microbial metabolites hold promise for future therapeutic applications, TMAO stands out as one of the most extensively studied (Descamps et al. 2019).

A comprehensive exploration of the microbiome in patients afflicted with conditions beyond metabolic syndrome could yield valuable insights into diagnosis and potential responses to conventional treatment approaches, irrespective of specific therapeutics. For instance, a significant challenge in modern medicine lies in the staging and diagnosis of non-alcoholic fatty liver disease (NAFLD). Growing evidence suggests that microbial analysis might lead to less invasive strategies for addressing this issue. Intriguing discoveries have linked dysbiosis to the clinical manifestation of NAFLD-related fibrosis, with fibrosis notably associated with an increase in gram-negative bacteria and a reduction in overall microbial diversity, potentially contributing to inflammatory endotoxin production (Sharpton et al. 2021).

References

Aasmets O, Lüll K, Lang JM, Pan C, Kuusisto J, Fischer K et al (2021) Machine learning reveals time-varying microbial predictors with complex effects on glucose regulation. Msystems 6(1):10–128

Alisi A, Bedogni G, Baviera G, Giorgio V, Porro E, Paris C et al (2014) Randomised clinical trial: the beneficial effects of VSL#3 in obese children with non-alcoholic steatohepatitis. Aliment Pharmacol Ther 39:1276–1285

Allegretti JR, Kassam Z, Mullish BH, Chiang A, Carrellas M, Hurtado J et al (2020) Effects of fecal microbiota transplantation with oral capsules in obese patients. Clin Gastroenterol Hepatol 18(4):855–863

Amon P, Sanderson I (2017) What is the microbiome? Arch Dis Childhood-Educ Pract 102(5):257–260

An D, Oh SF, Olszak T, Neves JF, Avci FY, Erturk-Hasdemir D et al (2014) Sphingolipids from a symbiotic microbe regulate homeostasis of host intestinal natural killer T cells. Cell 156(1):123–133

Asnicar F, Berry SE, Valdes AM, Nguyen LH, Piccinno G, Drew DA et al (2021) Microbiome connections with host metabolism and habitual diet from 1,098 deeply phenotyped individuals. Nat Med 27:321–332

Awoyemi A, Mayerhofer C, Felix AS, Hov JR, Moscavitch SD, Lappegard KT et al (2021) Rifaximin or saccharomyces boulardii in heart failure with reduced ejection fraction: results from the randomized GutHeart trial. EBioMedicine 70:103511

Basolo A, Hohenadel M, Ang QY, Piaggi P, Heinitz S, Walter M et al (2020) Effects of underfeeding and oral vancomycin on gut microbiome and nutrient absorption in humans. Nat Med 26:589–598

Becher T, Palanisamy S, Kramer DJ, Eljalby M, Marx SJ, Wibmer AG et al (2021) Brown adipose tissue is associated with cardiometabolic health. Nat Med 27:58–65

Bellerba F, Muzio V, Gnagnarella P, Facciotti F, Chiocca S, Bossi P et al (2021) The association between vitamin D and gut microbiota: a systematic review of human studies. Nutrients 13(10):3378

Bornfeldt KE, Tabas I (2011) Insulin resistance, hyperglycemia, and atherosclerosis. Cell Metab 14(5):575–585

Breton J, Tennoune N, Lucas N, Francois M, Legrand R, Jacquemot J et al (2016) coli proteins activate host satiety pathways following nutrient-induced bacterial growth. Cell Metab 23(2):324–334

Brown EM, Ke X, Hitchcock D, Jeanfavre S, Avila-Pacheco J, Nakata T et al (2019) Bacteroides-derived sphingolipids are critical for maintaining intestinal homeostasis and symbiosis. Cell Host Microbe 25(5):668–680

Brown EM, Clardy J, Xavier RJ (2023) Gut microbiome lipid metabolism and its impact on host physiology. Cell Host Microbe 31(2):173–186

Brunt EM, Kleiner DE, Wilson LA, Belt P, Neuschwander-Tetri BA, NASH Clinical Research Network (CRN) (2011) Nonalcoholic fatty liver disease (NAFLD) activity score and the histopathologic diagnosis in NAFLD: distinct clinicopathologic meanings. Hepatology 53(3):810–820

Byndloss MX, Olsan EE, Rivera-Chávez F, Tiffany CR, Cevallos SA, Lokken KL et al (2017) Microbiota-activated PPAR-γ signaling inhibits dysbiotic Enterobacteriaceae expansion. Science 357(6351):570–575

Cani PD, Amar J, Iglesias MA, Poggi M, Knauf C, Bastelica D et al (2007) Metabolic endotoxemia initiates obesity and insulin resistance. Diabetes 56(7):1761–1772

Cannon B, Nedergaard J (2004) Brown adipose tissue: function and physiological significance. Physiol Rev 84:277–359

Charbonneau MR, Blanton LV, DiGiulio DB, Relman DA, Lebrilla CB et al (2016) A microbial perspective of human developmental biology. Nature 535(7610):48–55

Chaudhari SN, Luo JN, Harris DA, Aliakbarian H, Yao L, Paik D et al (2021) A microbial metabolite remodels the gut-liver axis following bariatric surgery. Cell Host Microbe 29(408–424):e407

Chen T, Zheng X, Ma X, Bao Y, Ni Y, Hu C et al (2016) Tryptophan predicts the risk for future type 2 diabetes. PLoS One 11(9):e0162192

Chouchani ET, Kazak L, Spiegelman BM (2019) New advances in adaptive thermogenesis: UCP1 and beyond. Cell Metab 29:27–37

Christ A, Lauterbach M, Latz E (2019) Western diet and the immune system: an inflammatory connection. Immunity 51(5):794–811

Cui X, Ye L, Li J, Jin L, Wang W, Li S et al (2018) Metagenomic and metabolomic analyses unveil dysbiosis of gut microbiota in chronic heart failure patients. Sci Rep 8(1):635

de Medeiros IC, de Lima JG (2015) Is nonalcoholic fatty liver disease an endogenous alcoholic fatty liver disease?–A mechanistic hypothesis. Med Hypotheses 85(2):148–152

Del Chierico F, Nobili V, Vernocchi P, Russo A, De Stefanis C, Gnani D et al (2017) Gut microbiota profiling of pediatric nonalcoholic fatty liver disease and obese patients unveiled by an integrated meta-omics-based approach. Hepatology 65(2):451–464

Delannoy-Bruno O, Desai C, Raman AS, Chen RY, Hibberd MC, Cheng J et al (2021) Evaluating microbiome-directed fibre snacks in gnotobiotic mice and humans. Nature 595:91–95
Depommier C, Everard A, Druart C, Plovier H, Van Hul M, Vieira-Silva S et al (2019) Supplementation with Akkermansia muciniphila in overweight and obese human volunteers: a proof-of-concept exploratory study. Nat Med 25(7):1096–1103
Descamps HC, Herrmann B, Wiredu D, Thaiss CA (2019) The path toward using microbial metabolites as therapies. EBioMedicine 44:747–754
Deshpande AD, Harris-Hayes M, Schootman M (2008) Epidemiology of diabetes and diabetes-related complications. Phys Ther 88(11):1254–1264
Fluhr L, Mor U, Kolodziejczyk AA, Dori-Bachash M, Leshem A, Itav S et al (2021) Gut microbiota modulates weight gain in mice after discontinued smoke exposure. Nature 600:713–719
Fujisaka S, Avila-Pacheco J, Soto M, Kostic A, Dreyfuss JM, Pan H et al (2018) Diet, genetics, and the gut microbiome drive dynamic changes in plasma metabolites. Cell Rep 22(11):3072–3086
Fujisaka S, Watanabe Y, Tobe K (2023) The gut microbiome: a core regulator of metabolism. J Endocrinol 256(3):e220111
Gavin PG, Mullaney JA, Loo D, Cao KA, Gottlieb PA, Hill MM et al (2018) Intestinal metaproteomics reveals host-microbiota interactions in subjects at risk for type 1 diabetes. Diabetes Care 41(10):2178–2186
Gehrig JL, Venkatesh S, Chang HW, Hibberd MC, Kung VL, Cheng J et al (2019) Effects of microbiota-directed foods in gnotobiotic animals and undernourished children. Science 365(6449):eaau4732
Gérard P (2020) The crosstalk between the gut microbiota and lipids. OCL 27:70
Girdhar K, Huang Q, Chow IT, Vatanen T, Brady C, Raisingani A et al (2022) A gut microbial peptide and molecular mimicry in the pathogenesis of type 1 diabetes. Proc Natl Acad Sci 119(31):e2120028119
Gophna U, Konikoff T, Nielsen HB (2017) Oscillospira and related bacteria–from metagenomic species to metabolic features. Environ Microbiol 19(3):835–841
Haghikia A, Li XS, Liman TG, Bledau N, Schmidt D, Zimmermann F et al (2018) Gut microbiota-dependent trimethylamine N-Oxide predicts risk of cardiovascular events in patients with stroke and is related to proinflammatory monocytes. Arterioscler Thromb Vasc Biol 38:2225–2235
Hildebrand F, Nguyen TLA, Brinkman B, Yunta RG, Cauwe B, Vandenabeele P et al (2013) Inflammation-associated enterotypes, host genotype, cage and inter-individual effects drive gut microbiota variation in common laboratory mice. Genome Biol 14:R4
Huda MN, Ahmad SM, Kalanetra KM, Taft DH, Alam MJ, Khanam A et al (2019) Neonatal vitamin A supplementation and vitamin A status are associated with gut microbiome composition in Bangladeshi infants in early infancy and at 2 years of age. J Nutr 149(6):1075–1088
Jang C, Oh SF, Wada S, Rowe GC, Liu L, Chan MC et al (2016) A branched-chain amino acid metabolite drives vascular fatty acid transport and causes insulin resistance. Nat Med 22(4):421–426
Jensen AB, Sørensen TI, Pedersen O, Jess T, Brunak S, Allin KH (2018) Increase in clinically recorded type 2 diabetes after colectomy. elife 7:e37420
Jiang W, Wu N, Wang X, Chi Y, Zhang Y, Qiu X et al (2015) Dysbiosis gut microbiota associated with inflammation and impaired mucosal immune function in intestine of humans with non-alcoholic fatty liver disease. Sci Rep 5(1):8096
Jie Z, Xia H, Zhong SL, Feng Q, Li S, Liang S et al (2017) The gut microbiome in atherosclerotic cardiovascular disease. Nat Commun 8(1):845
Johnson EL, Heaver SL, Waters JL, Kim BI, Bretin A, Goodman AL et al (2020) Sphingolipids produced by gut bacteria enter host metabolic pathways impacting ceramide levels. Nat Commun 11(1):2471
Jones RB, Alderete TL, Martin AA, Geary BA, Hwang DH, Palmer SL et al (2018) Probiotic supplementation increases obesity with no detectable effects on liver fat or gut microbiota in obese Hispanic adolescents: a 16-week, randomized, placebo-controlled trial. Pediatr Obes 13:705–714

Karlsson FH, Tremaroli V, Nookaew I, Bergström G, Behre CJ, Fagerberg B et al (2013) Gut metagenome in European women with normal, impaired and diabetic glucose control. Nature 498(7452):99–103

Kelly D, Campbell JI, King TP, Grant G, Jansson EA, Coutts AG et al (2004) Commensal anaerobic gut bacteria attenuate inflammation by regulating nuclear-cytoplasmic shuttling of PPAR-γ and RelA. Nat Immunol 5(1):104–112

Kimber C, Zhang S, Johnson C, West RE 3rd, Prokopienko AJ, Mahnken JD et al (2020) Randomized, placebo-controlled trial of rifaximin therapy for lowering gut-derived cardiovascular toxins and inflammation in CKD. Kidney360 1:1206–1216

Kishino S, Takeuchi M, Park SB, Hirata A, Kitamura N, Kunisawa J et al (2013) Polyunsaturated fatty acid saturation by gut lactic acid bacteria affecting host lipid composition. Proc Natl Acad Sci 110(44):17808–17813

Knip M, Siljander H (2016) The role of the intestinal microbiota in type 1 diabetes mellitus. Nat Rev Endocrinol 12(3):154–167

Koh A, De Vadder F, Kovatcheva-Datchary P, Bäckhed F (2016) From dietary fiber to host physiology: short-chain fatty acids as key bacterial metabolites. Cell 165(6):1332–1345

Kostic AD, Gevers D, Siljander H, Vatanen T, Hyötyläinen T, Hämäläinen AM et al (2015) The dynamics of the human infant gut microbiome in development and in progression toward type 1 diabetes. Cell Host Microbe 17(2):260–273

Koutnikova H, Genser B, Monteiro-Sepulveda M, Faurie J-M, Rizkalla S, Schrezenmeir J et al (2019) Impact of bacterial probiotics on obesity, diabetes and non-alcoholic fatty liver disease related variables: a systematic review and meta-analysis of randomised controlled trials. BMJ Open 9:e017995

Krisko TI, Nicholls HT, Bare CJ, Holman CD, Putzel GG, Jansen RS et al (2020) Dissociation of adaptive thermogenesis from glucose homeostasis in microbiome-deficient mice. Cell Metab 31(592–604):e599

Laakso M, Kuusisto J (2014) Insulin resistance and hyperglycaemia in cardiovascular disease development. Nat Rev Endocrinol 10(5):293–302

Lee JH, Li X, O'Sullivan DJ (2011) Transcription analysis of a lantibiotic gene cluster from Bifidobacterium longum DJO10A. Appl Environ Microbiol 77(17):5879–5887

Li L, Krause L, Somerset S (2017) Associations between micronutrient intakes and gut microbiota in a group of adults with cystic fibrosis. Clin Nutr 36(4):1097–1104

Li B, Li L, Li M, Lam SM, Wang G, Wu Y et al (2019) Microbiota depletion impairs thermogenesis of brown adipose tissue and browning of white adipose tissue. Cell Rep 26(2720–2737):e2725

Liu R, Hong J, Xu X, Feng Q, Zhang D, Gu Y et al (2017) Gut microbiome and serum metabolome alterations in obesity and after weight-loss intervention. Nat Med 23(7):859–868

Long KZ, García C, Santos JI, Rosado JL, Hertzmark E, DuPont HL et al (2007) Vitamin A supplementation has divergent effects on norovirus infections and clinical symptoms among Mexican children. J Infect Dis 196(7):978–985

Louis P, Flint HJ (2017) Formation of propionate and butyrate by the human colonic microbiota. Environ Microbiol 19(1):29–41

Lozupone CA, Stombaugh JI, Gordon JI, Jansson JK, Knight R (2012) Diversity, stability and resilience of the human gut microbiota. Nature 489(7415):220–230

Lund J, Gerhart-Hines Z, Clemmensen C (2020) Role of energy excretion in human body weight regulation. Trends Endocrinol Metabol 31(10):705–708

Magnúsdóttir S, Ravcheev D, de Crécy-Lagard V, Thiele I (2015) Systematic genome assessment of B-vitamin biosynthesis suggests co-operation among gut microbes. Front Genet 6:148

Mariamenatu AH, Abdu EM (2021) Overconsumption of Omega-6 polyunsaturated fatty acids (PUFAs) versus deficiency of Omega-3 PUFAs in modern-day diets: the disturbing factor for their "balanced antagonistic metabolic functions" in the human body. J Lipids 2021(1):8848161

Mavros CF, Ronda C (2022) Microbiome and metabolism: advancements in microbiome engineering. Curr Opin Endocrine Metabol Res 27:100404

McAllister EJ, Dhurandhar NV, Keith SW, Aronne LJ, Barger J, Baskin M et al (2009) Ten putative contributors to the obesity epidemic. Crit Rev Food Sci Nutr 49(10):868–913

Meropol SB, Edwards A (2015) Development of the infant intestinal microbiome: a bird's eye view of a complex process. Birth Defects Res C Embryo Today 105(4):228–239
Michos ED, McEvoy JW, Blumenthal RS (2019) Lipid management for the prevention of atherosclerotic cardiovascular disease. N Engl J Med 381(16):1557–1567
Million M, Diallo A, Raoult D (2017) Gut microbiota and malnutrition. Microb Pathog 106:127–138
Missailidis C, Hallqvist J, Qureshi AR, Barany P, Heimburger O, Lindholm B et al (2016) Serum trimethylamine-N-oxide is strongly related to renal function and predicts outcome in chronic kidney disease. PLoS One 11:e0141738
Miyamoto J, Igarashi M, Watanabe K, Karaki SI, Mukouyama H, Kishino S et al (2019) Gut microbiota confers host resistance to obesity by metabolizing dietary polyunsaturated fatty acids. Nat Commun 10(1):4007
Nair S, Cope K, Terence RH, Diehl AM (2001) Obesity and female gender increase breath ethanol concentration: potential implications for the pathogenesis of nonalcoholic steatohepatitis. Off J Am Coll Gastroenterol 96(4):1200–1204
Napolitano M, Covasa M (2020) Microbiota transplant in the treatment of obesity and diabetes: current and future perspectives. Front Microbiol 11:590370
Pedersen HK, Gudmundsdottir V, Nielsen HB, Hyotylainen T, Nielsen T, Jensen BA et al (2016) Human gut microbes impact host serum metabolome and insulin sensitivity. Nature 535(7612):376–381
Perna S, Ilyas Z, Giacosa A, Gasparri C, Peroni G, Faliva MA et al (2021) Is probiotic supplementation useful for the management of body weight and other anthropometric measures in adults affected by overweight and obesity with metabolic related diseases? A systematic review and meta-analysis. Nutrients 13:666
Pham VT, Dold S, Rehman A, Bird JK, Steinert RE (2021) Vitamins, the gut microbiome and gastrointestinal health in humans. Nutr Res 95:35–53
Psichas A, Sleeth ML, Murphy KG, Brooks L, Bewick GA, Hanyaloglu AC et al (2015) The short chain fatty acid propionate stimulates GLP-1 and PYY secretion via free fatty acid receptor 2 in rodents. Int J Obes 39(3):424–429
Qin J, Li Y, Cai Z, Li S, Zhu J, Zhang F et al (2012) Metagenome-wide association study of gut microbiota in type 2 diabetes. Nature 490(7418):55–60
Rajkumar H, Mahmood N, Kumar M, Varikuti SR, Challa HR, Myakala SP (2014) Effect of probiotic (VSL#3) and omega-3 on lipid profile, insulin sensitivity, inflammatory markers, and gut colonization in overweight adults: a randomized, controlled trial. Mediat Inflamm 2014:348959
Rein M, Ben-Yacov O, Godneva A, Shilo S, Zmora N, Kolobkov D et al (2022) Effects of personalized diets by prediction of glycemic responses on glycemic control and metabolic health in newly diagnosed T2DM: a randomized dietary intervention pilot trial. BMC Med 20:56
Ridaura VK, Faith JJ, Rey FE, Cheng J, Duncan AE, Kau AL et al (2013) Gut microbiota from twins discordant for obesity modulate metabolism in mice. Science 341(6150):1241214
Rivera-Chávez F, Zhang LF, Faber F, Lopez CA, Byndloss MX, Olsan EE et al (2016) Depletion of butyrate-producing Clostridia from the gut microbiota drives an aerobic luminal expansion of Salmonella. Cell Host Microbe 19(4):443–454
Roberts AB, Gu X, Buffa JA, Hurd AG, Wang Z, Zhu W et al (2018) Development of a gut microbe-targeted nonlethal therapeutic to inhibit thrombosis potential. Nat Med 24:1407–1417
Saklayen MG (2018) The global epidemic of the metabolic syndrome. Curr Hypertens Rep 20(2):1–8
Sarafidis PA, Nilsson PM (2006) The metabolic syndrome: a glance at its history. J Hypertens 24(4):621–626
Sato Y, Atarashi K, Plichta DR, Arai Y, Sasajima S, Kearney SM et al (2021) Novel bile acid biosynthetic pathways are enriched in the microbiome of centenarians. Nature 599(7885):458–464
Schiattarella GG, Sannino A, Toscano E, Giugliano G, Gargiulo G, Franzone A et al (2017) Gut microbe-generated metabolite trimethylamine-N-oxide as cardiovascular risk biomarker: a systematic review and dose-response meta-analysis. Eur Heart J 38(39):2948–2956
Schwenger KJ, Clermont-Dejean N, Allard JP (2019) The role of the gut microbiome in chronic liver disease: the clinical evidence revised. JHEP Rep 1(3):214–226

Sharpton SR, Schnabl B, Knight R, Loomba R (2021) Current concepts, opportunities, and challenges of gut microbiome-based personalized medicine in nonalcoholic fatty liver disease. Cell Metab 33:21–32

Shilo S, Godneva A, Rachmiel M, Korem T, Bussi Y, Kolobkov D et al (2022) The gut microbiome of adults with type 1 diabetes and its association with the host glycemic control. Diabetes Care 45(3):555–563

Shimokawa C, Kato T, Takeuchi T, Ohshima N, Furuki T, Ohtsu Y et al (2020) CD8+ regulatory T cells are critical in prevention of autoimmune-mediated diabetes. Nat Commun 11(1):1922

Si J, Kang H, You HJ, Ko G (2022) Revisiting the role of Akkermansia muciniphila as a therapeutic bacterium. Gut Microbes 14(1):2078619

Smith MI, Yatsunenko T, Manary MJ, Trehan I, Mkakosya R, Cheng J et al (2013) Gut microbiomes of Malawian twin pairs discordant for kwashiorkor. Science 339(6119):548–554

Stecher B, Hardt WD (2008) The role of microbiota in infectious disease. Trends Microbiol 16(3):107–114

Steinert RE, Lee YK, Sybesma W (2020) Vitamins for the gut microbiome. Trends Mol Med 26(2):137–140

Suarez-Zamorano N, Fabbiano S, Chevalier C, Stojanovic O, Colin DJ, Stevanovic A et al (2015) Microbiota depletion promotes browning of white adipose tissue and reduces obesity. Nat Med 21:1497–1501

Suez J, Zmora N, Segal E, Elinav E (2019) The pros, cons, and many unknowns of probiotics. Nat Med 25:716–729

Tazoe H, Otomo Y, Karaki SI, Kato I, Fukami Y, Terasaki M et al (2009) Expression of short-chain fatty acid receptor GPR41 in the human colon. Biomed Res 30(3):149–156

Thaiss CA, Shapiro H, Elinav E (2017) Post-dieting weight gain: the role of persistent microbiome changes. Future Microbiol 12:555–559

Thaiss CA, Levy M, Grosheva I, Zheng D, Soffer E, Blacher E et al (2018) Hyperglycemia drives intestinal barrier dysfunction and risk for enteric infection. Science 359(6382):1376–1383

Thingholm LB, Rühlemann MC, Koch M, Fuqua B, Laucke G, Boehm R et al (2019) Obese individuals with and without type 2 diabetes show different gut microbial functional capacity and composition. Cell Host Microbe 26(2):252–264

Tims S, Derom C, Jonkers DM, Vlietinck R, Saris WH, Kleerebezem M et al (2013) Microbiota conservation and BMI signatures in adult monozygotic twins. ISME J 7(4):707–717

Turnbaugh PJ, Ley RE, Mahowald MA, Magrini V, Mardis ER, Gordon JI (2006) An obesity-associated gut microbiome with increased capacity for energy harvest. Nature 444(7122):1027–1031

Veiga P, Suez J, Derrien M, Elinav E (2020) Moving from probiotics to precision probiotics. Nat Microbiol 5:878–880

Vich Vila A, Collij V, Sanna S, Sinha T, Imhann F, Bourgonje AR et al (2020) Impact of commonly used drugs on the composition and metabolic function of the gut microbiota. Nat Commun 11(1):362

Vieira-Silva S, Falony G, Belda E, Nielsen T, Aron-Wisnewsky J, Chakaroun R et al (2020) Statin therapy is associated with lower prevalence of gut microbiota dysbiosis. Nature 581(7808):310–315

Vogt NM, Romano KA, Darst BF, Engelman CD, Johnson SC, Carlsson CM et al (2018) The gut microbiota-derived metabolite trimethylamine N-oxide is elevated in Alzheimer's disease. Alzheimers Res Ther 10:124

Wahlström A, Sayin SI, Marschall HU, Bäckhed F (2016) Intestinal crosstalk between bile acids and microbiota and its impact on host metabolism. Cell Metab 24(1):41–50

Wang Z, Klipfell E, Bennett BJ, Koeth R, Levison BS, DuGar B et al (2011) Gut flora metabolism of phosphatidylcholine promotes cardiovascular disease. Nature 472(7341):57–63

Wang Z, Roberts AB, Buffa JA, Levison BS, Zhu W, Org E et al (2015) Non-lethal inhibition of gut microbial trimethylamine production for the treatment of atherosclerosis. Cell 163(7):1585–1595

Wang J, Thingholm LB, Skiecevičienė J, Rausch P, Kummen M, Hov JR et al (2016) Genome-wide association analysis identifies variation in vitamin D receptor and other host factors influencing the gut microbiota. Nat Genet 48(11):1396–1406

Wastyk HC, Fragiadakis GK, Perelman D, Dahan D, Merrill BD, Yu FB et al (2021) Gut-microbiota-targeted diets modulate human immune status. Cell 184:4137–4153.e4114

Watanabe Y, Fujisaka S, Ikeda K, Ishikawa M, Yamada T, Nawaz A et al (2021) Gut microbiota, determined by dietary nutrients, drive modification of the plasma lipid profile and insulin resistance. Iscience 24(5):102445

Wen L, Ley RE, Volchkov PY, Stranges PB, Avanesyan L, Stonebraker AC et al (2008) Innate immunity and intestinal microbiota in the development of Type 1 diabetes. Nature 455(7216):1109–1113

Wiciński M, Gębalski J, Gołębiewski J, Malinowski B (2020) Probiotics for the treatment of overweight and obesity in humans-a review of clinical trials. Microorganisms 8:1148

Wu GD, Chen J, Hoffmann C, Bittinger K, Chen YY, Keilbaugh SA et al (2011) Linking long-term dietary patterns with gut microbial enterotypes. Science 334(6052):105–108

Wu S, Yoon S, Zhang YG, Lu R, Xia Y, Wan J et al (2015) Vitamin D receptor pathway is required for probiotic protection in colitis. Am J Physiol Gastrointest Liver Physiol 309(5):G341–G349

Wu H, Tremaroli V, Schmidt C, Lundqvist A, Olsson LM, Krämer M et al (2020) The gut microbiota in prediabetes and diabetes: a population-based cross-sectional study. Cell Metab 32(3):379–390

Xu J, Lai KK, Verlinsky A, Lugea A, French SW, Cooper MP et al (2011) Synergistic steatohepatitis by moderate obesity and alcohol in mice despite increased adiponectin and p-AMPK. J Hepatol 55(3):673–682

Yang S, Li X, Yang F, Zhao R, Pan X, Liang J et al (2019) Gut microbiotadependent marker TMAO in promoting cardiovascular disease: inflammation mechanism, clinical prognostic, and potential as a therapeutic target. Front Pharmacol 10:10

Yoon HS, Cho CH, Yun MS, Jang SJ, You HJ, Kim JH, Han D, Cha KH, Moon SH, Lee K, Kim YJ (2021) Akkermansia muciniphila secretes a glucagon-like peptide-1-inducing protein that improves glucose homeostasis and ameliorates metabolic disease in mice. Nat Microbiol 6(5):563–573

Yoshimoto S, Loo TM, Atarashi K, Kanda H, Sato S, Oyadomari S et al (2013) Obesity-induced gut microbial metabolite promotes liver cancer through senescence secretome. Nature 499(7456):97–101

Yuan J, Chen C, Cui J, Lu J, Yan C, Wei X et al (2019) Fatty liver disease caused by high-alcohol-producing Klebsiella pneumoniae. Cell Metab 30(4):675–688

Zeevi D, Korem T, Zmora N, Israeli D, Rothschild D, Weinberger A et al (2015) Personalized nutrition by prediction of glycemic responses. Cell 163:1079–1094

Zhang X, Zhong H, Li Y, Shi Z, Ren H, Zhang Z et al (2021) Sex- and agerelated trajectories of the adult human gut microbiota shared across populations of different ethnicities. Nat Aging 1:87–100

Zhong H, Ren H, Lu Y, Fang C, Hou G, Yang Z et al (2019) Distinct gut metagenomics and metaproteomics signatures in prediabetics and treatment-naïve type 2 diabetics. EBioMedicine 47:373–383

Zhu W, Gregory JC, Org E, Buffa JA, Gupta N, Wang Z et al (2016) Gut microbial metabolite TMAO enhances platelet hyperreactivity and thrombosis risk. Cell 165(1):111–124

Zhuang R, Ge X, Han L, Yu P, Gong X, Meng Q et al (2019) Gut microbegenerated metabolite trimethylamine N-oxide and the risk of diabetes: a systematic review and dose-response meta-analysis. Obes Rev 20:883–894

Zmora N, Zilberman-Schapira G, Suez J, Mor U, Dori-Bachash M, Bashiardes S et al (2018) Personalized gut mucosal colonization resistance to empiric probiotics is associated with unique host and microbiome features. Cell 174:1388–1405.e1321

Metabolic Pattern of Microbiome in Healthy Versus Patient Individuals

14

Ashraf El-Baz, Ashraf S. A. El-Sayed, Yousseria Shetaia, and Amera A. Abaza

Abstract

Microbiota is a complex and diverse collection of microorganisms, including bacteria, viruses, yeast, and fungi, that colonize various parts of the human body, such as the skin surface, oral cavity, intestine, and genital tract. These microorganisms are essential to maintaining a healthy immune system, digesting food, and producing essential vitamins and hormones. They also play a critical role in protecting against harmful pathogens and diseases by competing for resources and producing antimicrobial substances. In short, the microbiota is an integral part of our body's ecology and is crucial for our overall health and well-being. Depending on their location, microbiota can be categorized as gut, oral, respiratory, or skin microbiota. The imbalances or dysbiosis in the microbial communities can lead to disruptions in various bodily processes and procedures, potentially contributing to diseases. This chapter discusses the current perception of how microbiota impacts health and disease pathogenesis from the perspective of microbial community composition and interactions with the human host and its immune system. Finally, we discussed some clinical approaches that leverage microbiota for treating diseases, such as modulating the microbiota in probiotic and prebiotic approaches or using fecal microbiota transplantation to restore microbial balance.

A. El-Baz · A. A. Abaza
Department of Industrial Biotechnology, Genetic Engineering and Biotechnology Research Institute, University of Sadat City, Sadat City, Egypt

A. S. A. El-Sayed (✉)
Enzymology and Fungal Biotechnology Lab, Botany and Microbiology Department, Faculty of Science, Zagazig University, Zagazig, Egypt

Y. Shetaia
Department of Microbiology, Faculty of Science, Ain Shams University, Cairo, Egypt

S. S. M. Soliman, M. I. Husseiny (eds.), *Metabolic Dynamics in Host-Microbe Interaction*, https://doi.org/10.1007/978-981-96-1305-2_14

Keywords
Human microbiome · Dysbiosis · Chronic inflammation · Metabolic disorders · Gut-brain axis · Homeostasis · Probiotic · Prebiotic

14.1 Introduction

The human body is home to trillions of microbial cells (e.g., bacteria, viruses, and fungi) in different areas like the gastrointestinal tract, skin, oral cavity, and other mucosal surfaces. These microbial communities referred to as the human microbiome significantly impact human health and metabolism (Dong et al. 2018; Fujio-Vejar et al. 2017; Tang et al. 2022). The microbial communities within this system work together with the human body in a mutually beneficial relationship, providing specialized metabolic functions that benefit both parties. Through complex interactions, the microbiome aids in nutrient processing, energy harvesting, and the production of vital metabolites and vitamins. It also regulates immune system development and protects against pathogen colonization (Kasubuchi et al. 2015; McCarville et al. 2020; Rowland et al. 2018).

The gut microbiome, particularly in the large intestine, aids in the breakdown of indigestible dietary components through specialized enzyme repertoires and carbohydrate metabolic pathways. It converts complex plant fibers and resistant starches into short-chain fatty acids (SCFAs), amino acids, and other absorbable metabolites, enhancing energy extraction and nutrient assimilation (Kasubuchi et al. 2015; Rowland et al. 2018; El-Baz et al. 2018). The gut microbiota communicates bidirectionally with the immune system through metabolic signals and pathogen-associated molecular patterns, promoting immunotolerance against innocuous antigens. This differentiation between pathogenic and symbiotic microbes is essential for maintaining homeostasis and preventing inflammation (Abt and Pamer 2014).

Dysbiosis, which refers to any perturbations in the composition and functions of the microbiome, has been linked to various metabolic disorders such as obesity, diabetes, and heart disease (Singh et al. 2021a). Differences in the gut microbiota of obese individuals compared with lean individuals suggest its involvement in energy regulation and fat storage. Alterations associated with diabetes indicate modulation of host glycemic control pathways (Heiss and Olofsson 2017). Although studies have revealed taxonomic and functional changes in patient microbiomes, interactions between dysbiotic microbiota and human metabolism remain poorly understood.

Gut bacterial metabolites play an important role in communicating between microbes and the host, influencing various disease states. Key metabolites like SCFAs and methylamines, such as butyrate and trimethylamine (TMA), respectively, exhibit roles in multiple sclerosis, brain inflammation, cardiovascular risk, and certain cancers (Tomasova et al. 2021). Levels of these substances fluctuate in obesity, diabetes, and inflammatory bowel disease, where beneficial bacteria tied to butyrate production, for instance, decline. Furthermore, an imbalance in the gut microbiota, or gut bacteria, can disrupt the natural barriers in the gut and allow bacterial components such as lipopolysaccharides (LPS) to enter the bloodstream, causing chronic low-level inflammation. This drives insulin resistance, a key factor in metabolic syndrome like obesity (Vetrani et al. 2022). Also, bacterial metabolites can transmit signals systemically beyond the gut via absorption and inflammation-based impacts, which support concepts such as gut-lung and gut-brain axes (Lami et al. 2021; Zhang et al. 2020) or can regulate glucose and fat metabolism, thus influencing metabolic syndrome risk (Thomas et al. 2022). Consequently, maintaining a balanced gut microbiota may help prevent and treat metabolic syndrome.

Current therapies that modulate the microbiome composition and resultant metabolites through prebiotics, probiotics, fecal transplants, or supplementation show promise for addressing disorders linked to metabolic derangements and dysbiosis. Elucidating the metabolic profiles of healthy versus diseased microbiomes could provide novel insights into disease etiology. Comparative metagenomic and metabolomic analyses may uncover the signature pathways, genes, and metabolites that distinguish metabolic phenotypes. Such knowledge could facilitate the development of diagnostic biomarkers and guide microbiome-targeted therapeutic strategies (Federici 2019; Torres-Sánchez et al. 2023). This chapter explores the relationship between gut microbiota and host metabolism, focusing on how the gut microbiome breaks down nutrients, produces vitamins, and synthesizes metabolites that influence metabolic pathways. It explains how an imbalance in gut microbiota can disrupt metabolic homeostasis and potentially lead to disorders like obesity, diabetes, and non-alcoholic fatty liver disease (NAFLD). The chapter also examines the communication channels through which gut microbiota impacts metabolic processes in distant organs and tissues. Understanding microbiome-host metabolic crosstalk in health and disease may enable the management of metabolic diseases through systems-level approaches. It discusses how diet, lifestyle, and antibiotic usage can alter the microbiota-metabolism relationship. The chapter reviews therapeutic approaches, including probiotics, prebiotics, fetal microbiota transplantation, and microRNA therapies, to optimize gut microbiota and restore metabolic health.

14.2 Microbiome Profile and Its Metabolites Beneficial to Healthy Individuals

Research on healthy individuals' dominant phyla, genera, and species has revealed consistent patterns across different body sites. In the lungs and the respiratory tract, the predominant phyla are *Firmicutes, Proteobacteria, Bacteroidetes, Actinobacteria, and Fusobacteria,* with *Veillonella, Streptococcus*, *Prevotella, Neisseria*, and *Fusobacterium* being the most common bacterial genera. Ascomycota and Basidiomycota are dominant pulmonary fungal phyla in healthy individuals (Fig. 14.1), while bacteriophages are the predominant order in the virome (Tang et al. 2022).

Three phyla, *Bacteroidetes*, *Firmicutes*, and *Verrucomicrobia* are the dominant gut microbiome of healthy adults, which comprise over 90% of the bacterial population. At the genus level, prominent members include *Bacteroides*, *Prevotella*,

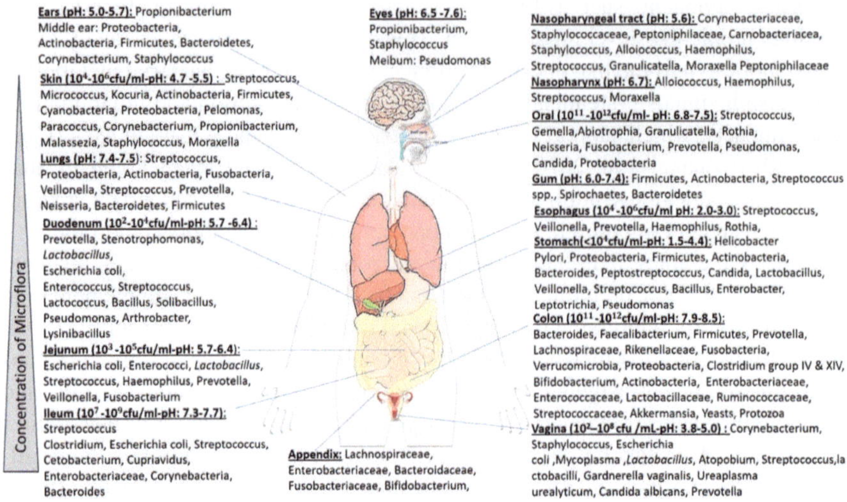

Fig. 14.1 Normal microbiota colonizes different human body organs

Ruminococcus, *Faecalibacterium*, *Roseburia*, *Coprococcus*, *Oscillospira*, and *Megamonas*. The most abundant species is *Bacteroides thetaiotaomicron*, closely followed by *Bacteroides vulgatus* and *Ruminococcus bromii* (Fujio-Vejar et al. 2017) (Fig. 14.1). The gut microbiota also contains a phylogenetically diverse core of nine genera, all known to contain short-chain fatty acid producers (Zhang et al. 2015). The most studied fungi in the gut are *Candida*, *Saccharomyces*, *Malassezia*, and *Cladosporium*. The gut also contains archaea, mainly *Methanobrevibacter smithii*.

The skin microbiota of healthy individuals is primarily composed of Gram-positive bacteria, with a significant diversity of Gram-negative taxa (Cosseau et al. 2016). The skin microbiome has a high degree of site-specificity and individual signatures. The skin microbiota is influenced by skin sites, age, and sex, with the most prevalent phyla on the skin being *Actinobacteria, Firmicutes, Cyanobacteria, and Proteobacteria*. The genus *Propionibacterium*, which includes *Propionibacterium acnes*, predominates in sebaceous areas. *Staphylococcus epidermidis* and *Corynebacterium* spp. are widespread skin commensals (Fig. 14.1). In addition, the cheeks facilitate the survival of *Betaproteobacteria* and *Alphaproteobacteria*, being prevalent on the volar forearms (Zhai et al. 2018).

Streptococci are the most abundant tongue and hard palate colonizers in the oral cavity. Species such as *Streptococcus oralis* and *Streptococcus mitis* account for over 50% of bacterial populations in these niches. While oral cavity and esophagus, the most abundant phyla are *Proteobacteria*, *Firmicutes*, *Bacteroidetes*, *Actinobacteria*, *Fusobacteria*, and *TM7*, with *Streptococcus*, *Neisseria*, *Prevotella*, *Actinobacillus*, and *Veillonella* being the most common genera (Fig. 14.1) (Dong et al. 2018).

These resident microbiotas play crucial metabolic roles through the degradation of complex carbohydrates and the production of SCFAs in the gut. These SCFAs, mainly acetate, butyrate, and propionate, are essential for host energy acquisition and metabolic regulation (Kasubuchi et al. 2015). The gut microbiota also produces a variety of nutrients, including B vitamins and vitamin K, and influences human metabolism through their interactions with the host (Ramakrishna 2013). On the skin, resident coagulase-negative Staphylococcus spp. Metabolize sebum triglycerides to produce antimicrobial fatty acids like trans-10-hydroxy-2-decenoic acid via β-oxidation pathways (Kasubuchi et al. 2015; Koh et al. 2016). Within the oral cavity, streptococci efficiently ferment dietary sugars to lactic acid through glycolysis, lowering the pH and preventing pathogen overgrowth. The resulting metabolic byproducts interact with host taste receptors, influencing appetite regulation (Abranches et al. 2018). Furthermore, colonization resistance against pathogens is another key function supported by the metabolic activities of healthy microbiota (Rowland et al. 2018; Ramakrishna 2013).

14.2.1 Functions and Metabolic Pathways of Major Microbiota

The gut microbiota plays a crucial role in human health by impacting host physiology through its metabolic activity (Krishnan et al. 2015). This diverse microbial community has a distinct metabolic repertoire that complements mammalian enzymes, particularly in the metabolism of dietary components and host-generated substances (Rowland et al. 2018). The microbiota's metabolites, small-molecule intermediates, regulate various physiological responses in the host, including immunity, inflammation, and metabolism (McCarville et al. 2020). However, dysbiosis in the microbiota composition can lead to metabolic imbalances and compromised health. The resident microbiota of healthy individuals performs indispensable metabolic functions through diverse catabolic and anabolic pathways. A prominent example is the saccharolytic fermentation of indigestible polysaccharides in the lower gut. Bacteroidetes such as *Bacteroides thetaiotaomicron* and *Prevotella copri* harbor an extensive array of polysaccharide utilization loci that encode hydrolytic enzymes for degrading complex plant fibers (Ravcheev et al. 2013). This enables the breakdown of otherwise indigestible carbohydrates into absorbable monosaccharides and SCFAs (Koh et al. 2016).

The gut microbiota breaks down foods into metabolites, including SCFAs, hydrogen sulfide (H_2S), and trimethylamine (TMA). That can be used for energy by cells or processed further after absorption into the bloodstream and transport to other organs like the liver. The majority of butyrate is used directly by colonic epithelial cells as fuel. It gets converted to acetyl-CoA, which enters the mitochondria. Also, butyrate in particular plays a trophic role through histone deacetylase inhibition and transcriptional regulation of host genes. Propionate and acetate can be partially metabolized in the gut. However, most propionate moves to the liver, where it is converted to succinyl-CoA and used for energy. In addition, propionate modulates host lipid and glucose metabolism via G-protein-coupled receptor signaling. Acetate

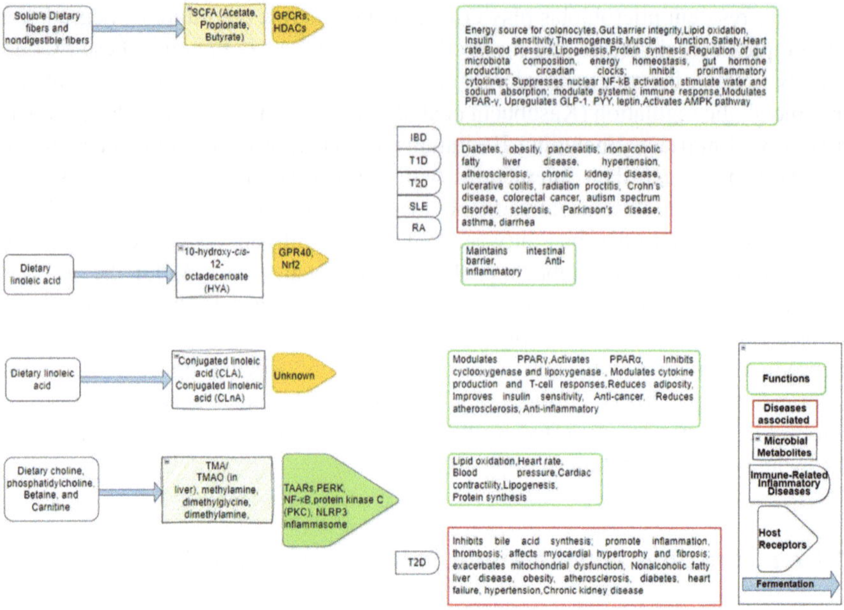

Fig. 14.2 Various metabolites produced by gut microbiota and their interaction with host receptors and signaling pathways. Short-chain fatty acids (SCFAs) produced during fermentation of dietary fibers bind G-protein-coupled receptors (GPRs) and inhibit histone deacetylases (HDACs). Choline is metabolized to trimethylamine (TMA) and then trimethylamine N-oxide (TMAO) by the liver. Trace amine-associated receptors (TAARs), protein kinase R-like endoplasmic reticulum kinase (PERK), nuclear factor kappa B (NF-κB), protein kinase c (PKC), NLRP3 inflammasome, inflammatory bowel diseases (IBD), rheumatoid arthritis (RA), systemic lupus erythematosus (SLE), type 2 diabetes(T2D), type 1 diabetes (T2D), peroxisome proliferator-activated receptor gamma (PPARγ), AMP-activated protein kinase (AMPK), peptide-1 (GLP-1) and peptide YY (PYY), Nrf2 nuclear factor (erythroid-derived 2)-like 2

acts as a substrate for cholesterol synthesis, and de novo lipogenesis in the liver reaches the highest levels in the blood and contributes significantly to energy production in other tissues around the body (Fig. 14.2). The gut lining also uses the enzyme sulfide quinone reductase (SQR) to oxidize or process hydrogen sulfide into thiosulfate. Thiosulfate can be recycled back to hydrogen sulfide in the gut or transported to the liver for further oxidation into sulfite and sulfate by other enzymes. In addition, any bacterial TMA transfer to the liver was converted to TMAO by the enzyme flavin-containing monooxygenase 3 (FMO3) (Tomasova et al. 2021).

14.2.2 Role in Nutrient Breakdown and Absorption

The gut microbiome is vital in maximizing nutrient extraction from the diet through specialized catabolic capabilities. The host lacks enzymes for degrading complex plant polysaccharides, so microbial fermentation is essential for liberating nutrients locked within dietary fibers. Bacteroidetes utilize polysaccharide utilization loci to hydrolyze indigestible starches, pectins, and glycans in the colon. This generates simple sugars that can be absorbed or further metabolized to SCFAs. Furthermore, intestinal bacteria aid in the synthesis of vitamins B and K and the metabolism of bile acids, other sterols, and xenobiotics (Fig. 14.3).

SCFAs lower colonic pH, facilitating mineral solubilization and absorption. Butyrate is the primary source of energy for colonocytes, stimulating epithelial cell proliferation and maintaining gut barrier integrity. This enhances the absorptive surface area and nutrient uptake. Propionate modulates host cholesterol synthesis and gluconeogenesis in the liver, improving lipid and glucose homeostasis. The microbiome also contributes to the bioavailability of micronutrients, such as vitamins and minerals, and can influence individual responses to dietary manipulation (Hadadi et al. 2021).

Inulin-type fructans, a type of prebiotic, have been shown to promote the proliferation of beneficial Bifidobacterium species and enhance the production of SCFAsin the gut (Birkeland et al. 2020; Toe et al. 2020). These SCFAs, particularly acetate and propionate, play a crucial role in metabolic health and can be increased through the consumption of inulin-type fructans (Birkeland et al. 2020). Furthermore, the use of a symbiotic supplement, which combines prebiotics and probiotics, has been found to enhance the beneficial effects of prebiotics on gut microbiota and associated metabolic activities (Singh et al. 2021b). The microbiome acts in synergy with the host to maximize the breakdown and absorption of nutrients from the diet, underscoring its importance in human nutrition and metabolic health.

14.2.3 Production of Vitamins and Metabolites Beneficial to the Host

Aside from aiding nutrient processing, the gut microbiota produces an array of vitamins, cofactors, and signaling metabolites that benefit the host. As mentioned earlier, *Bacteroides* and *Eubacterium* spp. Synthesize vitamin K_2 via the menaquinone pathway. Vit K_2 supports bone and cardiovascular health by facilitating calcium deposition and modulating inflammatory pathways. Gut microbes synthesize various B vitamins and vitamin K, which are essential cofactors for both bacterial and mammalian metabolism. Early investigations in germ-free rats found hemorrhaging attributable to deficient prothrombin levels due to a lack of vitamin K in the absence of gut colonization, highlighting the physiological importance of microbiota-derived nutrients. Metagenomic analyses have revealed microbial communities enriched biosynthesis pathways of precursors involved in the synthesis of thiamine (Vit B1) and pyridoxine (Vit B6) (Fig. 14.3).

Comprehensive in silico analyses mapping the biosynthetic capabilities of over 200 prevalent gut bacterial genomes identified complete pathways for producing eight B vitamins. Modeling estimates that the intestinal microbiota may account for over 25% of the recommended daily intake of specific B vitamins. Studies utilizing colonic preparations provide evidence that the colonic epithelium can absorb B

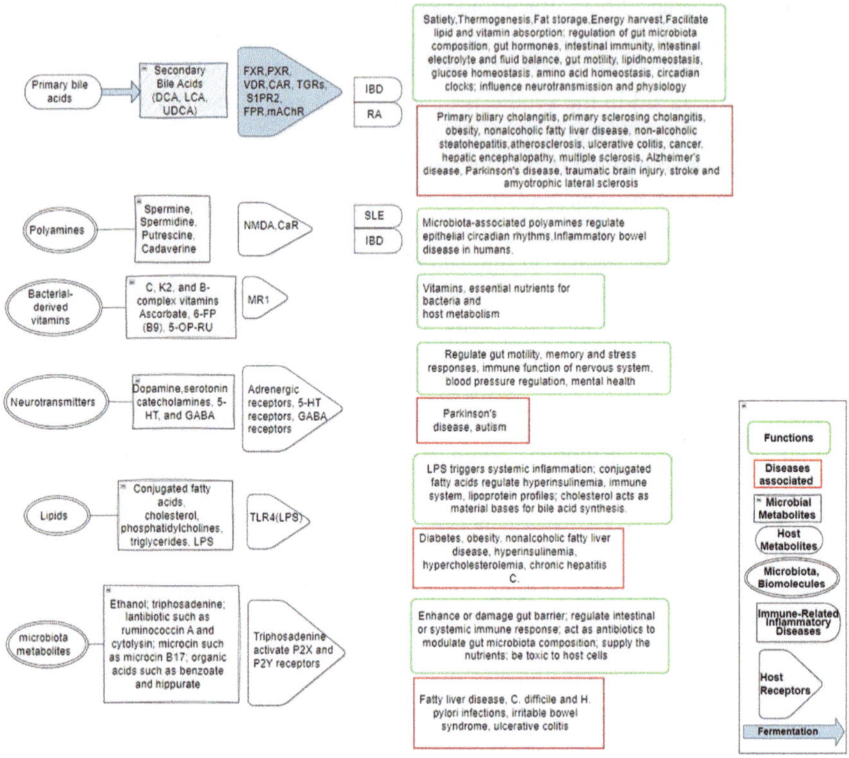

Fig. 14.3 Various metabolites produced by gut microbiota and their interaction with host receptors and signaling pathways. Bile acid metabolites like deoxycholate (DCA), lithocholic acid (LCA), and ursodeoxycholic acid (UDCA) regulate farnesoid X receptor (FXR) and G protein-coupled bile acid receptor 1 (TGR5) with high affinity in the gastrointestinal tract. Pregnane X receptor (PXR), vitamin D receptor (VDR), and constitutive androstane receptor (CAR) interact with bile acids to varying degrees and influence drug metabolism and transport and inflammatory and immune responses. In extra-intestinal tissues, bile acids signal through sphingosine 1-phosphate receptor 2 (S1PR2), formyl-peptide receptor (FPR), and muscarinic acetylcholine receptor (mAChR) to regulate inflammation and immune functions. Dysregulation of bile acid receptor signaling contributes to diseases such as inflammatory bowel diseases (IBD) and rheumatoid arthritis (RA). Commensal bacteria synthesize polyamines, which signal to host epithelial cells via activation of calcium-sensing receptors (CaR) and N-methyl-D-aspartate receptors (NMDA), shaping circadian programs with implications for immunity and disease susceptibility. Extracellular adenosine triphosphate (eATP) (Triphosadenine) released by gut microbiota interacts with the host through purinergic signaling pathways. ATP acts as an extracellular signaling molecule by binding to and activating specific P2X and P2Y receptor subtypes located on host cells such as enteric neurons, enteroendocrine cells, immune cells, and epithelial cells. Gamma-aminobutyric acid (GABA), systemic lupus erythematosus (SLE), toll-like receptor 4 (TLR4), and lipopolysaccharides (LPS)

vitamins synthesized by the intestinal microflora via specialized carrier-mediated uptake mechanisms. Collectively, these findings substantiate the integral role of the gut microbiota in meeting host vitamin demands and maintaining nutrient homeostasis through cooperative metabolic functions (Rowland et al. 2018).

SCFAs, including butyrate and propionate, have been shown to have a range of beneficial effects on host metabolism (Abaza et al. 2024). Butyrate, for example, acts as a histone deacetylase inhibitor, influencing gene expression and enhancing intestinal barrier integrity, which can protect against colitis (Canani et al. 2011). Conversely, propionate binds to GPR41/GPR43 receptors (Yoshida et al. 2019), influencing host metabolism by reducing hepatic gluconeogenesis and lowering cholesterol and triglyceride levels (Lin et al. 2012). These findings suggest that SCFAs play a crucial role in maintaining metabolic health (Fig. 14.2).

Neurotransmitters such as gamma-aminobutyric acid (GABA) and serotonin are produced by commensal bacteria, such as *Lactobacillus* and *Bifidobacterium*, while dopamine is produced by *Escherichia* species (Fig. 14.3), which can impact host behavior and mood (Strandwitz 2018; Sudo 2019). These neuroactive compounds are thought to mediate the gut-brain axis, influencing gut physiology and brain function (Margolis et al. 2021). For example, emotional behavior and central GABA receptor expression can be regulated by *Lactobacillus rhamnosus* in mice, suggesting a potential role in stress-related disorders (Bravo et al. 2011). However, further research is needed to fully understand these findings' implications and explore potential therapeutic applications.

Certain urease-positive taxa, such as *Eubacterium aerofaciens* and *Bifidobacterium infantis*, play a crucial role in maintaining gut homeostasis by hydrolyzing urea to release ammonia, which elevates luminal pH and supports the growth of acid-sensitive commensals (Suzuki et al. 1979). This process is particularly important in the gut, where the resident microbiota synthesizes bioactive factors and communicates with host cells to maintain homeostasis (Abt and Pamer 2014).

The gut microbiota plays a role in regulating host circadian rhythms by producing polyamines. Commensal bacteria synthesize polyamines such as putrescine, spermidine, and spermine, which signal to host epithelial cells via activation of calcium-sensing receptors (CaR) and N-methyl-D-aspartate receptors (NMDA). This microbiota-epithelial communication network modulates circadian expression patterns in host cells. Impairing circadian clock function has been linked to dysbiosis and disrupted polyamine rhythms. Since circadian disturbances are involved in inflammatory and autoimmune diseases, the microbiota-polyamine axis may influence conditions such as rheumatoid arthritis and systemic lupus erythematosus through effects on peripheral and central circadian clocks. Further research is warranted to understand better how polyamine-mediated microbiota signaling impacts the host circadian system and whether targeting this pathway could help manage certain immune-mediated diseases associated with circadian dysregulation (Di Martino et al. 2013) (Fig. 14.3).

Primary bile acids (BAs), such as chenodeoxycholic acid (CDCA) and cholic acid (CA), are synthesized in the liver from cholesterol by hepatocytes and make up over 75% of the body's total BA content. Secondary BAs, such as deoxycholic acid

(DCA), lithocholic acid (LCA), ursodeoxycholic acid (UDCA), and taurine, are produced by bacterial deconjugation, dehydrogenation, 7α-dehydroxylation, and epimerization in the small and large intestine, regulating farnesoid X receptor (FXR) and Takeda-G-protein-receptor-5 (TGR5) activity (Fig. 14.3). Bacteria with bile salt hydrolase activity, such as *Lactobacillus* and *Clostridium*, facilitate BA metabolism. Germ-free mice have higher BA levels, mostly ursodeoxycholic acid (UDCA), and larger gallbladders due to activating the transmembrane Takeda-G protein-coupled receptor 5 (TGR5) pathway. Recent studies have shown that UDCA improves insulin sensitivity, promotes weight loss in type 2 diabetes patients, and fights infection by inhibiting the growth of *Clostridium difficile* in fecal samples from patients with recurring *C. difficile* infections. Additionally, higher concentrations of deoxycholic acid (DCA) are associated with colonic crypt regeneration after wounding, as DCA utilizes FXR to hinder prostaglandin E2 synthesis, promoting wound repair. Studies have also focused on inhibiting secondary BAs as a treatment, as they have been linked to increased risk for various types of cancer and their ability to promote DNA mutations and cell death. Also, bacterial deconjugation of bile acids leads to increased levels of the amino acid taurine in the gut. Taurine has been shown to induce NLRP6 inflammasome signaling and may impact inflammatory bowel diseases. It is also being studied for protective effects against colorectal cancer and diabetes (Duszka 2022).

14.3 Imbalance in Metabolic Pathways and Loss of Beneficial Functions

Gut dysbiosis, a disruption in the balance of the gut microbiota, is a key factor in the development of metabolic disorders such as obesity, cardiovascular disease, diabetes, and fatty liver disease (Rajani and Jia 2018). This dysbiosis can lead to altered metabolic products, shifting the metabolic of the host balance toward chronic inflammation (Rajani and Jia 2018). The gut microbiota's role in regulating host immunity, energy supply, and pathogen prevention is crucial, and its dysbiosis is associated with gastrointestinal dysmotility and metabolic disorders (Singh et al. 2021a). The interaction of gut microbiota with host sensing and signaling pathways, as well as the role of gut microbial metabolites, particularly SCFAs, are implicated in the development of metabolic diseases (Li et al. 2017).

The gut microbiota comprises three phyla *Bacteroidetes*, *Firmicutes*, and *Actinobacteria*. Fermentation of non-digestible carbohydrates by the gut microbiota produces SCFAs. *Bacteroidetes* predominantly produce acetate and propionate, while *Firmicutes* specialize in butyrate production. These SCFAs influence various metabolic pathways and processes in the host. Butyrate, propionate, and acetate are essential compounds in the body, each playing a crucial role in various bodily functions, including colonic epithelial cell energy, gut barrier integrity, insulin sensitivity, and cholesterol synthesis. Given these diverse effects of microbiota-derived SCFAs on host metabolism and physiology, alterations in the gut microbiota composition can plausibly contribute to the pathogenesis of obesity. Reduced production of butyrate and propionate coupled with elevated acetate may promote adiposity by influencing energy extraction, nutrient absorption, lipid and cholesterol metabolism, appetite regulation, and low-grade inflammation. These findings provide evidence linking the human gut microbiota to obesity risk through modulation of SCFA profiles and various metabolic pathways.

The depletion of butyrate-producers like *Faecalibacterium prausnitzii* in inflammatory bowel disease (IBD) leads to decreased butyrate synthesis, crucial for gut barrier integrity and disease susceptibility (Zhang et al. 2019). Butyrate, produced by *F. prausnitzii*, inhibits histone deacetylase 3 (HDAC3) and the c-Myc-related metabolism pathway, decreasing T helper type 17 (Th17) cell differentiation and attenuating colitis (Zhang et al. 2019). Butyrate-producing bacteria, including *F. prausnitzii*, can enhance intestinal epithelial barrier integrity and increase butyrate production in IBD patients (Geirnaert et al. 2017). Butyrate also mediates the anti-inflammatory effects of *F. prausnitzii* in intestinal epithelial cells through Dact3, an epigenetic regulator of the Wnt/JNK pathway (Lenoir et al. 2020). Furthermore, butyrate maintains Th17/Treg (proinflammation /anti-inflammation) balance and ameliorates colorectal colitis by inhibiting HDAC1 and promoting Foxp3 (Zhou et al. 2018). These findings underscore the importance of butyrate and butyrate-producing bacteria in IBD management. Reduced representation of *Bacteroidetes* capable of complex polysaccharide breakdown limits fiber fermentation and SCFA yields. This diminishes energy extraction from non-digestible substrates in the diet. Dysbiotic states also demonstrate downregulation of amino acid metabolism and cofactor biosynthesis pathways.

14.4 Gut Microbiota and Specific Metabolic Disorders and Diseases

14.4.1 Relationship Between Gut Microbiota Composition and Metabolic Syndrome

The relationship between gut microbiota composition and metabolic syndrome has been the subject of extensive research in recent years. Metabolic syndrome (MetS) is a cluster of conditions that includes obesity, high blood pressure, high blood sugar levels, and abnormal cholesterol levels. It is associated with an increased risk of developing common chronic diseases such as cardiovascular diseases, cerebrovascular disease, NAFLD, and type 2 diabetes. The gut microbiota plays a role in metabolic syndrome (MetS) through several mechanisms:

I. Imbalance of Gut Microbiota: MetS is often accompanied by an imbalance of the gut microbiota. This imbalance can lead to a low-grade inflammatory response in the body, contributing to the progression of MetS. Gut microbiota imbalances are seen in conditions that make up MetS, like obesity, hyperglycemia, dyslipidemia, and hypertension. Specific bacteria are increased or decreased compared to healthy individuals.

II. Bacterial ethanol production: The gut microbiota can produce ethanol as a byproduct of fermentation. Studies found obese mice had higher levels of gut ethanol compared to lean mice and treating the obese mice with antibiotics significantly reduced their gut ethanol levels by 50%. Similarly, people with non-alcoholic steatohepatitis (NASH) disease were found to have elevated ethanol circulating in their blood. Scientists believe this extra ethanol comes from certain gut bacteria like *E. coli* which specializes in ethanol production. The ethanol made by gut bacteria could potentially be absorbed into the bloodstream and cause harm by traveling to the liver and triggering inflammation (Festi et al. 2014).

III. Gut barrier dysfunction and inflammation: An imbalance in the gut microbiota can disrupt the gut barrier, allowing bacterial metabolites and pathogens to enter the bloodstream and cause chronic low-grade inflammation known as metabolic endotoxemia. This is believed to be a significant driver of MetS, as it induces insulin resistance (Wang et al. 2020). Metabolic endotoxemia, characterized by elevated circulating lipopolysaccharide (LPS) levels, leads to chronic inflammation and insulin resistance. LPS is typically present in healthy individuals due to gut permeability, but in obese individuals, it is significantly higher. This is due to LPS binding to TLR4 receptors (Fig. 14.4), activating inflammatory cascades that impair insulin signaling. This suggests that obesity's abnormal gut barrier function allows for increased translocation of microbial antigens, leading to peripheral insulin resistance (Cai et al. 2005; Cani et al. 2007).

IV. Insulin Resistance: Metabolites produced by the imbalanced gut microbiota can affect host metabolism and hormone release, leading to insulin resistance. Insulin resistance is the state of the body's cells becoming less responsive to the effects of insulin, which causes increased blood sugar levels.

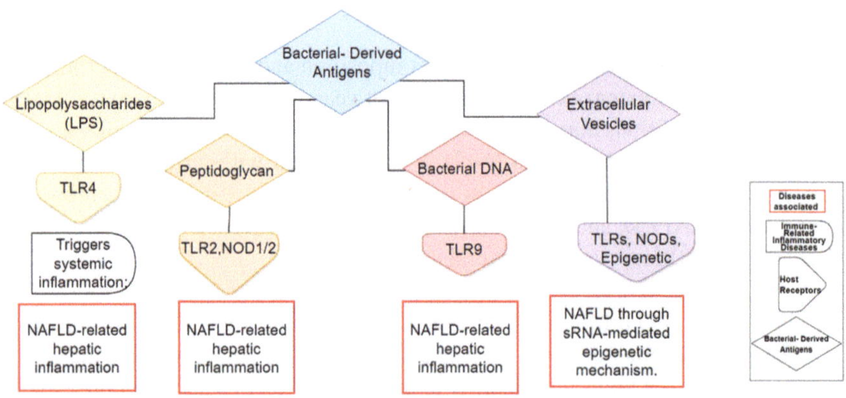

Fig. 14.4 Microbiota-derived antigens and their interaction with host receptors/signaling pathways involved in metabolic dysregulation. TLR Toll-like receptor, NAFLD non-alcoholic fatty liver disease, NOD Nucleotide Binding Oligomerization Domain Containing

V. Energy metabolism: Commensals in the gut ferment dietary fiber and produce SCFAs that help regulate glucose and lipid metabolism and improve insulin sensitivity. Gut microbes produce SCFAs, which activate G protein-coupled receptors like GPR41 and GPR43, which regulate host physiological processes. GPR41-deficient mice have an improved metabolic profile, increasing lean body mass and reducing total adiposity. GPR41 serves as a receptor for the microbial metabolites acetate and propionate, suggesting that activation by SCFAs may promote energy storage as fat (Thomas et al. 2022). The gut microbiota can indirectly influence an individual's susceptibility to weight gain and obesity by regulating SCFA sensing pathways involved in energy homeostasis. They also strengthen the gut barrier and reduce inflammation (Samuel et al. 2008). Conversely, the imbalanced gut microbiota harvests more energy from food; also, germ-free mice have been shown to absorb fewer calories from food and gain less weight than normal mice (Turnbaugh et al. 2006).

VI. Alterations of choline and bile acid metabolism: The gut microbiota alters bile acid composition and signaling through receptors like farnesol X (FXR) and G protein-coupled membrane receptor 5 (TGR5). This affects glucose and lipid metabolism in the liver and throughout the body (Molinaro et al. 2018). Intestinal bacteria break down choline-containing nutrients into trimethylamine (TMA) and convert it to trimethylamine N-oxide (TMAO) in the liver. TMAO is linked to atherosclerosis development and is associated with Western diet patterns. Mediterranean diets have lower plasma TMAO, indicating nutrition that can influence microbial-mediated host metabolism. The gut microbiota regulates host lipid profiles through secondary bile acid production and other metabolic functions. Dysbiosis disrupting these microbial activities can impact lipid processing and cardiometabolic risk, potentially promoting atherosclerosis through elevated TMAO. Also, alterations in diet and gut community structure may mediate cardiovascular implications (Thomas et al. 2022).

VII. Vicious Circle: The imbalanced gut microbiota, low-grade inflammation, gut barrier dysfunction, and insulin resistance form a vicious circle that promotes the continuous progression of MetS. The inflammatory response further disrupts the gut barrier, exacerbating insulin resistance and contributing to the metabolic imbalance associated with MetS (Wang et al. 2020).

It is important to note that the relationship between gut microbiota composition and MetS is complex and multifactorial. Various factors, including genetics, diet, lifestyle, and medication use, can influence the composition of the gut microbiota and its impact on metabolic health (Thomas et al. 2022).

14.4.2 Metabolically Healthy Versus Unhealthy Obesity

Metabolically healthy obesity refers to individuals who are obese but do not exhibit obesity-related comorbidities like insulin resistance, dyslipidemia, or hypertension. Research has found differences in the gut microbiota composition among obese individuals who are metabolically healthy versus obese individuals with metabolic complications. Specifically, studies have shown that metabolically healthy obese individuals tend to have higher levels of *Faecalibacterium prausnitzii* associated with reduced inflammation. They also exhibit a lower Firmicutes to Bacteroidetes ratio compared to obese individuals with metabolic issues. Thus, the gut microbiota profile may play a role in determining whether obesity leads to metabolic disturbances (Chakraborti 2015).

The gut microbiota of obese individuals tends to have a higher ratio of Firmicutes to Bacteroidetes compared to lean or metabolically healthy obese individuals (Chakraborti 2015; Iacobini et al. 2019). Obese individuals also tend to have a higher relative abundance of the phylum Actinobacteria in their gut microbiota than lean individuals. Conversely, obese individuals' gut microbiota tends to contain less of the phylum *Verrucomicrobia* than lean individuals. Furthermore, specific bacterial genera and species have also been linked to obesity or lean phenotypes in different studies, such as higher levels of bacteria from the genera *Ruminococcus, Lactobacillus*, and the species *Clostridium ramosum* being found in obese individuals versus lean controls.

Several factors contribute to the development of obesity; in addition to different composition patterns of gut microbiota, SCFAs produced by gut bacteria, especially butyrate, propionate, and acetate, play a critical role. Butyrate and propionate seem to be anti-obesogenic, while acetate is more obesogenic. Lipopolysaccharides (LPS) from gram-negative bacteria cell walls can induce inflammation and increase adiposity, contributing to obesity (Fig. 14.4). Also, a high-fat diet increases LPS absorption, potentially altering the gut microbiota composition. The gut microbiota and their metabolites, like SCFAs and LPS, can influence obesity through multiple mechanisms, including activating receptors, modifying cytokine/hormone functions, and regulating gene expression in ways that impact metabolism, inflammation, and energy storage/expenditure (Chakraborti 2015).

14.4.2.1 Gut Microbiota and Obesity-Related Inflammation

Obesity is a complex condition associated with various comorbidities, including low-grade inflammation. Recent research has highlighted the role of gut microbiota in regulating intestinal immunity and inflammation, which can contribute to obesity-related inflammation (Vetrani et al. 2022). The gut microbiota plays a crucial role in maintaining gut homeostasis and influencing various aspects of human health, including metabolism and immune function (Boulangé et al. 2016).

A summary of the relationship between gut microbiota and obesity-related inflammation, including (A) Dysbiosis, an imbalance in the composition and function of the gut microbiota, has been observed in individuals with obesity. Dysbiosis occurs when beneficial bacteria are decreased and harmful bacteria are increased (Boulangé et al. 2016). (B) Metabolic endotoxemia; dysbiosis can lead to increased gut permeability, allowing the translocation of bacterial components such as lipopolysaccharides (LPS) into the bloodstream, triggering an inflammatory response and contributing to systemic low-grade inflammation (Boulangé et al. 2016). (C) Inflammatory signaling: LPS in the bloodstream activates immune cells, such as macrophages, producing pro-inflammatory cytokines. These cytokines, including tumor necrosis factor-alpha (TNF-α) and interleukin-6 (IL-6), contribute to the development of obesity-related inflammation (Boulangé et al. 2016). (D) Adipose tissue inflammation in obesity, adipose tissue undergoes significant changes, including the infiltration of immune cells and the secretion of pro-inflammatory molecules. The gut microbiota can influence adipose tissue inflammation by modulating immune cell activation and the release of inflammatory mediators (Vetrani et al. 2022). (E) Short-chain fatty acids (SCFAs): SCFAs are metabolic byproducts of gut bacteria fermentation of dietary fiber. They have been shown to have anti-inflammatory effects and can help maintain gut barrier integrity. However, in obesity, the production of SCFAs may be reduced, contributing to gut dysbiosis and inflammation.

SCFAs are organic acids with less than six carbons produced by gut bacteria during the fermentation of dietary fibers and resistant starches that escape digestion in the small intestine. Acetate, propionate, and butyrate are the three most abundant SCFAs detected in feces, normally present in molar ratios ranging from 3:1:1 to 10:2:1. Butyrate forms the key energy source for human colonocytes and has potential anti-cancer activity. Propionate is also an energy source and plays a role in gluconeogenesis and satiety signaling. Acetate is used in cholesterol metabolism and lipogenesis. SCFA production involves cross-feeding between bacterial species, with some bacteria producing intermediates like lactate that are then converted to propionate or butyrate by other bacteria. The predominant butyrate producers are Firmicutes, like some *Lachnospiraceae* and *Faecalibacterium prausnitzii*. *Bacteroides* species, *Negativicutes*, and some *Clostridia* produce propionate. Acetate production is more widespread. Dietary factors like carbohydrates content can impact SCFA concentrations by affecting bacterial numbers and composition. Wheat bran and other resistant starches and fibers have been proven to increase levels of SCFAs. The key points are that SCFAs are essential bacterial fermentation products with varied roles in host health. Their production involves metabolic cross-feeding between distinct but complementary groups of gut bacteria (Rowland et al. 2018; Aj et al. 2015).

14.4.2.2 Gut Microbiota and Non-Alcoholic Fatty Liver Disease (NAFLD)

Gut microbiota has been found to play a significant role in the development and progression of NAFLD. NAFLD is a common liver disease characterized by the accumulation of fat in the liver; up to one-quarter of cases advance to NASH, characterized by cirrhosis, and hepatocellular carcinoma (Duarte et al. 2019). NAFLD represents a growing global health concern, afflicting over 25% of the world's population with an asymptomatic yet progressive metabolic disease. An accumulating body of evidence implicates alterations in the composition and function of the gut microbiota as a key environmental factor contributing to NAFLD pathogenesis. Dysbiosis leading to disruption of intestinal barrier integrity facilitates bacterial translocation and exposure of host tissues to microbial toxins such as lipopolysaccharides (LPS) (Fig. 14.4), priming hepatic and systemic inflammation hypothesized to drive NASH development (Bashiardes et al. 2016).

Metabolites generated through microbial cross-feeding in the gut lumen, such as SCFAs, bile acids, and ethanol, have also been scrutinized as mediators of liver pathology through diverse mechanisms (Figs. 14.3 and 14.4). Proposed mechanisms link gut microbial metabolites to NAFLD progression through hepatic immune modulation and epigenetic reprogramming, including permitting translocation of bacterial components such as lipopolysaccharides, peptidoglycan, DNA, and extracellular vesicles into the portal circulation (Fig. 14.4). These microbial factors interact with pathogen recognition receptors, including Toll-like receptors (TLRs) 2, 4, and 9, as well as Nod-like receptors (NLRs) 1 and 2 on hepatic resident macrophages (Kupffer cells) and other cell types like hepatic stellate cells and hepatocytes. This induces pro-inflammatory signaling cascades that contribute to NAFLD pathogenesis.

Additionally, short, regulatory RNAs within extracellular vesicles may function as epigenetic regulators of gene expression in the liver. Fermentation of dietary fibers and non-digestible carbohydrates yields SCFAs such as acetate, propionate, and butyrate. These SCFAs signal through G-protein-coupled receptors (GPCRs) and inhibit histone deacetylases (HDACs) in the liver, modulating the epigenetic landscape. Furthermore, tryptophan is metabolized to indole derivatives, including indole-3-aldehyde, indole-3-acetic acid, and indole-3-pyruvic acid, through microbial tryptophanase activity. These products activate aryl hydrocarbon receptor (AHR) and pregnane X receptor (PXR) signaling in hepatocytes via the portal vein (Fig. 14.5).

Choline is also metabolized to trimethylamine, subsequently converted to trimethylamine-N-oxide (TMAO) by hepatic flavin-containing monooxygenase 3 (FMO3) linked to cardiometabolic risk, potentially promoting atherosclerosis (Fig. 14.2). Dietary components additionally serve as substrates for microbiota-derived carotenoids and phenols impacting liver nuclear receptors and protect against NAFLD through antioxidant and anti-inflammatory (Fig. 14.6) (Ji et al. 2019). In addition, microbial production of ethanol can act as a hepatotoxin (Fig. 14.3). Consequently, the gut microbiome emerges as a potential therapeutic avenue worthy of intensive characterization, as modulating microbiota-derived factors may improve clinical management of this widespread disease by decelerating inflammatory progression to life-threatening stages of NASH (Tokuhara 2021).

14.4.2.3 Gut Microbiota and Type 2 Diabetes (T2DM)

Type 2 diabetes mellitus (T2DM) is a progressive metabolic disorder characterized by pancreatic β-cell dysfunction and peripheral insulin resistance, resulting in impaired glucose metabolism and low-grade chronic inflammation. Genetic and environmental factors contribute to T2DM development and progression, as highlighted by the continuing global surge in prevalence, which parallels trends in obesogenic diets, sedentary lifestyles, and urbanization. T2DM is expected to affect over 700 million individuals by 2045, and novel insights into pathogenic mechanisms and therapeutic targets are urgently needed. Much research implicates the gut

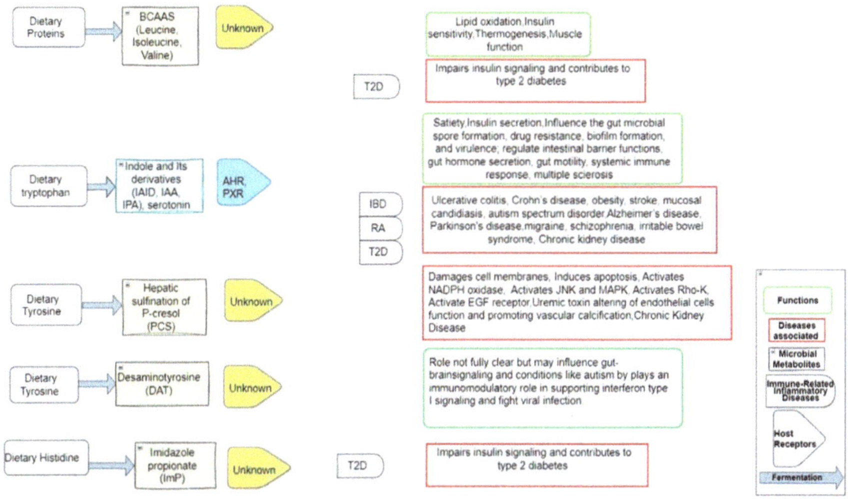

Fig. 14.5 Various metabolites produced by gut microbiota and their interaction with host receptors and signaling pathways. Branched-chain amino acids (BCAAs) are elevated in circulation through microbial protein fermentation in the gut. Higher BCAA levels are associated with insulin resistance and increased type 2(T2D) diabetes risk, possibly involving activation of rapamycin (mTOR) signaling. Tryptophan metabolites: indole-3-aldehyde (IAD), indole-3-acetic acid (IAA), and indole-3-pyruvic acid (IPA) activate aryl hydrocarbon receptor (AhR) and pregnane X receptor (PXR) in the liver. Para-cresyl sulfate (PCS) induces programmed cell death (apoptosis), activates the enzyme NADPH oxidase involved in oxidative stress and kinases linked to inflammation; Jun N-terminal protein kinase (JNK), mitogen-activated protein kinase (MAPK), rho-kinase (Rho-K) also activating the epidermal growth factor (EGF) receptor signaling pathway. Desaminotyrosine (DAT), a degradation product of the amino acid tyrosine, is dependent on the gut bacterium *Clostridium orbiscendens*. Imidazole propionate (ImP) is produced by urocanate reductase (urdA) in gut bacteria. Inflammatory bowel diseases (IBD) and rheumatoid arthritis (RA)

microbiota with its complex intestinal ecosystem composed of bacteria, archaea, fungi, and protozoa as a potentially modifiable factor linked to T2DM risk. Metagenomic studies have observed distinct intestinal microbiota profiles between Chinese individuals with obesity and T2DM compared to healthy controls, with specific bacterial taxa, including *Acidaminococcales*, *Bacteroides plebeius*, and *Phascolarctobacterium* identified as potential biomarkers, underscoring the role of the gut microbiota in T2DM pathogenesis (Zhou et al. 2022).

The gut microbiota influences the pathogenesis of T2DM through several metabolic pathways. SCFAs such as acetate, propionate, and butyrate produced during bacterial fermentation impact glucose homeostasis and insulin sensitivity by stimulating glucagon-like peptide-1 (GLP-1) and peptide YY (PYY) secretion (Fig. 14.3). Secondary bile acids metabolized from primary bile acids by gut bacteria stimulate fibroblast growth factor 19/15 (FGF19/15) release, leading to enhanced insulin sensitivity and glucose tolerance (Fig. 14.6). Conversely, lipopolysaccharide from gram-negative gut microbes promotes chronic low-grade inflammation and insulin resistance by interacting with CD14 and toll-like receptor 4 (TLR4) (Figs. 14.2 and

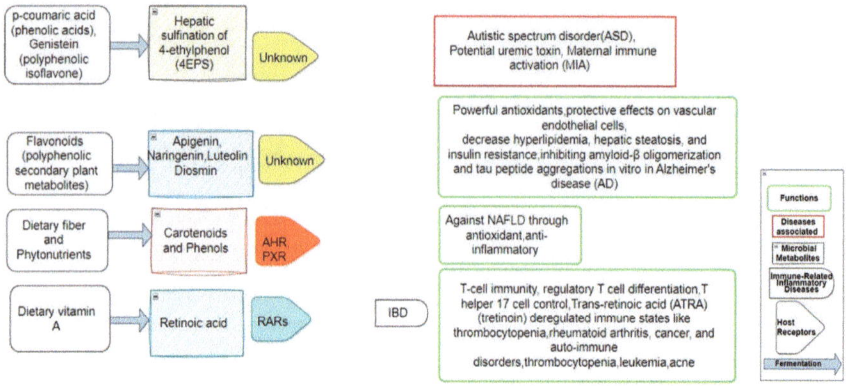

Fig. 14.6 Various metabolites produced by gut microbiota and their interaction with host receptors and signaling pathways. 4-ethylphenylsulfate (4-EPS) links circumstantially to autism and kidney dysfunction based on structural similarities to potentially pathogenic sulfated compounds. A probiotic bacterium Bacteroides *fragilis* helped reduce 4-EPS levels and behavioral abnormalities, suggesting 4-EPS and modulating the gut microbiome may be a potential therapeutic target for autism. Carotenoids and phenols protect against non-alcoholic fatty liver disease (NAFLD) through antioxidant, anti-inflammatory, and nuclear receptor-mediated activities; aryl hydrocarbon receptor (AhR) and pregnane X receptor (PXR) in the liver. Retinoic acid, a lipid metabolite derived from vitamin A, plays a crucial role in the immune system's development and regulation. It is maintained in the gut by Clostridia commensal bacteria by suppressing retinol dehydrogenase 7 in intestinal epithelial cells and retinoic acid receptors (RARs). Inflammatory bowel diseases (IBD)

14.6). Additionally, branched-chain amino acids (BCAAs) have been correlated with increased type 2 diabetes risk, potentially involving the mechanistic target of rapamycin (mTOR) signaling (Fig. 14.4). These diverse bacterial metabolites modulating host inflammation, immunity, and metabolic processes provide plausible ways for compositional and functional alterations in the gut microbiota that may influence type 2 diabetes mellitus development (Zhang et al. 2021).

14.4.3 Communication of Gut Microbiota with Organs-Axis

14.4.3.1 The Gut-Lung Axis

It has become more recognized that there is bidirectional communication between the microorganisms that inhabit the respiratory and gastrointestinal tracts, known as the "gut-lung axis," which has a significant impact on health and disease (Zhang et al. 2020). Research in animal models and humans has associated disruptions to this gut-lung relationship through microbial imbalances in the gut with elevated risk and severity of respiratory infections, asthma, chronic obstructive pulmonary disease (COPD), and other pulmonary conditions (Chunxi et al. 2020). Studies involving germ-free mice or antibiotic treatment found impaired immune defenses against lung viruses, while fecal transplants have aided acute lung injury in rats (Rastogi et al. 2022). The evidence confirms that maintaining a balanced gut microbiota composition impacts lung health by strengthening immunity and reducing inflammation across interconnected bodily systems. The gut microbiota and its metabolites, particularly SCFAs, have been identified as key factors in the gut-lung axis, influencing lung immunity and inflammation (Yang et al. 2020).

14.4.3.2 The Gut-Brain Axis

The gut-brain axis, a complex communication network between the gut microbiota and the central nervous system, plays a crucial role in various neurodegenerative and psychiatric conditions through multiple neurological and endocrine signaling systems linked between the gut microbiota and brain (Lami et al. 2021). Accumulating research implicates gut dysbiosis and disruptions along this brain-gut-microbiome axis in various neurodegenerative and psychiatric conditions through mechanisms like overactive immune responses in the brain. Diseases such as Alzheimer's and Parkinson's may involve abnormal protein aggregates or "proteopathy" like amyloid beta and alpha-synuclein. Also, gut microbiota secretes amyloids that may cross-react with human antigens and induce misfolding. Microbial amyloids could be induced through molecular mimicry, potentially spreading from the gut in a "prion-like" fashion (Manole et al. 2021). Age-related alterations to the gut microbiota composition contribute to systemic immune activation in older adults and disruptions to barriers separating the brain and gut, leading to neuroinflammation and degeneration. Modulating gut bacteria through dietary changes, prebiotics, or probiotics shows promise for alleviating neurological disorder symptoms, highlighting the microbiota's influential role in central nervous system functions via the gut-brain axis.

Commensal gut bacteria can produce important neurotransmitters like GABA, serotonin, and dopamine. *Lactobacillus* and *Bifidobacterium* species produce GABA and serotonin, while *Escherichia* species produce dopamine (Fig. 14.6). The production of these neuroactive compounds by gut microbes allows them to influence gut and brain function through the gut-brain axis. GABA, serotonin, and dopamine are known to regulate processes in the brain that impact mood, behavior, and stress responses. Studies in mice have shown that the probiotic *Lactobacillus rhamnosus* can modulate GABA receptor expression in the brain and reduce anxiety and depression-related behaviors. This suggests gut bacteria may play a role in neurological and mental health conditions involving stress and mood regulation (Margolis et al. 2021).

14.4.3.3 The Gut-Liver Axis

Through signaling pathways, the gut microbiota produces SCFAs that can affect the liver and brain. SCFAs enter the bloodstream via the portal vein and travel to the liver, where they regulate processes like glycolipid homeostasis through AMP-activated protein kinase (AMPK) and interact with peroxisome proliferator-activated receptor gamma (PPARγ) (Fig. 14.3). They also impact gluconeogenesis, AMPK activity, and insulin sensitivity within the liver. SCFAs induce a switch in lipid metabolism in the body from synthesis to utilization by promoting (PPARγ)-dependent pathway. When mice were given SCFA supplements in their high-fat diets, they prevented and reversed metabolic abnormalities like increased body weight and insulin resistance. Because SCFAs lowered PPARγ expression and activity in tissues like liver and adipose tissue, upregulating mitochondrial uncoupling protein 2 (UCP2) involved in oxidative metabolism and raising the AMP:ATP ratio, activating AMPK to stimulate fat and liver cell oxidation of fuels. Mice lacking PPARγ specifically in fat or liver tissue did not see the beneficial effects of SCFAs on weight, insulin sensitivity, or liver fat. Therefore, PPARγ activity in both fat and liver cells is critical for SCFAs to improve hepatic steatosis (den Besten et al. 2015).

Furthermore, gut microbiota plays a key role in choline metabolism by facilitating the conversion of choline to trimethylamine (TMA). Once absorbed by the host, TMA is metabolized in the liver to TMAO by hepatic flavin monooxygenases. TMAO exerts effects on the host through various molecular mechanisms and signaling pathways. TMAO can activate trace amine-associated receptors (TAARs), which are G protein-coupled receptors expressed in the gut and cardiovascular systems. Activation of TAARs by TMAO impacts inflammation and cardiovascular disease. TMAO also acts as an activator of protein kinase R-like endoplasmic reticulum kinase (PERK), triggering endoplasmic reticulum stress responses implicated in atherosclerosis. PERK signaling stimulates downstream nuclear factor kappa B (NF-κB) inflammatory pathways involved in vascular dysfunction. Additionally, TMAO may activate protein kinase c (PKC) isoforms, which induce NLRP3 inflammasome activation, exacerbating inflammation linked to cardiometabolic disorders (Fig. 14.3).

14.4.3.4 The Gut-Skin Axis

The gut-skin axis, a bidirectional relationship between the gut microbiome and skin health, has been implicated in various skin disorders, including psoriasis and atopic dermatitis (Lee et al. 2018; Thye et al. 2022). Changes in the composition of bacteria in the gut and skin have been linked to these conditions, implying that imbalances affecting one site may impact the other through this axis. Psoriasis patients specifically exhibit alterations to their gut and skin microbiomes, and gut dysbiosis can exacerbate inflammation and cardiovascular risks in these individuals (Thye et al. 2022). The gut microbiome is similar to an endocrine organ and produces hormone-like compounds that affect bodily systems, such as SCFAs; secondary bile acids; cortisol; and neurotransmitters such as gamma-aminobutyric acid (GABA), serotonin, dopamine, and tryptophan. Modulating the immune system releases them into the bloodstream and can affect distant organs and systems, such as the skin (De Pessemier et al. 2021).

Furthermore, stress, smoking, and poor diet may disrupt microbiota balances across this gut-skin communication pathway, influencing conditions. Additionally, acne pathogenesis appears tied to disruptions along the brain-gut-skin microbiota network, with gut microbe changes contributing to acne severity. Acne vulgaris patients have altered gut microbiota, with decreased Actinobacteria and increased Proteobacteria, as well as decreased levels of specific genera: *Bifidobacterium*, *Butyricicoccus*, *Coprobacillus*, *Lactobacillus*, and *Allobaculum* were all decreased

(Yan et al. 2018). Maintaining equilibrium between the resident microorganisms populating the gut and skin may aid certain skin ailments through this interaction pathway (Thye et al. 2022).

14.4.3.5 The Gut-Kidney Axes (Chronic Kidney Disease (CKD))

Evidence indicates that gut microbiota imbalances, or dysbiosis, play a role in the development and progression of chronic kidney disease (Armani et al. 2017). Studies have found differences in gut bacterial abundances between CKD patients and healthy controls, such as higher abundances of *Neisseria*, *Lachnoclostridium*, and *Enterobacteriaceae* and lower abundances of *Bifidobacteriaceae* and *Lactobacillaceae* in CKD patients. This dysbiosis can lead to the generation of excessive uremic toxins and increased cardiovascular risk (Armani et al. 2017), worsening cardiovascular comorbidities, and more severe CKD complications. Factors like dialysis further disrupt the gut microbiome. Impaired intestinal permeability and bacterial translocation into circulation are the causes of dysbiosis, which fuels inflammation and leads to further renal disability. Given the gut-kidney axes relationship, modulating the composition through prebiotics or probiotics may help reduce inflammation and potentially slow CKD progression. However, more research is still needed to explore this emerging therapeutic target.

14.5 Interactions Between Altered Microbiome and Host Metabolism

The intestinal microbiota influences metabolic homeostasis through several signaling metabolites. The fermentation of dietary fibers by gut microbes generates SCFAs such as butyrate, acetate, and propionate. These SCFAs interact with G-protein-coupled receptors and positively impact lipid oxidation, helping to improve metabolic control. Certain bacterial species also modulate amino acid metabolism, with organisms like *Prevotella copri* and *Bacteroides vulgatus* producing substrates that elevate circulating branched-chain amino acids (BCAAs) like isoleucine, leucine, and valine, which are linked to insulin resistance and type 2 diabetes (Fig. 14.4). Additionally, gut microbial catabolism of amino acids and the resultant aromatic acid metabolites like phenylacetic acid can impact pathways such as hepatic branched-chain amino acid utilization and steatosis development (Hoyles et al. 2018). Indole derivatives formed by intestinal bacteria signaling via the aryl hydrocarbon receptor (AHR) pathway are further connected to reduced inflammation and protection against diabetes (Brial et al. 2018). Through utilizing host nutrients, the intestinal microbiota generates an array of signaling molecules that regulate metabolic homeostasis via diverse host receptors and physiological routes (Figs. 14.3,14.4, and 14.5).

Gut microbiota dysbiosis, defined by decreased diversity and declines in fiber-fermenting and SCFA-producing bacteria along with rises in pro-inflammatory species such as adherent/invasive *Escherichia coli* and hydrogen sulfide generators, directly influences host metabolism and contributes to inflammatory bowel diseases

(IBD) through a complex interplay with the immune system. This pathological intestinal microbiota shifts the crosstalk between the microbiota and host in a manner that may exacerbate chronic intestinal inflammation. As such, researchers are pursuing probiotic and prebiotic therapies to modulate the gut microbiota composition to reduce inflammation and improve clinical outcomes for these conditions. Strategies focused on restoring beneficial microbiota metabolism can potentially complement medical therapies for IBD (Sultan et al. 2021). Reduced fiber fermentation lowers SCFA yields, diminishing substrate availability for intestinal epithelial cells. This compromises nutrient absorption and energy extraction from the diet. SCFAs, mainly acetate, butyrate, and propionate, play a crucial role in host energy regulation and metabolic health (Kasubuchi et al. 2015). The intestinal microbiota, through its metabolic capacity, can influence nutrient availability and absorption; therefore, dysbiosis of the intestinal microbiota has been implicated in conditions involving nutrient malabsorption and undernutrition, where disruptions to the microbiota-diet-GI function interplay may perpetuate deficiencies (Jonkers 2016). Furthermore, the microbiome is implicated in regulating energy balance and metabolic function, with altered microbial ecology potentially contributing to metabolic diseases (Nieuwdorp et al. 2014).

Butyrate, a SCFA, is a crucial energy source for over 60% of colonocytes and plays a significant role in maintaining colonic homeostasis. Butyrate has emerged as a key SCFA produced via the fermentation of dietary fiber by the gut microbiota. It exerts potent effects on numerous functions of the colonic mucosa through diverse mechanisms of action. Butyrate inhibits nuclear factor kappa B activation (NF-κB), suppressing pro-inflammatory signaling cascades involved in conditions like inflammatory bowel disease and colorectal cancer. Additionally, butyrate acts as a histone deacetylase inhibitor to regulate gene expression epigenetically. Butyrate reinforces various components of the colonic defense barrier through these molecular pathways, decreasing oxidative stress and inflammation. This supports the inhibition of carcinogenesis in the large intestine. Beyond its role in the colon, evidence indicates that butyrate may promote satiety through unclear central and peripheral signals. Some hypothesize this occurs via activation of gut-brain signaling via free fatty acid receptors in enteroendocrine L cells and related enteric neurons (Hamer et al. 2008). As seen in quiescent ulcerative colitis, its deficit can impair epithelial integrity and barrier function, negatively impacting mineral and vitamin uptake (Chapman et al. 1994). However, dietary intervention can enhance butyrate, leading to improved epithelial barrier function (Nielsen et al. 2018).

Furthermore, SCFAs act as signal transduction molecules through G-protein-coupled receptors (FFAR2, FFAR3) and as epigenetic regulators of gene expression through the inhibition of histone deacetylase (HDAC) (Kasubuchi et al. 2015). Reduced propionate and butyrate signaling through FFAR2/FFAR3 and HDAC, respectively, inhibition can dysregulate host genes involved in adipogenesis, gluconeogenesis, and lipogenesis. Propionate and butyrate are shown to activate intestinal gluconeogenesis (IGN) through complementary mechanisms. Butyrate stimulates IGN gene expression via a cAMP-dependent pathway, while propionate, a substrate for IGN itself, activates the pathway through a gut-brain neural circuit involving FFAR3 receptors. Moreover, succinate generated by the fiber-fermenting microbiota response to the fructo-oligosaccharides (FOS) diet improves glycemic control and energy metabolism in the liver and body weight through IGN. Critically, mice deficient in IGN fail to maintain body weight and glucose regulation induced by SCFAs or dietary fibers, revealing IGN activation to be required to elicit these effects. These findings emphasize the interlinked roles of the gut microbiota, its metabolites, and IGN in modulating glucose and energy homeostasis with relevance for preventing metabolic diseases (De Vadder et al. 2014). The dysregulation of these microbial metabolites is associated with the pathogenesis of NAFLD (Chen and Vitetta 2020). However, the protective effects of butyrate and propionate against diet-induced obesity and insulin resistance are independent of FFAR3, suggesting the involvement of additional mediators (Lin et al. 2012).

The endocannabinoid (eCB) system is a major signaling network in the body that plays a pivotal role in regulating glucose and lipid metabolism through the actions of endocannabinoids like anandamide (AEA) and 2-arachidonoylglycerol (2-AG) on receptors such as CB1 in the brain and CB2 in peripheral metabolic tissues mainly in the immune system and in the CNS, whereas CB1 mediates of the psychoactive effects, In contrast, the CB2 receptor is implicated in anti-inflammatory and immunosuppressive activity. This system becomes dysregulated during the development and progression of obesity, contributing to metabolic impairments through dysfunctional pathways involving hormones, inflammation, and lipids. Beyond just anandamide and 2-AG, the endocannabinoid system incorporates signaling lipids called endocannabinoid-like mediators that influence metabolism via additional receptors. Surprisingly, evidence indicates that the gut microbiota interacts bidirectionally with the endocannabinoid system to control metabolism cooperatively (Bourdeau-Julien et al. 2023). Therapeutically targeting the interplay between these interconnected systems could offer promise for obesity and related metabolic disorders (Iannotti and Marzo 2021).

14.6 Biomarkers for Distinguishing Metabolic Phenotypes

14.6.1 Potential for Diagnosis, Prognosis, and Monitoring of Disease Progression

The human gut microbiome is recognized as an essential host metabolism and physiology component, as distinct microbiome compositions, or enterotypes, have been associated with various metabolic phenotypes in healthy and diseased populations. Characterizing the microbiome-metabolic phenotype relationship could provide novel microbiome-based biomarkers to aid in the clinical assessment of metabolic disorders. Comparative metagenomic, metabolomic, or longitudinal profiling of fecal, mucosal, or luminal samples between healthy and patient cohorts with metabolic conditions may help distinguish microbiome enterotypes linked to fast versus slow metabolism, insulin sensitivity versus resistance, and inflammation, while fecal microbiota transplantation or probiotic administration offers opportunities to experimentally modulate the microbiome and assess resultant metabolic effects, as well-defined microbiome biomarkers may non-invasively complement traditional biomarkers in metabolic phenotyping with applications to precision diagnosis and personalized management of metabolic disorders.

14.6.2 Signature Taxa, Genes, and Metabolites in Disease Versus Health

Recent research has highlighted the potential of transcriptomic, metabolomic, and metagenomic studies in identifying biomarkers for metabolic disorders. By combining analyses of the genetic makeup of people's gut bacteria (through metagenomic sequencing) and the metabolic byproducts produced by those microbes (via metabolomic profiling), researchers believe they may be able to develop a unique "signature" made up of microbial and biochemical factors. This signature could help predict future disease risk among otherwise healthy individuals by identifying subtle perturbations in the gut microbiome and its function that predispose individuals to conditions like diabetes or heart disease in the future. The hope is that in the future, applying both metagenomics and metabolomics approaches may enable the creation of biomarker profiles for healthy people to recognize those at higher risk early on. This could allow for preventative lifestyle changes or medical interventions before any illnesses manifest, helping people stay well (Federici 2019).

Torres-Sánchez et al. (Torres-Sánchez et al. 2023) identified key microbial genera and metabolites associated with metabolic diseases; genera like *Faecalibacterium, Bifidobacterium, Ruminococcus, Parabacteroides, Roseburia, Akkermansia, Alistipes, Coprococcus, Oscillospira, Oscillibacter, Butyricicoccus, Odoribacter*, and *Paraprevotella* were commonly reported to be decreased in patients with conditions like obesity, diabetes, metabolic syndrome, NAFLD, etc. compared to healthy controls. Genera like *Escherichia, Blautia, Streptococcus, Klebsiella, Collinsella, Dorea,* and *Enterococcus* are seen to be increased in patients with metabolic diseases. Torres-Sánchez and his team also identified 63 metabolites related to health status, of which 35 metabolites were exclusively associated with health status. These included metabolites from fatty acid pathways like acetate and caproic acid, amino acid pathways like glycylvaline and isoleucine, and bile acid pathways like chenodeoxyglycocholate. Meanwhile, 171 metabolites were associated with metabolic diseases, of which 143 were exclusively associated (Torres-Sánchez et al. 2023). These included disease-related metabolites from fatty acid pathways like arachidonic acid and docosahexaenoic acid; amino acid pathways like asymmetric dimethylarginine and citrulline; and bile acid pathways like 7-dehydrocholic acid and lithocholic acid.

Furthermore, Surowiec et al. (2019) identified specific metabolite and lipid signatures connected to the metabolic syndrome. Using orthogonal projections to latent structures (OPLS), a model revealed several distinct classes of small-molecule biomarkers in plasma that were positively correlated with metabolic syndrome severity based on p (corr) values. These included multiple acylcarnitines like butyryl-carnitine and valeryl-carnitine, certain amino acids such as alanine and valine, and various organic acids such as lactic acid and alpha-ketoglutaric acid. In addition, lipid profiling by OPLS identified numerous triglycerides, phosphatidylcholines, phosphatidylinositols, and ceramides positively correlated with metabolic syndrome scores. One triglyceride, TG (53:2), showed a negative correlation. Thus, multivariate modeling of metabolomic and lipidomic datasets successfully identified distinct biochemical signatures reflecting worsening metabolic syndrome, providing insight into pathways that may modulate disease risk and severity. These studies collectively underscore the potential of these approaches in identifying novel diagnostic and prognostic biomarkers for metabolic disorders. Validated biomarker panels could enable non-invasive diagnosis, disease staging, and monitoring of treatment responses. Longitudinal studies are needed to establish biomarker stability and clinical utility for personalized disease management.

14.7 Therapeutic Approaches Targeting the Gut Microbiota for Metabolic Health

14.7.1 Probiotics and Prebiotics as Potential Interventions for Metabolic Disorders

Metabolic disorders such as obesity and type 2 diabetes are significant global health concerns, and the gut microbiota plays a crucial role in their development and management. Prebiotics, probiotics, and dietary modifications can improve metabolic health by selectively influencing the composition and functionality of the microbiome (Bock et al. 2021). For example, prebiotics like inulin and oligofructose can stimulate beneficial bacteria in the gut, leading to improvements in glucose regulation, lipid profiles, and weight management (O'Connor et al. 2017; Sharma and Puri 2015). They also support the growth of beneficial gut bacteria such as *Bifidobacterium*, which can increase the production of SCFAs and reduce lipid levels in the blood, thus lowering the risk of coronary heart disease (Sharma and Puri 2015). Certain types of dietary fibers, such as inulin, oligofructose, lactulose, and resistant starch, can be broken down by the gut microbiota through fermentation, resulting in the production of SCFAs with important physiological effects. Other fibers, including galactooligosaccharides, transgalactooligosaccharides, polydextrose, wheat dextrin, acacia gum, psyllium, banana, and whole grains, also demonstrate prebiotic effects by selectively increasing the populations of *Bifidobacteria* and *lactobacilli*.

Probiotic supplementation introduces exogenous bacteria to repopulate the gut with strains linked to significant health benefits for obesity and diabetes by modifying gut bacteria. *Lactobacillus fermentum* KC4b, a type of probiotic, has shown promise in improving cholesterol and fat processing. Probiotics can lower total cholesterol levels and enhance lipid metabolism. Some *Lactobacilli* can help break down bile salts in the intestines, affecting cholesterol absorption. Consuming dairy containing *Lactobacillus acidophilus* reduced total cholesterol by about 2.4% on average. Additionally, supplements with *Lactobacillus casei* and *Saccharomyces boulardii* showed significant drops in total and LDL cholesterol, 37% and 19%, respectively (Badr et al. 2021; El-Sayed et al. 2021). Probiotics may aid cholesterol elimination through the digestive tract by impacting bile salt composition and cholesterol absorption. Consequently, long-term probiotic use could support healthy cholesterol metabolism. A symbiotic combining prebiotics with probiotics may provide additional synergistic effects. Overall, probiotics hold promise for managing metabolic conditions through gut microbiota modulation (Aggarwal et al. 2013).

Multiple mice and human clinical trials report probiotic administration resulting in decreased weight gain, adipose tissue accrual, and serum biomarkers of obesity, including leptin, glucose, and triglycerides. Several mechanisms were hypothesized for these effects, such as elevated fat metabolism and energy expenditure potentially mediated by probiotic-induced factors like angiopoietin-like protein 4 and sympathetic nervous system activation. Furthermore, certain trials note the benefits of probiotics for reducing visceral and subcutaneous fat stores, attenuating gestational diabetes risk, and facilitating weight loss after bariatric procedures. Altogether, the evidence suggests a role for targeted probiotic supplementation in obesity treatment and prevention (Arora et al. 2013).

Furthermore, probiotics may help address metabolic conditions by influencing the intestinal mucosal barrier and immune system. This barrier that separates gut contents from the body consists of mucus, epithelial cells, tight junctions, and immune components and prevents endotoxins such as LPS from entering circulation. However, obesity and related disorders disrupt barrier integrity and increase permeability, contributing to chronic low-grade inflammation. Probiotics can restore the compromised barrier by elevating the production of tight junction proteins like occludin and claudins that seal epithelial cells. Reinforcing barrier function inhibits LPS passage into the bloodstream, decreasing metabolic endotoxemia. Probiotics also interact with intestinal dendritic cells and T-cells, stimulating anti-inflammatory regulatory responses and cytokine/IgA secretion to modulate immunity. These actions, which restore the mucosal barrier and temper inflammation, may help combat metabolic dysfunction (Barz et al. 2015). Specific probiotic species, including *Lactobacillus* spp., *Bifidobacterium* spp., and *Saccharomyces* spp., have been identified as promising in the prevention of degenerative diseases, including obesity and diabetes (Azad et al. 2018).

The gut microbiota plays a vital role in bidirectional signaling between the gastrointestinal and central nervous systems, known as the gut-brain axis. Specifically, intestinal bacteria can influence hunger and satiety cues by generating hormones and neurotransmitters that interact with the brain via the vagus nerve (Fig. 14.6). Metabolites resulting from microbial fermentation, such as SCFAs, also appear to modulate the release of factors modulating food intake within the brain (Fig. 14.3). The composition of gut bacteria is often altered in overweight/obese individuals. Restoring a balanced microbiota state through the consumption of probiotic foods or supplements represents a potential strategy for weight management by optimizing gut-brain communication and appetite regulation. The hypothalamus is key in maintaining energy balance and metabolism through neural and hormonal pathways. This homeostatic center directly communicates with the gastrointestinal system via the gut-brain axis. The gut contains the largest endocrine organ in the form of enteroendocrine cells (EEC) scattered along the intestinal tract. These cells secrete hormones and neurotransmitters in response to the presence of food that signals sensations like hunger. Hormones such as ghrelin, cholecystokinin, glucose-dependent insulinotropic peptide (GIP), glucagon-like peptide 1 (GLP-1), and peptide YY (PYY) mediate effects on the secretion of substances like insulin and gastric acid, as well as gut motility and food intake. This occurs through activation of the vagus nerve, which stimulates the digestive tract and links to the hypothalamus. Notably, the gut microbiota substantially influences this bidirectional signaling between the gastrointestinal and central nervous systems by regulating host energy expenditure, producing proteins activating satiety pathways, and ultimately impacting food intake and energy homeostasis (Gallegos-Inzunza et al. 2021).

14.7.2 Fecal Microbiota Transplantation (FMT) and Its Role in Metabolic Health

Fecal microbiota transplantation (FMT) has shown a potential to improve metabolic health, particularly with obesity and metabolic syndrome. In diabetes, some trials found FMT temporarily improved insulin resistance in obese individuals or those with type 2 diabetes using healthy donors. However, results have been mixed and may depend on donor selection as the donor's metabolic traits may transfer. Also, the microbiome profile and diet of the recipient may be affected. The autologous FMT did better than FMT from other people (allogenic) over a year by maintaining beta cell function in type 1 diabetes, possibly by reshaping the composition of gut bacteria, indicating it could slow disease progression. Furthermore, FMT from vegan donors in NAFLD noted decreased liver inflammation markers and genes and reduced gut permeability following FMT from healthy volunteers (Hanssen et al. 2021).

For recurring *Clostridium difficile* infections, several randomized controlled trials found FMT achieved high success rates of 85–90% compared to just 30–80% for standard medical therapies alone, leading to FMT now being recommended after two *C. difficile* recurrences (Xu et al. 2015). While early case studies of FMT for inflammatory bowel diseases like ulcerative colitis reported symptom improvements, the first proper clinical trials only saw remission in about 24–30% of patients versus placebo, indicating it is not as effective as for *C. difficile.* The pilot case series also suggests FMT may help around 70% of people with irritable bowel syndrome or constipation, but more rigorous studies are still needed. It briefly mentions early exploratory work on using FMT for other non-GI issues tied to gut imbalances like metabolic disorders, though the evidence remains very preliminary overall (Choi and Cho 2016).

FMT commonly produces only minor and temporary adverse effects like abdominal discomfort, bloating, diarrhea, and constipation; it also acknowledges a lack of information regarding potential long-term risks. Even though immediate side effects usually resolve within two days, very little is known about FMT's long-term immunological impacts or induction of latent infections. A key concern raised is that alterations to the gut microbiome following FMT could theoretically influence chronic disease development over the long run. The conditions listed as hypothetical long-term adverse events that warrant continued monitoring and research include obesity, diabetes, heart disease, inflammatory bowel disease, colon cancer, NAFLD, irritable bowel syndrome, asthma, and autism. So, while short-term complications tend to be mild and brief, the lack of data on immunological and microbiome changes underscores the need to evaluate FMT's long-term safety profile thoroughly (Choi and Cho 2016).

14.7.3 Metabolites Supplement Approach

Microbial metabolites, small molecules produced through microbial metabolism in the gut, play significant roles in signaling between gut bacteria and the host. Metabolites like TMAO, indoxyl sulfate (IndS), and para-cresyl sulfate (PCS) are linked to increased risk of cardiovascular disease, dysbiosis, and inflammation, potentially exacerbating conditions like diabetes or colitis. PCS has been linked to inflammation and oxidative stress in gut tissues. TMAO has been tied to increased cardiovascular risk. Levels of these microbial products are altered in obesity, diabetes, heart disease, cancer, neurological conditions, and more. On the other hand, metabolites like SCFAs, indole-3-propionate (IPA), conjugated linoleic acid (CLA), and 10-hydroxy-*cis*-12-octadecenoate (HYA) seem to confer protective effects. SCFAs like butyrate fuel colonic cells and support anti-inflammatory gut immune responses. At the same time, IPA, CLA, and HYA have antioxidant and anti-inflammatory properties that could reduce disease activity in conditions involving oxidative stress or chronic inflammation. Supplementation with specific metabolites has shown benefits in disease models, indicating that targeting microbial metabolites offers potential as a novel therapeutic approach. Metabolite profiling using metabolomics has enhanced understanding of their involvement in health and disease, aiding the identification of biomarkers and treatment targets (Descamps et al. 2019; Zhang and Davies 2016).

14.7.4 MicroRNAs Therapeutic Approach

MicroRNAs (miRNAs) are small molecules, non-coding RNAs that regulate gene expression post-transcriptionally. They are essential in mediating interactions between the gut microbiome and the host. They are synthesized primarily within intestinal epithelial cells, and miRNAs help regulate responses to commensal, pathogenic, and probiotic bacteria by modulating levels of various genes. (Staedel and Darfeuille 2013). The microbial community inhabiting the gut can distinctly influence the levels and makeup of miRNAs present in these gut cells in a species- and strain-specific manner. In turn, changes to the intestinal miRNA profile impact processes such as barrier function, immunity, and metabolism within the gastrointestinal tract. Dysregulated fecal miRNAs are potential biomarkers for intestinal diseases, including colorectal cancer and celiac disease. They also shape the gut microbiota and influence bacterial metabolism (Sarshar et al. 2020). Any deficiencies disrupting the average production of miRNAs by intestinal cells have also been linked to imbalances in the gut microbiota. Therefore, strategically manipulating miRNA compositions in response to modulating the microbiome could present a potential means of optimizing intestinal health.

14.8 Conclusions and Future Perspectives

Further research on the human microbiome is still needed to develop a more comprehensive understanding of how it interacts with the body and influences health and disease. While imbalances in the microbiome have been linked to various conditions, the specific biological mechanisms at play are still unclear. More studies are needed to elucidate these causal pathways and identify the specific bacterial strains or microbial metabolites involved (Cullen et al. 2020). There is also a need for more personalized approaches tailored to each individual's unique microbiome profile and other relevant factors. Future research should explore using biomarkers or other indicators to match patients with microbiota-based therapies based on their distinct microbiome characteristics.

Regulatory agencies continue to face challenges in regulating these novel living therapies due to their complexity. Clear guidelines are still required to facilitate development and market approval (Gebrayel et al. 2022). Additionally, longer-term follow-up studies are still needed to fully evaluate patient outcomes and safety over several years after microbial manipulation. This continuing collaborative research aims to strengthen evidence, drive adoption, and unlock the full therapeutic potential of microbiota-based strategies.

In conclusion, this emerging area of microbiome-metabolism research is promising to advance our understanding and management of complex metabolic diseases. Longitudinal studies using multi-omics approaches are needed to map dynamic changes in the human-microbial metabolic network over time and across diverse populations and environmental exposures. More comprehensive clinical trials are necessary to validate microbiome-based biomarkers and delineate their applications in precision diagnosis, risk stratification, and treatment effectiveness monitoring. Standardization of analytical methods will help translate findings between studies. Developing microbiome-modulating therapies is a key focus, with ongoing research evaluating prebiotic formulations, probiotic strains, personalized diets, and other gut ecosystem manipulations. Combinatorial interventions designed through systems-level network analyses may most effectively restore regulatory balance. Continuous digital monitoring platforms incorporating multi-omic, lifestyle, and physiological data promise to advance predictive and preventative approaches. The critical role of our inner microbial communities must be considered from a systems perspective of human health and resilience.

References

Abaza AA, Shetaia YM, Sorour NM, El-Sayed ASA, El-Baz AF (2024) Enhancing the biosynthesis of polyunsaturated fatty acids by Rhodotorula mucilaginosa and Lodderomyces elongisporus. Bioresour Bioprocess 11(1):39

Abranches J, Zeng L, Kajfasz JK, Palmer SR, Chakraborty B, Wen ZT et al (2018) Biology of oral Streptococci. Microbiol Spectr 6:5. https://doi.org/10.1128/microbiolspec.gpp3-0042-2018

Abt MC, Pamer EG (2014) Commensal bacteria mediated defenses against pathogens. Curr Opin Immunol 29:16–22

Aggarwal J, Swami G, Kumar M (2013) Probiotics and their effects on metabolic diseases: an update. J Clin Diagn Res 7(1):173–177

Aj C, Np W, Aw C (2015) Obesity, inflammation, and the gut microbiota. Lancet Diab Endocrinol [Internet]. 2015 Mar [cited 2023 Dec 13];3(3). Available from: https://pubmed.ncbi.nlm.nih.gov/25066177/

Armani RG, Ramezani A, Yasir A, Sharama S, Canziani MEF, Raj DS (2017) Gut microbiome in chronic kidney disease. Curr Hypertens Rep 19(4):29

Arora T, Singh S, Sharma RK (2013) Probiotics: interaction with gut microbiome and antiobesity potential. Nutrition 29(4):591–596

Azad MAK, Sarker M, Li T, Yin J (2018) Probiotic species in the modulation of gut microbiota: an overview. Biomed Res Int 2018:1–8

Badr H, El-Baz A, Mohamed I, Shetaia Y, El-Sayed ASA, Sorour N (2021) Bioprocess optimization of glutathione production by Saccharomyces boulardii: biochemical characterization of glutathione peroxidase. Arch Microbiol 203(10):6183–6196

Barz ML, Anhê FF, Varin TV, Desjardins Y, Levy E, Roy D et al (2015) Probiotics as complementary treatment for metabolic disorders. Diabetes Metab J 39(4):291–303

Bashiardes S, Shapiro H, Rozin S, Shibolet O, Elinav E (2016) Non-alcoholic fatty liver and the gut microbiota. Mol Metab 5(9):782–794

Birkeland E, Gharagozlian S, Birkeland KI, Valeur J, Måge I, Rud I et al (2020) Prebiotic effect of inulin-type fructans on faecal microbiota and short-chain fatty acids in type 2 diabetes: a randomised controlled trial. Eur J Nutr 59(7):3325–3338

Bock PM, Telo GH, Ramalho R, Sbaraini M, Leivas G, Martins AF et al (2021) The effect of probiotics, prebiotics or synbiotics on metabolic outcomes in individuals with diabetes: a systematic review and meta-analysis. Diabetologia 64(1):26–41

Boulangé CL, Neves AL, Chilloux J, Nicholson JK, Dumas ME (2016) Impact of the gut microbiota on inflammation, obesity, and metabolic disease. Genome Med 8(1):42

Bourdeau-Julien I, Castonguay-Paradis S, Rochefort G, Perron J, Lamarche B, Flamand N et al (2023) The diet rapidly and differentially affects the gut microbiota and host lipid mediators in a healthy population. Microbiome 11(1):26

Bravo JA, Forsythe P, Chew MV, Escaravage E, Savignac HM, Dinan TG et al (2011) Ingestion of Lactobacillus strain regulates emotional behavior and central GABA receptor expression in a mouse via the vagus nerve. Proc Natl Acad Sci 108(38):16050–16055

Brial F, Le Lay A, Dumas ME, Gauguier D (2018) Implication of gut microbiota metabolites in cardiovascular and metabolic diseases. Cell Mol Life Sci 75(21):3977–3990

Cai D, Yuan M, Frantz DF, Melendez PA, Hansen L, Lee J et al (2005) Local and systemic insulin resistance resulting from hepatic activation of IKK-β and NF-κB. Nat Med 11(2):183–190

Canani RB, Costanzo MD, Leone L, Pedata M, Meli R, Calignano A (2011) Potential beneficial effects of butyrate in intestinal and extraintestinal diseases. World J Gastroenterol 17(12):1519–1528

Cani PD, Neyrinck AM, Fava F, Knauf C, Burcelin RG, Tuohy KM et al (2007) Selective increases of bifidobacteria in gut microflora improve high-fat-diet-induced diabetes in mice through a mechanism associated with endotoxaemia. Diabetologia 50(11):2374–2383

Chakraborti CK (2015) New-found link between microbiota and obesity. World J Gastrointest Pathophysiol 6(4):110–119

Chapman MA, Grahn MF, Boyle MA, Hutton M, Rogers J, Williams NS (1994) Butyrate oxidation is impaired in the colonic mucosa of sufferers of quiescent ulcerative colitis. Gut 35(1):73–76
Chen J, Vitetta L (2020) Gut microbiota metabolites in NAFLD pathogenesis and therapeutic implications. Int J Mol Sci 21(15):5214
Choi HH, Cho YS (2016) Fecal microbiota transplantation: current applications, effectiveness, and future perspectives. Clin Endosc 49(3):257–265
Chunxi L, Haiyue L, Yanxia L, Jianbing P, Jin S (2020) The gut microbiota and respiratory diseases: new evidence. J Immunol Res 31:e2340670
Cosseau C, Romano-Bertrand S, Duplan H, Lucas O, Ingrassia I, Pigasse C et al (2016) Proteobacteria from the human skin microbiota: species-level diversity and hypotheses. One Health 2:33–41
Cullen CM, Aneja KK, Beyhan S, Cho CE, Woloszynek S, Convertino M et al (2020) Emerging priorities for microbiome research. Front Microbiol 11:136
De Pessemier B, Grine L, Debaere M, Maes A, Paetzold B, Callewaert C (2021) Gut–skin axis: current knowledge of the interrelationship between microbial dysbiosis and skin conditions. Microorganisms 9(2):353
De Vadder F, Kovatcheva-Datchary P, Goncalves D, Vinera J, Zitoun C, Duchampt A et al (2014) Microbiota-generated metabolites promote metabolic benefits via gut-brain neural circuits. Cell 156(1):84–96
den Besten G, Bleeker A, Gerding A, van Eunen K, Havinga R, van Dijk TH et al (2015) Short-chain fatty acids protect against high-fat diet–induced obesity via a PPARγ-dependent switch from lipogenesis to fat oxidation. Diabetes 64(7):2398–2408
Descamps HC, Herrmann B, Wiredu D, Thaiss CA (2019) The path toward using microbial metabolites as therapies. EBioMedicine 44:747–754
Di Martino ML, Campilongo R, Casalino M, Micheli G, Colonna B, Prosseda G (2013) Polyamines: emerging players in bacteria–host interactions. Int J Med Microbiol 303(8):484–491
Dong L, Yin J, Zhao J, Ma S, Rui WH, Wang M, et al (2018) Microbial similarity and preference for specific sites in healthy oral cavity and Esophagus. Front Microbiol [Internet]. 2018 [cited 2023 Dec 11];9. Available from: https://www.frontiersin.org/articles/10.3389/fmicb.2018.01603
Duarte SMB, Stefano JT, Oliveira CP (2019) Microbiota and nonalcoholic fatty liver disease/nonalcoholic steatohepatitis (NAFLD/NASH). Ann Hepatol 18(3):416–421
Duszka K (2022) Versatile triad alliance: bile acid, taurine and microbiota. Cells [Internet]. 2022 Aug [cited 2023 Dec 25];11(15). Available from: https://www.ncbi.nlm.nih.gov/pmc/articles/PMC9367564/
El-Baz AF, El-Enshasy HA, Shetaia YM, Mahrous H, Othman NZ, Yousef AE (2018) Semi-industrial scale production of a new yeast with probiotic traits, Cryptococcus sp. YMHS, isolated from the Red Sea. Prob Antimicrob Prot 10(1):77–88
El-Sayed ASA, Fathalla M, Shindia AA, Rady AM, El-Baz AF, Morsy Y et al (2021) Purification and biochemical characterization of Taxadiene synthase from Bacillus koreensis and Stenotrophomonas maltophilia. Sci Pharm 89(4):48
Federici M (2019) Gut microbiome and microbial metabolites: a new system affecting metabolic disorders. J Endocrinol Investig 42(9):1011–1018
Festi D, Schiumerini R, Eusebi LH, Marasco G, Taddia M, Colecchia A (2014) Gut microbiota and metabolic syndrome. World J Gastroenterol 20(43):16079–16094
Fujio-Vejar S, Vasquez Y, Morales P, Magne F, Vera-Wolf P, Ugalde JA, et al (2017) The gut microbiota of healthy Chilean subjects reveals a high abundance of the Phylum Verrucomicrobia. Front Microbiol [Internet]. 2017 [cited 2023 Dec 11];8. Available from: https://www.frontiersin.org/articles/10.3389/fmicb.2017.01221
Gallegos-Inzunza B, Prado-Ramírez J, Romo-Araiza A (2021) Probiotic foods as functional foods for modulating obesity. Proc Sci Res Univ Anáhuac Multidiscipl J Healthcare 1(2):46–57
Gebrayel P, Nicco C, Al Khodor S, Bilinski J, Caselli E, Comelli EM et al (2022) Microbiota medicine: towards clinical revolution. J Transl Med 20(1):111

Geirnaert A, Calatayud M, Grootaert C, Laukens D, Devriese S, Smagghe G et al (2017) Butyrate-producing bacteria supplemented in vitro to Crohn's disease patient microbiota increased butyrate production and enhanced intestinal epithelial barrier integrity. Sci Rep 7(1):11450

Hadadi N, Berweiler V, Wang H, Trajkovski M (2021) Intestinal microbiota as a route for micro-nutrient bioavailability. Curr Opin Endocr Metabol Res 20:100285

Hamer HM, Jonkers D, Venema K, Vanhoutvin S, Troost FJ, Brummer RJ (2008) Review article: the role of butyrate on colonic function. Aliment Pharmacol Ther 27(2):104–119

Hanssen NMJ, De Vos WM, Nieuwdorp M (2021) Fecal microbiota transplantation in human metabolic diseases: from a murky past to a bright future? Cell Metab 33(6):1098–1110

Heiss CN, Olofsson LE (2017) Gut microbiota-dependent modulation of energy metabolism. J Innate Immun 10(3):163–171

Hoyles L, Fernández-Real JM, Federici M, Serino M, Abbott J, Charpentier J et al (2018) Molecular phenomics and metagenomics of hepatic steatosis in non-diabetic obese women. Nat Med 24(7):1070–1080

Iacobini C, Pugliese G, Blasetti Fantauzzi C, Federici M, Menini S (2019) Metabolically healthy versus metabolically unhealthy obesity. Metabolism 92:51–60

Iannotti FA, Marzo VD (2021) The gut microbiome, endocannabinoids and metabolic disorders. J Endocrinol 248(2):R83–R97

Ji Y, Yin Y, Li Z, Zhang W (2019) Gut microbiota-derived components and metabolites in the progression of Non-Alcoholic Fatty Liver Disease (NAFLD). Nutrients 11(8):1712

Jonkers DMAE (2016) Microbial perturbations and modulation in conditions associated with malnutrition and malabsorption. Best Pract Res Clin Gastroenterol 30(2):161–172

Kasubuchi M, Hasegawa S, Hiramatsu T, Ichimura A, Kimura I (2015) Dietary gut microbial metabolites, short-chain fatty acids, and host metabolic regulation. Nutrients 7(4):2839–2849

Koh A, De Vadder F, Kovatcheva-Datchary P, Bäckhed F (2016) From dietary fiber to host physiology: short-chain fatty acids as key bacterial metabolites. Cell 165(6):1332–1345

Krishnan S, Alden N, Lee K (2015) Pathways and functions of gut microbiota metabolism impacting host physiology. Curr Opin Biotechnol 36:137–145

Lami KF, de Oliveira VF, Batista KZS (2021) Gut-brain axis and immunoneuroendocrine modulation in neurological and psychiatric disorders: a systematic review. Res Soc Dev 10(4):e28110414185–e28110414185

Lee SY, Lee E, Park YM, Hong SJ (2018) Microbiome in the gut-skin axis in atopic dermatitis. Allergy Asthma Immunol Res 10(4):354

Lenoir M, Martín R, Torres-Maravilla E, Chadi S, González-Dávila P, Sokol H et al (2020) Butyrate mediates anti-inflammatory effects of Faecalibacterium prausnitzii in intestinal epithelial cells through Dact3. Gut Microbes 12(1):1826748

Li X, Watanabe K, Kimura I (2017) Gut microbiota dysbiosis drives and implies novel therapeutic strategies for diabetes Mellitus and related metabolic diseases. Front Immunol [Internet]. 2017 [cited 2023 Dec 12];8. Available from: https://www.frontiersin.org/articles/10.3389/fimmu.2017.01882

Lin HV, Frassetto A Jr, Nawrocki AR, Lu MM, Kosinski JR et al (2012) Butyrate and propionate protect against diet-induced obesity and regulate gut hormones via free fatty acid receptor 3-independent mechanisms. PLoS One 7(4):e35240

Manole E, Dumitrescu L, Niculiţe C, Popescu B, Ceafalan L (2021) Potential roles of functional bacterial amyloid proteins, bacterial biosurfactants and other putative gut microbiota products in the etiopathogeny of Parkinson's disease. Biocell 45:1–16

Margolis KG, Cryan JF, Mayer EA (2021) The microbiota-gut-brain axis: from motility to mood. Gastroenterology 160(5):1486–1501

McCarville JL, Chen GY, Cuevas VD, Troha K, Ayres JS (2020) Microbiota metabolites in health and disease. Annu Rev Immunol 38(1):147–170

Molinaro A, Wahlström A, Marschall HU (2018) Role of bile acids in metabolic control. Trends Endocrinol Metab 29(1):31–41

Nielsen DSG, Jensen BB, Theil PK, Nielsen TS, Knudsen KEB, Purup S (2018) Effect of butyrate and fermentation products on epithelial integrity in a mucus-secreting human colon cell line. J Funct Foods 40:9–17
Nieuwdorp M, Gilijamse PW, Pai N, Kaplan LM (2014) Role of the microbiome in energy regulation and metabolism. Gastroenterology 146(6):1525–1533
O'Connor S, Chouinard-Castonguay S, Gagnon C, Rudkowska I (2017) Prebiotics in the management of components of the metabolic syndrome. Maturitas 104:11–18
Rajani C, Jia W (2018) Disruptions in gut microbial-host co-metabolism and the development of metabolic disorders. Clin Sci 132(7):791–811
Ramakrishna BS (2013) Role of the gut microbiota in human nutrition and metabolism. J Gastroenterol Hepatol 28(S4):9–17
Rastogi S, Mohanty S, Sharma S, Tripathi P (2022) Possible role of gut microbes and host's immune response in gut–lung homeostasis. Front Immunol [Internet]. 2022 [cited 2023 Dec 17];13. Available from: https://www.frontiersin.org/articles/10.3389/fimmu.2022.954339
Ravcheev DA, Godzik A, Osterman AL, Rodionov DA (2013) Polysaccharides utilization in human gut bacterium Bacteroides thetaiotaomicron: comparative genomics reconstruction of metabolic and regulatory networks. BMC Genomics 14(1):873
Rowland I, Gibson G, Heinken A, Scott K, Swann J, Thiele I et al (2018) Gut microbiota functions: metabolism of nutrients and other food components. Eur J Nutr 57(1):1–24
Samuel BS, Shaito A, Motoike T, Rey FE, Backhed F, Manchester JK et al (2008) Effects of the gut microbiota on host adiposity are modulated by the short-chain fatty-acid binding G protein-coupled receptor, Gpr41. Proc Natl Acad Sci USA 105(43):16767–16772
Sarshar M, Scribano D, Ambrosi C, Palamara AT, Masotti A (2020) Fecal microRNAs as innovative biomarkers of intestinal diseases and effective players in host-microbiome interactions. Cancers 12(8):2174
Sharma S, Puri S (2015) Prebiotics and lipid metabolism: a review. Altern Ther Health Med 21:34–42
Singh R, Zogg H, Wei L, Bartlett A, Ghoshal UC, Rajender S et al (2021a) Gut microbial dysbiosis in the pathogenesis of gastrointestinal dysmotility and metabolic disorders. J Neurogastroenterol Motil 27(1):19–34
Singh V, Muthuramalingam K, Kim YM, Park S, Kim SH, Lee J et al (2021b) Synbiotic supplementation with prebiotic Schizophyllum commune derived β-(1,3/1,6)-glucan and probiotic concoction benefits gut microbiota and its associated metabolic activities. Appl Biol Chem 64(1):7
Staedel C, Darfeuille F (2013) MicroRNAs and bacterial infection. Cell Microbiol 15(9):1496–1507
Strandwitz P (2018) Neurotransmitter modulation by the gut microbiota. Brain Res 1693:128–133
Sudo N (2019) Biogenic Amines: signals between commensal microbiota and gut physiology. Front Endocrinol [Internet]. 2019 [cited 2023 Dec 12];10. Available from: https://www.frontiersin.org/articles/10.3389/fendo.2019.00504
Sultan S, El-Mowafy M, Elgaml A, Ahmed TAE, Hassan H, Mottawea W (2021) Metabolic influences of gut microbiota dysbiosis on inflammatory bowel disease. Front Physiol [Internet]. 2021 [cited 2023 Dec 13];12. Available from: https://www.frontiersin.org/articles/10.3389/fphys.2021.715506
Surowiec I, Noordam R, Bennett K, Beekman M, Slagboom PE, Lundstedt T et al (2019) Metabolomic and lipidomic assessment of the metabolic syndrome in Dutch middle-aged individuals reveals novel biological signatures separating health and disease. Metabolomics 15(2):23
Suzuki K, Benno Y, Mitsuoka T, Takebe S, Kobashi K, Hase J (1979) Urease-producing species of intestinal anaerobes and their activities. Appl Environ Microbiol 37(3):379–382
Tang Z, Yang S, He Z (2022) Defining the baseline of pulmonary microbiota in healthy populations and influencing factors. HSET 11:38–48
Thomas MS, Blesso CN, Calle MC, Chun OK, Puglisi M, Fernandez ML (2022) Dietary influences on gut microbiota with a focus on metabolic syndrome. Metab Syndr Relat Disord 20(8):429–439

Thye AYK, Bah YR, Law JWF, Tan LTH, He YW, Wong SH et al (2022) Gut–skin axis: unravelling the connection between the gut microbiome and psoriasis. Biomedicines 10(5):1037

Toe LC, Kerckhof FM, De Bodt J, Morel FB, Ouedraogo JB, Kolsteren P et al (2020) A prebiotic-enhanced lipid-based nutrient supplement (LNSp) increases Bifidobacterium relative abundance and enhances short-chain fatty acid production in simulated colonic microbiota from undernourished infants. FEMS Microbiol Ecol 96(7):fiaa105

Tokuhara D (2021) Role of the gut microbiota in regulating non-alcoholic fatty liver disease in children and adolescents. Front Nutr [Internet]. 2021 [cited 2023 Dec 8];8. Available from: https://www.frontiersin.org/articles/10.3389/fnut.2021.700058

Tomasova L, Grman M, Ondrias K, Ufnal M (2021) The impact of gut microbiota metabolites on cellular bioenergetics and cardiometabolic health. Nutr Metab (Lond) 18(1):72

Torres-Sánchez A, Ruiz-Rodríguez A, Ortiz P, Aguilera M (2023) Key stratification of microbiota taxa and metabolites in the host metabolic health–disease balance. Int J Mol Sci 24(5):4519

Turnbaugh PJ, Ley RE, Mahowald MA, Magrini V, Mardis ER, Gordon JI (2006) An obesity-associated gut microbiome with increased capacity for energy harvest. Nature 444(7122):1027–1031

Vetrani C, Di Nisio A, Paschou SA, Barrea L, Muscogiuri G, Graziadio C et al (2022) From gut microbiota through low-grade inflammation to obesity: key players and potential targets. Nutrients 14(10):2103

Wang PX, Deng XR, Zhang CH, Yuan HJ (2020) Gut microbiota and metabolic syndrome. Chin Med J 133(7):808–816

Xu MQ, Cao HL, Wang WQ, Wang S, Cao XC, Yan F et al (2015) Fecal microbiota transplantation broadening its application beyond intestinal disorders. World J Gastroenterol 21(1):102–111

Yan HM, Zhao HJ, Guo DY, Zhu PQ, Zhang CL, Jiang W (2018) Gut microbiota alterations in moderate to severe acne vulgaris patients. J Dermatol 45(10):1166–1171

Yang D, Xing Y, Song X, Qian Y (2020) The impact of lung microbiota dysbiosis on inflammation. Immunology 159(2):156–166

Yoshida H, Ishii M, Akagawa M (2019) Propionate suppresses hepatic gluconeogenesis via GPR43/AMPK signaling pathway. Arch Biochem Biophys 672:108057

Zhai W, Huang Y, Zhang X, Fei W, Chang Y, Cheng S et al (2018) Profile of the skin microbiota in a healthy Chinese population. J Dermatol 45(11):1289–1300

Zhang LS, Davies SS (2016) Microbial metabolism of dietary components to bioactive metabolites: opportunities for new therapeutic interventions. Genome Med 8(1):46

Zhang J, Guo Z, Xue Z, Sun Z, Zhang M, Wang L et al (2015) A phylo-functional core of gut microbiota in healthy young Chinese cohorts across lifestyles, geography and ethnicities. ISME J 9(9):1979–1990

Zhang M, Zhou L, Wang Y, Dorfman RG, Tang D, Xu L et al (2019) Faecalibacterium prausnitzii produces butyrate to decrease c-Myc-related metabolism and Th17 differentiation by inhibiting histone deacetylase 3. Int Immunol 31(8):499–514

Zhang D, Li S, Wang N, Tan HY, Zhang Z, Feng Y (2020) The cross-talk between gut microbiota and lungs in common lung diseases. Front Microbiol [Internet]. 2020 [cited 2023 Dec 17];11. Available from: https://www.frontiersin.org/articles/10.3389/fmicb.2020.00301

Zhang L, Chu J, Hao W, Zhang J, Li H, Yang C et al (2021) Gut microbiota and type 2 diabetes mellitus: association, mechanism, and translational applications. Mediat Inflamm 2021:5110276

Zhou L, Zhang M, Wang Y, Dorfman RG, Liu H, Yu T et al (2018) Faecalibacterium prausnitzii produces Butyrate to maintain Th17/Treg balance and to Ameliorate colorectal colitis by inhibiting histone deacetylase 1. Inflamm Bowel Dis 24(9):1926–1940

Zhou Z, Sun B, Yu D, Zhu C (2022) Gut microbiota: an important player in type 2 diabetes Mellitus. Front Cell Infect Microbiol [Internet]. 2022 [cited 2023 Dec 8];12. Available from: https://www.frontiersin.org/articles/10.3389/fcimb.2022.834485

Role of Microbiome in Defense

15

Aya Al Ali and Walaa K. Mousa

Abstract

The human body hosts trillions of diverse microbial species, along with their secreted metabolites and genes, collectively known as the human microbiome. Advances in bioinformatics, particularly culture-independent omics analyses, have enhanced our ability to study the microbiome and elucidate its functions. Understanding the roles of these microbes is crucial for comprehending human biology and developing novel therapeutics. The microbiome's role in defending against invading pathogens is fundamental, starting with its critical contribution to the development and maturation of the immune system. This book chapter provides a comprehensive illustration of the microbiome's role in defense, from mechanistic insights to therapeutic applications. We detail the interactions between the microbiota and the immune system, discussing various mechanisms and signaling pathways that underpin the antipathogenic activities of the microbiota, including colonization resistance, suppression of virulence factors, and the production of secreted antimicrobials. We also highlight unique host-microbiome-pathogen interactions across different body sites. We conclude by discussion the recent advances in the development of microbiome-based therapeutics to control infectious diseases. Besides, we analyze opportunities and challenges in the field, and explore the potential of machine learning tools in the discovery of

A. Al Ali
College of Pharmacy, Al Ain University, Abu Dhabi, UAE

AAU Health and Biomedical Research Center, Al Ain University, Abu Dhabi, UAE

W. K. Mousa (✉)
College of Pharmacy, Al Ain University, Abu Dhabi, UAE

AAU Health and Biomedical Research Center, Al Ain University, Abu Dhabi, UAE

College of Pharmacy, Mansoura University, Mansoura, Egypt
e-mail: walaa.mousa@aau.ac.ae

S. S. M. Soliman, M. I. Husseiny (eds.), *Metabolic Dynamics in Host-Microbe Interaction*, https://doi.org/10.1007/978-981-96-1305-2_15

novel antimicrobial molecules and the promising avenue of designing engineered live therapeutics.

Keywords
Gut microbiota · Probiotics · Microbiome-based therapeutics · Antimicrobial peptides · Immune activation

15.1 Introduction

The human body is colonized by a magnitude of distinct microbial communities at different body sites, along with their genes and secreted substances, collectively forming the human microbiome (Salvadori and Rosso 2024). The gut, compared to other body sites, is a densely colonized microbial habitat. This microbial colonization begins immediately after birth, with the method of delivery playing a significant role in shaping the composition of the infant's microbiome. Vaginal delivery results in colonization by vaginal bacteria, while infants born via Cesarean section exhibit bacterial communities similar to those found on the skin surface (Zhang et al. 2021). Throughout our lives, multiple factors contribute to shaping the structure and diversity of the microbiome. These include host genetics, diet, lifestyle, cigarette smoking, medications, and stress. Each of these elements plays a significant role in the evolving composition of the microbiome (Imchen et al. 2023).

Early colonization of the gut microbiota is critical in shaping the development of the host's immune responses, directly impacting the host's ability to recognize and respond to invading pathogens (Campbell et al. 2023). To facilitate their colonization, commensal microbes initially establish immune tolerance by multiple mechanisms. For example, commensal microbes activate Group 3 innate lymphoid cells (ILC3), which selectively elevate microbiome-specific RORγt+ Tregs and prevent the expansion of Th17 cell responses by a transforming growth factor-β1 (TGFβ1)-αVβ3 integrin-dependent mechanism, thereby enforcing their tolerance (Lyu et al. 2022). In addition, some secondary metabolites from the gut microbes such as 3-oxolithocholic acid and isolithocholic acid exhibit an inhibitory effect on Th17 cell differentiation enabling immune tolerance (Paik et al. 2022). Other studies revealed the crucial role of immunoglobulin A (IgA) in mucosal colonization. A study on *Bacteroides thetaiotaomicron* reported that binding of IgA to surface glycan upregulates the expression of a subset of genes, namely, mucus-associated functional factors (MAFFs), which play a crucial role in a successful microbiome colonization of the mucous layer, thus regulating the microbial composition of gut microbiome (Nakajima et al. 2018). Another study revealed that *B. fragilis* manipulate commensal colonization factor (CCF) to help anchoring *B. fragilis* to the gut mucosa. CCF controls the expression of certain capsular polysaccharides that attract the binding to IgA (Donaldson et al. 2018).

Once established their colonization, commensal microbiota employs diverse mechanisms in fighting invading pathogens such as production of antimicrobial peptides (AMPs), colonization resistance through reinforcing the gut epithelial barrier by promoting the production of mucus and tight junction (TJ) proteins, and blocking access to nutrients (Horrocks et al. 2023). An expanding body of research confirms that richness in microbiome is associated with better defense, while dysbiosis impairs its crucial function, making the host more vulnerable to infectious diseases (Spragge et al. 2023). Therefore, therapeutic interventions that involve restoration of the balanced microbiome are gaining momentum as ecological therapies. These interventions include the use of probiotics, prebiotics, and microbiota transplant. In this chapter, we detail mechanisms with case studies that illustrate the role of microbiome at defense. We discuss the future directions to harness the power of microbiome in translational research and development of therapeutics.

15.2 Strategies of the Microbiome at Defense

15.2.1 Microbiome Modulation of the Host Immune Response

Exposure to the microbiome in early life is essential for the development and maturation of immune system components, establishing a foundation for the host's defense against pathogens. Figure 15.1 illustrates examples of the interactions between gut microbes and the immune system. The microbiota, or its secreted metabolites, interacts with cellular receptors to mediate cascades of signals that lead to control of both the innate and adaptive immune responses (Zhang and Frenette 2019). This interaction is mediated through pathogen-associated molecular patterns (PAMPs), which stimulate pattern recognition receptors (PRRs) such as Toll-like receptors (TLRs), C-type lectin receptors, and nucleotide-binding oligomerization domain (NOD)-like receptors that are expressed on immune cells such as dendritic cells (DCs), T cells, B cells, and macrophages (Negi et al. 2019). Within the gastrointestinal mucosa, macrophages are primarily responsible for engulfing and digesting foreign particles, microbes, and cellular debris. Reduction in total intestinal macrophages was reported in germ-free (GF) mice (Kang et al. 2020). This decline is due to the absence of microbiome that regulates the activity of C-C chemokine receptor type 2, a primary modulator for circulating monocytes to influx to the intestine, where it differentiates into mature macrophages (Bain et al. 2014). Gut microbiota induces the production of tumor necrosis factor (TNF) by monocytes and macrophages, which in turn stimulates the maturation of precursors of type 1 conventional DCs through the C1q and Wtn/β-catenin pathway (Köhler et al. 2021).

Microbiota-derived molecules such as lipopolysaccharide and peptidoglycan or other bacterial components such as flagella can bind to TLR present on the immature DC. This binding activates the DC and promotes its migration to the mesenteric lymph node where the differentiation of naïve T cells into functional T cells including T helper (Th) cells and T regulatory (Treg) cells occurs (Ivanov et al. 2022). Thus, microbiome maintains the immunity homeostasis through balancing between anti-inflammatory and pro-inflammatory cytokines through regulating T-cell differentiation. A study reported that the introduction of polysaccharides derived from commensal *B. fragilis* in mice promotes the maturation of the developing immune

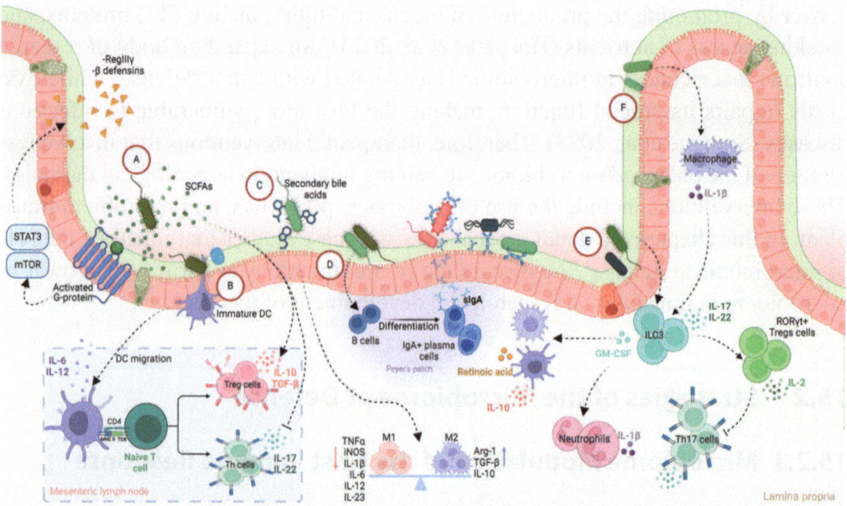

Fig. 15.1 Role of microbiome in immunomodulation. (**a**) Short-chain fatty acids (SCFAs) are types of microbiome metabolites that promote the anti-inflammatory status of the immune system by inhibiting inflammatory cytokines (e.g., IL-17 and IL-22). Moreover, SCFAs, through binding to G-protein-coupled receptor, activate mTOR and STAT3 pathways and prime the immune system by activating the production of antimicrobials (e.g., RegIIIγ and β-defensins). (**b**) Binding of microbiome to the immature dendritic cell induces the latter migration to the mesenteric lymph node where the differentiation of the naïve T cell initiates. (**c**) Secondary bile acids, another type of microbiome metabolite, control the macrophage polarization and thus mediate the equilibrium among the anti-inflammatory and inflammatory subtypes of macrophages (M2 and M1, respectively). (**d**) Microbiome is also an inducer of B-cell differentiation. The IgA secreted by IgA+ plasma cells selectively provides protection against pathogens and anchors the microbiome in the mucous layer. (**e, f**) Bacterial components such as lipopolysaccharide activate ILC3 and induce the release of GM-CSF that activates DC, macrophages, and neutrophils. Furthermore, activated ILC3 inhibits the inflammatory activity of Th17 cells through activating RORγt+ Tregs cells. Direct activation of macrophages by microbiome promotes the release of IL-1β that further participates in ILC3 activation

system, regulates the Th1/Th2 balance, and corrects the T-cell deficiencies (Mazmanian et al. 2005). Another study identified a bacterial cell-surface polysaccharide, named CSGG, from the *Bifidobacterium bifidum* strain PRI1. CSGG induces the generation of Treg cells. This stimulation is mediated through the CSGG-TLR2-DC axis (Verma et al. 2018). Similarly, the polysaccharide A derived from *B. fragilis* recognized by TLR2/TLR1 leads to the activation of the phosphoinositide 3-kinase (PI3K) pathway. This pathway triggers the deactivation of glycogen synthase kinase 3β (GSK3β), which consequently leads to the activation of cAMP response element-binding protein (CREB) that enhances the transcription of anti-inflammatory genes (Erturk-Hasdemir et al. 2019). Another example from young children population highlights the immune interaction of segmented filamentous bacteria (SFB), which is commonly known as inducers for interleukin (IL)-17 and IL-22 production. SFB colonization results in the production of serum amyloid

A (SAA) in the terminal ileum. SAA promotes Th17 cell differentiation from CD4+ T cells in vitro through its acting on lamina propria DCs (Chen et al. 2018). Th17 is a subgroup of CD4+ T cells known for their pro-inflammatory activity by secreting IL-17, which is crucial for defense against external pathogens (Schnell et al. 2023). The secretion of IL-17 also induces the production of TJ proteins and antimicrobials from gut epithelial cells (ECs), further enhancing the protective barrier function of the gut (Valeri and Raffatellu 2016).

15.2.2 Role of Microbiota-Derived Short-Chain Fatty Acids in Defense

Commensal gut bacteria generate short-chain fatty acids (SCFAs), including butyrate, propionate, and acetate, through the anaerobic fermentation of nondigestible substrates like dietary fibers (Portincasa et al. 2022). These SCFAs are crucial in maintaining intestinal homeostasis and resistance against pathogens (Fusco et al. 2023). The amount and the type of SCFAs produced vary with microbiota composition (Fusco et al. 2023). Within the gut, the gram-negative Bacteroidetes is the main producer of propionate (Zafar and Saier 2021), while gram-positive Firmicutes produce butyrate (Houtman et al. 2022). SCFAs enhance the intestinal barrier's function and integrity, and modulate host's inflammatory responses (Xiong et al. 2022) as illustrated in Fig. 15.2. In addition, maintaining a supply of SCFAs to colonocytes leads to the production of adenosine triphosphate (ATP) via mitochondrial β-oxidation of butyrate derived from bacterial metabolism. The utilization of oxygen in colonocytes during this process ensures the maintenance of an anaerobic milieu in the lumen (Litvak et al. 2018). A study shows that depletion of butyrate-producing microbiota results in pathogen expansion due to the shift toward anaerobic glycolysis by colonocytes leading to increased oxygen diffusion into the lumen and elevated epithelial oxygenation. This elevation, resulting from butyrate deficiency, facilitates the aerobic growth of pathogens such as *Salmonella* (Rivera-Chávez et al. 2016). Therefore, SCFA production is crucial in modulating susceptibility to bacterial infections in the gut.

15.2.2.1 SCFAs Modulate Host Immune Response

SCFAs suppress inflammation through various mechanisms. For example, acetate and butyrate inhibit nuclear factor-κB (NF-κB) by diminishing histone deacetylase (HDAC) activity, thereby restricting access of NF-κB to inflammatory gene promoter sites (Caetano-Silva et al. 2023). Butyrate directly modulates the immune responses of neutrophils through a reduced secretion of pro-inflammatory cytokines such as IFN-γ, IL-6, and TNF-α (Li et al. 2021). SCFAs balance the inflammatory response by promoting the differentiation of T cells into Treg cells. Studies show that mice fed with butyrate and propionate show an increase in extrathymic $Foxp3^+$ Treg cells (Arpaia et al. 2013). Another study reported that binding of SCFAs to G-protein-coupled receptor (GPCR)-43, expressed on colonic Treg cells, restores Treg levels and confers protection against colitis in murine models (Smith et al.

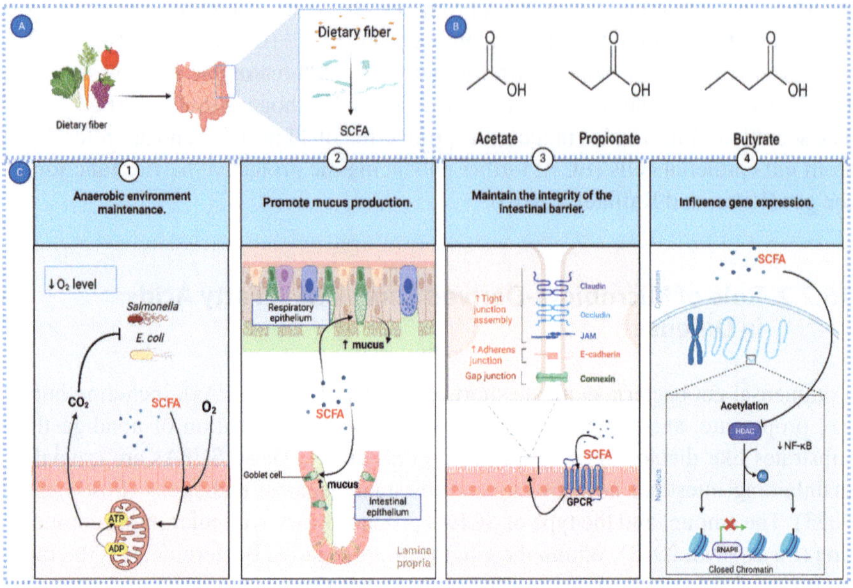

Fig. 15.2 Microbiome-derived short-chain fatty acid mediates the host's homeostasis. (**a**) Gut microbiota ferments the dietary fibers and produces SCFAs through anaerobic fermentation. (**b**) Chemical structure of the most common types of SCFAs. (**c**) (Salvadori and Rosso 2024; Zhang et al. 2021; Imchen et al. 2023; Campbell et al. 2023) Role of SCFAs in regulation of homeostasis (Salvadori and Rosso 2024). Butyrate is an energy source for colonic epithelial cells, and its β-oxidation produces CO_2 and consumes oxygen, which maintains the anaerobic environment within the gut and protects against aerobic pathogens (Zhang et al. 2021). SCFAs stimulate the production of mucus within many body sites, including the intestine and lungs (Imchen et al. 2023). SCFAs, through binding to G-protein-coupled receptors (GPCR), elevate the levels of tight junction expression, which protect the integrity of the intestinal barrier (Campbell et al. 2023). Intracellularly, SCFAs inhibit histone deacetylase (HDAC), which leads to inflammation inhibition and immune response modulation

2013). Additionally, SCFAs increase the acetylation of histones H3 at the Foxp3 gene locus, consequently promoting Treg differentiation (Furusawa et al. 2013). Another study reported that acetate-producing *B. longum* prevents death in *Escherichia coli* O157+ mice and was associated with lower serum concentration of Shiga toxin (Stx). Studying the metabolic profiles of mice revealed that higher acetate levels in the feces are linked to *E. coli* resistance, increased pathogen apoptosis, and decreased mucosal inflammation (Fukuda et al. 2011). Furthermore, the production of acetate induced the gene expression of C3, Pla2g2a, and Apoe, and decreased permeability of the epithelial monolayer (Fukuda et al. 2012).

Microbiota-derived acetate activates neutrophils and facilitates their migration to the site of inflammation via binding to GPCR43, known as free fatty acid receptor 2 (FFAR2) (Vinolo et al. 2011; Le Poul et al. 2003). This recognition initiated by acetate-FFAR2 interaction activates the NOD-like receptor protein 3 (NLRP3) inflammasome, facilitating the release of IL-1β from neutrophils (Fachi et al. 2020).

Inflammasome is a multiprotein complex that detects danger signals and triggers caspase-1 activation, an enzyme responsible for the maturation of pro-IL-1β into its active form, IL-1β (Zheng et al. 2020). However, the exact mechanism through which FFAR2 activation by acetate influences inflammasome production remains inadequately studied, presenting a pertinent area for further investigation. Additionally, acetate upregulates IL-1 receptor expression on ILC3 cells, enhancing IL-22 secretion in response to neutrophil-derived IL-1β. IL-22 supports epithelial wound healing and prevents bacterial translocation following *Clostridium difficile* infection (CDI) (Fachi et al. 2020).

15.2.2.2 SCFAs Maintain Integrity of the Intestinal Barrier

Another crucial function for SCFAs in the gastrointestinal tract (GIT) involves the promotion of intestinal cell functions and the maintenance of barrier integrity. SCFAs contribute to the enhancement of TJ production and function within the gut epithelial layer (Pérez-Reytor et al. 2021; Saleri et al. 2022). Tong et al. reported that propionate increases the expression of TJ-associated proteins, including occludin, E-cadherin, and Zonula occludens-1 (ZO-1) when given to mice with dextran sulfate sodium (DSS)-induced colitis (Tong et al. 2016). Although the primary mechanism underlying SCFAs' role in enhancing the expression of these proteins remains unclear, some studies suggest that upon binding to GPCR, SCFAs activate AMP-activated protein kinase (AMPK). AMPK is involved in the regulation of TJ formation, and its activation results in the phosphorylation of TJ proteins and their related proteins (Zhu et al. 2018). For example, in a Caco-2 cell, AMPK activation by butyrate promotes the redistribution of occludin and ZO-1, thereby strengthening intestinal TJs (Peng et al. 2009). Another study reported that *B. fragilis* exerts prophylactic activity against CDI by inhibiting the attachment of *C. difficile* to colon cells, reducing the cellular damage induced by *C. difficile* infection through decreasing apoptosis, and preventing the loss of MUC-2 and ZO-1 and consequently increasing the host survival rate (Deng et al. 2018).

15.2.3 Role of the Microbiome in Colonization Resistance

The human body possesses inherent physical barriers that serve as an impenetrable layer against a variety of external invaders. The gastrointestinal barrier consists of three interconnected layers that create an efficient physical and immunological defense against external microbial colonization. These layers include the epithelial layer, the internal layer of immune cells, and the mucous layer. The microbiota plays a vital role in preventing colonization by occupying available niches, blockage of nutrient access, enhancing production of mucin, and enhancing the TJs and intestinal barrier integrity as illustrated in Fig. 15.3. In addition, commensal microbes interfere in quorum sensing (QS), modulate gene expression, and prevent biofilm formation in invading pathogens.

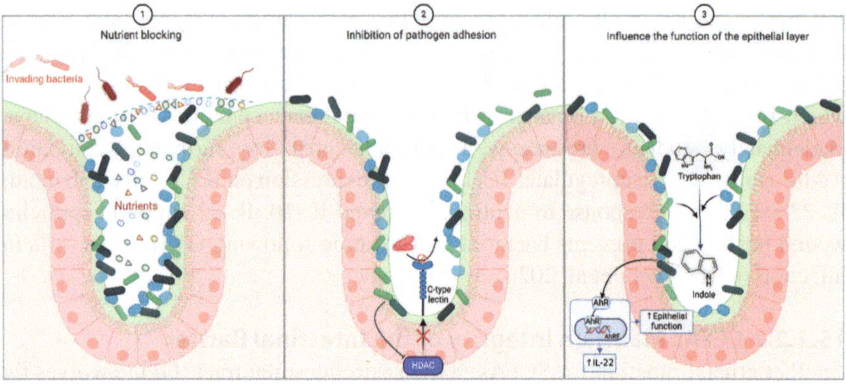

Fig. 15.3 Microbiome-mediated mechanisms of colonization resistance. The diverse microbial community of the gut resists pathogen colonization through a variety of mechanisms (Salvadori and Rosso 2024). Microbiome consumes the same nutrients needed by some pathogens, thus blocking the access of the pathogens to these nutrients (Zhang et al. 2021). By reducing histone acetylation and recruiting the histone deacetylase 3 (HDAC3), the microbiome suppresses the expression of cell-surface C-type lectin 2e (Clec2e), thereby inhibiting pathogen adhesion to intestinal epithelial cells (Imchen et al. 2023). Metabolites of tryptophan metabolism by microbiota activate aryl hydrocarbon receptor (AhR) within the cytosol. AhR regulates the transcription of genes responsible for enhancing epithelial barrier function and anti-inflammatory cytokine release

15.2.3.1 Colonization Resistance by Promoting Production of Mucin

To establish gut colonization, the pathogen must adhere to mucosal surfaces. However, a thick mucous layer covering the intestinal epithelium provides a primary line of defense hindering such attachment. The formation and maintenance of this mucous layer are influenced by the microbiota. This was proven by the distinct mucous properties, the thinner or even absent mucous layer in GF mice compared to conventionally raised rats (Johansson et al. 2015), and the restoration of the functional inner mucous layer after colonizing GF mice (Jakobsson et al. 2015). Microbiome, such as *Bifidobacterium* species, inhabits the outer mucous layer and utilizes MUC2 glycans as an energy source, maintaining the integrity of the inner mucous layer while enhancing mucous production by goblet cells. However, certain strains of *Bifidobacterium* can also degrade protective glycans present on mucin proteins (Engevik et al. 2019). Importantly, the equilibrium between bacterial degradation and mucous renewal is kept tightly controlled. Moreover, microbiome metabolites can also aid in maintaining the intestinal barrier integrity. For example, indole, a bacterial byproduct of tryptophan metabolism, enhances the barrier functions of ECs (Bansal et al. 2010; Shimada et al. 2013). The mechanism of action involves inducing the expression of genes critical to the functions of intestinal ECs and enhancing immune cell function through IL-22.

The outer mucin layer in the distal part of the intestine harbors bacteria possessing unique mucin-degrading abilities, such as *Akkermansia muciniphila* which represent 3–5% of the microbial community within host's gut (Cheng and Xie 2021). *A. muciniphila* utilizes mucin as a substrate, generating SCFAs which function as an energy source for goblet cells, facilitating increased mucous production (Kim et al. 2021a). Of note, *A. muciniphila* abundance within the intestine is negatively associated with inflammatory disorders like IBD (Earley et al. 2019). In line with this, treatment with *A. muciniphila* protects against gut damage initiated by methotrexate and radiation (Kim et al. 2021b).

15.2.3.2 Colonization Resistance by Nutrient Blockage

The gut microbiota has an essential role in conferring colonization resistance against enteric bacteria through nutrient competition (Spragge et al. 2023; Oliveira et al. 2020). Analysis of genomic and metabolic profiles of the most protective symbiont communities revealed significant similarities in protein composition and metabolic pathways with the pathogen, which therefore indicates similar nutrient demands. This suggests that the high diversity of the gut microbiome provides protection via consuming nutrients that pathogen needs (Spragge et al. 2023). Oliveira et al. reported that *Klebsiella michiganensis* resist the colonization of *E. coli* and *Salmonella enterica* by outcompeting for nutrients, leading to a delay in host mortality (Oliveira et al. 2020).

15.2.3.3 Colonization Resistance by Interfering in Virulence Gene Expression

Microbiota can also resist pathogen colonization through production of molecules that interfere in QS process in the pathogen which disables communications and impacts virulence gene expression such as those related to invasion and biofilm formation (Pena et al. 2019). For example, *Ruminococcus obeum* produces AI-2 autoinducer to restrict colonization of *Vibrio cholera* in mice. AI-2 molecules interfere with the expression of the *V. cholera* toxin-coregulated pilus operon, which is used by this pathogen to effectively colonize the intestinal tract (Hsiao et al. 2014). Another example is production of a lantibiotic peptide known as subtilosin by *Bacillus subtilis* that inhibits the QS signal production and biofilm formation in *Gardnerella vaginalis* (Algburi et al. 2017), an anaerobic bacterium linked to bacterial vaginosis (Morrill et al. 2020).

Staphylococci utilize the accessory gene regulator (*agr*) system as the main communication mechanism (Williams et al. 2019). Several commensal Staphylococci such as *Staphylococcus hominis* and *S. simulans* (Brown et al. 2020; Severn et al. 2022a) produce autoinducing peptides (AIPs) that repress pathogens' *agr* activity. This suppression inhibits the expression of phenol-soluble modulin (PSM)-α, a virulence factor produced by *S. aureus* that stimulates inflammation and epidermal barrier disruption (Nakagawa et al. 2017). A commensal isolate of S. *simulans* exhibited inhibitory activity on methicillin-resistant *S. aureus* (MRSA). Furthermore, this isolate protected the outermost layer of the skin from damage caused by the same pathogen (Brown et al. 2020). Similarly, several AIPs including those isolated from *S. warneri* have potential as therapeutic agents in atopic dermatitis (AD) (Severn et al. 2022b).

Another study reported that production of 3,4-dimethylbenzoic acid by commensals lowers the invasion of *S. enterica* serovar Typhimurium to HeLa cells by 82%.

This reduction was attributed to the suppression of the expression of the *hilA* gene affecting bacterial invasion (Peixoto et al. 2017). Likewise, De Sablet et al. reported that *B. thetaiotaomicron* releases molecules that can repress the expression of Stx2 which is the primary virulence factor of *E. coli* O157:H7 (De Sablet et al. 2009). Enteric bacteria such as *Flavonifractor plautii*, *C. orbiscindens*, *Lactiplantibacillus pentosus,* and *C. sporogenes* can produce anti-inflammatory and immune modulator molecules called desaminotyrosine (DAT) (Wang et al. 2023; Wei et al. 2020). DAT, also known as 4-hydroxyphenylpropionic acid, is a bacterial metabolite resulting from flavonoid metabolism. Steed et al. reported that DAT can provide protection against SARS-CoV-2 infection and influenza through enhancing the IFN-I signaling pathway to suppress virus replication (Steed et al. 2017).

Several studies reported the ability of microbiota to inhibit biofilm formation in pathogens which is a major virulence mechanism that pathogens employ to resist host immunity and antibiotic treatment (Peng et al. 2022). Microbiota members such as *S. epidermidis* suppress biofilm formation in *S. aureus.* Mechanistically, transcriptomic analysis revealed that cell-free conditioned media of *S. epidermidis* isolates could modulate *S. aureus* genes, such as *icaR* and *rsp*, that are involved in biofilm formation. These antibiofilm molecules are hydrophobic with a molecular weight range of 3–10 kDA, and tolerant to heat, sodium periodate, trypsin, proteinase K, and protease inhibitors. However, further investigations are required to define the chemical nature of these molecules and their mechanism of action (Glatthardt et al. 2020).

15.2.3.4 Colonization Resistance by Production of Bile Acids

Secondary bile acids (SBAs) are bacterial metabolites that contribute to colonization resistance. Several gut commensals, including Firmicutes (e.g., commensal *Clostridia* carrying the bile acid-inducible (*bai*) operon), are known producers of SBAs in intestinal tract and provide resistance to *C. difficile* (Reed et al. 2020; Theriot et al. 2016; Thanissery et al. 2017). *Bai* operon is conserved in some bacterial species, including human gut microbiome, and it encodes enzymes responsible for the transformation of liver-produced bile acid into SBA in a process named bile acid 7α-dehydroxylation (Kim et al. 2022). The production of the SBA lithocholate and its epimer isolithocholate inhibits the growth of *C. difficile* without affecting commensal gut microbes. This inhibition activity is due to the ability of SBA and its epimers to damage bacterial membranes and interfere with the expression and activity of toxins released by *C. difficile* (Kisthardt et al. 2023). *Additionally, SBAs can mediate the inflammatory state within GIT.* The activity of macrophages is partially triggered by the interaction between SBAs and the Takeda G-protein-coupled bile acid receptor-1 (TGR5) that is found on macrophage surface (Zeng et al. 2019a). This activation influences various physiological and immunological processes. Of note is that the activation of TGR5, by SBAs, can also elevate IL-10 and inhibit pro-inflammatory cytokines like IL-6 and TNF-α. This is mediated by the partial shift toward the anti-inflammatory M2-like phenotype while inhibiting the pro-inflammatory M1-like phenotype following TGR5 activation (Jia et al. 2018).

15.2.4 Production of Antimicrobial Molecules by the Human Microbiota

One of the strategies that microbiota employs to fight pathogens is production of antimicrobial molecules. Although the number of fully elucidated metabolites is limited, due to technical limitation of culturing these microbes, genome mining reveals a richness of potential encoded metabolites (Santos-Júnior et al. 2023; Ma et al. 2022). These molecules are gaining increasing interest especially with the escalation of antibiotic resistance and the need to find new antibiotics. One interesting class of antimicrobial that is widely produced by microbiota is AMPs which function by several mechanisms including pore formation in the bacterial membrane, double DNA breakage, and transcription inhibition or ATP synthase (Lei et al. 2019). Examples of elucidated AMPs are shown in Table 15.1.

Lactic acid bacteria are prolific producers of AMPs such as gassericin and plantaricins (Gallardo-Becerra et al. 2024). Another class of AMPs is microcins that outcompete pathogenic strains such as *Klebsiella, E. coli, Salmonella, Yersinia,* and *Shigella* especially upon exposure to stress conditions such as depletion in nutrients (Marković et al. 2022). A common type of AMPs is lantipeptides, which are produced by various bacterial species and disrupt the cell wall by pore formation or direct inhibition of its synthesis (Chattopadhyay et al. 2023). *R. gnavus* produce ruminococcin A, a lantibiotic that displays potent antibacterial properties against *C. perfringens* associated with food poisoning (Popoff 2014). Another lantipeptide that inhibits C. *perfringens* is nisin O produced by *Blautia obeum* (Hatziioanou et al. 2017). The oral mucosa is also considered a vital source for a common group of lantibiotics, namely, salivaricins. This group of AMP is produced by *Streptococcus salivarius and* contributes to the natural defense mechanisms in the oral cavity by inhibiting the colonization of oral pathogens such as *S. pyogenes* (Vertillo Aluisio et al. 2022). Of note, salivaricin has been purified from *L. salivarius* isolated from gut origin as well (Wayah and Philip 2018). Another class of AMPs is nonribosomal peptides (NRPs). *S. lugdunensis* IVK28, a common beneficial microbe found within the nasal cavity, produces one type of NRP called lugdunin. This AMP is a cyclic peptide containing thiazolidine, exhibiting inhibitory effects on the colonization of glycopeptide intermediate-resistant *S. aureus*, MRSA, and isolates of vancomycin-resistant *Enterococcus.* Interestingly, even following prolonged serial passaging of *S. aureus* in the presence of subinhibitory levels of lugdunin for 30 days, no resistance to lugdunin was observed (Zipperer et al. 2016). Trending research is the prediction of AMP through genome sequences using machine learning tools followed by their total synthesis and biological evaluation. Following these protocols, hundreds of new AMP have been developed and proved to be active against antibiotic-resistant pathogens as validated by in vitro and in vivo models. For instance, bioinformatics predictions have revealed new NRPs called humimycins that show activity against *Staphylococcus* and *Streptococcus* species by preventing the movement of peptidoglycan precursors across the pathogen membrane (Chu et al. 2016).

In addition to secretion of antimicrobial molecules, microbiota can prime the expression of host-secreted antimicrobials. For example, a study reported that

Table 15.1 Examples of antimicrobials produced by microbiota at different body sites along with their target pathogen and mechanism of action

Antimicrobial	Type	Body site	Producer	Target pathogen	Mechanism of action	Ref.
Gassericin A	Bacteriocin *class IIc*	GI	*L. gasseri* LA39	*L. monocytogenes, B. cereus*, and *S. aureus*	Incorporation of hydrophobic alpha helices into the cell membrane leading to pore development and death of the cell.	Selle and Klaenhammer (2013) and Pandey et al. (2013)
Lactomodulin	Bacteriocin	GI	*L. rhamnosus*	*E. coli, L. monocytogenes*, MRSA, VRE, B. *cereus*, and *C. difficile*	Unknown	Mousa et al. (2023)
Nisin O	Lantibiotic class Ia	GI	*B. obeum* strain A2-162	*C. perfringens*	Cell wall synthesis inhibition through disruption of lipid II and membrane pore formation.	Hatziioanou et al. (2017)
Thuricin CD	Class V bacteriocin (Sactibiotics)	GI	*B. thuringiensis* DPC 6431	*L. monocytogenes, C. tyrobutyricum, C. difficile, C. lithuseburense, C. perfringens, B. firmus,* and *C. indolis*	Bacterial membrane depolarization leads to pore formation.	Rea et al. (2010) and Mathur et al. (2017)
Humimycins A and B	Nonribosomal peptide	GI	*Rhodococcus equi* and *R. erythropolis*	*Various Streptococcus and Staphylococcus* species (including MRSA)	Inhibit the activity of lipid II flippase and promote β-lactam activity	Chu et al. (2016)
Ruminococcin A	Lantibiotics class IIA	GI	*R. gnavus*	*C. perfringens*	Unknown	Dabard et al. (2001)
Enterocin-7	Bacteriocin class III	GI	*E. faecium*	*L. monocytogenes*	Increases membrane permeabilization	Vasilchenko et al. (2018)
Salivaricins (e.g., A,B,D,9,10)	Lantibiotics	Oral GI	*S. salivarius*, and *L. salivarius*	*S. pyogenes* and *Micrococcus luteus*	Cell wall synthesis inhibition through disruption of lipid II and membrane pore formation.	Barbour et al. (2020)

Plantaricin NC8 α and β	Bacteriocin class IIb	GI Oral	*L. plantarum*	*S. epidermidis, Porphyromonas gingivalis*	Weaken the stability of the bacterial cell wall and membrane.	Bengtsson et al. (2020)
Mutacin 1140	Lantibiotic class Ia	Oral	*S. mutans*	Broad-spectrum gram-positive bacteria.	Cell wall synthesis inhibition through disruption of lipid II and membrane pore formation.	Pokhrel et al. (2019)
Iturin A	Cyclic antifungal lipopeptide	Oral	*Bacillus* sp.	*C. albicans*	Inhibition of filamentation and biofilm formation and cell membrane disruption.	Han et al. (2024)
Mutanobactins and mutanamide	Nonribosomal peptide/ polyketide hybrids	Oral	*S. mutans*	*C. albicans*	Inhibition of filamentation and biofilm formation	Mousa et al. (2017)
Lugdunin	Nonribosomal peptides	Nasal	*S. lugdunensis*	MRSA, GISA, and VRE isolates	Dispersion of cell membrane potential through forming hydrophilic nanotubes in the bacterial lipid membrane.	Zipperer et al. (2016) and Ruppelt et al. (2024)
Epifadin	Nonribosomal peptide	Nasal	*S. epidermidis IVK83*	*S. aureus, M. luteus* and *S. pyogenes*	Cell membrane disruption.	Torres Salazar et al. (2024)
Hh1-BLS	Bacteriocin-like substance	Respiratory	*Haemophilus haemolyticus*	*Haemophilus influenzae*	Unknown	Latham et al. (2017)
PSM	Amphipathic, α-helical peptides	Skin	*S. epidermidis*	*C. acnes,* and group A *Streptococcus*	Formation of membrane pores.	O'Neill et al. (2020) and Cogen et al. (2010)
Micrococcin P1	Thiopeptide	Skin	*S. hominis* S34-1	*S. aureus* infection	Inhibition of protein synthesis.	Liu et al. (2022)

(continued)

Table 15.1 (continued)

Antimicrobial	Type	Body site	Producer	Target pathogen	Mechanism of action	Ref.
Hominicin	Lantibiotics class Ia	Skin	*S. hominis*	Multiple strains of *S. aureus including MRSA*	Unknown	Kim et al. (2010)
Sh-lantibiotics	Lantibiotics	Skin	*S. epidermidis* and *S. hominis*	*S. aureus*	Cell membrane disruption.	Nakatsuji et al. (2017)
Capidermicin	Bacteriocin class II	Skin	*S. capitis* CIT060	Gram-positive (e.g., *S. aureus, M. luteus,* and *S. intermedius*)	Increases membrane permeabilization	Lynch et al. (2019)
Nisin J	Lantibiotic class Ia	Skin	*S. capitis APC 2923*	Gram-positive (e.g., MRSA, *L. monocytogenes, and Cutibacterium acnes*)	Membrane pore formation	O'Sullivan et al. (2020)
Gassericin E	Class IIb two-peptide bacteriocins	Vagina	*L. gasseri* EV1461	Pathogens associated with vaginosis (e.g., *A. vaginae*, *G. vaginalis*, *P. gingivalis* and *P. bivia*)	Unknown	Maldonado-Barragán et al. (2016)

RNR ribonucleotide reductase, *GISA* glycopeptide intermediate-resistant *S. aureus*, *VRE* vancomycin-resistant *Enterococcus*, *PSM* phenol-soluble modulin β-defensin production depends on the binding of microbial-derived SCFAs to GPR43, which activate mammalian target of rapamycin (mTOR) and signal transducers and activator of transcription 3 (STAT3) pathways leading to higher antimicrobial expression and efficient defense (Zhao et al. 2018). Besides, microbiota activates specific receptors such as ILC3 inducing the secretion of IL-22 and IL-17, which prompt AMP production (Zeng et al. 2019b).

15.3 Unique Activities of Microbiome in Defense at Defendant Body Niches

Each body site is a unique ecosystem that harbors distinctive consortia of microbiota. Many pieces of evidence suggest that these microbiomes confer specific antimicrobial activity against body site-specific pathogens. Here, we will discuss some examples of the interactions between commensal microbiota and invading pathogens at each body site.

15.3.1 Role of Skin Microbiota in Defense

Skin microbiome is a complex community that shows variation in composition between different skin niches. For example, *Streptococcus* species are enriched in dry sites, *Cutibacterium* and *Staphylococcus* mainly colonize oily sites, while moist areas are dominant with *Corynebacterium* and ß-Proteobacteria (Lee and Kim 2022; Byrd et al. 2018). Significant changes to the microbiome structure of the skin occur during puberty, which is marked by an increased prevalence of lipophilic microbes including *Corynebacterium and C. acnes* (Townsend and Kalan 2023), then stabilized by adulthood (Oh et al. 2016).

Skin microbiota directly inhibits pathogens' entry through physical hindrance, nutrient blockage, secretion of antimicrobial compounds, and interference with pathogen virulence factors (Bier and Schittek 2021) as illustrated in Fig. 15.4. Skin microbiota contributes to skin integrity through signaling on aryl hydrocarbon receptor (AHR) on the keratinocytes, which promotes skin barrier repair and self-renewal (Uberoi et al. 2021). A study reported that GF mice are more susceptible to skin damage and show delays in barrier recovery (Uberoi et al. 2021). Introducing human skin microbiome to GF mice restores AHR activity in keratinocytes and elevates the expression of differentiation genes (Uberoi et al. 2021).

In addition, skin commensals such as *C. acnes* and *Corynebacterium* sp. secrete lipase enzyme, which maintain the normal levels of free fatty acids (FFAs) in skin by hydrolyzing FFAs from sebum's triglyceride (Flowers and Grice 2020). This action maintains an acidic surface that restricts the colonization of bacteria. Moreover, skin microbiota enhances the skin's innate immunity by increasing the production of host-derived molecules. For example, FFAs released by microbiomes' lipase promote host innate immunity by enhancing the expression of β-defensin (hBD)-2 in human sebocytes with antimicrobial activity against *P. acnes* (Nakatsuji

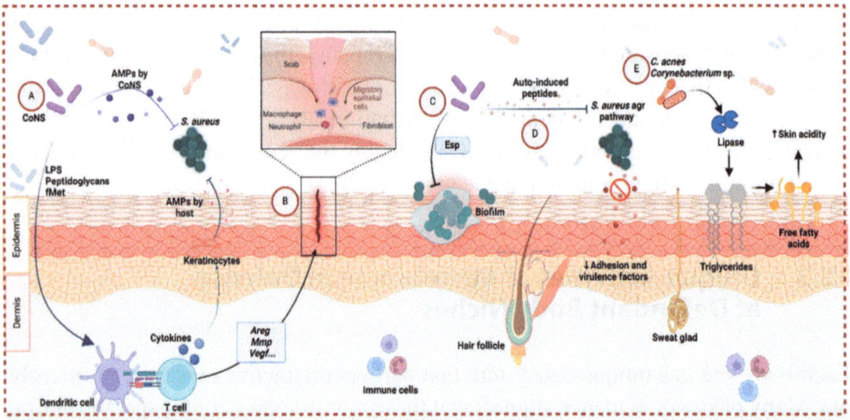

Fig. 15.4 Role of skin microbiome in protection. (**a**) Besides production of its own AMP, coagulase-negative Staphylococcus (CoNS) stimulates the production of host-related antimicrobials through activating T cells to produce several cytokines that induce the production of AMPs by keratinocytes. (**b**) Skin microbiome activates commensal-specific T cells that overexpress many genes related to wound healing and extracellular matrix production such as Areg, Mmp, and Vegf and thus accelerate wound closure. (**c**) Through secreting serine protease Esp, *S. epidermidis* restrains the development of biofilm by degrading many S. aureus-related proteins aided in biofilm construction. (**d**) CoNS produces autoinduced peptide that interferes with the virulence factors released by S. aureus, thus preventing infection and inflammation related to *S. aureus*. (**e**) CoNS secretes lipase enzymes that hydrolyze skin triglycerides, generating free fatty acids that maintain skin surface acidity

et al. 2010). Besides, hBD-2 protects keratinocyte barrier function from pathogen-derived proteases such as the V8, an extracellular protease secreted by *S. aureus* (Shelley et al. 2023).

One of the major microbial taxa among the skin is coagulase-negative *Staphylococcus* (CoNS) (Bier and Schittek 2021). Representative members of this group include *S. epidermidis*, *S. capitis*, *S. simulans*, *S. caprae*, *S. hominis*, *S. haemolyticus*, and *S. lugdunensis* (Flowers and Grice 2020). While traditionally considered opportunistic pathogens, current investigations reveal that particular strains of CoNS may play a protective role in the skin microbiome especially against *S. aureus* and *C. acnes* (Bier and Schittek 2021). *S. epidermidis* is among the most commonly bacterial species found in the human skin and showed ability to induce the expression of host-derived antimicrobials such as hBD-2 and hBD-3 in keratinocytes, thereby inhibiting the growth of *S. aureus* (Lai et al. 2010). Furthermore, *S. epidermidis* activates gamma delta T cells (γδ T) through upregulating the expression of perforin-2 (P-2) (Pastar et al. 2020), an AMP secreted by human immune cells (McCormack et al. 2015). *S. epidermidis* influences skin structure and lowers water loss through the skin by secreting a sphingomyelinase that helps in the production of ceramides, a key lipid component in the skin that regulates skin barrier function and permeability (Zheng et al. 2022). Of note is that *S. epidermidis* utilizes the components degraded by the sphingomyelinase enzyme as nutrients to enhance its colonization in the skin (Zheng et al. 2022), thus conferring a mutual benefit. Moreover, *S. epidermidis* and *S. hominis* also produce antimicrobial compounds such as Sh-lantibiotics and hominicin, which inhibit *S. aureus colonization* (Kim et al. 2010; Nakatsuji et al. 2017). *S. epidermidis* inhibits biofilm formation by *S. aureus* through secretion of serine protease Esp (Iwase et al. 2010). However, cysteine protease EcpA secreted by overcolonized *S. epidermidis* induces skin damage and inflammation among some patients with AD, and the expression of EcpA mRNA was directly linked to AD severity (Cau et al. 2021). Some studies reported that *S. epidermidis* enhances wound healing by promoting the accumulation of CD8+ T lymphocytes in the skin and inhibits the formation of cutaneous tumors through the secretion of 6-N-hydroxyaminopurine (Linehan et al. 2018; Nakatsuji et al. 2018). *Other members of CoNS such as S. capitis* secrete different PSM-β peptides that exhibit efficacy against various *C. acnes* strains including those isolated from both lesional and nonlesional skin of individuals with acne while being nonlethal to normal skin commensal (O'Neill et al. 2020).

15.3.2 Role of Oral Microbiota in Defense

The oral cavity hosts the largest number of organisms after the gut (Deo and Deshmukh 2019). Until now, 774 bacterial species have been identified within the human oral microbiome, according to the Human Oral Microbiome Database (Escapa et al. 2018). These species belong to more than 12 phyla including *Proteobacteria, Fusobacteria, Bacteroidetes, Firmicutes, Chlamydiae, Saccharibacteria* (TM7), *Actinobacteria, Spirochaetes, SR1, Synergistetes, Chloroflexi,* and *Gracilibacteria* (GN02) (Escapa et al. 2018). The abundance of bacteria within the mouth is site-specific, and these oral species assemble into biofilm communities on surfaces of the mouth (McLean et al. 2022; Lamont et al. 2018). The connectivity of the buccal cavity to the external environment and the provision of favorable conditions for bacterial survival, including pH and temperature, enhance the composition of the microbiome in the mouth. Many host-related factors such as smoking, oral hygiene, plaque levels, diet, and tooth loss can change the mucosal colonization structure (Li et al. 2022). Dysbiosis in the mouth is linked to oral, systemic, and autoimmune diseases such as Alzheimer's disease, diabetes, and IBD (Gao et al. 2018).

The main infectious disease in the mouth is dental caries, which is caused by tooth-adherent bacteria such as *S. mutans*, lactobacilli, *Actinomyces* spp., and fungi such as *C. albicans* (Baker et al. 2021; Mira et al. 2017; Menon et al. 2022). These microorganisms consume sugars and refined carbohydrates, thus producing acids that can erode the enamel (Featherstone 2008). Some oral commensal species such as *S. gordonii, S. sanguinis, S. mitis,* and *S. parasanguis* encodes arginine deiminase system (ADS), a pathway that metabolizes arginine to release ammonia, ornithine, and CO_2 (Liu et al. 2012) as shown in Fig. 15.5. This aids in maintaining pH homeostasis within the mouth, consequently protecting cells from harsh acidification and reducing the progression of dental caries and oral cancers (Nascimento et al. 2013).

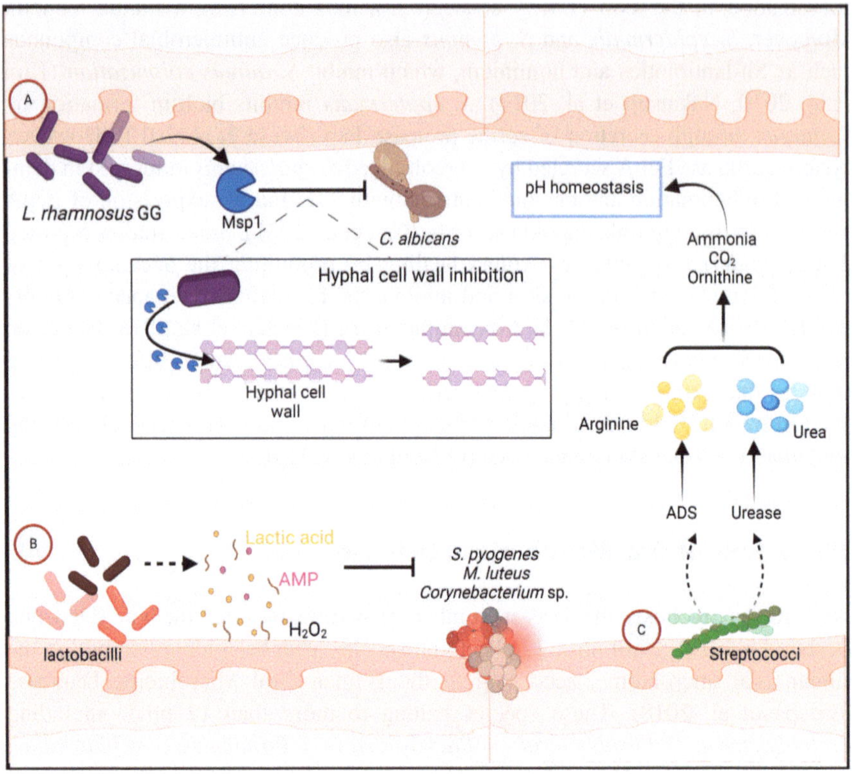

Fig. 15.5 Mechanisms of the oral microbiome in maintaining oral health and infection prevention. (**a**) *L. rhamnosus* GG protects against oral candida infection through the production of peptidoglycan hydrolase Msp1, which breakdowns chitin, a polymer in the hyphal cell wall. (**b**) Beneficial oral species such as *S. salivarius* are a source of various antimicrobial peptides, lactic acid, and H_2O_2 which inhibits the colonization of pathogens, such as *M. luteus* and *Corynebacterium* sp., within the oral cavity. (**c**) Oral Streptococci secrete urease and arginine deiminase system (ADS) that degrade urea and arginine, respectively. The degradation products such as ammonia and CO_2 maintain the optimal pH within the oral cavity, thus preventing caries

Individuals with active caries exhibit elevated levels of free arginine in their saliva and dental plaque compared to those without caries (Nascimento et al. 2013; Silveira et al. 2023), which is a consequence of lower ADS-positive bacteria in the mouth.

Likewise, salivary glands secrete urea in concentrations ranging from 3 to 10 mM in healthy individuals (Mora and Arioli 2014). Urea is hydrolyzed by bacterial urease to produce ammonia and CO_2. Many ureolytic organisms are present in the mouth such as *S. salivarius* and *A. naeslundii* (Dahlén et al. 2018). A *streptococcus* strain isolated from a dental plaque was shown to produce significant amounts of hydrogen peroxide (H_2O_2) suppressing the growth of *S. mutans* and disrupting its signaling pathways through deactivation of bacteriocin-inducing peptide signal (Huang et al. 2016). Oral commensals exhibit other defense mechanisms against oral bacterial invaders. For example, *S. mitis* directly inhibits the adhesion of *P.*

gingivalis to the mouth surfaces (Van Hoogmoed et al. 2008). *S. cristatus* represses FimA production and inhibits the biofilm formation of *P. gingivalis* (Wu and Xie 2010). *S. salivarius,* a key commensal organism found on the oral mucosal surfaces of healthy individuals, inhibits *S. pyogenes* and other pathogens including *M. luteus* and *Corynebacterium* sp. through the production of lantibiotics such as salivaricin B (Mousa et al. 2017).

Another crucial role of oral commensals in defense is protection against Candida infections such as *C. albicans*. Oral candidiasis (OC) is a common opportunistic infection affecting oral mucosa (Vila et al. 2020; Patil et al. 2015). Secondary to oral dysbiosis, *C. albicans* can become opportunistic and invade tissues causing mucosal infection (Vila et al. 2020; Allonsius et al. 2019). Oral commensals such as *Lactobacillus* species prevent OC by several mechanisms including the production of organic acids (such as lactic acid) and metabolites (such as bacteriocins, chitinases, and fatty acids) (Allonsius et al. 2019; Matsubara et al. 2016; Zangl et al. 2020). Peptidoglycan hydrolase Msp1 was identified in *L. rhamnosus* GG. Msp1 can degrade the main polymer in the hyphal cell wall of *C. albicans* (Allonsius et al. 2019). While *S. mutans* secretes mutanobactins and mutanamide, NRP/polyketide hybrids inhibit *C. albicans* hyphal transition, thus preventing oral thrush caused by *C. albicans* (Mousa et al. 2017). Some microbial interactions at the mouth are illustrated in Fig. 15.5.

15.3.3 Role of Vaginal Microbiota in Defense

Vaginal microbiota contributes to a woman's reproductive and general health as demonstrated in Fig. 15.6. A well-balanced vaginal microbiota usually exhibits low diversity, with a predominant presence of lactic acid producers *Lactobacillus* species such as *L. crispatus* or *L. iners* (Baud et al. 2023; Dols et al. 2016). Several conditions including pregnancy, antibiotics use, sexual activity, and hormonal change disrupt the balance in vaginal microbiome abundance and composition (Younes et al. 2018). Vaginal alteration is characterized by a reduction in lactobacillus presence, higher microbial biodiversity, and higher presence of pathogenic bacteria in the vagina, leading to vaginal diseases and infections such as bacterial vaginosis (BV) (Machado et al. 2022). *Lactobacillus* generates lactic acid, which contributes to the acidic pH in the vagina, as a metabolic byproduct of carbohydrate metabolism, specifically derived from the human alpha-amylase breakdown of glycogen present in vaginal fluids (Tester and Al-Ghazzewi 2018). Some mechanisms employed by vaginal microbiota in defense are illustrated in Fig. 15.6.

Besides lactic acid, *lactobacilli* also produce H_2O_2 and bacteriocins that exert important bactericidal effects. Lactobacilli can form coaggregates, particularly around multidrug-resistant urogenital pathogens (Scillato et al. 2021; Pino et al. 2019), creating interlinked structures that serve as a barrier, diminishing the adhesion of pathogens to the vaginal mucous layer. This coaggregation not only restricts pathogen adhesion but also exposes them to elevated levels of inhibitory compounds or metabolites produced by the commensal (Zakaria 2013). For example, Lactobacilli

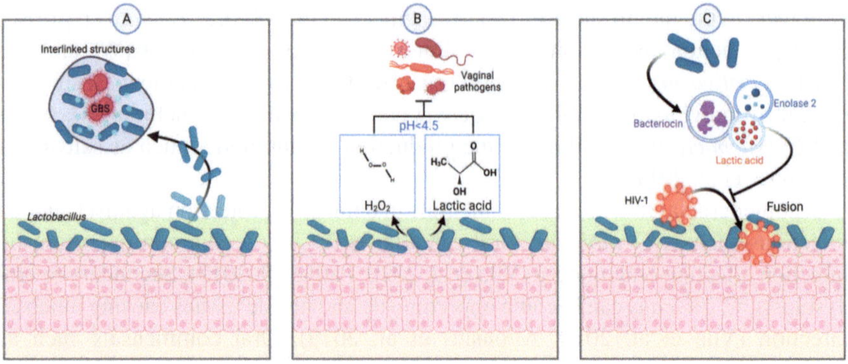

Fig. 15.6 *Lactobacillus* species are well-known contributors to vaginal health. (**a**) *Lactobacillus* prevents the vaginal colonization of pathogens by forming an interlinked structure that covers the pathogen and inhibits its adhesion to the vaginal surface. (**b**) *Lactobacillus* secretes several metabolites that protect the host against pathogens. While H_2O_2 has direct antimicrobial activity, lactic acid works by maintaining the low pH within the vagina, and the acidic environment prevents the growth of several harmful microorganisms. (**c**) Some Lactobacilli secrete membrane vesicles that contain several antimicrobials, metabolites, and enzymes such as enolase 2 that prevent the fusion of HIV-1 to the vaginal cells

offer protection against Group B streptococcus (GBS) colonization during pregnancy (Starc et al. 2022) mainly by production of lactic acid, H_2O_2, and coaggregation with the pathogen. GBS in pregnant women predisposes the mother and her infant to pregnancy-associated risks such as preterm birth, and invasive- and noninvasive maternal diseases (Shabayek and Spellerberg 2018). Reduced levels of GBS were observed in the vaginal microbiota of third-trimester pregnant women when lactobacilli predominated (Starc et al. 2022; Shabayek et al. 2022).

Interestingly, Lactobacilli use vehicles to deliver DNA, virulence factors, extracellular-matrix-degrading enzymes, and antimicrobial metabolites (Liu et al. 2018; Briaud and Carroll 2020). Bacterial vesicles, also known as membrane vesicles, are membrane-derived lipid bilayer structures, with an average diameter of 20–200 nm, that mediate the communications between bacteria. Recent research has emphasized the potential of bacterial vesicles as delivery vehicles, especially after findings showed that vesicles produced by probiotic bacteria can transport bacteriocin peptides to target opportunistic pathogens (Dean et al. 2020). These vesicles have been reported in various *Lactobacillus* species such as in *L. casei* (Domínguez Rubio et al. 2017), *L. reuteri* (Grande et al. 2017), *L. acidophilus* (Dean et al. 2020), *L. plantarum* (Kim et al. 2018), *L. gasseri BC12, and L. crispatus BC5* (Croatti et al. 2022). *L. acidophilus* secrete membrane vesicles loaded with bacteriocin to target and kill opportunistic pathogens such as *L. delbrueckii* (Dean et al. 2019, 2020). *L. crispatus* and *L. gasseri* exert their viral blockage activity against HIV via secretion of extracellular vesicles. These vesicles have nanosized diameters and exhibit concentration-dependent entry inhibition of HIV-1; this inhibition is mediated through the presence of lactic acid and various proteins, including enolase 2.

The latter diminishes the exposure of the viral envelope, thereby disrupting virus-cell interactions (Ñahui Palomino et al. 2019).

15.3.4 Role of Respiratory Microbiota in Defense

In healthy adults, the respiratory microbiota demonstrates a low density and high diversity, including members from taxa such as *Prevotella*, *Streptococcus*, *Veillonella*, and *Porphyromonas*. The upper respiratory tract, consisting of the nasal cavity, nasopharynx, and oropharynx, exhibits a higher microbial density with distinct microbiota in each anatomical section, in contrast to the lower respiratory tract (Natalini et al. 2023). Due to persistent exposure to exogenous pollutants, allergens, and microorganisms, the lungs and the respiratory system undergo continuous dynamic alterations and adaptations (Santacroce et al. 2020).

Prevotella is one of the core airway microbiota, and its enrichment has been associated with a state of respiratory health. Conversely, diminished abundance of this genus has been correlated with respiratory diseases such as pneumonia (De Steenhuijsen Piters et al. 2016). *Prevotella* has been shown to improve hosts' immunity and maintain the microbiome homeostasis within the airway through different mechanisms as illustrated in Fig. 15.7. For example, *P. melaninogenica,* an anaerobic bacterium prevalent in the oral cavity and airway, has demonstrated the capacity to promote early clearance of *S. pneumonia,* a predominant bacterial etiology of pneumonia. This is achieved via TLR2 signaling, resulting in increasing neutrophil

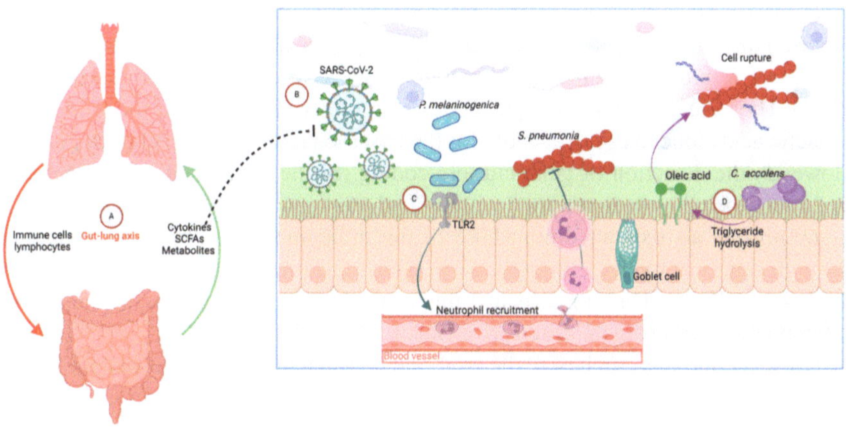

Fig. 15.7 Gut-lung crosstalk and role of respiratory microbiome in providing protection. (**a**) Gut and lung exchange microorganisms and immune components reciprocally. Lungs mediate the inflammatory response in gut disease through sharing inflammatory mediators and lymphocytes to the gut, (**b**) while the later provide cytokine and microbial metabolites that aid in protection against respiratory pathogens including SARS-CoV-2. (**c**) Detection of *P. melaninogenica* by TLR stimulates neutrophil recruitment from bloodstream that mediates the eradication of S. pneumonia. (**d**) Hydrolysis of cell-surface triglyceride by C. accolens-produced lipase (LipS1) produces oleic acid, a free fatty acid that depolarize S. pneumonia membrane resulting in cell rupture and death

recruitment to the lungs. Consequently, this pathway enhances the production of pro-inflammatory cytokines and augments serine protease-mediated clearance of *S. pneumonia* (Horn et al. 2022a). However, investigations using high-throughput 16S rRNA sequencing revealed that higher *Prevotella* abundance is linked to the severity COVID-19 infection (Hernández-Terán et al. 2021). This effect could be explained in line with Th17 induction ability of *Prevotella*, which increases the inflammation response.

In the upper airway, *Corynebacterium* and *Dolosigranulum* are known for mitigating pathogenic colonization, in particular against *S. aureus* (Brugger et al. 2020). Similar to their role in the skin, *C. accolens* hydrolyze triacylglycerols present in sebum and surfaces coating nostrils. This process, facilitated by the lipase LipS1, releases FFAs that inhibit pneumococcal growth. In vitro and in vivo studies alongside mass spectrometric analysis revealed that *C. accolens* releases free oleic acid from triolein (Bomar et al. 2016; Horn et al. 2022b). Oleic acid exerts inhibitory effects on *S. pneumonia* directly by inducing membrane depolarization leading to cell rupture, and indirectly by upregulating the expression of the AMP hBD-2 (Clementi et al. 2013; Lee et al. 2004). Moreover, Hardy et al. showed that *C. pseudodiphtheriticum* can disrupt *S. aureus* virulence through interfering with Agr QS system of the later by yet-to-be-identified released substances (Hardy et al. 2019). Aqr QS regulates the expression of several genes involved in virulence and colonization factors of *S. aureus* including adhesion agents and pore-forming toxins. Interestingly, *S. aureus* has developed resistance to the bactericidal activity of *C. pseudodiphtheriticum* through downregulation of Agr QS components (Hardy et al. 2019).

D. pigrum is another prevalent upper airway tract commensal. Several studies reported the role of *D. pigrum alone or in synergism with Corynebacterium* reduces growth of *S. pneumonia and S. aureus* (Brugger et al. 2020*;* Raya Tonetti et al. 2021). The exact mechanism of prevention is not fully elucidated, as the production of lactic acid alone does not account for the observed reduction in pathogen growth. However, the detection of gene clusters coding for the production of lantipeptide and bacteriocin has been reported within *D. pigrum* genomes (Brugger et al. 2020).

D. pigrum also shows diminishing of the inflammation induced by *S. aureus* through significantly decreasing the expression of pro-inflammatory cytokines such as IL-8 and IL-1β. This suggests that this commensal reduces the tissue-damaging inflammation caused by *S. aureus* (De Boeck et al. 2021).

Growing findings emphasize the crosstalk between gut microbiota and the respiratory system, particularly the lungs where the gut microbiome modulates the immunological activity of the lung. Mechanistically, LPS produced by gut commensals binds to TLR, and activates DC, which further activates various T cells. Besides, microbiome-produced SCFAs reduce cytokine production such as TNF-α and IFN-γ through downregulation of PRR. The contribution of the gut microbiome in protection from respiratory tract infections has been investigated in more depth following the COVID-19 pandemic (Santacroce et al. 2020). Fecal samples from COVID-19 patients displayed a reduction in beneficial microbiome species and an increase in opportunistic pathogens (Zuo et al. 2020). The gut microbiome has been found to have an anti-COVID-19 action, as it can elevate several interferon expression and downregulate both local and systemic inflammatory responses. This mechanism of defense is mediated by the gut-resident *B. dorei.* Interestingly, COVID-19 patients showed reduced abundance of *B. dorei,* which showed an inverse relationship with the SARS-CoV-2 viral load in fecal samples (Zuo et al. 2020). Moreover, influenza A virus infection initiates dysbiosis and reduces the levels of SCFAs-producing species in the gut. Therefore, the bactericidal activity of alveolar macrophages is depleted, resulting in higher chances of secondary bacterial superinfection with *S. pneumonia* (Sencio et al. 2020). These findings along with others demonstrated that microbiome-based therapeutics could be implemented against respiratory viral infections.

15.4 Conclusion and Future Directions

Much excitement exists in understanding the functional roles of the human microbiome and how they impact host biology. In this book chapter, we discussed the diverse strategies employed by the microbiome to control infections which include hindering colonization, blocking access to nutrients, suppression of virulence genes, production of antimicrobial compounds, and stimulation of efficient and balanced host immune response. We detailed examples of the distinct microbial communities at each body site and their role in combating site-specific pathogens. Translational research is crucially needed to transfer this fundamental knowledge into application to design and develop novel therapeutics. This fast-developing domain of research will reveal new opportunities to address challenging health threats such as infectious diseases and antibiotic resistance crisis. A key area of development involves creating innovative machine learning tools to mine multiomics data from the microbiome. These tools aim to identify antimicrobial molecules that could potentially be developed into commercial antibiotics. A recent study identified 243 AMPs from mining 2229 microbiome genomes followed by synthesis or heterologous expression in *E coli* for 70 peptides. The study revealed new antibiotics that selectively targets human pathogens at various body niches including against clinically resistant strains such as vancomycin-resistant *Enterococci* and MRSA (King et al. 2023). Total synthesis of predicted peptides is gaining momentum because it bypasses the challenges of culturing host-associated microbes under laboratory conditions. Previously, this strategy led to discovery of humimycins with activity against *S. aureus* (Chu et al. 2016). Humimycin impairs cell wall biosynthesis by targeting flippase MurJ. Further studies identified more bioactive analogues with increased potency (Chu et al. 2018). A recent study utilized a dataset of 87,920 genomes and 63,410 metagenomes to predict 863,498 unique AMPS from multiple microbial systems including the human microbiome, from which 50 AMPs have been fully synthesized and showed promising activity against human pathogens (Santos-Júnior et al. 2023). Another study employed functional metagenomic strategy to identify antibacterial compounds from the gut microbiota where the researchers constructed a metagenomic clone library with DNA from gut microbiome and screened against pathogenic *E. coli* leading to the discovery of a narrow-spectrum active bacteriocin (Cohen et al. 2018). In this context, several databases have been curated and developed for prediction of AMPs such as iAMPpred, AntiBP, and CAMP as reviewed previously (Wang et al. 2022). Another research employed deep learning strategies to predict AMPs specifically from the gut microbiome (Ma et al. 2022). In this research, 2349 sequences were predicted as AMPs, of which 216 were synthesized and more than 83% proven to be active as predicted, including against clinically antibiotic-resistant pathogens. Of these active AMPs, 11 were validated in animal models and shown to decrease the infection load by over 10-fold (Ma et al. 2022). In line with the advances in AMP prediction and development, several research lines have been developed to investigate nanotechnology delivery of AMPs and design more druggable candidates using machine learning (Sowers et al. 2023).

Another exciting venue of developing microbiome-based products to control infectious diseases is the use of these microbes as live therapeutics either as defined microbial consortia or as community transplant (Gulliver et al. 2022). The main challenge of this approach is the unpredictable interaction of these microbes once introduced to a well-established microbial community and in the context of the human host. One promising strategy is the development of engineered probiotic strains (Dosoky et al. 2018). Palmer et al. constructed an engineered *E. coli* strain that could sense the utilization of tetrathionate, an electron acceptor produced in the gut upon excessive inflammation caused by *Salmonella* infection, and then produce the AMP, microcin H47 (Palmer et al. 2018).

In conclusion, understanding the interplay between host and microbes and the molecular mechanisms mediating these interactions is essential to better understand human biology. This understanding will advance our appreciation of the ecology of these microbial communities and enable the designing and development of novel therapeutics for a wide array of human diseases. Besides, this venue of translational research may drastically change how we control and treat infectious diseases and lead to innovative microbiome-based interventions.

Author Contribution Both AA and WKM contributed equally to this book chapter.

Conflict of Interests The authors declare no conflict of interest. The authors declare that there is no copyright issue. All figures and data in this chapter are original and never taken from other sources. All figures were designed and created here for the first time using BioRender with a subscription and permission to publish.

References

Algburi A, Zehm S, Netrebov V, Bren AB, Chistyakov V, Chikindas ML (2017) Subtilosin prevents biofilm formation by inhibiting bacterial quorum sensing. Probiotics Antimicro 9(1):81–90

Allonsius CN, Vandenheuvel D, Oerlemans EFM, Petrova MI, Donders GGG, Cos P et al (2019) Inhibition of Candida albicans morphogenesis by chitinase from lactobacillus rhamnosus GG. Sci Rep 9(1):2900

Arpaia N, Campbell C, Fan X, Dikiy S, Van Der Veeken J, deRoos P et al (2013) Metabolites produced by commensal bacteria promote peripheral regulatory T-cell generation. Nature 504(7480):451–455
Bain CC, Bravo-Blas A, Scott CL, Gomez Perdiguero E, Geissmann F, Henri S et al (2014) Constant replenishment from circulating monocytes maintains the macrophage pool in the intestine of adult mice. Nat Immunol 15(10):929–937
Baker JL, Morton JT, Dinis M, Alvarez R, Tran NC, Knight R et al (2021) Deep metagenomics examines the oral microbiome during dental caries, revealing novel taxa and co-occurrences with host molecules. Genome Res 31(1):64–74
Bansal T, Alaniz RC, Wood TK, Jayaraman A (2010) The bacterial signal indole increases epithelial-cell tight-junction resistance and attenuates indicators of inflammation. Proc Natl Acad Sci USA 107(1):228–233
Barbour A, Wescombe P, Smith L (2020) Evolution of Lantibiotic Salivaricins: new weapons to fight infectious diseases. Trends Microbiol 28(7):578–593
Baud A, Hillion KH, Plainvert C, Tessier V, Tazi A, Mandelbrot L et al (2023) Microbial diversity in the vaginal microbiota and its link to pregnancy outcomes. Sci Rep 13(1):9061
Bengtsson T, Selegård R, Musa A, Hultenby K, Utterström J, Sivlér P et al (2020) Plantaricin NC8 αβ exerts potent antimicrobial activity against Staphylococcus spp. and enhances the effects of antibiotics. Sci Rep 10(1):3580
Bier K, Schittek B (2021) Beneficial effects of coagulase-negative Staphylococci on Staphylococcus aureus skin colonization. Exp Dermatol 30(10):1442–1452
Bomar L, Brugger SD, Yost BH, Davies SS, Lemon KP (2016) Corynebacterium accolens releases antipneumococcal free fatty acids from human nostril and skin surface triacylglycerols. MBio 7(1):e01725–e01715
Briaud P, Carroll RK (2020) Extracellular vesicle biogenesis and functions in gram-positive bacteria. Richardson AR (ed). Infect Immun 88(12):e00433–e00420
Brown MM, Kwiecinski JM, Cruz LM, Shahbandi A, Todd DA, Cech NB et al (2020) Novel peptide from commensal Staphylococcus simulans blocks methicillin-resistant Staphylococcus aureus quorum sensing and protects host skin from damage. Antimicrob Agents Chemother 64(6):e00172–e00120
Brugger SD, Eslami SM, Pettigrew MM, Escapa IF, Henke MT, Kong Y et al (2020) Dolosigranulum pigrum cooperation and competition in human nasal microbiota. D'Orazio SEF (ed). mSphere 5(5):e00852-20
Byrd AL, Belkaid Y, Segre JA (2018) The human skin microbiome. Nat Rev Microbiol 16(3):143–155
Caetano-Silva ME, Rund L, Hutchinson NT, Woods JA, Steelman AJ, Johnson RW (2023) Inhibition of inflammatory microglia by dietary fiber and short-chain fatty acids. Sci Rep 13(1):2819
Campbell C, Kandalgaonkar MR, Golonka RM, Yeoh BS, Vijay-Kumar M, Saha P (2023) Crosstalk between gut microbiota and host immunity: impact on inflammation and immunotherapy. Biomedicines 11(2):294
Cau L, Williams MR, Butcher AM, Nakatsuji T, Kavanaugh JS, Cheng JY et al (2021) Staphylococcus epidermidis protease EcpA can be a deleterious component of the skin microbiome in atopic dermatitis. J Allergy Clin Immunol 147(3):955–966.e16
Chattopadhyay A, Das N, Banerjee D (2023) Mechanism of action of the lantibiotics on the multidrug-resistant organisms. In: Lantibiotics as alternative therapeutics [Internet]. Elsevier. [cited 2024 Jan 15]:85–117. Available from: https://linkinghub.elsevier.com/retrieve/pii/B9780323991414000102
Chen B, Chen H, Shu X, Yin Y, Li J, Qin J et al (2018) Presence of segmented filamentous bacteria in human children and its potential role in the modulation of human gut immunity. Front Microbiol 9:1403
Cheng D, Xie MZ (2021) A review of a potential and promising probiotic candidate—*Akkermansia muciniphila*. J Appl Microbiol 130(6):1813–1822

Chu J, Vila-Farres X, Inoyama D, Ternei M, Cohen LJ, Gordon EA et al (2016) Discovery of MRSA active antibiotics using primary sequence from the human microbiome. Nat Chem Biol 12:1004–1006

Chu J, Vila-Farres X, Inoyama D, Gallardo-Macias R, Jaskowski M, Satish S et al (2018) Human microbiome inspired antibiotics with improved β-lactam synergy against MDR *Staphylococcus aureus*. ACS Infect Dis 4(1):33–38

Clementi EA, Wilhelm KR, Schleucher J, Morozova-Roche LA, Hakansson AP (2013) A complex of equine lysozyme and oleic acid with bactericidal activity against Streptococcus pneumoniae. Netto LES (ed). PLoS ONE 8(11):e80649

Cogen AL, Yamasaki K, Muto J, Sanchez KM, Crotty Alexander L, Tanios J et al (2010) Staphylococcus epidermidis antimicrobial δ-Toxin (Phenol-Soluble Modulin-γ) cooperates with host antimicrobial peptides to kill group A Streptococcus. DeLeo FR (ed). PLoS ONE 5(1):e8557

Cohen LJ, Han S, Huang YH, Brady SF (2018) Identification of the Colicin V Bacteriocin gene cluster by functional screening of a human microbiome metagenomic library. ACS Infect Dis 4(1):27–32

Croatti V, Parolin C, Giordani B, Foschi C, Fedi S, Vitali B (2022) Lactobacilli extracellular vesicles: potential postbiotics to support the vaginal microbiota homeostasis. Microb Cell Factories 21(1):237

Dabard J, Bridonneau C, Phillipe C, Anglade P, Molle D, Nardi M et al (2001) Ruminococcin A, a new lantibiotic produced by a Ruminococcus gnavus strain isolated from human feces. Appl Environ Microbiol 67(9):4111–4118

Dahlén G, Hassan H, Blomqvist S, Carlén A (2018) Rapid urease test (RUT) for evaluation of urease activity in oral bacteria in vitro and in supragingival dental plaque ex vivo. BMC Oral Health 18(1):89

De Boeck I, Wittouck S, Martens K, Spacova I, Cauwenberghs E, Allonsius CN et al (2021) The nasal mutualist Dolosigranulum pigrum AMBR11 supports homeostasis via multiple mechanisms. iScience 24(9):102978

De Sablet T, Chassard C, Bernalier-Donadille A, Vareille M, Gobert AP, Martin C (2009) Human microbiota-secreted factors inhibit Shiga toxin synthesis by enterohemorrhagic *Escherichia coli* O157:H7. Infect Immun 77(2):783–790

De Steenhuijsen Piters WAA, Huijskens EGW, Wyllie AL, Biesbroek G, Van Den Bergh MR, Veenhoven RH et al (2016) Dysbiosis of upper respiratory tract microbiota in elderly pneumonia patients. ISME J 10(1):97–108

Dean SN, Leary DH, Sullivan CJ, Oh E, Walper SA (2019) Isolation and characterization of lactobacillus-derived membrane vesicles. Sci Rep 9(1):877

Dean SN, Rimmer MA, Turner KB, Phillips DA, Caruana JC, Hervey WJ et al (2020) Lactobacillus acidophilus membrane vesicles as a vehicle of Bacteriocin delivery. Front Microbiol 11:710

Deng H, Yang S, Zhang Y, Qian K, Zhang Z, Liu Y et al (2018) Bacteroides fragilis prevents Clostridium difficile infection in a mouse model by restoring gut barrier and microbiome regulation. Front Microbiol 9:2976

Deo P, Deshmukh R (2019) Oral microbiome: unveiling the fundamentals. J Oral Maxillofac Pathol 23(1):122

Dols JAM, Molenaar D, Van Der Helm JJ, Caspers MPM, De Kat A-BA, Schuren FHJ et al (2016) Molecular assessment of bacterial vaginosis by lactobacillus abundance and species diversity. BMC Infect Dis 16(1):180

Domínguez Rubio AP, Martínez JH, Martínez Casillas DC, Coluccio Leskow F, Piuri M, Pérez OE (2017) Lactobacillus casei BL23 produces microvesicles carrying proteins that have been associated with its probiotic effect. Front Microbiol 8:1783

Donaldson GP, Ladinsky MS, Yu KB, Sanders JG, Yoo BB, Chou WC et al (2018) Gut microbiota utilize immunoglobulin A for mucosal colonization. Science 360(6390):795–800

Dosoky NS, Guo L, Chen Z, Feigley AV, Davies SS (2018) Dietary fatty acids control the species of N-acyl-phosphatidylethanolamines synthesized by therapeutically modified bacteria in the intestinal tract. ACS Infect Dis 4(1):3–13

Earley H, Lennon G, Balfe Á, Coffey JC, Winter DC, O'Connell PR (2019) The abundance of Akkermansia muciniphila and its relationship with sulphated colonic mucins in health and ulcerative colitis. Sci Rep 9(1):15683

Engevik MA, Luk B, Chang-Graham AL, Hall A, Herrmann B, Ruan W et al (2019) Bifidobacterium dentium fortifies the intestinal mucus layer via autophagy and calcium signaling pathways. Gorska S, Pujol N (eds). mBio 10(3):e01087-19

Erturk-Hasdemir D, Oh SF, Okan NA, Stefanetti G, Gazzaniga FS, Seeberger PH et al (2019) Symbionts exploit complex signaling to educate the immune system. Proc Natl Acad Sci USA 116(52):26157–26166

Escapa IF, Chen T, Huang Y, Gajare P, Dewhirst FE, Lemon KP (2018) New insights into human nostril microbiome from the expanded human oral microbiome database (eHOMD): a resource for the microbiome of the human aerodigestive tract. Xu J (ed). mSystems 3(6):e00187–e00118

Fachi JL, Sécca C, Rodrigues PB, FCPD M, Di Luccia B, JDS F et al (2020) Acetate coordinates neutrophil and ILC3 responses against C. difficile through FFAR2. J Exp Med 217(3):jem.20190489

Featherstone JDB (2008) Dental caries: a dynamic disease process. Aust Dent J 53(3):286–291

Flowers L, Grice EA (2020) The skin microbiota: balancing risk and reward. Cell Host Microbe 28(2):190–200

Fukuda S, Toh H, Hase K, Oshima K, Nakanishi Y, Yoshimura K et al (2011) Bifidobacteria can protect from enteropathogenic infection through production of acetate. Nature 469(7331):543–547

Fukuda S, Toh H, Taylor TD, Ohno H, Hattori M (2012) Acetate-producing bifidobacteria protect the host from enteropathogenic infection via carbohydrate transporters. Gut Microbes 3(5):449–454

Furusawa Y, Obata Y, Fukuda S, Endo TA, Nakato G, Takahashi D et al (2013) Commensal microbe-derived butyrate induces the differentiation of colonic regulatory T cells. Nature 504(7480):446–450

Fusco W, Lorenzo MB, Cintoni M, Porcari S, Rinninella E, Kaitsas F et al (2023) Short-chain fatty-acid-producing bacteria: key components of the human gut microbiota. Nutrients 15(9):2211

Gallardo-Becerra L, Cervantes-Echeverría M, Cornejo-Granados F, Vazquez-Morado LE, Ochoa-Leyva A (2024) Perspectives in searching antimicrobial peptides (AMPs) produced by the microbiota. Microb Ecol 87(1):8

Gao L, Xu T, Huang G, Jiang S, Gu Y, Chen F (2018) Oral microbiomes: more and more importance in oral cavity and whole body. Protein Cell 9(5):488–500

Glatthardt T, Campos JCDM, Chamon RC, De Sá Coimbra TF, Rocha GDA, De Melo MAF et al (2020) Small molecules produced by commensal Staphylococcus epidermidis disrupt formation of biofilms by Staphylococcus aureus. Stabb EV (ed). Appl Environ Microbiol 86(5):e02539–e02519

Grande R, Celia C, Mincione G, Stringaro A, Di Marzio L, Colone M et al (2017) Detection and physicochemical characterization of membrane vesicles (MVs) of Lactobacillus reuteri DSM 17938. Front Microbiol 8:1040

Gulliver EL, Young RB, Chonwerawong M, D'Adamo GL, Thomason T, Widdop JT et al (2022) Review article: the future of microbiome-based therapeutics. Aliment Pharmacol Ther 56(2):192–208

Han D, Ji Y, Yang S, Song P, Shi Y, Shao D et al (2024) Therapeutic effect of iturin A on Candida albicans oral infection and its pathogenic factors. Antimicrob Agents Chemother 68(1):e0094823

Hardy BL, Dickey SW, Plaut RD, Riggins DP, Stibitz S, Otto M et al (2019) Corynebacterium pseudodiphtheriticum exploits Staphylococcus aureus virulence components in a novel polymicrobial defense strategy. Salama NR (ed). mBio 10(1):e02491-18

Hatziioanou D, Gherghisan-Filip C, Saalbach G, Horn N, Wegmann U, Duncan SH et al (2017) Discovery of a novel lantibiotic nisin O from Blautia obeum A2-162, isolated from the human gastrointestinal tract. Microbiology 163(9):1292–1305

Hernández-Terán A, Mejía-Nepomuceno F, Herrera MT, Barreto O, García E, Castillejos M et al (2021) Dysbiosis and structural disruption of the respiratory microbiota in COVID-19 patients with severe and fatal outcomes. Sci Rep 11(1):21297

Horn KJ, Schopper MA, Drigot ZG, Clark SE (2022a) Airway Prevotella promote TLR2-dependent neutrophil activation and rapid clearance of Streptococcus pneumoniae from the lung. Nat Commun 13(1):3321

Horn KJ, Jaberi Vivar AC, Arenas V, Andani S, Janoff EN, Clark SE (2022b) Corynebacterium species inhibit Streptococcus pneumoniae colonization and infection of the mouse airway. Front Microbiol 12:804935

Horrocks V, King OG, Yip AYG, Marques IM, McDonald JAK (2023) Role of the gut microbiota in nutrient competition and protection against intestinal pathogen colonization. Microbiology [Internet]. [cited 2024 Jun 22];169(8). Available from: https://www.microbiologyresearch.org/content/journal/micro/10.1099/mic.0.001377

Houtman TA, Eckermann HA, Smidt H, De Weerth C (2022) Gut microbiota and BMI throughout childhood: the role of firmicutes, bacteroidetes, and short-chain fatty acid producers. Sci Rep 12(1):3140

Hsiao A, Ahmed AMS, Subramanian S, Griffin NW, Drewry LL, Petri WA et al (2014) Members of the human gut microbiota involved in recovery from Vibrio cholerae infection. Nature 515(7527):423–426

Huang X, Palmer SR, Ahn SJ, Richards VP, Williams ML, Nascimento MM et al (2016) A highly Arginolytic streptococcus species that potently antagonizes Streptococcus mutans. Schaffner DW (ed). Appl Environ Microbiol 82(7):2187–2201

Imchen M, Salim SA, Kumavath R, Busi S (2023) Factors affecting the composition of the human microbiome. In: Kothari V, Kumar P, Ray S (eds) Probiotics, prebiotics, synbiotics, and postbiotics [Internet]. Springer Nature Singapore, Singapore [cited 2024 Jun 22], pp 49–63. Available from: https://link.springer.com/10.1007/978-981-99-1463-0_3

Ivanov II, Tuganbaev T, Skelly AN, Honda K (2022) T cell responses to the microbiota. Annu Rev Immunol 40:559–587

Iwase T, Uehara Y, Shinji H, Tajima A, Seo H, Takada K et al (2010) Staphylococcus epidermidis Esp inhibits Staphylococcus aureus biofilm formation and nasal colonization. Nature 465(7296):346–349

Jakobsson HE, Rodríguez-Piñeiro AM, Schütte A, Ermund A, Boysen P, Bemark M et al (2015) The composition of the gut microbiota shapes the colon mucus barrier. EMBO Rep 16(2):164–177

Jia W, Xie G, Jia W (2018) Bile acid–microbiota crosstalk in gastrointestinal inflammation and carcinogenesis. Nat Rev Gastroenterol Hepatol 15(2):111–128

Johansson MEV, Jakobsson HE, Holmén-Larsson J, Schütte A, Ermund A, Rodríguez-Piñeiro AM et al (2015) Normalization of host intestinal mucus layers requires long-term microbial colonization. Cell Host Microbe 18(5):582–592

Kang B, Alvarado LJ, Kim T, Lehmann ML, Cho H, He J et al (2020) Commensal microbiota drive the functional diversification of colon macrophages. Mucosal Immunol 13(2):216–229

Kim PI, Sohng JK, Sung C, Joo HS, Kim EM, Yamaguchi T et al (2010) Characterization and structure identification of an antimicrobial peptide, hominicin, produced by Staphylococcus hominis MBBL 2–9. Biochem Biophys Res Commun 399(2):133–138

Kim MH, Choi SJ, Choi HI, Choi JP, Park HK, Kim EK et al (2018) *Lactobacillus plantarum* -derived extracellular vesicles protect atopic dermatitis induced by *Staphylococcus aureus* -derived extracellular vesicles. Allergy Asthma Immunol Res 10(5):516

Kim S, Shin YC, Kim TY, Kim Y, Lee YS, Lee SH et al (2021a) Mucin degrader Akkermansia muciniphila accelerates intestinal stem cell-mediated epithelial development. Gut Microbes 13(1):1–20

Kim S, Shin YC, Kim TY, Kim Y, Lee YS, Lee SH et al (2021b) Mucin degrader *Akkermansia muciniphila* accelerates intestinal stem cell-mediated epithelial development. Gut Microbes 13(1):1892441

Kim KH, Park D, Jia B, Baek JH, Hahn Y, Jeon CO (2022) Identification and characterization of major bile acid 7α-dehydroxylating bacteria in the human gut. Chu H (ed). mSystems 7(4):e00455-22

King AM, Zhang Z, Glassey E, Siuti P, Clardy J, Voigt CA (2023) Systematic mining of the human microbiome identifies antimicrobial peptides with diverse activity spectra. Nat Microbiol 8(12):2420–2434

Kisthardt SC, Thanissery R, Pike CM, Foley MH, Theriot CM (2023) The microbial-derived bile acid lithocholate and its epimers inhibit Clostridioides difficile growth and pathogenicity while sparing members of the gut microbiota. J Bacteriol 205(9):e0018023

Köhler A, Delbauve S, Smout J, Torres D, Flamand V (2021) Very early-life exposure to microbiota-induced TNF drives the maturation of neonatal pre-cDC1. Gut 70(3):511–521

Lai Y, Cogen AL, Radek KA, Park HJ, MacLeod DT, Leichtle A et al (2010) Activation of TLR2 by a small molecule produced by Staphylococcus epidermidis increases antimicrobial defense against bacterial skin infections. J Invest Dermatol 130(9):2211–2221

Lamont RJ, Koo H, Hajishengallis G (2018) The oral microbiota: dynamic communities and host interactions. Nat Rev Microbiol 16(12):745–759

Latham RD, Gell DA, Fairbairn RL, Lyons AB, Shukla SD, Cho KY et al (2017) An isolate of Haemophilus haemolyticus produces a bacteriocin-like substance that inhibits the growth of nontypeable Haemophilus influenzae. Int J Antimicrob Agents 49(4):503–506

Le Poul E, Loison C, Struyf S, Springael JY, Lannoy V, Decobecq ME et al (2003) Functional characterization of human receptors for short chain fatty acids and their role in polymorphonuclear cell activation. J Biol Chem 278(28):25481–25489

Lee HJ, Kim M (2022) Skin barrier function and the microbiome. Int J Mol Sci 23(21):13071

Lee HY, Andalibi A, Webster P, Moon SK, Teufert K, Kang SH et al (2004) Antimicrobial activity of innate immune molecules against Streptococcus pneumoniae, Moraxella catarrhalis and nontypeable Haemophilus influenzae. BMC Infect Dis 4(1):12

Lei J, Sun L, Huang S, Zhu C, Li P, He J et al (2019) The antimicrobial peptides and their potential clinical applications. Am J Transl Res 11(7):3919–3931

Li G, Lin J, Zhang C, Gao H, Lu H, Gao X et al (2021) Microbiota metabolite butyrate constrains neutrophil functions and ameliorates mucosal inflammation in inflammatory bowel disease. Gut Microbes 13(1):1968257

Li X, Liu Y, Yang X, Li C, Song Z (2022) The Oral microbiota: community composition, influencing factors, pathogenesis, and interventions. Front Microbiol 13:895537

Linehan JL, Harrison OJ, Han SJ, Byrd AL, Vujkovic-Cvijin I, Villarino AV et al (2018) Non-classical immunity controls microbiota impact on skin immunity and tissue repair. Cell 172(4):784–796.e18

Litvak Y, Byndloss MX, Bäumler AJ (2018) Colonocyte metabolism shapes the gut microbiota. Science 362(6418):eaat9076

Liu YL, Nascimento M, Burne RA (2012) Progress toward understanding the contribution of alkali generation in dental biofilms to inhibition of dental caries. Int J Oral Sci 4(3):135–140

Liu Y, Defourny KAY, Smid EJ, Abee T (2018) Gram-positive bacterial extracellular vesicles and their impact on health and disease. Front Microbiol 9:1502

Liu Y, Liu Q, Zhao L, Dickey SW, Wang H, Xu R et al (2022) Essential role of membrane vesicles for biological activity of the bacteriocin micrococcin P1. J Extracell Vesicles 11(4):e12212

Lynch D, O'Connor PM, Cotter PD, Hill C, Field D, Begley M (2019) Identification and characterisation of capidermicin, a novel bacteriocin produced by Staphylococcus capitis. Tse H, editor. PLoS ONE 14(10):e0223541

Lyu M, Suzuki H, Kang L, Gaspal F, Zhou W, Goc J et al (2022) ILC3s select microbiota-specific regulatory T cells to establish tolerance in the gut. Nature 610(7933):744–751

Ma Y, Guo Z, Xia B, Zhang Y, Liu X, Yu Y et al (2022) Identification of antimicrobial peptides from the human gut microbiome using deep learning. Nat Biotechnol 40(6):921–931

Machado A, Foschi C, Marangoni A (2022) Editorial: vaginal dysbiosis and biofilms. Front Cell Infect Microbiol 12:976057

Maldonado-Barragán A, Caballero-Guerrero B, Martín V, Ruiz-Barba JL, Rodríguez JM (2016) Purification and genetic characterization of gassericin E, a novel co-culture inducible bacteriocin from Lactobacillus gasseri EV1461 isolated from the vagina of a healthy woman. BMC Microbiol 16(1):37

Marković KG, Grujović MŽ, Koraćević MG, Nikodijević DD, Milutinović MG, Semedo-Lemsaddek T et al (2022) Colicins and Microcins produced by Enterobacteriaceae: characterization, mode of action, and putative applications. Int J Environ Res Public Health 19(18):11825

Mathur H, Fallico V, O'Connor PM, Rea MC, Cotter PD, Hill C et al (2017) Insights into the mode of action of the Sactibiotic Thuricin CD. Front Microbiol 8:696

Matsubara VH, Wang Y, Bandara HMHN, Mayer MPA, Samaranayake LP (2016) Probiotic lactobacilli inhibit early stages of Candida albicans biofilm development by reducing their growth, cell adhesion, and filamentation. Appl Microbiol Biotechnol 100(14):6415–6426

Mazmanian SK, Liu CH, Tzianabos AO, Kasper DL (2005) An immunomodulatory molecule of symbiotic bacteria directs maturation of the host immune system. Cell 122(1):107–118

McCormack RM, de Armas LR, Shiratsuchi M, Fiorentino DG, Olsson ML, Lichtenheld MG et al (2015) Perforin-2 is essential for intracellular defense of parenchymal cells and phagocytes against pathogenic bacteria. elife 4:e06508

McLean AR, Torres-Morales J, Dewhirst FE, Borisy GG, Mark Welch JL (2022) Site-tropism of streptococci in the oral microbiome. Mol Oral Microbiol 37(6):229–243

Menon LU, Scoffield JA, Jackson JG, Zhang P (2022) Candida albicans and early childhood caries. Front Dent Med 3:849274

Mira A, Simon-Soro A, Curtis MA (2017) Role of microbial communities in the pathogenesis of periodontal diseases and caries. J Clinic Periodontology [Internet]. [cited 2024 Jan 21];44(S18). Available from: https://onlinelibrary.wiley.com/doi/10.1111/jcpe.12671

Mora D, Arioli S (2014) Microbial Urease in health and disease. Miller V (ed). PLoS Pathog 10(12):e1004472

Morrill S, Gilbert NM, Lewis AL (2020) Gardnerella vaginalis as a cause of bacterial vaginosis: appraisal of the evidence from in vivo models. Front Cell Infect Microbiol 10:168

Mousa WK, Athar B, Merwin NJ, Magarvey NA (2017) Antibiotics and specialized metabolites from the human microbiota. Nat Prod Rep 34(11):1302–1331

Mousa WK, Ghemrawi R, Abu-Izneid T, Ramadan A, Al-Marzooq F (2023) Discovery of Lactomodulin, a unique microbiome-derived peptide that exhibits dual anti-inflammatory and antimicrobial activity against multidrug-resistant pathogens. IJMS 24(8):6901

Ñahui Palomino RA, Vanpouille C, Laghi L, Parolin C, Melikov K, Backlund P et al (2019) Extracellular vesicles from symbiotic vaginal lactobacilli inhibit HIV-1 infection of human tissues. Nat Commun 10(1):5656

Nakagawa S, Matsumoto M, Katayama Y, Oguma R, Wakabayashi S, Nygaard T et al (2017) Staphylococcus aureus virulent PSMα peptides induce keratinocyte Alarmin release to orchestrate IL-17-dependent skin inflammation. Cell Host Microbe 22(5):667–677.e5

Nakajima A, Vogelzang A, Maruya M, Miyajima M, Murata M, Son A et al (2018) IgA regulates the composition and metabolic function of gut microbiota by promoting symbiosis between bacteria. J Exp Med 215(8):2019–2034

Nakatsuji T, Kao MC, Zhang L, Zouboulis CC, Gallo RL, Huang CM (2010) Sebum free fatty acids enhance the innate immune defense of human sebocytes by upregulating beta-defensin-2 expression. J Invest Dermatol 130(4):985–994

Nakatsuji T, Chen TH, Narala S, Chun KA, Two AM, Yun T et al (2017) Antimicrobials from human skin commensal bacteria protect against Staphylococcus aureus and are deficient in atopic dermatitis. Sci Transl Med 9(378):eaah4680

Nakatsuji T, Chen TH, Butcher AM, Trzoss LL, Nam SJ, Shirakawa KT et al (2018) A commensal strain of Staphylococcus epidermidis protects against skin neoplasia. Sci Adv 4(2):eaao4502

Nascimento MM, Liu Y, Kalra R, Perry S, Adewumi A, Xu X et al (2013) Oral arginine metabolism may decrease the risk for dental caries in children. J Dent Res 92(7):604–608

Natalini JG, Singh S, Segal LN (2023) The dynamic lung microbiome in health and disease. Nat Rev Microbiol 21(4):222–235

Negi S, Das DK, Pahari S, Nadeem S, Agrewala JN (2019) Potential role of gut microbiota in induction and regulation of innate immune memory. Front Immunol 10:2441
O'Neill AM, Nakatsuji T, Hayachi A, Williams MR, Mills RH, Gonzalez DJ et al (2020) Identification of a human skin commensal bacterium that selectively kills Cutibacterium acnes. J Invest Dermatol 140(8):1619–1628.e2
O'Sullivan JN, O'Connor PM, Rea MC, O'Sullivan O, Walsh CJ, Healy B et al (2020) Nisin J, a novel natural nisin variant, is produced by Staphylococcus capitis sourced from the human skin microbiota. O'Toole G (ed). J Bacteriol [Internet]. [cited 2024 Jun 23];202(3). Available from: https://journals.asm.org/doi/10.1128/JB.00639-19
Oh J, Byrd AL, Park M, Comparative Sequencing Program NISC, Kong HH, Segre JA (2016) Temporal stability of the human skin microbiome. Cell 165(4):854–866
Oliveira RA, Ng KM, Correia MB, Cabral V, Shi H, Sonnenburg JL et al (2020) Klebsiella michiganensis transmission enhances resistance to Enterobacteriaceae gut invasion by nutrition competition. Nat Microbiol 5(4):630–641
Paik D, Yao L, Zhang Y, Bae S, D'Agostino GD, Zhang M et al (2022) Human gut bacteria produce TH17-modulating bile acid metabolites. Nature 603(7903):907–912
Palmer JD, Piattelli E, McCormick BA, Silby MW, Brigham CJ, Bucci V (2018) Engineered probiotic for the inhibition of salmonella via Tetrathionate-induced production of Microcin H47. ACS Infect Dis 4(1):39–45
Pandey N, Malik RK, Kaushik JK, Singroha G (2013) Gassericin A: a circular bacteriocin produced by lactic acid bacteria lactobacillus gasseri. World J Microbiol Biotechnol 29(11):1977–1987
Pastar I, O'Neill K, Padula L, Head CR, Burgess JL, Chen V et al (2020) Staphylococcus epidermidis boosts innate immune response by activation of Gamma Delta T cells and induction of Perforin-2 in human skin. Front Immunol 11:550946
Patil S, Rao RS, Majumdar B, Anil S (2015) Clinical appearance of Oral Candida infection and therapeutic strategies. Front Microbiol [Internet]. [cited 2024 Jan 21];6. Available from: http://journal.frontiersin.org/Article/10.3389/fmicb.2015.01391/abstract
Peixoto RJM, Alves ES, Wang M, Ferreira RBR, Granato A, Han J et al (2017) Repression of salmonella host cell invasion by aromatic small molecules from the human fecal metabolome. Müller V (ed). Appl Environ Microbiol 83(19):e01148–e01117
Pena RT, Blasco L, Ambroa A, González-Pedrajo B, Fernández-García L, López M et al (2019) Relationship between quorum sensing and secretion systems. Front Microbiol 10:1100
Peng L, Li ZR, Green RS, Holzman IR, Lin J (2009) Butyrate enhances the intestinal barrier by facilitating tight junction assembly via activation of AMP-activated protein kinase in Caco-2 cell monolayers. J Nutr 139(9):1619–1625
Peng Q, Tang X, Dong W, Sun N, Yuan W (2022) A review of biofilm formation of Staphylococcus aureus and its regulation mechanism. Antibiotics (Basel) 12(1):12
Pérez-Reytor D, Puebla C, Karahanian E, García K (2021) Use of short-chain fatty acids for the recovery of the intestinal epithelial barrier affected by bacterial toxins. Front Physiol 12:650313
Pino A, Bartolo E, Caggia C, Cianci A, Randazzo CL (2019) Detection of vaginal lactobacilli as probiotic candidates. Sci Rep 9(1):3355
Pokhrel R, Bhattarai N, Baral P, Gerstman BS, Park JH, Handfield M et al (2019) Molecular mechanisms of pore formation and membrane disruption by the antimicrobial lantibiotic peptide Mutacin 1140. Phys Chem Chem Phys 21(23):12530–12539
Popoff MR (2014) CLOSTRIDIUM | detection of enterotoxin of Clostridium perfringens. In: Encyclopedia of food microbiology [Internet]. Elsevier. [cited 2024 Jan 15], pp 474–480. Available from: https://linkinghub.elsevier.com/retrieve/pii/B9780123847300000690
Portincasa P, Bonfrate L, Vacca M, De Angelis M, Farella I, Lanza E et al (2022) Gut microbiota and short chain fatty acids: implications in glucose homeostasis. Int J Mol Sci 23(3):1105
Raya Tonetti F, Tomokiyo M, Ortiz Moyano R, Quilodrán-Vega S, Yamamuro H, Kanmani P et al (2021) The respiratory commensal bacterium Dolosigranulum pigrum 040417 improves the innate immune response to Streptococcus pneumoniae. Microorganisms 9(6):1324

Rea MC, Sit CS, Clayton E, O'Connor PM, Whittal RM, Zheng J et al (2010) Thuricin CD, a posttranslationally modified bacteriocin with a narrow spectrum of activity against Clostridium difficile. Proc Natl Acad Sci USA 107(20):9352–9357

Reed AD, Nethery MA, Stewart A, Barrangou R, Theriot CM (2020) Strain-dependent inhibition of Clostridioides difficile by commensal *Clostridia* carrying the bile acid-inducible (*bai*) operon. Comstock LE (ed). J Bacteriol [Internet]. [cited 2024 Jan 21];202(11). Available from: https://journals.asm.org/doi/10.1128/JB.00039-20

Rivera-Chávez F, Zhang LF, Faber F, Lopez CA, Byndloss MX, Olsan EE et al (2016) Depletion of butyrate-producing clostridia from the gut microbiota drives an aerobic luminal expansion of salmonella. Cell Host Microbe 19(4):443–454

Ruppelt D, Trollmann MFW, Dema T, Wirtz SN, Flegel H, Mönnikes S et al (2024) The antimicrobial fibupeptide lugdunin forms water-filled channel structures in lipid membranes. Nat Commun 15(1):3521

Saleri R, Borghetti P, Ravanetti F, Cavalli V, Ferrari L, De Angelis E et al (2022) Effects of different short-chain fatty acids (SCFA) on gene expression of proteins involved in barrier function in IPEC-J2. Porc Health Manag 8(1):21

Salvadori M, Rosso G (2024) Update on the gut microbiome in health and diseases. World J Methodol 14(1):89196

Santacroce L, Charitos IA, Ballini A, Inchingolo F, Luperto P, De Nitto E et al (2020) The human respiratory system and its microbiome at a glimpse. Biology 9(10):318

Santos-Júnior CD, Der Torossian TM, Duan Y, Del Río ÁR, Schmidt TSB, Chong H et al (2023) Computational exploration of the global microbiome for antibiotic discovery. bioRxiv. 2023.08.31.555663

Schnell A, Littman DR, Kuchroo VK (2023) TH17 cell heterogeneity and its role in tissue inflammation. Nat Immunol 24(1):19–29

Scillato M, Spitale A, Mongelli G, Privitera GF, Mangano K, Cianci A et al (2021) Antimicrobial properties of Lactobacillus cell-free supernatants against multidrug-resistant urogenital pathogens. Microbiology 10(2):e1173

Selle K, Klaenhammer TR (2013) Genomic and phenotypic evidence for probiotic influences of lactobacillus gasseri on human health. FEMS Microbiol Rev 37(6):915–935

Sencio V, Barthelemy A, Tavares LP, Machado MG, Soulard D, Cuinat C et al (2020) Gut dysbiosis during influenza contributes to pulmonary pneumococcal superinfection through altered short-chain fatty acid production. Cell Rep 30(9):2934–2947.e6

Severn MM, Williams MR, Shahbandi A, Bunch ZL, Lyon LM, Nguyen A et al (2022a) The ubiquitous human skin commensal Staphylococcus hominis protects against opportunistic pathogens. mBio 13(3):e0093022

Severn MM, Cho YSK, Manzer HS, Bunch ZL, Shahbandi A, Todd DA et al (2022b) The commensal Staphylococcus warneri makes peptide inhibitors of MRSA quorum sensing that protect skin from atopic or necrotic damage. J Invest Dermatol 142(12):3349–3352.e5

Shabayek S, Spellerberg B (2018) Group B streptococcal colonization, molecular characteristics, and epidemiology. Front Microbiol 9:437

Shabayek S, Abdellah AM, Salah M, Ramadan M, Fahmy N (2022) Alterations of the vaginal microbiome in healthy pregnant women positive for group B streptococcus colonization during the third trimester. BMC Microbiol 22(1):313

Shelley JR, McHugh BJ, Wills J, Dorin JR, Weller R, Clarke DJ et al (2023) A mechanistic evaluation of human beta defensin 2 mediated protection of human skin barrier in vitro. Sci Rep 13(1):2271

Shimada Y, Kinoshita M, Harada K, Mizutani M, Masahata K, Kayama H et al (2013) Commensal bacteria-dependent indole production enhances epithelial barrier function in the colon. Heimesaat MM (ed). PLoS ONE 8(11):e80604

Silveira EGD, Prato LS, Pilati SFM, Arthur RA (2023) Comparison of oral cavity protein abundance among caries-free and caries-affected individuals—a systematic review and meta-analysis. Front Oral Health 4:1265817

Smith PM, Howitt MR, Panikov N, Michaud M, Gallini CA, Bohlooly-Y M et al (2013) The microbial metabolites, short-chain fatty acids, regulate colonic Treg cell homeostasis. Science 341(6145):569–573

Sowers A, Wang G, Xing M, Li B (2023) Advances in antimicrobial peptide discovery via machine learning and delivery via nanotechnology. Microorganisms 11(5):1129

Spragge F, Bakkeren E, Jahn MT, Araujo EBN, Pearson CF, Wang X et al (2023) Microbiome diversity protects against pathogens by nutrient blocking. Science 382(6676):eadj3502

Starc M, Lučovnik M, Eržen Vrlič P, Jeverica S (2022) Protective effect of lactobacillus crispatus against vaginal colonization with group B streptococci in the third trimester of pregnancy. Pathogens 11(9):980

Steed AL, Christophi GP, Kaiko GE, Sun L, Goodwin VM, Jain U et al (2017) The microbial metabolite desaminotyrosine protects from influenza through type I interferon. Science 357(6350):498–502

Tester R, Al-Ghazzewi FH (2018) Intrinsic and extrinsic carbohydrates in the vagina: a short review on vaginal glycogen. Int J Biol Macromol 112:203–206

Thanissery R, Winston JA, Theriot CM (2017) Inhibition of spore germination, growth, and toxin activity of clinically relevant C. difficile strains by gut microbiota derived secondary bile acids. Anaerobe 45:86–100

Theriot CM, Bowman AA, Young VB (2016) Antibiotic-induced alterations of the gut microbiota alter secondary bile acid production and allow for Clostridium difficile spore germination and outgrowth in the large intestine. mSphere 1(1):e00045–e00015

Tong LC, Wang Y, Wang ZB, Liu WY, Sun S, Li L, Su DF, Zhang LC (2016) Propionate ameliorates dextran sodium sulfate-induced colitis by improving intestinal barrier function and reducing inflammation and oxidative stress. Front Pharmacol [Internet]. [cited 2024 Feb 14];7. Available from: http://journal.frontiersin.org/Article/10.3389/fphar.2016.00253/abstract

Torres Salazar BO, Dema T, Schilling NA, Janek D, Bornikoel J, Berscheid A et al (2024) Commensal production of a broad-spectrum and short-lived antimicrobial peptide polyene eliminates nasal Staphylococcus aureus. Nat Microbiol 9(1):200–213

Townsend EC, Kalan LR (2023) The dynamic balance of the skin microbiome across the lifespan. Biochem Soc Trans 51(1):71–86

Uberoi A, Bartow-McKenney C, Zheng Q, Flowers L, Campbell A, Knight SAB et al (2021) Commensal microbiota regulates skin barrier function and repair via signaling through the aryl hydrocarbon receptor. Cell Host Microbe 29(8):1235–1248.e8

Valeri M, Raffatellu M (2016) Cytokines IL-17 and IL-22 in the host response to infection. Napier B (ed). Pathogens and disease 74(9):ftw111

Van Hoogmoed CG, Geertsema-Doornbusch GI, Teughels W, Quirynen M, Busscher HJ, Van der Mei HC (2008) Reduction of periodontal pathogens adhesion by antagonistic strains. Oral Microbiol Immunol 23(1):43–48

Vasilchenko AS, Vasilchenko AV, Valyshev AV, Rogozhin EA (2018) A novel high-molecular-mass Bacteriocin produced by enterococcus faecium: biochemical features and mode of action. Probiotics Antimicro 10(3):427–434

Verma R, Lee C, Jeun EJ, Yi J, Kim KS, Ghosh A et al (2018) Cell surface polysaccharides of *Bifidobacterium bifidum* induce the generation of Foxp3 $^{+}$ regulatory T cells. Sci Immunol 3(28):eaat6975

Vertillo Aluisio G, Spitale A, Bonifacio L, Privitera GF, Stivala A, Stefani S et al (2022) Streptococcus salivarius 24SMBc genome analysis reveals new biosynthetic gene clusters involved in antimicrobial effects on Streptococcus pneumoniae and streptococcus pyogenes. Microorganisms 10(10):2042

Vila T, Sultan AS, Montelongo-Jauregui D, Jabra-Rizk MA (2020) Oral candidiasis: a disease of opportunity. JoF 6(1):15

Vinolo MAR, Ferguson GJ, Kulkarni S, Damoulakis G, Anderson K, Bohlooly-Y M et al (2011) SCFAs induce mouse neutrophil chemotaxis through the GPR43 receptor. Ko BCB (ed). PLoS ONE 6(6):e21205

Wang G, Vaisman II, van Hoek ML (2022) Machine learning prediction of antimicrobial peptides. Methods Mol Biol 2405:1–37

Wang Q, Fang Z, Xiao Y, Wang H, Zhang P, Lu W et al (2023) Lactiplantibacillus pentoses CCFM1227 produces Desaminotyrosine to protect against influenza virus H1N1 infection through the type I interferon in mice. Nutrients 15(16):3659

Wayah SB, Philip K (2018) Purification, characterization, mode of action, and enhanced production of Salivaricin mmaye1, a novel bacteriocin from Lactobacillus salivarius SPW1 of human gut origin. Electron J Biotechnol 35:39–47

Wei Y, Gao J, Kou Y, Liu M, Meng L, Zheng X et al (2020) The intestinal microbial metabolite desaminotyrosine is an anti-inflammatory molecule that modulates local and systemic immune homeostasis. FASEB J 34(12):16117–16128

Williams MR, Costa SK, Zaramela LS, Khalil S, Todd DA, Winter HL et al (2019) Quorum sensing between bacterial species on the skin protects against epidermal injury in atopic dermatitis. Sci Transl Med 11(490):eaat8329

Wu J, Xie H (2010) Role of arginine deiminase of Streptococcus cristatus in Porphyromonas gingivalis colonization. Antimicrob Agents Chemother 54(11):4694–4698

Xiong RG, Zhou DD, Wu SX, Huang SY, Saimaiti A, Yang ZJ et al (2022) Health benefits and side effects of short-chain fatty acids. Food Secur 11(18):2863

Younes JA, Lievens E, Hummelen R, Van Der Westen R, Reid G, Petrova MI (2018) Women and their microbes: the unexpected friendship. Trends Microbiol 26(1):16–32

Zafar H, Saier MH (2021) Gut *Bacteroides* species in health and disease. Gut Microbes 13(1):1848158

Zakaria GE (2013) Antimicrobial and anti-adhesive properties of biosurfactant produced by lactobacilli isolates, biofilm formation and aggregation ability. J Gen Appl Microbiol 59(6):425–436

Zangl I, Pap IJ, Aspöck C, Schüller C (2020) The role of lactobacillus species in the control of Candida via biotrophic interactions. Microb Cell 7(1):1–14

Zeng H, Umar S, Rust B, Lazarova D, Bordonaro M (2019a) Secondary bile acids and short chain fatty acids in the colon: a focus on colonic microbiome, cell proliferation, inflammation, and cancer. Int J Mol Sci 20(5):1214

Zeng B, Shi S, Ashworth G, Dong C, Liu J, Xing F (2019b) ILC3 function as a double-edged sword in inflammatory bowel diseases. Cell Death Dis 10(4):315

Zhang D, Frenette PS (2019) Cross talk between neutrophils and the microbiota. Blood 133(20):2168–2177

Zhang C, Li L, Jin B, Xu X, Zuo X, Li Y et al (2021) The effects of delivery mode on the gut microbiota and health: state of art. Front Microbiol 12:724449

Zhao Y, Chen F, Wu W, Sun M, Bilotta AJ, Yao S et al (2018) GPR43 mediates microbiota metabolite SCFA regulation of antimicrobial peptide expression in intestinal epithelial cells via activation of mTOR and STAT3. Mucosal Immunol 11(3):752–762

Zheng D, Liwinski T, Elinav E (2020) Inflammasome activation and regulation: toward a better understanding of complex mechanisms. Cell Discov 6(1):36

Zheng Y, Hunt RL, Villaruz AE, Fisher EL, Liu R, Liu Q et al (2022) Commensal Staphylococcus epidermidis contributes to skin barrier homeostasis by generating protective ceramides. Cell Host Microbe 30(3):301–313.e9

Zhu MJ, Sun X, Du M (2018) AMPK in regulation of apical junctions and barrier function of intestinal epithelium. Tissue Barriers 6(2):1–13

Zipperer A, Konnerth MC, Laux C, Berscheid A, Janek D, Weidenmaier C et al (2016) Human commensals producing a novel antibiotic impair pathogen colonization. Nature 535(7613):511–516

Zuo T, Zhang F, Lui GCY, Yeoh YK, Li AYL, Zhan H et al (2020) Alterations in gut microbiota of patients with COVID-19 during time of hospitalization. Gastroenterology 159(3):944–955.e8

Human Microbiota and Metabolic Alteration in Cancer Progression and Treatment

16

Sarra B. Shakartalla, Tasneem M. Elsayed, Fatimah Almsailem, Fatma Almousa, Aseel Aldaihani, and Sameh S. M. Soliman

Abstract

In recent years, there has been a growing interest in the gut microbiota and its role in human health and disease. This important role is influenced by the composition of microbiota. A normal microbiota composition plays an important role in maintaining human health, while an alteration in microbiota composition and their metabolism can lead to several diseases including cancer. Recent studies have focused on the role of gut microbiota in cancer development, progression, metastasis, and response to therapy. Certain microbial communities can promote cancer cell migration, invasion, and formation of premetastatic niche, which are crucial in metastasis. On the other side, some metabolites secreted by the microbiota were found to inhibit metastasis. Gut microbiota can have an impact on the efficacy and side effects of anticancer drugs including chemotherapy, target ther-

S. B. Shakartalla
Research Institute for Medical and Health Sciences (RIMHS), University of Sharjah, Sharjah, UAE

Faculty of Pharmacy, University of Gezira, Wad Madani, Sudan

T. M. Elsayed
Division of Cardiology, The Lundquist Institute at Harbor-UCLA Medical Center, Los Angeles, CA, USA

F. Almsailem · F. Almousa · A. Aldaihani
Department of Medicinal Chemistry, College of Pharmacy, University of Sharjah, Sharjah, UAE

S. S. M. Soliman (✉)
Department of Medicinal Chemistry, College of Pharmacy, University of Sharjah, Sharjah, UAE

Research Institute for Medical and Health Sciences (RIMHS), University of Sharjah, Sharjah, UAE
e-mail: ssoliman@sharjah.ac.ae

S. S. M. Soliman, M. I. Husseiny (eds.), *Metabolic Dynamics in Host-Microbe Interaction*, https://doi.org/10.1007/978-981-96-1305-2_16

apy, and immunotherapy. For instance, manipulation of gut microbiota could modulate metastasis and therapy-induced side effects and may be used as an adjuvant therapy to enhance therapeutic outcomes. In this chapter, we explore the complex role of gut microbiota in cancer, cancer metastasis, and their effects on specific anticancer drugs. Potential microbiota-modulating strategies to improve cancer treatment were also discussed.

Keywords

Gut microbiota, metabolism · Cancer development · Cancer metastasis · Anticancer drugs

16.1 Introduction

The human microbiota refer to the collective microorganisms that live in and on the human body including bacteria, viruses, fungi, archaea, and protozoa and their genome (Ursell et al. 2012; Li et al. 2019a). These microorganisms inhabit various body sites, including skin, oral cavity, gastrointestinal tract (GIT), respiratory tract, and urogenital tract (Ursell et al. 2012; Li et al. 2019a). Human microbiota play an important role in promoting health and influencing disease processes. More than 97% of the human microbiota is found in the GIT, particularly in the colon (Sender et al. 2016a, b). This population is known as the gut microbiota, and it has been extensively studied, revealing its role in regulating various physiological functions, including supporting the immune system by preserving the integrity of the mucosal barrier, supplying essential nutrients and vitamins such as thiamine (B-1), biotin (B-7), riboflavin (B-2), and folate, and protecting against pathogens (Adak and Khan 2019; Gensollen et al. 2016; McNeil 1984). Besides, gut microbiota provide pathogen resistance by competing for attachment sites or nutrient sources and producing antimicrobial substances (Virtue et al. 2019). The microbiota impact not only the GIT but also other vital organs including the pancreas, liver, and brain (Riquelme et al. 2019). Dysbiosis is defined as an imbalance in the composition and function of the gut microbiota that contribute to the development of various clinical conditions including neurodegenerative diseases (Maini Rekdal et al. 2019; Erny et al. 2015), obesity (Virtue et al. 2019; Foley et al. 2018; Sun et al. 2018), diabetes mellitus, and cancer (Koh et al. 2018; Brown et al. 2016; Livanos et al. 2016). Dysbiosis can result from factors such as medication use (e.g., antibiotics), stress, infection, diet, and environmental toxins (Wen and Duffy 2017).

Studies have found that there is a strong relationship between the microbiota and cancer, demonstrating its involvement in nearly all aspects of carcinogenesis, from cancer susceptibility and progression in response to anticancer therapies, such as chemotherapy, target therapy, and immunotherapy (Panebianco et al. 2018; Khan et al. 2012). Several types of cancer may benefit from the restoration of microbiota dysbiosis via the use of helpful probiotic bacteria. Furthermore, several bacteria have been discovered to possess anticancer properties and are now being researched for efficient anticancer treatment (Khan et al. 2021; Sedighi et al. 2019). The microbiota can contribute to tumor development in three primary ways: (i) disrupting the cell growth and death balance, (ii) altering the immune system and its responses, and (iii) affecting the metabolism of substances like host-secreted factors, nutrients, and medications (Garrett 2015).

Microbiota were also found to play a pivotal role in cancer metastasis by influencing various biological pathways that regulate tumor progression. Certain microbial communities can promote cancer cell invasion, migration, the epithelial-mesenchymal transition (EMT), and formation of premetastatic niche, which are crucial in metastasis (Ivleva and Grivennikov 2022). The gut and intratumoral microbiota have been found to interact with the immune system, modulate inflammation, and influence the tumor microenvironment. These interactions can either suppress or enhance the metastatic potential of cancer cells, highlighting the microbiota's dual role in cancer progression.

Additionally, growing evidence over the past 10 years has demonstrated the critical functions that microbiota play in the toxicity and effectiveness of anticancer medications through processes such as microbial enzymatic breakdown and immunomodulation (Bashiardes et al. 2017). Thus, the human microbiota offers an attractive opportunity for the development of interventional strategies to enhance therapeutic benefits and biomarkers for predicting treatment outcomes. In this chapter, we provide an in-depth analysis of the role of microbiota in cancer progression and metastasis, highlighting how specific microbial species influence these processes. Additionally, we explored various examples of anticancer drugs whose therapeutic efficacy or adverse effects are modulated by the microbiota, emphasizing the dual impact of these microorganisms on treatment outcomes.

16.2 Microbiota in Cancer Progression and Metastasis

Microbiota can profoundly impact cancer invasion and metastasis by modifying the tumor microenvironment, regulating immune responses, and affecting oncogenic signaling pathways. Additionally, microbiota influence key stages of the metastatic process, including EMT, matrix metalloproteinase (MMP) activity, and the development of the premetastatic niche (Yang et al. 2023).

Numerous research has shown that gut microbiota play a vital role in modulating host immune responses. However, the intratumoral microbiota can notably affect the tumor microenvironment immunological responses, thus influencing cancer progression (Yang et al. 2023). The origin of intratumoral microbiota can be classified into three main sources: (i) mucosal barrier sources that are found in cancers such as colorectal, pancreatic, lung, and cervical, where organs with external exposure harbor microorganisms on their mucosal surfaces. These microbes may invade tumors when disruption in the mucosa occurs during cancer development; (ii) adjacent normal tissues—studies indicate that bacteria are present in organs previously considered sterile, with microbial compositions in tumors closely resembling those in nearby healthy tissues. The tumor's immunosuppressive and hypoxic environment facilitates colonization of microbes, though the microbe's origin in normal tissues is unclear and could involve spread from tumors, warranting further investigation; and (iii) hematogenous spread—microbes may enter the bloodstream from the mouth, gut, and other regions and infiltrate tumors through damaged blood vessels (Xie et al. 2022) (Fig. 16.1).

16.2.1 Microbial Effect in Cancer Cell Migration and Invasion

It has been reported that intratumoral microbiota profiles are closely linked to the expression of genes associated with EMT (Rubinstein et al. 2013). Microbes initiate EMT via adhering to the surface of mucosa, leading to loss of cell adhesion between epithelial cells (Rubinstein et al. 2013). Bacterial adhesins attach to the E-cadherin/β-catenin complex in epithelial cells, thereby disrupting cell polarity and triggering downstream signaling pathways that eventually induce EMT (Rubinstein et al. 2013). The microbiota within the tumors can trigger the release of various cytokines and stimulate the NF-κB signaling pathway through the interaction pattern recognition receptors (PRRs). This creates a feedback loop that sustains chronic inflammation, ultimately contributing to tumor growth and progression (Yang et al. 2023) (Table 16.1 and Fig. 16.2).

Fusobacterium nucleatum (*F. nucleatum*) is a gram-negative, anaerobic bacterium commonly found in the oral cavity, historically identified as an opportunistic pathogen linked to periodontal disease (Chen et al. 2022). However, *F. nucleatum* was reported recently to present in the tissues of colon cancer (Castellarin et al. 2012), and it has been identified as a potential cause of colorectal cancer (CRC) (Rubinstein et al. 2013). *Fusobacterium* adhesin A (FadA) is a key virulence factor in *F. nucleatum,* which allows the microbe to adhere and invade CRC cells, triggering oncogenic and inflammatory responses that promote tumor growth (Rubinstein et al. 2013). FadA binds to E-cadherin, activates β-catenin signaling, and selectively controls inflammatory and cancer-promoting signaling pathways (Rubinstein et al. 2013). Further, *F. nucleatum* was reported to stimulate the TLR4/MYD88/NF-κB signaling pathway, leading to elevated levels of miR-21 and suppression of the expression of RASA1 in CRC cells. This subsequently activates the RAS signaling pathway, promoting the upregulation of genes involved in cancer growth and progression (Yang et al. 2017). In addition, *F. nucleatum* was reported to trigger TLR4 signaling and elevate the expression of the cellular cytochrome P450 enzyme CYP2J2. This enzyme facilitates the conversion of linoleic acid into 12,13-EpOME, which subsequently induces EMT, thereby contributing to the development and spread of CRC (Kong et al. 2021). In laryngeal squamous cell carcinoma (LSCC), *F. nucleatum* upregulates miR-155-5p and miR-205-5p, which downregulates alcohol dehydrogenase 1B (ADH1B) and transforms growth factor β receptor 2 (TGFBR2) through innate immune signaling activation. This leads to ethanol metabolism reprogramming, facilitating the accumulation of *F. nucleatum* and triggering the PI3K/AKT signaling pathway. As a result, EMT is promoted, further driving the progression and metastasis of LSCC (Hsueh et al. 2022). Several studies

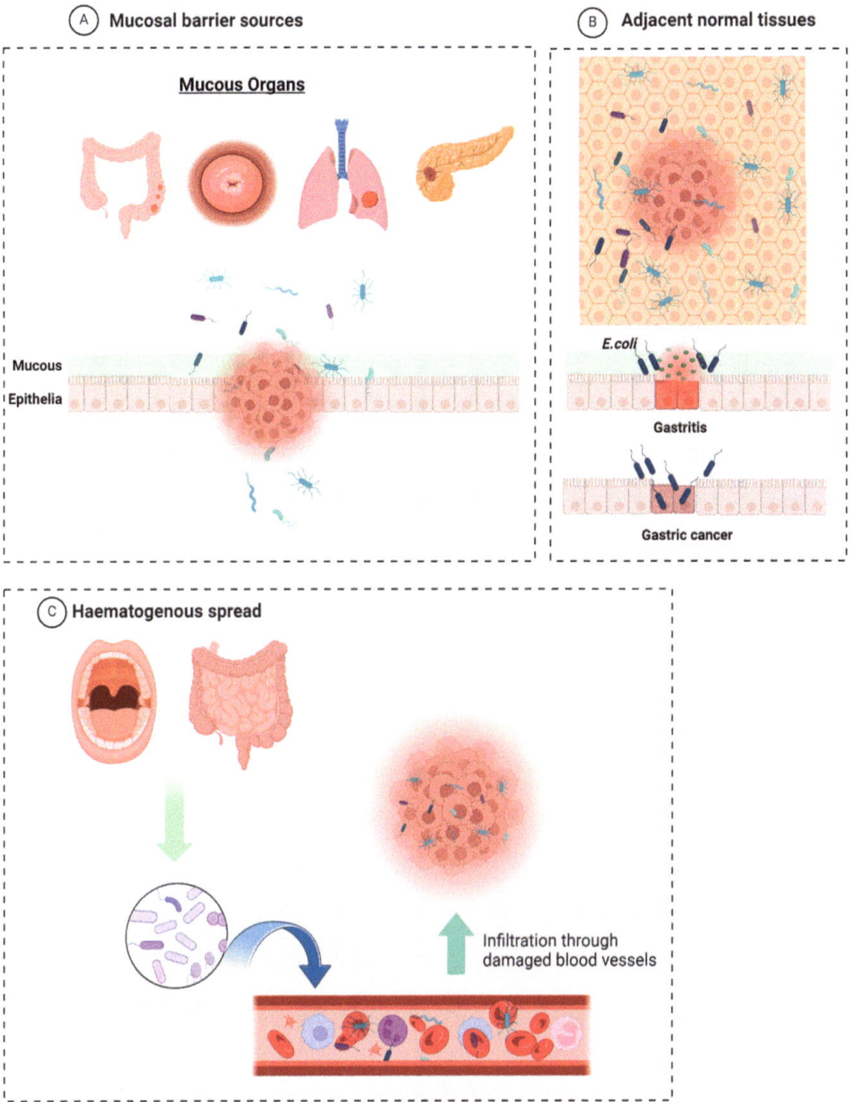

Fig. 16.1 Origin of intratumoral microbiota. The origin of intratumoral microbiota is categorized into three main sources: (**a**) mucosal surfaces, where microbes from organs exposed to external environments, like colorectal, pancreatic, lung, and cervical cancers, enter tumors due to mucosal disruption; (**b**) adjacent normal tissues, where microbes found in tumors resemble those in adjacent healthy tissues, potentially spreading due to the immunosuppressive and hypoxic conditions in tumors; and (**c**) hematogenous spread, where microbes from regions like the mouth and gut travel through the bloodstream and colonize tumors via damaged blood vessels

Table 16.1 Role of microbes in metastasis of different cancers

Microbe	Type of cancer	Effect on metastasis	Mechanism	Ref.
Fusobacterium nucleatum	Colorectal cancer (CRC)	Increase	Activates β-catenin signaling that selectively controls inflammatory and cancer-promoting signaling pathways leading to cancer progression	Rubinstein et al. (2013)
	CRC	Increase	Stimulates the TLR4/MYD88/NF-κB signaling pathway, leading to elevated levels of miR-21, and subsequent activation of the RAS signaling pathway, leading to cancer progression	Yang et al. (2017)
	CRC	Increase	Activates TLR4 signaling and increases the expression of CYP2J2 with conversion of linoleic acid into 12,13-EpOME resulting in EMT	Kong et al. (2021)
	Laryngeal squamous cell carcinoma (LSCC)	Increase	Upregulates specific microRNAs and reprogramming of ethanol metabolism leading to activation of PI3K/AKT signaling pathway and cancer aggressiveness	Hsueh et al. (2022)
	Breast cancer	Increase	Suppresses the accumulation of tumor-infiltrating T cells, through mechanism relying on Fap2-dependent binding of the bacterium to breast cancer tissues.	Parhi et al. (2020)
Bacteroides fragilis	CRC	Increase	Promotes E-cadherin cleavage and activates β-catenin signaling leading to cancer progression	Sears (2009)
Helicobacter pylori	Gastric cancer	Increase	Increases the expression of EMT markers like Snail and vimentin, while also promoting the release of HB-EGF, which is partly regulated by gastrin and MMP-7	Yin et al. (2010)
	Gastric cancer	Increase	Triggers the MAPK signaling pathway, enhancing gastrin expression and accelerating EMT progression	Gunawardhana et al. (2018)
	Gastric cancer	Increase	Deregulation of the β-catenin signaling pathway through CagA-dependent interaction of the microbe with E-cadherin.	Murata-Kamiya et al. (2007)

Escherichia coli	CRC	Increase	LPS produced by *E. coli* enhances the secretion of cathepsin K, leading to the polarization of macrophages toward an "M2-like" state	Li et al. (2019b)
	CRC	Increase	It travels to the liver through the portal vein, where it enhances the recruitment of innate immune cells and establishes an inflammatory, pro-metastatic environment	Bertocchi et al. (2021)
Porphyromonas gingivalis	Oral squamous cell carcinoma (OSCC)	Increase	Increases the expression of CD44 and CD133, activates MMP-1 and MMP-10, and increases the expression of EMT markers	Ha et al. (2015) and Sztukowska et al. (2016)
Trichomonas vaginalis	Prostate cancer	Increase	Polarization of macrophages toward an "M2-like" state	Han et al. (2020)
Lachnoclostridium Sutterella	Bladder cancer	Increase	Induction of inflammation and upregulation of EMT-related genes and components of the extracellular matrix	Li et al. (2024)
Dysbiotic oral microbiome (*Porphyromonas gingivalis*, *F. nucleatum*, *Streptococci*, *Peptostreptococci*, and *Prevotella*)	OSCC	Increase	Increases the expression of mesenchymal markers and decreases the expression of epithelial markers	Chakraborti and Das (2020)
Dysbiotic intestinal microbiome	Ovarian cancer	Increase	Activation of tumor-associated macrophages and elevation of IL-6 and TNF-α, which lead to EMT	Xu et al. (2019)
Intratumoral microbiota	Breast cancer	Increase	Helps in metastatic colonization by assisting in survival of CTCs	Fu et al. (2022)

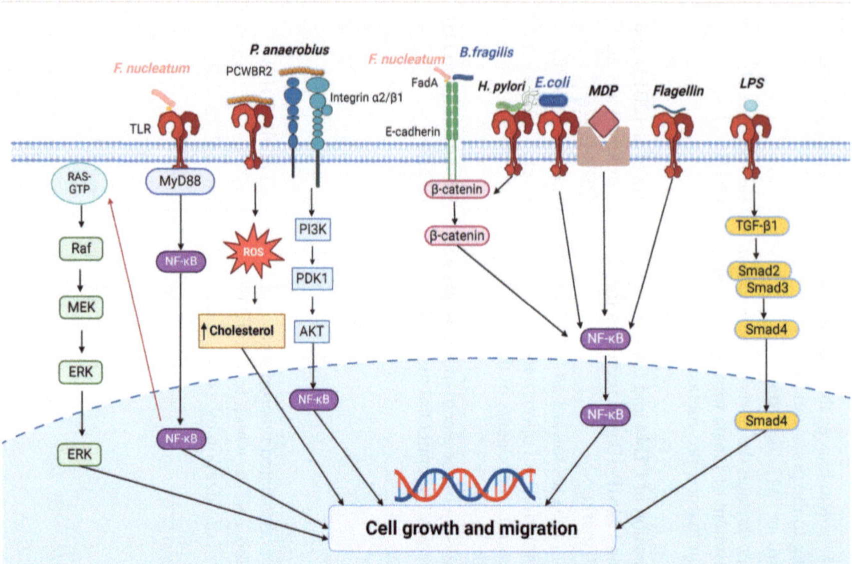

Fig. 16.2 Mechanisms by which microbiota induce cancer aggressiveness and metastasis. Microbiota can influence various signaling pathways, including β-catenin, PI3K/AKT, ERK, NF-κB, and TGF-β-Smad pathway. (i) β-catenin: *F. nucleatum, B. fragilis*, and *H. pylori* can either directly or indirectly initiate β-catenin signaling by phosphorylating E-cadherin. This process prompts β-catenin to move into the nucleus, where it activates T-cell factor (TCF), subsequently stimulating the transcription of downstream target genes and promoting cell proliferation and migration. (ii) PI3K-AKT: surface protein of *P. anaerobius*, known as PCWBR2, can directly bind to epithelial cells through the α2/β1 integrin triggering the PI3K-Akt signaling pathway. (iii) NF-κB: the microbiota can stimulate NF-κB activation via the β-catenin or TLR pathways, while *P. anaerobius*-induced PI3K signaling can also trigger NF-κB, leading to the production of pro-inflammatory cytokines and expression of EMT markers. (iv) ERK pathway: *F. nucleatum*-induced NF-κB signaling can cause activation of RAS-RAF-MEK-ERK signaling that promotes ERK to enter the nucleus and activates transcription of genes responsible for cell proliferation and aggressiveness. (v) TGF-β-Smad pathway: LPS-activated TLR can cause activation of TGF-β-regulated Smad proteins, particularly Smad2 and Smad3. These phosphorylated Smads form a complex with the co-Smad, Smad4, and this complex translocates into the nucleus and induces transcription of genes responsible for cancer cell migration

provide evidence indicating the role of *F. nucleatum* in suppression of immune cells (Kostic et al. 2013; Kim et al. 2021; Parhi et al. 2020). *F. nucleatum* was reported to suppress the activity of natural killer (NK) cells and lead to the buildup of myeloid cells (Kostic et al. 2013; Kim et al. 2021). Similarly, *F. nucleatum* suppresses the accumulation of tumor-infiltrating T cells, thereby increasing metastasis of breast cancer cells through mechanism dependent on fibroblast-associated protein (Fap2)-dependent binding of the bacterium to breast cancer (Parhi et al. 2020).

Bacteroides fragilis (*B. fragilis*) strains are opportunistic pathogens, frequently isolated as the dominant anaerobic bacteria in clinical samples, including cases of bloodstream infections and abdominal abscesses, although they make up less than 1–2% of the cultured fecal microbiota (Sears 2009). *B. fragilis* toxin (Bft) was reported to promote E-cadherin cleavage and activation of β-catenin signaling leading to cancer progression (Sears 2009).

Peptostreptococcus is a gram-positive anaerobic bacterium found in various parts of the human body, including the oral cavity, upper respiratory system, intestines, and female reproductive organs (Xu et al. 2020). This bacterium is reported to be implicated in CRC development (Long et al. 2019). It was found that the surface protein of *Peptostreptococcus anaerobius*, known as PCWBR2, can directly bind to colorectal epithelial cells through the α2/β1 integrin, which is often highly expressed in the tissues of CRC. Integrin α2/β1 and PCWBR2 interaction in CRC cells can trigger the PI3K/AKT pathway, which subsequently activates NF-κB, promoting increased cell proliferation and accelerating CRC progression (Long et al. 2019). Additionally, *P. anaerobius* has been shown to exacerbate intestinal dysplasia in azoxymethane-treated mice (Tsoi et al. 2017). This occurs as a result of binding of *P. anaerobius* with TLR2 and TLR4, leading to elevated intracellular reactive oxygen species (ROS) levels, which drive cholesterol synthesis and cell growth, ultimately accelerating cancer development (Tsoi et al. 2017). These findings suggest that *P. anaerobius* can thrive in the tumor microenvironment and contribute to CRC progression.

Helicobacter pylori (*H. pylori*) is a gram-negative microaerophilic bacterium found within or under the mucous layer covering the stomach epithelium (Amieva and Peek Jr 2016; Hooi et al. 2017; Mitchell 1999). *H. pylori* can persist asymptomatically in many individuals throughout life, particularly in those infected at an early age, without causing noticeable disease. Its low virulence in some cases and its ability to coexist within the human stomach indicate a commensal-like relationship; therefore, *H. pylori* is considered part of the normal microbiota, particularly in individuals who do not develop clinical symptoms (Reshetnyak et al. 2021). *H. pylori* is the primary contributor to gastric cancer and was designated as a class I carcinogen by the World Health Organization in 1994 (Amieva and Peek Jr 2016; Vogiatzi et al. 2007). *H. pylori* plays a critical role in driving EMT in gastric cancer through pathways involving soluble heparin-binding epidermal growth factor (HB-EGF), gastrin, and MMP-7 (Yin et al. 2010). The virulent strain of *H. pylori* boosts the expression of EMT markers like Snail and vimentin, while also promoting the release of HB-EGF, which is partly regulated by gastrin and MMP-7 (Yin et al. 2010). Furthermore, HB-EGF activation by *H. pylori* triggers the MAPK signaling pathway, enhancing gastrin expression and accelerating EMT progression (Gunawardhana et al. 2018). The main virulent factors of *H. pylori*, cytotoxin-associated gene A (CagA), and vacuolating cytotoxin (VacA) have been found to disrupt the function of epithelial cells. CagA affects the apical cell junction complex and induces reorganizations in the actin cytoskeleton, whereas VacA damages tight junctions, resulting in the loss of cell-to-cell adhesion and epithelial cell polarity, which facilitates cell migration (Papini et al. 1998; Amieva et al. 2003; Murata-Kamiya et al. 2007). In addition, *H. pylori* can also activate TLR2 and TLR5, leading to the activation of NF-κB and inducing pro-inflammatory responses that can lead to cancer aggressiveness (Smith et al. 2003). The involvement of *H. pylori* in various signaling pathways, including the activation of β-catenin pathway suppression of programmed cell death protein 4 (PDCD4) via the CagA protein, further illustrates its role in promoting EMT and the progression of gastric cancer (Yu et al. 2014; Franco et al. 2005).

Escherichia coli (*E. coli*) is a gram-negative bacterium that establishes itself in the GIT shortly after birth and generally maintains a harmonious coexistence with its human host (Murphy et al. 2021). In contrast, when junctional proteins are incorrectly localized, *E. coli* has the potential to cause diarrhea. This may result in EMT in chronic situations (König et al. 2016). Studies have demonstrated that diffusion-adherent *E. coli* can invade the epithelial cells of the intestine and facilitate EMT via stimulation of PI3K and Ras signaling pathways (Bétis et al. 2003). Additionally, *E. coli* promotes the overexpression of HIF-1α, which contributes to the reduction of E-cadherin and cytokeratin 18, alongside an increase in fibronectin, highlighting its potential involvement in EMT (Cane et al. 2010).

Staphylococcus aureus (*S. aureus*) is a gram-positive bacterium, a human microbiota particularly found within the nasal cavity, where it can exist as both a commensal and a pathogen (Laux et al. 2019). It was demonstrated that α-hemolysin, a cytotoxin secreted by *S. aureus* in human airway epithelial cells, has been found to enhance actin filament reorganization by disrupting cell-to-cell junctions and focal adhesion, thereby increasing epithelial layer permeability (Ziesemer et al. 2018). Moreover, *S. aureus* has been linked to nasal polyposis, where nasal polyps display reduced levels of E-cadherin and occludin, along with increased TGF-β and vimentin compared to healthy nasal tissues. This indicates the potential role of *S. aureus* in facilitating EMT.

Porphyromonas gingivalis (*P. gingivalis*) is a gram-negative anaerobic bacterium commonly found in individuals with periodontitis, and is widely recognized as a key pathogen responsible for the chronic, age-related inflammatory condition (Visentin et al. 2023; Ha et al. 2015). Persistent infection of oral cancer cells by *P. gingivalis* leads to the increased expression of cancer stem cell markers (CD44 and CD133) and release of IL-8 and gingival proteases, which subsequently activate MMPs including MMP-1 and MMP-10 and transcription factors like Slug, Snail, and Zeb1, driving EMT activity in oral squamous cell carcinoma (OSCC) (Ha et al. 2015; Sztukowska et al. 2016). This EMT process, induced by chronic infection, equips oral cells with enhanced invasive and migratory capabilities (Ha et al. 2015; Sztukowska et al. 2016).

Trichomonas vaginalis (*T. vaginalis*) is a protozoan parasite linked to sexually transmitted diseases and has been found in prostate-related conditions like prostatitis and prostate cancer (Han et al. 2020). It was demonstrated that IL-6 secreted from *T. vaginalis*-infected prostate cells significantly affects the tumor microenvironment, promoting EMT and prostate cancer progression through M2 macrophage polarization (Han et al. 2020).

The oral cavity harbors a diverse range of microorganisms, including *Porphyromonas gingivalis*, *Peptostreptococci, F. nucleatum*, *Prevotella*, *Peptostreptococci*, and *Streptococci* (Lee et al. 2017). Oral cancer arises in the mucosal tissues, with about 15% of cases associated with microbial imbalances in the oral microbiota (Zhao et al. 2017). Such dysbiosis can modulate the immune system, elevating the risk of periodontal diseases and OSCC (Nagy et al. 1998; Karpiński 2019). It was reported that oral microbiome is associated with lymph node metastasis in OSCC patients (Eun et al. 2021). Dysbiotic oral microbiota was reported to induce EMT by decreasing the expression of epithelial markers such as E-cadherin and β-catenin, while increasing mesenchymal markers like N-cadherin, fibronectin, and vimentin (Chakraborti and Das 2020). This shift promotes cancer cell invasion and metastasis.

In ovarian cancer, it has been shown that intestinal dysbiosis triggers the activation of tumor-associated macrophages resulting in elevations in the levels of circulating IL-6 and TNF-α, which are pro-inflammatory mediators playing a role in the onset of EMT (Xu et al. 2019).

In bladder cancer, it was shown that specific intratumoral bacteria, such as *Lachnoclostridium* and *Sutterella*, can induce inflammation, leading to changes in gene expression patterns that facilitate EMT and tumor progression (Li et al. 2024). Moreover, the presence of these microbes is linked to the upregulation of EMT-related genes and components of the extracellular matrix, establishing a strong connection between the microbiota and tumor progression (Li et al. 2024).

Several bacterial components, including lipopolysaccharide (LPS), flagellin, and muramyl dipeptides (MDP), have been the focus of many cancer-related studies. In gram-negative bacteria, LPS is an essential part of the outer membrane, interacting with TLR4 to create endotoxins (Park and Lee 2013). It has been shown that LPS decreases the expression of E-cadherin and increases the expression of S100A4 and α-SMA (mesenchymal markers) in intrahepatic biliary epithelial cells (Zhao et al. 2011). More significantly, it was demonstrated that LPS induces the overexpression of TGF-β1, a key driver of EMT through the Smad 2/3 pathway (Zhao et al. 2011). Similarly, both flagellin and MDP have been shown to induce EMT, primarily through the activation of NF-κB and MAPK signaling pathways (Honko and Mizel 2005). Additionally, flagellin promotes the synthesis of TGF-β and TGF-β1 that can induce EMT (Yang et al. 2011). Likewise, MDP promotes the gene expression of EMT markers in intestinal epithelium (Ferrand et al. 2019). LPS produced by *E. coli* enhances the secretion of cathepsin K, leading to the polarization of macrophages toward an "M2-like" state, which supports metastasis in CRC (Li et al. 2019b).

16.2.2 Microbiota in Cancer Metastatic Colonization

The role of microbiota in the formation of a premetastatic niche is increasingly recognized as a critical factor in cancer metastasis, particularly in CRC. Studies reported that gut microbiota can influence tumor microenvironments and facilitate the establishment of premetastatic niche in distant organs such as the liver. *E. coli* compromises the gut-vascular barrier and travels to the liver through the portal vein, where it enhances the recruitment of innate immune cells and establishes an inflammatory, pro-metastatic environment (Bertocchi et al. 2021). In cases of CRC, *E. coli* breaches the barrier between gut and blood vessels, enters the bloodstream, and settles in the liver, triggering the development of a premetastatic niche that encourages liver metastasis (Bertocchi et al. 2021). Administration of broad-spectrum antibiotics resulted in a reduction of tumor growth in liver metastasis models for both pancreatic and colon cancers (Sethi et al. 2018).

In breast cancer, during the process of metastatic colonization, bacteria residing within circulating tumor cells (CTCs) that travel through the bloodstream including *Staphylococcus, Lactobacillus*, and *Streptococcus* help these cells survive by increasing their resistance to fluid shear stress. This is achieved through the reorganization of the actin cytoskeleton, which reinforces the cells' structural integrity under mechanical pressure (Fu et al. 2022).

16.2.3 Role of Metabolites Secreted by Human Microbiota in Cancer Progression and Metastasis

Metabolism is a complex process shaped by gut microbiota, which react to environmental signals and dietary intake. The microbiota generate metabolites that interact and communicate with different organs and tissues in the body, such as the liver, adipose tissue, brain, cardiovascular system, and lungs (Schroeder and Bäckhed 2016). In cancer, these metabolites were found to play a critical role in tumor progression and metastasis (Fig. 16.3).

Spermine, a metabolite generated by gut microbiota, has been shown to enhance non-small-cell lung cancer (NSCLC) metastasis to the brain by driving M2 polarization of microglia through the STAT3 pathway, which leads to the suppression of immune responses (Liu et al. 2022). Similarly, several studies indicated that bile acids, generated by gut bacteria, are associated with the progression of CRC, showing carcinogenic potential that could enhance tumor growth (Wu et al. 2024). Metagenomic and metabolomic analyses of paired samples from CRC tumors, nearby normal colon tissue, liver metastases, and adjacent normal liver revealed that liver metastases were notably enriched with *Bacteroides* and *Fusobacteria*. Additionally, 12-hydroxy-12-octadecanoylcarnitine and 3-hydroxy-9Z-octadeceno yl-carnitine were linked to the presence of liver metastases and could contribute to the metabolic alterations seen in CRC patients (Wang et al. 2023a, b).

Short-chain fatty acids (SCFAs), such as butyrate, acetate, and propionate, are metabolic byproducts of dietary fiber fermentation by the gut microbiota in the colon (Tan et al. 2014; McNabney and Henagan 2017). These SCFAs play a role in regulating the resistance to insulin and inflammation in the colon (McNabney and Henagan 2017). The presence of *Lactobacillus, Bacteroides, Bifidobacterium*, and *Akkermansia* is linked to the increased production of SCFAs (Blaak et al. 2020). In colon cancer, it was found that propionate inhibits the growth of CRC cells by reducing the expression of protein arginine *N*-methyltransferase 1 (PRMT1), leading to cell apoptosis through modulation of the mTOR pathway (Ryu et al. 2019). Sodium butyrate selectively targets and kills cancer cells, while inhibiting migration in CRC. It suppresses cell growth and reduces Trx-1 protein expression in CRC cells without affecting normal colon epithelial cells. Additionally, sodium butyrate induces apoptosis, inhibits colony formation, and impairs migration and EMT in

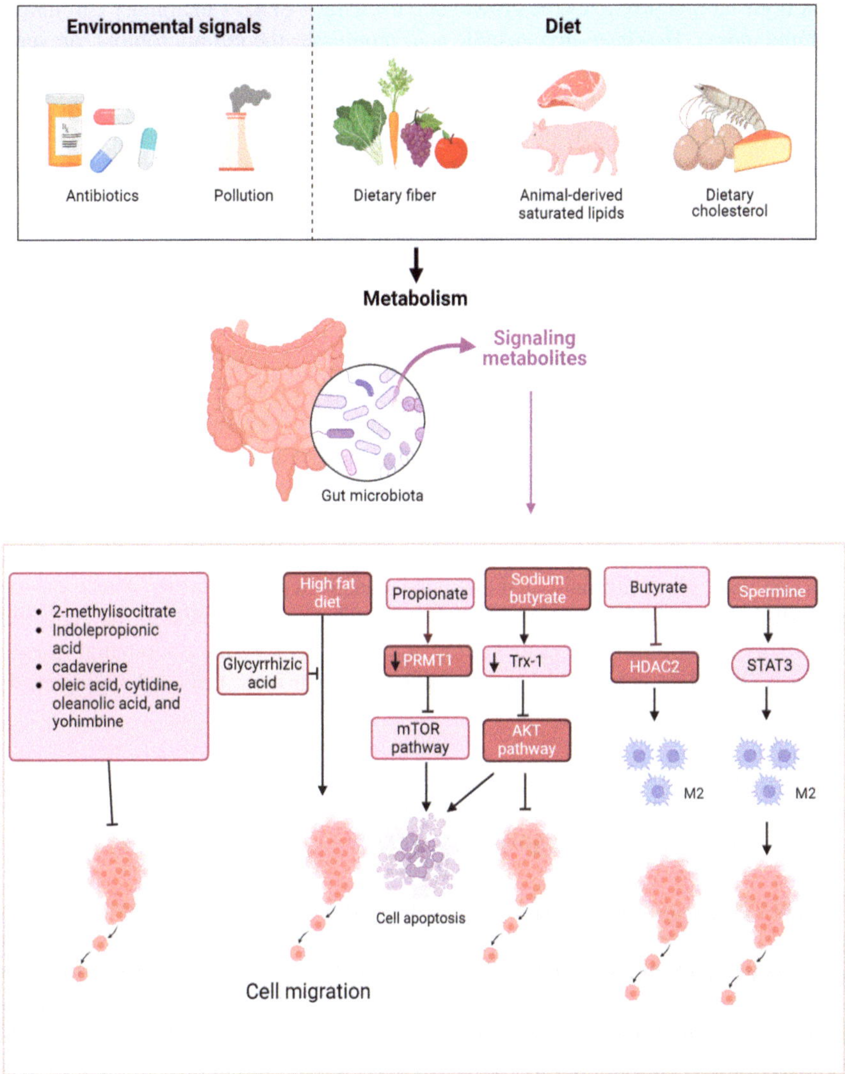

Fig. 16.3 Role of metabolites secreted by microbiota in cancer cell migration and metastasis. Gut microbiota interact with the environmental signals and diet, thereby shaping the host metabolism. The microbiota generates signaling metabolites that can either inhibit or promote cancer metastasis through different mechanisms including mTOR, AKT, and STAT signaling and immunomodulation

CRC cells (Wang et al. 2020). However, butyrate, produced by intratumoral microbiota, was reported to promote lung cancer metastasis by inducing H19 expression in tumor cells, inhibiting histone deacetylase (HDAC2), increasing H3K27 acetylation, and inducing M2 macrophage polarization (Ma et al. 2024).

A high-fat diet promotes the growth and metastasis of 4 T1 mammary carcinoma and melanoma. However, glycyrrhizic acid modulates the gut microbiota, preventing the formation of a premetastatic niche and inhibiting high-fat diet-enhanced metastasis to the lungs in cases of breast cancer and melanoma (Jung et al. 2015; Kim et al. 2011). Similarly, the microbiota-produced metabolite 2-methylisocitrate suppresses cancer metastasis by limiting tumor spread and enhancing survival rates, highlighting its promise as a supportive agent in cancer therapy (Martinez-Martinez et al. 2024). In breast cancer, indole propionic acid, which is a metabolite secreted by gut microbiota, was found to lower the levels of cancer stem cells and inhibit the growth, migration, and metastasis of cancer cells (Sári et al. 2020). This was accomplished by suppressing EMT, promoting oxidative and nitrosative stress, and enhancing the antitumor immune response (Sári et al. 2020). Likewise, cadaverine, which is also secreted by the gut microbiota, was reported to reverse the EMT, suppress cell movement and invasion, and reduce the stem-like properties of breast cancer cells by lowering mitochondrial oxidation. These effects were mediated through trace amino acid receptors, specifically TAAR1, TAAR8, and TAAR9 (Kovács et al. 2019).

Metabolites secreted from *Lactobacillus plantarum* YYC-3 including oleic acid, cytidine, and oleanolic acid demonstrated a significant inhibitory effect on the growth, invasion, and migration of CRC cells. They also suppressed the gene and protein expression of MMP-2 and MMP-9, as well as vascular endothelial growth factor A (VEGFA), which are key regulators of cancer metastasis and angiogenesis (Yue et al. 2020).

16.3 The Human Microbiota and Its Effect on Anticancer Agents

The microbiota plays a crucial role in modulating the efficacy of anticancer therapies, including chemotherapy, target therapy, and immunotherapy (Fig. 16.4). Microbial communities within the body can either enhance or reduce the effectiveness of cancer treatments (Zhao et al. 2023). On the positive side, specific microbiota can improve host immune responses, boosting the efficacy of immunotherapies such as immune checkpoint inhibitors (ICIs). However, certain microbiota may also diminish treatment outcomes by interfering with drug mechanisms or inducing resistance. In addition to efficacy, microbiota also affect the adverse effects of chemotherapy and immunotherapy, potentially exacerbating or mitigating toxicity levels, thus impacting overall treatment tolerability and success (Zhao et al. 2023).

Enterohepatic recycling plays a critical role in drug metabolism, particularly through its interaction with gut microbiota. Drugs excreted into bile can be metabolized by gut microbial enzymes such as oxidoreductase, hydrolases, and lyases, which may either inactivate the drug or convert it to more active or toxic metabolites. The drug or its metabolites can then be reabsorbed into the circulation, leading to prolonged drug activity and altered pharmacokinetics. This recycling process can influence bioavailability, duration of action, and overall efficacy of oral drugs.

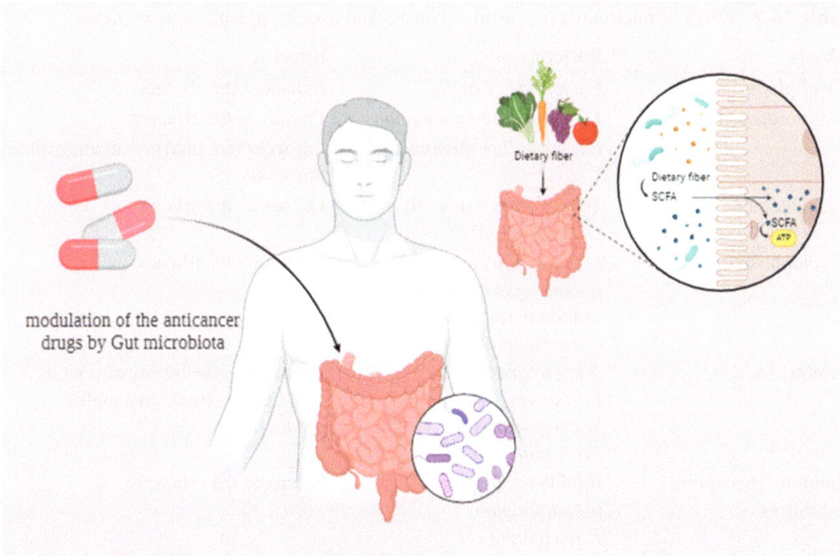

Fig. 16.4 Modulation of anticancer drugs by gut microbiota

For instance, *Eggerthella* is a gram-positive anaerobic bacterium that exists in the gut microbiota and responsible for the deactivation of some therapeutic medications that reduce the efficacy of the drugs (Dobkin et al. 1982; Lindenbaum et al. 1981). Moreover, gut microbiota can affect the drug by converting it to a toxic metabolite, which can increase the adverse reaction of the drug (Dobrinska 1989; Malik et al. 2016). Table 16.2 summarizes the role of microbiota in anticancer drug efficacy and toxicity.

16.3.1 Effect of Microbiota on the Efficacy of Chemotherapeutic Agents

Oxaliplatin is an alkylating agent that belongs to the class of platinum-based anticancer drugs that interfere with the replication of cancer cells by damaging their DNA, specifically through creation of platinum-DNA abducts (Chen et al. 2019). It is the first-line treatment for advanced colon and rectal cancer (Schoch et al. 2020). Oxaliplatin stimulates immune cells like $CD8^{+}$ T lymphocytes, which results in immunogenic cell death (Schoch et al. 2020). Butyrate, a gut microbiota metabolite, was reported to enhance the oxaliplatin's chemotherapeutic efficiency by controlling $CD8^{+}$ T-cell activity in the tumor microenvironment via IL-12 signaling (Schoch et al. 2020; He et al. 2021). Moreover, cancer patients who showed a positive response to oxaliplatin treatment had higher levels of butyrate in their serum compared to those who did not respond to the therapy (He et al. 2021). As a result, specific metabolites produced by gut microbiota could potentially be employed in

Table 16.2 Effect of microbiota on the effectiveness and toxicity of anticancer therapies

Drugs	Bacteria	Effect
Oxaliplatin	*Bacteroides fragilis*	Enhances the efficacy
5-Fluorouracil	*Fusobacterium nucleatum*	Decreases the efficacy
	Lactobacillus rhamnosus	lowers severe diarrhea and intestinal mucositis
Gemcitabine	*Mycoplasma hyorhinis* *Gammaproteobacteria*	Decreases the efficacy
Cyclophosphamide	*Lactobacillus johnsonii* *Lactobacillus murinus* *Enterococcus hirae* *Barnesiella intestinihominis*	Enhances the efficacy
Irinotecan	*Bacteroidetes* *Firmicutes* *Verrucomicrobia* *Proteobacteria*	Increases gastrointestinal toxicity and diarrhea (toxic metabolite production)
Immune checkpoint inhibitors	*Bifidobacterium pseudolongum* *F. nucleatum*	Enhances the efficacy
EGFR inhibitor	Commensal skin microbiota	Induces skin toxicities (toxic metabolite production)
Anti-CD47 antibodies	*Bifidobacterium*	Enhances the efficacy

the future as adjuncts to oxaliplatin, improving its anticancer effects. Additionally, it was discovered that ileal microbiota, such as *B. fragilis*, may strengthen the antitumor efficacy of oxaliplatin-based chemotherapy and stimulate local immunological responses by increasing the expression of immune-related genes and promoting the infiltration of the $CD45^+$ lymphoid cell (Picard et al. 2021). Significantly, reactive oxygen species (ROS) generated by the inflammatory response triggered by commensal bacteria are essential for maintaining the cytotoxic effectiveness of oxaliplatin (Iida et al. 2013).

Cyclophosphamide, a well-known alkylating anticancer drug, triggers immunogenic cell death in cancer cells, disrupts the activity of immunosuppressive T cells, and enhances the activation of T helper 1 (Th1) and T helper 17 (Th17) cells, which are crucial for inhibiting tumor progression (Sistigu et al. 2011; Ghiringhelli et al. 2004; Viaud et al. 2011). Studies on mouse models demonstrate that cyclophosphamide alters the gut microbiota, specifically in the small intestine resulting in enhancement of its anticancer effect (Viaud et al. 2013; Daillère et al. 2016). It allows certain gram-positive bacteria including *Lactobacillus johnsonii, Lactobacillus murinus*, and *Enterococcus hirae* to migrate to secondary lymphoid organs, where they stimulate the formation of "pathogenic" (pTh17) cells and memory Th1 immune responses (Viaud et al. 2013). In tumor-bearing mice that were either germ-free or treated with antibiotics to remove these gram-positive bacteria, the immune response was weakened, resulting in reduced pTh17 activity and resistance to cyclophosphamide therapy (Viaud et al. 2013). Following cyclophosphamide treatment, *Barnesiella intestinihominis* was reported to stimulate Th1 response, increase the number of cytotoxic CD8⁺ T cells, and encourage the infiltration of T cells that produce IFN-γ in the tumor microenvironment (Daillère et al. 2016). Remarkably, cyclophosphamide resistance is brought on by the depletion of gram-positive and gram-negative bacteria, respectively, *Enterococcus hirae* and *Barnesiella intestinihominis* (Daillère et al. 2016).

16.3.2 Role of Microbiota in Chemotherapeutic Resistance

5-Fluorouracil (5-FU) is an antimetabolite chemotherapeutic agent and one of the most used anticancer drugs. It has a crucial role in the treatment of CRC, breast cancer, and other cancers (Grem 2000). 5-FU has a similar structure to the pyrimidine molecules that exist in the DNA and RNA, which is inserted incorrectly in replacement of uracil or thymine. An incorrect replacement of DNA's uracil with the 5-FU in the targeted treated cells that will be accumulated in the genome leads to 5-FU cytotoxicity (An et al. 2007). Due to these structure similarities, 5-FU interferes with nucleoside metabolism and can be integrated into RNA and DNA; therefore, it may lead to cytotoxicity and cell death (Thomas and Zalcberg 1998; Noordhuis et al. 2004). In CRC, high abundance of *F. nucleatum* is correlated with resistance of 5-FU (Zhang et al. 2019; Yu et al. 2017). *F. nucleatum* enhances the chemotherapy resistance through diverse mechanisms, including increasing the expression of BIRC3, regulating TLR4-MYD88 signaling pathway, genomic loss of specific miRNAs, such as miR-18a and miR-4802, and upregulating key autophagy elements, including ULK1 and ATG7 (Zhang et al. 2019; Yu et al. 2017).

Gemcitabine (2′,2′-difluoro-2′-deoxycytidine) is a cytidine analog that is used as a first-line treatment for several solid cancers including pancreatic ductal adenocarcinoma (PDAC) (Jia and Xie 2015; Geller et al. 2017). The anticancer effects of gemcitabine involve inhibiting the passage of the cell cycle from the G1 to S phase and killing cells that are actively synthesizing DNA during the S phase (Jia and Xie 2015). Certain enzymes secreted from bacteria have the ability to convert gemcitabine into its inert form (2′,2′-difluorodeoxycytidine), which can lead to the development of drug resistance (Geller et al. 2017) (Fig. 16.5). Gemcitabine resistance was observed in colon cancer models due to *Mycoplasma hyorhinis* and other strains of bacteria, primarily belonging to the *Gammaproteobacteria* class. An isoform of bacterial cytidine deaminase (CDDL), primarily found in *Gammaproteobacteria*, is responsible for the metabolism of gemcitabine. However, the application of antibiotics reversed this impact suggesting that the anticancer activity of chemotherapeutic agents can be increased by combining them with antibiotic against *Gammaproteobacteria* (Geller et al. 2017). Moreover, intratumoral bacteria and their contributions to resistance against gemcitabine chemotherapy were confirmed by the discovery of bacterial DNA in human PDAC and the induction of gemcitabine resistance in human carcinoma cell lines by bacteria cultivated from fresh human pancreatic ductal adenocarcinoma tumors (Jia and Xie 2015; Geller et al. 2017). In contrast, butyrate was found to boost the effectiveness of

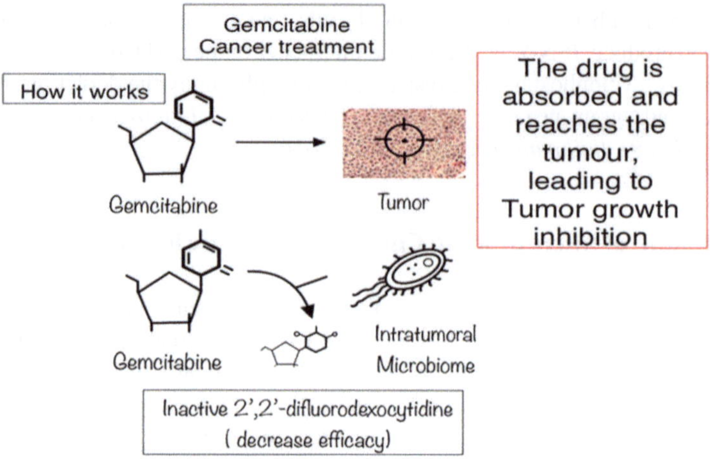

Fig. 16.5 Effect of microbiota on gemcitabine

gemcitabine in targeting cancer cells by promoting apoptosis (Panebianco et al. 2022).

16.3.3 Effect of Microbiota on the Efficacy of Immunotherapeutic Agents

ICIs are monoclonal antibodies used in cancer therapy. ICIs utilize the components of the immune system to fight against tumor cells by inhibiting key regulatory mechanisms known as checkpoints, which restrict the immune responses within the body (Sadeghi Rad et al. 2021). Programmed cell death (PD-1) and cytotoxic T lymphocyte antigen 4 (CTLA-4) are inhibitory coreceptors found on the surface of T cells suppressing the immune responses induced by T cells; nevertheless, these inhibitory components are utilized by cancer cells to promote tumor tolerance and exhaustion of T cells (Seidel et al. 2018). Consequently, ICIs such as anti-CTLA-4 (ipilimumab), anti-PD-1 (nivolumab, pembrolizumab), and anti-PD-L1 (atezolizumab, durvalumab) can bind to these receptors to boost anticancer immune responses against tumor cells (Seidel et al. 2018).

The microbiota exhibits a significant effect on the human immune system by regulating both innate and adaptive immunity. The microbiota trigger PRRs on intestinal epithelial cells, leading to downstream signaling that modulates the antimicrobial response. This process involves the expression of pro-inflammatory cytokines, including TNF-α, IL-1, and IL-6 (Thaiss et al. 2016). The activation of innate immunity by the microbiota and PRRs also triggers the activation of adaptive immunity. A prospective study by Nomura et al. investigated the concentrations of SCFAs in feces and serum, and their variations in response to anti-PD-1 antibodies. The findings revealed that elevated levels of SCFAs, including fecal acetic acid, propionic acid, butyric acid, and valeric acid, were associated with extended progression-free survival (Nomura et al. 2020). SCFAs can directly modulate immune responses, suppressing regulatory T cells (Treg) and promoting effector T-cell activity based on the availability of various cytokines (Park et al. 2015). *Bifidobacterium pseudolongum* was reported to improve the response to ICIs by generating the metabolite inosine, which subsequently activates anticancer response. It was demonstrated that elevated levels of *F. nucleatum* are associated with better responses to PD-L1 blockade therapy in CRC, leading to stronger antitumor effects and extended survival in mouse models (Gao et al. 2021). This bacterium triggers STING signaling, which upregulates the expression of PD-L1 and promotes the recruitment of $CD8^+$ tumor-infiltrating lymphocytes, further enhancing the response of tumor to immunotherapy (Gao et al. 2021).

CD47 blockade is an effective strategy for increasing the immune ability to target and eliminate tumors (Zhao et al. 2022). Anti-CD47 therapy targets the interaction between CD47 on tumor cells and SIRPα on macrophages, which normally sends a "don't eat me" signal to inhibit phagocytosis. By blocking CD47, this therapy allows macrophages to engulf and destroy cancer cells (Maute et al. 2022). Additionally, anti-CD47 antibodies can stimulate tumor cell apoptosis and promote antigen presentation, enhancing T-cell-mediated responses (Maute et al. 2022). Through the STING signaling pathway, intratumoral *Bifidobacterium* activates dendritic cells (DCs) that alter the tumor's response to anti-CD47 immunotherapy (Shi et al. 2020). Furthermore, after anti-CD47 therapy, local administration of *Bifidobacterium* enhances cross-priming of DCs by activating STING signaling, presenting a novel strategy for T-cell-targeted immunotherapy in conjunction with intratumoral bacteria (Shi et al. 2020). The exact relationship between microbiota and tumor immune microenvironment remains widely unclear and needs deep investigations. Nevertheless, research offers compelling support for developing a microbiota-focused strategy to improve and anticipate the therapeutic outcomes of immunotherapeutic treatments in patients.

16.3.4 Effect of Microbiota on Chemotherapy-Induced Side Effect

Microbiota can either increase or decrease the side effects of chemotherapeutic agents. Irinotecan is widely used in first-line treatment alongside 5-FU and folinic acid to treat CRC (Chamseddine et al. 2019). Irinotecan and its active metabolite 7-ethyl-10-hydroxycamptothecin (SN-38) form a complex with the DNA inhibiting DNA topoisomerase I and eventually preventing the reconnection of DNA strands, leading to DNA fragmentation and cellular apoptosis (Chamseddine et al. 2019). The gut microbiota can modify the side effects of irinotecan via microbial enzyme degradation (Bhatt et al. 2020). The irinotecan-induced gastrointestinal toxicity is mediated by gut bacterial β-glucuronidase (BGUS), which may lead to decrease in dosage and potential discontinuation of therapy. In the GIT, glucuronidase releases the active metabolite (SN-38) of irinotecan after cleaving the glucuronide from the inactive metabolite (SN-38G) of irinotecan, which causes intestinal epithelial damage and diarrhea (Bhatt et al. 2020). There are many microbial organisms that can produce BGUS, including *Bacteroidetes*, *Firmicutes*, *Verrucomicrobia*, and *Proteobacteria* found in the gut microbiota (Bhatt et al. 2020). It was demonstrated that combining irinotecan with BGUS inhibitors (e.g., ciprofloxacin, amoxapine) can reduce the irinotecan-induced gastrointestinal toxicity and diarrhea (Bhatt et al. 2020; Kodawara et al. 2016). Since most strains of *E. coli* have BGUS with up to 50% of the highly conserved amino acid sequence in common with human BGUS, direct targeting of *E. coli* BGUS appears to be a safe and effective way to prevent irinotecan-induced intestinal toxicity (Sun et al. 2020). High-throughput screening was used to identify several distinct *E. coli* BGUS inhibitors such as caffeic acid ethyl ester (CAEE) and acteoside (Li et al. 2020). These selective BGUS inhibitors decrease intestinal BGUS activity and the incidence of irinotecan-induced intestinal toxicity without eradicating the bacteria or interfering with host cell BGUS or affecting irinotecan's therapeutic efficacy (Li et al. 2020; Wang 2021). However, this is still an area of research that can aid in developing synthetically engineered selective intestinal bacterial BGUS inhibitors as a strategy to reduce irinotecan-induced GI toxicity.

Some microbiota were reported to reduce the side effects of chemotherapeutic agents. It has been demonstrated that administration of 5-FU results in alteration of the gut microbiota resulting in 5-FU induced intestinal mucositis (Li et al. 2017). These results suggest that 5-FU-induced mucositis may be alleviated by restoring the balance of gut microbiota (Li et al. 2017). Hence, reducing gastrointestinal mucosa damage by using probiotic supplements of *Lactobacillus* could reduce and prevent the mucositis that happens upon administration of 5-FU. Specifically in CRC patients, an oral-administered *Lactobacillus rhamnosus* was successful in lowering severe diarrhea and intestinal mucositis that results from 5-FU, by decreasing and regulating the expression of NF-κB, TNF-α, and IL-6 (Österlund et al. 2007; Chang et al. 2018) (Fig. 16.6).

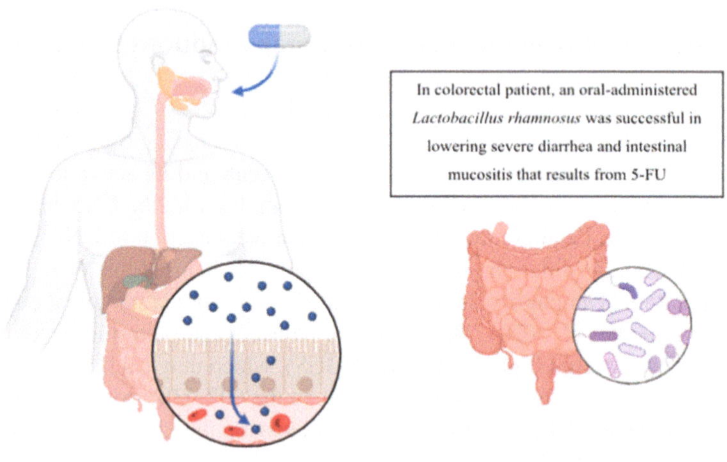

Fig. 16.6 Impact of oral-administering *Lactobacillus rhamnosus* on the toxicity of 5-FU

16.3.5 Effect of Microbiota on Immunotherapy-Induced Side Effects

Although ICI therapy is considered effective in treating cancer, it can induce some toxicity that is known as immune-related adverse events. With an increased immune system response, ICIs can cause inflammatory side effects on the skin, GIT, liver, and endocrine glands (Postow et al. 2018; Weber et al. 2017). The exact mechanism responsible for immune-related adverse events caused by ICIs is still not fully understood (Postow et al. 2018). Among these adverse events, ICI-induced colitis is the most frequent and severe one (Martins et al. 2019; Nielsen et al. 2022). This adverse effect can vary from mild diarrhea to severe colitis (Martins et al. 2019). The occurrence of ICI-induced colitis has been linked to a decreased diversity and alteration in gut microbiota composition. For instance, in melanoma patients receiving anti-CTLA-4 treatment, the presence of certain *Firmicutes* species (such as *Faecalibacterium prausnitzii* and *Gemmiger formicilis*) was linked to the development of ICI-induced colitis, whereas *Bacteroidetes* was associated with the absence of this condition (Martins et al. 2019). Before starting anti-CTLA-4 therapy, researchers investigated the baseline microbiota of patients with metastatic melanoma. They found that resistance to ICI-induced colitis was connected to a higher abundance of the *Bacteroidetes* phylum and three of its families (*Bacteroidaceae*, *Rikenellaceae*, and *Barnesiellaceae*) (Martins et al. 2019). These results propose that based on the composition of gut microbiota, we can assess the risk of ICI-induced colitis before treatment (Martins et al. 2019).

16.3.6 Effect of Microbiota on Target Therapy-Induced Side Effects

Apart from conventional chemotherapies, several targeted medicines have been made accessible to treat cancer. For instance, certain forms of cancer can be treated with epidermal growth factor receptor (EGFR) inhibitors like cetuximab or panitumumab (Klufa et al. 2019). Antibodies that cause internalization and degradation of EGFR, such as cetuximab and panitumumab, are commonly used to treat NSCLC and CRC (Klufa et al. 2019). Commensal skin microbiota may exacerbate skin toxicities such as itching, inflammation, and papulopustular skin rash when applying EGFR inhibitors. Anti-EGFR therapy-induced disruption of the skin barrier may facilitate the proliferation of microbiota, aggravating immunological responses including an increase in TNF-α, Th2 cytokines IL-4 and IL-33 and resulting in skin inflammation (Klufa et al. 2019).

16.4 Microbiota-Modulating Strategies in Cancer Therapy

Potential microbiota-modulating strategies for improving cancer therapeutic effects including fecal microbiota transplantation (FMT) and probiotics administration were used to improve the effect of anticancer agents (Fig. 16.7). FMT is a medical procedure used to transfer a small sample of stool from a healthy donor to a diseased recipient (Wang et al. 2019a). The stool sample contains a variety of beneficial microbiota, and it has been proven by recent studies that the combination of FMT and anti-PD-1 restored and altered the tumor microenvironment, as well as increased activation of CD 8^+ T cell. In addition, FMT can reduce the adverse effects of cancer therapy by restoring the reduced diversity and dysbiosis of microbiota. It has been shown that ICI-related colitis can be successfully treated by FMT as it is highly effective against chemotherapy-induced *Clostridium difficile* infection (Baruch et al. 2021). For these reasons, FMT is considered a potential strategy to improve cancer therapy.

Probiotics comprise various bacteria, including *Lactobacillus* and *Bifidobacterium*, which are live microorganisms capable of altering the bacterial composition in the GIT to address various conditions, such as inflammatory bowel disease (Kechagia et al. 2013). Recent studies have shown that natural probiotics and genetically engineered probiotics have great potential to both improve efficacy and alleviate adverse effects of cancer chemotherapy. It was reported that the adverse effects of irinotecan can be mitigated by using specific probiotics (Wang et al. 2019b). This probiotic helps to improve the integrity of the intestinal barrier, alleviate gut dysbiosis, and consequently reduce irinotecan-induced intestinal complications like diarrhea (Wang et al. 2019b).

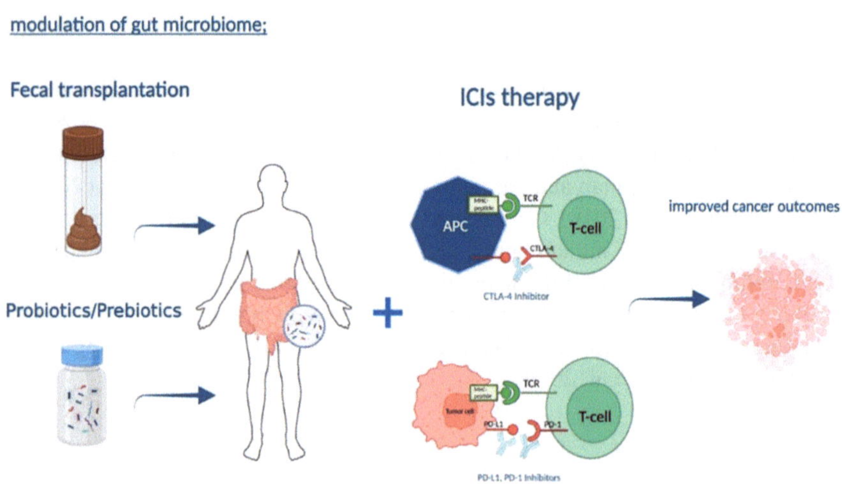

Fig. 16.7 Strategies for modulation of gut microbiota in cancer therapy

16.5 Conclusions

The complex interactions between the gut microbiota, cancer development and metastasis, and anticancer medications are increasingly recognized as pivotal in shaping therapeutic outcomes. Microbiota plays a dual role, both in enhancing and inhibiting the efficacy of anticancer therapies and in influencing metastatic progression through immune modulation and metabolic alterations. These microorganisms can metabolize drugs, modulate immune responses, and influence the tumor microenvironment, contributing to drug resistance or potentiation. Understanding these interactions opens new avenues for personalized cancer therapies, where modulation of gut microbiota could optimize drug efficacy, reduce toxicity, and potentially inhibit metastasis. Future research should focus on the identification of key microbial players and their mechanisms to enhance therapeutic strategies, potentially improving patient outcomes in oncology.

Acknowledgement This study was funded by Al Jalila Foundation Research Grant (AJF2023-078) to SSMS.

References

Adak A, Khan MR (2019) An insight into gut microbiota and its functionalities. Cell Mol Life Sci 76:473–493

Amieva M, Peek RM Jr (2016) Pathobiology of Helicobacter pylori–induced gastric cancer. Gastroenterology 150(1):64–78

Amieva MR, Vogelmann R, Covacci A, Tompkins LS, Nelson WJ, Falkow S (2003) Disruption of the epithelial apical-junctional complex by Helicobacter pylori CagA. Science 300(5624):1430–1434

An Q, Robins P, Lindahl T, Barnes DE (2007) 5-Fluorouracil incorporated into DNA is excised by the Smug1 DNA glycosylase to reduce drug cytotoxicity. Cancer Res 67(3):940–945

Baruch EN, Youngster I, Ben-Betzalel G, Ortenberg R, Lahat A, Katz L et al (2021) Fecal microbiota transplant promotes response in immunotherapy-refractory melanoma patients. Science 371(6529):602–609

Bashiardes S, Tuganbaev T, Federici S, Elinav E (eds) (2017) The microbiome in anti-cancer therapy. In: Seminars in immunology. Elsevier

Bertocchi A, Carloni S, Ravenda PS, Bertalot G, Spadoni I, Cascio AL et al (2021) Gut vascular barrier impairment leads to intestinal bacteria dissemination and colorectal cancer metastasis to liver. Cancer Cell 39(5):708–724. e11

Bétis F, Brest P, Hofman V, Guignot J, Bernet-Camard M-F, Rossi B et al (2003) The Afa/Dr adhesins of diffusely adhering Escherichia coli stimulate interleukin-8 secretion, activate mitogen-activated protein kinases, and promote polymorphonuclear transepithelial migration in T84 polarized epithelial cells. Infect Immun 71(3):1068–1074

Bhatt AP, Pellock SJ, Biernat KA, Walton WG, Wallace BD, Creekmore BC et al (2020) Targeted inhibition of gut bacterial β-glucuronidase activity enhances anticancer drug efficacy. Proc Natl Acad Sci 117(13):7374–7381

Blaak E, Canfora E, Theis S, Frost G, Groen A, Mithieux G et al (2020) Short chain fatty acids in human gut and metabolic health. Benefic Microbes 11(5):411–455

Brown K, Godovannyi A, Ma C, Zhang Y, Ahmadi-Vand Z, Dai C et al (2016) Prolonged antibiotic treatment induces a diabetogenic intestinal microbiome that accelerates diabetes in NOD mice. ISME J 10(2):321–332

Cane G, Ginouvès A, Marchetti S, Buscà R, Pouysségur J, Berra E et al (2010) HIF-1α mediates the induction of IL-8 and VEGF expression on infection with Afa/Dr diffusely adhering E. coli and promotes EMT-like behaviour. Cell Microbiol 12(5):640–653

Castellarin M, Warren RL, Freeman JD, Dreolini L, Krzywinski M, Strauss J et al (2012) Fusobacterium nucleatum infection is prevalent in human colorectal carcinoma. Genome Res 22(2):299–306

Chakraborti S, Das RK (2020) Dysbiosis of Oral microbiota and its effect on epithelial-mesenchymal transition: a review. SN Compr Clin Med 2(11):2324–2335

Chamseddine AN, Ducreux M, Armand J-P, Paoletti X, Satar T, Paci A et al (2019) Intestinal bacterial β-glucuronidase as a possible predictive biomarker of irinotecan-induced diarrhea severity. Pharmacol Ther 199:1–15

Chang C-W, Liu C-Y, Lee H-C, Huang Y-H, Li L-H, Chiau J-SC et al (2018) Lactobacillus casei variety rhamnosus probiotic preventively attenuates 5-fluorouracil/oxaliplatin-induced intestinal injury in a syngeneic colorectal cancer model. Front Microbiol 9:983

Chen W, Lian W, Yuan Y, Li M (2019) The synergistic effects of oxaliplatin and piperlongumine on colorectal cancer are mediated by oxidative stress. Cell Death Dis 10(8):600

Chen Y, Shi T, Li Y, Huang L, Yin D (2022) Fusobacterium nucleatum: the opportunistic pathogen of periodontal and peri-implant diseases. Front Microbiol 13:860149

Daillère R, Vétizou M, Waldschmitt N, Yamazaki T, Isnard C, Poirier-Colame V et al (2016) Enterococcus hirae and Barnesiella intestinihominis facilitate cyclophosphamide-induced therapeutic immunomodulatory effects. Immunity 45(4):931–943

Dobkin J, Saha J, Butler V Jr, Neu H, Lindenbaum J (1982) Inactivation of digoxin by Eubacterium lentum, an anaerobe of the human gut flora. Trans Assoc Am Phys 95:22–29

Dobrinska M (1989) Enterohepatic circulation of drugs. J Clin Pharmacol 29(7):577–580

Erny D, Hrabě de Angelis AL, Jaitin D, Wieghofer P, Staszewski O, David E et al (2015) Host microbiota constantly control maturation and function of microglia in the CNS. Nat Neurosci 18(7):965–977

Eun Y-G, Lee J-W, Kim SW, Hyun D-W, Bae J-W, Lee YC (2021) Oral microbiome associated with lymph node metastasis in oral squamous cell carcinoma. Sci Rep 11(1):23176

Ferrand A, Al Nabhani Z, Tapias NS, Mas E, Hugot J-P, Barreau F (2019) NOD2 expression in intestinal epithelial cells protects toward the development of inflammation and associated carcinogenesis. Cell Mol Gastroenterol Hepatol 7(2):357–369

Foley KP, Zlitni S, Denou E, Duggan BM, Chan RW, Stearns JC et al (2018) Long term but not short term exposure to obesity related microbiota promotes host insulin resistance. Nat Commun 9(1):4681

Franco AT, Israel DA, Washington MK, Krishna U, Fox JG, Rogers AB et al (2005) Activation of β-catenin by carcinogenic Helicobacter pylori. Proc Natl Acad Sci 102(30):10646–10651

Fu A, Yao B, Dong T, Chen Y, Yao J, Liu Y et al (2022) Tumor-resident intracellular microbiota promotes metastatic colonization in breast cancer. Cell 185(8):1356–1372. e26

Gao Y, Bi D, Xie R, Li M, Guo J, Liu H et al (2021) Fusobacterium nucleatum enhances the efficacy of PD-L1 blockade in colorectal cancer. Signal Transduct Target Ther 6(1):398

Garrett WS (2015) Cancer and the microbiota. Science 348(6230):80–86

Geller LT, Barzily-Rokni M, Danino T, Jonas OH, Shental N, Nejman D et al (2017) Potential role of intratumor bacteria in mediating tumor resistance to the chemotherapeutic drug gemcitabine. Science 357(6356):1156–1160

Gensollen T, Iyer SS, Kasper DL, Blumberg RS (2016) How colonization by microbiota in early life shapes the immune system. Science 352(6285):539–544

Ghiringhelli F, Larmonier N, Schmitt E, Parcellier A, Cathelin D, Garrido C et al (2004) CD4+ CD25+ regulatory T cells suppress tumor immunity but are sensitive to cyclophosphamide which allows immunotherapy of established tumors to be curative. Eur J Immunol 34(2):336–344

Grem JL (2000) 5-Fluorouracil: forty-plus and still ticking. A review of its preclinical and clinical development. Investig New Drugs 18:299–313

Gunawardhana N, Jang S, Choi YH, Hong YA, Jeon Y-E, Kim A et al (2018) Helicobacter pylori-induced HB-EGF upregulates gastrin expression via the EGF receptor, C-Raf, Mek1, and Erk2 in the MAPK pathway. Front Cell Infect Microbiol 7:541

Ha NH, Woo BH, Kim DJ, Ha ES, Choi JI, Kim SJ et al (2015) Prolonged and repetitive exposure to Porphyromonas gingivalis increases aggressiveness of oral cancer cells by promoting acquisition of cancer stem cell properties. Tumour Biol 36:9947–9960

Han I-H, Song H-O, Ryu J-S (2020) IL-6 produced by prostate epithelial cells stimulated with Trichomonas vaginalis promotes proliferation of prostate cancer cells by inducing M2 polarization of THP-1-derived macrophages. PLoS Negl Trop Dis 14(3):e0008126

He Y, Fu L, Li Y, Wang W, Gong M, Zhang J et al (2021) Gut microbial metabolites facilitate anticancer therapy efficacy by modulating cytotoxic CD8+ T cell immunity. Cell Metab 33(5):988–1000. e7

Honko AN, Mizel SB (2005) Effects of flagellin on innate and adaptive immunity. Immunol Res 33:83–101

Hooi JK, Lai WY, Ng WK, Suen MM, Underwood FE, Tanyingoh D et al (2017) Global prevalence of Helicobacter pylori infection: systematic review and meta-analysis. Gastroenterology 153(2):420–429

Hsueh C-Y, Huang Q, Gong H, Shen Y, Sun J, Lau H-C et al (2022) A positive feed-forward loop between Fusobacterium nucleatum and ethanol metabolism reprogramming drives laryngeal cancer progression and metastasis. Iscience 25(2):103829

Iida N, Dzutsev A, Stewart CA, Smith L, Bouladoux N, Weingarten RA et al (2013) Commensal bacteria control cancer response to therapy by modulating the tumor microenvironment. Science 342(6161):967–970

Ivleva EA, Grivennikov SI (2022) Microbiota-driven mechanisms at different stages of cancer development. Neoplasia 32:100829

Jia Y, Xie J (2015) Promising molecular mechanisms responsible for gemcitabine resistance in cancer. Genes Dis 2(4):299–306

Jung JI, Cho HJ, Jung YJ, Kwon SH, Her S, Choi SS et al (2015) High-fat diet-induced obesity increases lymphangiogenesis and lymph node metastasis in the B16F10 melanoma allograft model: roles of adipocytes and M2-macrophages. Int J Cancer 136(2):258–270

Karpiński TM (2019) Role of oral microbiota in cancer development. Microorganisms 7(1):20

Kechagia M, Basoulis D, Konstantopoulou S, Dimitriadi D, Gyftopoulou K, Skarmoutsou N et al (2013) Health benefits of probiotics: a review. Int Sch Res Notices 2013:1

Khan AA, Shrivastava A, Khurshid M (2012) Normal to cancer microbiome transformation and its implication in cancer diagnosis. Biochimica et Biophysica Acta (BBA)-Reviews on. Cancer 1826(2):331–337

Khan AA, Nema V, Khan Z (2021) Current status of probiotics for prevention and management of gastrointestinal cancers. Expert Opin Biol Ther 21(3):413–422

Kim EJ, Choi M-R, Park H, Kim M, Hong JE, Lee J-Y et al (2011) Dietary fat increases solid tumor growth and metastasis of 4T1 murine mammary carcinoma cells and mortality in obesity-resistant BALB/c mice. Breast Cancer Res 13:1–13

Kim YJ, Kim BK, Park SJ, Kim JH (2021) Impact of Fusobacterium nucleatum in the gastrointestinal tract on natural killer cells. World J Gastroenterol 27(29):4879

Klufa J, Bauer T, Hanson B, Herbold C, Starkl P, Lichtenberger B et al (2019) Hair eruption initiates and commensal skin microbiota aggravate adverse events of anti-EGFR therapy. Sci Transl Med 11(522):eaax2693

Kodawara T, Higashi T, Negoro Y, Kamitani Y, Igarashi T, Watanabe K et al (2016) The inhibitory effect of ciprofloxacin on the β-glucuronidase-mediated deconjugation of the irinotecan metabolite SN-38-G. Basic Clin Pharmacol Toxicol 118(5):333–337

Koh A, Molinaro A, Ståhlman M, Khan MT, Schmidt C, Mannerås-Holm L et al (2018) Microbially produced imidazole propionate impairs insulin signaling through mTORC1. Cell 175(4):947–961. e17

Kong C, Yan X, Zhu Y, Zhu H, Luo Y, Liu P et al (2021) Fusobacterium nucleatum promotes the development of colorectal cancer by activating a cytochrome P450/epoxyoctadecenoic acid axis via TLR4/Keap1/NRF2 signaling. Cancer Res 81(17):4485–4498

König J, Wells J, Cani PD, García-Ródenas CL, MacDonald T, Mercenier A et al (2016) Human intestinal barrier function in health and disease. Clin Transl Gastroenterol 7(10):e196

Kostic AD, Chun E, Robertson L, Glickman JN, Gallini CA, Michaud M et al (2013) Fusobacterium nucleatum potentiates intestinal tumorigenesis and modulates the tumor-immune microenvironment. Cell Host Microbe 14(2):207–215

Kovács T, Mikó E, Vida A, Sebő É, Toth J, Csonka T et al (2019) Cadaverine, a metabolite of the microbiome, reduces breast cancer aggressiveness through trace amino acid receptors. Sci Rep 9(1):1300

Laux C, Peschel A, Krismer B (2019) Staphylococcus aureus colonization of the human nose and interaction with other microbiome members. Microbiol Spectr 7(2). https://doi.org/10.1128/microbiolspec.gpp3-0029-2018

Lee W-H, Chen H-M, Yang S-F, Liang C, Peng C-Y, Lin F-M et al (2017) Bacterial alterations in salivary microbiota and their association in oral cancer. Sci Rep 7(1):1–11

Li H-L, Lu L, Wang X-S, Qin L-Y, Wang P, Qiu S-P et al (2017) Alteration of gut microbiota and inflammatory cytokine/chemokine profiles in 5-fluorouracil induced intestinal mucositis. Front Cell Infect Microbiol 7:455

Li W, Deng Y, Chu Q, Zhang P (2019a) Gut microbiome and cancer immunotherapy. Cancer Lett 447:41–47

Li R, Zhou R, Wang H, Li W, Pan M, Yao X et al (2019b) Gut microbiota-stimulated cathepsin K secretion mediates TLR4-dependent M2 macrophage polarization and promotes tumor metastasis in colorectal cancer. Cell Death Differ 26(11):2447–2463

Li X-N, Hua L-X, Zhou T-S, Wang K-B, Wu Y-Y, Emam M et al (2020) Cinnamic acid derivatives: inhibitory activity against Escherichia coli β-glucuronidase and structure–activity relationships. J Enzyme Inhib Med Chem 35(1):1372–1378

Li Q, Sun Y, Zhai K, Geng B, Dong Z, Ji L et al (2024) Microbiota-induced inflammatory responses in bladder tumors promote epithelial-mesenchymal transition and enhanced immune infiltration. Physiol Genomics 56:544

Lindenbaum J, Rund DG, Butler VP Jr, Tse-Eng D, Saha JR (1981) Inactivation of digoxin by the gut flora: reversal by antibiotic therapy. N Engl J Med 305(14):789–794

Liu L, Bin Y, Zeng H, Wang J, Zhang R, Tong F et al (2022) Microbiota-modulated spermine promotes brain metastasis in non-small cell lung cancer by regulating microglia M2 polarization via the STAT3 pathway. https://doi.org/10.21203/rs.3.rs-2259805/v1

Livanos AE, Greiner TU, Vangay P, Pathmasiri W, Stewart D, McRitchie S et al (2016) Antibiotic-mediated gut microbiome perturbation accelerates development of type 1 diabetes in mice. Nat Microbiol 1(11):1–13

Long X, Wong CC, Tong L, Chu ESH, Ho Szeto C, Go MYY et al (2019) Peptostreptococcus anaerobius promotes colorectal carcinogenesis and modulates tumour immunity. Nat Microbiol 4(12):2319–2330

Ma Y, Chen H, Li H, Zheng M, Zuo X, Wang W et al (2024) Intratumor microbiome-derived butyrate promotes lung cancer metastasis. Cell Rep Med 5(4):101488

Maini Rekdal V, Bess EN, Bisanz JE, Turnbaugh PJ, Balskus EP (2019) Discovery and inhibition of an interspecies gut bacterial pathway for Levodopa metabolism. Science 364(6445):eaau6323

Malik MY, Jaiswal S, Sharma A, Shukla M, Lal J (2016) Role of enterohepatic recirculation in drug disposition: cooperation and complications. Drug Metab Rev 48(2):281–327

Martinez-Martinez D, Peres TV, Gehling K, Quintaneiro L, Cabrera C, Cutty SJ et al (2024) Chemotherapy modulation by a cancer-associated microbiota metabolite. https://doi.org/10.2139/ssrn.4832225

Martins F, Sofiya L, Sykiotis GP, Lamine F, Maillard M, Fraga M et al (2019) Adverse effects of immune-checkpoint inhibitors: epidemiology, management and surveillance. Nat Rev Clin Oncol 16(9):563–580

Maute R, Xu J, Weissman IL (2022) CD47-SIRPα-targeted therapeutics: status and prospects. Immunooncol Technol 13:100070
McNabney SM, Henagan TM (2017) Short chain fatty acids in the colon and peripheral tissues: a focus on butyrate, colon cancer, obesity and insulin resistance. Nutrients 9(12):1348
McNeil N (1984) The contribution of the large intestine to energy supplies in man. Am J Clin Nutr 39(2):338–342
Mitchell H (1999) The epidemiology of Helicobacter pylori. In: Gastroduodenal disease and Helicobacter pylori: pathophysiology, diagnosis and treatment, pp 11–30
Murata-Kamiya N, Kurashima Y, Teishikata Y, Yamahashi Y, Saito Y, Higashi H et al (2007) Helicobacter pylori CagA interacts with E-cadherin and deregulates the β-catenin signal that promotes intestinal transdifferentiation in gastric epithelial cells. Oncogene 26(32):4617–4626
Murphy R, Palm M, Mustonen V, Warringer J, Farewell A, Parts L et al (2021) Genomic epidemiology and evolution of Escherichia coli in wild animals in Mexico. Msphere 6(1). https://doi.org/10.1128/msphere.00738-20
Nagy K, Sonkodi I, Szöke I, Nagy E, Newman H (1998) The microflora associated with human oral carcinomas. Oral Oncol 34(4):304–308
Nielsen DL, Juhl CB, Chen IM, Kellermann L, Nielsen OH (2022) Immune checkpoint Inhibitor–Induced diarrhea and Colitis: incidence and management. A systematic review and meta-analysis. Cancer Treat Rev 109:102440
Nomura M, Nagatomo R, Doi K, Shimizu J, Baba K, Saito T et al (2020) Association of short-chain fatty acids in the gut microbiome with clinical response to treatment with nivolumab or pembrolizumab in patients with solid cancer tumors. JAMA Netw Open 3(4):e202895
Noordhuis P, Holwerda U, Van der Wilt C, Van Groeningen C, Smid K, Meijer S et al (2004) 5-Fluorouracil incorporation into RNA and DNA in relation to thymidylate synthase inhibition of human colorectal cancers. Ann Oncol 15(7):1025–1032
Österlund P, Ruotsalainen T, Korpela R, Saxelin M, Ollus A, Valta P et al (2007) Lactobacillus supplementation for diarrhoea related to chemotherapy of colorectal cancer: a randomised study. Br J Cancer 97(8):1028–1034
Panebianco C, Andriulli A, Pazienza V (2018) Pharmacomicrobiomics: exploiting the drug-microbiota interactions in anticancer therapies. Microbiome 6:1–13
Panebianco C, Villani A, Pisati F, Orsenigo F, Ulaszewska M, Latiano TP et al (2022) Butyrate, a postbiotic of intestinal bacteria, affects pancreatic cancer and gemcitabine response in in vitro and in vivo models. Biomed Pharmacother 151:113163
Papini E, Satin B, Norais N, de Bernard M, Telford JL, Rappuoli R et al (1998) Selective increase of the permeability of polarized epithelial cell monolayers by Helicobacter pylori vacuolating toxin. J Clin Invest 102(4):813–820
Parhi L, Alon-Maimon T, Sol A, Nejman D, Shhadeh A, Fainsod-Levi T et al (2020) Breast cancer colonization by Fusobacterium nucleatum accelerates tumor growth and metastatic progression. Nat Commun 11(1):3259
Park BS, Lee J-O (2013) Recognition of lipopolysaccharide pattern by TLR4 complexes. Exp Mol Med 45(12):e66
Park J, Kim M, Kang SG, Jannasch AH, Cooper B, Patterson J et al (2015) Short-chain fatty acids induce both effector and regulatory T cells by suppression of histone deacetylases and regulation of the mTOR–S6K pathway. Mucosal Immunol 8(1):80–93
Picard M, Yonekura S, Slowicka K, Petta I, Rauber C, Routy B et al (2021) Ileal immune tonus is a prognosis marker of proximal colon cancer in mice and patients. Cell Death Differ 28(5):1532–1547
Postow MA, Sidlow R, Hellmann MD (2018) Immune-related adverse events associated with immune checkpoint blockade. N Engl J Med 378(2):158–168
Reshetnyak VI, Burmistrov AI, Maev IV (2021) Helicobacter pylori: commensal, symbiont or pathogen? World J Gastroenterol 27(7):545–560
Riquelme E, Zhang Y, Zhang L, Montiel M, Zoltan M, Dong W et al (2019) Tumor microbiome diversity and composition influence pancreatic cancer outcomes. Cell 178(4):795–806. e12

Rubinstein MR, Wang X, Liu W, Hao Y, Cai G, Han YW (2013) Fusobacterium nucleatum promotes colorectal carcinogenesis by modulating E-cadherin/β-catenin signaling via its FadA adhesin. Cell Host Microbe 14(2):195–206
Ryu TY, Kim K, Son M-Y, Min J-K, Kim J, Han T-S et al (2019) Downregulation of PRMT1, a histone arginine methyltransferase, by sodium propionate induces cell apoptosis in colon cancer. Oncol Rep 41(3):1691–1699
Sadeghi Rad H, Monkman J, Warkiani ME, Ladwa R, O'Byrne K, Rezaei N et al (2021) Understanding the tumor microenvironment for effective immunotherapy. Med Res Rev 41(3):1474–1498
Sári Z, Mikó E, Kovács T, Jankó L, Csonka T, Lente G et al (2020) Indolepropionic acid, a metabolite of the microbiome, has cytostatic properties in breast cancer by activating AHR and PXR receptors and inducing oxidative stress. Cancers 12(9):2411
Schoch S, Gajewski S, Rothfuß J, Hartwig A, Köberle B (2020) Comparative study of the mode of action of clinically approved platinum-based chemotherapeutics. Int J Mol Sci 21(18):6928
Schroeder BO, Bäckhed F (2016) Signals from the gut microbiota to distant organs in physiology and disease. Nat Med 22(10):1079–1089
Sears CL (2009) Enterotoxigenic Bacteroides fragilis: a rogue among symbiotes. Clin Microbiol Rev 22(2):349–369
Sedighi M, Zahedi Bialvaei A, Hamblin MR, Ohadi E, Asadi A, Halajzadeh M et al (2019) Therapeutic bacteria to combat cancer; current advances, challenges, and opportunities. Cancer Med 8(6):3167–3181
Seidel JA, Otsuka A, Kabashima K (2018) Anti-PD-1 and anti-CTLA-4 therapies in cancer: mechanisms of action, efficacy, and limitations. Front Oncol 8:86
Sender R, Fuchs S, Milo R (2016a) Revised estimates for the number of human and bacteria cells in the body. PLoS Biol 14(8):e1002533
Sender R, Fuchs S, Milo R (2016b) Are we really vastly outnumbered? Revisiting the ratio of bacterial to host cells in humans. Cell 164(3):337–340
Sethi V, Kurtom S, Tarique M, Lavania S, Malchiodi Z, Hellmund L et al (2018) Gut microbiota promotes tumor growth in mice by modulating immune response. Gastroenterology 155(1):33–37. e6
Shi Y, Zheng W, Yang K, Harris KG, Ni K, Xue L et al (2020) Intratumoral accumulation of gut microbiota facilitates CD47-based immunotherapy via STING signaling. J Exp Med 217(5):e20192282
Sistigu A, Viaud S, Chaput N, Bracci L, Proietti E, Zitvogel L (eds) (2011) Immunomodulatory effects of cyclophosphamide and implementations for vaccine design. In: Seminars in immunopathology. Springer
Smith MF, Mitchell A, Li G, Ding S, Fitzmaurice AM, Ryan K et al (2003) Toll-like receptor (TLR) 2 and TLR5, but not TLR4, are required for Helicobacter pylori-induced NF-κB activation and chemokine expression by epithelial cells. J Biol Chem 278(35):32552–32560
Sun L, Ma L, Ma Y, Zhang F, Zhao C, Nie Y (2018) Insights into the role of gut microbiota in obesity: pathogenesis, mechanisms, and therapeutic perspectives. Protein Cell 9(5):397–403
Sun R, Zhu L, Li L, Song W, Gong X, Qi X et al (2020) Irinotecan-mediated diarrhea is mainly correlated with intestinal exposure to SN-38: critical role of gut Ugt. Toxicol Appl Pharmacol 398:115032
Sztukowska MN, Ojo A, Ahmed S, Carenbauer AL, Wang Q, Shumway B et al (2016) Porphyromonas gingivalis initiates a mesenchymal-like transition through ZEB1 in gingival epithelial cells. Cell Microbiol 18(6):844–858
Tan J, McKenzie C, Potamitis M, Thorburn AN, Mackay CR, Macia L (2014) The role of short-chain fatty acids in health and disease. Adv Immunol 121:91–119
Thaiss CA, Zmora N, Levy M, Elinav E (2016) The microbiome and innate immunity. Nature 535(7610):65–74
Thomas D, Zalcberg J (1998) 5-Fluorouracil: a pharmacological paradigm in the use of cytotoxics. Clin Exp Pharmacol Physiol 25(11):887–895

Tsoi H, Chu ESH, Zhang X, Sheng J, Nakatsu G, Ng SC et al (2017) Peptostreptococcus anaerobius induces intracellular cholesterol biosynthesis in colon cells to induce proliferation and causes dysplasia in mice. Gastroenterology 152(6):1419–33.e5

Ursell LK, Metcalf JL, Parfrey LW, Knight R (2012) Defining the human microbiome. Nutr Rev 70(suppl_1):S38–S44

Viaud S, Flament C, Zoubir M, Pautier P, LeCesne A, Ribrag V et al (2011) Cyclophosphamide induces differentiation of Th17 cells in cancer patients. Cancer Res 71(3):661–665

Viaud S, Saccheri F, Mignot G, Yamazaki T, Daillère R, Hannani D et al (2013) The intestinal microbiota modulates the anticancer immune effects of cyclophosphamide. Science 342(6161):971–976

Virtue AT, McCright SJ, Wright JM, Jimenez MT, Mowel WK, Kotzin JJ et al (2019) The gut microbiota regulates white adipose tissue inflammation and obesity via a family of microRNAs. Sci Transl Med 11(496):eaav1892

Visentin D, Gobin I, Maglica Ž (2023) Periodontal pathogens and their links to neuroinflammation and neurodegeneration. Microorganisms 11(7):1832

Vogiatzi P, Cassone M, Luzzi I, Lucchetti C, Otvos L Jr, Giordano A (2007) Helicobacter pylori as a class I carcinogen: physiopathology and management strategies. J Cell Biochem 102(2):264–273

Wang P (2021) Discovery of bacterial β-Glucuronidase inhibitors to attenuate irinotecan toxicity, University of Macau

Wang J-W, Kuo C-H, Kuo F-C, Wang Y-K, Hsu W-H, Yu F-J et al (2019a) Fecal microbiota transplantation: review and update. J Formos Med Assoc 118:S23–S31

Wang Y, Sun L, Chen S, Guo S, Yue T, Hou Q et al (2019b) The administration of Escherichia coli Nissle 1917 ameliorates irinotecan-induced intestinal barrier dysfunction and gut microbial dysbiosis in mice. Life Sci 231:116529

Wang W, Fang D, Zhang H, Xue J, Wangchuk D, Du J et al (2020) Sodium butyrate selectively kills cancer cells and inhibits migration in colorectal cancer by targeting thioredoxin-1. Onco Targets Ther 13:4691–4704

Wang X, Zhang J, Hu X, Fan B, Wong C-C, Hu Y et al (2023a) IDDF2023-ABS-0250 Comprehensive analyses of microbiome and metabolism reveal liver metastasis-specific signature in colorectal cancer. Gut 72(Suppl 1):A116

Wang X, Zhang J, Hu X, Fan B, Wong C-C, Hu Y et al (2023b) IDDF2023-ABS-0250 Comprehensive analyses of microbiome and metabolism reveal liver metastasis-specific signature in colorectal cancer. BMJ Publishing Group

Weber JS, Hodi FS, Wolchok JD, Topalian SL, Schadendorf D, Larkin J et al (2017) Safety profile of nivolumab monotherapy: a pooled analysis of patients with advanced melanoma, vol 35, p 785

Wen L, Duffy A (2017) Factors influencing the gut microbiota, inflammation, and type 2 diabetes. J Nutr 147(7):1468S–1475S

Wu H, Ma W, Wang Y, Wang Y, Sun X, Zheng Q (2024) Gut microbiome-metabolites axis: a friend or foe to colorectal cancer progression. Biomed Pharmacother 173:116410

Xie Y, Xie F, Zhou X, Zhang L, Yang B, Huang J et al (2022) Microbiota in tumors: from understanding to application. Adv Sci 9(21):2200470

Xu S, Liu Z, Lv M, Chen Y, Liu Y (2019) Intestinal dysbiosis promotes epithelial-mesenchymal transition by activating tumor-associated macrophages in ovarian cancer. Pathog Dis 77(2):ftz019

Xu S, Yin W, Zhang Y, Lv Q, Yang Y, He J (2020) Foes or friends? Bacteria enriched in the tumor microenvironment of colorectal cancer. Cancers 12(2):372

Yang J-j, Wang D-d, Sun T-y (2011) Flagellin of Pseudomonas aeruginosa induces transforming growth factor beta 1 expression in normal bronchial epithelial cells through mitogen activated protein kinase cascades. Chin Med J 124(4):599–605

Yang Y, Weng W, Peng J, Hong L, Yang L, Toiyama Y et al (2017) Fusobacterium nucleatum increases proliferation of colorectal cancer cells and tumor development in mice by activating toll-like receptor 4 signaling to nuclear factor-κB, and up-regulating expression of microRNA-21. Gastroenterology 152(4):851–66.e24

Yang L, Li A, Wang Y, Zhang Y (2023) Intratumoral microbiota: roles in cancer initiation, development and therapeutic efficacy. Signal Transduct Target Ther 8(1):35

Yin Y, Grabowska AM, Clarke PA, Whelband E, Robinson K, Argent RH et al (2010) Helicobacter pylori potentiates epithelial: mesenchymal transition in gastric cancer: links to soluble HB-EGF, gastrin and matrix metalloproteinase-7. Gut 59(8):1037–1045

Yu H, Zeng J, Liang X, Wang W, Zhou Y, Sun Y et al (2014) Helicobacter pylori promotes epithelial–mesenchymal transition in gastric cancer by downregulating programmed cell death protein 4 (PDCD4). PLoS One 9(8):e105306

Yu T, Guo F, Yu Y, Sun T, Ma D, Han J et al (2017) Fusobacterium nucleatum promotes chemoresistance to colorectal cancer by modulating autophagy. Cell 170(3):548–563. e16

Yue Y-C, Yang B-Y, Lu J, Zhang S-W, Liu L, Nassar K et al (2020) Metabolite secretions of Lactobacillus plantarum YYC-3 may inhibit colon cancer cell metastasis by suppressing the VEGF-MMP2/9 signaling pathway. Microb Cell Factories 19:1–12

Zhang S, Yang Y, Weng W, Guo B, Cai G, Ma Y et al (2019) Fusobacterium nucleatum promotes chemoresistance to 5-fluorouracil by upregulation of BIRC3 expression in colorectal cancer. J Exp Clin Cancer Res 38:1–13

Zhao L, Yang R, Cheng L, Wang M, Jiang Y, Wang S (2011) LPS-induced epithelial-mesenchymal transition of intrahepatic biliary epithelial cells. J Surg Res 171(2):819–825

Zhao H, Chu M, Huang Z, Yang X, Ran S, Hu B et al (2017) Variations in oral microbiota associated with oral cancer. Sci Rep 7(1):11773

Zhao H, Song S, Ma J, Yan Z, Xie H, Feng Y et al (2022) CD47 as a promising therapeutic target in oncology. Front Immunol 13:757480

Zhao L-Y, Mei J-X, Yu G, Lei L, Zhang W-H, Liu K et al (2023) Role of the gut microbiota in anticancer therapy: from molecular mechanisms to clinical applications. Signal Transduct Target Ther 8(1):201

Ziesemer S, Eiffler I, Schönberg A, Müller C, Hochgräfe F, Beule AG et al (2018) Staphylococcus aureus α-toxin induces actin filament remodeling in human airway epithelial model cells. Am J Respir Cell Mol Biol 58(4):482–491

The Microbial and Metabolic Link Between Gut and Brain

17

Rand Salamalek, Yasmin Shareef, Farah Abdulridha, and Sameh S. M. Soliman

Abstract

The gut-brain connection plays a critical role in the pathophysiology of numerous neurological disorders. This book chapter explores the role of the gut microbiome in this complex interaction through key mechanisms, including the endocrine, immune, and vagus nerve pathways and metabolic activities. Dysbiosis, a gut microbiota imbalance, contributes to neuroinflammation, altered neurotransmitter levels, and increased blood-brain barrier (BBB) permeability, which are associated with the development and progression of neurological conditions such as Alzheimer's disease, Parkinson's disease, anxiety, depression, and autism spectrum disorder. Gut microbiome-derived metabolites such as short-chain fatty acids (SCFAs), including butyrate, acetate, and propionate, are known to cross the BBB and modulate the neuroinflammatory pathways, synaptic plasticity, and cognitive functions. Additionally, microbial production of neurotransmitters like gamma-aminobutyric acid (GABA), serotonin, and dopamine directly affects CNS signaling, with implications in mood disorders and neurodegenerative diseases. Dysregulated gut microbiota can impair the gut-brain communication via the vagus nerve, which further contributes to neurodegenerative processes. The gut microbiome showed also significant impact on the pharmacokinetics and pharmacodynamics of neurological drugs. Emerging therapeutic strategies can include the development of tailored psychomicrobiot-

R. Salamalek · Y. Shareef · F. Abdulridha
College of Pharmacy, University of Sharjah, Sharjah, UAE

S. S. M. Soliman (✉)
Department of Medicinal Chemistry, College of Pharmacy, University of Sharjah, Sharjah, UAE

Research Institute for Medical and Health Sciences (RIMHS), University of Sharjah, Sharjah, UAE
e-mail: ssoliman@sharjah.ac.ae

S. S. M. Soliman, M. I. Husseiny (eds.), *Metabolic Dynamics in Host-Microbe Interaction*, https://doi.org/10.1007/978-981-96-1305-2_17

ics, which aim to modulate the gut microbiome composition by enhancing the drug efficacy and restoration of neuroimmune homeostasis. Dietary interventions, including prebiotics and probiotics, are also proposed as adjunct therapies to modulate gut microbiota and improve brain health. Overall, the gut microbiome offers novel choices for next-generation treatments aimed at improving brain function and limiting disease progression.

Keywords

Gut microbiota · Brain · Neurological disorders · Neurological drugs · Neurological metabolites

17.1 Introduction

The link between the human microbiome and brain functioning has recently become a hotspot in scientific research (Cryan et al. 2019). Gut-brain link mainly represented in the microbiome apparently contributes to almost all neurological disorders. The gut microbiome serves as an ecosystem rich in fungi, bacteria, archaea, and viruses making more than 90% of our body and consists of around 150 times more genes than the human (Cryan et al. 2019). The microbiome aids in digestion, absorption, and immune defense. Several studies and research have discovered the role of microbiome in regulating the central nervous system function (Nicholson et al. 2012; Belkaid and Hand 2014; Zhu et al. 2010; Bull and Plummer 2014; Holzer 2017). Dysbiosis was noticed when the microbiome levels were altered by factors such as inflammation, infections, eating high levels of sugar, or taking antibiotics (Hrncir 2022). The human gut has ~100 million nerve cells enabling them to continuously communicate with the human brain (Mayer 2011).

The discovery of this connection started in the eighteenth century, when the vitalists search for the existence of the gut-brain connection (Haigh 1975). In the nineteenth century, some physicians proposed that the gastrointestinal system fully controls the nerve endings, and thus, the gut was named as the "great abdominal brain" or the "great nervous center" with clear advice on a healthy eating habit (Miller 2018). In contrast, in the twentieth century, scientists believed that it is the brain that controls the gut and is responsible for all health abnormalities (Lewandowska-Pietruszka et al. 2022). The British psychiatrist Peter Down hypothesized that all patients with gastrointestinal abnormalities should undergo a mental health test (Moore et al. 2019). Investigations on the gut-brain connection continued to take place after the twentieth century. In the twenty-first century, the term "human microbiome" has been coined (Alexander 1934; Nature 2012; Health Gmf 2012). In 2007, the human microbiome project was developed to know the role of each microorganism (Chaplin et al. 2017). In 2008, a new international consortium called "Meta HIT" was introduced, which aimed to understand and analyze the gut genomes to further determine the species and type of these microorganisms (Nature 2012; Health Gmf 2012; Davenport et al. 2017). The human microbiome was then shown to be a promising therapeutic agent (Lewandowska-Pietruszka et al. 2022). Specifically, the microbiome that treats neurodegenerative disorders is called the "psychobiotics." Earlier research has proven that it had a positive influence on patients with several neurological diseases such as depression, anxiety, Alzheimer's disease (AD), Parkinson's disease (PD), and sleep disturbances (Del Toro-Barbosa et al. 2020). In 2011, the genuine idea of the influence of the gut microbiome on human brain health was well established by experiments performed on germ-free (GF) mice, which showed an altered cognitive function with a decrease in anxiety-related factors (Damiani et al. 2023).

The gut-brain connection involves several processes including the endocrine system in which it releases specific hormones and metabolites that regulate the CNS (Canfora et al. 2015). For example, the gut microbiome produces short-chain fatty acids (SCFAs) that pass through the blood-brain barrier (BBB) to interact with specific receptors and then provide several beneficial activities on the brain such as reducing inflammation, enhancing mood and cognitive function (Canfora et al. 2015; Dinan and Cryan 2017; Sun et al. 2020). This book chapter highlights the connection between the human gut microbiome and the brain. The influence of metabolites of the human gut microbiome on neurological disorders such as Alzheimer's disease, Parkinson's disease, anxiety, depression, and autism will be discussed. The significant effect on shaping the brain health will also be highlighted. Additionally, the effect of metabolites of the gut microbiome on the neurological drug's efficacy and their therapeutic concentrations will be addressed. A future prospective on using the microbiome in treatment will be proposed.

17.2 Gut Microbiota and Brain Connection

There are many ways that the gut is connected to the brain; it has been shown that brain plasticity (neuroplasticity) can be altered by gut microbiota via regulating gene expression, and production of neuroactive substances (Damiani et al. 2023). For instance, SCFAs produced by microbiota can cross the BBB and interfere with the function of epigenetic marks including histone acetylation, microRNA regulation, and DNA methylation, which alter gene expression (Damiani et al. 2023). Furthermore, microbiota can produce neurotransmitters including gamma-aminobutyric acid (GABA), serotonin (5-HT), tryptophan, histamine, and dopamine, or their precursors such as tyrosine, tryptophan, choline, and acetate, which can be converted later in the brain (Damiani et al. 2023). Consequently, it can cause a change in the neurotransmitter concentration in the brain, leading to cognitive-associated activity (Damiani et al. 2023).

A recent study showed that these metabolites can modulate the BBB permeability, stimulation of brain microglia, and integrity of BBB (Kasarello et al. 2023). GF mice revealed a significant decrease in the tight junctions that line the endothelium of the BBB, which in turn increased the permeability of molecules with toxic effects (Kasarello et al. 2023). On the contrary, re-establishment of a dominant SCFA-rich bacterial species in GF mice retains BBB integrity (Kasarello et al. 2023). SCFAs exhibit an indirect effect on the CNS by binding to enteroendocrine cells (EEC) in the gastrointestinal tract and pancreas, resulting in an increase in the release of gastrointestinal hormones such as glucagon-like peptide 1 (GLP-1) and peptide YY (PYY), which regulate the body metabolism and appetite (Kasarello et al. 2023). SCFAs influence the memory and learning abilities; for instance, they stimulate the expression of brain-derived neurotrophic factor (BDNF), serotonin transporter, and others, in addition to regulating the circadian rhythm (Kasarello et al. 2023).

17.3 Microflora Affecting the Brain Performance

The CNS and enteric nervous system (ENS) exhibit a communication line using various chemical signaling pathways, including endocrine, immune, and direct neuronal pathways (Yoo and Mazmanian 2017). It is a two-way communication between the brain and the gut, in which it is crucial in maintaining neurological, microbial, and gastrointestinal system functions (Martin et al. 2018). Microflora is a key component responsible for maturation, aging, maintaining homeostasis, and development of the nervous system. For example, microflora oppose changes to the expression of *N*-methyl-*D*-aspartate (NMDA) receptors, thus altering the neurotrophic factors (Bercik et al. 2011a; Diaz Heijtz et al. 2011).

17.3.1 Endocrine Pathway

GIT enteroendocrine cells play a major role in its functioning, including digestion, motility, and hormone production in the gut (Gribble and Reimann 2019). The enteroendocrine cells sense the changes in the gut and accordingly release several hormones. These hormones can act locally and systemically (Buffa et al. 1978; Sternini et al. 2008). There are different types of enteroendocrine cells according to their secretion. The main types of enteroendocrine cells are the enterochromaffin, L cells, and the D cells (Gunawardene et al. 2011). The endocrine system is linked to the nervous system with the term "neuroendocrine" and regulates various physiological processes in the body. The neuroendocrine system plays a major role in forming the tissues (Cussotto et al. 2018). Microflora releases metabolites that can change the function of the gut-brain link by producing specific gut hormones at different levels (Tolhurst et al. 2012). These gut hormones are produced by activating the G-protein-coupled receptors (GPCRs) by SCFAs in the large intestine (Psichas et al. 2015). This process promotes the production of PYY and GLP1 in the enteroendocrine L cells. These hormones have a great effect on people's mood, cognitive function, and memory (Dalile et al. 2019). It was shown that GLP1 in mice enhanced both memory and learning skills, as well as the neuroprotection at the hippocampus (Isacson et al. 2011; McClean et al. 2011) (Fig. 17.1).

On Alzheimer's disease, animal models showed lower amounts of beta-amyloid plaques (McClean et al. 2011). GLP1 has many receptors in the body, mostly in the brain that it can affect the function of the human brain by both hormonal and

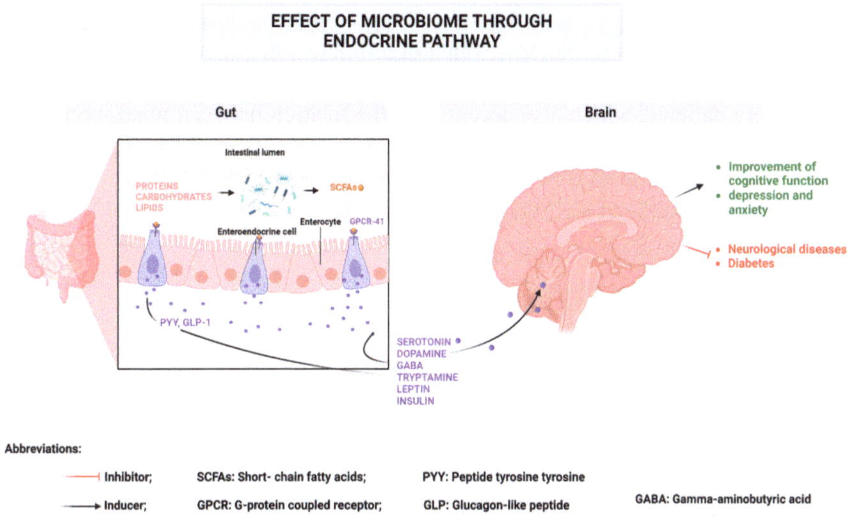

Fig. 17.1 Effect of microbiome through the endocrine pathway. Proteins, carbohydrates, and lipids feed the gut microbiota to produce SCFAs, which activates GPCR-41 in the large intestine to enhance the production of several hormones like PYY, GLP-1, serotonin, tryptamine, dopamine, GABA, leptin, and insulin. Activating these hormones results in an improvement in the brain function and inhibits them from causing neurological diseases and diabetes

neuronal pathways (Alvarez et al. 2005; Katsurada and Yada 2016). The other neuropeptide called PYY reduces the appetite significantly and gastric motility by passing BBB or by the vagal afferent routes (Nonaka et al. 2003). This PYY is produced by the L cells present in the ileum and colon and stored in the hypothalamus and the pituitary gland at the highest levels (Morimoto et al. 2008).

There are other hormones that play a role in affecting the human brain, and they are regulated by SCFAs such as leptin, insulin, and ghrelin (Hube et al. 1996). The hormone leptin mainly has a weight loss impact and has effects on regulating the energy balance in the body (Obradovic et al. 2021). Although serotonin is one of the vital neuromodulators in the brain, 90% of serotonin presents in the human intestine, with a possible effect on the human behavior (Nakatani et al. 2008). Moreover, bacterial metabolites increase the expression of the gene called tryptophan hydrolase 1, leading to a subsequent increase in the synthesis of serotonin by the endocrine cells (Whiley et al. 2021). GF mice showed lower blood concentrations of serotonin than the normal mice (Yano et al. 2015). The serum bioavailability of serotonin is linked to a number of neurological disorders because it hardly crosses the BBB. For example, serotonin serum levels have been found to be decreased in AD (Whiley et al. 2021) and PD patients (Tong et al. 2015). Furthermore, it was found that dopamine was also produced in some types of bacteria such as *Bacillus* and *Serratia* (Strandwitz 2018). GF mice showed lower concentration of dopamine compared to the normal mice; this highlights a relation between the intestinal microbiome and Parkinson's disease (Lyte and Freestone 2010; Asano et al. 2012;

Dobbs et al. 2012). Additionally, GABA, a suppressing neurotransmitter, was found to be supplied by the microbiome, which in turn alters the human manners (Bravo et al. 2011). Alteration in GABA receptor expression was reported in depression and anxiety patients, which also accentuates the relation between microbiome and brain-related diseases (Li et al. 2008). Giving mice a *Lactobacillus rhamnosus* showed a change in the expression of GABA receptors, which in turn reduced anxiety and depression (Li et al. 2008; Bravo et al. 2011).

17.3.2 The Vagal Nerve Pathway

Microbiome affects the CNS, directly or indirectly, by affecting the vagus nerve, which constitutes a major part of parasympathetic nervous system, resulting in significant control of digestion, heart rate, and immune response of visceral organs (Claudino dos Santos et al. 2023; Liu and Forsythe 2021). It is the most well-studied direct connection between the gut and the brain, in which the bacterial metabolites communicate with the human brain (Fülling et al. 2019). 80–90% of the vagus nerves are delivering signals from the gut to the brain, and they are called the afferent fibers, while the rest 10–20% are efferent fibers that transfer the signals from the brain to the gut (Claudino dos Santos et al. 2023). In dysbiosis, an intestinal permeability is increased, which in turn leads to a cholinergic anti-inflammatory effect by the vagus nerve via reducing the tumor necrosis factor-α (TNF-α) in the GIT (Claudino dos Santos et al. 2023) (Fig. 17.2).

The PYY hormone also directly affects the signaling of the vagus nerve (Bonaz et al. 2018). The vago-vagal reflex, which basically translates signals to and from the brain, plays an important role in obese people (Noor et al. 2023). Transplanting microbiome from healthy individuals to obese people helped in regulating the appetite and decreasing the adiposity (Noor et al. 2023). The use of *Lactobacillus* influenced the vagal sensory signaling in anxiety, depression, and stress (Bravo et al. 2011).

Microbiome induces signals that affected the levels of the neurotransmitters in the brain including the hypothalamus and parabrachial nucleus (Zubcevic and Raizada 2023). It was noticed that gut microbiota can send direct signals to the neurons, and hence increase its neuronal activity (Brun et al. 2013). GF mice showed a reduced neuronal innervation to the colonic lining, which is regained with the restoration of gut microbiota (De Vadder et al. 2018). Isolated polysaccharides A and B from *Bacteroides fragilis*, and *Lactobacillus rhamnosus* have all been shown to stimulate afferent neurons in vivo (Mao et al. 2013). The long-term treatment with *Bifidobacterium longum* induced reduction in anxiety caused by gut inflammation (Bercik et al. 2011b). This effect is completely deleted when the integrity of the vagus nerve is reduced by undertaking a vagotomy (Bravo et al. 2011). Moreover, the neurons of the gut were also shown to be influenced by chemical reactions; for example, a recent study showed that switching on aryl hydrocarbon receptors (AHRs) in mice had an effect on the mice's gut motility by affecting the ENS (Obata et al. 2020).

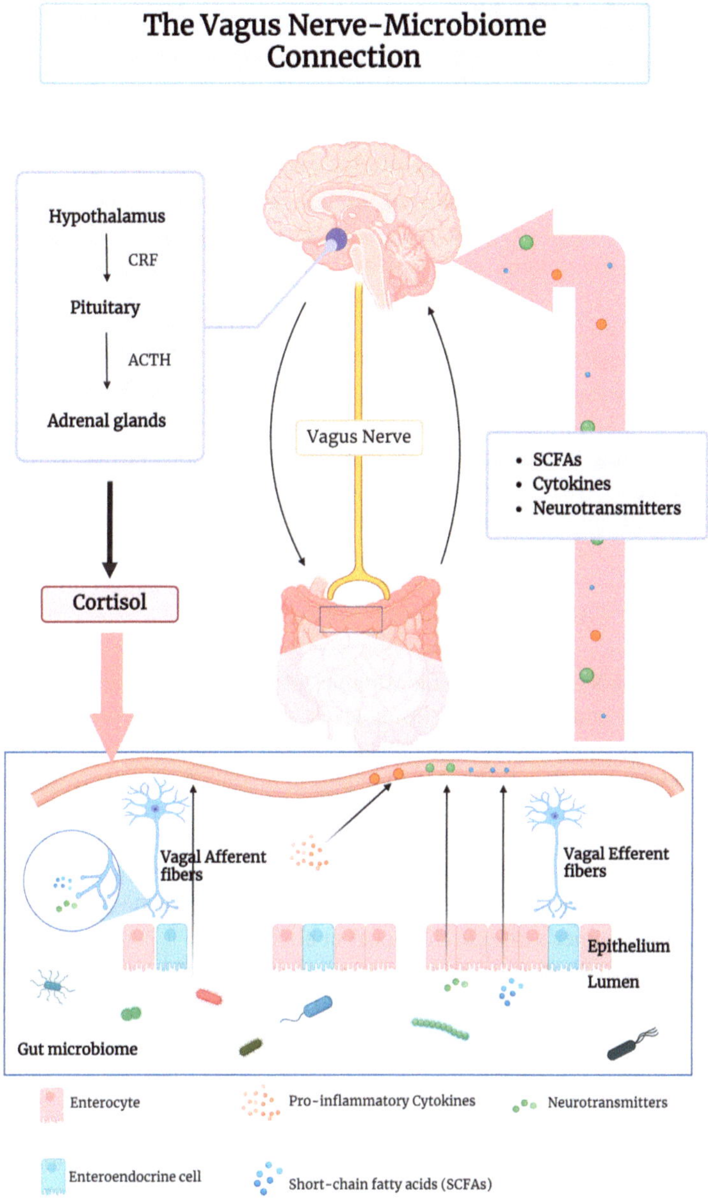

Fig. 17.2 Vagus nerve-microbiome connection. There is a two-way effect between the gut microbiome and the vagus nerve. The bacteria secrete SCFAs and neurotransmitters; it may also induce the release of pro-inflammatory cytokines which travel to the brain through circulation inducing its effects, for example, on the hypothalamus which results in release of cortisol, the stress hormone, and cortisol in return produces changes in the microbiome

17.3.3 Immune Pathway

The immune system is essential to facilitate the communication between the gut microbiota and the brain (Fung 2020). The gut microbiota activates the immune cells in both the gut and the brain, and can lead to neuroinflammation, which is a factor in the development of neurological disorders (Sampson and Mazmanian 2015). Microbes are first encountered by the immune system during birth, either through the birth canal or by maternal milk (Perez-Muñoz et al. 2017). This early activity teaches the immune system to distinguish between harmful and beneficial microbes (Hooper et al. 2012). Newborns have an underdeveloped immune system; therefore, early microbial encounter aids in the establishment of immunological tolerance (Siegrist 2001). Less inflammatory cytokines are produced by the body during this early stage of life, which promote the growth of regulatory T and B cells that are essential for preserving immunological homeostasis (PrabhuDas et al. 2011). This setting facilitates the entry of beneficial microorganisms, which support healthy immune system development. Toll-like receptors (TLRs), which are in charge of identifying microbial-associated molecular patterns, are important participants in this process. These receptors aid in the recognition of infections by the gut's mucosal immune system, particularly those with significant impact on young ages, like respiratory syncytial virus (RSV) and group B *Streptococcus* (Levy 2007) (Fig. 17.3).

Interferon (IFN)-α and IL-12 are examples of inflammatory signals that are produced by the immune system at lower levels during infancy, and increased with time, thereby boosting the immunological response (PrabhuDas et al. 2011). As the gut microbiota continues to develop, it begins to modulate immune responses more directly. Microbial components that trigger the release of pro-inflammatory cytokines such as TNF-α, IL-6, and IL-1β by TLRs can pass the BBB and induce neuroinflammation (Sampson and Mazmanian 2015). Neuroinflammation has been connected to the onset of neurological conditions such as autism spectrum disorder (ASD) and Alzheimer's disease (Lammert et al. 2018).

Studies have indicated that certain gut microorganisms have the ability to stimulate or inhibit inflammation (Braniste et al. 2014). For example, segmented filamentous bacteria (SFB) are known to cause a pro-inflammatory response, increasing symptoms in disorders including ASD by altering Th17 cells in the gut, and balancing antibodies by inhibiting IL-17A (Lee et al. 2011). On the other hand, some bacteria, such as *Prevotella histicola* and *Bacteroides fragilis*, enhance the production of regulatory T cells, which aid in the suppression of inflammation and the defense against neuroinflammatory disorders (Mangalam et al. 2017). Another key pathway in neuroinflammation includes the inflammasome pathway that activates cytokines like IL-1β and IL-18, leading to the maturation of caspase-1 (Yang et al. 2019). Inflammasome pathways and the generation of pro-inflammatory cytokines can be triggered by microbe-associated molecular patterns (MAMPs) (Yang et al. 2019). These processes have been linked to a variety of neurological disorders (Wong et al. 2016). Furthermore, after prolonged stress, animals with a hereditary

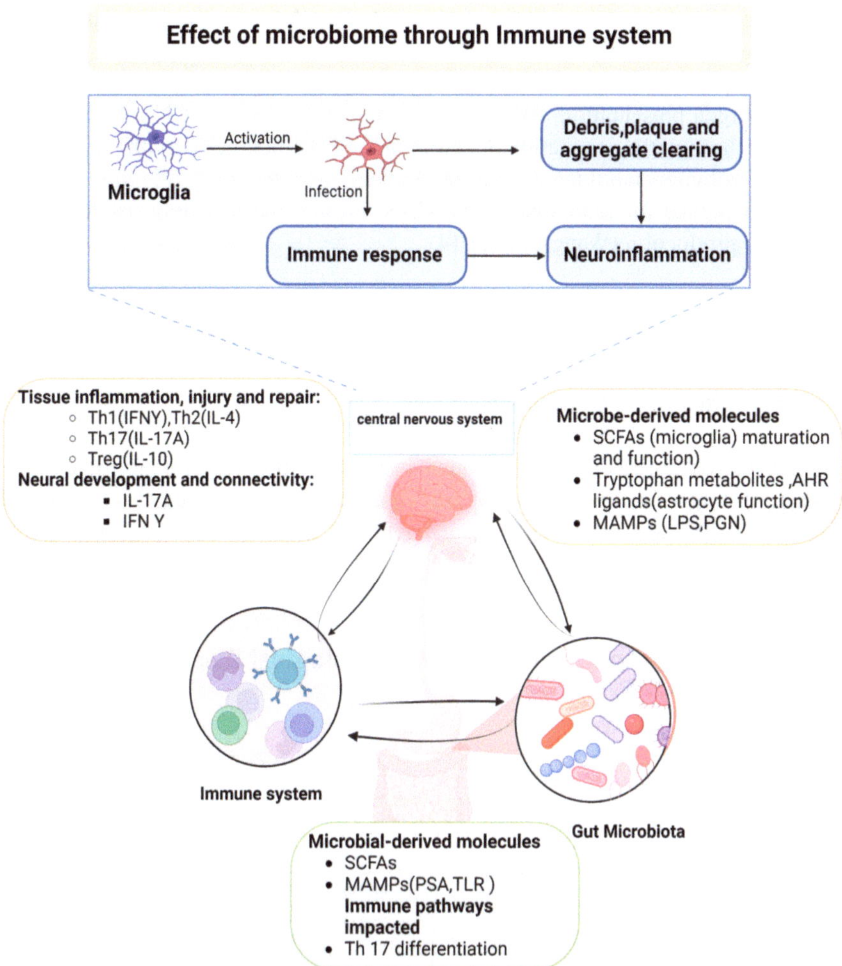

Fig. 17.3 Gut microbiome influences the immune system through microbial metabolites and immune signaling pathways. These interactions promote microglial maturation, tissue repair, and regulation of inflammation, showing how gut-derived molecules impact immune responses throughout the body

caspase-1 defect exhibit fewer anxiety- and depression-like behaviors (Wong et al. 2016).

Beyond the immune system's impact on inflammation, gut microbiota-derived molecules that are able to pass BBB may have an impact on the development and activation of brain immune cells, as well as modulation of peripheral disorders including microglia and astrocytes (Fung 2020). Microglia have a crucial role in homeostasis, development, and defense against infections, in addition to preserving brain health by releasing cytokines, phagocytosis, and complement activation (Salter and Stevens 2017). Gut bacteria are essential for healthy microglial function since microglia show signs of immaturity and altered expression of inflammatory genes in gut microbiota-deficient mice (Erny et al. 2015). Similarly, astrocytes, another class of brain cells, have a variety of roles in the nervous system. These include antigen presentation, BBB control, and management of CNS development and repair through the production of cytokines and chemokines (Rothhammer et al. 2016). Microbial metabolites, such as tryptophan, mediate type 1 IFN in astrocytes, which can reduce inflammation in the CNS through the activation of AHR and metabolite production (Wang et al. 2014).

17.4 Microbiome Effects on Neurological Disorders

17.4.1 Alzheimer's Disease

The gut microbiome as a key contributor in the development of neurological diseases including Alzheimer's disease has significantly highlighted (Fig. 17.4 and Table 17.1). Alzheimer's disease is a huge global health problem that affects

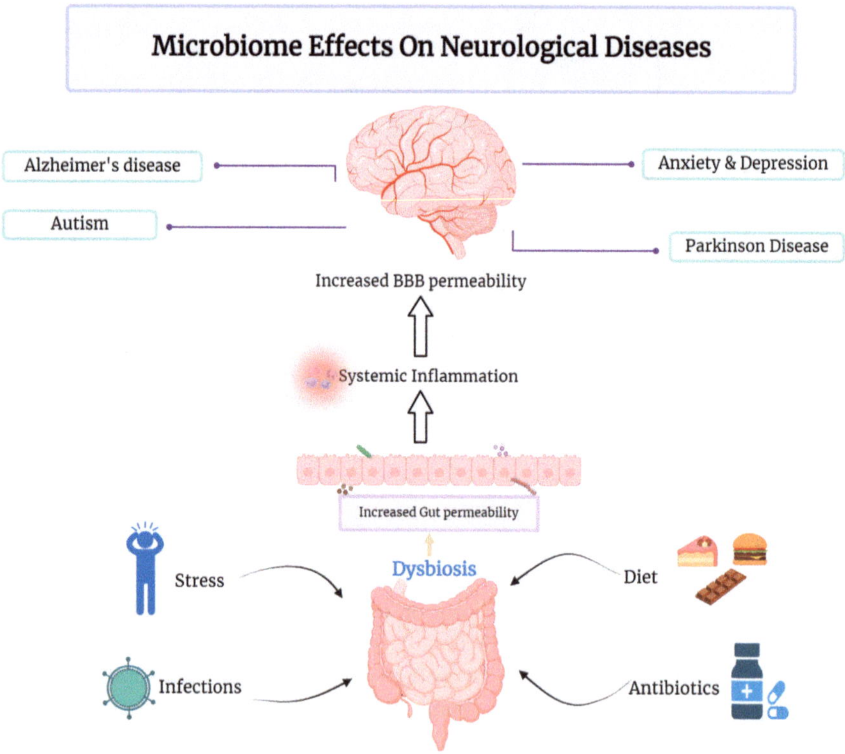

Fig. 17.4 Environmental and physical factors of a person's lifestyle affect the microbiome and may lead to dysbiosis. This may contribute to the pathology of several neurological diseases

Table 17.1 Microbiome composition alteration in neurological disorders

Neurological disorder	Model type	Microbiota altered	Supplements
Alzheimer's disease	Human	Increase in pro-inflammatory *Faecalibacterium prausnitzii, Eubacterium rectale, & Eubacterium hallii*	*Bifidobacterium bifidum, Lactobacillus fermentum, Lactobacillus casei, Lactobacillus acidophilus* (Akbari et al. 2016)
Parkinson's disease	Human	*Decrease in short-chain fatty acid (SCFA)-producing bacteria, such as genera Butyricicoccus and Coprococcus & an increase in pro-inflammatory bacteria, such as the genera Akkermansia* (Keshavarzian et al. 2015)	*S. thermophilus CRL 807, S. thermophilus CRL 808, Bifidobacterium animalis lactis, & Lactobacillus acidophilus* (Zhu et al. 2022)
Autism disorder	Human (child)	*Increase In Enterobacteriaceae Faecalibacterium* and *Ruminococcaceae* & decrease in *Alistipes*, *Bilophila*, and *Veillonella* (De Angelis et al. 2013)	*L. plantarum* WCSF1, *Lactobacillus plantarum* WCSF1, *Bifidobacterium*, & *Streptococcus* (Parracho et al. 2010)
Anxiety & depression	Human	*Increase in Oscillibacter, Alistipes, Campylobacter jejuni*, & *Clostridium* spp. (Mayer et al. 2014; Naseribafrouei et al. 2014)	*L. helveticus* and *B. longum* (Messaoudi et al. 2011)

~50,000,000 patients worldwide (Hodson 2018). This disease is mainly characterized by the progressive plaques deposing β-amyloid and hyperphosphorylated tau neurofibril tangles, resulting in an inflammation of the neurons with an ongoing cognitive deterioration (Mattson 2004). The neuroinflammation or a brain trauma eventually results in neuron death and disabled synaptic function leading to disease progression. Recently, the gut microbiome effects on brain are at the center of research and are recommended as a treatment option for neurological disorders, such as Alzheimer's disease (Table 17.1).

As age increases, the composition of the gut microbiome gets altered leading to enhancing the pro-inflammatory bacteria such as *Bacillus fragilis*, *Faecalibacterium prausnitzii*, *Eubacterium rectale*, and *Eubacterium hallii*, causing neuroinflammation (Nagpal et al. 2018). The presence of such pro-inflammatory bacteria in Alzheimer's disease patients compared to healthy people was observed by Cattaneo et al. (2017). Moreover, the amyloids of bacteria were shown to be a contributing causative factor in the aggregation of amyloid β-peptide in AD (Kesika et al. 2021). For instance, bacterial components such as curli (*E. coli*), TasA (*Bacillus subtilis*), CsgA (*S. Typhimurium*), FapC (*Pseudomonas fluorescens*), and phenol-soluble modulins (*Staphylococcus aureus*) have been proven to increase the development of AD, specifically by assisting in fibril and Aβ oligomer formation (Kesika et al. 2021).

17.4.2 Parkinson's Disease

Parkinson's disease is one of the most common neurodegenerative diseases that is characterized by α-synuclein aggregation, which functions in regulating the stability of neuronal membranes. These aggregates are present in both the CNS and the periphery (Atik et al. 2016). In Parkinson's disease, there are two types of symptoms: the motor, which includes bradykinesia, tremor, and rigidity; and the non-motor, which includes hyposmia, cognitive alterations, rapid eye movement, sleep behavior disorder (RBD), orthostatic hypotension, and some gastrointestinal difficulties like constipation (Goetze et al. 2005; Litvan et al. 2003). It was theorized that Parkinson's disease either presents as body-first or brain-first depending on when RBD develops, either before PD diagnosis or after. In other words, in the brain-first phenotype patients, PD is initiated prior to the development of RBD, while in body-first phenotype patients, RBD comes first. This happens because of the α-synuclein aggregation, which is initiated from the entry of a pathogen into the gastrointestinal tract; these aggregates are then taken up by the vagus nerve to the CNS, eventually to the substantia nigra and brain, leading to PD (Borghammer et al. 2021; Braak et al. 2003; Hawkes et al. 2009) (Table 17.1).

The microbiota-gut-brain axis plays a crucial role in the body-first phenotype due to its relation to the ENS (Fig. 17.4). Pathogens that enter through the gastrointestinal tract tend to cause inflammation; this increases the permeability of the intestines, leading to neuroinflammation as another mechanism for PD progression (Forsyth et al. 2011; Schwiertz et al. 2018). It was suggested that decreasing the levels of SCFAs in the gut contributes to the etiology of the disease, as these SCFAs help maintain the integrity of the intestinal wall and their anti-inflammatory impacts (Nuzum et al. 2020; Tan et al. 2014). Certain types of bacteria, like *Faecalibacterium prausnitzii* and members of the *Lachnospiraceae* family (such as *Roseburia*), are less common in people with PD compared to those without the condition (Romano et al. 2021; Shen et al. 2021). On the other hand, PD patients had an increase in *Enterobacteriaceae*, which was associated with postural instability (Scheperjans et al. 2015). Supporting evidence of autopsy studies of PD patients showed the presence of α-synuclein aggregates in both the substantia nigra and the ENS, emphasizing this connection (Elfil et al. 2020). In addition, a majority of PD patients are infected with *H. pylori*, a gastric ulcer-causing bacteria, which has been connected to PD. Moreover, it was shown that PD patients with this bacteria had severe manifestations and a decrease in motor function (Tan et al. 2015).

17.4.3 Anxiety and Depression

While depression is a common problem, major depressive disorder (MDD) is currently recognized as the most serious mental illness in the world. Pro-inflammatory cytokine levels have been found to rise in MDD patients, and the composition of their gut microbiota has been shown to be altered (Park et al. 2013). Patients with depression showed lower levels of *Firmicutes* and higher levels of *Bacteroidetes* and *Proteobacteria* (Valles-Colomer et al. 2019). People with depression had lower levels of *Dialister* and *Coprococcus* spp. in a recent large microbiome population cohort research (Valles-Colomer et al. 2019). Additionally, researchers noted that in depressed individuals, the gut microbiota can synthesize the neurotransmitter dopamine (3,4-dihydroxyphenylacetic acid, or DOPAC) (Chen et al. 2018). Researchers observed clear changes in the abundances of 16 bacterial families and the composition of the fecal microbiota in four phyla between healthy people and patients with MDD (Braak et al. 2003). Another study found a relative reduction in the alpha-diversity of gut microbiota in patients with current depressive episodes (Hawkes et al. 2009) (Table 17.1).

Changes associated with depression and anxiety may include a large number of gut microbes, involving 15 genera and 18 groupings (Ritchie et al. 2023). In generalized anxiety disorder (GAD) patients, there are a decrease in bacteria that produce SCFAs and an increase in *E. coli*, *Fusobacterium*, and *Ruminococcus gnavus*. Even after GAD subsided, these alterations persisted (Jiang et al. 2018). The amount of *E. coli* and *Bacteroides* has an impact on the intensity of anxiety in people with active GAD (Chen et al. 2019). C-reactive protein (CRP) has been found to be correlated with dysbiosis in *Bacteroidales, Selenomonadales, Clostridiales*, and *Holdemanella* in GAD patients. Differences in the genus *Holdemanella, Desulfovibrio, Barnesiella*, and the family *Coriobacteriaceae* were shown in both GAD and depression (Chen et al. 2021). Gut dysbiosis shares many characteristics with GAD and depression, but it also offers a way to distinguish these conditions based on the composition of the gut microbiome. The population of *Firmicutes* decreases, while *Bacteroides* increases in GAD, which is opposite to what happens in depression (Chen et al. 2019; Dong et al. 2021). Furthermore, *Proteobacteria* are more abundant in GAD. There are discernible variations in the quantity of bacteria present. Compared to the control, there is a decrease in the number of *Christensenellaceae* R7 group and *Fusicatenibacter* in GAD. In contrast, there are a drop in *Proteobacteria*, *Actinobacteria*, and *Firmicutes* and an increase in *Fusobacteria*, *Tenericutes*, *Verrucomicrobia*, and *Bacteroidetes* when comparing the amount in GAD and depression. The potential to develop targeted probiotic therapy is proposed based on these differences (Dong et al. 2021).

17.4.4 Autism Spectrum Disorder (ASD)

ASD is a diverse neurodevelopmental illness (Wasilewska and Klukowski 2015). It has revealed differences in the gut microbiota's abundance and composition between those with ASD and those without the condition (Dilmore et al. 2021). Research has linked dysbiotic gut microbiota to autism by demonstrating that autistic individuals exhibited a change in the gut microbiota, which can affect the immune system and trigger the release of its metabolites (Haba et al. 2012). Moreover, there may be a connection between the severity of ASD and the gut microbiome. It shows, for example, that children with ASD (who also had sleep disorders) exhibited lower levels of *Agathobacter* and *Faecalibacterium* and higher intensity core symptoms of

ASD. These bacterial strains showed a negative correlation with serotonin levels but a favorable correlation with 3-hydroxybutyric acid and melatonin levels (Hua et al. 2020). According to one study, children with Pitt-Hopkins syndrome, a severe ASD case, exhibited a higher relative abundance of *Clostridium bolteae* than their family members, who were not affected (Xiao et al. 2021). Furthermore, it was reported that only children with ASD have *Clostridium difficile* and *Clostridium clostridioforme*, whereas normal children only had *Clostridium tertium* (Kandeel et al. 2020). Further evidence showed that lower populations of *Akkermansia muciniphila* and elevated levels of TNF-α and Iba1 were present in a mouse model exhibiting autistic-like behavior (Goo et al. 2020). Additionally, a cross-sectional study showed increased quantities of *Bacilli, Actinobacteria*, and *Proteobacteria* in ADS (Plaza-Díaz et al. 2019) (Table 17.1).

17.5 Effect of Microbiome on Neurological Drug Efficacy

Medicines that are given orally pass through several challenges before they are systemically distributed; one of these challenges is the stability of the drug in the GIT (El Aidy et al. 2015). Gut microbiota influences the metabolism of these drugs significantly, thus altering drug's stability and efficacy (Enright et al. 2016). These alterations vary significantly from patient to patient and even between twins because of microbiota diversity (Turnbaugh et al. 2010). This is because of the variation in the interindividual and intraindividual microbiota composition (Enright et al. 2016). Reports have demonstrated the relation of microbiota and the pharmacokinetic profile of neurological drugs (Sousa et al. 2008).

17.5.1 Effect on Drugs Used in Alzheimer's Disease

Medicines such as donepezil, rivastigmine, galantamine are usually given to Alzheimer's patients to alleviate the inflammation and symptoms (Briggs et al. 2016). There is a bidirectional relationship between the drug used for AD and gut microbiota (Koppel et al. 2017). Drug consumption may promote changes in the microbial density by changing the environment in the intestine, affecting the growth of bacteria and their function. In addition, the microbiota can directly alter the patient's response to specific drugs by enzymatic reactions, thus influencing the drug's efficacy, bioavailability, and safety (Koppel et al. 2017). The acetylcholinesterase inhibitor, donepezil, has been used widely in the treatment of AD. Using a cognitive impaired amyloid β (Aβ)-induced animal models, it was shown in several studies that treatment with donepezil and Aβ influences the gut microbiota (Jo et al. 2022). Moreover, genotyping of fecal samples using 16S rRNA was employed to identify the effect of donepezil treatment on the gut microbiota (Jo et al. 2022). The AB+donepezil group showed high levels of *Cyanobacteria* and *Verrucomicrobia* than the AB group alone. The results also demonstrated that there is an increase in the number of *Blautia* and *Lachnospiraceae in the control group. However, there was a noticeable dominance of Prevotellaceae, Parabacteroides, Tannerellaceae, Ruminococcaceae, Akkermansia, Gastranaerophilales*, and *Verrucomicrobiales* in the Aβ + donepezil group (Segata et al. 2011). There is increased evidence on the relationship of microbiota on therapeutic efficacy and safety, which highlights the potential of increasing the drug's efficacy by modulating the gut microbiota (Li et al. 2016).

17.5.2 Effect on Drugs Used in Parkinson's Disease

Drugs used in the management of Parkinson's disease include levodopa-carbidopa combination, dopamine-receptor agonists, monoamine oxidase inhibitors (MAOI), catechol-o-methyltransferase inhibitors, and others (Weersma et al. 2020). Levodopa-carbidopa combination is the most widely used therapy for symptomatic PD; the effects of levodopa depend mainly on the concentration that enters the brain. It is highly affected by its conversion in the periphery rather the brain (Maini Rekdal et al. 2019). The major conversion that happens outside the brain is in the gut, either by the human aromatic amino acid decarboxylase (AADC) or by the gut microbiota. This alteration decreases the amount of dopamine in the brain and produces unwanted side effects (Mulroy and Bhatia 2019). Following oral levodopa administration, different patients showed variable concentrations, which can be explained by the fact that some bacteria metabolize the drug in the gut in two steps, first is decarboxylation to dopamine followed by dehydroxylation to *m*-tyramine (Maini Rekdal et al. 2019). *Enterococcus faecalis* is the major l-dopa decarboxylator using the tyrosine decarboxylases (tyrDCs) (Maini Rekdal et al. 2019). It was demonstrated that carbidopa and other human aromatic amino acid decarboxylase inhibitors are not able to block the bacterial decarboxylation. On the other hand, alpha-fluoromethyltyrosine, which is a selective inhibitor of TyrDC but not AADC, was able to block l-dopa decarboxylation that happened in the gut by microbiome (Maini Rekdal et al. 2019). This discovery allows for tailored therapy for PD patients, who can be given alpha-fluoromethyltyrosine plus l-dopa-carbidopa combination to achieve higher concentrations in the brain.

17.5.3 Effect on Drugs Used in Anxiety and Depression

FDA-approved antidepressants include tricyclic antidepressants (TCAs), antihistamines, selective serotonin reuptake inhibitors (SSRIs), benzodiazepines (BZDs), and MAOI (Murrough et al. 2015). Ketamine has been utilized as a fast-acting antidepressant to treat depression in recent years (Hess et al. 2022). In LPS-induced depressed mice, ketamine may raise *Lactobacillus johnsonii* levels, which could reduce depressive-like behaviors through the hypothalamic-pituitary-adrenal axis. Enhancing the quantity of microbiota that produces SCFAs, such *as Butyricimonas, Turicibacter*, and *Clostridiales*, may also contribute to the antidepressant properties of ketamine and its by-products (Hua et al. 2022). It was discovered that SSRIs fluoxetine (Flu) and escitalopram decrease the amount of intestinal microbiota, particularly those of *Ruminococcus, Adlercreutzia*, and an unidentified *Alphaproteobacterium*; this was also confirmed for two SNRIs, venlafaxine and duloxetine (Lukic et al. 2019). Potential adverse effects could arise from a reduction in the richness of the gut bacteria. One type of *Ruminococcus, R. flavefaciens*, was found to be able to reduce the effects of antidepressants by altering the expression of genes in the mitochondria and synapses and by changing the amounts of monoamine neurotransmitters (Lukic et al. 2019). A prior study that used a mouse model to investigate the antidepressant qualities of neferine (Nef) showed similar effects to those of an antidepressant with a rise in the relative abundance of *Lactobacillus*. This finding suggests that Nef might alleviate depression condition through controlling *Lactobacillus* levels, which may affect the triple reuptake of dopamine, serotonin, and norepinephrine (Dong et al. 2022). Supplemental diets high in omega-3 and omega-6 polyunsaturated fatty acids support the growth of *Bifidobacterium* and *Lactobacillus* and regulate the metabolism of microorganisms, especially in times of early life stress (Robertson et al. 2017; Costantini et al. 2017).

It has also been demonstrated that omega-3 fatty acids, such as eicosapentaenoic and docosahexaenoic acid, enhance adult cognition and lessen stress and sadness (Robertson et al. 2017). When the probiotic *L. rhamnosus* was given to stressed mice, it reduced the expression of gamma-aminobutyric acid 2 and corticosterone gene levels, without affecting the hippocampal expression of gamma-aminobutyric acid 2 (Bravo et al. 2011; Cheung et al. 2019). In one study, probiotics such as Bifidum, *Lactococcus lactis*, Bifidum, *L. casei*, *L. salivarius*, *L. brevis*, and *L. acidophilus* showed encouraging benefits in lowering negative thoughts and behavior (Karen et al. 2021). Probiotics, such as *Bifidobacterium longum*, were administered to restore BDNF levels and reduce anxiety-like behavior caused by inflammation (Cheung et al. 2019; Pinto-Sanchez et al. 2017). *B. longum* was able to lower the daily self-reported stress and cortisol levels. A randomized study found that synbiotic FOS, GOS, and inulin mixed with a probiotic mixture decreased depression and raised serum levels of a brain-derived neurotrophic factor in people with depression in comparison with controls (Haghighat et al. 2021). On the other hand, FMT improves gut diversity by introducing healthy microorganisms to the patient's gut. Like probiotics, FMT was developed to promote healthy gut bacteria composition and activity. Transferring the fecal microbiota of healthy donors to anxious mice reduced anxiety and depression effects. FMT is currently among the most widely utilized treatments for GIT and neuropsychiatric disorders (Chinna Meyyappan et al. 2020). However, transferring fecal microbiota from someone with a disease might result in adverse effects, such as depression. For instance, a study found that mice exhibiting depressive-like behaviors were given fecal microbiota from rheumatoid arthritis patients, who also had depression (Pu et al. 2022).

17.5.4 Effect on Drugs Used in Autism Spectrum Disorder

A growing number of preclinical studies demonstrates the significance of gut microorganisms in modulating host physiology, including endophenotypes linked to ASD (Zhang et al. 2024). More precisely, in a mother high-fat diet mouse model for neurodevelopmental problems, it is accidentally found that the commensal bacterial species *Limosilactobacillus reuteri* (L.) preferentially restored the social deficiencies (Zhang et al. 2024). *L. reuteri* targets the oxytocin-dopaminergic reward circuit, a crucial pathway involved in social behaviors, to boost social reward and restore the social deficits in ASD models via the vagus nerve. This is a way that shows how bacteria can affect social behavior. Furthermore, it has been discovered recently that *L. reuteri* repaired social issues in the absence of a fully developed adaptive immune system, suggesting that B and T cells are probably not necessary for *L. reuteri's* prosocial effects (Dooling et al. 2022). Also, the favorable effects of *L. reuteri* seem to be limited to social behavior, as the bacteria are unable to treat other behavioral impairments in ASD mouse models, such as repetitive behavior and anxiety-related behavior (Sharon et al. 2019). FDA has designated *L. reuteri* as "generally recognized as safe" (GRAS) for use in humans. In a clinical trial, *L. reuteri* has already been safely administered to neonates within the first 3 months of life (Sgritta et al. 2019).

Although research indicates that early antibiotic exposure may cause autism, there have also been cases where antibiotics have been used to cure autism. For example, aminoglycosides have been shown to partially alleviate autism symptoms by modulating a premature stop codon mutation in a polymorphic gene associated with autism (Manev and Manev 2001). Following an 8-week therapy regimen consisting of vancomycin, an antibiotic mostly used to treat Gram-positive bacteria like *Clostridium*, improvements in communication and behavior were noted in 11 children with ASD (Johnson et al. 2020). It recently became known that treatment therapies such as *Bifidobacterium longum* or fecal transplantation could improve gut dysbiosis and restore the levels of propionic acid producing *Butyricicoccus* in rat model for ASD (Happé and Frith 2020). Tryptophan metabolism was found to have changed, and Bifidobacteria modifications were linked to metabolites that differed in abundance from tryptophan metabolism derivatives (Xiao et al. 2021). Additionally, it has been demonstrated that dietary therapies containing *Bifidobacterium* strains, other bacterial strains, and/or substances that raise the amount of gut Bifidobacteria improve both intestinal outcomes and the severity of ASD (Happé and Frith 2020; Shaaban et al. 2018).

17.6 Futuristic Therapy

17.6.1 Tailored Psychomicrobiotics

Various studies were performed to investigate the effect of the drugs on the human microbiota structurally and functionally, and the effect of microbiota on drug efficacy. Investigations were highlighted to limit the use of drugs and the use of tailored microbiome instead to enhance the brain function (Li et al. 2020). Unlike human genetics, the gut microbiome is modifiable, making it an attractive therapeutic target to optimize therapy (Koppel et al. 2017). The term "tailored psych postbiotic" was pointed out to demonstrate the necessity of individualizing the dose and the content of the probiotic to efficiently target neurological disease in the brain (Lee et al. 2021). The microbiome-targeted treatment is being studied thoroughly. For example, administration of yeast β-glucans in vitro to mouse models with AD was effectively balanced the inflammatory and the non-inflammatory gut microbiota, reducing the neuroinflammation and enhancing the production of SCFAs (Xu et al. 2020) (Fig. 17.5).

It also suggests the idea of tailoring treatment uniquely to fit the patient's genetic code. This can be implemented in real life by testing for microbiome composition and specific bacterial deficiencies to formulate a specialized treatment for this patient. For instance, urine and feces samples could be collected from an Alzheimer's patient, followed by analysis of specific microbiota metabolites to know the identity of the downregulated microbiota. Accordingly, a tailored treatment can be prepared to effectively prevent disease progression and promote disease treatment; the term for such process is "postbiotic." Amino acid derivatives make up a good candidate for postbiotics. Tryptophan also decreases inflammation by significantly enhancing the SCFA's production (Claesson et al. 2012).

17.6.2 Probiotic Lifestyle

The human microbiome is easily affected by a lot of factors, which can be genetic or environmental. Diet habit is a major factor; nowadays, our fast-paced and busy lives have impacted our eating and general habits. To achieve a healthy microbiome, one should make sure to eat a balanced diet, prioritize sleep, and reduce stress. Dietary habits have proposed to be closely connected to our microbiome. Studies have shown that there is a link between weight control and the consumption of dairy-fermented foods, in addition to reduced risk of cardiovascular disease, type 2 diabetes, and better metabolism of glucose (Tapsell 2015; Soedamah-Muthu et al. 2013; Eussen et al. 2016; Chen et al. 2014). Moreover, fermented foods have antioxidant properties, which are crucial for both our cells and microbiome (Kamal-Eldin 2012). It was shown that fermented milk was able to increase the levels of SCFAs, which are essential for maintaining gut health (Veiga et al. 2014). Moreover, protein source is important; it was demonstrated that people who got their protein from meats and eggs had a higher gut microbiome than who got protein from vegetarian sources (Ren et al. 2023). Consuming prebiotics and probiotics are vital for the gut health (Table 17.2).

Probiotics can enhance human health by strengthening the immune system by fortifying the gut lining, preventing harmful bacteria and their by-products from entering the body, increasing antibody production, improving the gut's protective layer, and directly fighting off harmful microorganisms (Crittenden et al. 2005).

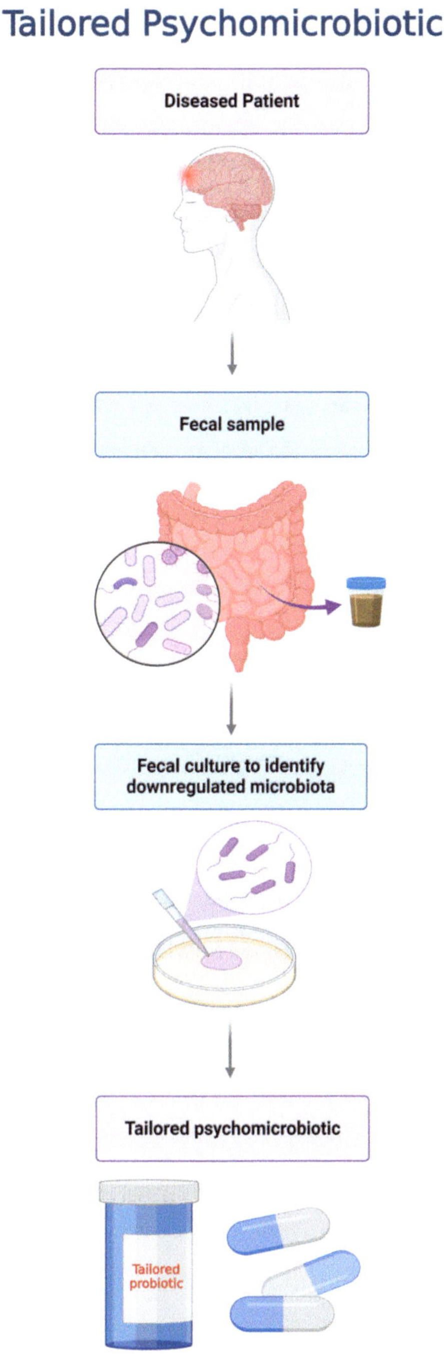

Fig. 17.5 Tailored psychomicrobiotics strategy. A futuristic therapy demonstrates the use of fecal samples taken from a neurological diseased patient and cultured to identify specific downregulated microbiota

Table 17.2 Studies on animal models to demonstrate the effect of probiotics on neurological disorders

Type of disease	Study objective	Animal model	Intervention	Outcome
Alzheimer's disease (AD) (Rezaeiasl et al. 2019)	To analyze the effect of probiotic supplements on for 6 weeks in AD rats	Beta-amyloid-induced rats	500 mg probiotic daily by gavage for 6 weeks	Probiotic supplementation improved learning but not memory impairment; increased levels of PPF ratios were seen compared to the placebo group
Autism spectrum disorder (ASD) (Hsiao et al. 2013)	To study the microbiota effect on ASD mouse model	Mice	*Bacteroides fragilis* for 6 days	*Bacteroides fragilis* oral administration corrected gut permeability, and ameliorated defects in anxiety and sensorimotor behavior in offspring mice
Depression (Liu et al. 2016)	To study the effect of *Bifidobacterium infantis* in depression	Rats	Chronic treatment of Bifidobacteria and then subjected to forced swim test (FST)	Probiotic treatment normalized immune response, reversed behavioral deficits, and returned the norepinephrine levels in brains to normal
Anxiety disorder (Liang et al. 2015)	To study the effect of *Lactobacillus helveticus* on stress levels in rats	Rats	*Lactobacillus helveticus* for 21 days	*L. helveticus* NS8 improved cognitive dysfunction and behavior anxiety and depression with a decrease in CPRT levels and ACTH levels, and an increase in IL-10 levels, and restored levels of 5-HT and NE
Parkinson's disease (Liao et al. 2020; Wang et al. 2022)	Study the effect of *L. plantarum* DP189 & PS128	MPTP-induced PD mice	0.2 mL of *L. plantarum* DP189 orally for 14 days *L. plantarum PS128* was administered as 10^9 CFU in 200 μL saline daily for 28 days	*DP189* reduced alpha-synuclein aggravation via regulating oxidative damage, inflammation, and gut microbiota dysfunction. *PS128* can suppress glial cell hyperactivation, get rid of MPTP-induced oxidative stress and neuroinflammation, but it also promotes intestinal motility and mucin production, and relieves motor deficits and neurotoxicity in mouse models of PD

Moving to smoking, it was shown that smoking nicotine has caused decreased microbiome in mice (Tam et al. 2020). Cigarette smoking increased the mucosal permeability and pH of the intestine, leading to overgrowth of certain microbes and an imbalance in the microflora (Bai et al. 2022; Berkowitz et al. 2018).

17.7 Conclusion

The connection between human gut microbiota and human brain functions reveals a relationship that is crucial for the understanding of mental and neurological health. The intricate ways in which gut bacteria communicate with the brain through hormonal, immune, and neural pathways demonstrate a fascinating bidirectional interaction. The implications of this relationship are particularly significant for patients affected by neurological disorders such as Alzheimer's disease, Parkinson's disease, anxiety, depression, and autism. The evidence suggests that imbalances in gut microbiota can exacerbate these conditions and can be the cause of them, highlighting the potential for personalized treatment strategies that consider an individual's unique microbiome. Emerging therapies, such as tailored psychomicrobiotics and dietary adjustments, offer exciting possibilities for improving treatment outcomes. By focusing on gut health, we can not only enhance cognitive function but also determine the root causes of various disorders. As we advance our research in this field, we set a new understanding of healthcare that embraces the complexity of the microbiome and its essential role in the overall human well-being. This holistic perspective may ultimately lead to more effective interventions and a better quality of life for those affected by neurological challenges.

References

Akbari E, Asemi Z, Daneshvar Kakhaki R, Bahmani F, Kouchaki E, Tamtaji OR et al (2016) Effect of probiotic supplementation on cognitive function and metabolic status in Alzheimer's disease: a randomized, double-blind and controlled trial. Front Aging Neurosci 8:256

Alexander F (1934) The influence of psychologic factors upon gastro-intestinal disturbances: a symposium. Psychoanal Q 3(4):501–539

Alvarez E, Martínez MD, Roncero I, Chowen JA, García-Cuartero B, Gispert JD et al (2005) The expression of GLP-1 receptor mRNA and protein allows the effect of GLP-1 on glucose metabolism in the human hypothalamus and brainstem. J Neurochem 92(4):798–806

Asano Y, Hiramoto T, Nishino R, Aiba Y, Kimura T, Yoshihara K et al (2012) Critical role of gut microbiota in the production of biologically active, free catecholamines in the gut lumen of mice. Am J Physiol Gastrointest Liver Physiology 303(11):G1288–G1G95

Atik A, Stewart T, Zhang J (2016) Alpha-synuclein as a biomarker for Parkinson's disease. Brain Pathol 26(3):410–418

Bai X, Wei H, Liu W, Coker OO, Gou H, Liu C et al (2022) Cigarette smoke promotes colorectal cancer through modulation of gut microbiota and related metabolites. Gut 71(12):2439–2450

Belkaid Y, Hand TW (2014) Role of the microbiota in immunity and inflammation. Cell 157(1):121–141

Bercik P, Denou E, Collins J, Jackson W, Lu J, Jury J et al (2011a) The intestinal microbiota affect central levels of brain-derived neurotropic factor and behavior in mice. Gastroenterology 141(2):599–609.e1-3

Bercik P, Park AJ, Sinclair D, Khoshdel A, Lu J, Huang X et al (2011b) The anxiolytic effect of Bifidobacterium longum NCC3001 involves vagal pathways for gut-brain communication. Neurogastroenterol Motil 23(12):1132–1139

Berkowitz L, Schultz BM, Salazar GA, Pardo-Roa C, Sebastián VP, Álvarez-Lobos MM et al (2018) Impact of cigarette smoking on the gastrointestinal tract inflammation: opposing effects in Crohn's disease and ulcerative colitis. Front Immunol 9:74

Bonaz B, Bazin T, Pellissier S (2018) The vagus nerve at the Interface of the microbiota-gut-brain axis. Front Neurosci 12:49

Borghammer P, Horsager J, Andersen K, Van Den Berge N, Raunio A, Murayama S et al (2021) Neuropathological evidence of body-first vs. brain-first Lewy body disease. Neurobiol Dis 161:105557

Braak H, Del Tredici K, Rüb U, De Vos RA, Steur ENJ, Braak E (2003) Staging of brain pathology related to sporadic Parkinson's disease. Neurobiol Aging 24(2):197–211

Braniste V, Al-Asmakh M, Kowal C, Anuar F, Abbaspour A, Tóth M et al (2014) The gut microbiota influences blood-brain barrier permeability in mice. Sci Transl Med 6(263):263ra158-263ra158

Bravo JA, Forsythe P, Chew MV, Escaravage E, Savignac HM, Dinan TG et al (2011) Ingestion of Lactobacillus strain regulates emotional behavior and central GABA receptor expression in a mouse via the vagus nerve. Proc Natl Acad Sci USA 108(38):16050–16055

Briggs R, Kennelly SP, O'Neill D (2016) Drug treatments in Alzheimer's disease. Clin Med (Lond) 16(3):247–253

Brun P, Giron MC, Qesari M, Porzionato A, Caputi V, Zoppellaro C et al (2013) Toll-like receptor 2 regulates intestinal inflammation by controlling integrity of the enteric nervous system. Gastroenterology 145(6):1323–1333

Buffa R, Capella C, Fontana P, Usellini L, Solcia E (1978) Types of endocrine cells in the human colon and rectum. Cell Tissue Res 192(2):227–240

Bull MJ, Plummer NT (2014) Part 1: The human gut microbiome in health and disease. Integr Med (Encinitas) 13(6):17–22

Canfora EE, Jocken JW, Blaak EE (2015) Short-chain fatty acids in control of body weight and insulin sensitivity. Nat Rev Endocrinol 11(10):577–591

Cattaneo A, Cattane N, Galluzzi S, Provasi S, Lopizzo N, Festari C et al (2017) Association of brain amyloidosis with pro-inflammatory gut bacterial taxa and peripheral inflammation markers in cognitively impaired elderly. Neurobiol Aging 49:60–68

Chaplin A, Rebrikov D, Boldyreva M (2017) The human microbiome. Bull Russ State Med Univ 2:5–13

Chen M, Sun Q, Giovannucci E, Mozaffarian D, Manson JE, Willett WC et al (2014) Dairy consumption and risk of type 2 diabetes: 3 cohorts of US adults and an updated meta-analysis. BMC Med 12:1–14

Chen Z, Li J, Gui S, Zhou C, Chen J, Yang C et al (2018) Comparative metaproteomics analysis shows altered fecal microbiota signatures in patients with major depressive disorder. Neuroreport 29(5):417–425

Chen Y-h, Bai J, Wu D, Yu S-f, Qiang X-l, Bai H et al (2019) Association between fecal microbiota and generalized anxiety disorder: severity and early treatment response. J Affect Disord 259:56–66

Chen Y, Meng P, Cheng S, Jia Y, Wen Y, Yang X et al (2021) Assessing the effect of interaction between C-reactive protein and gut microbiome on the risks of anxiety and depression. Mol Brain 14:1–8

Cheung SG, Goldenthal AR, Uhlemann AC, Mann JJ, Miller JM, Sublette ME (2019) Systematic review of gut microbiota and major depression. Front Psych 10:34

Chinna Meyyappan A, Forth E, Wallace CJ, Milev R (2020) Effect of fecal microbiota transplant on symptoms of psychiatric disorders: a systematic review. BMC Psychiatry 20:1–19

Claesson MJ, Jeffery IB, Conde S, Power SE, O'Connor EM, Cusack S et al (2012) Gut microbiota composition correlates with diet and health in the elderly. Nature 488(7410):178–184
Claudino dos Santos JC, Oliveira LF, Noleto FM, Gusmão CTP, Brito GAC, Viana GSB (2023) Gut-microbiome-brain axis: the crosstalk between the vagus nerve, alpha-synuclein and the brain in Parkinson's disease. Neural Regen Res 18(12):2611–2614
Costantini L, Molinari R, Farinon B, Merendino N (2017) Impact of Omega-3 fatty acids on the gut microbiota. Int J Mol Sci 18(12)
Crittenden R, Bird AR, Gopal P, Henriksson A, Lee Y, Playne M (2005) Probiotic research in Australia, New Zealand and the Asia-Pacific region. Curr Pharm Des 11(1):37–53
Cryan JF, O'Riordan KJ, Cowan CSM, Sandhu KV, Bastiaanssen TFS, Boehme M et al (2019) The microbiota-gut-brain axis. Physiol Rev 99(4):1877–2013
Cussotto S, Sandhu KV, Dinan TG, Cryan JF (2018) The neuroendocrinology of the microbiota-gut-brain axis: a behavioural perspective. Front Neuroendocrinol 51:80–101
Dalile B, Van Oudenhove L, Vervliet B, Verbeke K (2019) The role of short-chain fatty acids in microbiota-gut-brain communication. Nat Rev Gastroenterol Hepatol 16(8):461–478
Damiani F, Cornuti S, Tognini P (2023) The gut-brain connection: exploring the influence of the gut microbiota on neuroplasticity and neurodevelopmental disorders. Neuropharmacology 231:109491
Davenport ER, Sanders JG, Song SJ, Amato KR, Clark AG, Knight R (2017) The human microbiome in evolution. BMC Biol 15(1):127
De Angelis M, Piccolo M, Vannini L, Siragusa S, De Giacomo A, Serrazzanetti DI et al (2013) Fecal microbiota and metabolome of children with autism and pervasive developmental disorder not otherwise specified. PLoS One 8(10):e76993
De Vadder F, Grasset E, Mannerås Holm L, Karsenty G, Macpherson AJ, Olofsson LE et al (2018) Gut microbiota regulates maturation of the adult enteric nervous system via enteric serotonin networks. Proc Natl Acad Sci USA 115(25):6458–6463
Del Toro-Barbosa M, Hurtado-Romero A, Garcia-Amezquita LE, García-Cayuela T (2020) Psychobiotics: mechanisms of action, evaluation methods and effectiveness in applications with food products. Nutrients 12(12)
Diaz Heijtz R, Wang S, Anuar F, Qian Y, Björkholm B, Samuelsson A et al (2011) Normal gut microbiota modulates brain development and behavior. Proc Natl Acad Sci USA 108(7):3047–3052
Dilmore AH, McDonald D, Nguyen TT, Adams JB, Krajmalnik-Brown R, Elijah E et al (2021) The fecal microbiome and metabolome of Pitt Hopkins Syndrome, a severe autism spectrum disorder. mSystems 6(6):e0100621
Dinan TG, Cryan JF (2017) The microbiome-gut-brain axis in health and disease. Gastroenterol Clin N Am 46(1):77–89
Dobbs RJ, Charlett A, Dobbs SM, Weller C, Iguodala O, Smee C et al (2012) Towards defining a rigidity-associated pathogenic pathway in idiopathic parkinsonism. Neurodegener Dis 10(1–4):183–186
Dong Z, Shen X, Hao Y, Li J, Li H, Xu H et al (2021) Gut microbiome: a potential indicator for differential diagnosis of major depressive disorder and general anxiety disorder. Front Psych 12:651536
Dong Z, Xie Q, Xu F, Shen X, Hao Y, Li J et al (2022) Neferine alleviates chronic stress-induced depression by regulating monoamine neurotransmitter secretion and gut microbiota structure. Front Pharmacol 13:974949
Dooling SW, Sgritta M, Wang IC, Duque A, Costa-Mattioli M (2022) The effect of Limosilactobacillus reuteri on social behavior is independent of the adaptive immune system. mSystems 7(6):e0035822
El Aidy S, van den Bogert B, Kleerebezem M (2015) The small intestine microbiota, nutritional modulation and relevance for health. Curr Opin Biotechnol 32:14–20
Elfil M, Kamel S, Kandil M, Koo BB, Schaefer SM (2020) Implications of the gut microbiome in Parkinson's disease. Mov Disord 35(6):921–933
Enright EF, Gahan CG, Joyce SA, Griffin BT (2016) The impact of the gut microbiota on drug metabolism and clinical outcome. Yale J Biol Med 89(3):375–382

Erny D, Hrabě de Angelis AL, Jaitin D, Wieghofer P, Staszewski O, David E et al (2015) Host microbiota constantly control maturation and function of microglia in the CNS. Nat Neurosci 18(7):965–977

Eussen SJ, van Dongen MC, Wijckmans N, den Biggelaar L, Elferink SJO, Singh-Povel CM et al (2016) Consumption of dairy foods in relation to impaired glucose metabolism and type 2 diabetes mellitus: the Maastricht Study. Br J Nutr 115(8):1453–1461

Forsyth CB, Shannon KM, Kordower JH, Voigt RM, Shaikh M, Jaglin JA et al (2011) Increased intestinal permeability correlates with sigmoid mucosa alpha-synuclein staining and endotoxin exposure markers in early Parkinson's disease. PLoS One 6(12):e28032

Fülling C, Dinan TG, Cryan JF (2019) Gut microbe to brain signaling: what happens in vagus…. Neuron 101(6):998–1002

Fung TC (2020) The microbiota-immune axis as a central mediator of gut-brain communication. Neurobiol Dis 136:104714

Goetze O, Wieczorek J, Mueller T, Przuntek H, Schmidt WE, Woitalla D (2005) Impaired gastric emptying of a solid test meal in patients with Parkinson's disease using 13C-sodium octanoate breath test. Neurosci Lett 375(3):170–173

Goo N, Bae HJ, Park K, Kim J, Jeong Y, Cai M et al (2020) The effect of fecal microbiota transplantation on autistic-like behaviors in Fmr1 KO mice. Life Sci 262:118497

Gribble FM, Reimann F (2019) Function and mechanisms of enteroendocrine cells and gut hormones in metabolism. Nat Rev Endocrinol 15(4):226–237

Gunawardene AR, Corfe BM, Staton CA (2011) Classification and functions of enteroendocrine cells of the lower gastrointestinal tract. Int J Exp Pathol 92(4):219–231

Haba R, Shintani N, Onaka Y, Wang H, Takenaga R, Hayata A et al (2012) Lipopolysaccharide affects exploratory behaviors toward novel objects by impairing cognition and/or motivation in mice: possible role of activation of the central amygdala. Behav Brain Res 228(2):423–431

Haghighat N, Rajabi S, Mohammadshahi M (2021) Effect of synbiotic and probiotic supplementation on serum brain-derived neurotrophic factor level, depression and anxiety symptoms in hemodialysis patients: a randomized, double-blinded, clinical trial. Nutr Neurosci 24(6):490–499

Haigh E (1975) The roots of the vitalism of Xavier Bichat. Bull Hist Med 49(1):72–86

Happé F, Frith U (2020) Annual research review: looking back to look forward – changes in the concept of autism and implications for future research. J Child Psychol Psychiatry 61(3):218–232

Hawkes CH, Del Tredici K, Braak H (2009) Parkinson's disease: the dual hit theory revisited. Ann N Y Acad Sci 1170(1):615–622

Health Gmf (2012) MetaHIT: GMFH Editing Team. Available from: https://www.gutmicrobiotaforhealth.com/metahit/

Hess EM, Riggs LM, Michaelides M, Gould TD (2022) Mechanisms of ketamine and its metabolites as antidepressants. Biochem Pharmacol 197:114892

Hodson R (2018) Alzheimer's disease. Nature 559(7715):S1

Holzer P (2017) Interoception and gut feelings: unconscious body signals' impact on brain function, behavior and belief processes. In: Angel H-F, Oviedo L, Paloutzian RF, Runehov ALC, Seitz RJ (eds) Processes of believing: the acquisition, maintenance, and change in creditions. Springer, Cham, pp 435–442

Hooper LV, Littman DR, Macpherson AJ (2012) Interactions between the microbiota and the immune system. Science 336(6086):1268–1273

Hrncir T (2022) Gut microbiota dysbiosis: triggers, consequences, diagnostic and therapeutic options. Microorganisms 10(3)

Hsiao EY, McBride SW, Hsien S, Sharon G, Hyde ER, McCue T et al (2013) Microbiota modulate behavioral and physiological abnormalities associated with neurodevelopmental disorders. Cell 155(7):1451–1463

Hua X, Zhu J, Yang T, Guo M, Li Q, Chen J et al (2020) The gut microbiota and associated metabolites are altered in sleep disorder of children with autism spectrum disorders. Front Psych 11:855

Hua H, Huang C, Liu H, Xu X, Xu X, Wu Z et al (2022) Depression and antidepressant effects of ketamine and its metabolites: the pivotal role of gut microbiota. Neuropharmacology 220:109272

Hube F, Lietz U, Igel M, Jensen PB, Tornqvist H, Joost HG et al (1996) Difference in leptin mRNA levels between omental and subcutaneous abdominal adipose tissue from obese humans. Horm Metab Res 28(12):690–693

Isacson R, Nielsen E, Dannaeus K, Bertilsson G, Patrone C, Zachrisson O et al (2011) The glucagon-like peptide 1 receptor agonist exendin-4 improves reference memory performance and decreases immobility in the forced swim test. Eur J Pharmacol 650(1):249–255

Jiang H-y, Zhang X, Yu Z-h, Zhang Z, Deng M, Zhao J-h et al (2018) Altered gut microbiota profile in patients with generalized anxiety disorder. J Psychiatr Res 104:130–136

Jo JK, Lee G, Nguyen CD, Park SE, Kim EJ, Kim HW et al (2022) Effects of donepezil treatment on brain metabolites, gut microbiota, and gut metabolites in an amyloid beta-induced cognitive impairment mouse pilot model. Molecules 27(19)

Johnson D, Letchumanan V, Thurairajasingam S, Lee L-H (2020) A revolutionizing approach to autism spectrum disorder using the microbiome. Nutrients 12(7):1983

Kamal-Eldin A (2012) The role of fermentation in providing biologically active compounds for the human organism. In: Fermentation: effects on food properties, pp 151–168

Kandeel WA, Meguid NA, Bjørklund G, Eid EM, Farid M, Mohamed SK et al (2020) Impact of Clostridium bacteria in children with autism spectrum disorder and their anthropometric measurements. J Mol Neurosci 70(6):897–907

Karen C, Shyu DJ, Rajan KE (2021) Lactobacillus paracasei supplementation prevents early life stress-induced anxiety and depressive-like behavior in maternal separation model-possible involvement of microbiota-gut-brain axis in differential regulation of microrna124a/132 and glutamate receptors. Front Neurosci 15:719933

Kasarello K, Cudnoch-Jedrzejewska A, Czarzasta K (2023) Communication of gut microbiota and brain via immune and neuroendocrine signaling. Front Microbiol 14

Katsurada K, Yada T (2016) Neural effects of gut- and brain-derived glucagon-like peptide-1 and its receptor agonist. J Diabetes Investig 7 Suppl 1(Suppl 1):64–69

Keshavarzian A, Green SJ, Engen PA, Voigt RM, Naqib A, Forsyth CB et al (2015) Colonic bacterial composition in Parkinson's disease. Mov Disord 30(10):1351–1360

Kesika P, Suganthy N, Sivamaruthi BS, Chaiyasut C (2021) Role of gut-brain axis, gut microbial composition, and probiotic intervention in Alzheimer's disease. Life Sci 264:118627

Koppel N, Maini Rekdal V, Balskus EP (2017) Chemical transformation of xenobiotics by the human gut microbiota. Science 356(6344):eaag2770

Lammert CR, Frost EL, Bolte AC, Paysour MJ, Shaw ME, Bellinger CE et al (2018) Cutting edge: critical roles for microbiota-mediated regulation of the immune system in a prenatal immune activation model of autism. J Immunol 201(3):845–850

Lee YK, Menezes JS, Umesaki Y, Mazmanian SK (2011) Proinflammatory T-cell responses to gut microbiota promote experimental autoimmune encephalomyelitis. Proc Natl Acad Sci 108(supplement_1):4615–4622

Lee YS, Lai DM, Huang HJ, Lee-Chen GJ, Chang CH, Hsieh-Li HM et al (2021) Prebiotic lactulose ameliorates the cognitive deficit in Alzheimer's disease mouse model through macroautophagy and chaperone-mediated autophagy pathways. J Agric Food Chem 69(8):2422–2437

Levy O (2007) Innate immunity of the newborn: basic mechanisms and clinical correlates. Nat Rev Immunol 7(5):379–390

Lewandowska-Pietruszka Z, Figlerowicz M, Mazur-Melewska K (2022) The history of the intestinal microbiota and the gut-brain axis. Pathogens 11(12)

Li H, Gao D, Cao Y, Xu H (2008) A high γ-aminobutyric acid-producing Lactobacillus brevis isolated from Chinese traditional paocai. Ann Microbiol 58:649–653

Li H, He J, Jia W (2016) The influence of gut microbiota on drug metabolism and toxicity. Expert Opin Drug Metab Toxicol 12(1):31–40

Li L, Ning Z, Zhang X, Mayne J, Cheng K, Stintzi A et al (2020) RapidAIM: a culture- and metaproteomics-based rapid assay of individual microbiome responses to drugs. Microbiome 8(1):33

Liang S, Wang T, Hu X, Luo J, Li W, Wu X et al (2015) Administration of Lactobacillus helveticus NS8 improves behavioral, cognitive, and biochemical aberrations caused by chronic restraint stress. Neuroscience 310:561–577

Liao J-F, Cheng Y-F, You S-T, Kuo W-C, Huang C-W, Chiou J-J et al (2020) Lactobacillus plantarum PS128 alleviates neurodegenerative progression in 1-methyl-4-phenyl-1,2,3,6-tetrahydropyridine-induced mouse models of Parkinson's disease. Brain Behav Immun 90:26–46

Litvan I, Bhatia KP, Burn DJ, Goetz CG, Lang AE, McKeith I et al (2003) Movement disorders society scientific issues committee report: SIC task force appraisal of clinical diagnostic criteria for parkinsonian disorders. Mov Disord 18(5):467–486

Liu Y, Forsythe P (2021) Vagotomy and insights into the microbiota-gut-brain axis. Neurosci Res 168:20–27

Liu WH, Chuang HL, Huang YT, Wu CC, Chou GT, Wang S et al (2016) Alteration of behavior and monoamine levels attributable to Lactobacillus plantarum PS128 in germ-free mice. Behav Brain Res 298(Pt B):202–209

Lukic I, Getselter D, Ziv O, Oron O, Reuveni E, Koren O et al (2019) Antidepressants affect gut microbiota and Ruminococcus flavefaciens is able to abolish their effects on depressive-like behavior. Transl Psychiatry 9(1):133

Lyte M, Freestone PP (2010) Microbial endocrinology: interkingdom signaling in infectious disease and health. Springer

Maini Rekdal V, Bess EN, Bisanz JE, Turnbaugh PJ, Balskus EP (2019) Discovery and inhibition of an interspecies gut bacterial pathway for Levodopa metabolism. Science 364(6445):eaau6323

Manev R, Manev H (2001) Aminoglycoside antibiotics and autism: a speculative hypothesis. BMC Psychiatry 1(1):5

Mangalam A, Shahi SK, Luckey D, Karau M, Marietta E, Luo N et al (2017) Human gut-derived commensal bacteria suppress CNS inflammatory and demyelinating disease. Cell Rep 20(6):1269–1277

Mao YK, Kasper DL, Wang B, Forsythe P, Bienenstock J, Kunze WA (2013) Bacteroides fragilis polysaccharide A is necessary and sufficient for acute activation of intestinal sensory neurons. Nat Commun 4:1465

Martin CR, Osadchiy V, Kalani A, Mayer EA (2018) The brain-gut-microbiome axis. Cell Mol Gastroenterol Hepatol 6(2):133–148

Mattson MP (2004) Pathways towards and away from Alzheimer's disease. Nature 430(7000):631–639

Mayer EA (2011) Gut feelings: the emerging biology of gut–brain communication. Nat Rev Neurosci 12(8):453–466

Mayer EA, Knight R, Mazmanian SK, Cryan JF, Tillisch K (2014) Gut microbes and the brain: paradigm shift in neuroscience. J Neurosci 34(46):15490–15496

McClean PL, Parthsarathy V, Faivre E, Hölscher C (2011) The diabetes drug liraglutide prevents degenerative processes in a mouse model of Alzheimer's disease. J Neurosci 31(17):6587–6594

Messaoudi M, Violle N, Bisson JF, Desor D, Javelot H, Rougeot C (2011) Beneficial psychological effects of a probiotic formulation (Lactobacillus helveticus R0052 and Bifidobacterium longum R0175) in healthy human volunteers. Gut Microbes 2(4):256–261

Miller I (2018) The gut-brain axis: historical reflections. Microb Ecol Health Dis 29(1):1542921

Moore AM, Mathias M, Valeur J (2019) Contextualising the microbiota-gut-brain axis in history and culture. Microb Ecol Health Dis 30(Suppl 1):1546267

Morimoto R, Satoh F, Murakami O, Totsune K, Saruta M, Suzuki T et al (2008) Expression of peptide YY in human brain and pituitary tissues. Nutrition 24(9):878–884

Mulroy E, Bhatia KP (2019) The gut microbiome: a therapeutically targetable site of peripheral levodopa metabolism. Mov Disord Clin Pract 6(7):547–548

Murrough JW, Yaqubi S, Sayed S, Charney DS (2015) Emerging drugs for the treatment of anxiety. Expert Opin Emerg Drugs 20(3):393–406

Nagpal R, Mainali R, Ahmadi S, Wang S, Singh R, Kavanagh K et al (2018) Gut microbiome and aging: physiological and mechanistic insights. Nutr Healthy Aging 4(4):267–285

Nakatani Y, Sato-Suzuki I, Tsujino N, Nakasato A, Seki Y, Fumoto M et al (2008) Augmented brain 5-HT crosses the blood–brain barrier through the 5-HT transporter in rat. Eur J Neurosci 27(9):2466–2472

Naseribafrouei A, Hestad K, Avershina E, Sekelja M, Linløkken A, Wilson R et al (2014) Correlation between the human fecal microbiota and depression. Neurogastroenterol Motil 26(8):1155–1162

Nicholson JK, Holmes E, Kinross J, Burcelin R, Gibson G, Jia W et al (2012) Host-gut microbiota metabolic interactions. Science 336(6086):1262–1267

Nonaka N, Shioda S, Niehoff ML, Banks WA (2003) Characterization of blood-brain barrier permeability to PYY3-36 in the mouse. J Pharmacol Exp Ther 306(3):948–953

Noor J, Chaudhry A, Batool S, Noor R, Fatima G (2023) Exploring the impact of the gut microbiome on obesity and weight loss: a review article. Cureus 15(6):e40948

Nuzum ND, Loughman A, Szymlek-Gay EA, Hendy A, Teo W-P, Macpherson H (2020) Gut microbiota differences between healthy older adults and individuals with Parkinson's disease: a systematic review. Neurosci Biobehav Rev 112:227–241

Obata Y, Castaño Á, Boeing S, Bon-Frauches AC, Fung C, Fallesen T et al (2020) Neuronal programming by microbiota regulates intestinal physiology. Nature 578(7794):284–289

Obradovic M, Sudar-Milovanovic E, Soskic S, Essack M, Arya S, Stewart AJ et al (2021) Leptin and obesity: role and clinical implication. Front Endocrinol (Lausanne) 12:585887

Park A, Collins J, Blennerhassett P, Ghia J, Verdu E, Bercik P et al (2013) Altered colonic function and microbiota profile in a mouse model of chronic depression. Neurogastroenterol Motil 25(9):733–e575

Parracho HM, Gibson GR, Knott F, Bosscher D, Kleerebezem M, McCartney AL (2010) A double-blind, placebo-controlled, crossover-designed probiotic feeding study in children diagnosed with autistic spectrum disorders. Int J Probiot Prebiot 5(2):69

Perez-Muñoz ME, Arrieta M-C, Ramer-Tait AE, Walter J (2017) A critical assessment of the "sterile womb" and "in utero colonization" hypotheses: implications for research on the pioneer infant microbiome. Microbiome 5:1–19

Pinto-Sanchez MI, Hall GB, Ghajar K, Nardelli A, Bolino C, Lau JT et al (2017) Probiotic Bifidobacterium longum NCC3001 reduces depression scores and alters brain activity: a pilot study in patients with irritable bowel syndrome. Gastroenterology 153(2):448–459. e8

Plaza-Díaz J, Gómez-Fernández A, Chueca N, Torre-Aguilar MJ, Gil Á, Perez-Navero JL et al (2019) Autism spectrum disorder (ASD) with and without mental regression is associated with changes in the fecal microbiota. Nutrients 11(2)

PrabhuDas M, Adkins B, Gans H, King C, Levy O, Ramilo O et al (2011) Challenges in infant immunity: implications for responses to infection and vaccines. Nat Immunol 12(3):189–194

Psichas A, Sleeth ML, Murphy KG, Brooks L, Bewick GA, Hanyaloglu AC et al (2015) The short chain fatty acid propionate stimulates GLP-1 and PYY secretion via free fatty acid receptor 2 in rodents. Int J Obes 39(3):424–429

Pu Y, Zhang Q, Tang Z, Lu C, Wu L, Zhong Y et al (2022) Fecal microbiota transplantation from patients with rheumatoid arthritis causes depression-like behaviors in mice through abnormal T cells activation. Transl Psychiatry 12(1):223

Ren Y, Wu J, Wang Y, Zhang L, Ren J, Zhang Z et al (2023) Lifestyle patterns influence the composition of the gut microbiome in a healthy Chinese population. Sci Rep 13(1):14425

Rezaeiasl Z, Salami M, Sepehri G (2019) The effects of probiotic lactobacillus and Bifidobacterium strains on memory and learning behavior, long-term potentiation (LTP), and some biochemical parameters in β-amyloid-induced Rat's model of Alzheimer's disease. Prev Nutr Food Sci 24(3):265–273

Ritchie G, Strodl E, Parham S, Bambling M, Cramb S, Vitetta L (2023) An exploratory study of the gut microbiota in major depression with anxious distress. J Affect Disord 320:595–604

Robertson RC, Oriach CS, Murphy K, Moloney GM, Cryan JF, Dinan TG et al (2017) Omega-3 polyunsaturated fatty acids critically regulate behaviour and gut microbiota development in adolescence and adulthood. Brain Behav Immun 59:21–37

Romano S, Savva GM, Bedarf JR, Charles IG, Hildebrand F, Narbad A (2021) Meta-analysis of the Parkinson's disease gut microbiome suggests alterations linked to intestinal inflammation. npj Parkinson's Dis 7(1):27

Rothhammer V, Mascanfroni ID, Bunse L, Takenaka MC, Kenison JE, Mayo L et al (2016) Type I interferons and microbial metabolites of tryptophan modulate astrocyte activity and central nervous system inflammation via the aryl hydrocarbon receptor. Nat Med 22(6):586–597

Salter MW, Stevens B (2017) Microglia emerge as central players in brain disease. Nat Med 23(9):1018–1027

Sampson TR, Mazmanian SK (2015) Control of brain development, function, and behavior by the microbiome. Cell Host Microbe 17(5):565–576

Scheperjans F, Aho V, Pereira PAB, Koskinen K, Paulin L, Pekkonen E et al (2015) Gut microbiota are related to Parkinson's disease and clinical phenotype. Mov Disord 30(3):350–358

Schwiertz A, Spiegel J, Dillmann U, Grundmann D, Bürmann J, Faßbender K et al (2018) Fecal markers of intestinal inflammation and intestinal permeability are elevated in Parkinson's disease. Parkinsonism Relat Disord 50:104–107

Segata N, Izard J, Waldron L, Gevers D, Miropolsky L, Garrett WS et al (2011) Metagenomic biomarker discovery and explanation. Genome Biol 12(6):R60

Sgritta M, Dooling SW, Buffington SA, Momin EN, Francis MB, Britton RA et al (2019) Mechanisms underlying microbial-mediated changes in social behavior in mouse models of autism spectrum disorder. Neuron 101(2):246–59.e6

Shaaban SY, El Gendy YG, Mehanna NS, El-Senousy WM, El-Feki HSA, Saad K et al (2018) The role of probiotics in children with autism spectrum disorder: a prospective, open-label study. Nutr Neurosci 21(9):676–681

Sharon G, Cruz NJ, Kang DW, Gandal MJ, Wang B, Kim YM et al (2019) Human gut microbiota from autism spectrum disorder promote behavioral symptoms in mice. Cell 177(6):1600–18.e17

Shen T, Yue Y, He T, Huang C, Qu B, Lv W et al (2021) The association between the gut microbiota and Parkinson's disease, a meta-analysis. Front Aging Neurosci 13:636545

Siegrist C-A (2001) Neonatal and early life vaccinology. Vaccine 19(25–26):3331–3346

Soedamah-Muthu SS, Masset G, Verberne L, Geleijnse JM, Brunner EJ (2013) Consumption of dairy products and associations with incident diabetes, CHD and mortality in the Whitehall II study. Br J Nutr 109(4):718–726

Sousa T, Paterson R, Moore V, Carlsson A, Abrahamsson B, Basit AW (2008) The gastrointestinal microbiota as a site for the biotransformation of drugs. Int J Pharm 363(1–2):1–25

Sternini C, Anselmi L, Rozengurt E (2008) Enteroendocrine cells: a site of 'taste' in gastrointestinal chemosensing. Curr Opin Endocrinol Diabetes Obes 15(1):73–78

Strandwitz P (2018) Neurotransmitter modulation by the gut microbiota. Brain Res 1693(Pt B):128–133

Structure, function and diversity of the healthy human microbiome (2012) Nature 486(7402):207–214

Sun LJ, Li JN, Nie YZ (2020) Gut hormones in microbiota-gut-brain cross-talk. Chin Med J 133(7):826–833

Tam A, Filho FSL, Ra SW, Yang J, Leung JM, Churg A et al (2020) Effects of sex and chronic cigarette smoke exposure on the mouse cecal microbiome. PLoS One 15(4):e0230932

Tan J, McKenzie C, Potamitis M, Thorburn AN, Mackay CR, Macia L (2014) The role of short-chain fatty acids in health and disease. Adv Immunol 121:91–119

Tan AH, Mahadeva S, Marras C, Thalha AM, Kiew CK, Yeat CM et al (2015) Helicobacter pylori infection is associated with worse severity of Parkinson's disease. Parkinsonism Relat Disord 21(3):221–225

Tapsell LC (2015) Fermented dairy food and CVD risk. Br J Nutr 113(S2):S131–S1S5

Tolhurst G, Heffron H, Lam YS, Parker HE, Habib AM, Diakogiannaki E et al (2012) Short-chain fatty acids stimulate glucagon-like peptide-1 secretion via the G-protein-coupled receptor FFAR2. Diabetes 61(2):364–371

Tong Q, Zhang L, Yuan Y, Jiang S, Zhang R, Xu Q et al (2015) Reduced plasma serotonin and 5-hydroxyindoleacetic acid levels in Parkinson's disease are associated with nonmotor symptoms. Parkinsonism Relat Disord 21(8):882–887

Turnbaugh PJ, Quince C, Faith JJ, McHardy AC, Yatsunenko T, Niazi F et al (2010) Organismal, genetic, and transcriptional variation in the deeply sequenced gut microbiomes of identical twins. Proc Natl Acad Sci USA 107(16):7503–7508

Valles-Colomer M, Falony G, Darzi Y, Tigchelaar EF, Wang J, Tito RY et al (2019) The neuroactive potential of the human gut microbiota in quality of life and depression. Nat Microbiol 4(4):623–632

Veiga P, Pons N, Agrawal A, Oozeer R, Guyonnet D, Brazeilles R et al (2014) Changes of the human gut microbiome induced by a fermented milk product. Sci Rep 4(1):6328

Wang Y, Telesford KM, Ochoa-Repáraz J, Haque-Begum S, Christy M, Kasper EJ et al (2014) An intestinal commensal symbiosis factor controls neuroinflammation via TLR2-mediated CD39 signalling. Nat Commun 5:4432

Wang L, Zhao Z, Zhao L, Zhao Y, Yang G, Wang C et al (2022) Lactobacillus plantarum DP189 reduces α-SYN aggravation in MPTP-induced Parkinson's disease mice via regulating oxidative damage, inflammation, and gut microbiota disorder. J Agric Food Chem 70(4):1163–1173

Wasilewska J, Klukowski M (2015) Gastrointestinal symptoms and autism spectrum disorder: links and risks–a possible new overlap syndrome. Pediatric Health Med Ther 6:153–166

Weersma RK, Zhernakova A, Fu J (2020) Interaction between drugs and the gut microbiome. Gut 69(8):1510–1519

Whiley L, Chappell KE, D'Hondt E, Lewis MR, Jiménez B, Snowden SG et al (2021) Metabolic phenotyping reveals a reduction in the bioavailability of serotonin and kynurenine pathway metabolites in both the urine and serum of individuals living with Alzheimer's disease. Alzheimers Res Ther 13(1):20

Wong ML, Inserra A, Lewis MD, Mastronardi CA, Leong L, Choo J et al (2016) Inflammasome signaling affects anxiety- and depressive-like behavior and gut microbiome composition. Mol Psychiatry 21(6):797–805

Xiao L, Yan J, Yang T, Zhu J, Li T, Wei H et al (2021) Fecal microbiome transplantation from children with autism spectrum disorder modulates tryptophan and serotonergic synapse metabolism and induces altered behaviors in germ-free mice. mSystems 6(2)

Xu M, Mo X, Huang H, Chen X, Liu H, Peng Z et al (2020) Yeast β-glucan alleviates cognitive deficit by regulating gut microbiota and metabolites in Aβ(1)(-)(42)-induced AD-like mice. Int J Biol Macromol 161:258–270

Yang Y, Wang H, Kouadir M, Song H, Shi F (2019) Recent advances in the mechanisms of NLRP3 inflammasome activation and its inhibitors. Cell Death Dis 10(2):128

Yano JM, Yu K, Donaldson GP, Shastri GG, Ann P, Ma L et al (2015) Indigenous bacteria from the gut microbiota regulate host serotonin biosynthesis. Cell 161(2):264–276

Yoo BB, Mazmanian SK (2017) The enteric network: interactions between the immune and nervous systems of the gut. Immunity 46(6):910–926

Zhang L, Zhang S, Jiang M, Ni X, Du M, Jiang H et al (2024) Limosilactobacillus reuteri alleviates anxiety-like behavior and intestinal symptoms in two stressed mouse models. Nutrients 16(18):3209

Zhu B, Wang X, Li L (2010) Human gut microbiome: the second genome of human body. Protein Cell 1(8):718–725

Zhu M, Liu X, Ye Y, Yan X, Cheng Y, Zhao L et al (2022) Gut microbiota: a novel therapeutic target for Parkinson's disease. Front Immunol 13:937555

Zubcevic J, Raizada MK (2023) Chapter 140 – Autonomic nervous system-gut-microbiome axis in chronic diseases. In: Biaggioni I, Browning K, Fink G, Jordan J, Low PA, Paton JFR (eds) Primer on the autonomic nervous system, 4th edn. Academic, pp 815–819

Tong Q, Zhang L, Yuan Y, Jiang S, Zhang R, Xu Q, et al. (2015) Reduced plasma serotonin and 5-hydroxyindoleacetic acid levels in Parkinson's disease are associated with nonmotor symptoms. Parkinsonism Relat Disord 21(8):882–887
Turnbaugh PJ, Quince C, Faith JJ, McHardy AC, Yatsunenko T, Niazi F et al (2010) Organismal, genetic, and transcriptional variation in the deeply sequenced gut microbiomes of identical twins. Proc Natl Acad Sci USA 107(16):7503–7508
[illegible]

Machine Learning Approaches in Metabolic Pathway Predictions and Drug-Target Interactions: Advancing Drug Discovery

18

Mohamed E. Hasan, Rawhia R. Allam, Alaa A. Hemeida, and Mahmoud ElHefnawi

Abstract

The integration of machine learning (ML) and prediction of metabolic pathways has enhanced the identification of drug-target interactions (DTIs) and hence effective drug discovery. Traditional drug discovery methods, which are often time-consuming and costly, are being supplemented by ML algorithms capable of analyzing large biological datasets and revealing complex patterns. Deep learning (DL), a subset of ML, has emerged as a powerful tool for modeling DTIs, leveraging neural networks to predict potential drug candidates based on structural and chemical properties (such as DeepDTIs, DeepACTION, DeepPurpose, and NeoDTI). We emphasize the integration of diverse data sources, including genomic, proteomic, and chemical information, utilizing platforms like DeepChem and AutoML to train robust predictive models. Furthermore, we discuss the implications of accurate DTI and metabolic pathways predictions on the efficiency of drug design and development, significantly reducing the time and costs associated with traditional methods. Our findings underscore the potential of machine learning to accelerate the discovery of novel drugs and optimize existing therapeutic strategies.

Keywords

Machine learning · Deep learning · Drug discovery · Drug-target interactions · Metabolic pathways

M. E. Hasan · R. R. Allam · A. A. Hemeida
Bioinformatics Department, Genetic Engineering and Biotechnology Research Institute, University of Sadat City, Sadat City, Egypt

M. ElHefnawi (✉)
Biomedical Informatics and Chemoinformatics Group (BICG), Informatics and Systems Department, National Research Centre, Cairo, Egypt
e-mail: mahef@aucegypt.edu

S. S. M. Soliman, M. I. Husseiny (eds.), *Metabolic Dynamics in Host-Microbe Interaction*, https://doi.org/10.1007/978-981-96-1305-2_18

18.1 Introduction

Artificial intelligence (AI) is a rapidly evolving concept that has the potential to greatly improve human life. The collaboration between AI and biological research is increasing, with ongoing progress in extracting and utilizing data from living organisms. As AI develops more advanced algorithms, its potential applications in fields such as epidemiology, studies of host-pathogen interactions, and drug design are expanding. AI is increasingly integrated into drug discovery, personalized medicine, gene editing, radiography, image analysis, and medication management. In the near future, AI technologies are expected to facilitate more accurate diagnoses and cost-effective treatments. Machine learning (ML) and deep learning, both subsets of AI, enable adjustments to metabolic pathways in living organisms, optimizing outputs while minimizing inputs. These innovations have the potential to enhance microbial strains for better production in bio-based industries. This article highlights the diverse applications of AI in biological fields such as pharmacy, medicine, and industrial biotechnology (Bhardwaj et al. 2022).

AI assists researchers in addressing complex challenges like quantitative and predictive epidemiology, precision medicine, and host-pathogen interactions. It aids in disease detection and diagnosis while also making programming more accessible for non-experts. Machine and deep learning techniques contribute to improved disease diagnosis, enhanced identification of individual cases, increased prediction accuracy, reduced errors, expedited decision-making, and better risk assessment. The growing number of tissue biomarkers and their intricate evaluations have further accelerated the implementation of AI-based techniques. AI-driven biomarkers help physicians predict and analyze diagnoses, patient reactions to treatments, and survival rates. Additionally, AI fosters the development of more accurate models of complex socio-biological systems through advanced techniques in knowledge representation and reasoning (Bhardwaj et al. 2022).

In precision medicine, it is crucial to consider individual differences in a person's metabolic, physical, physiological, and genetic characteristics that influence their body's response to medications. However, a generalized approach is currently being used, where all patients receive the same treatment regardless of their unique conditions (Krzyszczyk et al. 2018). Advancements in AI are ushering in a new era of personalized medicine, enabling the fitting of medications to meet individual needs and responses. Although this transition may seem straightforward, it involves substantial data collection, processing, management, and implementation. Additionally, the process includes millions of predictive analyses to identify the most suitable therapeutic molecules for specific cases. This approach enables healthcare providers to more accurately predict which treatment and prevention strategies will be most effective for distinct patient groups (Fig. 18.1).

AI can also enhance research in DNA, RNA, and protein studies, facilitating better visualization of drug dose effects on living tissues over time and the reorganization of signaling networks during treatment. For instance, IBM Watson uses AI to formulate appropriate treatment plans based on a patient's medical history and personal data, including genetic information. An AI-driven personalized medicine

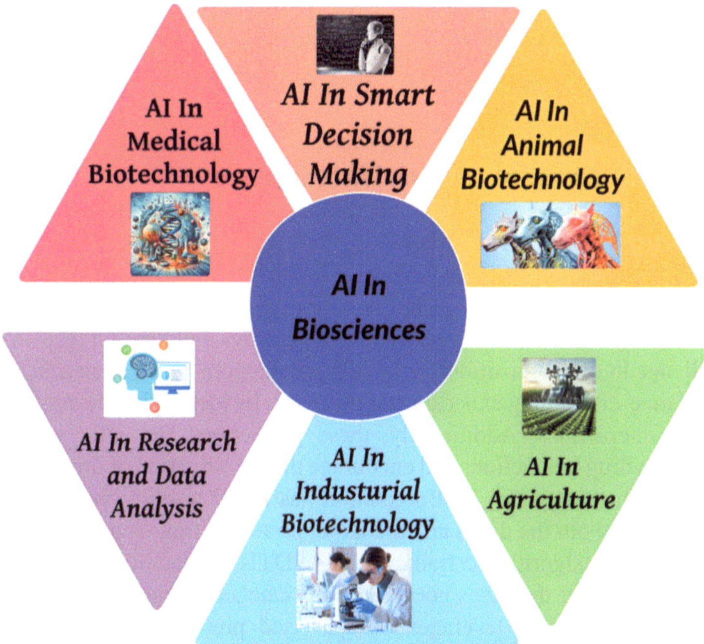

Fig. 18.1 This timeline outlines significant milestones in the development of artificial intelligence and its applications across various domains, particularly in the biosciences

system not only aims to reduce treatment costs but also to lessen the side effects of medications. Furthermore, AI can streamline processes in gene editing, radiography, and drug management planning, thereby enhancing efficiency and improving patient care with evidence-based clinical decision support systems (Johnson et al. 2021).

AI utilizes immense processing capabilities (via supercomputers), algorithms that learn rapidly (deep learning), and strategies that leverage physicians' cognitive abilities. This methodology can aid in developing innovative theoretical models of disease pathophysiology and in predicting significant adverse effects of long-term medications. Recent studies have demonstrated that AI-based methods are highly effective for the early identification, diagnosis, prognosis, and treatment of conditions like myopia. In fields such as cardiology, dermatology, and oncology, deep learning algorithms have been shown to surpass physicians in diagnostic accuracy. For instance, computational algorithms are capable of identifying metastatic breast cancer in sentinel lymph node biopsies from whole-slide images with an accuracy exceeding 91%, which improves to 99.5% when incorporating input from physicians. AI has also proven valuable in risk analysis, particularly for diagnosing heart issues through cardiovascular imaging. This involves automated monitoring for deviations from normal conditions, assessing myocardial function, and detecting and analyzing coronary atherosclerotic plaques (Davenport and Kalakota 2019). AI can efficiently analyze various variables, such as identifying which conditions are resistant to specific antibiotics. Such analyses can assist physicians while significantly reducing unnecessary testing and healthcare costs (Davenport and Kalakota 2019).

18.2 Applications of Artificial Intelligence

18.2.1 Drug-Target Interaction

One application of AI in drug design is "drug-target interactions" (DTIs), which involves identifying interactions between known drugs and their targets—a significant challenge in drug repositioning. In silico predictions of DTI can accelerate the costly and time-consuming experimental processes by identifying the most effective drug-target interactions. Additionally, these predictions can offer insights into potential drug-drug interactions and encourage the investigation of drug side effects. Traditionally, the success of DTI prediction has relied heavily on the descriptors used to represent both the drugs and the target proteins (Zeng et al. 2024).

The DeepDTIs algorithmic framework for DTIs is intended to create a deep learning-based model that does not require the categorization of targets into separate classes. This model leverages unsupervised pretraining for representation extraction and uses known label pairs for building a classification model, demonstrating performance that meets or exceeds that of existing state-of-the-art methods. Furthermore, DeepDTIs can be utilized to predict whether a novel drug targets existing proteins or whether a new target interacts with established drugs (Ching et al. 2018). Predicting interactions between drugs and targets is crucial in the drug discovery process. To avoid expensive and labor-intensive experiments that may not always yield conclusive results, there is a need to develop new and efficient prediction methods for determining DTIs. These methods should be able to identify potential DTIs quickly and effectively (Khojasteh et al. 2023).

A crucial step in the drug discovery process is identifying interactions between drugs and targets, such as genes, which can be reliably achieved through in vitro experiments. To minimize both time and costs, in silico approaches are becoming increasingly popular. Rather than conducting exhaustive in vitro searches, virtual screening is performed first to identify potential candidates, which are then experimentally validated. Generally, two main methods are used for in silico prediction of DTIs, also known as compound-protein interactions: docking simulations and machine learning techniques (Ren et al. 2023).

In docking simulations, the 3D structures of drug molecules and their targets are analyzed to identify potential binding sites. While this method is well accepted in biological contexts, it can be time-consuming and cannot be applied if the 3D structure of the protein is unknown. For example, only a limited number of structures have been crystallized for a class of proteins known as G-protein-coupled receptors (GPCRs), which means docking simulations cannot be utilized for orphan GPCRs. To address this limitation, chemogenomics was introduced to explore the entire chemical space for interactions with the biological space rather than examining each protein target in isolation from others (Liao et al. 2022). Chemogenomics approaches are typically classified as ligand-based, target-based, and target-ligand, all of which rely on similarities among proteins and targets. The goal of chemogenomics research is to connect the chemical space of potential compounds with the genomic space to identify potentially valuable compounds such as imaging probes and drug leads.

This similarity-based perspective in chemogenomics has made machine learning methods well-suited for predicting DTIs. In machine learning techniques, information about drugs, targets, and previously confirmed DTIs is converted into features that are used to train a predictive model, which is then applied to forecast interactions between new drugs and/or new targets (Rognan 2007). In practice, DTIs can be categorized into four classes based on the available information about interacting drug compounds and target proteins: (i) known drug versus known target, (ii) known drug versus new target candidate, (iii) new drug candidate versus known target, and (iv) new drug candidate versus new target candidate. Although the primary aim of machine learning methods is to predict interactions for new drugs and target candidates, most methods are confined to the first three categories (Wang et al. 2023). Although the primary focus of this work was not specifically on drug discovery, the aim was to create a ranked list of molecular ligands that bind to each orphan GPCR, where docking simulations could not be utilized due to the absence of crystallized 3D structures. In this context, machine learning approaches have been categorized into six groups (Fig. 18.2) with a list of corresponding methods (Atas Guvenilir and Doğan 2023).

18.3 Deep Learning

Deep learning is gaining popularity due to its impressive performance across various fields, including speech recognition, image recognition, and natural language processing. In recent years, the application of deep learning methods in drug discovery has been steadily rising (Chakraborty et al. 2024).

Deep learning approaches seem to address certain limitations by minimizing the loss of feature information when predicting DTIs. However, one challenge with deep learning methods is that there is not always enough information available to effectively implement them.

Most deep learning-based methods for predicting DTIs involve two primary steps: first, generating feature vectors, and second, applying deep learning to known DTIs. Typically, three types of properties—biological, topological, and physicochemical information—are utilized to create the feature vectors or matrices for these DTIs methods. Recently published studies have employed techniques such as deep belief neural networks, convolutional neural networks, and multilayer (Wu et al. 2024). Initially, compounds and proteins are encoded using their respective features. Subsequently, the feature embeddings of these compounds and proteins serve as inputs for the deep learning methods. Following this strategy, models such as deep belief neural networks, CNNs, and multilayer perceptrons have been

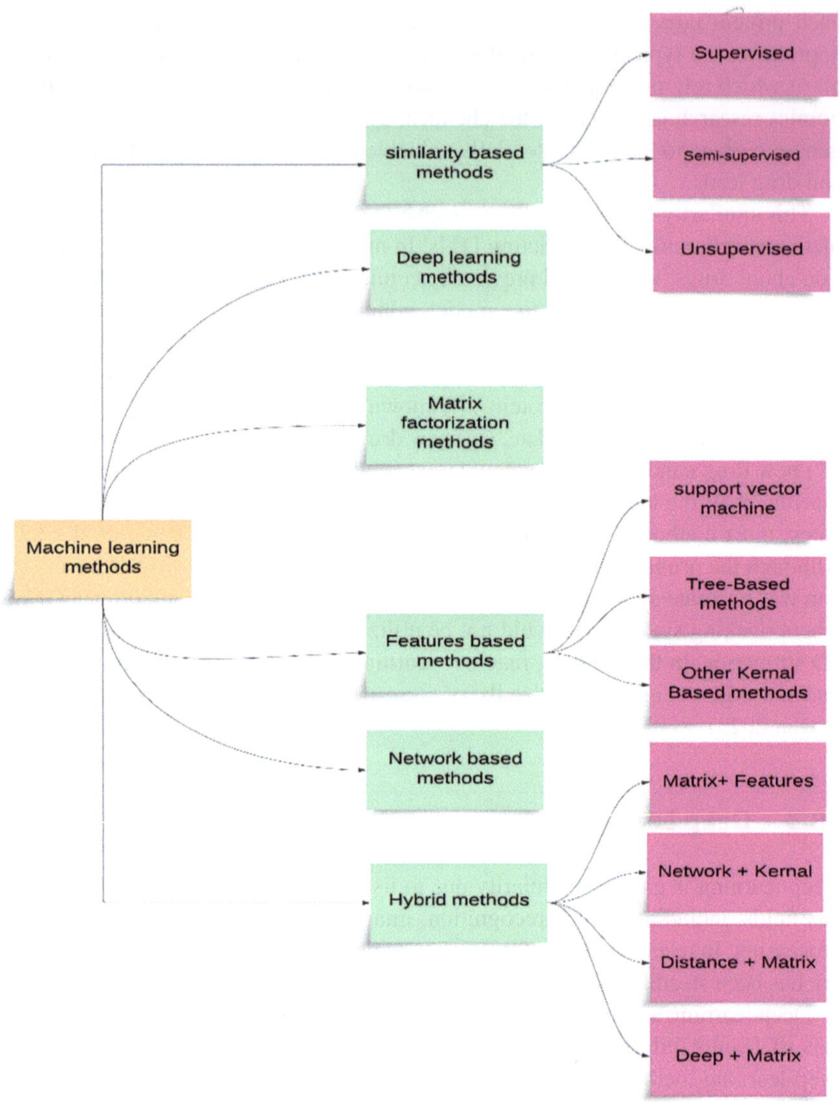

Fig. 18.2 Machine learning approaches comprise various methodologies, techniques, and algorithms that empower computers to learn from data and make predictions. These methodologies can primarily be categorized into three main types: supervised learning, unsupervised learning, and semi- supervised learning. Below is an overview of these categories along with additional approaches and techniques employed in machine learning

developed for predicting drug-protein interactions, significantly advancing the drug discovery process (Wu et al. 2024).

In reality, many diseases do not have clearly defined targets, making it challenging to find drugs for these conditions using traditional methods. Recently, Zhu et al.

introduced a deep learning-based efficacy prediction system (DLEPS) that identifies drug candidates based on changes in gene expression profiles rather than specific targets. Initially, compounds were encoded using SMILES and input into a convolutional neural network (CNN) to model gene expression changes. Then, the potential efficacy of these compounds against diseases was assessed based on gene signatures unique to particular diseases and ranked using a method akin to gene set enrichment analysis. DLEPS offers innovative perspectives for discovering new drugs for complex diseases (Zhu et al. 2021).

Typically, DTI prediction is viewed as a binary classification problem, often overlooking the binding affinity between a drug and its target. Binding affinity indicates the strength of interactions between drug-target pairs and provides valuable insights for drug discovery. While binding affinity can be experimentally measured through the determination of dissociation and inhibition constants, these procedures are time-consuming and costly. Consequently, there is a need to develop computational methods to predict binding affinity (Tang et al. 2024).

18.3.1 Deep Learning Methods

While deep learning methods demonstrate strong performance, there remain several areas for enhancement. First, generating robust negative datasets for supervised deep learning approaches poses a significant challenge. Most existing deep learning-based drug-target interaction (DTI) prediction programs depend on supervised machine learning, thereby necessitating the creation of unbiased negative DTI datasets for model training and testing as a crucial step. Furthermore, since DTI prediction seeks to uncover novel interactions, the selection of authentic non-interacting drug-target pairs can be particularly complex (Wang et al. 2023).

Second, as more diverse types of drug and target data become available, integrating heterogeneous data into high-dimensional features for deep learning methods presents another challenge. Finally, achieving excellent performance on a testing dataset does not guarantee that these deep learning methods will perform equally well in actual drug discovery scenarios (Table 18.1).

18.4 Types of AI Techniques

Currently, ML algorithms are typically categorized into four primary types: supervised, unsupervised, semi-supervised, and reinforcement learning (Haldorai et al. 2020). Supervised learning (SL) algorithms are utilized in scenarios where the training dataset includes labeled examples (Fabris et al. 2017). On the other hand, unsupervised learning focuses on developing models from unlabeled data, expecting the algorithm to uncover patterns and relationships within the data independently of predetermined labels. Semi-supervised learning merges elements of supervised and unsupervised learning by employing a small set of labeled data alongside a larger quantity of unlabeled data (Yang et al. 2023).

Table 18.1 Deep learning methods

Method	Description	Key features	Datasets used
DeepSurv (Katzman et al. 2018)	Deep learning model for survival analysis in drug response	Utilizes neural networks to predict survival outcomes, focusing on the Cox proportional hazards model	Various clinical datasets
PiDeeL (Kaynar et al. 2023)	Pathway-informed multilayer perceptron model for DTI prediction	Integrates biomarker metabolite concentrations at the metabolic pathway level, improving predictive performance	KEGG pathways
DeepACTION (Hasan Mahmud et al. 2020)	Deep learning model designed to predict drug-target interactions	Combines chemical structure representation and protein sequence features, using CNN for prediction	DrugBank, KEGG
DeepPurpose (Huang et al. 2021a)	A library for encoding and predicting DTIs using deep learning	Implements over 50 deep learning models, supports rapid prototyping, and provides modularity for DTI tasks	Large binding affinity datasets
NeoDTI (Wan et al. 2019)	Nonlinear end-to-end learning model that integrates heterogeneous data for DTI prediction	Focuses on topology-preserving representations and integrates neighborhood information for enhanced predictive power	Varied public datasets
MolTrans (Huang et al. 2021b)	Molecular Interaction Transformer designed for interpretable DTI prediction	Utilizes sub-structural pattern mining and transformer encoders for semantic relation extraction	Real-world biomedical data
Molecule Transformer DTI (MT-DTI) (Shin et al. 2019)	Self-attention-based model that learns high-dimensional molecular structures	Pre-trained on extensive chemical compound datasets for effective feature extraction and transfer to DTI tasks	Kiba, Davis benchmark datasets
iNGNN-DTI (Sun et al. 2024)	Interpretable nested graph neural network for DTI prediction	Employs a cross-attention mechanism to capture interactions between drug and target substructures based on 3D graphs	Benchmark datasets

SL algorithms have broader applications and are more widely favored. Some of the most frequently used SL algorithms include logistic regression (LR), decision trees (DT), support vector machines (SVMs), random forests (RF), and neural networks (Obaido et al. 2022). Each of these algorithms has its own advantages and drawbacks, with the selection of an algorithm largely depending on the specific nature of the problem. The significance and timeliness of this survey stems from the growing implementation of ML across various fields and the necessity of a thorough understanding of its potential and limitations, along with their mathematical underpinnings. By comprehending the strengths and weaknesses of the various algorithms, both researchers and practitioners can make well-informed decisions regarding the most suitable approach for addressing particular challenges.

Unsupervised learning involves the use of machine learning algorithms to analyze and categorize datasets that do not have labels. These algorithms identify hidden patterns within the data without requiring human guidance, which is why they are classified as "unsupervised." Typically, unsupervised learning models focus on three primary tasks (Polanski 2022): clustering, association, and dimensionality reduction, defined as follows:

- *Clustering* is a method for organizing unlabeled data based on their similarities or differences. For instance, K-means clustering algorithms group similar data points together, with the K value denoting the size and granularity of the clusters. This approach is particularly useful for categorizing travel behaviors, among other applications.
- *Dimensionality reduction* is a technique employed when a dataset contains an excessive number of features or dimensions. It condenses the number of data inputs to a more manageable level while maintaining data integrity. This method is frequently applied during data preprocessing, such as when autoencoders eliminate noise from visual data to enhance image quality.

A clear distinction exists between supervised and unsupervised learning, which can be bridged by semi-supervised learning. This approach shares similarities with supervised learning but leverages both labeled and unlabeled data. Labeled data includes information with meaningful tags that help the algorithm interpret the data, whereas unlabeled data lacks such information. By combining these data types, machine learning algorithms can learn to label the previously unlabeled data.

18.5 Metabolic Pathways

Metabolic pathways consist of a connected sequence of chemical reactions within a cell, where the products of one reaction serve as substrates for the subsequent reaction. In each pathway, catalytic enzymes facilitate the conversion of substrates into structurally similar products. Recognizing the processes and structural relationships among molecules and pathway classes is essential for designing reaction predictors aimed at synthesizing new molecules or optimizing drug metabolism. Additionally, knowledge of metabolic pathways aims to develop and evaluate a high-fidelity model that can accurately predict the associations of a given chemical structure representation with its corresponding pathway classes (Judge and Dodd 2020).

Various approaches have been employed to link protein annotations to pathway templates, facilitating the derivation of organism-specific metabolic pathways. These methods encompass data retrieval strategies, machine learning techniques, molecular fragment representations, and network integration methods.

As a result, numerous widely adopted tools for metabolic pathway analysis have been developed, including PathComp, PathPred, Pathway Tools, the UM-BBD

Pathway Prediction System, the MRE biosynthesis pathway finding tool, and TrackSM. Several methods have been specifically designed to classify compounds into distinct metabolic pathway categories, with validation occurring against publicly available metabolic pathway databases (Goshisht 2024). Notable among these databases are the Kyoto Encyclopedia of Genes and Genomes (KEGG), EcoCyc/MetaCyc, the Expert Protein Analysis System (ExPASy), the Cell-Signaling Networks Database (CSNDB), PathDB, the UM-BBD, and the Signaling Pathway Database (SPAD). KEGG, in particular, is frequently utilized to benchmark the classification performance of pathway prediction methods. It is a manually curated database that provides comprehensive pathway maps linked to detailed information regarding compounds, enzymes, and genes. Many pathways within KEGG are characterized by the chemical structures of their primary compounds, which include carbohydrates, lipids, polyketides, and amino acids. The representation of molecules in KEGG includes their names, chemical and structural formulas, associated metabolic pathways, and the enzymes catalyzing reactions involving those molecules (Baranwal et al. 2024).

To classify small molecules into metabolic pathway classes based on functional group composition, a nearest-neighbor (NN) algorithm has been introduced. This analysis utilized a dataset of 2764 compounds sourced from the KEGG database, with each compound assigned exclusively to one of 11 identified pathway classes. The NN classifier achieved an overall accuracy of 73.3% in predicting metabolic pathway classes. However, a limitation of this approach is its inability to directly classify compounds that belong to multiple pathway classes. To overcome this challenge, a random forest (RF) classifier was employed, using 32 physicochemical and topological descriptors to predict the association of 681 molecules with seven manually identified KEGG pathway classes. This approach yielded an average Matthews correlation coefficient of 0.73, demonstrating its effectiveness in pathway classification.

A multi-class model has been developed to predict the association of query compounds with one or more previously identified KEGG pathway classes. In the realm of single-class prediction, where compounds are assigned to only one pathway class, the model achieved an overall average accuracy of 77.97% using five-fold cross-validation on a benchmark dataset of 3137 compounds. This effort was further extended to a dataset of 3348 small molecules, yielding an average prediction accuracy of 77.12% through a leave-one-out cross-validation (LOOCV) methodology (Baranwal et al. 2024). However, a significant limitation in both studies arises from the models' reliance on known interactions among the compounds. This dependency led to the exclusion of 1229 small molecules due to insufficient interaction data (Baranwal et al. 2024). While these models demonstrate potential in accurately classifying compounds into pathways, the absence of comprehensive interaction data may limit their applicability and restrict the inclusion of a wider array of compounds in metabolic pathway predictions.

To address this limitation, the researchers proposed that structurally similar compounds often share common scaffolds, which are associated with biochemically related pathways. To validate this hypothesis, they created a tool named TrackSM, which extracts scaffolds from compounds within the same KEGG pathway classes. TrackSM demonstrated an impressive average accuracy of 84.92% on a dataset of 3190 small molecules using LOOCV. When a query compound's metabolic pathway class is unknown, the scaffolds of the query compound are compared to those of compounds with known pathway associations, and the classifier assigns the query compound to the class that displays the highest match score (Lopez-Ibañez et al. 2021).

Over recent decades, biomolecular databases have significantly expanded, yet leveraging these resources to predict the properties of novel biomolecular compounds remains a major challenge. To tackle this issue, machine learning—particularly deep learning—has been employed across various domains in computational chemistry, including drug discovery, toxicity prediction, genomic prediction, protein-protein interaction prediction, enzymatic function prediction, biological reaction energy prediction, quantitative structure-activity relationship (QSAR) modeling, and predicting the outcomes of biological assays. Given that the input data, such as chemical compounds, often have a highly structured nature, it is suggested that specialized machine learning techniques may be better suited to extracting relevant shape features compared to general-purpose deep neural networks. Moreover, advancements in these methods have also found applications in the generation and optimization of chemical structures (Wu et al. 2024).

In this context, six distinct machine learning models are compared for the prediction task:

1. A random forest (RF) classifier using local graph features, which inputs concatenated initial Graph Convolutional Network (GCN) node embedding vectors representing the shapes of two-radius subgraphs of the nodes.
2. An RF classifier utilizing global molecular features, incorporating 166-dimensional MACCS strings alongside seven additional molecular descriptors well-established in drug discovery. These descriptors assess the bioavailability and activity of small molecule compounds based on various molecular characteristics such as size (molecular weight), rigidity (rotatable bonds and ring counts), lipophilicity (logP and aromaticity), and polarizability (molar refractivity) (Lopez-Ibañez et al. 2021). Together, these 173 features are collectively referred to as global molecular features.
3. An RF classifier that employs GCN embeddings, using the output node embedding vectors (i.e., learned shape features) from the trained GCN as input.
4. A GCN that utilizes only the chemical structure as input, represented in SMILES format.
5. A GCN that integrates both chemical structure and global molecular features.
6. A GCN designed for multi-class classification that takes both SMILES and the aforementioned global molecular features as inputs.

By systematically comparing various approaches, the study aims to identify the most effective strategies for predicting the associations between query compounds

Table 18.2 Performance analysis of several machine learning methods

Methods	Accuracy		
	Top1	Top2	Top3
Hu et al. (2011)	77.97	NA	NA
Cai et al. (2008)	73.30	NA	NA
Gao et al. (2012)	77	79	85
Hamdalla et al. (2015)	84.92	92.82	95.39
RF w/local graph features	21.47	39.96	59.76
RF w/global features	88.01	95.05	96.70
RF w/GCN embeddings	95.16	98.20	98.99
GCN	88.79	93.49	95.44
GCN + global features	90.21	94.73	96.70

and KEGG pathway classes. This refined understanding of compound-pathway relationships is essential for advancing biochemical research.

Among the identified 4539 KEGG compounds categorized under a single pathway class, 3631 compounds (80%) were randomly selected for model training. The remaining 908 compounds (20%) were evenly split into cross-validation and test sets to allow for rigorous assessment of model performance. Crucially, the test samples are kept separate, ensuring that the evaluation of model performance is unbiased and occurs strictly after the completion of the training phase.

To enhance the robustness of the findings, this comprehensive training and testing protocol is executed 10 times, with a random selection of model training, testing, and cross-validation sets for each iteration. This method helps to reduce variations that could otherwise affect the results and ensures that the performance evaluations are reliable.

For each experiment, performance metrics are presented in Table 18.2, with a particular emphasis on top-n accuracy (where n can be 1, 2, or 3). A classifier is considered effective if the true class of a query compound ranks among the top n predicted classes. This metric is particularly valuable in multi-class classification scenarios, where achieving accurate class predictions is vital (Shin et al. 2019). The detailed analysis of results will shed light on the implications of these approaches, thereby enhancing predictive modeling in biochemical research. Moreover, these insights could lead to significant advancements in applications such as drug discovery and metabolic pathway analysis, facilitating more accurate predictions and a deeper understanding of biochemical interactions.

It modifies the output layer in our GCN model and replaces the SoftMax layer with a layer of element-wise sigmoid activation functions. Recall that the output of a sigmoid unit is restricted between 0 and 1 and, therefore, can be used to represent probabilities of association to pathway classes. The GCN is trained to minimize the sum of the binary cross-entropy losses at the sigmoid units (Baranwal et al. 2024). The performance of our multi-class GCN model is depicted in Table 18.3. For the multi-class classification problem, accuracy is defined as follows:

$$\text{Accuracy} = \sum_{i=1}^{N}\sum_{c=1}^{11}\frac{(\text{Correct predictions})_{i,c}}{N \times 11} \times 100\%,$$

In evaluating the performance of a classifier predicting compound-pathway associations, the accuracy is calculated as the fraction of correctly predicted labels across all compounds and pathway classes. While overall accuracy is important, the evaluation also considers precision and recall.

Precision measures the proportion of positively predicted associations that are correct, while recall reflects the proportion of true positive associations identified by the model. To assess these metrics, a compound's true and predicted pathway associations are compared. For example, if a compound is correctly predicted to belong to some pathway classes and not to others, true positives (TPs), true negatives (TNs), false positives (FPs), and false negatives (FNs) are identified. This comparison is repeated for all compounds to calculate the overall performance in terms of precision and recall, offering a detailed understanding of the classifier's strengths and weaknesses (Hicks et al. 2022). The cumulative statistics for TPs, TNs, FPs, and FNs are used to evaluate precision and recall as:

$$\text{Precision} = \frac{\text{TP}}{\text{TP}+\text{FP}},\ \text{Recall} = \frac{\text{TP}}{\text{TP}+\text{FN}}.$$

The performance of the proposed GCN-based multi-class classifier was evaluated against several traditional approaches, including k-nearest neighbors (k-NN), ensemble logistic regression, and 11 random forest (RF) classifiers trained separately for each class. These classifiers used global molecular features of query compounds as input (Barhate et al. 2024). The performance metrics presented in the columns of Table 18.3 serve as useful summaries of overall accuracy in the multi-class prediction task. However, there is still a chance that our classifier may underperform in certain less-represented pathway classes (Aguilar-Ruiz and Michalak 2024). To investigate this possibility, we present in Fig. 18.3 the accuracy, precision, recall, and Matthew's correlation coefficient (MCC) for our method across each pathway class. Our findings indicate that there is no pathway class where our method demonstrates notably poor performance.

Table 18.3 Performance analysis of multi-class classification

	Scores		
Methods	Accuracy	Precision	Recall
Hu et al. (2011)	94.64	77.97	67.83
K NN classifier	89.84	43.40	42.07
Ensemble logistic regression	89.20	14.07	13.06
Independent RFs	95.47	75.40	72.08
GCN + additional features	95.62	81.09	79.30

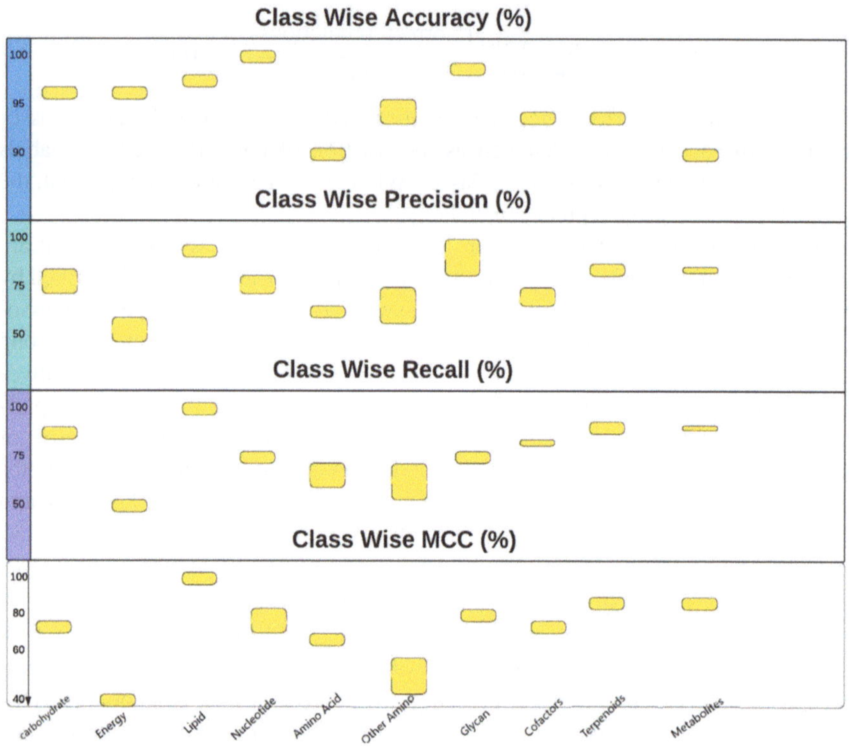

Fig. 18.3 Class-wise performance statistics for the multi-class GCN classifier

18.5.1 Metabolic Pathway Prediction Tools

Recent advancements in deep learning have led to the development of tools that effectively predict metabolic pathways, addressing the challenge of reconstructing an organism's metabolic network from genomic sequences in systems biology. These innovative methods have shown promising results, performing comparably to the traditional rule-based approach, PathoLogic. However, a notable limitation in past evaluations of PathoLogic is the lack of taxonomic pruning, which adversely affects its performance (Huckvale and Moseley 2024).

18.5.1.1 DeepRF: A Hybrid Model for Glioma Classification and Survival Prediction

Overview: DeepRF is a hybrid model that combines the strengths of deep learning (DL) and random forests (RF) to enhance glioma classification and survival prediction. While deep learning excels at feature extraction and processing high-dimensional data, random forests provide robust decision-making capabilities and help mitigate overfitting, particularly with smaller datasets. By merging these two approaches, DeepRF aims to improve both prediction accuracy and interpretability specific to glioma-related tasks (Wang et al. 2024a).

Key features

(a) *Performance on Classification and Survival Tasks.*
Enhanced Accuracy: DeepRF outperformed traditional methods in classifying glioma tissues pathologically and predicting patient survival outcomes. The integration of DL and RF allowed the model to capture complex patterns found in high-resolution magic angle spinning (HRMAS) nuclear magnetic resonance (NMR) spectra while maintaining strong generalization to unseen data (Karabacak et al. 2024).

(b) *Accuracy and Robustness.*
High Concordance Index: The hybrid model achieved a higher concordance index (c-index) compared to baseline models such as DeepSurv and RF alone. Its improved predictive capability was evident in both pathological grading and survival analysis (Yuan et al. 2024).
Mitigation of Overfitting: DeepRF effectively balanced accuracy and robustness, addressing common overfitting challenges seen in purely deep learning frameworks, particularly when utilizing a limited dataset of 568 glioma tissue samples (Sreedhar and David 2024).

(c) *Feature Importance.*
Insights into Critical Features: The random forest component offered significant insights into the most important features that influenced decision-making. These insights included the identification of key metabolites and their associated metabolic pathways, shedding light on the biological mechanisms underlying glioma progression (Malhotra et al. 2024).

Comparative Performance

Versus PiDeeL: While PiDeeL employs pathway-informed architectures for its predictions, DeepRF focuses on the synergy between deep feature extraction and the decision-making strengths of random forests. This focus makes DeepRF particularly effective in scenarios with smaller datasets.

Concordance Index Comparison: DeepRF achieved a high c-index for survival prediction, which was comparable to PiDeeL's performance (68.7%), yet demonstrated additional robustness against overfitting (Bjørkeli et al. 2024).

18.5.1.2 PiDeeL: Metabolic Pathway-Informed Deep Learning

Overview: PiDeeL (Metabolic Pathway-Informed Deep Learning) is an innovative framework that merges deep learning techniques with metabolic pathway information to enhance predictions related to drug-target interactions (DTIs) and metabolic processes. Unlike conventional deep learning methodologies that often focus solely on molecular features, PiDeeL leverages biological insights derived from established metabolic pathways. This incorporation significantly enhances the model's interpretability and accuracy, allowing it to better capture complex biological relationships and interactions that are often overlooked by purely data-driven approaches (Ren et al. 2023).

Key Evaluation Outcomes

(a) *GCN-Based Multi-Class Classifier.*
Model Performance: The GCN-based classifier demonstrated superior performance compared to traditional machine learning methods, such as k-nearest neighbors (k-NN), ensemble logistic regression, and individual random forest classifiers. The GCN's ability to utilize r-radii molecular subgraphs, rather than traditional feature representations like MACCS bits, contributed to its enhanced accuracy.
Metrics: The GCN achieved impressive results, exemplified by an average bitwise Hamming loss of 0.045 and a high exact match score of 0.717, indicating precise and reliable pathway predictions (Huynh et al. 2024).
(b) *Comparison with Traditional Classifiers.*
Performance Metrics: Although random forest classifiers provided decent results, they did not reach the performance levels of the GCN, with an exact match score of 0.648. Moreover, traditional classifiers such as k-NN and logistic regression generally lagged behind in terms of precision, recall, and overall accuracy (Johannesdottir et al. 2022).

Implications and Applications Overall, the incorporation of metabolic pathway information by PiDeeL enhances its prediction accuracy and interpretability, making it a valuable tool for various applications. This includes:

Intraoperative Feedback: Facilitating real-time decision-making during surgical procedures, particularly in glioma treatment, by providing insights based on predicted drug-target interactions and their metabolic implications.
Personalized Surgical Decision-Making: Assisting in tailoring treatments based on individual patient metabolic profiles and potential drug responses, ultimately leading to improved patient outcomes (Kaynar et al. 2023).

18.6 Drug-Target Interaction Prediction Tools

18.6.1 DeepACTION

Overview: Identifying DTIs is crucial in drug development and discovery, as it allows researchers to uncover new target proteins for existing drugs and develop new therapeutic agents. Traditional wet lab methods for predicting DTIs can be time-consuming, costly, and labor-intensive. As a response, computational approaches like deep learning have emerged to streamline the identification of novel interactions and facilitate drug repurposing. Among these computational tools, deepACTION stands out as a powerful deep learning-based method designed to predict potential or previously unknown DTIs.

To support these computational approaches, various public databases have been established to compile verified drug-target interaction information. Notable examples include KEGG, DrugBank, ChEMBL, STITCH, and TTD. These databases provide experimentally validated interaction data, serving as essential resources for training and validating DTI prediction algorithms.

18.6.1.1 Categories of Computational DTI Prediction Methods

Computational DTI prediction methods can be broadly categorized into three main approaches, each with unique strengths and limitations:

Ligand-Based Methods These methods leverage the similarity between target proteins to forecast interactions between drug chemical structures and protein sequences. This approach often operates within a quantitative structure-activity relationship (QSAR) framework.

Docking-Based Methods Docking methods perform dynamic simulations on target proteins to identify potential interactions. However, they require the 3D structures of proteins, which can be challenging to obtain experimentally. This method is thus often complex and resource-intensive.

Chemogenomic Methods These methods have gained considerable traction due to their ability to identify DTIs without relying on 3D protein structures. They commonly utilize simpler protein sequence formats, such as FASTA, to predict interactions, addressing some of the limitations inherent in other approaches.

18.6.1.2 Machine Learning Techniques in DTI Prediction

Recent reviews of machine learning techniques for DTI prediction classify these approaches into two primary categories:

Similarity-Based Methods These techniques, which include matrix factorization, kernel-based methods, and graph-based approaches, measure similarities between drug and target profiles and have been adapted for various types of data.

Feature-Based Methods This approach involves transforming input data into feature vectors and labeling drug-target pairs with binary values to indicate interactions. The models are trained to predict interactions based on these feature representations.

One noteworthy advancement is the use of a unified space technique that extracts both drug and protein features via a bipartite graph, facilitating the prediction of potential drug-target pairs based on pharmacological effects within an integrated framework. Moreover, a boosting-based method known as iDTi-CSsmoteB has been proposed, addressing the challenge of imbalanced datasets through both undersampling and oversampling techniques, further enhancing DTI prediction performance.

18.6.1.3 The deepACTION Model

To fill existing gaps in DTI prediction methodologies, the deepACTION model was developed as a convolutional neural network (CNN)-based approach, leveraging drug chemical structures and protein sequence information. The DTI data for deepACTION were sourced from the DrugBank database, along with the corresponding chemical structures and protein sequences gathered from the KEGG database.

18.6.1.4 Workflow and Feature Representation

1. Chemical Structure Transformation.

   The chemical structures of drugs are converted into multiple representations, including topological, constitutional, and geometrical forms, capturing the full scope of their molecular properties.
2. Protein Sequence Descriptors.
   - Multiple descriptors are employed to represent protein sequences, including:
   - Dipeptide composition.
   - Amino acid composition (AAC).
   - Pseudo-amino acid composition (PAAC).
   - Autocorrelation.
   - Quasi-sequence order.
   - Composition, Transition, and Distribution (CTD).

By utilizing these diverse representations, deepACTION is positioned to effectively predict potential DTIs, harnessing the strengths of deep learning to explore both complex chemical structures and protein sequences in a unified manner. This innovative approach aims to accelerate the identification of new drug-target interactions and contribute to the broader field of drug discovery and development (Nguyen-Vo et al. 2024).

18.6.2 DeepPurpose: A Deep Learning Library for Drug Discovery

Overview: DeepPurpose is a cutting-edge deep learning library specifically designed for encoding and predicting interactions between proteins and compounds. It provides a versatile programming framework that facilitates the rapid prototyping of over 50 deep learning models, alongside seven protein encoders and eight compound encoders. This makes DeepPurpose an essential tool for researchers seeking to advance drug discovery and bioinformatics research, as empirical evaluations indicate that its models achieve state-of-the-art performance on benchmark datasets for drug-target interactions (DTIs) (Huang et al. 2021a).

User-Friendly Design: Many current drug discovery tools are tailored for experienced interdisciplinary researchers, often posing challenges for newcomers, such as computer scientists entering the biomedical field or bioinformaticians with limited experience in deploying deep learning models. Additionally, open-source tools commonly feature distinct programming interfaces and code structures, complicating efforts to integrate outputs for model ensembles.

DeepPurpose addresses these challenges by providing a simplified interface and consolidated functionality that allows researchers of varying expertise to effectively experiment with deep learning models.

18.6.2.1 Key Features of DeepPurpose

Input Data Format: DeepPurpose takes two primary inputs:

- Compound Input: Specified as a Simplified Molecular Input Line Entry System (SMILES) string.
- Protein Input: Given as an amino acid sequence.

These inputs are subsequently fed into molecular encoders, which map the compounds and proteins to vector representations through advanced transformation functions (Huang et al. 2021a).

Compound Encoders

DeepPurpose offers eight different encoders for compounds, allowing users to select the most appropriate method based on their specific application:

1. Multilayer Perceptrons (MLP) applied to various 2D Fingerprints (Morgan, PubChem, Daylight, RDKit).
2. Convolutional Neural Networks (CNN) on SMILES strings.
3. Recurrent Neural Network (RNN) built on top of CNN representations.
4. Transformer Encoders that operate on substructure fingerprints.
5. Message Passing Graph Neural Network designed specifically for molecular graphs (Huang et al. 2021b).

Protein Encoders

For encoding protein sequences, DeepPurpose incorporates seven different approaches, including:

1. MLP on various amino acid representations (AAC, Pseudo AAC, Conjoint Triad, Quasi-Sequence).
2. CNN on amino acid sequences.
3. RNN built on CNN architectures.
4. Transformer Encoder on substructure fingerprints (Wu et al. 2023).

Functional Workflow

DeepPurpose provides a straightforward workflow that can be executed in six simple steps, each easily invoked with a single line of code, enhancing the user experience for rapid model development:

1. Load the Dataset: Users can input datasets whether from local files or pre-selected benchmark datasets provided by DeepPurpose.
2. Specify Encoders: Users choose which compound and protein encoders they wish to utilize based on their specific research needs.
3. Split the Dataset: The built-in data_process function enables users to divide the dataset into training, validation, and testing sets using various data-splitting strategies.
4. Create Configuration File: Users can create configuration files to define model parameters, and DeepPurpose offers the functionality to automatically search for optimal hyper-parameter values.
5. Initialize the Model: Models can be initialized based on the configuration, with options to load pre-trained models or previously saved models.
6. Train the Model: The train function allows users to train models while closely monitoring their progress and performance metrics (Huang et al. 2021a).

Additional Functionalities

- Drug Repurposing and Virtual Screening: With only a few lines of code, users can specify a library of compounds for virtual screening. DeepPurpose will train five different deep learning models, aggregate their predictions, and produce a ranked list of compound candidates based on predicted binding scores.
- Pre-trained Models: If a training dataset is not provided, DeepPurpose can still utilize pre-trained models to make predictions effectively. This flexibility benefits users who may not have extensive datasets available (Huang et al. 2021a).

Compatibility and Accessibility.

DeepPurpose is OS-agnostic and designed to operate smoothly within a Jupyter Notebook environment, allowing it to be executed both in the cloud and on local machines. All necessary resources—including datasets, models, documentation, installation instructions, and tutorials—are easily accessible via the DeepPurpose GitHub repository (Huang et al. 2021a).

18.6.3 NeoDTI: A Comprehensive Framework for Drug-Target Interaction Prediction

Overview: NeoDTI is an advanced framework for predicting drug-target interactions (DTIs), bolstered by its capability to integrate heterogeneous datasets that encompass chemical structures, protein sequences, functional roles, protein-protein interactions, and drug-disease associations. This multifaceted integration allows NeoDTI to significantly enhance predictive accuracy, as demonstrated in empirical evaluations against various baseline methods (Wan et al. 2019).

18.6.3.1 Key Features of NeoDTI

- *Integration of Heterogeneous Data.*
  NeoDTI excels in synthesizing diverse types of biological and chemical data, including compound-protein binding affinities, thereby improving its predictive reliability. The framework incorporates information from multiple sources to create a holistic representation of the relationship between drugs and targets (Wan et al. 2019).
- *End-to-End Learning Model.*
  Unlike existing models that may target either homogeneous or heterogeneous datasets in isolation, NeoDTI employs neural network architectures tailored to learn node-level embeddings from heterogeneous networks. This unique focus allows the model to accurately capture and reflect the complexities of DTI relationships (Wang et al. 2024b).
- *Innovative Neural Information Passing.*
  NeoDTI is distinct in its use of neural information passing combined with aggregation techniques within a topology-preserving optimization framework. This methodological innovation facilitates the extraction of latent features from the heterogeneous network, supporting precise DTI predictions (Wan et al. 2019).

Empirical Validation

Comprehensive evaluations indicate that NeoDTI not only outperforms traditional baseline models but also predicts novel DTIs, further validated by supporting literature. This underscores its potential as a robust tool for drug discovery and repositioning (Wan et al. 2019).

18.6.4 DeepLSTM: Leveraging Long Short-Term Memory for DTI Prediction

Overview: DeepLSTM presents an effective computational approach for predicting potential drug-target interactions, addressing the limitations of traditional experimental methods, which are often constrained by throughput, precision, and cost considerations (Wang et al. 2020a).

18.6.4.1 Methodology

- *Feature Extraction.*
  DeepLSTM employs Position Specific Scoring Matrix (PSSM) and Legendre Moment (LM) to extract evolutionary features from protein sequences. These features are then combined with drug molecular substructure fingerprints to create comprehensive feature vectors representing drug-target pairs (Wang et al. 2020b).
- *Sparse Principal Component Analysis (SPCA).*
  To optimize the representation of these feature vectors, SPCA is utilized, compressing the attributes of both drugs and proteins into a unified vector space that can be efficiently processed (Wang et al. 2020b).

18.6.4.2 Deep LSTM Architecture

Leveraging the unique capabilities of deep long short-term memory networks, DeepLSTM is constructed to predict drug-target interactions effectively, capturing both temporal dependencies and intrinsic feature relationships within the data (Wang et al. 2020b).

18.6.5 SRX-DTI: Importance in Drug Discovery

Overview: The identification of DTIs is pivotal in the drug discovery process, as it directly influences the development of new therapeutics and the repurposing of existing drugs (Sun et al. 2024). SRX-DTI offers a holistic approach to DTI identification, highlighting its implications in various aspects of drug development.

18.6.5.1 Key Points

- *Drug Development and Therapeutic Efficacy.*
  A deep understanding of DTIs enables the design of new drug compounds that effectively bind to specific biological targets, enhancing therapeutic efficacy while minimizing side effects (Su et al. 2023).
- *Drug Repurposing Opportunities.*
  Identifying new DTIs can uncover alternative therapeutic uses for existing drugs, potentially accelerating development timelines and reducing associated costs (Low et al. 2020).
- *Validation Through In Vitro Experiments.*
  While computational predictions provide valuable insights, in vitro validation remains essential for confirming the interactions observed through computational methods (Grassmann et al. 2024).
- *High Throughput Screening (HTS).*
  In vitro techniques such as HTS allow for the rapid evaluation of extensive drug-target pairs, facilitating the identification of promising candidates for further investigation (Ayon 2023).

18.6.5.2 Biological Insights

Understanding DTIs contributes significantly to elucidating biological pathways and mechanisms of action, enabling the development of targeted therapies that address specific disease mechanisms (Rintala et al. 2022).

18.6.6 GraphDTA: Graph-Based DTA Prediction

Overview: GraphDTA introduces a novel methodology for predicting drug-target affinity by representing drugs as molecular graphs. This approach addresses the inefficiencies of traditional string-based representations of drug structures, enhancing prediction capabilities and application efficiency (Nguyen et al. 2021).

18.6.6.1 Key Features

- *Graph Representation of Drugs.*
  By modeling drugs as graphs, GraphDTA captures the intricate relationships between molecular structures more naturally, facilitating more informed predictions of drug-target interactions (Yazdani-Jahromi et al. 2022).
- *Graph Neural Networks.*
  The use of graph neural networks allows GraphDTA to outperform both non-deep learning models and competing deep learning approaches in predicting drug-target affinity across multiple benchmarks (Nguyen et al. 2021).
- *Alternative Computational Methods.*
  Traditional DTA prediction approaches include molecular docking and collaborative filtering; however, GraphDTA's graph-based technique provides a more holistic picture, avoiding the limitations of requiring protein crystallization data and capturing localized molecular interactions (Luchi et al. 2022).
- *Performance Validation.*
  GraphDTA has shown superior performance on drug-target affinity prediction benchmarks compared to existing methodologies, indicating its potential for effective drug repurposing and targeted experimental validation (Tang et al. 2024).

18.7 Conclusion

These innovative tools and frameworks significantly enhance the landscape prediction of metabolic pathways and drug-target interaction. By integrating sophisticated machine learning techniques and extensive biological insights, they accelerate the processes of drug discovery, development, and repurposing, ultimately contributing to more effective therapeutic strategies. Through empirical validation, these tools underscore the importance of computational methods in modern drug development and repositioning, ultimately contributing to more efficient therapeutic interventions and improved patient outcomes. As computational methods continue to evolve, they hold immense promise for transforming the future of biomedicine.

References

Aguilar-Ruiz JS, Michalak M (2024) Classification performance assessment for imbalanced multiclass data. Sci Rep 14(1):10759

Atas Guvenilir H, Doğan T (2023) How to approach machine learning-based prediction of drug/compound–target interactions. J Cheminform 15(1):16

Ayon NJ (2023) High-throughput screening of natural product and synthetic molecule libraries for antibacterial drug discovery. Meta 13(5):625

Baranwal M, Magner A, Elvati P, Saldinger J, Violi A, Hero AO (2024) A deep learning architecture for metabolic pathway prediction. Bioinformatics 40(7)

Barhate D, Pathak S, Singh BK, Jain A, Dubey AK (2024) A systematic review of machine learning and deep learning approaches in plant species detection. Smart Agric Technol 9:100605

Bhardwaj A, Kishore S, Pandey DK (2022) Artificial intelligence in biological sciences. Life (Basel) 12(9)

Bjørkeli EB, Johannessen K, Geitung JT, Karlberg AM, Eikenes L, Esmaeili M (2024) Deep neural network modeling for brain tumor classification using magnetic resonance spectroscopic imaging

Cai YD, Qian Z, Lu L, Feng KY, Meng X, Niu B, Zhao GD, Lu WC (2008) Prediction of compounds' biological function (metabolic pathways) based on functional group composition. Mol div 12:131–137.

Chakraborty C, Bhattacharya M, Pal S, Lee SS (2024) From machine learning to deep learning: advances of the recent data-driven paradigm shift in medicine and healthcare. Curr Res Biotechnol 7:100164

Ching T, Himmelstein DS, Beaulieu-Jones BK, Kalinin AA, Do BT, Way GP et al (2018) Opportunities and obstacles for deep learning in biology and medicine. J R Soc Interface 15(141):20170387

Davenport T, Kalakota R (2019) The potential for artificial intelligence in healthcare. Future Healthc J 6(2):94–98

Fabris F, de Magalhães JP, Freitas AA (2017) A review of supervised machine learning applied to ageing research. Biogerontology 18(2):171–188

Gao YF, Chen L, Cai YD, Feng KY, Huang T, Jiang Y (2012) Predicting metabolic pathways of small molecules and enzymes based on interaction information of chemicals and proteins

Goshisht MK (2024) Machine learning and deep learning in synthetic biology: key architectures, applications, and challenges. ACS Omega 9(9):9921–9945

Grassmann G, Miotto M, Desantis F, Di Rienzo L, Tartaglia GG, Pastore A et al (2024) Computational approaches to predict protein-protein interactions in crowded cellular environments. Chem Rev 124(7):3932–3977

Haldorai A, Ramu A, Suriya M (2020) Organization Internet of Things (IoTs): supervised, unsupervised, and reinforcement learning, pp 27–53

Hamdalla MA, Rajasekaran S, Grant DF, Mandoiu II (2015) Metabolic pathway predictions for metabolomics: a molecular structure matching approach. J Chem Inf Model 23;55(3):709–718

Hasan Mahmud SM, Chen W, Jahan H, Dai B, Din SU, Dzisoo AM (2020) DeepACTION: a deep learning-based method for predicting novel drug-target interactions. Anal Biochem 610:113978

Hicks SA, Strümke I, Thambawita V, Hammou M, Riegler MA, Halvorsen P et al (2022) On evaluation metrics for medical applications of artificial intelligence. Sci Rep 12(1):5979

Hu LL, Chen C, Huang T, Cai YD, Chou KC (2011) Predicting biological functions of compounds based on chemical-chemical interactions. PloS one 6(12):e29491

Huang K, Fu T, Glass LM, Zitnik M, Xiao C, Sun J (2021a) DeepPurpose: a deep learning library for drug–target interaction prediction. Bioinformatics 36(22–23):5545–5547

Huang K, Xiao C, Glass LM, Sun J (2021b) MolTrans: Molecular Interaction Transformer for drug–target interaction prediction. Bioinformatics 37(6):830–836

Huckvale ED, Moseley HNB (2024) Predicting the pathway involvement of metabolites based on combined metabolite and pathway features. Metabolites 14(5):266

Huynh N, Yan D, Ma Y, Wu S, Long C, Sami MT et al (2024) The use of generative adversarial network and graph convolution network for neuroimaging-based diagnostic classification. Brain Sci 14(5)

Johannesdottir KB, Kehlet H, Petersen PB, Aasvang EK, Sørensen HBD, Jørgensen CC et al (2022) Machine learning classifiers do not improve prediction of hospitalization >2 days after fast-track hip and knee arthroplasty compared with a classical statistical risk model. Acta Orthop 93:117–123

Johnson KB, Wei W, Weeraratne D, Frisse ME, Misulis K, Rhee K et al (2021) Precision medicine, AI, and the future of personalized health care. Clin Transl Sci 14(1):86–93

Judge A, Dodd MS (2020) Metabolism. Essays Biochem 64(4):607–647

Karabacak M, Jagtiani P, Di L, Shah AH, Komotar RJ, Margetis K (2024) Advancing precision prognostication in neuro-oncology: machine learning models for data-driven personalized survival predictions in IDH-wildtype glioblastoma. Neurooncol Adv 6(1)

Katzman JL, Shaham U, Cloninger A, Bates J, Jiang T, Kluger Y (2018) DeepSurv: personalized treatment recommender system using a Cox proportional hazards deep neural network. BMC Med Res Methodol [Internet] 18(1):24. https://doi.org/10.1186/s12874-018-0482-1

Kaynar G, Cakmakci D, Bund C, Todeschi J, Namer IJ, Cicek AE (2023) PiDeeL: metabolic pathway-informed deep learning model for survival analysis and pathological classification of gliomas. Bioinformatics 39(11)

Khojasteh H, Pirgazi J, Ghanbari SA (2023) Improving prediction of drug-target interactions based on fusing multiple features with data balancing and feature selection techniques. PLoS One 18(8):e0288173

Krzyszczyk P, Schloss R, Palmer A, Berthiaume F (2018) The role of macrophages in acute and chronic wound healing and interventions to promote pro-wound healing phenotypes. Front Physiol 9:9

Liao J, Wang Q, Wu F, Huang Z (2022) In silico methods for identification of potential active sites of therapeutic targets. Molecules 27(20):7103

Lopez-Ibañez J, Pazos F, Chagoyen M (2021) Predicting biological pathways of chemical compounds with a profile-inspired approach. BMC Bioinformatics 22(1):320

Low ZY, Farouk IA, Lal SK (2020) Drug repositioning: new approaches and future prospects for life-debilitating diseases and the COVID-19 pandemic outbreak. Viruses 12(9):1058

Luchi AM, Gomez Chávez JL, Villafañe RN, Conti GA, Pérez ER, Angelina E et al (2022) Graph neural networks and molecular docking as two complementary approaches for virtual screening: a case study on Cruzain

Malhotra R, Singh Saini B, Gupta S (2024) Survival classification of Gliomas through a novel enhancement-based strategy for class overlap of radiomics features. Expert Syst Appl 240:122320

Nguyen T, Le H, Quinn TP, Nguyen T, Le TD, Venkatesh S (2021) GraphDTA: predicting drug–target binding affinity with graph neural networks. Bioinformatics 37(8):1140–1147

Nguyen-Vo TH, Teesdale-Spittle P, Harvey JE, Nguyen BP (2024) Molecular representations in bio-cheminformatics. Memet Comput 16(3):519–536

Obaido G, Ogbuokiri B, Swart TG, Ayawei N, Kasongo SM, Aruleba K et al (2022) An interpretable machine learning approach for hepatitis B diagnosis. Appl Sci 12(21):11127

Polanski J (2022) Unsupervised learning in drug design from self-organization to deep chemistry. Int J Mol Sci 23(5):2797

Ren ZH, You ZH, Zou Q, Yu CQ, Ma YF, Guan YJ et al (2023) DeepMPF: deep learning framework for predicting drug-target interactions based on multi-modal representation with meta-path semantic analysis. J Transl Med 21(1):48

Rintala TJ, Ghosh A, Fortino V (2022) Network approaches for modeling the effect of drugs and diseases. Brief Bioinform 23(4)

Rognan D (2007) Chemogenomic approaches to rational drug design. Br J Pharmacol 152(1):38–52

Shin B, Park S, Kang K, Ho JC (2019) Self-attention based molecule representation for predicting drug-target interaction. In: Doshi-Velez F, Fackler J, Jung K, Kale D, Ranganath R, Wallace B et al (eds) Proceedings of the 4th machine learning for healthcare conference [Internet], Proceedings of machine learning research, vol 106. PMLR, pp 230–248. Available from: https://proceedings.mlr.press/v106/shin19a.html

Sreedhar D, David A (2024) A deep learning framework for diagnosis and survival prognosis of central nervous system tumors. J Stud Res 13(1)

Su Y, Hu Z, Wang F, Bin Y, Zheng C, Li H et al (2023) AMGDTI: drug–target interaction prediction based on adaptive meta-graph learning in heterogeneous network. Brief Bioinform 25(1)

Sun Y, Li YY, Leung CK, Hu P (2024) iNGNN-DTI: prediction of drug–target interaction with interpretable nested graph neural network and pretrained molecule models. Bioinformatics 40(3)

Tang X, Lei X, Zhang Y (2024) Prediction of drug-target affinity using attention neural network. Int J Mol Sci 25(10):5126

Wan F, Hong L, Xiao A, Jiang T, Zeng J (2019) NeoDTI: neural integration of neighbor information from a heterogeneous network for discovering new drug–target interactions. Bioinformatics 35(1):104–111

Wang YB, You ZH, Yang S, Yi HC, Chen ZH, Zheng K (2020a) A deep learning-based method for drug-target interaction prediction based on long short-term memory neural network. BMC Med Inform Decis Mak 20(S2):49

Wang YB, You ZH, Yang S, Yi HC, Chen ZH, Zheng K (2020b) A deep learning-based method for drug-target interaction prediction based on long short-term memory neural network. BMC Med Inform Decis Mak [Internet] 20(2):49. https://doi.org/10.1186/s12911-020-1052-0

Wang L, Zhou Y, Chen Q (2023) AMMVF-DTI: a novel model predicting drug–target interactions based on attention mechanism and multi-view fusion. Int J Mol Sci 24(18):14142

Wang B, Cai B, Sheng J, Jiao W (2024a) AAGCN: a graph convolutional neural network with adaptive feature and topology learning. Sci Rep 14(1):10134

Wang Z, He C, Hu Y, Luo H, Li C, Wu X et al (2024b) A hybrid deep learning scheme for MRI-based preliminary multiclassification diagnosis of primary brain tumors. Front Oncol 14:1363756

Wu X, Li Z, Chen G, Yin Y, Chen CYC (2023) Hybrid neural network approaches to predict drug–target binding affinity for drug repurposing: screening for potential leads for Alzheimer's disease. Front Mol Biosci 10:10

Wu H, Liu J, Zhang R, Lu Y, Cui G, Cui Z et al (2024) A review of deep learning methods for ligand based drug virtual screening. Fundam Res 4(4):715–737

Yang X, Song Z, King I, Xu Z (2023) A survey on deep semi-supervised learning. IEEE Trans Knowl Data Eng 35(9):8934–8954

Yazdani-Jahromi M, Yousefi N, Tayebi A, Kolanthai E, Neal CJ, Seal S et al (2022) AttentionSiteDTI: an interpretable graph-based model for drug-target interaction prediction using NLP sentence-level relation classification. Brief Bioinform 23(4)

Yuan Y, Zhang X, Wang Y, Li H, Qi Z, Du Z et al (2024) Multimodal data integration using deep learning predicts overall survival of patients with glioma. VIEW 5(5)

Zeng X, Li SJ, Lv SQ, Wen ML, Li Y (2024) A comprehensive review of the recent advances on predicting drug-target affinity based on deep learning. Front Pharmacol 15:15

Zhu J, Wang J, Wang X, Gao M, Guo B, Gao M et al (2021) Prediction of drug efficacy from transcriptional profiles with deep learning. Nat Biotechnol 39(11):1444–1452

MIX
Papier aus verantwortungsvollen Quellen
Paper from responsible sources
FSC® C105338

If you have any concerns about our products,
you can contact us on
ProductSafety@springernature.com

In case Publisher is established outside the EU,
the EU authorized representative is:
Springer Nature Customer Service Center GmbH
Europaplatz 3, 69115 Heidelberg, Germany

Printed by Libri Plureos GmbH
in Hamburg, Germany